Prealgebra

CROSSMONT COLLEGE
USED BOOK

Prealgebra

Elaine Hubbard
Kennesaw State University

Ronald D. Robinson

HOUGHTON MIFFLIN COMPANY Boston New York

Editor-in-Chief: Jack Shira
Assistant Editor: Yen Tieu
Senior Project Editor: Nancy Blodget
Editorial Assistants: Celeste Ng, Sean McGann
Senior Production/Design Coordinator: Carol Merrigan
Senior Manufacturing Coordinator: Marie Barnes
Senior Marketing Manager: Ben Rivera
Marketing Assistant: Lisa Lawler

Cover art: © 2002 Dale Chihuly.

Photo Credits: p. 1: © George H. H. Huey/Corbis; p. 77: © Dave Bartruff/Corbis; p. 185: Paul Buck, © AFP/Corbis; p. 271: © Mark E. Gibson/Corbis; p. 347: © Tom Stewart/Corbis; p. 471: © Bob Daemmrich/The Image Works; p. 573: © A. Ramey/Photo Edit; p. 673: © Jim Cummins/Corbis; p. 773: Sculpture by Menash Kadishman, © 1977, at Storm King Art Center, Mountainville, New York. Photograph © Margot Granitsas/The Image Works.

Copyright © 2005 by Houghton Mifflin Company. All rights reserved.

No part of this work may be reproduced or transmitted in any form or by any means, electronic or mechanical, including photocopying and recording, or by any information storage or retrieval system without the prior written permission of Houghton Mifflin Company unless such copying is expressly permitted by federal copyright law. Address inquiries to College Permissions, Houghton Mifflin Company, 222 Berkeley Street, Boston, MA 02116-3764.

Printed in the U.S.A.

Library of Congress Control Number: 2002109478

ISBN
Student Text: 0-618-01732-1
Instructor's Annotated Edition: 0-618-01733-X

23456789-WEB-08 07 06

Contents

Preface

The Audience

Prealgebra was written with three categories of students in mind:

- Students who have come directly from high school and whose placement tests indicate a need for remedial work
- Young adults who have entered or returned to college a few years after leaving high school
- Older, more mature students who have families and jobs and are seeking to improve their circumstances

Although it is true that these three groups of students have varied backgrounds and levels of experience, they also often tend to share several common traits. Many have weak study habits or a fear of the subject, or have lost whatever knowledge they once had. This book was written to help the student overcome these obstacles, to engage the student, and, most importantly, to give the student the best possible chance to succeed.

General Organization

Algebra Content Because programs of study vary from school to school, this book is designed to provide maximum flexibility in the choice of topics to include. All algebraic content is presented in separate sections so that those topics can optionally be included or omitted.

Equation solving is introduced early and intuitively. A more formal treatment is presented in Chapter 4. Also included in the beginning chapters are the topics of evaluating and simplifying algebraic expressions. Although all these topics follow naturally from the arithmetic content of the chapters, their presentation in separate sections allows for them to be omitted or deferred if the instructor so chooses.

Geometry Content Some sections are devoted to introducing geometry topics, which we treat as applications of the preceding arithmetic. For example, perimeter follows addition, area and volume follow multiplication, and so on. To reinforce these topics, we integrate geometry material in some other sections.

Sections Each section consists of three parts: the exposition, a Quick Reference, and a set of exercises. All three elements are referenced and connected by clearly identified subsections. This allows the student and the instructor to locate and cross-reference examples, summaries, and exercise groups quickly and easily.

The Exposition

Section Openers Each section opens with a feature entitled *Suggestions for Success,* which offers advice on study habits, ways to approach the coming material, and general observations on ways to think about mathematics.

Developing the Concept In addition to *how-to,* the authors emphasize *why.* Many sections include *Developing the Concept,* which provides background and reasoning for the definitions, rules, and procedures that follow.

Examples and Your Turns All the basic skills and concepts are illustrated with numerous examples. The steps in the solutions are annotated to explain what is being done and why.

Every example has a corresponding *Your Turn,* which is a similar example for the student to work. The answers to *Your Turns* are given (upside-down) at the bottom of the box, and worked-out solutions are included in the Answers to Selected Features section in the back of the book.

Learning Tips Scattered throughout the text are *Learning Tips* that relate to some specific explanation or example. These tips offer cautions, alternative methods, and suggestions for better understanding.

Keys to the Calculator *Prealgebra* is not a calculator-based book. However, many instructors permit, suggest, or require the use of at least a scientific calculator. Some instructors use graphing calculators to lay the groundwork for future courses.

Keys to the Calculator is a set-aside feature that assists the student with keystrokes for various arithmetic operations. Typical keystrokes are given for both scientific and graphing calculators, often with a sample resulting display. A brief set of calculator practice exercises is included.

This feature is designed to be visible and accessible enough for those who want to use it, but unobtrusive enough for those who do not. Calculator exercises in the exercise sets are clearly identified.

Art The authors have included abundant art, including bar graphs, geometric figures, assistance art for applications, and decorative art to please the student and add to the overall attractiveness of the book.

Quick Reference The *Quick Reference* feature is a summary of all the vocabulary, definitions, rules, procedures, and important concepts presented in the section. Where appropriate, brief refresher examples are given. Entries in the *Quick Reference* are keyed to the subsections in the exposition.

Exercise Sets

Instructors, not authors, assign homework. In this text, we have tried to provide an abundance and variety of exercises from which to construct assignments. Like the exposition and the *Quick Reference,* exercise sets are formatted by subsection for easy reference.

The bulk of the exercises are designed to give the student ample practice with the basic concepts and skills of the subsection. Included are mini-applications. Some blocks of exercises are identified with an icon as shown:

Usually, the exercise sets for each subsection begin with a pair of fill-in-the-blank exercises that focus on vocabulary in context.

≈ These exercise groups offer practice in the important skill of estimating the results of various arithmetic operations.

Many sourced, real-data applications are included. Because students in prealgebra classes come from diverse backgrounds, we have tried to design our applications with a wide variety of student interests and experiences in mind.

On a somewhat larger scale, *Explorations with Real-World Data* exercises are particularly suitable for class discussion and other group or collaborative activities. Some of these exercises require slightly extended skills and reasoning. The use of a calculator should be considered.

Bringing It Together exercises appear in selected, appropriate sections. This group of exercises cycles back and mixes in previous and related skills.

The Writing and Concept Extension exercises include writing exercises to encourage students to express themselves about ideas and concepts in their own words. These exercises can also be used as the basis for class discussion.

Other exercises in this section are mildly more challenging and extend the fundamental concepts of the section.

Chapter Organization

Chapter Openers Each chapter begins with a practical problem that is directly related to the material to be presented, along with a *Chapter Snapshot,* which describes the content of the chapter. This is followed by a feature entitled *Some Friendly Advice,* which informally suggests ways in which the student can succeed in the course. The chapter opener concludes with *Warm-Up Skills,* a brief set of review exercises targeted specifically to the content of the chapter.

Chapter Review Exercises Organized by section, these exercises review the material of the chapter.

Chapter Tests Every chapter includes a sample test that covers all sections.

Cumulative Tests A cumulative test appears at the end of Chapters 3, 5, 7, and 9.

The answers to the Review Exercises, the Chapter Tests, and the Cumulative Tests all cite the pertinent subsection so that students can easily refer back to the related material.

Supplements

For Instructors

Instructor's Annotated Edition This edition contains a replica of the student text and additional items just for the instructor. There are answers to the odd- and even-numbered exercises for the end-of-section activities, review problems, Chapter Review Exercises, Chapter Tests, Cumulative Tests, and selected features including Warm-Up Skills.

Instructor's Solutions Manual The *Instructor's Solutions Manual* contains worked-out solutions for all exercises in the text. It also includes a printed test bank, which provides a variety of test items for each chapter.

HM ClassPrep™ with HM Testing™ CD-ROM HM ClassPrep contains a multitude of text-specific resources for instructors to use to enhance the classroom experience. These resources can be easily accessed by chapter or resource type and can also link you to the text's web site. HM Testing is our computerized test generator and contains a database of algorithmic test items.

Instructor Text-Specific Web Site and PowerPoint Slides The resources available on the HM ClassPrep CD are also available on the instructor web site at *http://math.college.hmco.com/instructors.* Appropriate items are password-protected. Instructors also have access to the student part of the text's web site.

For Students

Student Solutions Manual The *Student Solutions Manual* contains complete solutions to all odd-numbered exercises in the text.

***Math Study Skills Workbook* by Paul D. Nolting** This workbook is designed to reinforce skills and minimize frustration for students in any math class, lab, or study skills course. It offers a wealth of study tips and sound advice on note taking, time management, and reducing math anxiety. In addition, numerous opportunities for self-assessment enable students to track their own progress.

HM mathSpace® Student CD-ROM This new tutorial CD-ROM allows you to practice skills and review concepts as many times as necessary, by algorithmically generating exercises and step-by-step solutions for you to practice on.

Eduspace® Online Learning Environment Eduspace is a text-specific online learning environment that combines an algorithmic tutorial program with homework capabilities. Specific content is available 24 hours a day to help you further understand your textbook.

SMARTHINKING™ Live, Online Tutoring Houghton Mifflin has partnered with SMARTHINKING to provide an easy-to-use, effective, online tutorial service. **Whiteboard Simulations** and **Practice Area** promote real-time visual interaction.

Three levels of service are offered.

- **Text-Specific Tutoring** provides real-time, one-on-one instruction with a specially qualified "instructor."
- **Questions Any Time** allows students to submit questions to the tutor outside the scheduled hours and receive a reply within 24 hours.
- **Independent Study Resources** connect students with around-the-clock access to additional educational services, including interactive web sites, diagnostic tests, and Frequently Asked Questions posed to SMARTHINKING e-structors.

Houghton Mifflin Instructional Videos and DVDs This text offers text-specific videos and DVDs, hosted by Dana Mosely, covering all sections of the text and providing a valuable resource for further instruction and review. At the beginning of each section, the serves as a reminder that the section is covered in a video/DVD lesson.

Student Text-Specific Web Site Online student resources can be found at this text's web site at *http://math.college.hmco.com/students.*

Acknowledgments

We would like to thank the following colleagues who reviewed the first edition manuscript and made helpful suggestions.

Pamela Arrindell	*Montgomery College, MD*
Kathleen Bavelas	*Manchester Community College, CT*
Linda K. Berg	*University of Great Falls, MT*
Karen Bingham	*Clarion University, PA*
Debra Bryant	*Tennessee Technological University, TN*
Mark Burtch	*Scottsdale Community College, AZ*
Deann Christianson	*University of the Pacific, CA*
James A. Cochran	*Kirkwood Community College, IA*
Pat C. Cook	*Weatherford College, TX*
Vincent A. Daniele	*National Technical Institute for the Deaf, NY*
Cheryl Taylor Dickson	*Columbus State Community College, OH*
Gail P. Figa	*Florence-Darlington Technical College, SC*
David J. French	*Tidewater Community College, VA*
Roy D. Frysinger	*Harrisburg Area Community College, PA*
Kay Haralson	*Austin Peay State University, TN*
Amy D. Hastings	*Johnson County Community College, KS*
Don Haussler	*University of Wisconsin at Sheboygan, WI*
Celeste Hernandez	*Richland College, TX*

Joyce Marie Hibbs	*St. Philip's College, TX*
Gail Johnston	*University of Alaska Anchorage, Mat-Su College, AK*
Vijay S. Joshi	*Virginia Intermont College, VA*
Maryann E. Justinger	*Erie Community College, NY*
Steven Kahn	*Anne Arundel Community College, MD*
Martha Kuklinski	*Salem State College, MA*
Patricia R. Lanz	*Erie Community College, NY*
David Longshore	*Victor Valley College, CA*
Patricia Marquis	*Central Community College, NE*
Mary Ann McCurdy	*Cambria County Area Community College, PA*
Timothy McLendon	*East Central College, MO*
Debra Packer	*New Mexico State University at Carlsbad, NM*
Peg Pankowski	*Community College of Allegheny County, PA*
Frank Pecchioni	*Jefferson Community College, KY*
Rick Ponticelli	*North Shore Community College, MA*
Melody Shipley	*North Central Missouri College, MO*
Robert Thomas	*Eastern Kentucky University, KY*
Sam Tinsley	*Richland College, TX*
Danny Whited	*Virginia Intermont College, VA*
Mary L. Wolyniak	*Broome Community College, NY*
Jill C. Zimmerman	*Manchester Community-Technical College, CT*

We are especially grateful to Daniel T. Paulsen and John Searcy, who took great care to ensure the accuracy of the text and answers.

Finally, we thank the staff at Houghton Mifflin for their assistance and contributions to the project. Particular thanks go to Jack Shira, Publisher; Marika Hoe and Yen Tieu, Assistant Editors; Nancy Blodget, Senior Project Editor; Celeste Ng and Sean McGann, Editorial Assistants; Carol Merrigan, Senior Production/Design Coordinator; Marie Barnes, Senior Manufacturing Coordinator; and Ben Rivera, Senior Marketing Manager.

To the Student: How To Use Your Book

Maybe someone has told you that your background in math is not strong enough and that you need more practice. Maybe you know that you need additional math skills and so you volunteered to take this course. Maybe you do not plan to major in mathematics or even in a math-related field but are required to take a mathematics course, regardless of your major. Whatever your reason, your goal now is to succeed!

This textbook will be a major resource in your learning. How you use it and how well you use it will be critical factors in reaching your goals. Here are a few tips:

Reading Your Text

Many students feel that they can't read a math textbook. However, if you go about it the right way, you definitely can. Reading mathematics is not the same as reading a novel. You will need to go slowly, pay attention to detail, and make notes about questions you want to ask.

Some students prefer to read material before it has been presented. Others prefer to do their reading afterwards. Choose the style that works best for you, but make a resolution to read the text!

Developing the Concept

Many sections of the book include a feature called ***Developing the Concept.*** The purpose of this feature is to explain not only *how to* but *why.* This feature encourages visualization, pattern recognition, and informal deduction, stepping you through the logic behind definitions, rules, and procedures.

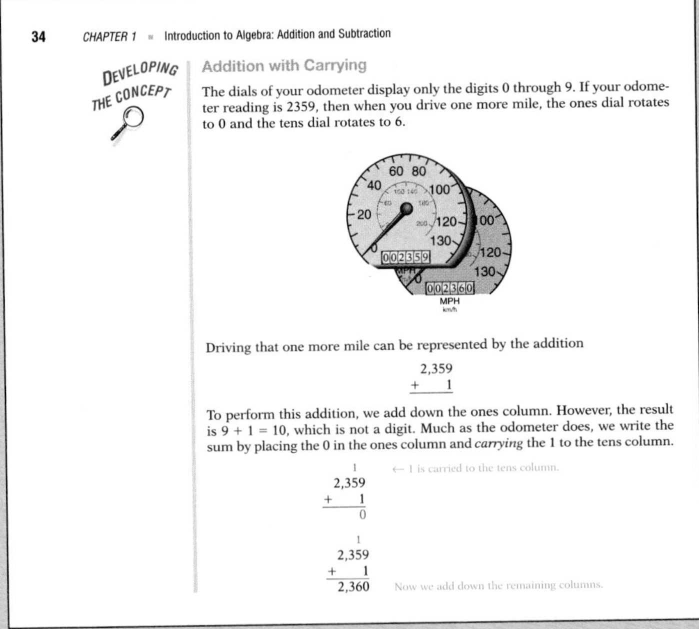
34 *CHAPTER 1* ▪ Introduction to Algebra: Addition and Subtraction

DEVELOPING THE CONCEPT

Addition with Carrying

The dials of your odometer display only the digits 0 through 9. If your odometer reading is 2359, then when you drive one more mile, the ones dial rotates to 0 and the tens dial rotates to 6.

Driving that one more mile can be represented by the addition

$$\begin{array}{r} 2{,}359 \\ +\quad 1 \\ \hline \end{array}$$

To perform this addition, we add down the ones column. However, the result is $9 + 1 = 10$, which is not a digit. Much as the odometer does, we write the sum by placing the 0 in the ones column and *carrying* the 1 to the tens column.

$$\begin{array}{r} 1 \\ 2{,}359 \\ +\quad 1 \\ \hline 0 \end{array} \quad \leftarrow \text{1 is carried to the tens column.}$$

$$\begin{array}{r} 1 \\ 2{,}359 \\ +\quad 1 \\ \hline 2{,}360 \end{array} \quad \text{Now we add down the remaining columns.}$$

page 34

Examples and Your Turns

Your book has hundreds of ***Examples*** that give you guided practice with concepts and skills. It is not enough to simply read these examples. You should actually work the examples with paper and pencil in hand. See how far you can go on your own before you need to peek at the worked-out solution.

There is also an abundance of ***Real-Data Examples and Exercises.*** These examples and exercises use familiar situations so that you can see the relevance of algebra to the world around you.

Matched with the examples are exercises called ***Your Turn.*** This is your chance to try working a problem that is similar to the example. The answers appear upside-down at the bottom of the box. If you have difficulties, worked-out solutions to selected Your Turn problems can be found at the back of the book. The best approach is to work through the example and then immediately do the Your Turn.

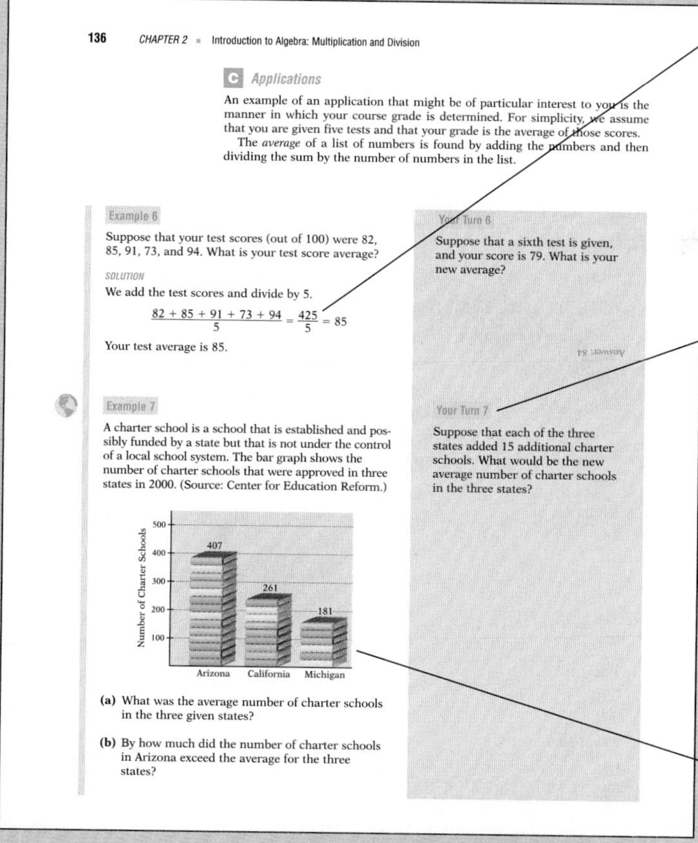

136 *CHAPTER 2* ▪ Introduction to Algebra: Multiplication and Division

C *Applications*

An example of an application that might be of particular interest to you is the manner in which your course grade is determined. For simplicity, we assume that you are given five tests and that your grade is the average of those scores.

The *average* of a list of numbers is found by adding the numbers and then dividing the sum by the number of numbers in the list.

Example 6

Suppose that your test scores (out of 100) were 82, 85, 91, 73, and 94. What is your test score average?

SOLUTION

We add the test scores and divide by 5.

$$\frac{82 + 85 + 91 + 73 + 94}{5} = \frac{425}{5} = 85$$

Your test average is 85.

Your Turn 6

Suppose that a sixth test is given, and your score is 79. What is your new average?

Answer: 84

Example 7

A charter school is a school that is established and possibly funded by a state but that is not under the control of a local school system. The bar graph shows the number of charter schools that were approved in three states in 2000. (Source: Center for Education Reform.)

(a) What was the average number of charter schools in the three given states?

(b) By how much did the number of charter schools in Arizona exceed the average for the three states?

Your Turn 7

Suppose that each of the three states added 15 additional charter schools. What would be the new average number of charter schools in the three states?

page 136

All the basic skills and concepts are illustrated with numerous examples. The steps in the solutions are annotated to explain what is being done and why.

Every example has a corresponding Your Turn, which is a similar example for you to work out. The answers to the Your Turns are given upside down at the bottom of the box, and worked-out solutions are included in the Answers to Selected Features at the back of the book.

A Real-Data Example teaches you how to organize and interpret data.

Keys to the Calculator

Your instructor will tell you whether a calculator is permitted or encouraged in this course. There are two different kinds (scientific and graphing) and many different models. Make sure you know what calculator your instructor recommends.

Scattered throughout this book are small sections called ***Keys to the Calculator.*** These sections give you the typical keystrokes for performing certain operations, show the typical resulting displays, and offer some practice exercises. If calculators are to be part of your course of study, you will want to take advantage of this information.

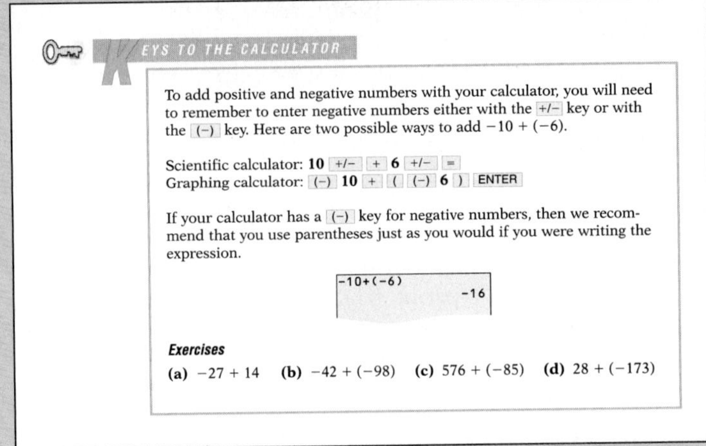

KEYS TO THE CALCULATOR

To add positive and negative numbers with your calculator, you will need to remember to enter negative numbers either with the +/− key or with the (−) key. Here are two possible ways to add $-10 + (-6)$.

Scientific calculator: **10** +/− + **6** +/− =
Graphing calculator: (−) **10** + ((−) **6**) ENTER

If your calculator has a (−) key for negative numbers, then we recommend that you use parentheses just as you would if you were writing the expression.

Exercises

(a) $-27 + 14$ **(b)** $-42 + (-98)$ **(c)** $576 + (-85)$ **(d)** $28 + (-173)$

page 202

Applying Yourself

In addition to the regular exercises, many exercise sets include features called ***Exploring with Real-World Data: Collaborative Activities*** and ***Writing and Concept Extension.*** These exercises challenge you to put your newly acquired knowledge into words by working with others and writing about algebraic concepts. We provide the answers to the odd-numbered exercises in the back of the book so that you can gauge your understanding of the concepts.

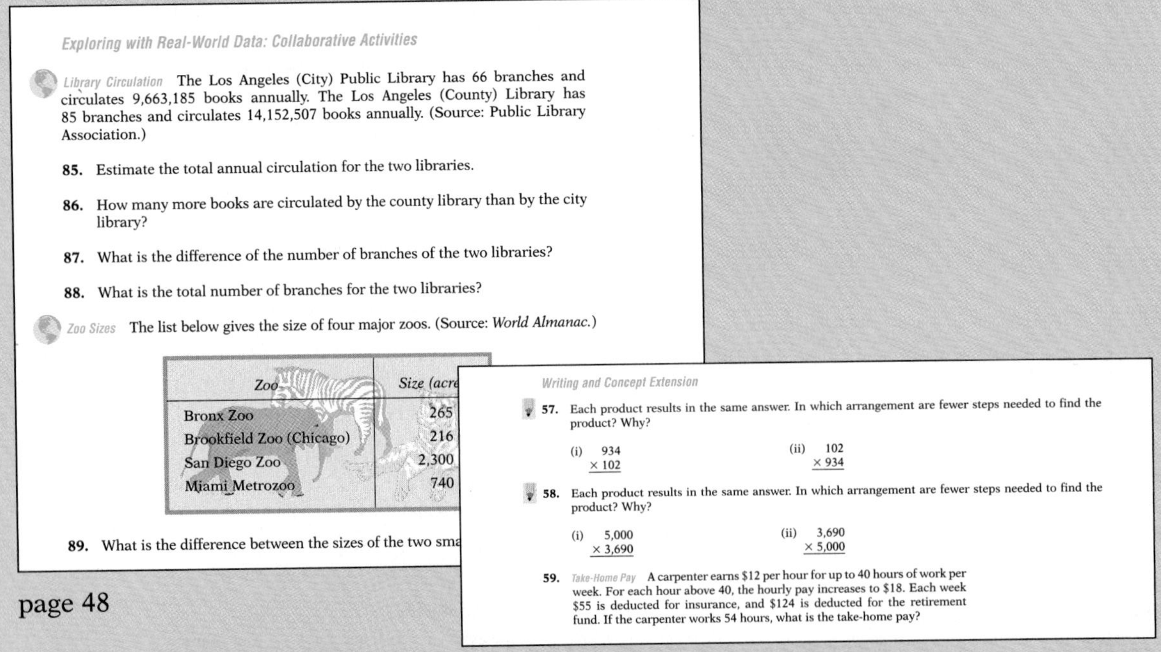

Exploring with Real-World Data: Collaborative Activities

Library Circulation The Los Angeles (City) Public Library has 66 branches and circulates 9,663,185 books annually. The Los Angeles (County) Library has 85 branches and circulates 14,152,507 books annually. (Source: Public Library Association.)

85. Estimate the total annual circulation for the two libraries.

86. How many more books are circulated by the county library than by the city library?

87. What is the difference of the number of branches of the two libraries?

88. What is the total number of branches for the two libraries?

Zoo Sizes The list below gives the size of four major zoos. (Source: *World Almanac.*)

Zoo	Size (acr
Bronx Zoo	265
Brookfield Zoo (Chicago)	216
San Diego Zoo	2,300
Miami Metrozoo	740

89. What is the difference between the sizes of the two sma

page 48

Writing and Concept Extension

57. Each product results in the same answer. In which arrangement are fewer steps needed to find the product? Why?

(i) 934×102 (ii) 102×934

58. Each product results in the same answer. In which arrangement are fewer steps needed to find the product? Why?

(i) $5{,}000 \times 3{,}690$ (ii) $3{,}690 \times 5{,}000$

59. *Take-Home Pay* A carpenter earns \$12 per hour for up to 40 hours of work per week. For each hour above 40, the hourly pay increases to \$18. Each week \$55 is deducted for insurance, and \$124 is deducted for the retirement fund. If the carpenter works 54 hours, what is the take-home pay?

page 103

Strengthen Your Knowledge

Included at the end of nearly every section is a ***Quick Reference,*** which is a summary of the key concepts and rules of the section. One excellent use of this feature is in your preparation for exams. Work through the summary and check off those things that you feel confident that you know. If you are unsure about a topic, go back and review, work examples, and work related exercises.

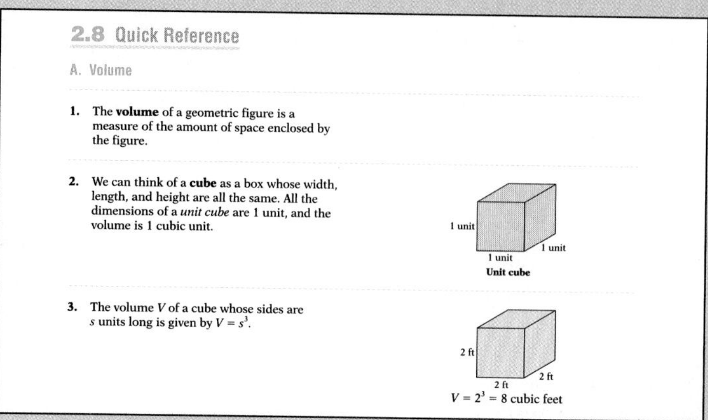

2.8 Quick Reference

A. Volume

1. The **volume** of a geometric figure is a measure of the amount of space enclosed by the figure.
2. We can think of a **cube** as a box whose width, length, and height are all the same. All the dimensions of a *unit cube* are 1 unit, and the volume is 1 cubic unit.
3. The volume V of a cube whose sides are s units long is given by $V = s^3$.

page 171

Every chapter concludes with a set of ***Review Exercises*** and a ***Chapter Test.*** Some chapters also have a ***Cumulative Test,*** which includes material from preceding chapters. The answers to the Review Exercises, Chapter Tests, and Cumulative Tests all cite the pertinent subsection so that students can easily refer back to the related material.

Chapter Review Exercises are organized by section. These exercises allow you to practice concepts learned in the chapter.

CHAPTER 8 REVIEW EXERCISES

Section 8.1

1. The word ______ means "per 100" or "for every 100."

In Exercises 2–4, write the percent as a simplified fraction.

2. *Reading for Pleasure* *USA Today* reported that 30% of the people in the United States said that reading was their favorite leisure activity.

page 761

Chapter Tests provide you with a self-assessment tool to check your learning progress.

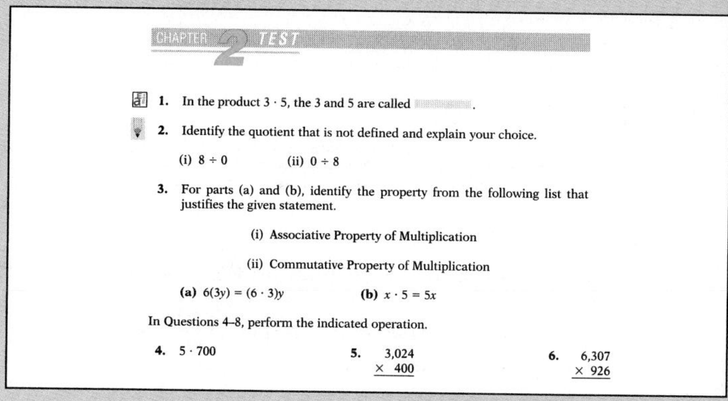

CHAPTER 2 TEST

1. In the product $3 \cdot 5$, the 3 and 5 are called ______.
2. Identify the quotient that is not defined and explain your choice.

   (i) $8 \div 0$ (ii) $0 \div 8$

3. For parts (a) and (b), identify the property from the following list that justifies the given statement.

   (i) Associative Property of Multiplication

   (ii) Commutative Property of Multiplication

   (a) $6(3y) = (6 \cdot 3)y$ **(b)** $x \cdot 5 = 5x$

In Questions 4–8, perform the indicated operation.

4. $5 \cdot 700$
5. $3{,}024 \times 400$
6. $6{,}307 \times 926$

page 183

Cumulative Tests appear at the ends of Chapters 3, 5, 7, and 9 and help tie together concepts from several chapters.

CHAPTERS 8–9 CUMULATIVE TEST

1. Write the percent as a simplified fraction and as a decimal.

   (a) 80% **(b)** 0.5% **(c)** 160%

2. Write the fraction as a percent.

   (a) $\frac{7}{40}$ **(b)** $\frac{2}{3}$

3. Write the decimal as a percent.

   (a) 0.12 **(b)** 3.2 **(c)** 0.008

page 819

Ongoing Help

Throughout the text, the authors offer hints about how to approach the material in the book, gained from their many years of teaching experience.

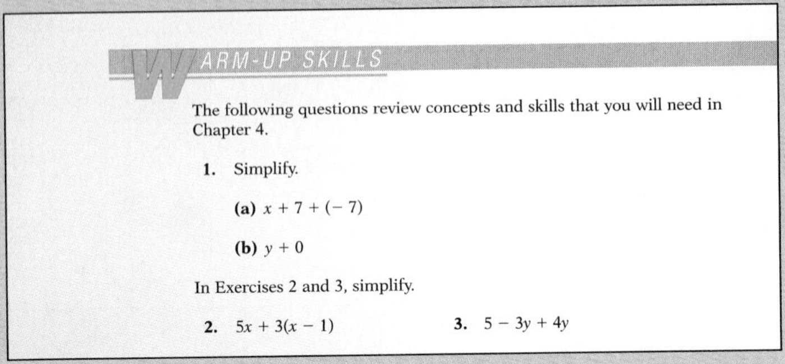

WARM-UP SKILLS

The following questions review concepts and skills that you will need in Chapter 4.

1. Simplify.

(a) $x + 7 + (-7)$

(b) $y + 0$

In Exercises 2 and 3, simplify.

2. $5x + 3(x - 1)$ **3.** $5 - 3y + 4y$

page 272

Every chapter (except Chapter 1) begins with a feature called ***Warm-Up Skills***—a set of 10–12 exercises that you can use to test your knowledge on the material you should have mastered before you proceed to the new chapter.

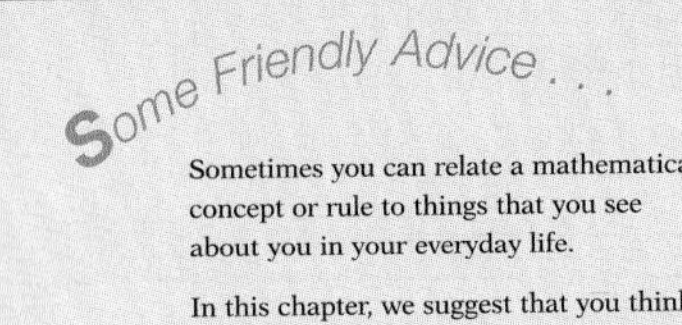

Some Friendly Advice . . .

Sometimes you can relate a mathematical concept or rule to things that you see about you in your everyday life.

In this chapter, we suggest that you think of an equation as being like a pan balance. If we add the same weight to both sides, the pans remain in balance. In the same way, if we add the same quantity to both sides of an equation, the equation remains in "balance." We also suggest the idea of a street map to think about moving around in a coordinate system.

Let your imagination run free. The more you can relate mathematical concepts to familiar things that you can visualize, the clearer the concepts will be to you.

page 272

Some Friendly Advice... offers suggestions about how to approach the chapter and how to improve your study methods.

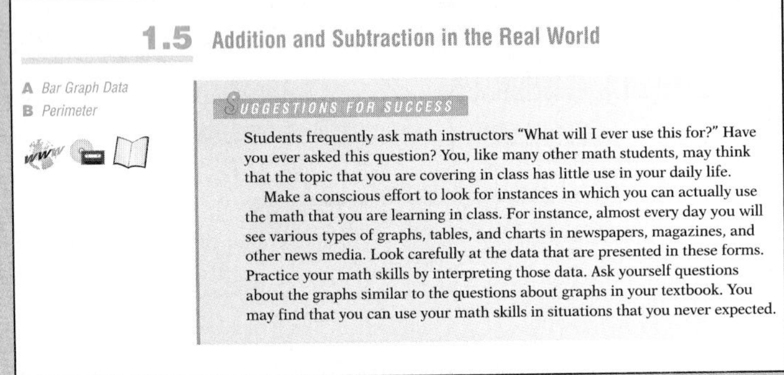

1.5 Addition and Subtraction in the Real World

A Bar Graph Data
B Perimeter

SUGGESTIONS FOR SUCCESS

Students frequently ask math instructors "What will I ever use this for?" Have you ever asked this question? You, like many other math students, may think that the topic that you are covering in class has little use in your daily life.

Make a conscious effort to look for instances in which you can actually use the math that you are learning in class. For instance, almost every day you will see various types of graphs, tables, and charts in newspapers, magazines, and other news media. Look carefully at the data that are presented in these forms. Practice your math skills by interpreting those data. Ask yourself questions about the graphs similar to the questions about graphs in your textbook. You may find that you can use your math skills in situations that you never expected.

page 49

Suggestions for Success offers advice on study habits, ways to approach the coming material, and general observations on ways to think about mathematics.

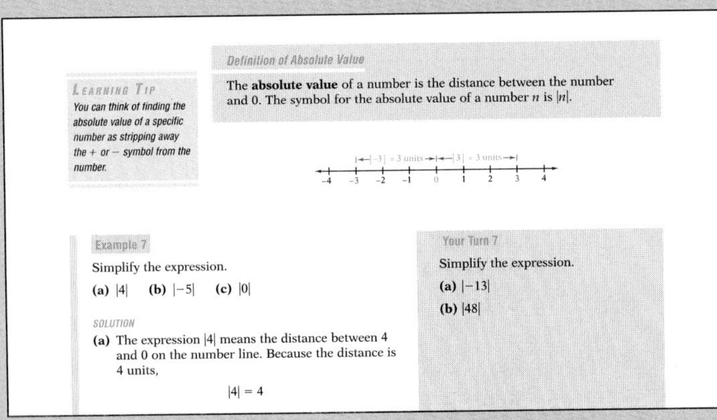

LEARNING TIP

You can think of finding the absolute value of a specific number as stripping away the + or − symbol from the number.

Definition of Absolute Value

The **absolute value** of a number is the distance between the number and 0. The symbol for the absolute value of a number n is $|n|$.

Example 7

Simplify the expression.

(a) $|4|$ **(b)** $|-5|$ **(c)** $|0|$

SOLUTION

(a) The expression $|4|$ means the distance between 4 and 0 on the number line. Because the distance is 4 units,

$$|4| = 4$$

Your Turn 7

Simplify the expression.

(a) $|-13|$

(b) $|48|$

page 192

Learning Tips are scattered throughout the text. Each one relates to a specific explanation or example. These tips offer cautions, alternative methods, and suggestions for a better understanding.

CHAPTER 1

Introduction to Algebra: Addition and Subtraction

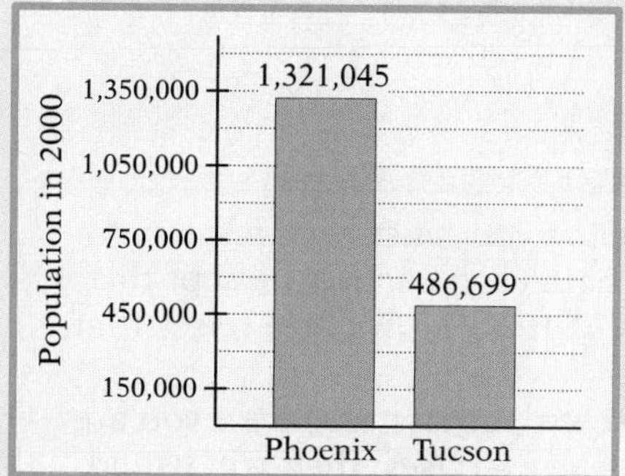

The bar graph shows the populations of the two largest cities in Arizona in 2000.

1. What was the total population of these two cities?
2. What was the difference in the populations of these two cities?

Chapter Snapshot

The answers to the questions on the left can be found by addition (Question 1) and by subtraction (Question 2). In this chapter, we will discuss both of those operations for the whole numbers 0, 1, 2, 3, 4, ... Along the way, we show you how to round numbers and estimate results, how to evaluate expressions, and how to solve simple equations. This chapter is an important beginning in your study of prealgebra.

Copyright © Houghton Mifflin Company. All rights reserved.

For online resources, visit the web site **math.college.hmco.com/students** and follow the links to Hubbard/Robinson, *Prealgebra*.

Some Friendly Advice . . .

The following is an actual conversation between one of the authors and a student.

STUDENT: *Do you know what we will be doing in class today?*

INSTRUCTOR: *Yes, I do. Do you?*

STUDENT: *Not a clue.*

INSTRUCTOR: *Did you look at the syllabus? It tells you what we are doing today and every other day. It also has your homework assignments.*

STUDENT: *Oh, I lost that a long time ago.*

We know that you aren't like this student. You usually receive your syllabus on the first day of class. It should be right at the front of your notebook so that you can refer to it every day. The syllabus usually lists the topics for every class, the homework that you are expected to do, and the dates on which tests will be given. Many instructors also use the syllabus to explain the objectives and requirements of the course, to make suggestions for success, and to provide other information that you really do need to know.

Reading your syllabus carefully, asking questions about it, and referring to it daily are excellent first steps in being aggressively involved in your studies.

WARM-UP SKILLS

To be successful with the material in a new chapter, you must have mastered certain skills from the preceding chapters. We begin each chapter (except this one) with a feature called Warm-Up Skills, a set of 10 to 12 brief exercises that you can use to test yourself.

If you can do all of the Warm-Up Skills exercises successfully, then you are well prepared. If you have difficulty with any of the exercises, then you would be wise to review that material before you proceed to the new chapter. You will find the answers to all the Warm-Up Skills in the back of the book.

There is nothing to review for Chapter 1, so instead of Warm-Up Skills, we have created a list of important questions that will help you to get organized and to guide you in preparing for this course. You will find these questions at our web site.

Copyright © Houghton Mifflin Company. All rights reserved.

1.1 Whole Numbers and Place Value

A *The Whole Numbers and Their Order*
B *Digits and Place Value*
C *Word Names for Numbers*

SUGGESTIONS FOR SUCCESS

Like any other subject that you study, mathematics has its own vocabulary. Throughout this book, you will find words in **bold face** print. We highlight words in this way to signal vocabulary that you should know. ("Thingy" and "whatcha-callit" are not good substitutes.)

As you go through the book, consider making a list of important words along with their meanings and reviewing it frequently. Getting together with fellow students and quizzing each other on the meanings of those words is a great study idea.

Much of your success will depend on your ability to use the correct words as you think, speak, and write about mathematics.

A *The Whole Numbers and Their Order*

A child learns to count at an early age. Learning to count begins with associating a *number* with a group of objects. For example, if you ask a three-year-old child how old he or she is, the child may hold up three fingers.

The familiar numbers

$$1, 2, 3, 4, 5, 6, \ldots$$

are called the **counting numbers** or, more formally, the **natural numbers.** The three dots at the end of the list indicate that the list of numbers continues forever.

If we combine 0 and the natural numbers, we have a new list of numbers called the **whole numbers.**

$$0, 1, 2, 3, 4, 5, 6, \ldots$$

A useful way to picture the whole numbers is with a **number line.**

0 1 2 3 4 5 6 7 8

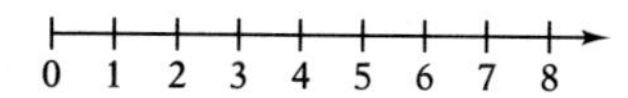

You can think of a number line as a row of hooks on which we hang the whole numbers in order. The arrow on the right indicates that the number line continues forever.

To **graph** a number on the number line, we highlight the number with a dot.

Example 1

Graph the whole numbers 2 and 5.

SOLUTION

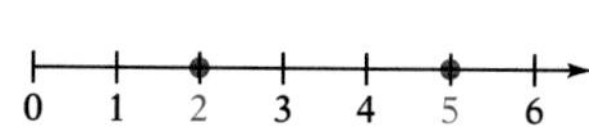

Your Turn 1

Graph the whole numbers 1 and 3.

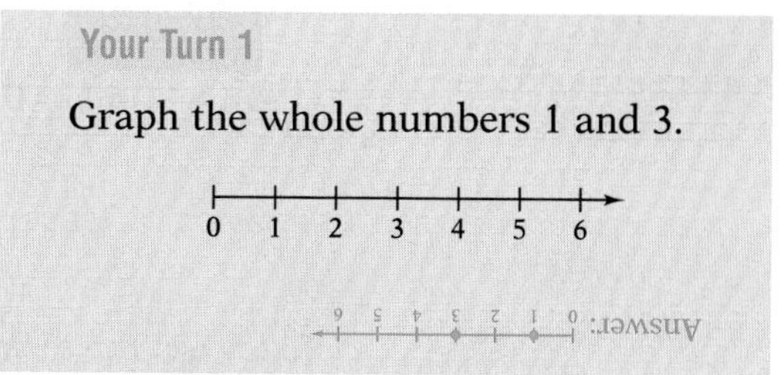

An important use of the number line is to help us see the **order** of numbers. As we move to the right along the number line, the numbers are larger and larger.

Copyright © Houghton Mifflin Company. All rights reserved.

As we move to the left, the numbers are smaller and smaller. Therefore, if we select any two numbers on the number line, we can see which is the smaller and which is the larger.

In the figure for Example 1, we see that 2 is to the left of 5. To write this fact, we use the symbol $<$.

LEARNING TIP

Symbols are just shorthand for words. When you see the symbol, say the words.

$$2 < 5$$ Read as "2 *is less than* 5."

We can also indicate that 5 is to the right of 2 with the symbol $>$.

$$5 > 2$$ Read as "5 *is greater than* 2."

The statements $2 < 5$ and $5 > 2$ are called **inequalities.**

Example 2

Insert $<$ or $>$ to make the statement true.

(a) 7 ▭ 12 **(b)** 10 ▭ 3

SOLUTION

(a)

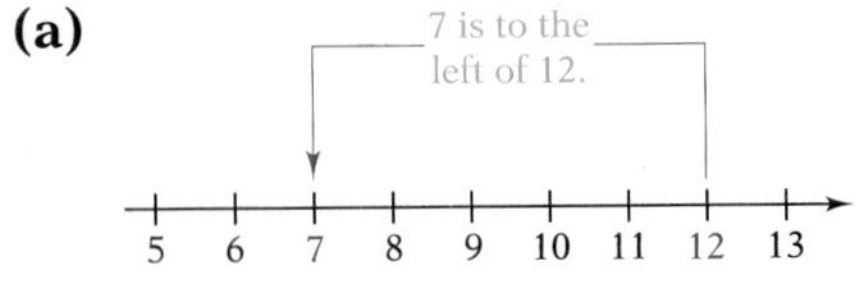

$$7 < 12$$

(b)

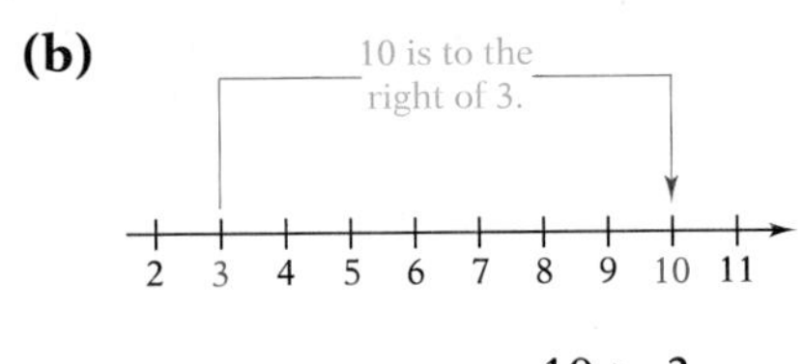

$$10 > 3$$

Your Turn 2

Insert $<$ or $>$ to make the statement true.

(a) 9 ▭ 1

(b) 6 ▭ 15

LEARNING TIP

The symbols $<$ and $>$ point to the smaller number.

Answers: (a) $9 > 1$; (b) $6 < 15$

Example 3

Arrange the following numbers from smallest to largest: 17, 2, 29, 31, 0, 8.

SOLUTION

The order is 0, 2, 8, 17, 29, 31.

Your Turn 3

Arrange the following numbers from *largest to smallest:*

1, 11, 6, 17, 32, 14

Answer: 32, 17, 14, 11, 6, 1

On a number line, the *distance* between 0 and 1 is 1 unit.

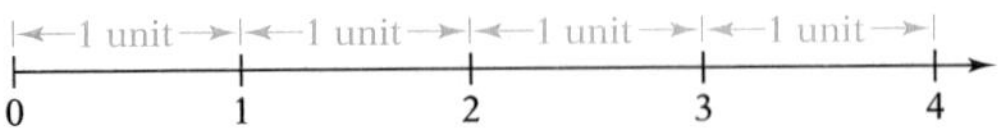

Likewise, the distance between 1 and 2, between 2 and 3, and so on, is 1 unit. To find the distance between any two whole numbers, we count the number of units from one to the other.

Copyright © Houghton Mifflin Company. All rights reserved.

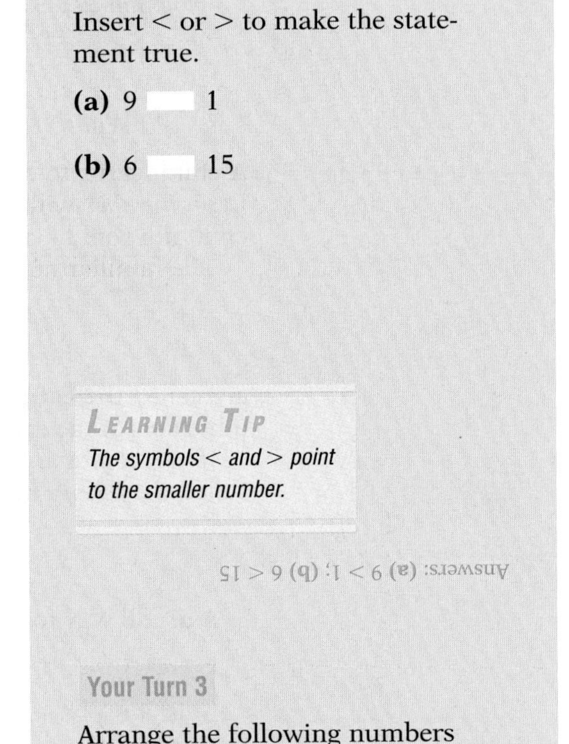

Example 4

On the number line, what is the distance between 3 and 8?

9 10

8 to 3, we find

Your Turn 4

On a number line, what is the distance between 9 and 1?

Answer: 8

nd Place Value

er in your car uses the following **digits** to tell you the number of
e car has been driven.

0, 1, 2, 3, 4, 5, 6, 7, 8, 9

lisplayed by your odometer has a **place value.** For example, if your
eading is

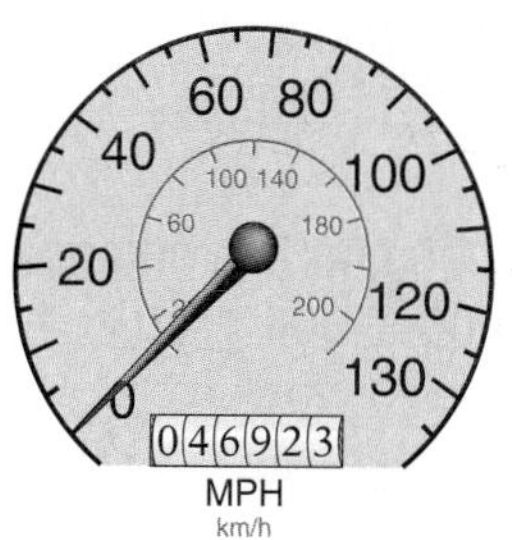

then the digits have the place values shown in the following chart:

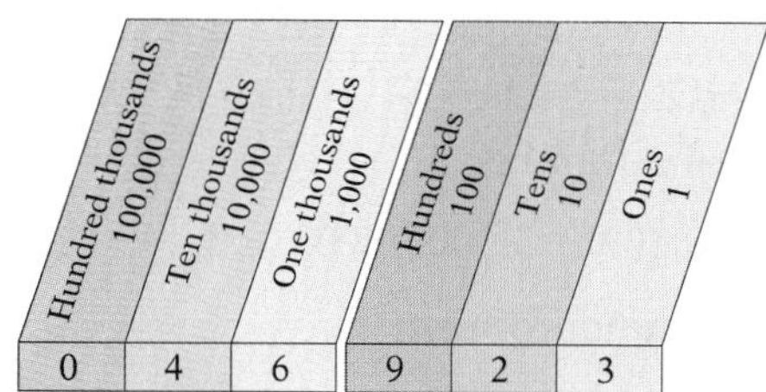

Digit	*Place Value*	*Number*
0	Hundred thousands	—
4	Ten thousands	40,000
6	One thousands	6,000
9	Hundreds	900
2	Tens	20
3	Ones	3

To make large numbers easier to read, we use commas to separate the digits into groups of three. For example, the odometer reading would be written as

Copyright © Houghton Mifflin Company. All rights reserved.

46,923. This **standard form** of the number also helps us to see the place values of the digits.

Referring to the place-value chart, we can also write the number in **expanded form.**

$$46{,}923 = 40{,}000 + 6{,}000 + 900 + 20 + 3$$

A place-value chart can be expanded for larger numbers. For example, the following chart shows the estimated population of the United States in 2002.

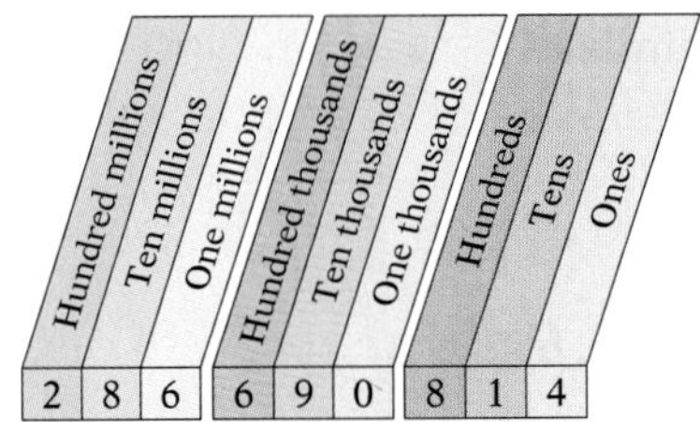

Example 5

Use the preceding chart to write the population (a) in standard form and (b) in expanded form.

SOLUTION

(a) The standard form is 286,690,814.

(b) The expanded form is

$$200{,}000{,}000 + 80{,}000{,}000 + 6{,}000{,}000 + 600{,}000 + 90{,}000 + 800 + 10 + 4$$

Your Turn 5

Suppose that your odometer reading is

Write this number (a) in standard form and (b) in expanded form.

Answers: (a) 102,596; (b) 100,000 + 2,000 + 500 + 90 + 6

Example 6

The expanded form of a number is

$$400{,}000 + 70{,}000 + 5{,}000 + 900 + 3$$

Write this number in standard form.

SOLUTION

The standard form is 475,903.

Your Turn 6

The expanded form of a number is

$$70{,}000 + 2{,}000 + 60 + 9$$

Write this number in standard form.

Answer: 72,069

C *Word Names for Numbers*

When you write a check, you must state the amount of the check both in standard form and in words. (The amount is in dollars and cents, but we will ignore the cents for now.)

Copyright © Houghton Mifflin Company. All rights reserved.

Jean Jones
10 Elm Street
Akron, OH 44260
23-125/03
517
0123
DATE May 16, 2003
PAY TO THE ORDER OF Monique's Boutique $ 28
Twenty-eight DOLLARS
USA BANK, N.A.
ALLIANCE, OH 02434 1-800-USA-BANK
FOR Jean Jones MP
I:045201291: 29 1294 8II: 0123

Note that the words for the number of tens and the number of ones are separated by a hyphen. For three-digit numbers, we start the word name with the number of hundreds.

Example 7

Write the given number in words.

(a) 32 **(b)** 741

SOLUTION

(a) Thirty-two

(b) Seven hundred forty-one

Your Turn 7

Write the given number in words.

(a) 83 **(b)** 485

Answers: (a) Eighty-three; (b) Four hundred eighty-five

Example 8

Write the following numbers in standard form.

(a) Fifty-eight

(b) Three hundred six

SOLUTION

(a) 58 **(b)** 306

Your Turn 8

Write the following numbers in standard form.

(a) Seventy-nine

(b) Nine hundred thirty-four

Answers: (a) 79; (b) 934

Now let's consider larger numbers such as 6,247 and 38,135 and 762,109. We start the word names for these numbers with the number of thousands.

Example 9

Write the given number in words.

(a) 6,247 **(b)** 38,135 **(c)** 762,109

SOLUTION

(a) Six thousand two hundred forty-seven

(b) Thirty-eight thousand one hundred thirty-five

(c) Seven hundred sixty-two thousand one hundred nine

Your Turn 9

Write the given number in words.

(a) 2,154

(b) 59,783

(c) 102,660

Answers: (a) Two thousand one hundred fifty-four; (b) Fifty-nine thousand seven hundred eighty-three; (c) One hundred two thousand six hundred sixty

Copyright © Houghton Mifflin Company. All rights reserved.

Example 10

Write the number one thousand twenty-five in standard form.

SOLUTION

The standard form is 1,025.

Your Turn 10

Write the number twenty-six thousand five hundred three in standard form.

Answer: 26,503

Word names for even larger numbers can be quite long, but the same scheme is followed. In writing the word names of such numbers, you may find it helpful to refer to the following place-value chart:

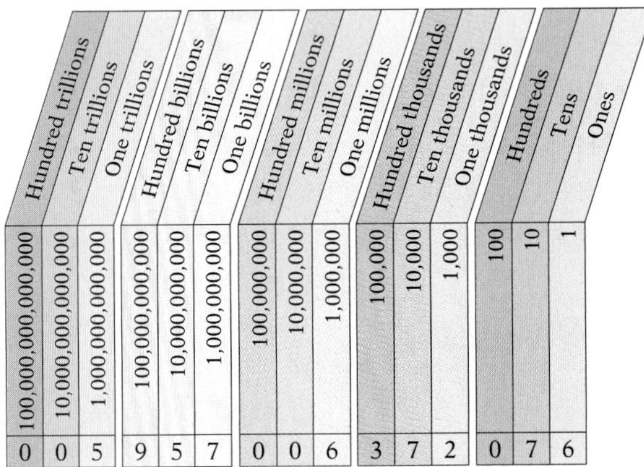

NOTE In case you are wondering, the number shown in the preceding chart was the U.S. national debt in March 2002.

Example 11

Write the given number in words.

(a) 6,284,713

(b) 539,401,287

(c) 32,893,741,225

SOLUTION

(a) Six million two hundred eighty-four thousand seven hundred thirteen

(b) Five hundred thirty-nine million four hundred one thousand two hundred eighty-seven

(c) Thirty-two billion eight hundred ninety-three million seven hundred forty-one thousand two hundred twenty-five

Your Turn 11

Write the given number in words.

(a) 73,591,280

(b) 2,046,500,739

(c) 800,000,000,000

Answers: **(a)** Seventy-three million five hundred ninety-one thousand two hundred eighty; **(b)** Two billion forty-six million five hundred thousand seven hundred thirty-nine; **(c)** Eight hundred billion

Copyright © Houghton Mifflin Company. All rights reserved.

Example 12

Write the number eight million five thousand eight hundred twelve in standard form.

SOLUTION

The standard form is 8,005,812.

Your Turn 12

Write the number fifteen million eighteen thousand forty-three in standard form.

Answer: 15,018,043

1.1 Quick Reference

A. The Whole Numbers and Their Order

1. The **natural numbers** are 1, 2, 3, 4, 5, ...
2. The **whole numbers** are 0, 1, 2, 3, 4, ...

3. We can **graph** a whole number on the **number line.**

Graph of 2 on a number line

4. A number line can be used to see the **order** of numbers.

The order is written as an **inequality.**

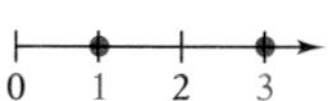

$1 < 3$ (1 *is less than* 3)

$3 > 1$ (3 *is greater than* 1)

5. The *distance* between two numbers on a number line is found by counting the number of units between the numbers.

Distance between 1 and 4

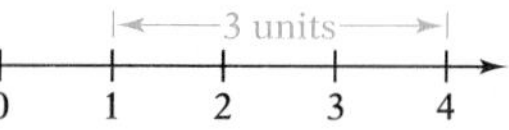

B. Digits and Place Value

1. The **digits** are

0, 1, 2, 3, 4, 5, 6, 7, 8, 9

In a whole number, digits have a **place value.**

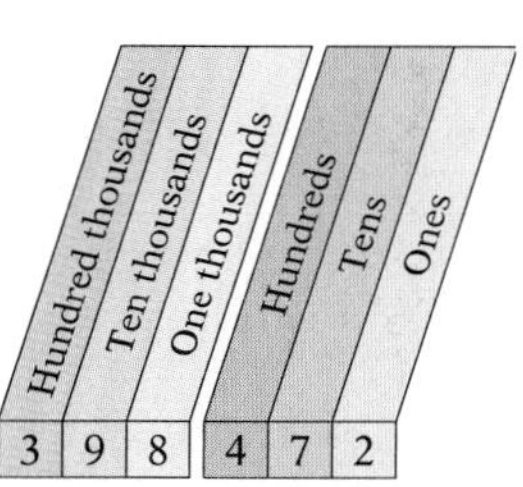

2. A whole number is usually written in **standard form,** but it can also be written in **expanded form.**

Standard form: 6,943

Expanded form:

6,000 + 900 + 40 + 3

Copyright © Houghton Mifflin Company. All rights reserved.

C. Word Names for Numbers

1. Whole numbers can be written in words.

53: fifty-three

425: four hundred twenty-five

6,943: six thousand nine hundred forty-three

Copyright © Houghton Mifflin Company. All rights reserved.

1.1 Exercises

A. The Whole Numbers and Their Order

1. A useful way to picture the whole numbers is with a(n) ______.

2. For $4 < 9$, the symbol $<$ is read "______," whereas for $9 > 4$, the symbol $>$ is read "______."

In Exercises 3–6, graph the numbers on the number line.

3. 3, 7

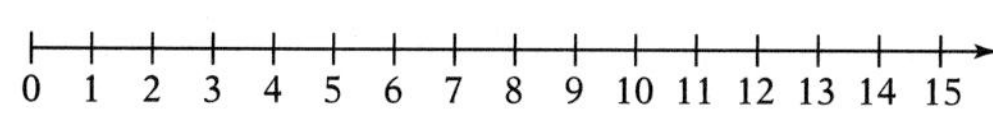

4. 2, 5

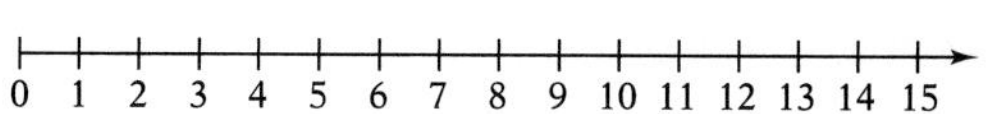

5. 6, 0, 13, 9

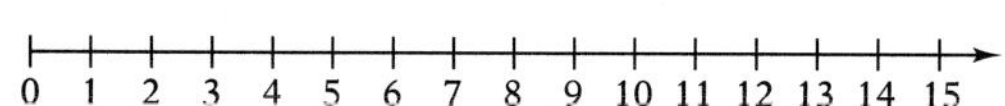

6. 8, 4, 7, 11

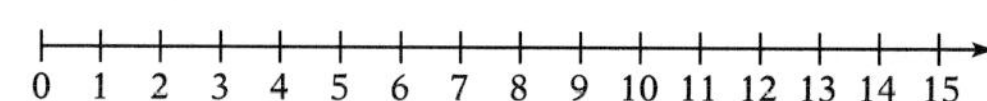

In Exercises 7–14, insert $<$ or $>$ to make the statement true.

7. 0 ______ 6
8. 4 ______ 12
9. 8 ______ 5
10. 3 ______ 0
11. 12 ______ 15
12. 23 ______ 22
13. 310 ______ 307
14. 403 ______ 405

In Exercises 15–18, arrange the numbers in the indicated order.

15. 32, 41, 21, 12, 3, 20
largest to smallest

16. 25, 22, 76, 9, 15, 52
smallest to largest

17. 213, 210, 245, 117, 312, 101
smallest to largest

18. 96, 149, 502, 490, 69, 43
largest to smallest

In Exercises 19–22, determine the distance between the two numbers on the number line.

19. 6, 10
20. 12, 18
21. 14, 9
22. 20, 17

B. Digits and Place Value

23. The number 1,527 is in the ______ form, and the ______ form is $1{,}000 + 500 + 20 + 7$.

24. In 3,000,000, the 3 represents ______, whereas in 3,000,000,000, the 3 represents ______.

Copyright © Houghton Mifflin Company. All rights reserved.

In Exercises 25–32, write the number in expanded form.

25. 536

26. 3,207

27. 40,690

28. 26,003

29. 796,000

30. 402,436

31. 16,200,037

32. 396,904,370

In Exercises 33–40, write the number in standard form.

33. 600 + 40 + 3

34. 1,000 + 300 + 20 + 6

35. 10,000 + 4,000 + 6

36. 30,000 + 9,000 + 400 + 20

37. 600,000 + 20,000 + 40 + 9

38. 100,000 + 3,000 + 800 + 30 + 2

39. 600,000,000 + 60,000 + 2,000 + 500 + 10 + 3

40. 90,000,000 + 7,000,000 + 200,000 + 60 + 5

C. Word Names for Numbers

In Exercises 41–48, write the number in words.

41. 54

42. 78

43. 407

44. 816

45. 8,014

46. 570,402

47. 3,290,700

48. 6,042,403,800

In Exercises 49–52, write the number in each sentence in words.

49. A mile is 5,280 feet.

50. The price of a car is 18,390 dollars.

51. The area of Alaska is 579,833 square miles.

Copyright © Houghton Mifflin Company. All rights reserved.

52. The vice president's salary is 186,300 dollars.

In Exercises 53–60, write the number in standard form.

53. Ninety-three

54. Thirty-six

55. Six hundred two

56. Five hundred thirteen

57. Nine hundred ninety

58. Eight thousand seven

59. Forty-two million eight

60. Seventeen billion two hundred million forty thousand two hundred

In Exercises 61–66, write the number in each sentence in standard form.

61. *TV Viewing* The average American family watches TV four hundred sixty-two minutes each day.

62. *Record Temperature* The highest temperature recorded in Death Valley is one hundred thirty-four degrees.

63. *Cable TV Subscribers* Fifty million four hundred fifty-five thousand people subscribe to cable TV.

64. *The Planet Pluto* The distance between the Sun and Pluto is three billion six hundred seventy-five million miles.

65. *Bank Check* A person writes a check for two thousand three hundred fifty-seven dollars.

66. *Area of Pennsylvania* The area of Pennsylvania is forty-four thousand three hundred thirty-three square miles.

Copyright © Houghton Mifflin Company. All rights reserved.

Writing and Concept Extension

67. Explain why there are two numbers that are 3 units from 9. What are the two numbers?

68. Explain why there is only one whole number that is 3 units from 1.

69. What number is 2 units to the right of 5?

70. What number is 4 units to the left of 10?

71. What is the smallest three-digit number?

72. What is the largest five-digit number?

Copyright © Houghton Mifflin Company. All rights reserved.

1.2 Data Interpretation and Presentation

A *Rounding Numbers*
B *Bar Graphs*

You will not please most instructors by asking, "Do we have to know this stuff?" or "Will this be on the test?" Even so, learning what your instructor's expectations are is an important part of your success in the course. There are many ways to do this, but here is the most important: Attend classes without fail!

We hope and expect that you will read your textbook, but cutting class means that you can't know what material the instructor has discussed and emphasized. (Additional hint: If you do cut a class, do not ask your instructor whether you missed anything important!)

A *Rounding Numbers*

On August 28, 1963, Dr. Martin Luther King, Jr., delivered his famous "I have a dream" speech during the March on Washington. About 250,000 people gathered at the Lincoln Memorial to hear the speech. Were there *exactly* 250,000 people in attendance? No, the reported number is only an **estimate.**

We use estimates, particularly of large numbers, when the exact number is not necessary. Typically, we estimate a number by **rounding** it to some selected place value.

Rounding Numbers

Look at the locations of 33 and 38 on the number line.

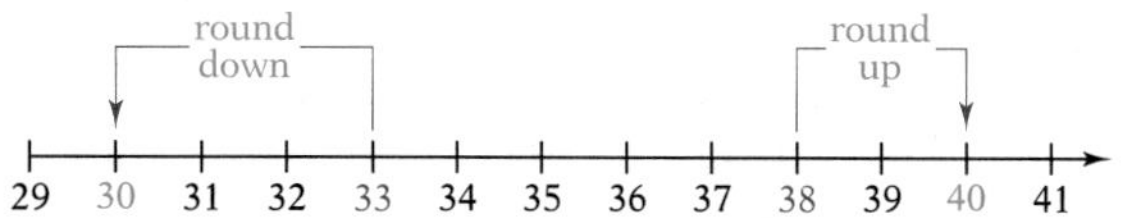

Suppose that we want to round these numbers to the nearest ten. We can see that 33 is closer to 30 than to 40, so we round 33 *down* to 30. On the other hand, 38 is closer to 40 than to 30, so we round 38 *up* to 40.

Note that 35 is exactly halfway between 30 and 40, so should we round 35 up or down? We simply agree to round 35 up to 40.

We can round numbers without having to draw a number line. Let's round an odometer reading of 46,753 to the nearest 10 miles.

Look here at the tens place.
↓
46,753
↑
Then look at the number to the right.

Because the 3 in the ones column is less than 5, we keep the 5 in the tens column and replace every digit to the right (in this case, the 3) with 0. The rounded number is 46,750.

Copyright © Houghton Mifflin Company. All rights reserved.

Now we round 46,753 to the nearest 100 miles.

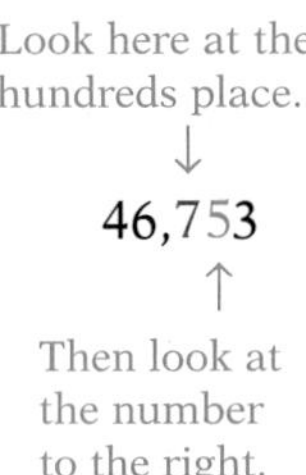

Because of the 5 in the tens column, we increase the 7 to 8 and replace every digit to the right (in this case, the 5 and 3) with 0. The rounded number is 46,800.

Next, we round 46,753 to the nearest 1,000 miles.

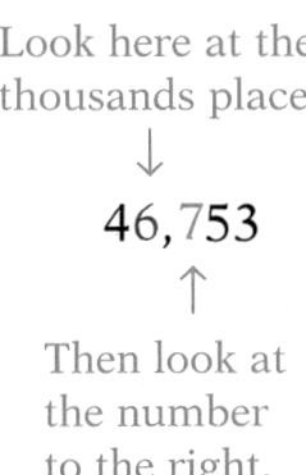

Because $7 > 5$, we increase the 6 to 7 and replace every digit to the right (in this case, the 7, 5, and 3) with 0. The rounded number is 47,000.

Finally, we round 46,753 to the nearest 10,000 miles.

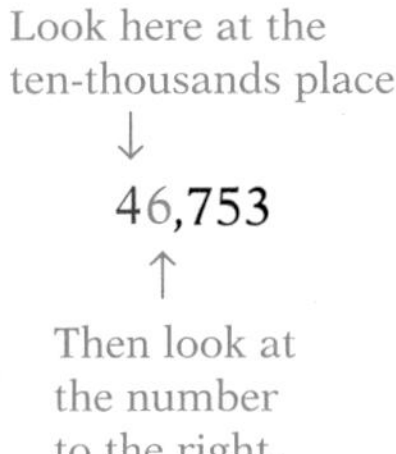

Because $6 > 5$, we increase the 4 to 5 and replace every digit to the right (in this case, the 6, 7, 5, and 3) with 0. The rounded number is 50,000.

We can summarize this method for rounding a number.

Rounding Numbers to a Given Place Value

To round a number to a given place value, locate the digit of the given place value. Then look at the digit to the right.

1. If the digit to the right is less than 5, replace it and every other digit to the right with 0.
2. If the digit to the right is 5 or greater, increase the digit of the given place value by 1. Then replace every digit to its right with 0.

Example 1

Round 195,499 to the nearest hundred.

Your Turn 1

Round 47,598 to the nearest thousand.

Copyright © Houghton Mifflin Company. All rights reserved.

SOLUTION

Here is the
hundreds place.
↓
195,499
↑
Here is the number
to the right.

Because 9 > 5, we increase the 4 to 5 and replace every other digit to the right with 0. The rounded number is 195,500.

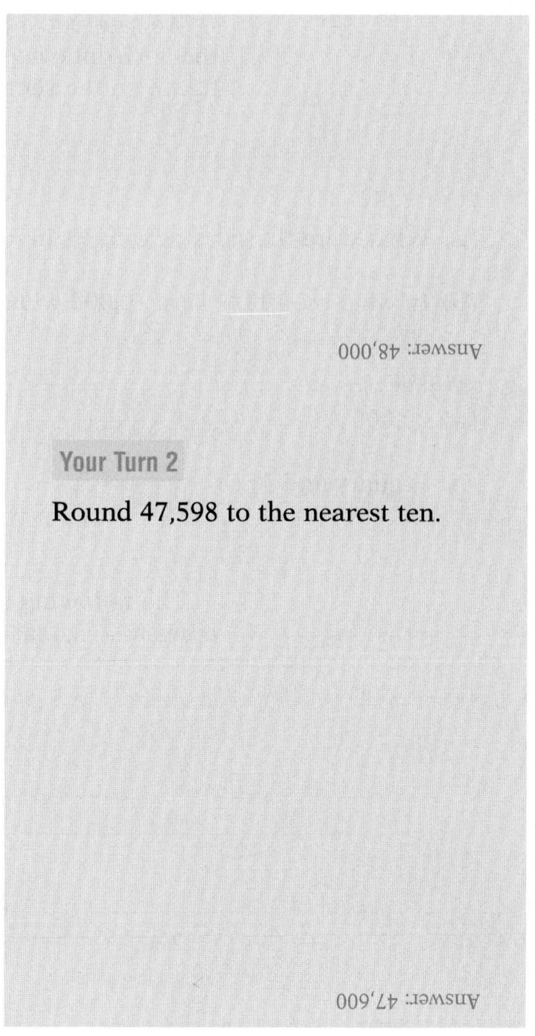

Example 2

Round 195,499 to the nearest ten thousand.

SOLUTION

Here is the
ten-thousands place.
↓
195,499
↑
Here is the number
to the right.

Be careful, this one is tricky. Because the number to the right of 9 is 5, we should increase the 9 by 1. But 9 is the largest digit, so it can't be increased. Instead, we increase the 19 to 20. The rounded number is 200,000.

Your Turn 2

Round 47,598 to the nearest ten.

Answer: 47,600

B Bar Graphs

A *bar graph* is a visual way of presenting information. The following bar graph shows the number of widgets sold by the ABC Corporation during the first five months of the year:

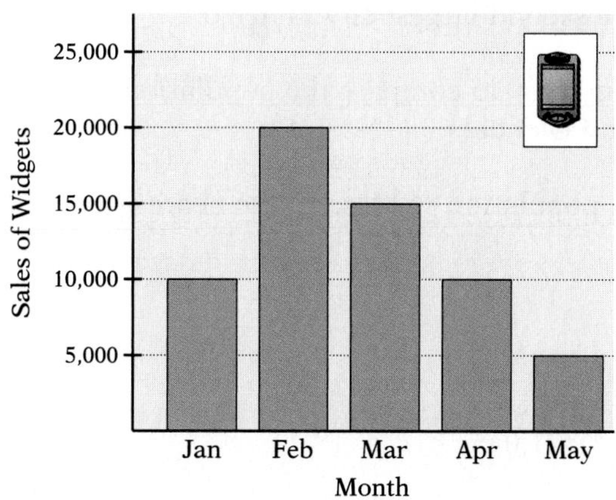

Note that the months are listed along the horizontal line (called the *horizontal axis*) and the sales of widgets are read from the vertical line (called the *vertical axis*).

Copyright © Houghton Mifflin Company. All rights reserved.

We can read a bar graph in either direction; that is, we can select a month on the horizontal axis and read the sales on the vertical axis, or we can select a sales figure on the vertical axis and read the month on the horizontal axis.

Example 3

(a) What were the sales of widgets in February?

(b) In what month(s) were 10,000 widgets sold?

SOLUTION

(a) 20,000

(b) January and April

Your Turn 3

(a) In what month(s) were 15,000 widgets sold?

(b) What were the sales of widgets in May?

Answers: (a) March; (b) 5,000

The following bar graph shows the year 2000 populations of the four largest cities in Arizona:

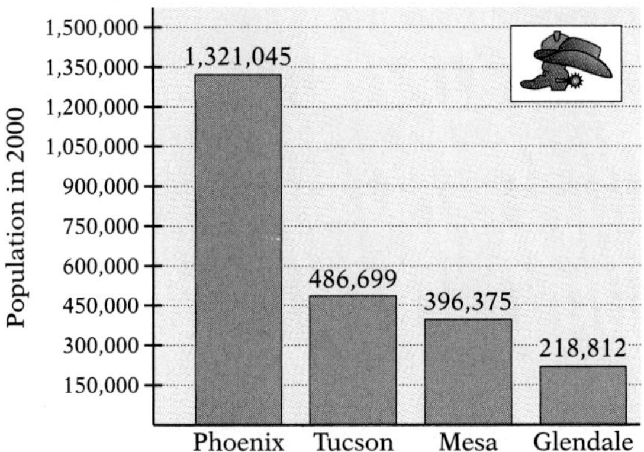

Bar graphs are especially helpful in making comparisons. For example, we can easily see from the heights of the bars that the largest city in Arizona is Phoenix.

Example 4

(a) What is the second-largest city in Arizona?

(b) Use the symbol $<$ to compare the populations of Phoenix and Glendale.

(c) Round the population of Mesa to the nearest ten thousand.

SOLUTION

(a) Tucson

(b) $218{,}812 < 1{,}321{,}045$

(c) 400,000

Your Turn 4

(a) What is the third-largest city in Arizona?

(b) Use the symbol $>$ to compare the populations of Mesa and Tucson.

(c) Round the population of Phoenix to the nearest thousand.

Answers: (a) Mesa; (b) $486{,}699 > 396{,}375$; (c) 1,321,000

Copyright © Houghton Mifflin Company. All rights reserved.

Bar graphs are also useful in showing trends. In a survey of 100 people in each of four different age groups, people were asked whether they believed in "love at first sight." The following bar graph illustrates the findings. (Source: Gallup Poll.)

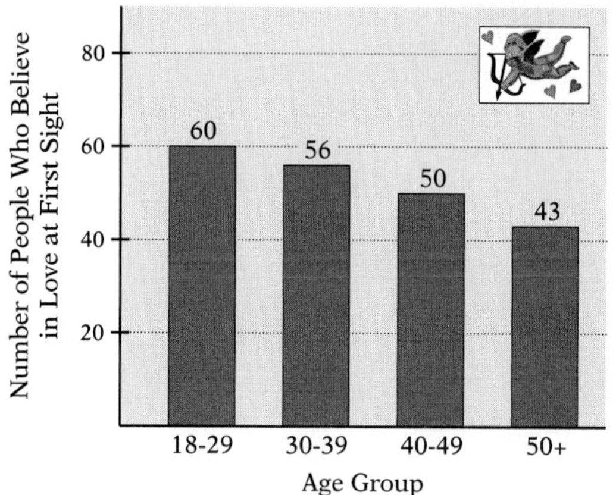

The bar graph clearly suggests that as people grow older, they are less likely to be romantically idealistic. When the bars of bar graphs continuously increase or decrease, such trends are easily visible.

To provide interest, a bar graph is sometimes displayed with horizontal bars rather than vertical bars. To do this in the bar graph above, the age groups would be shown on the vertical axis and the number of people would be shown on the horizontal axis. The disadvantage of such a presentation is that trends are not as evident. Bar graphs with horizontal bars can be used effectively when the data are meant for comparison, such as the ratings for a number of television shows.

A bar graph is not the only way in which data can be organized and visually presented. You may have seen *line graphs* that show the performance of the stock market and *pie graphs* that show how funds are allocated in the federal budget. The type of graph that you select will depend on the nature of the data and the purpose for which the graph is to be used.

1.2 Quick Reference

A. Rounding Numbers

1. A number is **estimated** by **rounding** the number to some selected place value.

2. To round a number to a given place value, locate the digit of the given place value. Then look at the digit to the right.

   a. If the digit to the right is less than 5, replace it and every other digit to the right with 0.

   b. If the digit to the right is 5 or greater, increase the digit of the given place value by 1. Then replace every digit to its right with 0.

85,934 to the nearest:

Hundred: 85,900
Thousand: 86,000
Ten thousand: 90,000

Copyright © Houghton Mifflin Company. All rights reserved.

B. Bar Graphs

1. A *bar graph* is a visual way of presenting information. Bar graphs are particularly useful in making comparisons.

2. A bar graph can be read from the *horizontal axis* to the *vertical axis* or from the vertical axis to the horizontal axis.

Copyright © Houghton Mifflin Company. All rights reserved.

1.2 Exercises

A. Rounding Numbers

1. To round 73,612 to the nearest thousand, we replace 3 with 4, and we replace the 6, 1, and 2 with ____.

2. We say that 8,600 is 8,592 rounded to the nearest ________.

In Exercises 3–14, round the number to the indicated place value.

3.	783	Hundreds	**4.**	342	Tens
5.	4,384	Tens	**6.**	3,643	Thousands
7.	15,503	Thousands	**8.**	12,490	Thousands
9.	93,958	Hundreds	**10.**	60,963	Hundreds
11.	537,206	Ten thousands	**12.**	145,329	Ten thousands
13.	8,340,217	Millions	**14.**	99,499,347	Millions

B. Bar Graphs

Average Low Temperatures The bar graph shows the average low temperatures (in °F) for the summer months in a certain city.

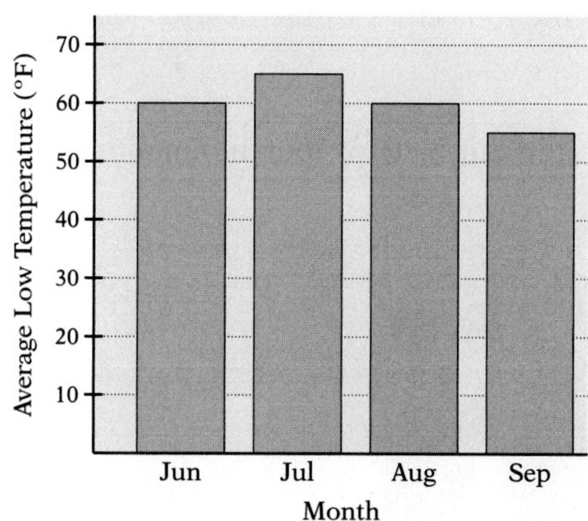

15. For what month is the average low temperature the highest?

16. What is the average low temperature for September?

Copyright © Houghton Mifflin Company. All rights reserved.

17. For what months are the average low temperatures the same?

18. For what months is the average low temperature at least 60°?

Louisiana Cities The bar graph shows the populations of four cities in Louisiana in 2000. (Source: *World Almanac*.)

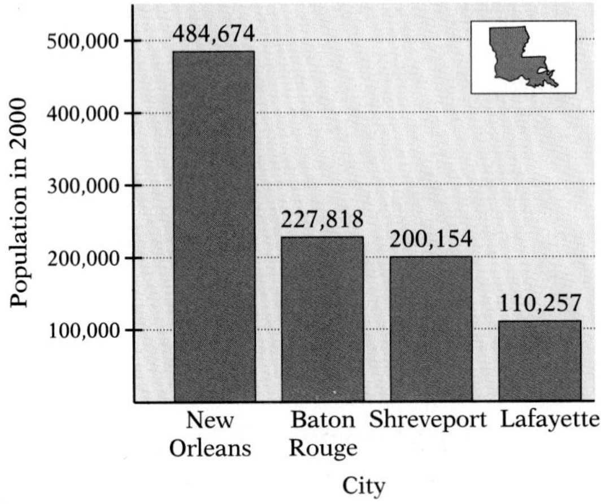

19. Use the symbol < to compare the populations of Baton Rouge and Shreveport.

20. For what city (cities) is the population less than 200,000?

21. Round the population of the largest of the four cities to the nearest hundred thousand.

22. Round the population of the smallest of the four cities to the nearest ten thousand.

Alaska Mountains The bar graph shows the heights (in feet) of four mountains in Alaska.

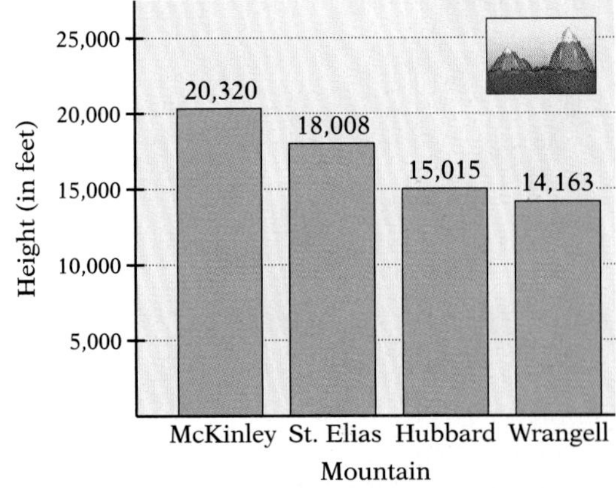

Copyright © Houghton Mifflin Company. All rights reserved.

23. When the heights are rounded to the nearest ten thousand, which mountains have the same height?

24. Which mountains are at least 14,000 feet high?

25. Use the symbol > to compare the heights of Hubbard and St. Elias.

26. Round the height of St. Elias to the nearest thousand.

Writing and Concept Extension

27. Which number would be more reasonable to round to the nearest million: the distance between Earth and the Sun or the number of students at your college? Why?

28. An item sells for $549 at one store and for $451 at another store. Would a wise shopper round these prices to the nearest hundred dollars in order to decide where to buy the item? Why?

29. Round 29,998 to the nearest (a) ten, (b) hundred, (c) thousand, and (d) ten thousand.

Exploring with Real-World Data: Collaborative Activities

Employment Opportunities The bar graph shows the four occupations that are projected to be growing the fastest by the year 2005. (Source: U.S. Bureau of Census.)

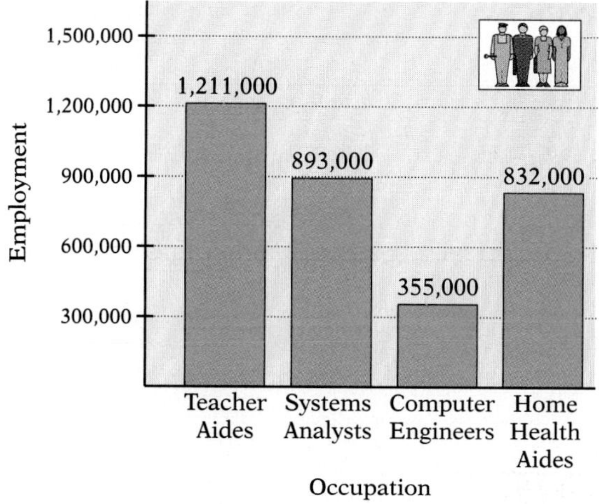

30. What is the projected employment for home health aides?

Copyright © Houghton Mifflin Company. All rights reserved.

31. For what occupation is the projected employment the highest?

32. For what occupations are the projected employment numbers at least 850,000?

33. To what place value do the projected employment numbers appear to be rounded?

34. Use the symbol < to compare the projected employment numbers for systems analysts and home health aides.

35. The graph shows that computer engineering has the lowest projected employment number of the four occupations shown. Would you conclude, therefore, that computer engineering would not be a promising career? Why or why not?

Copyright © Houghton Mifflin Company. All rights reserved.

1.3 Algebra of Addition

A *Addition Concepts and Vocabulary*
B *Variables and Properties of Addition*

SUGGESTIONS FOR SUCCESS

In this section, you will add some words to your vocabulary list and you will learn the first of many rules. These words and rules should be an important part of your notes, but to assist you, we provide a Quick Reference at the end of every section.

Each Quick Reference is a summary of the key points presented in the section, along with examples to illustrate those points. You should carefully work through the Quick Reference before you try the exercises. This will help you to spot any concepts or procedures that are not clear to you. Remember: If you don't understand something, it will almost certainly be on the test!

A Addition Concepts and Vocabulary

At a gardening store, bags of pine bark nuggets and mulch are displayed on pallets. There are currently 38 bags of nuggets and 47 bags of mulch.

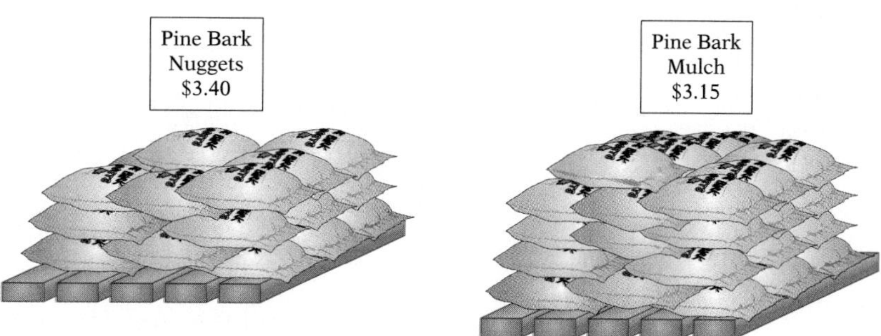

One way to determine the total number of bags on the two pallets is to count the bags. A faster way is to *add* the number of bags on one pallet to the number of bags on the other pallet: 38 + 47. **Addition** is the process of finding the total of two or more numbers.

These numbers are called **addends.**

$$38 + 47 = 85$$

This number is called the **sum.**

The numbers being added are called **addends.** The addition itself and the result of the addition are both called the **sum.** For example, in the addition 5 + 3 = 8, the 5 + 3 and the 8 are both called a sum.

Addition is important because we can't always find a total by counting. For example, if you put 5 gallons of gasoline in your car and then decide to put in 7 more gallons, you can't crawl into the gas tank and count the total number of gallons. Instead, you can perform the addition 5 + 7 = 12.

The number line can help us to see addition. Here is how the number line can be used to find the sum 3 + 5.

Copyright © Houghton Mifflin Company. All rights reserved.

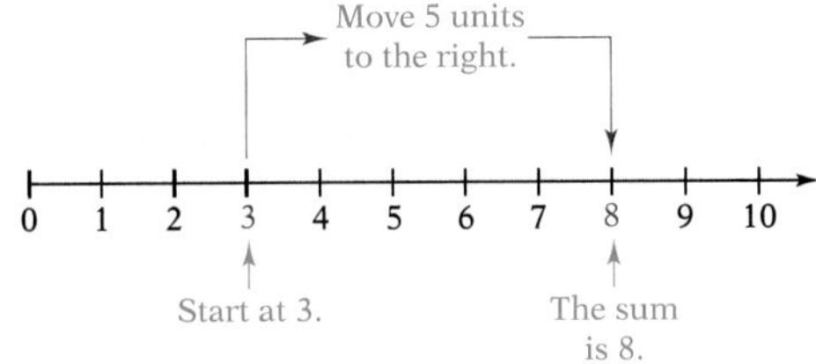

We call the sum of any two digits a *basic addition fact*. The following table summarizes the 100 basic addition facts:

+	0	1	2	3	4	5	6	7	8	9
0	0	1	2	3	4	5	6	7	8	9
1	1	2	3	4	5	6	7	8	9	10
2	2	3	4	5	6	7	8	9	10	11
3	3	4	5	6	7	8	9	10	11	12
4	4	5	6	7	8	9	10	11	12	13
5	5	6	7	8	9	10	11	12	13	14
6	6	7	8	9	10	11	12	13	14	15
7	7	8	9	10	11	12	13	14	15	16
8	8	9	10	11	12	13	14	15	16	17
9	9	10	11	12	13	14	15	16	17	18

LEARNING TIP

You must know the basic addition facts so well that you can state them with hardly any thought at all. If you need practice, try getting someone to help you with flash cards.

This table can be used to find the sum of any two digits. For example, to add 6 and 8, read across from the 6 in the left column and down from the 8 in the top row and find the entry 14: $6 + 8 = 14$.

To be a good problem solver, you must be able to recognize certain words and phrases that indicate addition. In the following list, the words in italics are examples of such words and phrases:

Word Description	*Sum*
The sum of 9 and 2	$9 + 2$
5 *plus* 8	$5 + 8$
6 *added to* 3	$3 + 6$
The total of 4 and 1	$4 + 1$
7 *more than* 8	$8 + 7$
2 *increased by* 5	$2 + 5$

Example 1

The original price of \$35 was increased by \$4. Write a sum to represent the new price.

SOLUTION

In symbols, "35 increased by 4" is

$$35 + 4$$

Your Turn 1

Today's price is \$6 more than yesterday's price of \$17. Write a sum to represent today's price.

Answer: 17 + 6

Copyright © Houghton Mifflin Company. All rights reserved.

B Variables and Properties of Addition

We turn now to one of the most important concepts in mathematics, the idea of a *variable.*

DEVELOPING THE CONCEPT

Variables

Consider the following conversation:

"She is a student at the college."
"Who?"
"Mary."
"Oh, yes, that's true."

In the first sentence, the word *she* is a pronoun, which is used to represent some person. The listener could not tell whether the first statement was true until "she" was replaced with a specific name, Mary.

In mathematics, a variable is used in a similar way. A **variable** is a symbol, usually a letter, that represents some unknown number.

For the sum $x + 3$, the symbol x is a variable that represents some number. Until we know what the number is, we can't perform the addition. However, if we *replace* the variable x with a specific number, then we can carry out the addition.

$x + 3 = ?$ We can't add because we don't know what number x represents.
$\downarrow$
$5 + 3 = 8$ If we replace x with 5, then we can add.

One important way in which variables are used is to make general statements about numbers. Here is an example of what we mean.

In the table of 100 basic addition facts, 20 of the sums involve the number 0.

$0 + 0 = 0$	$0 + 5 = 5$	$0 + 0 = 0$	$5 + 0 = 5$
$0 + 1 = 1$	$0 + 6 = 6$	$1 + 0 = 1$	$6 + 0 = 6$
$0 + 2 = 2$	$0 + 7 = 7$	$2 + 0 = 2$	$7 + 0 = 7$
$0 + 3 = 3$	$0 + 8 = 8$	$3 + 0 = 3$	$8 + 0 = 8$
$0 + 4 = 4$	$0 + 9 = 9$	$4 + 0 = 4$	$9 + 0 = 9$

In every case, we find that whenever we add a number and 0, the sum is the original number. We call this general fact a *property* of addition involving the number 0. Discovering a property such as this gives us a better understanding of numbers and reduces the amount of memorizing we must do.

The Addition Property of 0

If the variable a is used to represent any number, then

$$a + 0 = a$$
and
$$0 + a = a$$

In words, the sum of 0 and any number is the number.

Copyright © Houghton Mifflin Company. All rights reserved.

Example 2

Find the sum.

(a) $9 + 0$ **(b)** $0 + 4$ **(c)** $x + 0$

SOLUTION

(a) $9 + 0 = 9$ **(b)** $0 + 4 = 4$ **(c)** $x + 0 = x$

Your Turn 2

Find the sum.

(a) $0 + 7$

(b) $0 + 0$

(c) $0 + y$

Answers: (a) 7; (b) 0; (c) y

We can use the table of basic addition facts to discover another important property of addition. Consider the following sums:

$$5 + 8 = 13 \quad \text{and} \quad 8 + 5 = 13$$
$$9 + 2 = 11 \quad \text{and} \quad 2 + 9 = 11$$
$$6 + 4 = 10 \quad \text{and} \quad 4 + 6 = 10$$

Do you see that the order in which we add the two numbers has no effect on the sum? We can state this fact, in general, as a property.

LEARNING TIP

A commuter is a person who travels back and forth. Think of this as a way of remembering the Commutative Property of Addition.

The Commutative Property of Addition

If the variables a and b are used to represent any two numbers, then

$$a + b = b + a$$

In words, two numbers can be added in either order and the sum is the same.

One practical use of the Commutative Property of Addition is that we need to memorize only half of the basic addition facts. For example, if you know that $9 + 7 = 16$, then you also know that $7 + 9 = 16$.

Example 3

Use the Commutative Property of Addition to rewrite the sum.

(a) $8 + 4$

(b) $x + 3$

(c) $y + z$

SOLUTION

The Commutative Property of Addition allows us to change the order of the addends.

(a) $8 + 4 = 4 + 8$

(b) $x + 3 = 3 + x$

(c) $y + z = z + y$

Your Turn 3

Use the Commutative Property of Addition to rewrite the sum.

(a) $9 + 3$

(b) $5 + y$

(c) $x + a$

Answers: (a) $3 + 9$; (b) $y + 5$; (c) $a + x$

Copyright © Houghton Mifflin Company. All rights reserved.

So far, we have added only two digits at a time. Even if we need to add three or more digits, our brains are designed to add only two at a time. But with which two numbers should we begin the addition?

DEVELOPING THE CONCEPT

Sums of Three or More Digits

Suppose we need to perform the addition $1 + 5 + 2$. We use parentheses (called *grouping symbols*) to indicate the two digits we will add first.

Begin with the first two digits.	Begin with the last two digits.
$(1 + 5) + 2$	$1 + (5 + 2)$
$= 6 + 2$	$= 1 + 7$
$= 8$	$= 8$

We see that the grouping of the numbers has no effect on the sum. In both groupings, the sum is 8.

In general, we can group (or *associate*) addends any way we want without affecting the sum.

The Associative Property of Addition

If the variables a, b, and c are used to represent any three numbers, then

$$(a + b) + c = a + (b + c)$$

In words, we can group the addends of a sum in any way and the sum is the same.

Example 4

Use the Associative Property of Addition to rewrite the sum.

(a) $(5 + 7) + 2$ **(b)** $3 + (6 + 1)$

(c) $(x + 4) + 4$ **(d)** $5 + (a + y)$

SOLUTION

The Associative Property of Addition allows us to change the grouping of the addends.

(a) $(5 + 7) + 2 = 5 + (7 + 2)$

(b) $3 + (6 + 1) = (3 + 6) + 1$

(c) $(x + 4) + 4 = x + (4 + 4)$

(d) $5 + (a + y) = (5 + a) + y$

Your Turn 4

Use the Associative Property of Addition to rewrite the sum.

(a) $4 + (6 + 3)$

(b) $(8 + 1) + 5$

(c) $(3 + x) + 7$

(d) $a + (b + 2)$

Answers: **(a)** $(4 + 6) + 3$; **(b)** $8 + (1 + 5)$; **(c)** $3 + (x + 7)$; **(d)** $(a + b) + 2$

Copyright © Houghton Mifflin Company. All rights reserved.

1.3 Quick Reference

A. Addition Concepts and Vocabulary

1. In addition, the numbers being added are called **addends;** the indicated addition and the result are called the **sum.**	For $7 + 5 = 12$, the 7 and 5 are addends; both the $7 + 5$ and the 12 are called sums.
2. A number line can be used to represent addition.	
3. The sum of any two digits is called a *basic addition fact.*	
4. Phrases such as "increased by" and "more than" indicate addition.	"8 increased by 3": $8 + 3$ "4 more than 9": $9 + 4$

B. Variables and Properties of Addition

1. A **variable** is a symbol, usually a letter, that represents some unknown number.	
2. If a sum contains a variable, we cannot add until the variable is replaced with a specific value.	
3. Variables can be used to state a general rule or *property* of numbers.	
4. The Addition Property of 0 If a is any number, then $a + 0 = 0 + a = a$.	$3 + 0 = 3$ and $0 + 8 = 8$
5. The Commutative Property of Addition If a and b are any numbers, then $a + b = b + a$	$4 + 7 = 7 + 4$
6. The Associative Property of Addition If a, b, and c are any numbers, then $(a + b) + c = a + (b + c)$	$(3 + 5) + 4 = 3 + (5 + 4)$

Copyright © Houghton Mifflin Company. All rights reserved.

1.3 Exercises

A. Addition Concepts and Vocabulary

1. We call 5 + 9 a(n) sum, and the numbers 5 and 9 are called addends

2. The phrases "the total of" and "the sum of" indicate addition

In Exercises 3–8, write the information as a sum.

3. 5 more than 6

4. y plus 8

5. The sum of x and 4

6. The total of 12 and 5

7. 3 increased by 2

8. 4 added to n

In Exercises 9–12, write the information as a sum (omit units).

9. *Medication Dosage* A patient receives 7 milligrams of a medication. After the dose is increased by 4 milligrams, what is the dosage?

10. *Hourly Wage* A worker earns $6 per hour. After the amount is increased by $2 per hour, what is the hourly wage?

11. *Low Temperatures* Today's low temperature is 9 degrees more than yesterday's low temperature of 36 degrees. What is today's low temperature?

12. *Ticket Cost* What is the total cost of a $20 ticket plus a $3 service charge?

B. Variables and Properties of Addition

13. A letter that represents an unknown number is called a(n) variable

14. A rule that is true for all whole numbers is called a(n) ______ of the whole numbers.

15. The Commutative Property of Addition states that the result of adding two numbers is the same regardless of the order in which they are added.

16. The Associative Property of Addition states that the result of adding three numbers is the same regardless of which two numbers are added first.

In Exercises 17–22, use the indicated property to fill in the blank.

17. Commutative Property of Addition: $x + 6 = 6 + x$

18. Addition Property of 0: $8 = 8 + 0$

19. Associative Property of Addition: $z + (4 + y) = (z + 4) + y$

20. Commutative Property of Addition: $y + 8 = 8 + y$

21. Addition Property of 0: $c + 0 = c$

22. Associative Property of Addition: $(x + 3) + 6 = x + (3 + 6)$

Copyright © Houghton Mifflin Company. All rights reserved.

In Exercises 23–26, use the Addition Property of 0 to rewrite the expression.

23. $9 + 0$ **24.** $0 + 3$ **25.** $0 + z$ **26.** $w + 0$

In Exercises 27–30, use the Commutative Property of Addition to rewrite the expression.

27. $9 + 2$ **28.** $5 + 8$ **29.** $c + n$ **30.** $b + a$

In Exercises 31–34, use the Associative Property of Addition to rewrite the expression.

31. $(3 + 8) + 7$

32. $4 + (3 + 9)$

33. $4 + (x + y)$

34. $(a + 3) + b$

In Exercises 35–40, identify the property that justifies the given statement.

35. $z + 3 = 3 + z$

36. $z = z + 0$

37. $0 + y = y$

38. $(n + 4) + m = n + (4 + m)$

39. $7 + (p + 9) = (7 + p) + 9$

40. $x + 5 = 5 + x$

Writing and Concept Extension

41. In your own words, describe what the Commutative Property of Addition allows you to do.

42. In your own words, describe what the Associative Property of Addition allows you to do.

43. Identify the property of addition that is used to rewrite the sum.

(a) $(x + 3) + 7 = 7 + (x + 3)$ Both Commutative

(b) $(4 + a) + 5 = (a + 4) + 5$

44. Identify the properties that are used to rewrite the following sum.

$$7 + (a + 4) = 7 + (4 + a)$$ (a) ~~Associative~~ Com.

$$= (7 + 4) + a$$ (b) Assoc.

$$= 11 + a$$

In Exercises 45–48, use the Associative and Commutative Properties of Addition to rewrite the sum so that only numbers appear inside the parentheses.

45. $(4 + x) + 5$

46. $7 + (a + 6)$

47. $9 + (b + 2)$

48. $(1 + y) + 9$

Copyright © Houghton Mifflin Company. All rights reserved.

1.4 Addition and Subtraction of Whole Numbers

A *Addition of Whole Numbers*
B *Subtraction of Whole Numbers*
C *Estimating Sums and Differences*
D *Applications*

SUGGESTIONS FOR SUCCESS

Let's be realistic. Whether you are at home balancing your checkbook or on the job adding columns of numbers, you are likely to reach for your calculator.

Obviously, a calculator is a time- and labor-saving device. However, brushing aside the concepts and methods discussed in this section is unwise. Remember that one purpose of your study is to *understand* mathematics, not simply to *do* mathematics.

You may be tempted to fudge just a little by using your calculator in the exercises. But you will be cheating yourself if you do that. Here is a compromise: Do the exercises the way they are meant to be done. Then use your calculator as a check.

A Addition of Whole Numbers

The usual method for adding whole numbers other than single digits is to write the numbers in a vertical arrangement.

As you see in Example 1, one advantage of the vertical arrangement is that place values are aligned in columns. This gives us an easy way to perform the addition simply by adding the digits in each column.

Example 1

$$\begin{array}{r} 1{,}612 \\ +\ 5{,}127 \\ \hline \end{array}$$

SOLUTION

To find the sum, we add down the columns.

Ones	Tens	Hundreds	Thousands
↓	↓	↓	↓
1,612	1,612	1,612	1,612
+ 5,127	+ 5,127	+ 5,127	+ 5,127
9	39	739	6,739

Your Turn 1

$$\begin{array}{r} 4{,}123 \\ +\ 5{,}251 \\ \hline \end{array}$$

Answer: 9,374

In Example 1, adding down a column always resulted in a digit. However, this will not always happen.

$$\begin{array}{r} 928 \\ +\ 275 \\ \hline \end{array}$$

Adding down the ones column, we have $8 + 5 = 13$. What do we do when the sum of a column is not a digit?

Copyright © Houghton Mifflin Company. All rights reserved.

DEVELOPING THE CONCEPT

Addition with Carrying

The dials of your odometer display only the digits 0 through 9. If your odometer reading is 2359, then when you drive one more mile, the ones dial rotates to 0 and the tens dial rotates to 6.

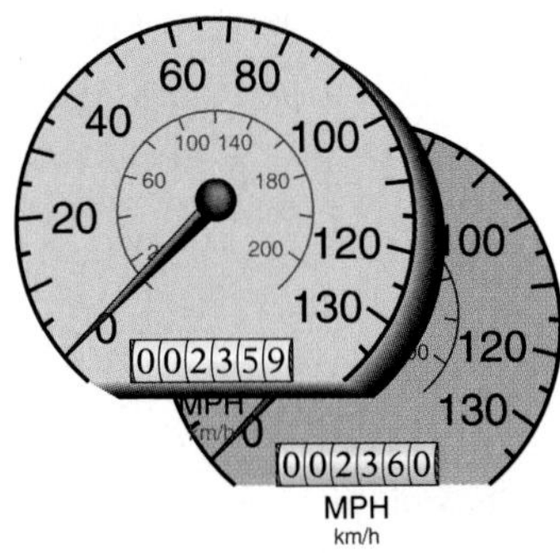

Driving that one more mile can be represented by the addition

$$\begin{array}{r} 2{,}359 \\ +\quad 1 \\ \hline \end{array}$$

To perform this addition, we add down the ones column. However, the result is 9 + 1 = 10, which is not a digit. Much as the odometer does, we write the sum by placing the 0 in the ones column and *carrying* the 1 to the tens column.

$$\begin{array}{r} 1 \\ 2{,}359 \\ +\quad 1 \\ \hline 0 \end{array}$$ ← 1 is carried to the tens column.

$$\begin{array}{r} 1 \\ 2{,}359 \\ +\quad 1 \\ \hline 2{,}360 \end{array}$$ Now we add down the remaining columns.

It is important to keep in mind that the value of the carried number is the place value of the column to which it is carried.

This carried 1 represents 10. ↓

$$\begin{array}{r} 1 \\ 237 \\ +\quad 8 \\ \hline 245 \end{array}$$

This carried 1 represents 100. ↓

$$\begin{array}{r} 1 \\ 237 \\ +\ 91 \\ \hline 328 \end{array}$$

LEARNING TIP

Eventually, you should be able to carry mentally. However, you should keep writing the carried number until you feel comfortable doing it in your head.

The need to carry a number may arise in any column, and we may need to carry more than once.

Example 2

$$\begin{array}{r} 928 \\ +\ 275 \\ \hline \end{array}$$

Your Turn 2

$$\begin{array}{r} 476 \\ +\ 528 \\ \hline \end{array}$$

Copyright © Houghton Mifflin Company. All rights reserved.

SOLUTION

$$\begin{array}{r} {\scriptstyle 1} \\ 928 \\ +\ 275 \\ \hline 3 \end{array}$$

When we add down the ones column, we have 8 + 5 = 13, but 13 is not a digit. Place the 3 in the ones column and carry the 1 to the tens column.

$$\begin{array}{r} {\scriptstyle 11} \\ 928 \\ +\ 275 \\ \hline 03 \end{array}$$

When we add down the tens column, we have 1 + 2 + 7 = 10, but 10 is not a digit. Place the 0 in the tens column and carry the 1 to the hundreds column.

$$\begin{array}{r} {\scriptstyle 111} \\ 928 \\ +\ 275 \\ \hline 203 \end{array}$$

When we add down the hundreds column, we have 1 + 9 + 2 = 12, but 12 is not a digit. Place the 2 in the hundreds column and carry the 1 to the thousands column.

$$\begin{array}{r} {\scriptstyle 111} \\ 928 \\ +\ 275 \\ \hline 1{,}203 \end{array}$$

Finally, add down the thousands column.

KEYS TO THE CALCULATOR

Scientific calculators usually have an = key, whereas graphing calculators have an ENTER key. These keys are pressed to obtain the result of the operation you are performing.

To perform the addition in Example 2, try these keystrokes:

928 + **275** = (or ENTER)

The figure shows how the sum would be displayed on a graphing calculator.

```
928+275
                1203
```

Exercises

(a) $\begin{array}{r} 27 \\ +\ 32 \\ \hline \end{array}$ **(b)** $\begin{array}{r} 736 \\ +\ 51 \\ \hline \end{array}$ **(c)** $\begin{array}{r} 147 \\ +\ 622 \\ \hline \end{array}$ **(d)** $\begin{array}{r} 3{,}925 \\ +\ 5{,}013 \\ \hline \end{array}$

B *Subtraction of Whole Numbers*

Just as addition involves an increase of a quantity, subtraction indicates a decrease of a quantity.

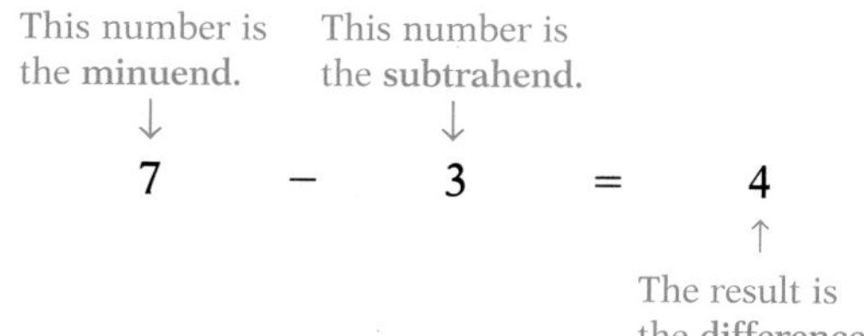

Both the subtraction, 7 − 3, and the result, 4, are called a **difference.**

Copyright © Houghton Mifflin Company. All rights reserved.

Subtraction

We can use a number line to represent subtraction. Here is how we can see the difference 7 − 3. Recall that adding a number involves moving to the right. Subtracting a number involves moving to the left.

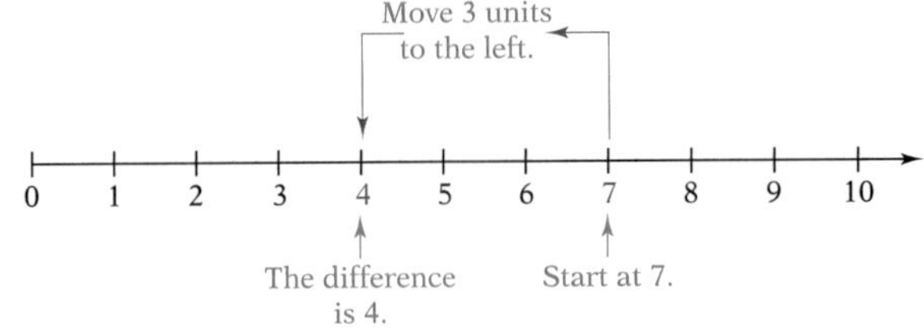

Another way to subtract 7 − 3 is to ask, "What number added to 3 gives us 7?"

Subtraction		*Addition*
$7 - 3 = 4$	because	$4 + 3 = 7$

The fact that subtraction and addition are related in this way allows us to check our work.

LEARNING TIP
You might be tempted to put addition away and treat subtraction as a completely new topic. However, you will often need to decide whether a problem requires addition or subtraction. As you work with subtraction, keep reviewing addition. (See Bringing It Together in the exercise sets.)

NOTE Subtraction is not commutative because the order in which we subtract numbers affects the result. Also, subtraction is not associative. For example,

$$(9 - 5) - 1 = 4 - 1 = 3$$
$$9 - (5 - 1) = 9 - 4 = 5$$

We can see that the result is affected by the way we group the numbers.

LEARNING TIP
In addition, the order of the addends does not affect the sum: 5 + 3 = 3 + 5. However, the order is *important in subtraction. The difference 5 − 3 is* not *the same as 3 − 5. Be very careful to use the correct order when you translate words into differences.*

As you solve problems, you need to be aware of other words and phrases that translate into subtraction.

The difference of 10 and 4	$10 - 4$
The difference of 4 and 10	$4 - 10$
9 *minus* 5	$9 - 5$
12 *decreased by* 8	$12 - 8$
2 *subtracted from* 6	$6 - 2$
7 *less than* 13	$13 - 7$

NOTE We translate "7 less than 13" into the difference 13 − 7. Compare this to "7 *is* less than 13," which translates into the inequality $7 < 13$. A small word can make a big difference in how you translate words into symbols.

Although we can think about subtractions in terms of related additions, a more efficient method is to subtract in columns.

Example 3

$$\begin{array}{r} 946 \\ -\ 432 \\ \hline \end{array}$$

Your Turn 3

$$\begin{array}{r} 6{,}859 \\ -\ 2{,}736 \\ \hline \end{array}$$

Copyright © Houghton Mifflin Company. All rights reserved.

SOLUTION

Subtract down the ones column.	Subtract down the tens column.	Subtract down the hundreds column.
↓	↓	↓
$\begin{array}{r} 946 \\ -\ 432 \\ \hline 4 \end{array}$	$\begin{array}{r} 946 \\ -\ 432 \\ \hline 14 \end{array}$	$\begin{array}{r} 946 \\ -\ 432 \\ \hline 514 \end{array}$

In Example 3, the lower digit was always less than the digit above it, so we were able to perform the subtraction right away. What do we do when the lower digit is greater than the digit above it?

DEVELOPING THE CONCEPT

Borrowing

Consider the following subtraction problem.

$$\begin{array}{r} 34 \\ -\ 6 \\ \hline \end{array}$$

We can't subtract 6 in the ones column because 34 has only 4 ones.

Take 10 from the tens column ↓ and add it to the ones column. ↓

$$\begin{array}{rclcl} 34 & = & 3 \text{ tens} + 4 \text{ ones} & = & 2 \text{ tens} + 14 \text{ ones} \\ -\ 6 & = & 6 \text{ ones} & = & 6 \text{ ones} \\ \hline & & & & 2 \text{ tens} + 8 \text{ ones} = 28 \end{array}$$

Now we can subtract.

The act of taking a 10 from the tens column and adding it to the ones column is called *borrowing.*

We can perform subtractions that require borrowing without writing out the numbers in words. Sometimes we need to borrow more than once.

Example 4

Subtract.

$$\begin{array}{r} 7{,}253 \\ -\ 675 \\ \hline \end{array}$$

SOLUTION

$$\begin{array}{r} {\scriptstyle 4\ 13} \\ 7{,}2\cancel{5}\cancel{3} \\ -\ 675 \\ \hline 8 \end{array}$$

Borrow a 10 and add it to the ones column.
In the ones column, subtract 13 − 5.

$$\begin{array}{r} {\scriptstyle 1\ 14\ 13} \\ 7{,}\cancel{2}\cancel{5}\cancel{3} \\ -\ 675 \\ \hline 78 \end{array}$$

Borrow a 100 and add it to the tens column.
In the tens column, subtract 14 − 7.

Your Turn 4

Subtract.

$$\begin{array}{r} 826 \\ -\ 139 \\ \hline \end{array}$$

continued

Copyright © Houghton Mifflin Company. All rights reserved.

```
  6 11 14 13
  7 , 2 5 3    Borrow a 1,000 and add it to the hundreds column.
−     6 7 5    In the hundreds column, subtract 11 − 6.
      5 7 8
```

```
  6 11 14 13
  7 , 2 5 3
−     6 7 5
  6 , 5 7 8    In the thousands column, subtract 6 − 0.
```

Answer: 687

Example 5

Subtract.

$$\begin{array}{r} 9{,}004 \\ -\ 2{,}359 \\ \hline \end{array}$$

SOLUTION

We need to borrow a 10, but the 0s in 9,004 indicate that we have neither tens nor hundreds that we can borrow. Therefore, we begin our borrowing from the thousands column.

```
  8 10        Borrow 1,000 (10 hundreds) and add it
  9 , 0 0 4   to the hundreds column.
− 2 , 3 5 9   Now we have 8 thousands and 10
              hundreds.
```

Now we can borrow from the hundreds column.

```
      9       Borrow 100 (10 tens) and add it to the tens
  8 10 10     column.
  9 , 0 0 4   Now we have 9 hundreds and 10 tens.
− 2 , 3 5 9
```

Finally, we can now borrow from the tens column.

```
      9 9
  8 10 10 14  Borrow 10 and add it to the ones column.
  9 , 0 0 4   Now we have 9 tens and 14 ones.
− 2 , 3 5 9
```

Now we can subtract down the columns.

```
      9 9
  8 10 10 14
  9 , 0 0 4
− 2 , 3 5 9
  6 , 6 4 5
```

Your Turn 5

Subtract.

$$\begin{array}{r} 8{,}050 \\ -\ 3{,}697 \\ \hline \end{array}$$

Answer: 4,353

LEARNING TIP

As with carrying in addition, you can borrow mentally. However, showing the result of borrowing is the wise thing to do until you are confident.

Copyright © Houghton Mifflin Company. All rights reserved.

KEYS TO THE CALCULATOR

The typical keystrokes for the subtraction in Example 5 are

9004 [−] 2359 [=] (or [ENTER])

Here is how the result might look on a graphing calculator.

```
9004-2359
                6645
```

Exercises

(a) $58 - 39$ **(b)** $819 - 283$ **(c)** $5{,}000 - 2{,}999$ **(d)** $63{,}041 - 8{,}762$

C Estimating Sums and Differences

In arithmetic problems, when you see words such as *about* or *approximately,* only an estimate of the answer is needed. Usually, we round off the given information to the highest place value before we add.

Even when an exact sum is required, estimating the sum can be helpful in preventing errors and making sure your answer is reasonable. For example, if you add 23, 58, and 95 with a calculator and the result is 235,895, you would realize that the sum is not reasonable and that you probably forgot to press the plus key.

We use the symbol ≈ to indicate that a number is an approximation. The symbol is read as "is approximately equal to."

Example 6

(a) Estimate the sum of 23, 58, and 95.

(b) Estimate the difference of 889 and 72.

SOLUTION

(a)

Given Number		*Rounded Number*
23	→	20
58	→	60
95	→	100
		180

We can write $23 + 58 + 95 \approx 180$ to indicate that the sum is approximately equal to 180. Note that the exact sum of the given numbers is 176.

Your Turn 6

(a) Estimate the sum of 111, 159, and 290.

(b) Estimate the difference of 1,198 and 653.

continued

Copyright © Houghton Mifflin Company. All rights reserved.

(b) We round the two numbers to the highest place value of the smaller number.

Given Number		*Rounded Number*
889	→	890
72	→	− 70
		820

Think: 89 tens minus 7 tens is 82 tens, or 820.

The estimated answer, 820, is close to the exact answer, which is 817.

D Applications

You will often need to decide whether the information that you are being asked to report must be exact or may be approximate. For example, a sales accountant usually works with exact amounts, but a sales manager may be satisfied with estimates. Knowing how to provide exact sums or approximate sums will increase your value as an employee.

Example 7

A college drama department produced a play that was performed three times. The attendance for the three shows was 892, 743, and 956. What were the approximate total attendance and the exact total attendance?

SOLUTION

Attendance		*Rounded*
892	→	900
743	→	700
956	→	1,000
2,591		2,600

The total attendance was about 2,600. The exact total attendance was 2,591.

Your Turn 7

A college book store had the following sales of a history textbook.

Fall: 395
Winter: 276
Spring: 144

What were (a) the approximate total sales and (b) the exact total sales?

Answers: (a) 800; (b) 815

In Example 7, if you were a reporter for the college newspaper, your story about the play would probably state that the total attendance for the three shows was approximately 2,600. However, if you were responsible for balancing the box office receipts with the ticket sales, you would need to know the exact attendance. As you learn to organize and interpret data, think about the purpose of your work and how the information will be used by others.

Questions that begin with "How much more . . . ?" or "How many more . . . ?" signal subtraction.

Copyright © Houghton Mifflin Company. All rights reserved.

Example 8

Two sweaters cost \$56 and \$49. How much more is the more expensive sweater?

SOLUTION

The question "How much more . . . ?" refers to the difference of the two costs.

$$\begin{array}{r} 56 \\ -\ 49 \\ \hline 7 \end{array}$$

The more expensive sweater costs \$7 more than the less expensive sweater.

Your Turn 8

A chat room on the Internet had 14 women and 5 men. How many more women than men were chatting?

Answer: 9

1.4 Quick Reference

A. Addition of Whole Numbers

1. When sums are arranged vertically, we add down the columns.

Ones ↓ $\begin{array}{r} 43 \\ +\ 16 \\ \hline 9 \end{array}$ Tens ↓ $\begin{array}{r} 43 \\ +\ 16 \\ \hline 59 \end{array}$

2. If the sum in any column is not a digit, we write the ones digit and carry the tens digit to the next column.

$$\begin{array}{r} \scriptstyle 1 \\ 517 \\ +\ 236 \\ \hline 753 \end{array}$$

3. *Carrying* may be needed in any column and may be needed more than once.

$$\begin{array}{r} \scriptstyle 1\ 1 \\ 809 \\ +\ 513 \\ \hline 1,322 \end{array}$$

B. Subtraction of Whole Numbers

1. For $9 - 6 = 3$, we call 9 the **minuend** and 6 the **subtrahend.** Both $9 - 6$ and 3 are called the **difference.**

2. For every subtraction we can write a related addition.

Subtraction: $8 - 2 = 6$

Addition: $6 + 2 = 8$

Copyright © Houghton Mifflin Company. All rights reserved.

3. A related addition can be used to check the result of a subtraction.

$$\begin{array}{r} 58 \\ -\ 23 \\ \hline 35 \end{array} \quad \text{because} \quad \begin{array}{r} 35 \\ +\ 23 \\ \hline 58 \end{array}$$

4. When we write a subtraction in a vertical arrangement, we subtract down the columns.

$$\begin{array}{r} 958 \\ -\ 613 \\ \hline 345 \end{array}$$

Subtract down the columns.

5. If a digit is less than the digit above it, we must *borrow* before we can subtract. We may need to borrow more than once in a subtraction.

$$\begin{array}{r} 52 \\ -\ 17 \\ \hline \end{array} \qquad \begin{array}{r} {}^{4\ 12} \\ \not{5}\not{2} \\ -\ 17 \\ \hline 35 \end{array}$$

C. Estimating Sums and Differences

1. We estimate a sum or difference by rounding the numbers before adding or subtracting.

$$\begin{array}{r} 41 \\ +\ 28 \\ \hline \end{array} \begin{array}{c} \rightarrow \\ \rightarrow \\ \\ \end{array} \begin{array}{r} 40 \\ +\ 30 \\ \hline 70 \end{array}$$

$$\begin{array}{r} 804 \\ -\ 298 \\ \hline \end{array} \begin{array}{c} \rightarrow \\ \rightarrow \\ \\ \end{array} \begin{array}{r} 800 \\ -\ 300 \\ \hline 500 \end{array}$$

Copyright © Houghton Mifflin Company. All rights reserved.

1.4 Exercises

A. Addition of Whole Numbers

1. The ________ Property of Addition allows us to add a column of numbers from top to bottom or from bottom to top.

2. To add 36 + 29, first add 6 and 9 to obtain 15. Place 5 in the ________ column and ________ 1 to the ________ column.

In Exercises 3–10, add.

3. 27 + 52

4. 532 + 267

5. $\begin{array}{r} 7{,}391 \\ +\ 2{,}306 \\ \hline \end{array}$

6. $\begin{array}{r} 3{,}466 \\ +\ \ \ 212 \\ \hline \end{array}$

7. 32 + 21 + 44

8. 102 + 410 + 476

9. $\begin{array}{r} 340 \\ 235 \\ +\ 412 \\ \hline \end{array}$

10. $\begin{array}{r} 3{,}210 \\ 403 \\ 1{,}124 \\ +\ 2{,}231 \\ \hline \end{array}$

In Exercises 11–20, add.

11. 769 + 635

12. 999 + 111

13. $\begin{array}{r} 7{,}384 \\ +\ 1{,}693 \\ \hline \end{array}$

14. $\begin{array}{r} 34{,}296 \\ +\ 6{,}985 \\ \hline \end{array}$

15. $\begin{array}{r} 647 \\ +\ 5{,}643 \\ \hline \end{array}$

16. $\begin{array}{r} 6{,}007 \\ +\ \ \ 996 \\ \hline \end{array}$

17. $\begin{array}{r} 25 \\ 76 \\ 43 \\ +\ 69 \\ \hline \end{array}$

18. $\begin{array}{r} 596 \\ 459 \\ +\ 387 \\ \hline \end{array}$

19. $\begin{array}{r} 6{,}246 \\ 3{,}078 \\ 4{,}357 \\ +\ 8{,}205 \\ \hline \end{array}$

20. $\begin{array}{r} 74 \\ 4{,}099 \\ 436 \\ 1{,}349 \\ +\ \ \ 231 \\ \hline \end{array}$

B. Subtraction of Whole Numbers

21. In the expression 12 − 5, 5 is called the ________ and 12 is called the ________.

22. Both 9 − 6 and the result 3 are called a(n) ________.

Copyright © Houghton Mifflin Company. All rights reserved.

In Exercises 23–28, write the information as a difference.

23. The difference of 12 and 3

24. y subtracted from 10

25. 9 less than x

26. The difference of b and 8

27. 2 subtracted from 11

28. x decreased by 4

29. *Hotel Prices* During the off-season, the price of a hotel room at a resort was $28 less than the peak-season price of $110. Write a difference that represents the off-season price.

30. *Meeting Attendance* The attendance at a civic club meeting in November was 12 less than the October attendance of 38. Write a difference that represents the attendance at the November meeting.

31. *Election Results* In a local election, the Democratic candidate received 1,830 votes and the Republican candidate received 1,497 votes. Write a difference that represents the number of votes by which the Democratic candidate won.

32. *Test Scores* A student scored 85 on the first test and 97 on the second test. Write a difference that represents the number of points by which the student improved.

In Exercises 33–36, subtract.

33. $\begin{array}{r} 37 \\ -\ 14 \\ \hline \end{array}$

34. $\begin{array}{r} 693 \\ -\ 491 \\ \hline \end{array}$

35. $\begin{array}{r} 3{,}782 \\ -\ 1{,}410 \\ \hline \end{array}$

36. $\begin{array}{r} 74{,}319 \\ -\ 4{,}116 \\ \hline \end{array}$

In Exercises 37–46, subtract.

37. $\begin{array}{r} 36 \\ -\ 18 \\ \hline \end{array}$

38. $\begin{array}{r} 83 \\ -\ 44 \\ \hline \end{array}$

39. $\begin{array}{r} 327 \\ -\ 219 \\ \hline \end{array}$

40. $\begin{array}{r} 913 \\ -\ 97 \\ \hline \end{array}$

41. $\begin{array}{r} 367 \\ -\ 238 \\ \hline \end{array}$

42. $\begin{array}{r} 584 \\ -\ 179 \\ \hline \end{array}$

43. $\begin{array}{r} 5{,}683 \\ -\ 2{,}491 \\ \hline \end{array}$

44. $\begin{array}{r} 8{,}543 \\ -\ 3{,}816 \\ \hline \end{array}$

45. $\begin{array}{r} 3{,}587 \\ -\ 2{,}837 \\ \hline \end{array}$

46. $\begin{array}{r} 4{,}932 \\ -\ 976 \\ \hline \end{array}$

C. Estimating Sums and Differences

47. To estimate a sum, we begin by ________ each number to the highest place value.

48. The symbol $\approx$ is read as ________ .

Copyright © Houghton Mifflin Company. All rights reserved.

In Exercises 49–52, estimate the sum or difference.

49. 109
93
+ 178

50. 540
− 190

51. 8,341
− 5,876

52. 1,230
3,829
+ 2,114

Bringing It Together

In Exercises 53–56, perform the indicated operations.

53. 871
296
+ 55

54. 631
− 58

55. 1,239
− 582

56. 57,106
+ 23,999

57. What is the difference of 28 and 17?

58. What is 28 increased by 17?

Calculator Exercises

In Exercises 59–62, use a calculator to determine the sum or difference.

59. 54,795 + 4,513 + 2,861

60. 7,946 − 815

61. 839,006
− 97,149

62. 13,647
+ 87,009

D. Applications

63. *Family Budget* For one month, a family paid $234 for utilities, $346 for food, $187 for a car payment, and $48 for gasoline. What were the total expenses for the month?

64. *College Costs* For one semester, a student paid $1,230 for tuition, $247 for books, $90 for a student activity fee, and $40 for parking. What was the total cost for the semester?

65. *Stock Loss* What is the loss on stock purchased for $3,245 and sold for $2,890?

66. *Television Sale* A television set, regularly priced for $210, is on sale for $178. By how much is the price reduced?

Copyright © Houghton Mifflin Company. All rights reserved.

67. *Junk Food Calories* What is the total number of calories for a meal consisting of hamburger, 455 calories; potato chips, 210 calories; soda, 369 calories; and cookies, 185 calories?

68. *Theater Attendance* A movie complex has five theaters. On one day the attendance at the theaters was 128, 210, 170, 145, and 97. What was the total attendance at all five theaters?

69. *Tuition Increase* At a certain college in 1970, the tuition for one semester was $530. Now the tuition is $2,179. How much more is tuition now than it was in 1970?

70. *College Enrollment* A college with an enrollment of 5,878 has 2,795 male students. How many female students are enrolled?

71. *Clothing Purchases* A shopper purchased a sweater for $58, a shirt for $34, slacks for $43, and a jacket for $76. What was the total bill?

72. *Breakfast Calories* What is the total calorie content of breakfast consisting of orange juice (110 calories), one fried egg (90 calories), fried potatoes (120 calories), and sausage (185 calories)?

73. *Seating Capacity* The seating capacity of the Gator Bowl is 76,000, and of the Fiesta Bowl 73,243. How many more people does the Gator Bowl accommodate than the Fiesta Bowl?

74. *Super Bowl* The Green Bay Packers and the New England Patriots played in Super Bowl XXXI. Green Bay passed for 208 yards and rushed for 115 yards. New England passed for 214 yards and rushed for 43 yards. For the two teams combined, what was the total number of yards of offense for the game?

Copyright © Houghton Mifflin Company. All rights reserved.

75. *Checking Account Balance* Suppose that the beginning balance of a checking account was $1,493. Complete the following table to show the balance after each transaction:

	Amount	*Balance*
Deposit	$290	**(a)**
Rent payment	$495	**(b)**
Car payment	$357	**(c)**
Deposit	$152	**(d)**

76. *Endangered Species* The United States has 337 animals on the endangered species list. The list for the rest of the world includes those same animals but has 520 additional endangered animals. The United States has 542 plants on the list, and the rest of the world has those plants plus 1 other. (Source: Fish and Wildlife Service.)

(a) What is the total number of plants and animals in the United States that are on the endangered species list?

(b) What is the total number of animals in the world that are on the endangered species list?

Writing and Concept Extension

77. Sums of 10 are easy to add. Explain how to use the Commutative and Associative Properties of Addition to add $3 + 6 + 8 + 7 + 4 + 2$ easily.

78. Explain why $n - n = 0$, where n represents any whole number.

79. For any whole number n, what is $n - 0$? Why?

80. How do “4 is less than 9” and “4 less than 9” differ?

81. *Auto Lease* A one-year auto lease agreement requires that a person pay a per-mile charge for all miles over 12,000 that a person drives during a calendar year. The odometer reading was 26,874 on January 1 and 45,851 on December 31. For how many miles was the person charged?

Copyright © Houghton Mifflin Company. All rights reserved.

Determining the information that is needed to solve a problem is an essential skill. In Exercises 82–84, not all of the given information is needed.

82. ***Credit Card Charges*** A shopper has $4,970 remaining on her credit card limit. She charges purchases of $120 at a department store, $97 at a grocery store, $351 at a home furnishing store, and $248 at an electronics store. What were the total charges to her credit card?

83. ***Wages and Expenses*** A mother earns $360 per week at a full-time job and $135 per week at a part-time job. She pays $95 per week for child care. What is the total weekly income from her two jobs?

84. ***Concert Attendance*** A concert was held at a college auditorium with a seating capacity of 520 people. The attendance at three evening performances was 410, 480, and 510. The attendance at two matinees was 390 and 425. What was the total attendance at the evening performances?

Exploring with Real-World Data: Collaborative Activities

Library Circulation The Los Angeles (City) Public Library has 66 branches and circulates 9,663,185 books annually. The Los Angeles (County) Library has 85 branches and circulates 14,152,507 books annually. (Source: Public Library Association.)

85. Estimate the total annual circulation for the two libraries.

86. How many more books are circulated by the county library than by the city library?

87. What is the difference of the number of branches of the two libraries?

88. What is the total number of branches for the two libraries?

Zoo Sizes The list below gives the size of four major zoos. (Source: *World Almanac.*)

Zoo	*Size (acres)*
Bronx Zoo	265
Brookfield Zoo (Chicago)	216
San Diego Zoo	2,300
Miami Metrozoo	740

89. What is the difference between the sizes of the two smallest zoos ?

90. What is the total number of acres in the four zoos?

91. What is the combined size of the three smallest zoos in the list?

92. How much larger is the San Diego Zoo than the combined size of the other three zoos?

Copyright © Houghton Mifflin Company. All rights reserved.

1.5 Addition and Subtraction in the Real World

A *Bar Graph Data*
B *Perimeter*

SUGGESTIONS FOR SUCCESS

Students frequently ask math instructors "What will I ever use this for?" Have you ever asked this question? You, like many other math students, may think that the topic that you are covering in class has little use in your daily life.

Make a conscious effort to look for instances in which you can actually use the math that you are learning in class. For instance, almost every day you will see various types of graphs, tables, and charts in newspapers, magazines, and other news media. Look carefully at the data that are presented in these forms. Practice your math skills by interpreting those data. Ask yourself questions about the graphs similar to the questions about graphs in your textbook. You may find that you can use your math skills in situations that you never expected.

A Bar Graph Data

Earlier in this chapter, we saw how a bar graph can be used to present information visually. You learned how to read a bar graph and how to use it to make comparisons of data.

Sometimes we need to add or subtract all or some of the data. We begin with a bar graph that you have seen before.

Example 1

The bar graph shows four occupations that are projected to be growing the fastest by the year 2005. (Source: U.S. Bureau of Census.)

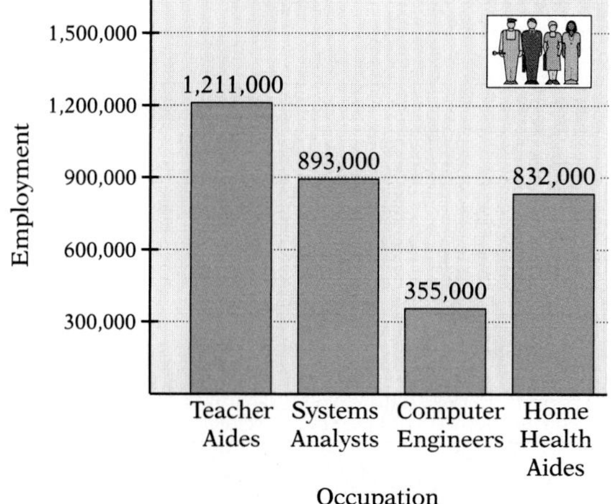

According to the data shown in the bar graph, what total employment is expected in these four occupations in 2005?

Your Turn 1

What is the expected total employment of systems analysts and computer engineers in the year 2005?

continued

Copyright © Houghton Mifflin Company. All rights reserved.

SOLUTION

$$\begin{array}{r} 832{,}000 \\ 893{,}000 \\ 355{,}000 \\ +\ 1{,}211{,}000 \\ \hline 3{,}291{,}000 \end{array}$$

For 2005, the expected employment in these four occupations is 3,291,000.

When we compare data, we often ask questions such as "What is the difference between . . . ?" and "How many more . . . ?" Such questions indicate that certain data are to be subtracted.

Example 2

In 1998, the New York Yankees set a record for the most games won in a season. The bar graph shows the number of games won by the teams in the American League East.

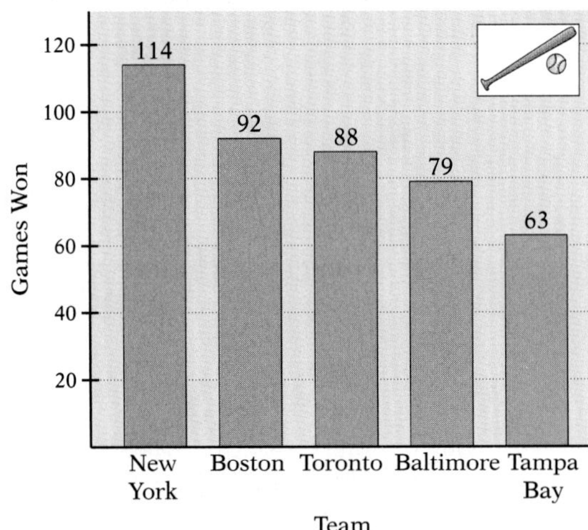

How many more games did New York win than Boston?

SOLUTION

The question asks for the difference between the number of wins for the two teams.

$$\begin{array}{r} 114 \\ -\ 92 \\ \hline 22 \end{array}$$

New York won 22 more games than Boston.

Your Turn 2

How many more games did the first-place team win than the last-place team?

Answer: 51

Copyright © Houghton Mifflin Company. All rights reserved.

B Perimeter

Squares, rectangles, and triangles are examples of familiar geometric figures. There are many other geometric figures whose names you may not know.

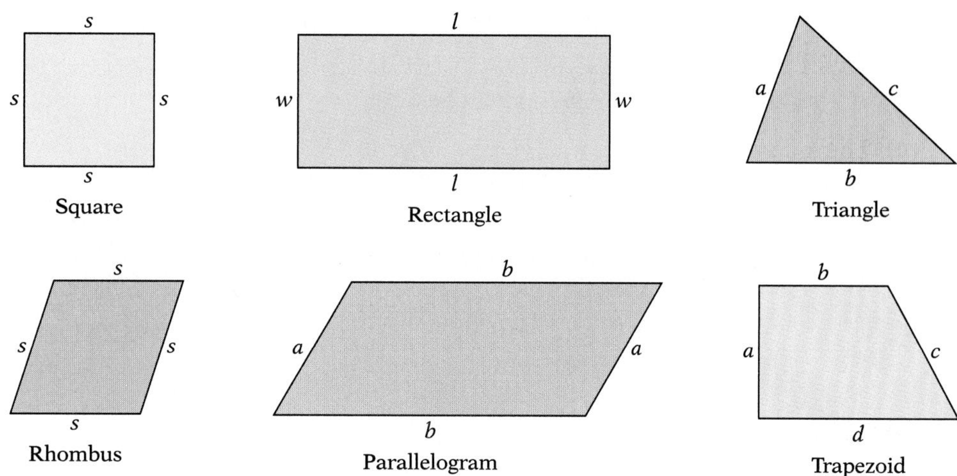

Each of the figures shown is a **polygon.** The sides of a polygon are line segments that completely enclose the figure.

The **perimeter** of a polygon is the sum of the lengths of the sides. Typically, we represent the perimeter with the letter P.

Example 3

Find the perimeter of the given geometric figure.

(a)

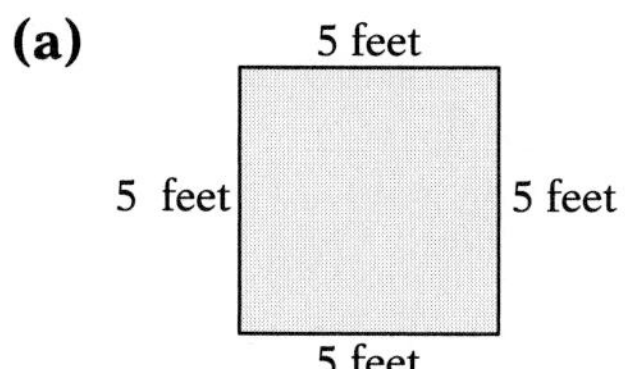

(b)

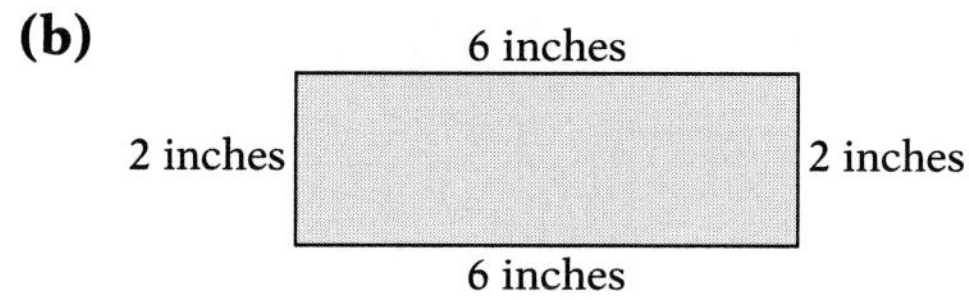

(c)

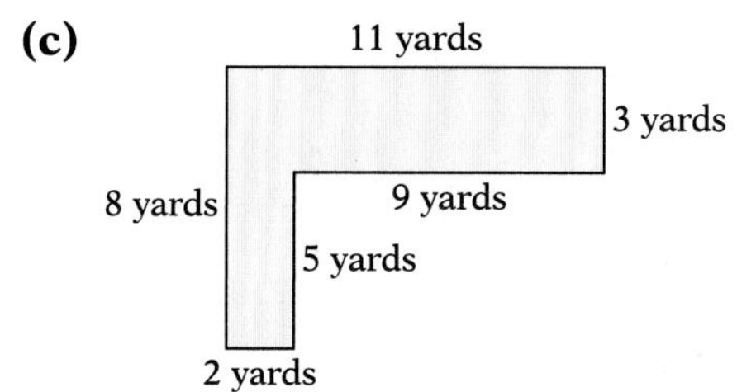

Your Turn 3

Find the perimeter of the given figure.

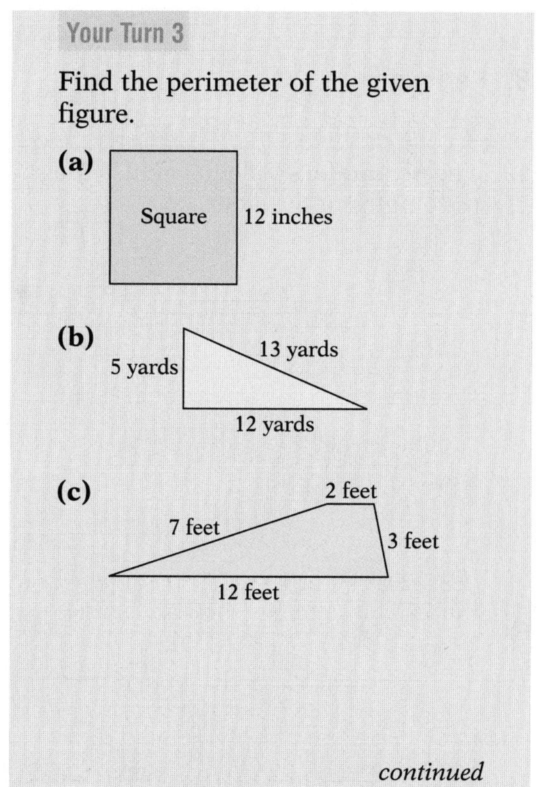

continued

Copyright © Houghton Mifflin Company. All rights reserved.

SOLUTION

(a) The four sides of a square are all the same length.

$$P = 5 + 5 + 5 + 5 = 20 \text{ feet}$$

(b) For a rectangle, the opposite sides have the same length.

$$P = 2 + 6 + 2 + 6 = 16 \text{ inches}$$

(c) The sides of this polygon all have different lengths. To find the perimeter, we add the lengths of the six sides.

$$P = 11 + 8 + 2 + 5 + 9 + 3 = 38 \text{ yards}$$

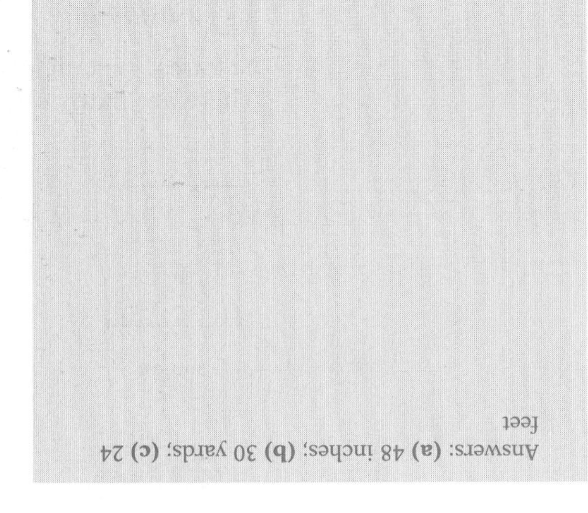

1.5 Quick Reference

A. Bar Graph Data

1. To obtain information from a bar graph, we may need to add or subtract displayed data.

B. Perimeter

1. The **perimeter** *P* of a polygon is the sum of the lengths of the sides.

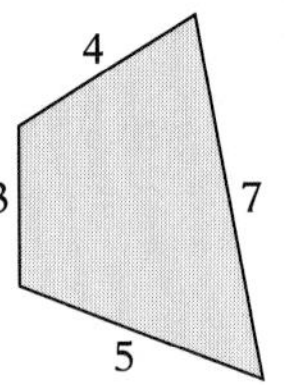

$$P = 5 + 3 + 4 + 7$$
$$= 19 \text{ units}$$

Copyright © Houghton Mifflin Company. All rights reserved.

1.5 Exercises

A. Bar Graph Data

Stadium Capacities The bar graph shows the seating capacity for five major-league baseball stadiums in California. (Source: *World Almanac.*)

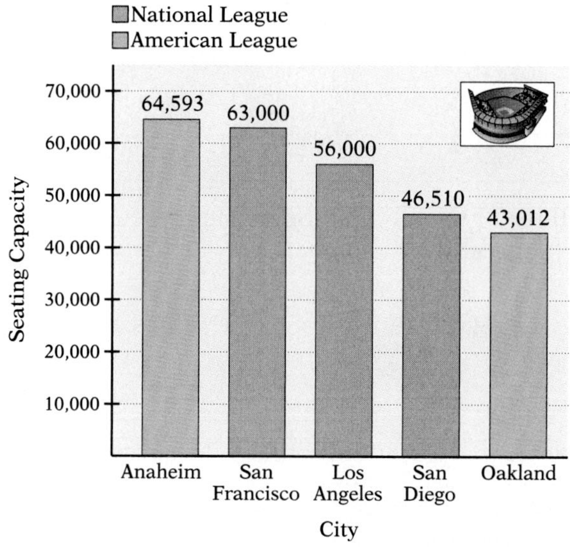

1. What is the total seating capacity of the five stadiums?

2. What is the total seating capacity of the three National League stadiums?

3. What is the total seating capacity of the two American League stadiums?

4. What is the total seating capacity of the three largest stadiums?

Top-Rated Television Programs The following bar graph shows the all-time top three television programs and the number of households that viewed each program. (Source: *World Almanac.*)

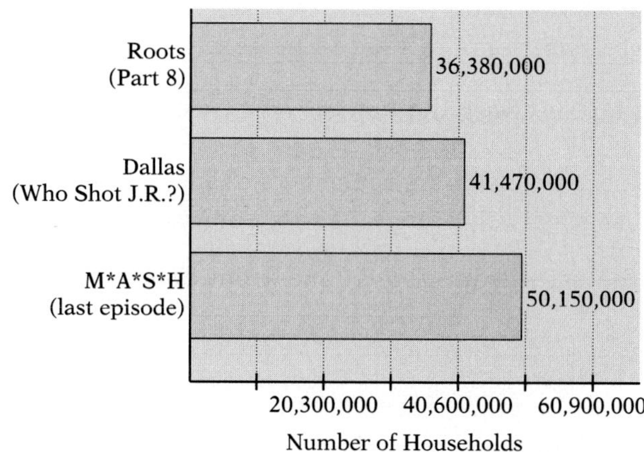

5. By how much did the number of households viewing *M*A*S*H* exceed the number viewing *Roots?*

Copyright © Houghton Mifflin Company. All rights reserved.

6. What was the difference between the number of households viewing *Dallas* and of those viewing *Roots?*

7. How many more households viewed *M*A*S*H* than *Dallas?*

8. What was the total number of households that viewed *M*A*S*H* and *Roots?*

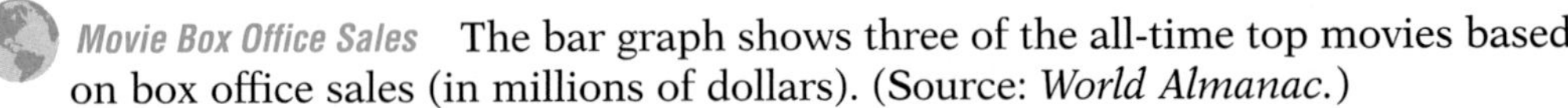

Movie Box Office Sales The bar graph shows three of the all-time top movies based on box office sales (in millions of dollars). (Source: *World Almanac.*)

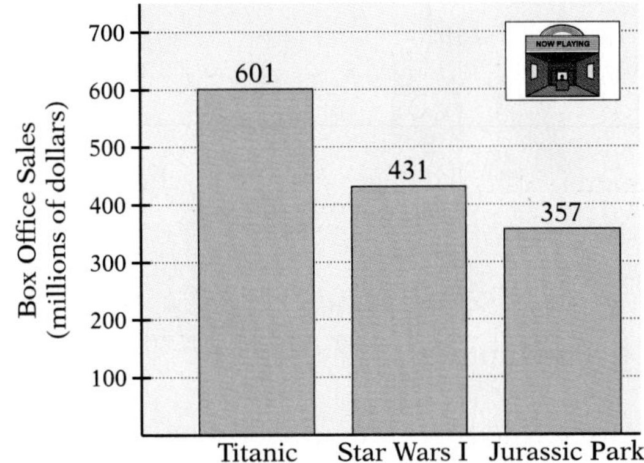

9. What was the difference between the box office sales for *Titanic* and for *Jurassic Park?*

10. How much more box office sales was earned from *Star Wars* than from *Jurassic Park?*

11. By how much did the combined sales for *Star Wars* and *Jurassic Park* exceed the sales for *Titanic?*

12. What was the difference between the sales from *Titanic* and from *Star Wars?*

13. What were the combined box office sales for the three movies?

14. By how much did the combined box office sales for the three movies exceed \$1 billion?

Copyright © Houghton Mifflin Company. All rights reserved.

Newspaper Reading Time The average adult spends 18 minutes a day reading a newspaper. The bar graph shows the average reading time for two age groups. (Source: Pew Research Center.)

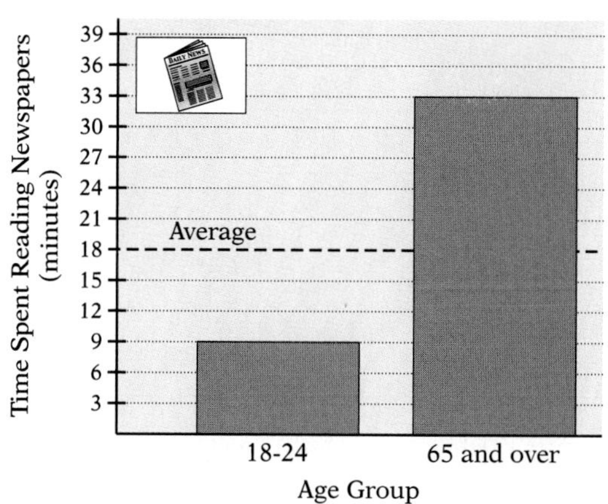

15. How many minutes below the overall adult average is the average reading time for the 18–24 age group?

16. How many minutes above the overall adult average is the average reading time for the 65 and over age group?

Military Personnel The bar graph shows the number of officers and enlisted personnel in the U.S. Navy and Army in 2000. (Source: U.S. Department of Defense.)

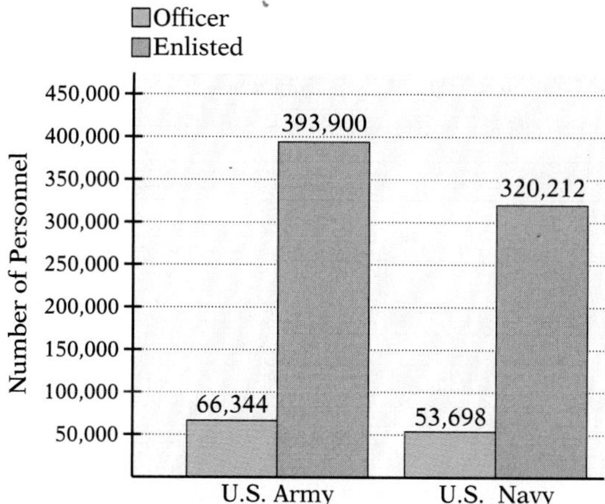

17. How many more officers were there in the Army than in the Navy?

18. How many more enlisted personnel than officers served in the Navy?

19. What was the difference of the number of enlisted personnel in the Army and the number in the Navy?

20. How many more personnel were in the Army than in the Navy?

B. Perimeter

21. The distance around a polygon is called the ________.

22. A rectangle has ________ sides and a(n) ________ has three sides.

Copyright © Houghton Mifflin Company. All rights reserved.

In Exercises 23–32, determine the perimeter of the geometric figure.

23.

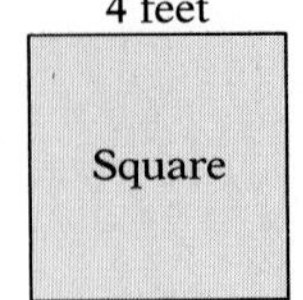

24.

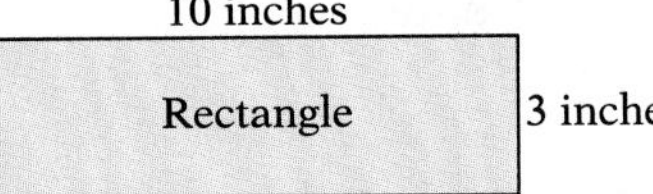

25.

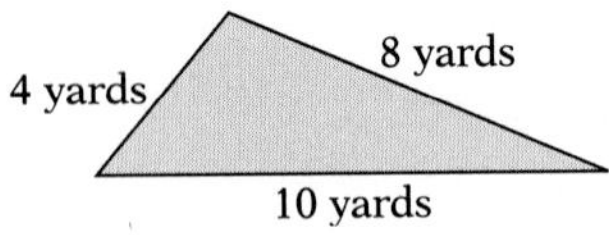

26.

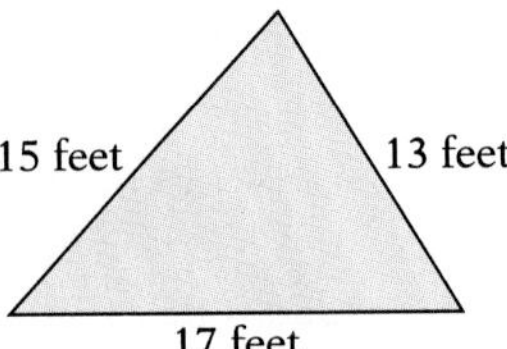

27.

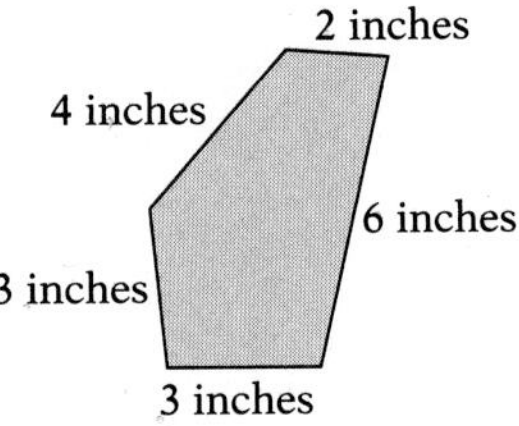

28.

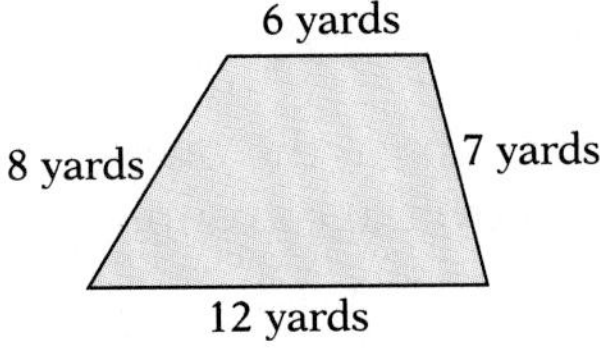

29.

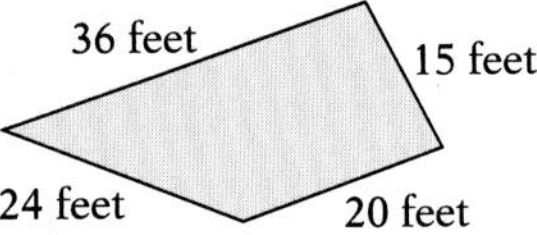

30.

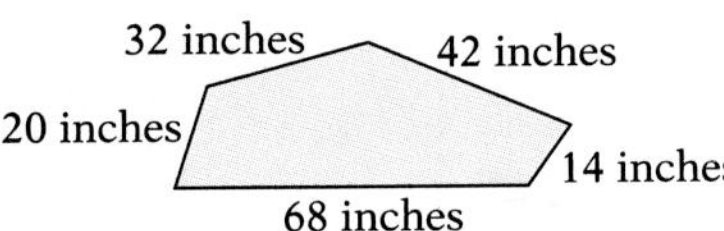

31.

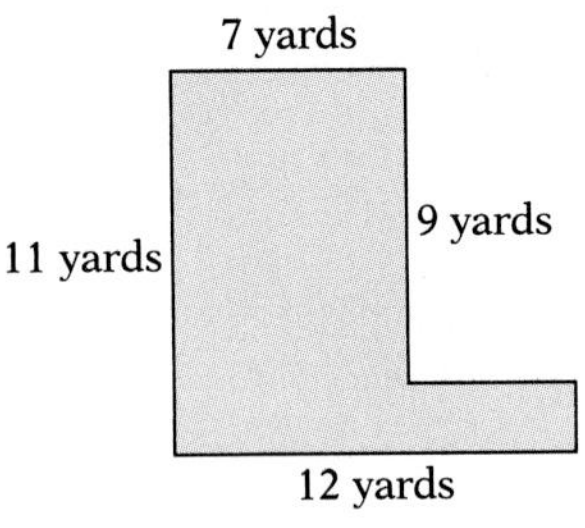

32.

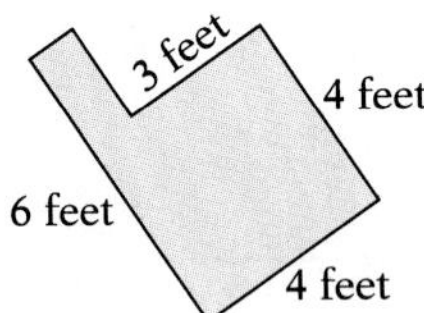

33. *Athletic Field Fencing* How many feet of fencing are needed to enclose a rectangular athletic field that is 350 feet long and 175 feet wide?

34. *Picture Frame* How many inches of framing are needed to construct a frame for a rectangular photo whose width is 8 inches and whose length is 10 inches?

Copyright © Houghton Mifflin Company. All rights reserved.

35. *Fitness Walk* A fitness group walks one lap along the outer boundary of a college parking lot. How many yards do they walk?

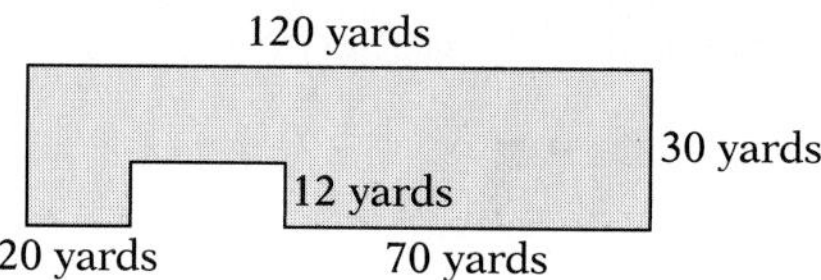

36. *Baseboard* How many feet of material are needed to install baseboard all the way around the room shown in the figure?

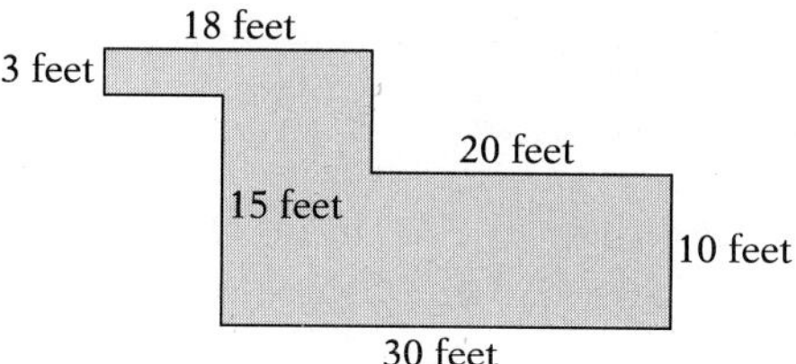

Writing and Concept Extension

37. Suppose that the length of a side of a square is a whole number. Can the perimeter be 6? Why or why not?

38. *Park Fencing* A rectangular city park is 120 feet long and 90 feet wide. The park is to be enclosed with a fence that is 4 feet high and costs $8 per linear foot. How many feet of fencing are needed?

39. Suppose that a rectangle has a perimeter of 10 inches and that the length and width are whole numbers. What are the possible dimensions?

40. Why would it be incorrect to say that a triangle whose sides are 5, 7, and 13 has a perimeter of 25?

41. If the width of a rectangle is 3 inches and the length is 4 inches, then a diagonal of the rectangle is 5 inches. What is the sum of the perimeter and the lengths of the two diagonals?

Copyright © Houghton Mifflin Company. All rights reserved.

Exploring with Real-World Data: Collaborative Activities

Health Care Expenditures The table shows the projected expenditures (in billions of dollars) for health care in 2005. (Source: U.S. Department of Commerce.)

	Expenditures (in billions of dollars)
Drugs	226
Physicians	348
Dentists	79
Hospitals	560
Insurance	90

42. Suppose that you construct a bar graph for the data in the table. On the basis of the data, which of the following would be the best choice for the scale on the vertical axis?
(i) 10, 20, . . . , 600 **(ii)** 100, 200, . . . , 600 **(iii)** 1, 2, . . . , 600

43. Use the scale that you chose in Question 1 to construct a bar graph for the data.

44. What are the total expenditures for health care?

45. What are the total expenditures for the two highest categories?

46. Which two categories have the greatest difference in expenditures?

47. By how much do the total expenditures for drugs and physicians exceed the expenditures for hospitals?

48. What are the total expenditures for physicians and dentists?

49. What are the total expenditures for the three lowest categories?

50. What is the difference between the expenditures for physicians and for dentists?

51. Should we conclude from the data that the average charge by dentists is less than the average charge by physicians? Why or why not?

Copyright © Houghton Mifflin Company. All rights reserved.

1.6 Expressions and Equations

A *Expressions*
B *Equations*

SUGGESTIONS FOR SUCCESS

Television has been blamed for a decrease in our listening skills. Even in social conversations, some people are so busy thinking about what they will say next that they never hear what anyone else says.

Research shows that the average attention span is about 10 minutes! Unless your classes are only 10 minutes long, you will need to try to do better than that. Here are some tips.

1. Take good notes so that you are doing something rather than just sitting around.
2. Join in the discussion. Ask questions. Answer questions.
3. Keep your eye off the clock. Watching it won't make it go any faster.

You can't always be as good as you sometimes are, but if you are aware of what your attention span is, you can work on improving it.

A Expressions

A **numerical expression** is any combination of numbers and operations. The sum $23 + 58$ and the difference $96 - 47$ are examples of numerical expressions. To **evaluate** numerical expressions, we simply perform the indicated operations.

$$23 + 58 = 81 \qquad 96 - 47 = 49$$

A **variable expression** is any combination of numbers, *variables,* and operations. Recall that a variable is a symbol, usually a letter, that represents some unknown number.

Unlike a numerical expression, which can always be evaluated, a variable expression cannot be evaluated until we are given a replacement for the variable. Suppose that we wish to evaluate $x + 6$ for $x = 9$.

$x + 6$	$x + 6$ is a variable expression.
$\downarrow$	Replace x with 9.
$9 + 6 = 15$	The numerical expression $9 + 6$ can be evaluated.

Example 1

Evaluate each expression for $y = 7$.

(a) $4 + y + 12$ **(b)** $18 - y$

SOLUTION

(a) $4 + y + 12$	The variable expression
$\downarrow$	Replace y with 7.
$4 + 7 + 12 = 23$	

(b) $18 - y$	The variable expression
$\downarrow$	Replace y with 7.
$18 - 7 = 11$	

Your Turn 1

Evaluate each expression for $x = 9$.

(a) $3 + 15 + x$

(b) $x - 6$

Answers: (a) 27; (b) 3

A variable expression can have more than one variable. To evaluate such an expression, we must have a replacement for each variable.

Copyright © Houghton Mifflin Company. All rights reserved.

Example 2

Evaluate each expression for $a = 5$ and $b = 7$.

(a) $a + 9 + b$ **(b)** $b - a$

SOLUTION

(a) $a + 9 + b$ The variable expression

$\downarrow \quad \downarrow$ Replace a with 5 and b with 7.

$5 + 9 + 7 = 21$

(b) $b - a$ The variable expression

$\downarrow \quad \downarrow$ Replace a with 5 and b with 7.

$7 - 5 = 2$

Your Turn 2

Evaluate each expression for $m = 9$ and $n = 4$.

(a) $m - n$

(b) $n + m + 1$

Answers: (a) 5; (b) 14

Sometimes the replacements for the variables are large enough that column addition (with carrying) or subtraction (with borrowing) is needed. Of course, you can also use a calculator if you are permitted to do so.

Example 3

Evaluate $x + y + 429$ for $x = 378$ and $y = 116$.

SOLUTION

$x + y + 429$ The variable expression

$\downarrow \quad \downarrow$ Replace x with 378 and y with 116.

$378 + 116 + 429$

$$\begin{array}{r} {}^{12} \\ 378 \\ 116 \\ +\ 429 \\ \hline 923 \end{array}$$

LEARNING TIP

Mentally replacing the variables with their values is risky. Even if you use a calculator, you should write the first two lines shown in this example.

Your Turn 3

Evaluate $a + 28 + b$ for $a = 37$ and $b = 86$.

Answer: 151

We **simplify** an expression by performing any operations that we can. Remember that grouping symbols indicate the operation that is to be performed first.

Example 4

Simplify the expression $(x + 9) + 3$.

SOLUTION

$(x + 9) + 3$ We can't add $x + 9$ because we don't know the value of x.

$= x + (9 + 3)$ Associative Property of Addition

$= x + 12$ Replace $(9 + 3)$ with 12.

Because we don't know what number x represents, we cannot simplify further.

Your Turn 4

Simplify the expression $(y + 5) + 8$.

Answer: $y + 13$

Copyright © Houghton Mifflin Company. All rights reserved.

B Equations

We can illustrate the Commutative Property of Addition with an *equation:* $5 + 3 = 3 + 5$. The expression $5 + 3$ is the *left side* of the equation, and the expression $3 + 5$ is the *right side.*

An **equation** is a statement that two expressions have the same value. An important point to keep in mind is that an equation may be true or false.

LEARNING TIP

The symbol = appears in an equation but not in an expression. Make sure that you know the difference between an equation and an expression.

$5 + 3 = 8$ A true equation

$4 + 6 = 7$ A false equation

If the two sides of an equation are numerical expressions, then we can decide whether the equation is true by evaluating the expressions to see whether they have the same value. If the expressions do not have the same value, we can indicate this with the symbol ≠ (read as "is not equal to").

$$4 + 6 \neq 7 \quad 9 \neq 8 + 0$$

Example 5

Determine whether the equation is true.

(a) $5 + (2 + 1) = 4 + 3$

(b) $6 + 7 = (1 + 3) + 9$

SOLUTION

(a)

Left Side	*Right Side*
$5 + (2 + 1)$	$4 + 3$
$5 + \quad 3$	
8	7

Because $8 \neq 7$, the equation is false.

(b)

Left Side	*Right Side*
$6 + 7$	$(1 + 3) + 9$
	$4 \quad + 9$
13	13

Because $13 = 13$, the equation is true.

Your Turn 5

Determine whether the equation is true.

(a) $8 + 1 = (2 + 3) + 3$

(b) $6 + (5 + 4) = 7 + 8$

Answers: (a) False; (b) True

If one side of an equation is a variable expression, then we can't know whether the equation is true until we replace the variable with a specific number.

$x + 4 = 9$ We don't know whether the equation is true or false.
↓
$3 + 4 = 9$ If $x = 3$, the equation is false.

$x + 4 = 9$
↓
$5 + 4 = 9$ If $x = 5$, the equation is true.

A **solution** of an equation is a replacement for the variable that makes the equation true. For $x + 4 = 9$, the only solution is 5. Any other replacement for x would make the equation false. Finding a solution is called **solving** the equation.

Copyright © Houghton Mifflin Company. All rights reserved.

One method for solving simple equations is called solving *by inspection.* This is just a fancy way of saying that we figure out the solution just by looking at the equation and asking, "What replacement for the variable would make the equation true?"

Example 6

Solve the equation $5 + x = 11$.

SOLUTION

The equation states that the sum of some unknown number and 5 is 11.

"By inspection," we see that the solution is 6 because $5 + 6 = 11$ is true.

Your Turn 6

Solve the equation $x + 4 = 12$.

Answer: 8

When a simple equation involves subtraction, one approach is to rewrite the equation as a related addition and then solve that equation by inspection.

Example 7

Solve each equation.

(a) $x - 7 = 2$ **(b)** $13 - x = 8$

SOLUTION

(a) *Subtraction* *Addition*

$x - 7 = 2$ $2 + 7 = x$

From the related addition, we see that $x = 9$. Therefore, the solution is 9.

(b) *Subtraction* *Addition*

$13 - x = 8$ $8 + x = 13$

From the related addition, we see that x must be 5. Therefore, the solution is 5.

Your Turn 7

Solve each equation.

(a) $y - 5 = 3$

(b) $15 - a = 9$

Answers: (a) 8; (b) 6

Sometimes we need to simplify one or both sides of an equation before we can solve the equation by inspection.

Example 8

Solve the equation $x + 4 = 9 + 2$.

SOLUTION

$$x + 4 = 9 + 2$$
$$x + 4 = 11 \quad \text{Simplify the right side.}$$

The solution is 7 because $7 + 4 = 11$.

Your Turn 8

Solve the equation $3 + y = 6 + 2$.

3+y = 8

8 - 3 = 5

Answer: 5

Copyright © Houghton Mifflin Company. All rights reserved.

Example 9

Solve the equation $(x + 3) + 5 = 15$.

SOLUTION

$(x + 3) + 5 = 15$

$x + (3 + 5) = 15$ — The Associative Property of Addition allows us to change the grouping.

$x + 8 = 15$ — Simplify the left side.

The solution is 7 because $7 + 8 = 15$.

Your Turn 9

Solve the equation $2 + (4 + x) = 11$.

Answer: 5

1.6 Quick Reference

A. Expressions

1. A **numerical expression** is any combination of numbers and operations. A **variable expression** is any combination of numbers, *variables,* and operations.	Numerical expression: $2 + 6$ Variable expression: $x + 1$
2. We **evaluate** a numerical expression by performing the indicated operations. We evaluate a variable expression by replacing each variable with a specific value and performing the indicated operations.	Evaluate $82 + 13$. $82 + 13 = 95$ Evaluate $b - a$ for $a = 3$ and $b = 11$. $b - a$ ↓ ↓ $11 - 3 = 8$
3. We **simplify** an expression by performing any operations that we can.	$(x + 8) + 2 = x + (8 + 2)$ $= x + 10$

B. Equations

1. An **equation** is a statement that two expressions (the *left side* and the *right side*) have the same value.	The equation $x + 5 = 7 + 2$ means that $x + 5$ and $7 + 2$ have the same value.
2. An equation may be true or false.	True: $8 + 4 = 12$ False: $9 + 5 = 4$
3. The symbol ≠ is read as "is not equal to."	$3 + 6 \neq 36$

Copyright © Houghton Mifflin Company. All rights reserved.

4. A **solution** of an equation is a replacement for the variable that makes the equation true.

The solution of $x + 4 = 9$ is 5 because $5 + 4 = 9$.

5. We solve simple equations such as $x + 7 = 13$ *by inspection* by asking, "What number plus 7 equals 13?"

$$x + 7 = 13$$
$$\downarrow$$
$$6 + 7 = 13$$

The solution is 6.

6. To solve a simple equation involving subtraction, one approach is to write the related addition and solve it by inspection.

Solve $11 - x = 4$.
Write the related addition:

$$4 + x = 11$$

Solve by inspection.
The solution is 7.

7. Sometimes we need to simplify one or both sides of an equation before we can solve the equation by inspection.

$$(x + 3) + 8 = 11$$
$$x + (3 + 8) = 11$$
$$x + 11 = 11$$

The solution is 0.

Copyright © Houghton Mifflin Company. All rights reserved.

1.6 Exercises

A. Expressions

1. We call $x + 3$ a(n) variable expression, whereas we call $4 + 3$ a(n) numerical expression.

2. We evaluate a variable expression by replacing each variable with a specific value and performing the indicated operations.

In Exercises 3–10, evaluate the expression for the indicated value of the variable.

3. $x + 7$ for $x = 3$

4. $y + 9$ for $y = 2$

5. $b - 8$ for $b = 12$

6. $w - 3$ for $w = 9$

7. $x + 5 + 9$ for $x = 2$

8. $10 + c + 3$ for $c = 7$

9. $6 - a$ for $a = 1$

10. $12 - b$ for $b = 5$

In Exercises 11–18, evaluate the expression for the given values of the variables.

11. $a + b$ for $a = 5$ and $b = 7$

12. $x + y$ for $x = 2$ and $y = 8$

13. $w - z$ for $w = 8$ and $z = 3$

14. $m - n$ for $m = 8$ and $n = 7$

15. $m + n + 9$ for $m = 3$ and $n = 6$

16. $8 + w + z$ for $w = 1$ and $z = 4$

17. $x + y + z$ for $x = 5, y = 1$, and $z = 2$

18. $b + c + a$ for $a = 3, b = 2$, and $c = 0$

In Exercises 19–26, evaluate the expression for the given values of the variables.

19. $a + b + 439; a = 836, b = 987$

20. $m + 159 + n; m = 402, n = 645$

21. $x - y; x = 1{,}399$ and $y = 599$

22. $b - c; b = 5{,}000$ and $c = 3{,}340$

23. $a + b + c; a = 29, b = 36, c = 98$

24. $x + y + z; x = 115, y = 103, z = 114$

25. $x + y - c; x = 30, y = 48, c = 12$

30+48 − 12 −30 52
78

26. $p - q + r; p = 23, q = 15, r = 19$

23 − 15
8+

Copyright © Houghton Mifflin Company. All rights reserved.

In Exercises 27–34, simplify the expression.

27. $(b + 7) + 5$ **28.** $(y + 2) + 6$ **29.** $3 + (8 + y)$ **30.** $1 + (9 + x)$

31. $(z + 2) + 10$ **32.** $(x + 3) + 12$ **33.** $4 + (5 + x)$ **34.** $2 + (4 + a)$

B. Equations

35. A(n) ______ is a statement that two expressions have the same value.

36. A(n) ______ of an equation is a replacement for the variable that makes the equation true.

In Exercises 37–40, determine whether the given number is a solution of the equation.

37. $4 + y = 12$; 8 **38.** $a + 8 = 17$; 7 **39.** $8 + 14 = 5 + b$; 6 **40.** $4 + 7 = x + 1$; 10

In Exercises 41–52, solve the given equation.

41. $a + 4 = 13$ **42.** $t + 7 = 9$ **43.** $x - 3 = 5$ **44.** $x - 2 = 2$

45. $8 = x + 6$ **46.** $6 = c + 2$ **47.** $5 - x = 1$ **48.** $6 - x = 0$

49. $7 + y = 11$ **50.** $16 = w + 8$ **51.** $x - 8 = 6$ **52.** $16 - x = 7$

In Exercises 53–62, simplify and solve the given equation.

53. $b + 9 = 8 + 6$ **54.** $9 + 6 = x + 8$ **55.** $x - 8 = 3 - 1$ **56.** $4 + 6 - x = 3$

57. $9 + 4 - x = 8 - 5$ **58.** $7 - 2 - x = 1 + 2$ **59.** $(n + 5) + 1 = 10$ **60.** $(a + 4) + 2 = 14$

61. $11 = 6 + (2 + x)$ **62.** $9 = 7 + (1 + t)$

Writing and Concept Extension

63. Explain why *any* number is a solution of $x + 5 = 5 + x$.

Copyright © Houghton Mifflin Company. All rights reserved.

64. Explain why *no* number is a solution of $x + 5 = x$.

In Exercises 65 and 66, solve the given equation.

65. $x + 0 = x$

66. $x + 5 = x + 1$

In Exercises 67–70, use the Associative and Commutative Properties of Addition to simplify the expression.

67. $(4 + x) + 5$

68. $7 + (a + 6)$

69. $9 + (b + 2)$

70. $(1 + y) + 9$

Copyright © Houghton Mifflin Company. All rights reserved.

CHAPTER 1 REVIEW EXERCISES

Section 1.1

1. The numbers 1, 2, 3, 4, . . . are called the ______ numbers. If we include 0, the numbers are called the ______ numbers.

2. The statement 2 < 5 is called a(n) ______.

3. Graph the numbers 8 and 5 on the number line.

0 1 2 3 4 5 6 7 8

In Exercises 4 and 5, insert < or > to make the statement true.

4. 7 ___ 0

5. 25 ___ 41

6. Arrange the numbers 132, 21, 12, and 139 from largest to smallest.

7. Determine the distance on a number line between 12 and 5.

8. In the number 7,382, what does the digit 3 represent?

9. Write 7,045 in expanded form.

10. Write 500 + 60 + 3 in standard form.

11. *Population of Pittsburgh* In 2002, the population of Pittsburgh was 334,563. Write this number in words.

12. *College Enrollment* The enrollment at Dalton College is three thousand six. Write this number in standard form.

Section 1.2

13. If the price of an item is $39, we say that $40 is a(n) ______ of the price.

14. The way that we round a number to the nearest hundred is determined by the digit in the ______ place.

15. Round 6,987 to the nearest hundred.

16. Round 4,342,199 to the nearest million.

17. *Area of Kansas* The area of Kansas is 81,823 square miles. Round this number to the nearest thousand.

18. *Car Lease* A car leases for $449 per month. Round this number to the nearest hundred.

Copyright © Houghton Mifflin Company. All rights reserved.

Restaurant Spending For Exercises 19–24, the bar graph shows the average amount per year that a person spent on meals at restaurants (including fast-food restaurants) in three countries. (Source: *Euromonitor.*)

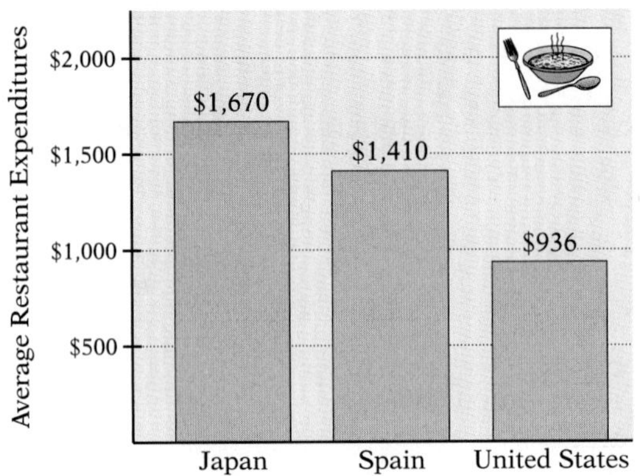

19. In which country is spending lowest?

20. In which country is spending highest?

21. Use the symbol > to compare spending in Japan to that in the United States.

22. Use the symbol < to compare spending in Spain to that in the United States.

23. Round each amount to the nearest thousand.

24. Round each amount to the nearest hundred.

Section 1.3

25. We call both 7 + 9 and the result 16 a(n) __________, and we call 7 and 9 __________.

26. Write the phrase "8 more than n" as a sum.

27. *Cost of a Sweater* A sweater sells for $32 plus $5 for monogramming. Write an expression for the total cost of a monogrammed sweater.

28. What is a variable?

In Exercises 29–31, use the indicated property to rewrite the sum.

29. Associative Property of Addition: $(a + 7) + b$

30. Commutative Property of Addition: $m + n$

31. Addition Property of 0: $y + 0$

In Exercises 32–34, identify the property that justifies the statement.

32. $7 + a = a + 7$

Copyright © Houghton Mifflin Company. All rights reserved.

33. $b = 0 + b$

34. $(x + 6) + 9 = x + (6 + 9)$

Section 1.4

In Exercises 35 and 36, determine the sum.

35.
$$\begin{array}{r} 4{,}093 \\ +\ \ 998 \\ \hline \end{array}$$

36.
$$\begin{array}{r} 49 \\ 8 \\ 67 \\ +\ 324 \\ \hline \end{array}$$

37. Explain how to estimate the sum of 193 and 512.

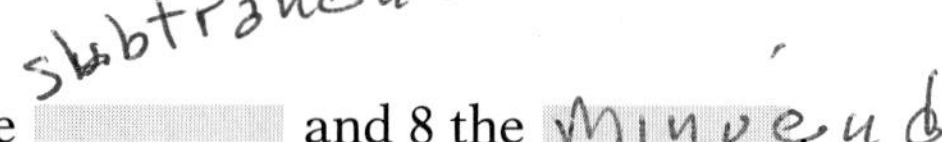

38. In the difference of 8 and 2, we call 2 the ________ and 8 the ________.

39. Write the phrase "7 less than n" as a difference.

In Exercises 40 and 41, determine the difference.

40.
$$\begin{array}{r} 638 \\ -\ 432 \\ \hline \end{array}$$

41.
$$\begin{array}{r} 7{,}003 \\ -\ 4{,}397 \\ \hline \end{array}$$

42. *Household Expenses* For a certain month, a couple's major expenses were \$680 for rent, \$346 for a car payment, \$178 for utilities, and \$530 for food. What was the total of the major expenses?

43. *Best-Selling Cars* In 2000, the three top-selling cars in the United States were the Ford Taurus, the Honda Accord, and the Toyota Camry. The sales (in thousands) were 382, 405, and 423, respectively. (Source: American Automobile Manufacturers Association.) What were the total sales for the three models?

44. *Registered Dogs* The Labrador retriever and the Rottweiler are the most popular registered dogs. The numbers registered are 149,505 and 89,867, respectively. (Source: American Kennel Club.) How many more Labrador retrievers than Rottweilers are registered?

Copyright © Houghton Mifflin Company. All rights reserved.

Section 1.5

State Populations For Exercises 45–50, the bar graph shows the 2000 populations of four states. (Source: *World Almanac.*)

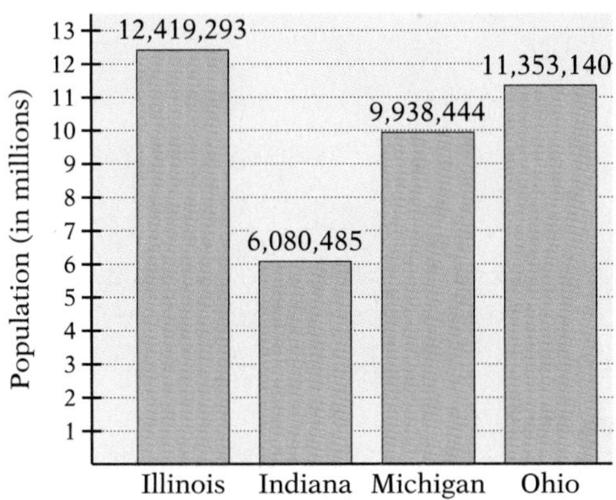

45. Estimate the total population of the four states shown in the bar graph.

46. Estimate the difference of the populations of Ohio and Michigan.

47. What is the total population of the two least populous states shown in the bar graph?

48. What is the total population of Michigan and Ohio?

49. What is the difference of the populations of the most and least populous states shown in the bar graph?

50. By how much does the total population of Indiana and Michigan exceed the population of Illinois?

51. The sum of the lengths of the sides of a polygon is called the ______.

52. A polygon with three sides is called a(n) ______.

Copyright © Houghton Mifflin Company. All rights reserved.

53. Determine the perimeter of the geometric figure.

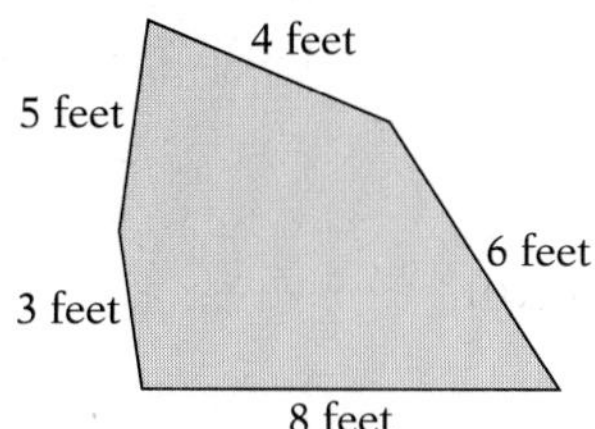

54. *Field Fencing* How many yards of chain-link fence are needed to enclose a triangular field if the lengths of the sides of the field are 124 yards, 92 yards, and 78 yards?

Section 1.6

55. A combination of numbers, variables, and operations is called a(n) variable expression.

In Exercises 56–58, evaluate the expression for the indicated value of the variable.

56. $y + 8$ for $y = 3$

57. $a + b + 2$ for $a = 5$ and $b = 1$

58. $n - m$ for $m = 45$ and $n = 134$

59. Simplify the expression $4 + (7 + a)$.

60. A statement that two expressions have the same value is called a(n) _____.

In Exercises 61–64, solve the given equation.

61. $5 + x = 14$

62. $17 - n = 9$

63. $12 - 7 = 5 + y$

64. $(n + 3) + 5 = 10 + 4$

Copyright © Houghton Mifflin Company. All rights reserved.

CHAPTER 1 TEST

 **1.** In the sum $5 + 3$, the 5 and 3 are both called ________, whereas in the difference $5 - 3$, the 5 is called the ________ and the 3 is called the ________.

 **2.** *Airport Traffic* Chicago's O'Hare Airport had a total of 72,136,000 passengers depart and arrive in 2000. (Source: *World Almanac.*)
(a) Write this number in words.

(b) Round this number to the nearest ten million.

3. In 1927, Charles Lindbergh flew three thousand six hundred ten miles from New York to Paris. Write this number of miles in standard form.

4. Insert $>$ or $<$ to make the statement true.

(a) 0 ____ 37 **(b)** 26 ____ 21

5. In parts (a) and (b), identify the property from the following list that justifies the given statement.
(i) Associative Property of Addition
(ii) Commutative Property of Addition

(a) $3 + x = x + 3$ **(b)** $(y + 2) + 4 = y + (2 + 4)$

In Questions 6 and 7, perform the indicated operation.

6. $\begin{array}{r} 987 \\ +\ 313 \\ \hline \end{array}$ **7.** $\begin{array}{r} 3{,}436 \\ -\ 1{,}368 \\ \hline \end{array}$

 **8.** The phrase "5 increased by n" indicates ________, whereas the phrase "17 less than 30" indicates ________.

9. Write each phrase in Question 8 as an expression. Indicate whether the expression is a numerical expression or a variable expression.

10. Find the sum of 324, 79, and 1,040.

11. *Cost of a VCR* A VCR costs $123, and the charge for an extended warranty is $15. Write a sum for the cost of a VCR with an extended warranty.

12. *Emergency Room* During a holiday weekend, an emergency room treated 78 people on Friday, 107 on Saturday, and 96 on Sunday. How many people were treated at the emergency room that weekend?

13. *Telephone Bill* Before a second telephone line was installed, a person's monthly bill was $37. After the second line was installed, the monthly bill was $61. What was the monthly cost of the second line?

Copyright © Houghton Mifflin Company. All rights reserved.

14. *Airline Passengers* In 2000, Delta Airlines carried 106 million passengers and United Airlines carried 84 million passengers.
(a) What is the total number of passengers that the two airlines carried?

(b) How many more passengers flew Delta than United?

Library Spending The bar graph shows the amounts (in millions of dollars) that libraries spent during a year for various types of books. (Source: Book Industry Study Group.) Use the information in the bar graph to answer Questions 15–18.

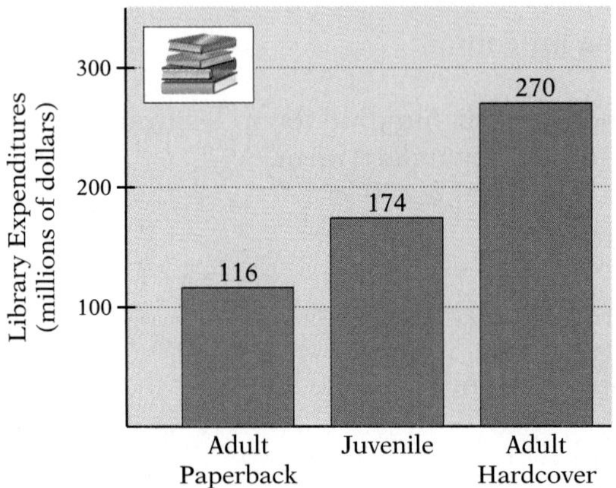

15. For which category is spending lowest?

16. Use the symbol < to compare the amounts spent for the two categories of adult books.

17. What is the total amount spent for books in the three categories?

18. By how much does the total spending for adult books exceed spending for juvenile books?

19. How many inches of trim are needed for a border of a rectangular tablecloth that is 64 inches long and 42 inches wide?

20. Determine the perimeter of the geometric figure.

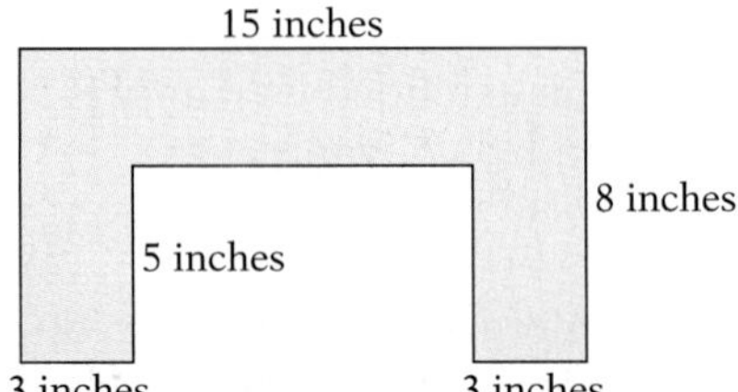

21. Simplify $(x + 3) + 7$.

22. Evaluate $x + 3 + y$ for $x = 4$ and $y = 9$.

23. Solve $n + 7 = 15$.

24. Solve $15 - y = 9$.

Copyright © Houghton Mifflin Company. All rights reserved.

CHAPTER 2

Introduction to Algebra: Multiplication and Division

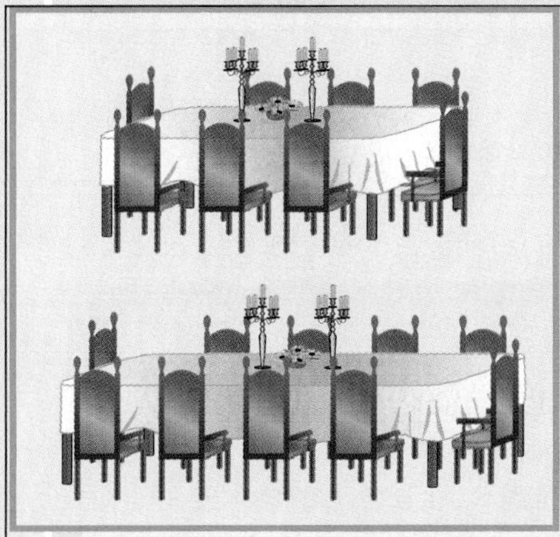

At a banquet, 10 tables that seat 8 people are set up.

1. If all the tables are filled, how many people attended the banquet?
2. If tables that seat 10 people are used instead, how many tables are needed?

The first question can be answered with multiplication:

(10 tables) $\times$ (8 people per table) = 80 people

The second question requires division:

(80 people) $\div$ (10 people per table) = 8 tables

Chapter Snapshot

In this chapter, we present methods for multiplying and dividing whole numbers. We introduce the use of an exponent, which represents repeated multiplication. Because expressions can involve addition, subtraction, multiplication, division, exponents, and grouping symbols, we present the agreement on the order in which operations are performed.

To give you a start toward algebra, we discuss methods for evaluating expressions and solving equations that involve multiplication, division, and exponents. Finally, we present formulas for finding the area, volume, and surface area of various geometric figures.

The first two chapters of this book give you a solid foundation in the algebra of whole numbers. Mastering these topics is essential to your success in algebra.

Copyright © Houghton Mifflin Company. All rights reserved.

For online resources, visit the web site **math.college.hmco.com/students** and follow the links to Hubbard/Robinson, *Prealgebra.*

Some Friendly Advice . . .

Have you ever wished that you could be an instructor for a day?

An excellent way to study a section of material is to pretend that you will be teaching it the next day. If you want to lower your sights and just pretend that you need to help a fellow student, that will work, too. The point is that you can't teach a class or tutor a friend unless you know the material inside and out.

Here are some pointers.

1. Read short passages of the text and then see whether you can restate what you've read in your own words, as though you were explaining it to someone else.
2. After you have worked an example, try making up an example of your own and think about how you would present it.
3. Select a few exercises from each group and consider how you would explain them to another student who is having trouble with them.
4. Finally, the best fantasy of all, make up a test. Think about what has been discussed and look back through homework assignments. What would *you* include in the test? (Of course, you should be able to answer your own questions.)

WARM-UP SKILLS

The following questions review concepts and skills that you will need in Chapter 2.

Perform the indicated operation.

1. $9 + 9 + 9 + 9$
2. $100 + 100 + 100$
3. $3 + (8 + 4)$
4. $\begin{array}{r} 32 \\ +\ 280 \\ \hline \end{array}$
5. $\begin{array}{r} 99 \\ -\ 84 \\ \hline \end{array}$
6. $\begin{array}{r} 151 \\ -\ 112 \\ \hline \end{array}$
7. Identify the property of addition that is illustrated.
   (a) $x + 8 = 8 + x$ (b) $(a + 4) + b = a + (4 + b)$
8. What is the perimeter of a square whose sides are 3 inches long?
9. Evaluate $x + y$ for $x = 9$ and $y = 22$.
10. Solve $y + 4 = 12$.

Copyright © Houghton Mifflin Company. All rights reserved.

2.1 Algebra of Multiplication

SUGGESTIONS FOR SUCCESS

Take charge of your own learning!

An excellent way to complete your reading and exercise assignments is to make a detailed list of the questions that you have. The list should include any examples or text passages that were unclear to you. It should also include specific exercises that you were unable to work.

In this way, when you seek help in or outside of class, you can ask your questions in an organized way. Instructors appreciate such a businesslike approach and will be impressed with your determination to succeed.

A Multiplication Concepts and Vocabulary

During each of the past 5 business days, an electronics store sold 2 computers. What total number of computers were sold during this period? One approach to the problem is to add the number of computers sold each day.

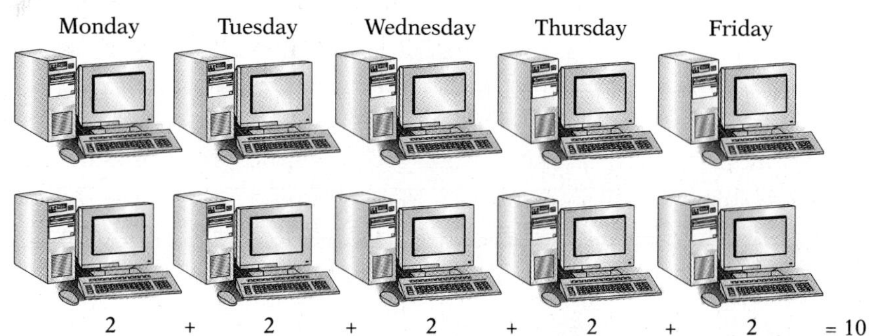

The repeated addition of the digit 2 can be written in words as "5 times 2." This more compact way of describing the total is called **multiplication,** which is the repeated addition of the same number.

We can represent "5 times 2" in any of the following ways:

These numbers are called **factors.** (pointing to 5 and 2)

$$5 \times 2 = 10 \qquad 5(2) \qquad (5)2 \qquad (5)(2) \qquad 5 \cdot 2$$

This number is called the **product.** (pointing to 10)

The numbers 5 and 2 are both called **factors** of 10. The multiplication, 5×2, and the result, 10, are both called **products.**

NOTE In algebra, the multiplication symbol $\times$ could be confused with the variable symbol x. Therefore, we will usually represent multiplication with one of the other forms.

When one or more of the factors is a variable, we have a variety of ways to write the product:

$$5 \cdot x \qquad 5x \qquad 5(x) \qquad a(b) \qquad (a)(b) \qquad ab$$

Copyright © Houghton Mifflin Company. All rights reserved.

Example 1

In each part, identify the products and factors.

(a) $7 \cdot 3 = 21$ **(b)** $8 = 4x$

SOLUTION

(a) The products are $7 \cdot 3$ and 21. The factors are 7 and 3.

(b) The products are 8 and $4x$. The factors are 4 and x.

Your Turn 1

In each part, identify the products and factors.

(a) $2 \cdot 3 \cdot 4 = 24$

(b) $9y = 45$

Answers: (a) Products: $2 \cdot 3 \cdot 4$ and 24 Factors: 2, 3, 4; (b) Products: $9y$ and 45 Factors: 9 and y

NOTE There will be times when it will be helpful to be aware that 1 is a factor of every number. For example, $6 = 1 \cdot 6$. More generally, we can say that two factors of any number x are 1 and x because $x = 1 \cdot x$. However, when we list the factors of a number, we usually omit the 1.

The product of two digits is called a *basic multiplication fact*.

×	*0*	*1*	*2*	*3*	*4*	*5*	*6*	*7*	*8*	*9*
0	0	0	0	0	0	0	0	0	0	0
1	0	1	2	3	4	5	6	7	8	9
2	0	2	4	6	8	10	12	14	16	18
3	0	3	6	9	12	15	18	21	24	27
4	0	4	8	12	16	20	24	28	32	36
5	0	5	10	15	20	25	30	35	40	45
6	0	6	12	18	24	30	36	42	48	54
7	0	7	14	21	28	35	42	49	56	63
8	0	8	16	24	32	40	48	56	64	72
9	0	9	18	27	36	45	54	63	72	81

LEARNING TIP

Once you understand how the entries in the table were determined, you need to memorize all the basic multiplication facts. As with addition, flash cards can be very helpful.

This table is read in the same way that the table of basic addition facts is read. Each entry is based on the idea of repeated addition. For example,

$$4 \cdot 9 = 9 + 9 + 9 + 9 = 36$$

When you are working with applications, you should be aware of the following words and phrases that indicate multiplication.

Word Description	*Product*
3 *times* 8	$3 \cdot 8$
Twice some number x	$2x$
10 *multiplied by* 5	$10(5)$
The *product* of 4 and 8	$(4)(8)$

In an applied problem, the words *factor* and *product* do not always appear. However, if you realize that the problem involves repeated addition, then you can represent the information with multiplication.

Copyright © Houghton Mifflin Company. All rights reserved.

Example 2

A person earns \$7 per hour and works an 8-hour day. Write a product that represents the person's income that day.

SOLUTION

The income can be found by adding the hourly wage 8 times. An easier approach is to use multiplication: $8 \cdot 7$.

Your Turn 2

A part-timer works 4 hours per day and is on the job 5 days per week. Write a product that represents the number of hours per week that this person works.

Answer: $4 \cdot 5$

B Properties of Multiplication

The product $5 \cdot 0 = 0 + 0 + 0 + 0 + 0 = 0$. In fact, no matter how many times we repeatedly add 0, the result will always be 0.

The Multiplication Property of 0

If the variable a is used to represent any number, then

$$a \cdot 0 = 0 \qquad \text{and} \qquad 0 \cdot a = 0$$

In words, if one factor is 0, then the product is 0.

The table of basic multiplication facts reveals two other properties of multiplication. First, whenever we multiply a number and 1, the product is the original number. For example, $3 \cdot 1 = 3$ and $1 \cdot 8 = 8$.

The Multiplication Property of 1

If the variable a is used to represent any number, then

$$a \cdot 1 = a \qquad \text{and} \qquad 1 \cdot a = a$$

In words, the product of 1 and any number is the number.

The other property revealed by the table is that the order of the factors has no effect on the product. For example, $4 \cdot 7 = 28$ and $7 \cdot 4 = 28$.

The Commutative Property of Multiplication

If the variables a and b are used to represent any two numbers, then

$$a \cdot b = b \cdot a$$

In words, two numbers can be multiplied in either order and the result will be the same.

Copyright © Houghton Mifflin Company. All rights reserved.

As with addition, the Commutative Property of Multiplication allows us to get by with memorizing only half of the basic multiplication facts. If you know that $9 \cdot 5 = 45$, then you also know that $5 \cdot 9 = 45$.

Example 3

Use the Commutative Property of Multiplication to rewrite the expression.

(a) $9 \cdot 7$

(b) $x \cdot 4$

(c) xy

SOLUTION

The Commutative Property of Multiplication allows us to change the order of the factors.

(a) $9 \cdot 7 = 7 \cdot 9$

(b) $x \cdot 4 = 4 \cdot x = 4x$

(c) $xy = yx$

Your Turn 3

Use the Commutative Property of Multiplication to rewrite the expression.

(a) $3 \cdot 8$

(b) $y \cdot 9$

(c) zx

Answers: (a) $8 \cdot 3$; (b) $9 \cdot y = 9y$; (c) xz

The Associative Property of Addition allows us to group the addends in any way we wish. Is there a corresponding property for multiplication?

$(3 \cdot 2) \cdot 4 = 6 \cdot 4 = 24$ Group the first two factors.

$3 \cdot (2 \cdot 4) = 3 \cdot 8 = 24$ Group the last two factors.

We see that changing the grouping of the factors has no effect on the product. This is true in general.

The Associative Property of Multiplication

If the variables a, b, and c are used to represent any three numbers, then

$$(a \cdot b) \cdot c = a \cdot (b \cdot c)$$

In words, we can group the factors of a product in any way and the product is the same.

The Commutative and Associative Properties of Multiplication have practical advantages when we need to multiply more than two numbers. Consider the following two multiplication problems:

$2 \cdot 7 \cdot 8 \cdot 5$ $\quad\quad$ $5 \cdot 637 \cdot 2$

If we simply multiply from left to right, the arithmetic becomes messy. However, our two properties allow us to reorder the factors and group them however we wish. Here are some good choices.

$(2 \cdot 5) \cdot (7 \cdot 8) = 10 \cdot 56$ $\quad\quad$ $(5 \cdot 2) \cdot 637 = 10 \cdot 637$

As we will see in the next subsection, multiplying by 10 is easy.

Copyright © Houghton Mifflin Company. All rights reserved.

Example 4

Use the Associative Property of Multiplication to rewrite the expression.

(a) $5 \cdot (2 \cdot 7)$

(b) $8(2x)$

(c) $(3y) \cdot z$

SOLUTION

The Associative Property of Multiplication allows us to change the grouping of the factors.

(a) $5 \cdot (2 \cdot 7) = (5 \cdot 2) \cdot 7$

(b) $8(2x) = (8 \cdot 2)x$

(c) $(3y) \cdot z = 3(yz)$

Your Turn 4

Use the Associative Property of Multiplication to rewrite the expression.

(a) $6 \cdot (1 \cdot 3)$

(b) $7(4y)$

(c) $(ax)y$

Answers: (a) $(6 \cdot 1)3$; (b) $(7 \cdot 4)y$; (c) $a(xy)$

The only purpose of Example 4 is to show how the Associative Property of Multiplication can be used to regroup the factors. There are differing reasons for wanting to do such regrouping.

We can continue part (a) by writing $5 \cdot (2 \cdot 7) = (5 \cdot 2) \cdot 7 = 10 \cdot 7$. As we will see, multiplying 7 by 10 is easier than multiplying 14 by 5, which is what $5 \cdot (2 \cdot 7)$ means.

In part (b), we can extend our work to $8(2x) = (8 \cdot 2)x = 16x$. The product $16x$ has a simpler look than the product $8(2x)$. Any time we can simplify things, we will want to.

Finally, the Associative Property of Multiplication allows us to place parentheses in a product wherever we want them, but the property also allows us to remove them. In part (c), we can write $(3y) \cdot z = 3(yz) = 3yz$. By eliminating the parentheses, we have again obtained a simpler look. Simplifying is an important aspect of algebra.

C Multiples of 10

We can write the number 60 as a repeated addition and as a product.

$$60 = 10 + 10 + 10 + 10 + 10 + 10 = 6 \cdot 10$$

Reversing this equation, we see that $6 \cdot 10 = 60$. This suggests that we can multiply $6 \cdot 10$ by writing 6 followed by one 0, which is the number of 0s in 10. This idea can be extended to multiplying any number by 10, 100, 1,000, 10,000, and so on.

Multiplying Any Number by 10, 100, 1,000, . . .

To multiply any number n by 10, 100, 1,000, . . . , follow these steps.

1. Count the number of 0s in the factor 10, 100, 1,000, ...
2. Write the number n followed by the number of 0s that you counted in step 1.

Copyright © Houghton Mifflin Company. All rights reserved.

Example 5

Multiply.

(a) $8 \cdot 100$ **(b)** $58 \cdot 10$ **(c)** $632 \cdot 1{,}000$

SOLUTION

(a) There are two 0s in 100.

$8 \cdot 100 = 800$ Write 8 followed by two 0s.

(b) There is one 0 in 10.

$58 \cdot 10 = 580$ Write 58 followed by one 0.

(c) There are three 0s in 1,000.

$632 \cdot 1{,}000 = 632{,}000$ Write 632 followed by three 0s.

Your Turn 5

Multiply

(a) $9 \cdot 1{,}000$

(b) $36 \cdot 100$

(c) $425 \cdot 10$

Answers: (a) 9,000; (b) 3,600; (c) 4,250

We give the following numbers special names:

10, 20, 30, 40, . . .	Multiples of 10
100, 200, 300, 400, . . .	Multiples of 100
1,000, 2,000, 3,000, 4,000, . . .	Multiples of 1,000
And so on	

We can extend the procedure for multiplying by 10, 100, 1,000, and so on, to other products in which one of these multiples is a factor.

Example 6

Multiply $7 \cdot 30$.

SOLUTION

$7 \cdot 30 = 7 \cdot (3 \cdot 10)$	
$= (7 \cdot 3) \cdot 10$	Associative Property of Addition
$= 21 \cdot 10$	$7 \cdot 3 = 21$
$= 210$	Write 21 followed by one 0.

Your Turn 6

Multiply $6 \cdot 70$.

Answer: 420

Example 7

Multiply $4 \cdot 800$.

SOLUTION

$4 \cdot 800 = 4 \cdot (8 \cdot 100)$	
$= (4 \cdot 8) \cdot 100$	Associative Property of Addition
$= 32 \cdot 100$	$4 \cdot 8 = 32$
$= 3{,}200$	Write 32 followed by two 0s.

Your Turn 7

Multiply $9 \cdot 500$.

Answer: 4,500

Copyright © Houghton Mifflin Company. All rights reserved.

A shortcut to finding products such as those in Examples 6 and 7 is to write the product of the leading digits followed by the number of 0s in the multiple of 10, 100, and so on.

Example 8

Multiply $3 \cdot 6{,}000$.

SOLUTION

Multiply the leading digits: $3 \cdot 6 = 18$. The factor 6,000 has three 0s.

$3 \cdot 6{,}000 = 18{,}000$ Write 18 followed by three 0s.

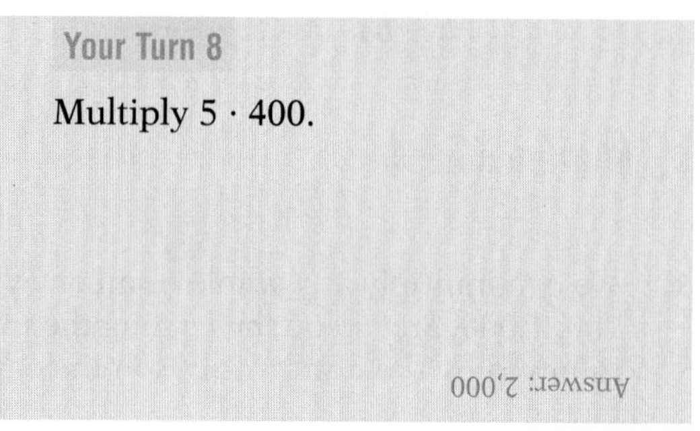

Your Turn 8

Multiply $5 \cdot 400$.

Answer: 2,000

2.1 Quick Reference

A. Multiplication Concepts and Vocabulary

1. **Multiplication** is the repeated addition of the same number.

   We can write $5 + 5 + 5$ as $3 \cdot 5$ or $3(5)$.

2. Numbers that are being multiplied are called **factors;** the multiplication and the result are called a **product.**

   For $8 \cdot 2$, the factors are 8 and 2; the product is $8 \cdot 2$, or 16.

3. The *basic multiplication facts* are all possible products of two digits.

B. Properties of Multiplication

1. The Multiplication Property of 0

   If a is any number, then $a \cdot 0 = 0$.

   $4 \cdot 0 = 0$ and $0 \cdot 9 = 0$

2. The Multiplication Property of 1

   If a is any number, then $a \cdot 1 = a$.

   $5 \cdot 1 = 5$ and $1 \cdot 7 = 7$

3. The Commutative Property of Multiplication

   If a and b are any numbers, then

   $$a \cdot b = b \cdot a$$

   $6 \cdot 8 = 8 \cdot 6$

   $x \cdot 3 = 3x$

Copyright © Houghton Mifflin Company. All rights reserved.

4. The Associative Property of Multiplication

If a, b, and c are any numbers, then

$$(a \cdot b) \cdot c = a \cdot (b \cdot c)$$

$(4 \cdot 5) \cdot 2 = 4 \cdot (5 \cdot 2)$

$2(7x) = (2 \cdot 7)x$

C. Multiples of 10

1. We can multiply any whole number by 10, 100, 1,000, and so on, by writing the number followed by the number of 0s in the other factor.

$2 \cdot 10 = 20$

$62 \cdot 100 = 6{,}200$

$45 \cdot 1{,}000 = 45{,}000$

2. Special names:

Multiples of 10: 10, 20, 30, 40, . . .
Multiples of 100: 100, 200, 300, . . .
Multiples of 1,000: 1,000, 2,000, . . .

3. To multiply a number by a multiple of 10, 100, 1,000, and so on, write the product of the leading digits followed by the number of 0s in the multiple.

To multiply $7 \cdot 400$, multiply $7 \cdot 4 = 28$, followed by two 0s: $7 \cdot 400 = 2{,}800$.

Copyright © Houghton Mifflin Company. All rights reserved.

2.1 Exercises

A. Multiplication Concepts and Vocabulary

1. In the expression $3 \cdot 7$, the numbers 3 and 7 are called ________.

2. The ________ of 9 and 5 can be written $9 \cdot 5$.

In Exercises 3–8, identify the factors in the expression.

3. $5 \cdot 11$
4. $7 \cdot 3$
5. $7x$
6. $3y$
7. $2ab$
8. $5mn$

In Exercises 9–14, write the word description as a product.

9. 9 multiplied by 7
10. The product of 6 and 0
11. Three times 5
12. Twice 8
13. The product of 4 and some number y
14. Some number a multiplied by 2
15. *Total Cost* Acme Supplies bought 58 gizmos at $6 each. Write a product that represents the total cost.
16. *Grocery Purchase* Round roast beef sells for $2 per pound. Write a product that represents the cost of a 6-pound roast.
17. *Class Makeup* A class had 12 male students, and there were 3 times as many female students. Write a product that represents the number of female students in the class.
18. *First-Year Student Enrollment* A college had 1,200 first-year students in 1995, but that number doubled by the year 2000. Write a product that represents the number of first-year students in 2000.
19. *Board Lengths* A board was cut into two pieces. One piece was x feet long, and the other piece was twice as long as the first piece. Write a product that represents the length of the second piece.
20. *Photo Enlargement* A photo that is w inches wide was enlarged to 4 times its original dimensions. Write a product that represents the width of the enlargement.

B. Properties of Multiplication

21. The ________ Property of Multiplication states that the result of multiplying two numbers is the same regardless of the order in which they are multiplied.

22. The ________ Property of Multiplication states that the result of multiplying three numbers is the same regardless of which two numbers are multiplied first.

Copyright © Houghton Mifflin Company. All rights reserved.

In Exercises 23–26, use the Commutative Property of Multiplication to rewrite the expression.

23. $5 \cdot 12$ **24.** $23 \cdot 4$ **25.** $c \cdot 8$ **26.** $5y$

In Exercises 27–30, use the Associative Property of Multiplication to rewrite the expression.

27. $5 \cdot (6 \cdot 9)$ **28.** $(3 \cdot 7) \cdot 8$ **29.** $7(3b)$ **30.** $(8w)z$

In Exercises 31–34, use the indicated property to fill in the blank.

31. Multiplication Property of 0: $\_\_\_ \cdot y = 0$

32. Multiplication Property of 1: $\_\_\_ \cdot c = c$

33. Commutative Property of Multiplication: $x \cdot 7 = \_\_\_ \cdot x$

34. Associative Property of Multiplication: $\_\_\_(yz) = (5 \cdot y)z$

In Exercises 35–38, identify the property that justifies the given statement.

35. $d = d \cdot 1$ **36.** $0 = 0 \cdot z$

37. $c \cdot 10 = 10c$ **38.** $5(3y) = (5 \cdot 3)y$

C. Multiples of 10

39. Numbers such as 30, 40, and 50 are ______ of 10.

40. To multiply $45 \cdot 1{,}000$, write 45 followed by three ______.

In Exercises 41–52, multiply.

41. $8 \cdot 100$ **42.** $5 \cdot 1{,}000$ **43.** $32 \cdot 10$ **44.** $49 \cdot 100$

45. $497 \cdot 1{,}000$ **46.** $302 \cdot 10$ **47.** $8 \cdot 600$ **48.** $5 \cdot 30$

49. $3{,}000 \cdot 7$ **50.** $4{,}000 \cdot 9$ **51.** $4 \cdot 50$ **52.** $5 \cdot 6{,}000$

Applications

53. *Child Care Costs* A mother pays $100 per week for child care. What is the cost of child care for 52 weeks?

54. *Study Time* You should study 2 hours for each hour of class time. How many hours per week should you study for a course that meets 1 hour each day and 5 days per week?

Copyright © Houghton Mifflin Company. All rights reserved.

55. *Car Rental* A rental car costs \$30 per day. What is the cost of renting the car for 8 days?

56. *Coffee Purchase* The owner of a specialty coffee shop purchased 50 pounds of coffee that sells for \$7 per pound. What was the total cost?

57. *Walking and Calories* A person burns 125 calories by walking 1 mile. How many calories does the person burn on a 10-mile walk?

58. *Calorie Content of Milk* One 8-ounce serving of low-fat milk contains 90 calories. How many calories do four 8-ounce servings contain?

59. *Chair Arrangement* A meeting room is arranged with 14 rows of chairs and 10 chairs per row. How many chairs are in the room?

60. *Checkerboard* A checkerboard has 8 rows of squares with 8 squares in each row. What is the total number of squares on the board?

61. *Orchard* An orchard has 100 rows of trees with 65 trees in each row. How many trees are in the orchard?

62. *Used-Car Lot* A used-car lot has 6 rows of cars with 20 cars in each row. How many cars are in the lot?

Fund Raiser Use the following information in Exercises 63 and 64. As a fund raiser, a civic club sold 1-pound blocks of cheese for \$4 and fruitcakes for \$5 each.

63. What was the income from selling 80 pounds of cheese?

64. If the club sold 30 cakes, how much money did the club earn?

65. *Charter Bus* A charter bus company charges \$8 per person for the first 30 passengers and \$6 per person for each passenger over 30. How much does the company charge for a group of 38 people?

66. *Hourly Wages* A person makes \$7 per hour for up to 40 hours of work and \$11 per hour for additional hours over 40 hours. How much is the person paid for 50 hours of work?

Copyright © Houghton Mifflin Company. All rights reserved.

67. *Landscaping Cost* A landscaper plans a rock wall to enclose a rectangular garden that is 30 feet long and 20 feet wide. Materials and labor cost $12 per foot. What is the total cost of building the wall?

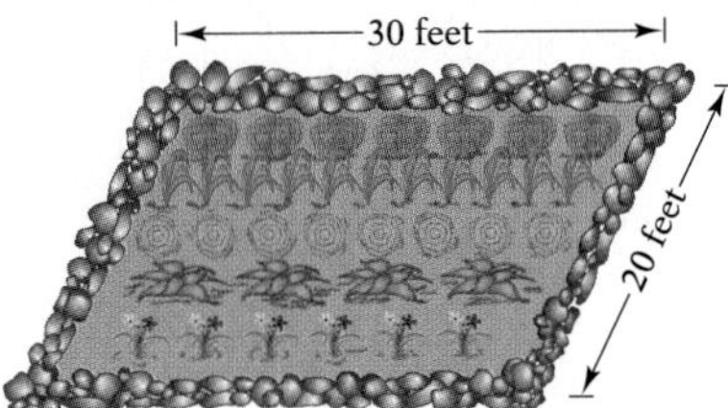

68. *Cost of Fencing* Fencing costs $8 per foot. What is the total cost of fencing for a rectangular yard that is 100 feet long and 50 feet wide?

Writing and Concept Extension

69. In terms of addition, what is the meaning of 6(13)?

70. A wall has 18 rows of bricks with 35 bricks in each row. Explain why 18(35) and 35(18) both represent the total number of bricks in the wall.

In Exercises 71–74, find two whole numbers whose product is the first given number and whose sum is the second given number.

	Product	*Sum*
71.	12	8
72.	8	9
73.	20	12
74.	15	8

75. What is the product of $8 \cdot 9 \cdot 0 \cdot 5 \cdot 3 \cdot 15$?

76. Suppose the product of several numbers is 0. What do you know about one of the numbers?

Copyright © Houghton Mifflin Company. All rights reserved.

2.2 Multiplication of Whole Numbers

A *The Distributive Property*
B *Multiplying Large Numbers*
C *Estimation and Applications*

SUGGESTIONS FOR SUCCESS

When, if at all, may you use a calculator?

Should you buy one? If so, what kind? Remember that a Wal-Mart sales clerk wants to sell you whatever is in stock. You could end up with a calculator that is too sophisticated or too basic. Get some very specific advice from your instructor.

Instructors vary in their opinions about the use of calculators in their courses. When are you allowed to use one? For homework? For tests? Make sure that you understand your instructor's rules. Your calculator can be a faithful ally, but it should not be a substitute for understanding what you are doing.

A *The Distributive Property*

All of the properties that we have presented so far are properties of either addition or multiplication. One property, called the *Distributive Property,* ties these two operations together.

DEVELOPING THE CONCEPT

Multiplication and Addition

Sugar can be purchased in 1-pound bags and in 5-pound bags. Suppose that you buy 3 bags of each size. How many pounds of sugar have you bought?

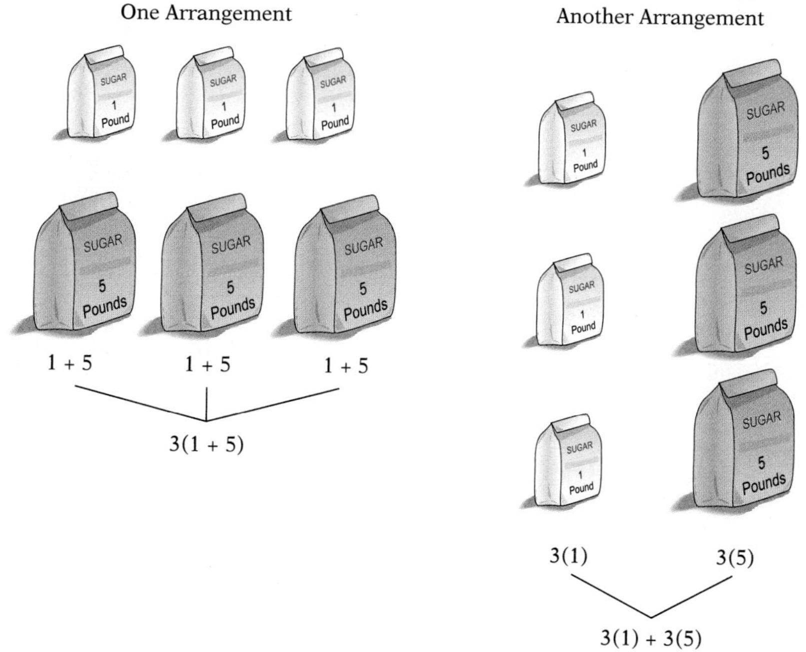

Both arrangements should give us the same total number of pounds. We can verify this by evaluating the two numerical expressions.

$$3(1 + 5) = 3(6) = 18 \qquad 3(1) + 3(5) = 3 + 15 = 18$$

Copyright © Houghton Mifflin Company. All rights reserved.

The total amount of sugar that you bought was 18 pounds.

Because $3(1 + 5)$ and $3(1) + 3(5)$ have the same value, we can write

$$3(1 + 5) = 3(1) + 3(5)$$

Note that the expression on the left is a *product:* "3 *times* (1 + 5)." The expression on the right is a *sum:* "3(1) *plus* 3(5)."

We can convert the product on the left to the sum on the right by multiplying the numbers inside the parentheses by 3.

$$3(1 + 5) = 3(1) + 3(5)$$

Multiplying in this way is called *distributing* the 3.

The property illustrated by the sugar purchase can be written for all numbers.

The Distributive Property

If the variables a, b, and c are used to represent any three numbers, then

$$a(b + c) = a(b) + a(c)$$

Example 1

Use the Distributive Property to write $8(3 + 6)$ as a sum. Then simplify the result.

SOLUTION

$8(3 + 6) = 8(3) + 8(6)$	Distributive Property
$= 24 + 48$	Multiply.
$= 72$	Add.

Your Turn 1

Use the Distributive Property to write $4(7 + 9)$ as a sum. Then simplify the result.

Answer: 28 + 36; 64

B Multiplying Large Numbers

Suppose that one floor of a building has 8 classrooms and that each classroom has 32 seats. What is the total number of students that can be seated?

Multiplying Large Numbers

The answer to the preceding question can be found by multiplying 8 times 32. The Distributive Property can be used to help us perform the operation.

$8(32) = 8(30 + 2)$	$32 = 30 + 2$
$= 8(30) + 8(2)$	Distributive Property
$= 240 + 16$	Multiply.
$= 256$	Add.

This same calculation can be written vertically. Note that when we use this form, we begin the multiplication with 8 times 2 and then multiply 8 times 30.

Copyright © Houghton Mifflin Company. All rights reserved.

$$\begin{array}{r} 30 + 2 \\ \times \quad 8 \\ \hline 16 \\ +\ 240 \\ \hline 256 \end{array}$$

Multiply 8(2) = 16.
Multiply 8(30) = 240.
Add.

To make the calculation even easier to write, we can simply write 32 rather than 30 + 2.

$$\begin{array}{r} 32 \\ \times\ 8 \\ \hline \end{array}$$

Now we will multiply and add at the same time. As before, we begin by multiplying 8 times 2. Because the result, 16, is not a digit, we *carry* just as we do in addition.

$$\begin{array}{r} {\scriptstyle 1}\ \\ 32 \\ \times\ 8 \\ \hline 6 \end{array}$$

Multiply 8(2) = 16.
Write the 6 and carry the 1.

$$\begin{array}{r} {\scriptstyle 1}\ \\ 32 \\ \times\ 8 \\ \hline 256 \end{array}$$

Multiply 8(3) = 24.
Add 24 and the carried 1; 24 + 1 = 25.

The Distributive Property helps us to understand how large numbers are multiplied. However, multiplying and adding at the same time is the more common method for performing multiplication.

Example 2

Multiply 8(357).

SOLUTION

For clarity, we show the solution in three steps. However, you should be able to combine the steps into one and perform the operations mentally.

$$\begin{array}{r} {\scriptstyle 5}\ \\ 357 \\ \times \quad 8 \\ \hline 6 \end{array}$$

Multiply 8(7) = 56.
Write the 6 and carry the 5.

$$\begin{array}{r} {\scriptstyle 4\,5}\ \\ 357 \\ \times \quad 8 \\ \hline 56 \end{array}$$

Multiply 8(5) = 40.
Add 40 and the carried 5; 40 + 5 = 45.
Write the 5 and carry the 4.

$$\begin{array}{r} {\scriptstyle 4\,5}\ \\ 357 \\ \times \quad 8 \\ \hline 2856 \end{array}$$

Multiply 8(3) = 24.
Add 24 and the carried 4: 24 + 4 = 28.
Write the 28.

The product is 2,856.

Your Turn 2

Multiply 6(239).

Answer: 1,434

Copyright © Houghton Mifflin Company. All rights reserved.

When both factors have two or more digits, the preceding methods are still used, but they are repeated for each digit.

Example 3

Multiply 72(40).

SOLUTION

Because 40 is a multiple of 10, we can multiply 72(4) = 288 followed by one 0.

$$72(40) = 2{,}880$$

The vertical form looks like this.

```
   72
×  40
 2880    Multiply 0(72) = 0.
         Multiply 4(72) = 288.
```

Your Turn 3

Multiply 13(70).

Answer: 910

Example 4

Multiply 28(53).

SOLUTION

We can use the Distributive Property to write the product as follows:

$$28(53) = 28(50 + 3) \quad 53 = 50 + 3$$
$$= 28(50) + 28(3) \quad \text{Distributive Property}$$

```
   28       28
×  50     ×  3
 1400       84
```

$$28(53) = 1{,}400 + 84 = 1{,}484$$

If we use the vertical arrangement, the problem looks like this:

```
     28
×    53
     84     3(28) = 84
+  1400     50(28) = 1,400
   1484
```

The following is a shortcut method.

```
    28
×   53
    84     Multiply 3(28); place the product under the 3.
+  140     Multiply 5(28); place the product under the 5.
  1484
```

Your Turn 4

Multiply 37(42).

LEARNING TIP

A carry is needed for each product, so writing the carries starts to become confusing. You can write the carries for one product and then erase them and write the carries for the next product. The sooner you can carry mentally, the better.

Answer: 1,554

Copyright © Houghton Mifflin Company. All rights reserved.

Example 5

Multiply 632(475).

SOLUTION

$$\begin{array}{rll} & 632 & \\ \times & 475 & \\ \hline & 3160 & 5(632) = 3160 \\ & 4424 & 7(632) = 4424 \\ + & 2528 & 4(632) = 2528 \\ \hline & 300200 & \end{array}$$

The product is 300,200.

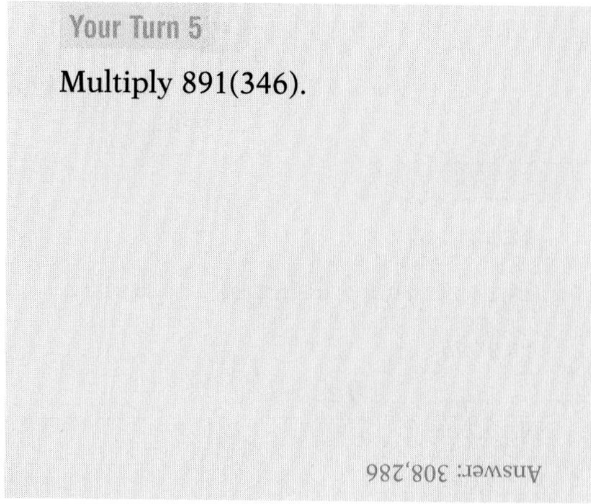

Your Turn 5

Multiply 891(346).

Answer: 308,286

We did not try to show the carries in Example 5 because there are too many of them. All three products involve at least one carry, as does the sum. Ideally, when you are multiplying by hand, carries should be managed mentally.

One advantage of using a calculator, especially for multiplying large numbers, is that carries are performed for you.

KEYS TO THE CALCULATOR

Understanding the concept of multiplication is important, and you should not have to rely on a calculator to perform simple multiplications. However, finding products of large numbers can be time-consuming, and there are many opportunities for errors.

Here are the typical keystrokes for the product in Example 5.

632 [×] **475** [=] (or [ENTER])

As shown in the following graphing calculator screen, your calculator may display the symbol * rather than ×.

```
632*475
                 300200
```

Exercises

(a) $\begin{array}{r} 38 \\ \times\ 92 \\ \hline \end{array}$ **(b)** $\begin{array}{r} 672 \\ \times\ 84 \\ \hline \end{array}$

(c) $\begin{array}{r} 734 \\ \times\ 981 \\ \hline \end{array}$ **(d)** $\begin{array}{r} 5{,}309 \\ \times\ 276 \\ \hline \end{array}$

Copyright © Houghton Mifflin Company. All rights reserved.

C Estimation and Applications

As with addition and subtraction, estimating a product begins with rounding the factors.

Example 6

Estimate the product of 498 and 723.

SOLUTION

Given Number		*Rounded Number*
498	→	500
723	→	700

We multiply the leading digits: 7(5) = 35. The two rounded numbers have a total of four 0s, so the estimated product is 35 followed by four 0s: 350,000.

Your Turn 6

Estimate the product of 79 and 92.

Answer: 7,200

In real-life situations, problems may involve two or more products.

Example 7

A truck driver earns $9 per hour for a 40-hour work week and $15 for every additional hour. What was the week's pay for a driver who worked 46 hours?

SOLUTION

	Hours ×	*Hourly Pay* =	*Total*
Regular	40	$ 9	$360
Overtime	6	$15	$ 90
Total pay			$450

Your Turn 7

What would be the driver's earnings if the overtime pay were double the regular pay?

Answer: $468

Example 8

For a graduation ceremony, a school set up folding chairs in the gymnasium. There were 23 rows with 36 chairs in each row. If 796 people attended the ceremony, how many empty chairs were there?

SOLUTION

The total number of chairs that were set up was

$$23(36) = 828$$

Therefore, 828 − 796 = 32 chairs were empty.

Your Turn 8

If one less row of chairs had been set up, how many people would have had to stand?

Answer: 4

Copyright © Houghton Mifflin Company. All rights reserved.

Example 9

The bar graph compares the number of births and the number of deaths that occur worldwide every minute. (Source: Population Reference Bureau.)

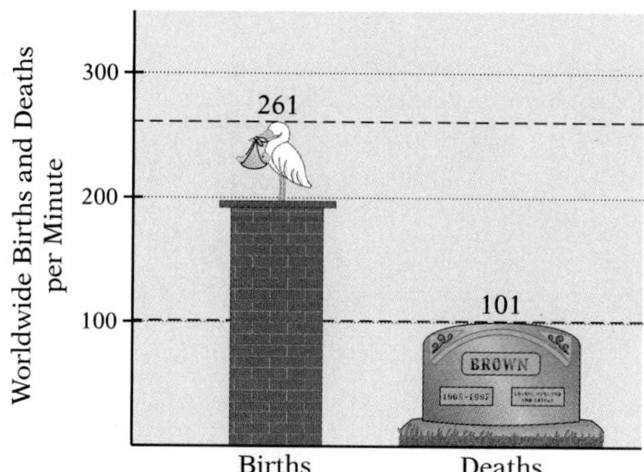

Find the change in the world's population every hour.

SOLUTION

The difference per minute between the number of births and the number of deaths is

$$261 - 101 = 160$$

Because there are 60 minutes in an hour, the hourly increase in the population is

$$60(160) = 9{,}600$$

Your Turn 9

What is the daily increase in the world's population?

Answer: 230,400

2.2 Quick Reference

A. The Distributive Property

1. If a, b, and c are any numbers, then
$$a(b + c) = a(b) + a(c)$$

$$9(2 + 5) = 9(2) + 9(5)$$

B. Multiplying Large Numbers

1. The Distributive Property can be used to multiply large numbers.

$$\begin{aligned} 4(51) &= 4(50 + 1) \\ &= 4(50) + 4(1) \\ &= 200 + 4 \\ &= 204 \end{aligned}$$

2. We can multiply large numbers with a vertical format.

$$\begin{array}{r} 51 \\ \times \quad 4 \\ \hline 204 \end{array}$$

Copyright © Houghton Mifflin Company. All rights reserved.

3. Multiplication often requires *carrying*.

$$\begin{array}{r} {\scriptstyle 1} \\ 72 \\ \times \quad 9 \\ \hline 648 \end{array}$$

4. For products of larger numbers, we usually need to carry more than once.

$$\begin{array}{r} 497 \\ \times \quad 26 \\ \hline 2982 \\ 994 \\ \hline 12922 \end{array}$$

C. Estimation and Applications

1. We estimate a product by rounding the factors before we multiply.

$$\begin{array}{rcr} 304 & \rightarrow & 300 \\ \times\ 695 & \rightarrow & \times \quad 700 \\ \hline & & 210{,}000 \end{array}$$

Copyright © Houghton Mifflin Company. All rights reserved.

2.2 Exercises

A. The Distributive Property

1. To write a product such as $3(x + 2)$ as a sum $3x + 6$, we apply the Distributive Property.

2. The Distributive Property ties the operations of addition and multiplication together.

In Exercises 3–8, use the Distributive Property to write the expression as a sum. Then simplify the result.

3. $5(2 + 7)$ **4.** $7(6 + 1)$ **5.** $4(20 + 3)$

6. $2(40 + 7)$ **7.** $6(300 + 7)$ **8.** $3(200 + 2)$

B. Multiplying Large Numbers

In Exercises 9–26, multiply.

9. 67×7 **10.** 215×2 **11.** $3{,}206 \times 6$ **12.** $9{,}003 \times 4$

13. 41×86 **14.** 54×26 **15.** 27×30 **16.** 69×50

17. 507×72 **18.** 989×42 **19.** 629×135 **20.** 448×272

21. 692×520 **22.** 396×650 **23.** 341×603 **24.** $1{,}020 \times 508$

25. $1{,}945 \times 300$ **26.** 876×500

Copyright © Houghton Mifflin Company. All rights reserved.

C. Estimation and Applications

≈ In Exercises 27–30, estimate the product.

27. $\begin{array}{r} 687 \\ \times\ 93 \\ \hline \end{array}$

28. $\begin{array}{r} 5{,}890 \\ \times\ 609 \\ \hline \end{array}$

29. $\begin{array}{r} 9{,}999 \\ \times\ 587 \\ \hline \end{array}$

30. $\begin{array}{r} 48{,}940 \\ \times\ 6{,}021 \\ \hline \end{array}$

Calculator Exercises

In Exercises 31–34, use a calculator to determine the product.

31. $\begin{array}{r} 53{,}090 \\ \times\ 4{,}836 \\ \hline \end{array}$

32. $\begin{array}{r} 6{,}888 \\ \times\ 5{,}047 \\ \hline \end{array}$

33. $\begin{array}{r} 41{,}322 \\ \times\ 17{,}989 \\ \hline \end{array}$

34. $\begin{array}{r} 421{,}390 \\ \times\ 7{,}503 \\ \hline \end{array}$

Applications

35. *Window Panes* An office building has 35 windows. Each window has 4 rows of panes with 5 panes per row. How many window panes are in the building?

36. *Book Arrangements* A discount store places stacks of books on a table. The books are arranged in 4 rows with 6 stacks per row. Each stack has 12 books. How many books are on the table?

37. *Parking Spaces* A parking lot has 6 rows of spaces that are reserved for faculty, and each row has 37 spaces. The lot also has 18 rows designated for students, and there are 34 spaces per row. How many cars can park in the lot?

38. *Marching Bands* At halftime, two marching bands lined up on a football field. One band was in a formation with 17 rows and 14 people per row. The other band was in a formation with 12 rows and 21 people per row. How many band members were on the field?

Copyright © Houghton Mifflin Company. All rights reserved.

39. *College Enrollment* This year the enrollment at a college is 5,436. The enrollment is projected to increase by 220 students per year for the next 4 years. What is the expected enrollment in 4 years?

40. *Cash Register Contents* A cash register contains 17 twenty-dollar bills, 14 ten-dollar bills, 32 five-dollar bills, and 26 one-dollar bills. What is the value of the money in the cash register?

41. *RV Travel* A recreational vehicle averages 11 miles per gallon. On a vacation, the owner purchased 175 gallons of gasoline. How far was the vehicle able to travel?

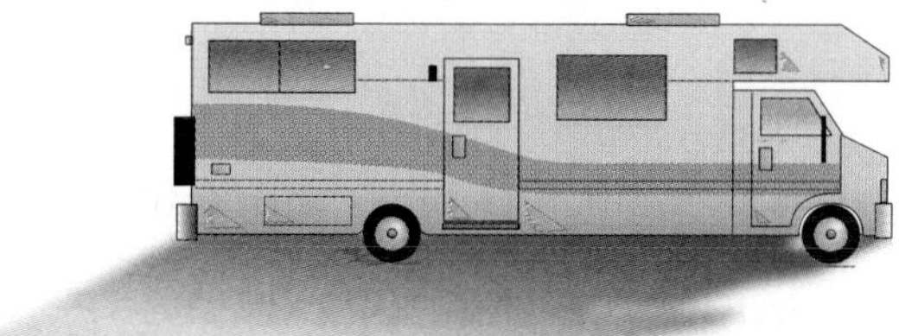

42. *Miles per Gallon* A car averages 23 miles per gallon. Starting with a full tank of 17 gallons, how far can the car travel?

Peanut Butter Use the following information in Exercises 43 and 44. A tablespoon of peanut butter has 95 calories, of which 70 calories are from fat.

43. How many calories are there in 4 tablespoons of peanut butter?

44. In a serving of 3 tablespoons of peanut butter, how many calories are *not* from fat?

45. *Food Stamps* From 1980 to 2000, the cost of the food stamp program approximately doubled. If the cost in 1980 was $9 billion, what was the approximate cost in 2000? (Source: U.S. Department of Agriculture.)

46. *State Populations* In 2000 the population of West Virginia was approximately 3 times the population of Alaska. If the population of Alaska was 627,000, what was the approximate population of West Virginia? (Source: U.S. Census Bureau.)

Copyright © Houghton Mifflin Company. All rights reserved.

Snack Bar Sales Use the following information in Exercises 47 and 48. A snack bar sells sandwiches for $4 and individual pizzas for $5 each.

47. During one day, the snack bar sold 56 sandwiches and 32 pizzas. How much money was collected that day?

48. During one week the snack bar sold 347 sandwiches and 210 pizzas. How much money was collected that week?

49. *College Costs* Each semester a student pays a $1,457 tuition fee, a $48 parking fee, and a $198 student activity fee. If the costs remain the same, what is the total cost for 4 years (8 semesters) of college?

50. *Vending Machines* A college employee stocks 15 vending machines with soft drinks. Each machine requires 4 cases of drinks per day, and each case holds 24 cans. How many cans are needed per week to stock all the machines?

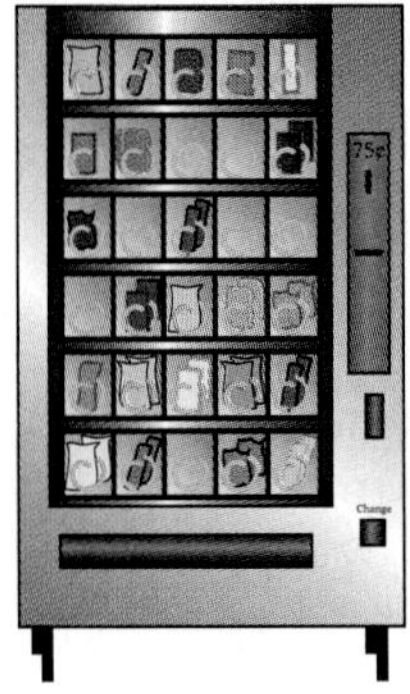

Bringing It Together

In Exercises 51–56, perform the indicated operation.

51. $\begin{array}{r} 8{,}761 \\ 43 \\ +\ 659 \\ \hline \end{array}$

52. $\begin{array}{r} 7{,}024 \\ \times\ 424 \\ \hline \end{array}$

53. $\begin{array}{r} 712 \\ -\ 67 \\ \hline \end{array}$

54. $\begin{array}{r} 53{,}284 \\ +\ 77{,}041 \\ \hline \end{array}$

55. $\begin{array}{r} 5{,}699 \\ \times\ 614 \\ \hline \end{array}$

56. $\begin{array}{r} 2{,}003 \\ -\ 957 \\ \hline \end{array}$

Copyright © Houghton Mifflin Company. All rights reserved.

Writing and Concept Extension

57. Each product results in the same answer. In which arrangement are fewer steps needed to find the product? Why?

(i) $\begin{array}{r} 934 \\ \times\ 102 \\ \hline \end{array}$ (ii) $\begin{array}{r} 102 \\ \times\ 934 \\ \hline \end{array}$

58. Each product results in the same answer. In which arrangement are fewer steps needed to find the product? Why?

(i) $\begin{array}{r} 5{,}000 \\ \times\ 3{,}690 \\ \hline \end{array}$ (ii) $\begin{array}{r} 3{,}690 \\ \times\ 5{,}000 \\ \hline \end{array}$

59. *Take-Home Pay* A carpenter earns $12 per hour for up to 40 hours of work per week. For each hour above 40, the hourly pay increases to $18. Each week $55 is deducted for insurance, and $124 is deducted for the retirement fund. If the carpenter works 54 hours, what is the take-home pay?

60. *Charter Bus Costs* A charter bus company charges a tour company $35 per hour for a bus. The company also charges $14 per hour for a driver for up to 8 hours and $18 per hour for hours over 8 hours. A tour group of 26 people paid $42 each for an 11-hour sightseeing trip. How much more than the cost of the bus did the group pay?

Exploring with Real-World Data: Collaborative Activities

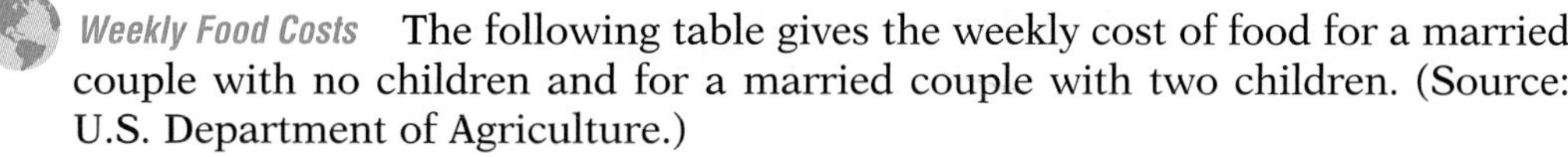

Weekly Food Costs The following table gives the weekly cost of food for a married couple with no children and for a married couple with two children. (Source: U.S. Department of Agriculture.)

Family Size		Weekly Food Cost
Family of 2	Age 20 - 50	$104
	Age 51 and over	$97
Family of 4	Children ages 1 - 5	$147
	Children ages 6 - 11	$173

Copyright © Houghton Mifflin Company. All rights reserved.

61. In the space below, construct a bar graph to illustrate the information for a family of 2.

62. In the space above, construct a bar graph to illustrate the information for a family of 4.

63. What is the food cost for a year (52 weeks) for a family with children ages 2 and 5?

64. What is the food cost for a month (4 weeks) for a family with children ages 7 and 10?

65. For married couples with no children, what is the difference in the monthly (4 weeks) food cost for a couple of age 25 and a couple of age 55?

66. For families of 4, what is the difference in the yearly (52 weeks) food cost for a family with children ages 6 and 11 and a family with children ages 2 and 3?

67. On the basis of the data in the table, can you determine the weekly food cost for a family of 4 with children ages 4 and 9?

68. On the basis of the information in the table, which of the following would be a reasonable estimate of the weekly food cost for a family of 4 with children ages 4 and 9?

(i) \$135 (ii) \$160 (iii) \$175

Copyright © Houghton Mifflin Company. All rights reserved.

2.3 Division of Whole Numbers

A *Division Concepts and Vocabulary*
B *Properties of Division*
C *Dividing Whole Numbers*
D *Remainders*
E *Applications*

SUGGESTIONS FOR SUCCESS

Procrastinate: to put off doing something until a future time.

Many people tend to wait until the last minute to do things, such as holiday shopping or filing tax returns. They may later wish that they hadn't, but more often than not, they do it again the next time.

For students, procrastinating is especially deadly. If you put off doing assignments or studying for exams, there is a good chance that you will end up not doing those things at all. Here are some tips that will help you avoid this problem.

1. Refer to your syllabus daily so that you will know whether you are on schedule.
2. Do assignments for your least favorite course first. Otherwise, you will be tempted to put them off and then never get around to doing them.
3. Read ahead. Your syllabus will tell you the next topic to be covered. This will help you to have a forward mindset.
4. Finally, catching up is very hard to do. Much of your success will depend on your never falling behind.

A Division Concepts and Vocabulary

In our everyday use of the word, *divide* means to separate into parts. You might divide a pizza into 6 slices, or the floor of an office building might have dividers that form cubicles.

Suppose that an instructor decides to divide a class of 28 students into 7 discussion groups.

 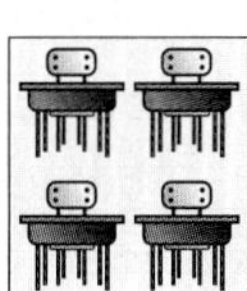 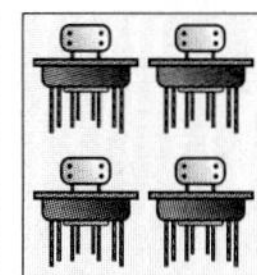 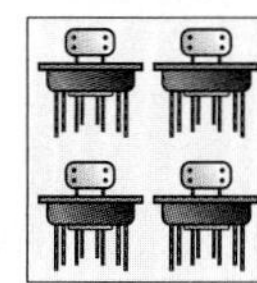 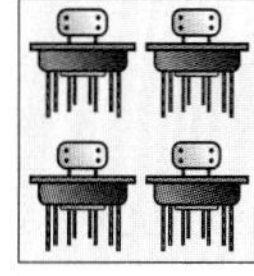 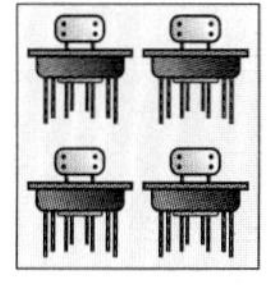 

Note that each group has 4 students. In this case, we say that "28 divided by 7 equals 4."

Recall that multiplication can be regarded as repeated addition. Another way to think about division is to regard it as repeated subtraction. When we divide the class of 28 students into 7 groups, we can find the number of students in each group by asking how many times we can subtract 7 from 28.

$$\begin{array}{r} 28 \\ -\ 7 \\ \hline 21 \\ -\ 7 \\ \hline 14 \\ -\ 7 \\ \hline 7 \\ -\ 7 \\ \hline 0 \end{array}$$

We see that 7 can be subtracted from 28 a total of 4 times. Therefore, each group has 4 students.

Copyright © Houghton Mifflin Company. All rights reserved.

One way to represent the statement "28 divided by 7 equals 4" is with the division symbol ÷.

This number is the **dividend.**		This number is the **divisor.**		
↓		↓		
28	÷	7	=	4
				↑
				The result is the **quotient.**

Other ways to represent $28 \div 7$ are 28/7 and $\frac{28}{7}$. (Although we usually think of $\frac{28}{7}$ as a fraction, for now we will regard the expression as representing division.) Both the division, $28 \div 7$, and the result, 4, are called a **quotient.** The first number given is the **dividend,** and the second number is the **divisor.**

NOTE A phrase such as "the quotient of 15 and 3" refers to the division of 15 by 3, that is, $15 \div 3$ or 15/3 or $\frac{15}{3}$.

Example 1

For the statement $16 \div 2 = 8$, identify the quotient, the dividend, and the divisor.

SOLUTION

The expression $16 \div 2$ and 8 are both called a quotient. The dividend is 16, and the divisor is 8.

Your Turn 1

For the statement $3 = 15 \div 5$, identify the quotient, the dividend, and the divisor.

Answers: Quotient: $15 \div 5$, 3; dividend: 15; divisor: 5

Do you remember that subtraction can be checked with addition?

$$13 - 5 = 8 \quad \text{because} \quad 8 + 5 = 13$$

In a similar way, division can be checked with multiplication.

$$28 \div 7 = 4 \quad \text{because} \quad 4 \cdot 7 = 28$$
$$16 \div 2 = 8 \quad \text{because} \quad 8 \cdot 2 = 16$$

B Properties of Division

For division, we need to pay special attention to the number 0.

The Role of 0 in Division

Consider the following two problems:

(a) $5 \div 0 =$ ____ because ____ $\cdot\, 0 = 5$
(b) $0 \div 0 =$ ____ because ____ $\cdot\, 0 = 0$

In problem (a), there is *no* number that we can multiply by 0 to obtain 5. In problem (b), *every* number multiplied by 0 is 0. In other words, neither of these problems has a unique (one and only one) answer. We simply cannot divide *by* 0. We say that division by 0 is *undefined.*

Copyright © Houghton Mifflin Company. All rights reserved.

Now consider the quotient $0 \div 3$.

$$0 \div 3 = \underline{\quad} \quad \text{because} \quad \underline{\quad} \cdot 3 = 0$$

Because 0 is the one and only number that we can multiply by 3 to obtain 0, $0 \div 3 = 0$.

We can summarize our findings with the following properties:

Division Properties of 0

If the variable a is used to represent any number, then:

1. $a \div 0$ is undefined.
   In words, we can never divide by 0.
2. $0 \div a = 0$ if $a \neq 0$.
   In words, if we divide 0 by any number except 0, the result is 0.

NOTE Recall that $a \neq 0$ is read as "a is not equal to 0." For property 2 in the preceding box, we say "if $a \neq 0$" because division by 0 is not defined.

Example 2

Find each quotient.

(a) $0 \div 53$ **(b)** $24 \div 0$

SOLUTION

(a) $0 \div 53 = 0$ $\quad 0 \div a = 0$ if $a \neq 0$

(b) $24 \div 0$ is undefined.

Your Turn 2

Find each quotient.

(a) $2 \div 0$

(b) $0 \div 2$

Answers: (a) Undefined; (b) 0

Two other properties of division are results of the Multiplication Property of 1. The following statements illustrate these properties:

$$6 \div 1 = 6 \quad \text{because} \quad 6 \cdot 1 = 6$$
$$8 \div 8 = 1 \quad \text{because} \quad 1 \cdot 8 = 8$$

Division Properties of 1

If the variable a is used to represent any number, then:

1. $a \div 1 = a$
   In words, if we divide any number by 1, the result is the number.
2. $a \div a = 1$ if $a \neq 0$
   In words, if we divide any number except 0 by itself, the result is 1.

Copyright © Houghton Mifflin Company. All rights reserved.

Example 3

Find each quotient.

(a) $17 \div 17$ **(b)** $35 \div 1$

SOLUTION

(a) $17 \div 17 = 1$ $\quad a \div a = 1$ if $a \neq 0$

(b) $35 \div 1 = 35$ $\quad a \div 1 = a$

Your Turn 3

Find each quotient.

(a) $19 \div 1$

(b) $100 \div 100$

Answers: (a) 19; (b) 1

NOTE There is no commutative property for division. For example, $12 \div 3 \neq 3 \div 12$. This means that we need to be careful about dividing numbers in the proper order. Also, division is not associative. For example,

$$(20 \div 10) \div 2 = 2 \div 2 = 1$$
$$20 \div (10 \div 2) = 20 \div 5 = 4$$

C Dividing Whole Numbers

All of the quotients that we have discussed so far can be determined from the basic multiplication facts. For example, we know that $20 \div 5 = 4$ because $4 \cdot 5 = 20$. Now we turn to divisions that involve larger whole numbers.

Long Division

Here is a convenient way to organize the problem $115 \div 5$. We call this arrangement *long division.*

$$\begin{array}{rl} \square & \leftarrow \text{We will write the quotient up here.} \\ \text{Divisor} \rightarrow \quad 5\overline{)115} & \\ \uparrow & \\ \text{Dividend} & \end{array}$$

We can think of the division $115 \div 5$ as the number of times that 5 can be subtracted from 115. Because we don't know, we just make the best guess that we can. We know that $5 \cdot 20 = 100$, so 20 seems like a reasonable guess.

$$\begin{array}{rl} 20 & \leftarrow \text{Write the first guess here.} \\ 5\overline{)115} & \\ -\,100 & \leftarrow \text{Multiply } 20(5) = 100. \text{ This shows that we are subtracting 20 fives.} \\ \hline 15 & \leftarrow \text{After we subtract 20 fives, 15 still remains.} \end{array}$$

Now we ask how many times 5 can be subtracted from the 15 that remains. Because $3 \cdot 5 = 15$, we know that the answer is 3.

$$\begin{array}{rl} 3 & \leftarrow \text{Write the second guess here.} \\ 20 & \\ 5\overline{)115} & \\ -\,100 & \\ \hline 15 & \\ -\,15 & \leftarrow \text{Multiply } 3(5) = 15. \text{ This shows that we are subtracting 3 more fives.} \\ \hline 0 & \leftarrow \text{There are no more fives that can be subtracted.} \end{array}$$

Copyright © Houghton Mifflin Company. All rights reserved.

We have subtracted 5 from 115 a total of 20 + 3 = 23 times. Therefore, the quotient is 23: 115 ÷ 5 = 23. We can verify this result with multiplication.

$$\begin{array}{r} 23 \\ \times \ \ 5 \\ \hline 115 \end{array}$$

Example 4 shows how we can shortcut this division process somewhat. However, until you are completely sure that you know what you are doing, the method we have just shown may be the safer way to go.

Example 4

Divide: 252 ÷ 6.

SOLUTION

A good way to begin is to look at 25, the first two digits of 252. Because 6 · 4 = 24 is close to 25, we use 4 as our first guess.

$$\begin{array}{r} 4 \\ 6\overline{)252} \\ -24 \\ \hline 12 \end{array}$$

← Rather than write 40, we write 4 in the tens column.
← Write 4 · 6 = 24 rather than 40 · 6 = 240.
← Sometimes we say, "Bring down the 2," but we are really subtracting.

Because 6(2) = 12, we complete the quotient.

$$\begin{array}{r} 42 \\ 6\overline{)252} \\ -24 \\ \hline 12 \\ -12 \\ \hline 0 \end{array}$$

← Write 2 in the ones column.
← Write 2 · 6 = 12 and subtract.
← There are no more sixes that can be subtracted.

Therefore, 252 ÷ 6 = 42. We can verify the result by multiplying: 42 · 6 = 252.

Your Turn 4

Divide: 332 ÷ 4.

Answer: 83

Sometimes arithmetic looks a little mysterious because mathematicians tend not to write more than is needed. There is no law against writing the 0s in a long division if you find that doing so helps to keep the problem organized. In the first part of Example 4, we could have written

$$\begin{array}{r} 40 \\ 6\overline{)252} \\ -240 \\ \hline 12 \end{array}$$

← 40 · 6 = 240

The long division process applies even when the divisor has two or more digits.

Copyright © Houghton Mifflin Company. All rights reserved.

Example 5

Divide: 9,912 ÷ 28.

SOLUTION

Note that the first two digits of 9,912 are 99. The following shows some possible first guesses.

$$\begin{array}{r} 2 \\ 28\overline{)99} \\ -56 \\ \hline 43 \end{array}$$ ← Bad guess because this remainder is larger than 28.

$$\begin{array}{r} 4 \\ 28\overline{)99} \\ 112 \\ \hline \end{array}$$ ← Bad guess because this product is larger than 99.

$$\begin{array}{r} 3 \\ 28\overline{)99} \\ -84 \\ \hline 15 \end{array}$$ Good guess because ← this product is less than 99 and ← this remainder is less than 28.

Now we can write the results of our first guess.

$$\begin{array}{r} 3 \\ 28\overline{)9912} \\ -84 \\ \hline 151 \end{array}$$
← Write 3 in the hundreds column.
← Write 3(28) = 84 and subtract.
← Bring down the 1.

Our next guess involves the number of times that 28 can be divided into 151.

$$\begin{array}{r} 34 \\ 28\overline{)9912} \\ -84 \\ \hline 151 \\ -112 \\ \hline 39 \end{array}$$
← Our guess is 4.
Bad guess because ← this remainder is larger than 28.

$$\begin{array}{r} 36 \\ 28\overline{)9912} \\ -84 \\ \hline 151 \\ -168 \\ \hline \end{array}$$
← Our guess is 6.
Bad guess because ← this product is larger than 151.

$$\begin{array}{r} 35 \\ 28\overline{)9912} \\ -84 \\ \hline 151 \\ -140 \\ \hline 11 \end{array}$$
← Our guess is 5.
Good guess because ← this product is less than 151 and ← this remainder is less than 28.

Now we write the results of our second guess.

Your Turn 5

Divide: 8,676 ÷ 36.

LEARNING TIP

Guessing becomes a little harder when the numbers are large. Help yourself by going off to the side and experimenting with some multiplications until you are satisfied that you have found the best guess.

Copyright © Houghton Mifflin Company. All rights reserved.

```
     35      ← Write 5 in the tens column.
 28)9912
   -84
    151
   -140      ← Write 5(28) = 140 and subtract.
    112      ← Bring down the 2.
```

Our last guess is based on the fact that $4 \cdot 28 = 112$, so the last digit in the quotient is 4.

```
    354      ← Write 4 in the ones column.
 28)9912
   -84
    151
   -140
    112
   -112      ← Write 4(28) = 112 and subtract.
      0      ← We can't subtract 28 any more.
```

We find that $9{,}912 \div 28 = 354$. We can check this answer with multiplication.

```
   354
 ×  28
  2832
 + 708
  9912
```

Example 5 illustrates how tedious and time-consuming the long division process can be. Realistically, if you have a calculator available, you will probably use it for such problems. However, anyone can push buttons and get results. If you are serious about *understanding* mathematics, we urge you to work the exercises the old-fashioned way.

KEYS TO THE CALCULATOR

Typically, the ÷ key is used to perform division with a calculator. You should be able to compute the quotient in Example 5 with these keystrokes.

9912 ÷ **28** = (or ENTER)

The figure shows a typical display on a graphing calculator. Note that your calculator may display the symbol / when you press the ÷ key.

```
9912/28
            354
```

Exercises

(a) $234 \div 9$ **(b)** 4,524/87 **(c)** $\frac{21{,}692}{29}$ **(d)** $66{,}198 \div 187$

Copyright © Houghton Mifflin Company. All rights reserved.

D Remainders

At the beginning of this section, an instructor divided a class of 28 students into discussion groups of 4 students per group. What would be the situation if there were 30 students in the class?

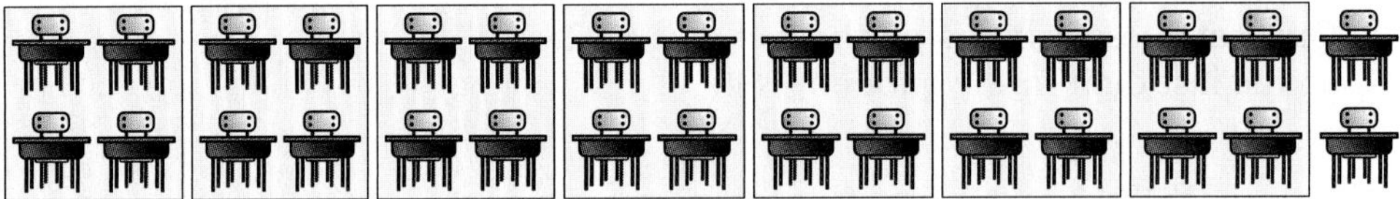

Two students are left over.

We can represent this situation with division, which results in a **remainder** of 2.

$$\begin{array}{r} 7 \\ 4\overline{)30} \\ -28 \\ \hline 2 \end{array}$$ ← This number is called the **remainder.**

Even though the remainder is not 0, we still can't subtract any more fours from 2. We end the long division process when the remainder is less than the divisor. We will use the following notation to indicate such answers:

$30 \div 4 = 7 \text{ R } 2$ The quotient is 7 with a remainder of 2.

To check the answer, we begin as we always do: Multiply $7 \cdot 4 = 28$. Now we add the remainder to the result: $28 + 2 = 30$.

When we divide a dividend a by a divisor b and the remainder is 0, we say that a is *divisible* by b. For example, 14 is divisible by 7 because $14 \div 7 = 2$ and the remainder is 0.

Example 6

Divide 1,617 by 43.

SOLUTION

$$\begin{array}{r} 37 \\ 43\overline{)1617} \\ -129 \\ \hline 327 \\ -301 \\ \hline 26 \end{array}$$ ← Remainder

Because $26 < 43$, we stop dividing.

$$1{,}617 \div 43 = 37 \text{ R } 26$$

Now we check the answer.

$$\begin{array}{r} 43 \\ \times\ 37 \\ \hline 301 \\ 129 \\ \hline 1591 \\ +\ 26 \\ \hline 1617 \end{array}$$

43 ← Divisor; 37 ← Quotient; 26 ← Add the remainder.; 1617 ← Dividend

Your Turn 6

Divide 3,685 by 92.

Answers: 40 R 5

Copyright © Houghton Mifflin Company. All rights reserved.

E Applications

When we perform division in real life, we will often find that the dividend is not divisible by the divisor—that is, the remainder is not often 0.

Interpreting Quotients and Remainders

Suppose you go to the post office to buy stamps. How many 37-cent stamps can you buy with $10? How much change will you receive? The answers to these questions can be found by dividing the amount of money you have by the value of each stamp.

$$\begin{array}{r} 27 \\ 37\overline{)1000} \\ \underline{74} \\ 260 \\ \underline{259} \\ 1 \end{array}$$

← We write $10 as 1000 cents.

The quotient indicates that you can buy 27 stamps, and the remainder indicates that your change will be 1 cent.

Solving real-life problems may involve only division, but sometimes one or more other operations must also be performed. In fact, long division itself requires both multiplication and subtraction.

Example 7

A jumbo cereal box contains 32 ounces of cereal. How many members of the family can have a 5-ounce serving, and how much cereal will be left over?

SOLUTION

The problem involves the number of times that 5 ounces can be served (subtracted) from the 32 ounces in the box. We solve the problem with division.

$$\begin{array}{r} 6 \\ 5\overline{)32} \\ \underline{-30} \\ 2 \end{array}$$

Six family members can have a 5-ounce serving, and 2 ounces of cereal will remain.

Your Turn 7

How many family members can have a 6-ounce serving? How much cereal will be left over?

Answers: 5; 2 ounces

Example 7 illustrates an important point to keep in mind when you work application problems. If the divisor is increased, then the quotient is decreased, and vice versa. Increasing the serving size decreases the number of family members who can be served. Increasing the number of servings decreases the serving size.

Copyright © Houghton Mifflin Company. All rights reserved.

Example 8

Suppose that you have determined that you need a salary of $3,000 per month in order to support your family. Would you take a job that pays an annual salary of $34,200?

SOLUTION

One approach is to divide the annual salary by 12 to determine the monthly salary.

$$\begin{array}{r} 2850 \\ 12\overline{)34200} \\ -24 \\ \hline 102 \\ -\ 96 \\ \hline 60 \\ -60 \\ \hline 00 \\ -\ 0 \\ \hline 0 \end{array}$$

← Even though the remainder is 0 here, we still have one more 0 to bring down.

The monthly salary is $2,850, which is not enough to meet your needs.

A second approach is to multiply your needed monthly salary by 12 to determine the annual salary that you need.

$$12(3{,}000) = 36{,}000$$

This shows that you need more than the $34,200 that is being offered.

Your Turn 8

If the annual salary offer were $38,400, how much should you be able to save each month?

Answer: $200

2.3 Quick Reference

A. Division Concepts and Vocabulary

1. We can think of division as repeated subtraction.	$15 \div 3$ is the number of times that 3 can be subtracted from 15.
2. There are several ways to represent division.	$12 \div 4$ 12/4 $\frac{12}{4}$
3. For $18 \div 9 = 2$, we call 18 the **dividend** and 9 the **divisor.** Both $18 \div 9$ and 2 are called the **quotient.**	

Copyright © Houghton Mifflin Company. All rights reserved.

4. For every division $a \div b = c$, we can write a related multiplication: $c \cdot b = a$. The related multiplication can be used to check the results of a division.

Division: $21 \div 7 = 3$
Multiplication: $3 \cdot 7 = 21$

B. Properties of Division

1. Division Properties of 0

If the variable a is used to represent any number, then:
(a) $a \div 0$ is undefined.
In words, we can never divide by 0.
(b) $0 \div a = 0$ if $a \neq 0$.
In words, if we divide 0 by any number except 0, the result is 0.

$7 \div 0$, $4/0$, and $\frac{53}{0}$ are all undefined.

$0 \div 5 = 0$

$0/13 = 0$

$\frac{0}{26} = 0$

2. Division Properties of 1

If the variable a is used to represent any number, then:
(a) $a \div 1 = a$
In words, if we divide any number by 1, the result is the number.

$23 \div 1 = 23$

$167/1 = 167$

$\frac{99}{1} = 99$

(b) $a \div a = 1$ if $a \neq 0$
In words, if we divide any number except 0 by itself, the result is 1.

$14 \div 14 = 1$

$37/37 = 1$

$\frac{52}{52} = 1$

C. Dividing Whole Numbers

1. The vertical arrangement for performing division is called *long division.*

$$\begin{array}{r} 24 \\ 4\overline{)96} \\ \underline{-8} \\ 16 \\ \underline{-16} \\ 0 \end{array}$$

2. In long division, we guess each digit of the quotient and check the guesses with multiplication.

$$\begin{array}{r} 17 \\ 29\overline{)493} \\ \underline{-29} \\ 203 \\ \underline{-203} \\ 0 \end{array}$$

We check results with multiplication. $17 \cdot 29 = 493$

Copyright © Houghton Mifflin Company. All rights reserved.

D. Remainders

1. The number at the bottom of a long division is called the **remainder.** The result of the division is written with the quotient and the remainder.

$$\begin{array}{r} 2 \\ 7\overline{)18} \\ -14 \\ \hline 4 \end{array} \leftarrow \text{Remainder}$$

$$18 \div 7 = 2 \text{ R } 4$$

We check the result by multiplying the quotient and the divisor and then adding the remainder.

Check:

$$\begin{array}{rl} 7 & \leftarrow \text{Divisor} \\ \times\ 2 & \leftarrow \text{Quotient} \\ \hline 14 & \\ +\ 4 & \leftarrow \text{Remainder} \\ \hline 18 & \leftarrow \text{Dividend} \end{array}$$

2. We say that a is *divisible* by b if $a \div b$ has a remainder of 0.

18 is divisible by 6 because $\frac{18}{6} = 3 \text{ R } 0$.

3. We stop dividing when the remainder is less than the divisor.

$$\begin{array}{r} 25 \\ 14\overline{)357} \\ -28 \\ \hline 77 \\ -70 \\ \hline 7 \end{array} \leftarrow \text{Remainder}$$

$$\frac{357}{14} = 25 \text{ R } 7$$

Copyright © Houghton Mifflin Company. All rights reserved.

2.3 Exercises

A. Division Concepts and Vocabulary

1. In the quotient 18 ÷ 6, the ______ is 6 and the ______ is 18.

2. Both 10 ÷ 5 and the result 2 are called a(n) ______.

In Exercises 3–6, identify (a) the quotient, (b) the dividend, and (c) the divisor.

3. 15 ÷ 3 = 5
4. 12 ÷ 6 = 2
5. $\frac{8}{4} = 2$
6. $\frac{20}{5} = 4$

In Exercises 7–10, fill in the blanks to show how the given division is related to multiplication.

7. 21 ÷ 3 = ____ because ____ · 3 = 21.
8. 30 ÷ 5 = ____ because ____ · 5 = 30.
9. 32 ÷ 8 = ____ because ____ · 8 = 32.
10. 18 ÷ 6 = ____ because ____ · 6 = 18.

B. Properties of Division

11. We say that division by 0 is ______.

12. The ______ of any number (except 0) and itself is 1.

In Exercises 13–20, determine the quotient.

13. 0 ÷ 5
14. 4 ÷ 0
15. 21 ÷ 21
16. 13 ÷ 1
17. 52 ÷ 1
18. 35 ÷ 35
19. 12 ÷ 0
20. 0 ÷ 9

21. Identify the quotient that is defined.
(i) 8 ÷ 0 (ii) 8/0 (iii) 0 ÷ 8 (iv) $\frac{8}{0}$

22. Identify the quotient that is *not* the same as the others in the list.
(i) The quotient of 5 and 1 (ii) $\frac{5}{1}$ (iii) 5 divided by 1 (iv) 1/5

C. Dividing Whole Numbers

In Exercises 23–32, determine the quotient.

23. 186 ÷ 6
24. 92 ÷ 4
25. 476 ÷ 7
26. 1,545 ÷ 5

Copyright © Houghton Mifflin Company. All rights reserved.

27. $21\overline{)1{,}218}$ **28.** $19\overline{)646}$ **29.** $85\overline{)6{,}035}$ **30.** $48\overline{)3{,}312}$

31. $41\overline{)12{,}587}$ **32.** $115\overline{)9{,}430}$

D. Remainders

33. When the remainder is 0, we say that the dividend is ______ by the divisor.

34. We continue the long division process until the ______ is less than the divisor.

In Exercises 35–44, determine the quotient and the remainder.

35. $52 \div 6$ **36.** $121 \div 5$ **37.** $1{,}690 \div 4$ **38.** $1{,}461 \div 7$

39. $42\overline{)2{,}582}$ **40.** $73\overline{)3{,}237}$ **41.** $67\overline{)4{,}711}$ **42.** $45\overline{)2{,}953}$

43. $96\overline{)44{,}456}$ **44.** $218\overline{)24{,}030}$

Calculator Exercises

In Exercises 45–48, use a calculator to determine the quotient.

45. $350{,}325 \div 865$ **46.** $358{,}308 \div 807$ **47.** $324{,}345 \div 3{,}089$ **48.** $2{,}223{,}111 \div 1{,}111$

Copyright © Houghton Mifflin Company. All rights reserved.

Applications

49. *Milk Servings* A half-gallon of milk contains 64 ounces. How many 6-ounce glasses can be filled from it? How much milk will be left over?

50. *Marigold Planting* A tray of marigolds contains 72 plants. A gardener wants to plant the flowers in rows of 5 plants. How many rows can be planted? How many plants will be left over?

51. *Unloading Times* A crew at an auto parts warehouse can unload a shipment of 616 cases of fan belts in 4 hours. How many cases can the crew unload in 1 hour?

52. *Orientation Packets* In 2 hours, a copier can copy, collate, and staple 360 copies of a student orientation packet. How many copies of the packet can the machine produce per minute?

53. *Furry Mutt Sales* A toy store sold \$3,696 worth of Furry Mutts at \$24 each. How many Furry Mutts were sold?

54. *Bottle Weight* A case of 18 bottles of kitchen cleaning solution weighs 396 ounces. What is the weight of one bottle?

55. *Hourly Wage* A part-time clerk at Discount City worked 28 hours one week. After \$30 was deducted for taxes, the take-home pay was \$138. How much per hour (before taxes) did Discount City pay the clerk?

Copyright © Houghton Mifflin Company. All rights reserved.

56. *Paper Cost* A company purchased 56 cases of copier paper. The cost, including a $12 delivery charge, was $2,532. What was the cost of a case of paper without the delivery charge?

57. *Juice Servings* Each morning, a daycare center serves each child a 5-ounce cup of orange juice. How many 32-ounce bottles of juice does the operator need for 44 children? How many ounces will be left over?

58. *Picnic Hamburgers* For an office picnic, organizers use 6 ounces of meat for each hamburger. How many 16-ounce packages of meat should they purchase to make 50 hamburgers? How many ounces of meat will be left over?

59. *Sizes of States* Connecticut is approximately 4,800 square miles in size, and Kansas is approximately 81,600 square miles. How many times as large as Connecticut is Kansas?

60. *Population Density* The population density of Rhode Island is approximately 960 people per square mile, and the population density of Michigan is approximately 160 people per square mile. How many times as dense as the population of Michigan is the population of Rhode Island?

61. *Monthly Income* The average annual income for a couple with both working is $56,196. (Source: U.S. Bureau of Census.) What is the average monthly income?

62. *Federal Tax* An average American worker pays $8,760 in federal tax each year. (Source: Internal Revenue Service.) Approximately how much tax per day (365 days per year) does an average worker pay?

Copyright © Houghton Mifflin Company. All rights reserved.

Bringing It Together

In Exercises 63–70, perform the indicated operations.

63. 729(53)

64. 8,061 − 437

65. $\begin{array}{r} 265 \\ 395 \\ +\ 84 \\ \hline \end{array}$

66. $71\overline{)1{,}988}$

67. $\begin{array}{r} 63{,}010 \\ -\ 57{,}892 \\ \hline \end{array}$

68. 382,791 + 4,546

69. 14,881 ÷ 23

70. $\begin{array}{r} 609 \\ \times\ 48 \\ \hline \end{array}$

Writing and Concept Extension

71. Explain whether there is any difference in the following phrases.

(i) 3 divided into 15 (ii) 15 divided by 3

72. If the quotient is 3 and the dividend is 21, what is the divisor?

In Exercises 73 and 74, insert parentheses to make the statement true.

73. 24 ÷ 6 ÷ 2 = 8

74. 7 + 8 ÷ 4 + 1 = 3

Exploring with Real-World Data: Collaborative Activities

The *population density* of a country is an indicator of how crowded the residents of the country are. Population density can be found by dividing the population of the country by the size of the country (in square miles). The result is the number of people per square mile.

Country Populations and Sizes The table shows the 2001 population and size of four countries. (Source: *World Almanac*.)

	Population	*Size (square miles)*
Canada	31,593,000	3,849,674
Germany	83,030,000	137,830
Japan	126,772,000	145,850
United States	278,059,000	3,675,031

In Exercises 75–84, ignore remainders when you divide.

75. Determine the population density of Canada.

Copyright © Houghton Mifflin Company. All rights reserved.

76. Determine the population density of Germany.

77. Determine the population density of Japan.

78. Determine the population density of the United States.

79. Use a bar graph to illustrate the population densities of the countries in the table.

80. How many times as crowded as Canada is the United States?

81. How many times as crowded as the United States is Germany?

82. In terms of size, how many times as large as Germany is Canada?

83. How many times as large as the population of Canada is the population of Japan?

84. How many times as great as the population density of Canada is the population density of Japan?

Copyright © Houghton Mifflin Company. All rights reserved.

2.4 Exponents

A *Introduction to Exponents*
B *The Zero Exponent*

SUGGESTIONS FOR SUCCESS

Have you ever returned to class and told your instructor "It looks so easy when I watch you do it, but when I try it myself, it seems much harder"? Most instructors have heard this many times.

The same statement could apply to music. Watching a pianist may make playing a piano look easy. However, just watching a pianist perform does not result in your being able to play the piano. In reality, study and practice are required to make playing a musical instrument look easy.

Mathematics requires similar active participation in order to be performed with ease. If your instructor provides an opportunity for you to try a problem on your own in class, take advantage of the chance to practice. Even if you don't complete the entire problem on your own, the practice increases your insight into the solution procedure and raises your level of understanding of the process involved.

A Introduction to Exponents

Recall that in a product such as $7 \cdot 9$, the 7 and the 9 are called *factors*. When the factors of a product are the same, we can represent the repeated multiplication with a special notation.

This number is called the **exponent.**

$$4 \cdot 4 = 4^2$$

This number is called the **base.**

The **base,** 4, is the factor that is being multiplied repeatedly. The **exponent,** 2, tells us the number of times that 4 is a factor. The expression is read as "4 to the second power" or "4 squared."

Similarly, here is how we can represent $2 \cdot 2 \cdot 2$.

The exponent means three factors of 2.

$$2 \cdot 2 \cdot 2 = 2^3$$

We read the expression 2^3 as "2 to the third power" or "2 cubed."

NOTE The use of the words *squared* and *cubed* will make more sense to you in the later sections on area and volume. These are the only two exponents that have special names.

We say that an expression such as 5^4 is in *exponential form* and that $5 \cdot 5 \cdot 5 \cdot 5$ is the *expanded form.*

Exponential Form	*Read As*	*Expanded Form*
3^5	3 to the 5th power	$3 \cdot 3 \cdot 3 \cdot 3 \cdot 3$
10^8	10 to the 8th power	$10 \cdot 10 \cdot 10 \cdot 10 \cdot 10 \cdot 10 \cdot 10 \cdot 10$

Because 7^1 means that 7 is a factor only once, $7^1 = 7$. In general, if the variable a represents any number, then $a^1 = a$.

Copyright © Houghton Mifflin Company. All rights reserved.

Example 1

Identify the base and the exponent, and state how the expression is read in words.

(a) 6^2 **(b)** 8^3 **(c)** 2^7

SOLUTION

(a) The base is 6 and the exponent is 2. The expression is read "6 to the second power" or "6 squared."

(b) The base is 8 and the exponent is 3. The expression is read "8 to the third power" or "8 cubed."

(c) The base is 2 and the exponent is 7. The expression is read "2 to the seventh power."

Your Turn 1

Identify the base and the exponent, and state how the expression is read in words.

(a) 7^3

(b) 4^6

(c) 1^2

Answers: **(a)** Base 7, exponent 3, "7 to the third power" or "7 cubed"; **(b)** Base 4, exponent 6, "4 to the sixth power"; **(c)** Base 1, exponent 2, "1 to the second power" or "1 squared."

Example 2

Write the expression in exponential form.

(a) $5 \cdot 5$

(b) $9(9)(9)$

(c) $3 \cdot 3 \cdot 3 \cdot 3 \cdot 3$

SOLUTION

(a) $5 \cdot 5 = 5^2$

(b) $9(9)(9) = 9^3$

(c) $3 \cdot 3 \cdot 3 \cdot 3 \cdot 3 = 3^5$

Your Turn 2

Write the expression in exponential form.

(a) $2 \cdot 2 \cdot 2$

(b) $5(5)(5)(5)(5)(5)$

(c) $7 \cdot 7 \cdot 7 \cdot 7$

Answers: **(a)** 2^3; **(b)** 5^6; **(c)** 7^4

Example 3

Translate the phrase into exponential form.

(a) 2 cubed

(b) 6 to the fourth power

(c) 8 squared

SOLUTION

(a) "2 cubed" means 2^3.

(b) "6 to the fourth power" means 6^4.

(c) "8 squared" means 8^2.

Your Turn 3

Translate the phrase into exponential form.

(a) 5 to the third power

(b) 4 to the first power

(c) 9 cubed

Answers: **(a)** 5^3; **(b)** 4^1; **(c)** 9^3

Copyright © Houghton Mifflin Company. All rights reserved.

Example 4

Write the expression in expanded form.

(a) 6^2

(b) 3^3

(c) 10^1

(d) 2^5

SOLUTION

(a) $6^2 = 6 \cdot 6$

(b) $3^3 = 3 \cdot 3 \cdot 3$

(c) $10^1 = 10$

(d) $2^5 = 2 \cdot 2 \cdot 2 \cdot 2 \cdot 2$

Your Turn 4

Write the expression in expanded form.

(a) 4^4

(b) 25^2

(c) 8^3

(d) 0^1

Answers: (a) $4 \cdot 4 \cdot 4 \cdot 4$; (b) $25 \cdot 25$; (c) $8 \cdot 8 \cdot 8$; (d) 0

We evaluate an expression in exponential form by performing the repeated multiplication.

Example 5

Evaluate.

(a) 7^2

(b) 10^3

(c) 2^4

(d) 1^{19}

SOLUTION

(a) $7^2 = 7 \cdot 7 = 49$

(b) $10^3 = 10 \cdot 10 \cdot 10 = 100 \cdot 10 = 1{,}000$

(c) $2^4 = 2 \cdot 2 \cdot 2 \cdot 2$

$= (2 \cdot 2) \cdot (2 \cdot 2)$ Associative Property of Multiplication

$= 4 \cdot 4$

$= 16$

(d) Fortunately, we don't have to write 1^{19} in expanded form. Because all the factors are 1, the product is 1.

Your Turn 5

Evaluate.

(a) 3^3

(b) 0^{23}

(c) 9^2

(d) 2^5

Answers: (a) 27; (b) 0; (c) 81; (d) 32

In part (b) of Example 5, we evaluated $10^3 = 1{,}000$. Did you notice that the exponent is 3 and that the answer has three 0s? We can use this observation to evaluate 10 raised to any power: Write 1 followed by the number of 0s indicated by the exponent.

Copyright © Houghton Mifflin Company. All rights reserved.

Example 6

Evaluate 10^4.

SOLUTION

The exponent, 4, tells us that the answer is 1 followed by four 0s.

$$10^4 = 10{,}000$$

Your Turn 6

Evaluate 10^5.

Answer: 100,000

Example 7

Write "one million" in exponential form.

SOLUTION

We write "one million" as 1,000,000, which is 1 followed by six 0s.

$$1{,}000{,}000 = 10^6$$

Your Turn 7

Write "one thousand" in exponential form.

Answer: 10^3

Example 8

Evaluate $10 \cdot 10 \cdot 10 \cdot 10 \cdot 10 \cdot 10 \cdot 10$ without performing any multiplication.

SOLUTION

$10 \cdot 10 \cdot 10 \cdot 10 \cdot 10 \cdot 10 \cdot 10 = 10^7$.
The answer is 1 followed by seven 0s.

$$10{,}000{,}000$$

Your Turn 8

Evaluate $10 \cdot 10 \cdot 10 \cdot 10$ without performing any multiplication.

Answer: 10,000

KEYS TO THE CALCULATOR

Depending on your calculator model, you may have a x^y key or a ^ key for exponents. Here are two typical keystroke sequences for evaluating 6^3.

Scientific calculator: **6** x^y **3** =

Graphing calculator: **6** ^ **3** ENTER

```
6^3
                216
```

Exercises

(a) 5^2 **(b)** 12^3 **(c)** 3^6 **(d)** 57^4

B The Zero Exponent

We have seen that when the natural numbers 1, 2, 3, . . . , are used as exponents, the exponent indicates the number of factors of the base. What would an expression such as 2^0 mean? No factors of 2?

Copyright © Houghton Mifflin Company. All rights reserved.

The Zero Exponent

In the following pattern, we begin with 32 and then keep dividing by 2. Observe how the exponent decreases by 1 each time.

$$
\begin{aligned}
32 &= 2 \cdot 2 \cdot 2 \cdot 2 \cdot 2 = 2^5 \\
\frac{32}{2} &= 16 = 2 \cdot 2 \cdot 2 \cdot 2 \quad = 2^4 \\
\frac{16}{2} &= 8 \quad = 2 \cdot 2 \cdot 2 \quad = 2^3 \\
\frac{8}{2} &= 4 \quad = 2 \cdot 2 \quad = 2^2 \\
\frac{4}{2} &= 2 \quad = 2^1 \\
\frac{2}{2} &= 1 \quad = 2^0
\end{aligned}
$$

The last line of the pattern indicates that 2^0 and 1 have the same value.

This experiment leads to the following definition:

Definition of the Zero Exponent

If b represents any base except 0, then

$$b^0 = 1$$

NOTE The definition states that *any* number except 0 raised to the 0 power is 1. The expression 0^0 is not defined.

Example 9

Evaluate.

(a) 12^0

(b) $8{,}679^0$

(c) $x^0, x \neq 0$

SOLUTION

(a) $12^0 = 1$

(b) $8{,}679^0 = 1$

(c) Even though we don't know what number x represents, we know that $x^0 = 1$ as long as x is not 0.

Your Turn 9

Evaluate.

(a) 63^0

(b) $7{,}423{,}598^0$

(c) $z^0, z \neq 0$

Answers: (a) 1; (b) 1; (c) 1

Asking what the meaning of a zero exponent might be took us into uncharted waters. We chose to define the zero exponent as we did because the definition is logically consistent with the pattern that we observed at the top of the page. However, the natural number exponents still have the same meaning: 5^3 still means $5 \cdot 5 \cdot 5$.

Copyright © Houghton Mifflin Company. All rights reserved.

2.4 Quick Reference

A. Introduction to Exponents

1. Exponents are used to indicate repeated multiplication.	Exponent ↓ $5^3 = 5 \cdot 5 \cdot 5$ ↑ Base
2. If a is any number, then $a^1 = a$.	$9^1 = 9$
3. An expression with an exponent is in *exponential form;* written as a repeated multiplication, the expression is in *expanded form.*	Exponential form: 7^2 Expanded form: $7 \cdot 7$
4. Exponents indicate the power to which the base is raised. Special names are often used for second and third powers.	4^2: "4 to the second power" or "4 squared" 8^3: "8 to the third power" or "8 cubed"
5. We evaluate an exponential expression by performing the repeated multiplication.	$2^4 = 2 \cdot 2 \cdot 2 \cdot 2 = 16$
6. To raise 10 to a power, we write 1 followed by the number of 0s given by the exponent.	$10^4 = 10{,}000$ ↑ Exponent indicates four 0s.

B. The Zero Exponent

1. If b represents any base except 0, then $b^0 = 1$. The expression 0^0 is not defined.	$27^0 = 1$ $x^0 = 1$ if $x \neq 0$

Copyright © Houghton Mifflin Company. All rights reserved.

2.4 Exercises

A. Introduction to Exponents

1. In the expression 5^3, we call 3 the ________ and we call 5 the ________.

2. The ________ form 3^2 can be written in ________ form as $3 \cdot 3$.

In Exercises 3–6, identify the base and the exponent. Then write the expression in words.

3. 5^3 **4.** 4^2 **5.** 7^5 **6.** 11^6

In Exercises 7–10, write the expression in exponential form.

7. $7 \cdot 7 \cdot 7$ **8.** $12 \cdot 12 \cdot 12 \cdot 12 \cdot 12$ **9.** $8 \cdot 8 \cdot 8 \cdot 8$ **10.** $21 \cdot 21$

In Exercises 11–14, write an exponential expression that corresponds to the given phrase.

11. 3 to the ninth power **12.** 5 squared **13.** 15 cubed **14.** 7 to the fifth power

In Exercises 15–18, write the exponential expression in expanded form.

15. 4^3 **16.** 14^2 **17.** 7^5 **18.** 1^8

In Exercises 19–24, evaluate the exponential expression.

19. 2^6 **20.** 12^2 **21.** 9^3 **22.** 3^5

23. 1^{20} **24.** 0^{33}

In Exercises 25–28, evaluate the exponential expression.

25. 10^6 **26.** 10^7 **27.** 10^8 **28.** 10^3

In Exercises 29–32, write the number in exponential form.

29. Ten thousand **30.** Ten million **31.** One billion **32.** One hundred thousand

B. The Zero Exponent

In Exercises 33–40, evaluate the exponential expression. Assume that no variable represents 0.

33. 10^0 **34.** 100^0 **35.** 0^1 **36.** 1^0

Copyright © Houghton Mifflin Company. All rights reserved.

37. $2{,}735^0$ **38.** $0^{2{,}375}$ **39.** c^0 **40.** y^0

Calculator Exercises

In Exercises 41–46, use a calculator to evaluate the exponential expression.

41. 15^3 **42.** 124^2 **43.** 32^6 **44.** 42^5

45. 21^0 **46.** 835^0

Bringing It Together

In Exercises 47–56, perform the indicated operations. For quotients, give the remainder, if any.

47. $405 \div 21$ **48.** 4^3 **49.** $1{,}101 - 996$ **50.** $91 \cdot 87$

51. $6{,}732 + 988$ **52.** $\frac{6{,}408}{89}$ **53.** 3^4 **54.** $312 + 497 + 504$

55. $13(274)$ **56.** $12{,}213 - 11{,}876$

Writing and Concept Extension

57. What are two common ways in which 7^2 can be read?

58. Technically, what is wrong with the following statement? "Any number raised to the 0 power is 1."

59. Which of the following expressions has a value that is different from all the others?

(i) a^0 (ii) 17^0 (iii) $\frac{x}{x}, x \neq 0$ (iv) $\frac{0}{x}, x \neq 0$

60. Consider the expression $x^3 \cdot x^5$.

(a) Replace x^3 and x^5 with their expanded forms.

(b) Write your result in part (a) in exponential form.

(c) Compare the exponent in your result in part (b) with the exponents in the original expression.

Copyright © Houghton Mifflin Company. All rights reserved.

2.5 The Order of Operations

A *Evaluating Numerical Expressions*
B *Translations*
C *Applications*

SUGGESTIONS FOR SUCCESS

Fear and trembling. I had to miss class today. Now what?
There are things that you can and should do to get caught up.

1. Does your instructor allow you to turn in assignments when you return to class? Be sure to do that if you can.
2. If you missed a quiz, can you make it up?
3. Find a classmate whose notes you can actually read. People worry about never getting their notes back, so photocopy them.
4. Many math labs have video tapes and computerized tutorials. Take advantage of them.
5. Visit the math lab for help, but make sure that you have done everything you can first, and have your questions organized.

We say again: Take charge of your own learning.

A Evaluating Numerical Expressions

A numerical expression may contain more than one operation. If the operations are all the same, then working from left to right seems to us the natural way to perform the operations, and that's exactly what we do.

$$6 + 3 + 2 = 9 + 2 = 11$$
$$9 - 5 - 2 = 4 - 2 = 2$$
$$2 \cdot 3 \cdot 5 = 6 \cdot 5 = 30$$
$$18 \div 3 \div 2 = 6 \div 2 = 3$$

If two or more different operations are indicated, then we have the question of which operation to perform first. For example, consider the expression $5 + 3 \cdot 2$.

Add first, then multiply.	*Multiply first, then add.*
$5 + 3 \cdot 2 = 8 \cdot 2 = 16$	$5 + 3 \cdot 2 = 5 + 6 = 11$

Depending on which operation we perform first, even a simple expression like $5 + 3 \cdot 2$ could have two different values. To avoid such confusion, we simply make agreements about the order in which we perform operations.

The Order of Operations

To evaluate a numerical expression, perform the indicated operations in the following order:

1. If the expression has grouping symbols, such as parentheses ( ), brackets [ ], or braces { }, then begin by performing all operations inside the grouping symbols.
2. Next, evaluate numbers that are in exponential form.
3. Working from left to right, perform all multiplications and divisions.
4. Working from left to right, perform all additions and subtractions.

PEMDAS

Copyright © Houghton Mifflin Company. All rights reserved.

Now we can return to the expression $5 + 3 \cdot 2$. Because multiplication ranks higher than addition in the Order of Operations, we perform the multiplication first.

$$5 + 3 \cdot 2 = 5 + 6 = 11$$

Grouping symbols are often used to change the normal order of operations. For example, in the expression $5 + 3 \cdot 2$, if we want the addition to be performed first, we write the expression as $(5 + 3) \cdot 2 = 8 \cdot 2 = 16$.

NOTE There are other grouping symbols besides parentheses, brackets, and braces. For example, a fraction bar indicates division, but it is also a grouping symbol. The expression above the bar is one group, and the expression below the bar is another group.

$$\frac{5 + 9}{6 - 4} \quad \text{means} \quad \frac{(5 + 9)}{(6 - 4)}$$

Example 1

Evaluate $12 - 15 \div 3$.

SOLUTION

$12 - 15 \div 3 = 12 - 5$ Division ranks higher than subtraction.

$= 7$

Your Turn 1

$6 + 24 \div 8$

Answer: 9

Example 2

Evaluate $14 \div (2 + 5) \cdot 4$.

SOLUTION

$14 \div (2 + 5) \cdot 4$ Grouping symbols first

$= 14 \div 7 \cdot 4$ Division and multiplication have the same rank. Work from left to right.

$= 2 \cdot 4$

$= 8$

Your Turn 2

Evaluate $15 - 2(5 - 3)$.

Answer: 11

Example 3

Evaluate $20 - 2(3^2 + 1)$.

SOLUTION

$20 - 2(3^2 + 1)$ Start inside the grouping symbols. Exponent first

$= 20 - 2(9 + 1)$ Then addition

$= 20 - 2(10)$ Then multiplication

$= 20 - 20$ Finally, subtraction

$= 0$

Your Turn 3

Evaluate $2^3 + \frac{8}{5 - 3}$.

Answer: 12

Copyright © Houghton Mifflin Company. All rights reserved.

EYS TO THE CALCULATOR

All calculators evaluate expressions according to the Order of Operations. However, the sequence used for entering the numbers differs among calculator models. Here are typical keystrokes for the expression in Example 3.

Scientific calculator:

20 [−] **2** [×] [( ] **3** [x^y] **2** [+] **1** [ )] [=]

Graphing calculator:

20 [−] **2** [×] [( ] **3** [^] **2** [+] **1** [ )] [ENTER]

```
20-2(3^2+1)
                    0
```

As shown in the screen display, some calculators may not require the [×] key in front of grouping symbols.

Sometimes you will need to insert grouping symbols yourself. For example, to evaluate $\frac{12 - 9}{2 + 1}$, if you enter $12 - 9 \div 2 + 1$, your calculator will follow the Order of Operations and evaluate $12 - (9 \div 2) + 1$. Instead, enter

[( ] **12** [−] **9** [ )] [÷] [( ] **2** [+] **1** [ )] [=] (or [ENTER])

```
(12-9)/(2+1)
                    1
```

Exercises

(a) $5 + 2(5 + 2)$

(b) $(5 + 2)(5 + 2)$

(c) $26 - 3^2 \div (10 - 1)$

(d) $\frac{7 + 5(4)}{2^3 - 5}$

LEARNING TIP

When you evaluate an expression, go just one step at a time. Trying to do too many steps at once increases the risk of errors.

Sometimes an expression has grouping symbols within grouping symbols. To evaluate such expressions, begin with the innermost grouping symbols and work outward.

$20 - [15 - (5 - 2)]$ Start with the innermost grouping symbols.

$= 20 - [15 - (3)]$ Now work inside the brackets.

$= 20 - [12]$

$= 8$

Copyright © Houghton Mifflin Company. All rights reserved.

Example 4

Evaluate $15 - [3^2 - (8 - 6)]$.

SOLUTION

The parentheses are the innermost grouping symbols, so we begin there.

$$
\begin{aligned}
&15 - [3^2 - (8 - 6)] \\
&= 15 - [3^2 - 2] && \text{Now work inside the brackets. Exponent first} \\
&= 15 - [9 - 2] && \text{Still inside the brackets} \\
&= 15 - 7 && \text{Now the final subtraction} \\
&= 8
\end{aligned}
$$

Your Turn 4

Evaluate

$$2[(3 + 9) \div 4]$$

Answer: 6

NOTE We stop writing grouping symbols once we have performed the indicated operations inside them. In Example 4, we began with $8 - 6$ inside the parentheses. After we performed that operation to obtain 2, we no longer needed the parentheses. Similarly, the brackets around $9 - 2$ were no longer needed after we performed that operation to obtain 7.

Special care must be taken with fraction bars, which represent division but are also grouping symbols.

$$18 - \frac{10 + 2^3}{3 + 6} \quad \text{means} \quad 18 - \frac{(10 + 2^3)}{(3 + 6)}$$

Therefore, the quantities above and below the fraction bar are understood to be in the innermost grouping symbols and are to be evaluated before dividing.

$$
\begin{aligned}
18 - \frac{10 + 2^3}{3 + 6} &= 18 - \frac{10 + 8}{3 + 6} && \text{Start with the exponent} \\
&= 18 - \frac{18}{9} && \text{Evaluate above and below the fraction bar.} \\
&= 18 - 2 && \text{Division ranks higher than subtraction.} \\
&= 16
\end{aligned}
$$

B Translations

In many real-life applications, information is given in words. To solve such problems, we need to be able to translate from the English language into mathematical language.

In preceding sections, we saw how a short phrase such as "6 increased by 2" is translated into the mathematical expression $6 + 2$. Sometimes we must translate longer descriptions into expressions that involve more than one operation.

Copyright © Houghton Mifflin Company. All rights reserved.

Suppose that we need to translate "the square of the sum of 7 and 1" into mathematical symbols. Rather than trying to translate the entire description all at once, a better approach is to build the expression a few words at a time.

The square of	$(\ )^2$	We are going to square some quantity, but we do not know what it is yet.
The sum of	$(\ +\ )^2$	The quantity being squared is a sum, but we do not know what we are adding yet.
7 and 1	$(7 + 1)^2$	The numbers being added are 7 and 1.

Translating from the English language to mathematical language takes a lot of practice and experience. Being patient and building expressions slowly are the keys to success.

Example 5

Translate into a mathematical expression.

(a) 6 increased by the cube of 2

(b) The product of 4 and the difference of 7 and 2

SOLUTION

(a) 6 increased by the cube of 2

6	6	Begin with 6.
increased by	$6 +$	Add some quantity.
the cube of	$6 + (\ )^3$	The quantity being added is the cube of something.
2	$6 + 2^3$	The quantity being cubed is 2.

The expression is $6 + 2^3$.

(b) The product of 4 and the difference of 7 and 2

The product of	$(\ ) \cdot (\ )$	We are multiplying, but we do not know what the factors are yet.
4	$4 \cdot (\ )$	The first factor is 4.
and the difference of	$4 \cdot (\ -\ )$	The second factor involves subtraction.
7 and 2	$4 \cdot (7 - 2)$	The numbers being subtracted are 7 and 2.

The expression is $4 \cdot (7 - 2)$, or simply $4(7 - 2)$.

Your Turn 5

Translate into a mathematical expression: 4 more than the product of 5 and 3.

Answer: $4 + (5)(3)$

Copyright © Houghton Mifflin Company. All rights reserved.

C Applications

An example of an application that might be of particular interest to you is the manner in which your course grade is determined. For simplicity, we assume that you are given five tests and that your grade is the average of those scores.

The *average* of a list of numbers is found by adding the numbers and then dividing the sum by the number of numbers in the list.

Example 6

Suppose that your test scores (out of 100) were 82, 85, 91, 73, and 94. What is your test score average?

SOLUTION

We add the test scores and divide by 5.

$$\frac{82 + 85 + 91 + 73 + 94}{5} = \frac{425}{5} = 85$$

Your test average is 85.

Your Turn 6

Suppose that a sixth test is given, and your score is 79. What is your new average?

Answer: 84

Example 7

A charter school is a school that is established and possibly funded by a state but that is not under the control of a local school system. The bar graph shows the number of charter schools that were approved in three states in 2000. (Source: Center for Education Reform.)

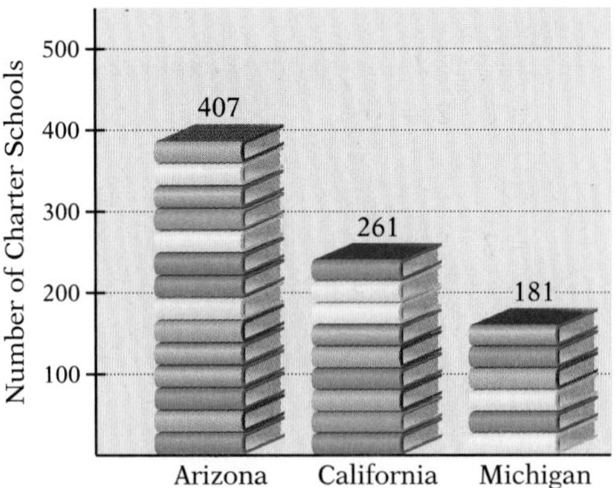

(a) What was the average number of charter schools in the three given states?

(b) By how much did the number of charter schools in Arizona exceed the average for the three states?

Your Turn 7

Suppose that each of the three states added 15 additional charter schools. What would be the new average number of charter schools in the three states?

Copyright © Houghton Mifflin Company. All rights reserved.

SOLUTION

(a) The average is found by adding the number of charter schools for all three states and dividing the sum by 3.

$$\frac{407 + 261 + 181}{3} = \frac{849}{3} = 283$$

(b) The question asks for the difference between the number of charter schools in Arizona and the average.

$$407 - 283 = 124$$

The number of charter schools in Arizona exceeded the average for the three states by 124.

2.5 Quick Reference

A. Evaluating Numerical Expressions

1. When an expression has two or more operations, we perform them according to an agreement called the Order of Operations.

(a) If the expression has grouping symbols, such as parentheses ( ), brackets [ ], or braces { }, then begin by performing all operations inside the grouping symbols.

(b) Next, evaluate numbers that are in exponential form.

(c) Working from left to right, perform all multiplications and divisions.

(d) Working from left to right, perform all additions and subtractions.

$7 + 2(6)$ $= 7 + 12 = 19$	Multiply first.
$11 - 3^2$ $= 11 - 9 = 2$	Exponent first
$15 - 12 \div 4$ $= 15 - 3 = 12$	Divide first.
$(5 + 3) \cdot 2$ $= 8 \cdot 2 = 16$	Parentheses first
$7 \cdot 2^3$ $= 7 \cdot 8 = 56$	Exponent first
$8 \div 4(3)$ $= 2(3) = 6$	Left to right
$7 - 2 + 9$ $= 5 + 9 = 14$	Left to right
$\frac{5 + 3}{4 - 2} = \frac{8}{2} = 4$	Numerator and denominator first
$2[5 - (3 - 1)]$ $= 2[5 - 2]$ $= 2[3]$ $= 6$	Start with innermost grouping symbols.

Copyright © Houghton Mifflin Company. All rights reserved.

2. To translate a word description into a mathematical expression, a good approach is to build the expression a few words at a time.

9 decreased by the quotient of 12 and 4

9	9
$9 - (\ )$	decreased by
$9 - \frac{(\ )}{(\ )}$	the quotient of
$9 - \frac{12}{4}$	12 and 4

Copyright © Houghton Mifflin Company. All rights reserved.

2.5 Exercises

A. Evaluating Numerical Expressions

1. In the expression $3 + [10 - (3 - 1)]$, the brackets and parentheses are called ________.

2. To evaluate an expression containing more than one operation, we follow the ________.

In Exercises 3–32, evaluate the expression.

3. $25 - 3 \cdot 7$

4. $3 + 4 \cdot 5$

5. $6 + 24 \div 3$

6. $15 - 12 \div 2$

7. $20 - 3(8 - 3)$

8. $1 + 4(5 - 2)$

9. $24 \div 6 \cdot 2$

10. $5 \cdot 6 \div 3$

11. $45 \div (3^2 - 4) \cdot 8$

12. $63 \div (2 + 5) - 3^2$

13. $28 - 3(15 - 2^3)$

14. $4 + 2(7 + 3 \cdot 5)$

15. $5 \cdot 2^3$

16. $2 \cdot 7^2$

17. $\dfrac{1 + 2 \cdot 5}{4^2 - 5}$

18. $\dfrac{3 + 3 \cdot 7}{2^3}$

19. $3 + 2(5^2 - 4)$

20. $19 - 4(2^3 - 5)$

21. $7^2 + \dfrac{3 \cdot 4}{5 - 2}$

22. $4^3 - \dfrac{3^2 + 11}{6 - 1}$

23. $8^2 - 2 \cdot 3 \cdot 5$

24. $(3 - 1)^2 + (5 + 1)^2$

25. $7 + 5 - 2 \cdot 6 \div 4$

26. $6 - (11 - 3) \div 4$

27. $5[30 \div (10 \div 2)]$

28. $12 \div 2 + 4[8 - 3(2 - 1)]$

29. $18 - [5^2 - (10 - 3)]$

Copyright © Houghton Mifflin Company. All rights reserved.

30. $6^2 + 2[8 - (3^2 - 1)]$

31. $7 - 3[5(6 - 2) \div 10]$

32. $7 + 2[5^2 - 2(4 - 1)^2]$

Calculator Exercises

In Exercises 33–36, use a calculator to evaluate the numerical expression.

33. $6^4 - 54(51 - 46)$

34. $48^2 - 4 \cdot 12 \cdot 35$

35. $\dfrac{27^2 + 2 \cdot 107}{5 \cdot 8 + 1}$

36. $\dfrac{1{,}286 - 564}{7 \cdot 9 - 5^2}$

B. Translations

In Exercises 37–42, write the given phrase as a numerical expression and evaluate it.

37. Five less than the product of 4 and 8

38. The square of the difference of 12 and 8

39. The quotient of 10 and the difference of 7 and 2

40. Twice the difference of 3 squared and 5

41. The difference of 8 squared and the product of 7 and 4

42. The quotient of 20 and the difference of 3 squared and 5

C. Applications

43. *Test Average* A student's scores on six 100-point tests were 95, 79, 86, 98, 82, and 94. What was the average score?

Copyright © Houghton Mifflin Company. All rights reserved.

44. *Miles per Gallon* A person drove 556 miles from Atlanta to Richmond and then returned by a longer route of 620 miles. The entire trip required 56 gallons of gasoline. How many miles per gallon did the car average?

45. *Grade-Point Average* In the calculation of a grade-point average, an A is worth 4 points, a B is worth 3 points, and a C is worth 2 points. One semester a student took four 3-hour courses and received one A, two B's, and one C. What was the student's grade-point average?

46. *Fund Raiser* A club with 120 members sold $34,440 worth of gift baskets to raise funds for its service projects. What was the average amount that each member sold?

47. *Grocery Bills* During a 6-week period, the weekly grocery bills for a family of four were $198, $162, $157, $237, $190, and $214. What was the average weekly grocery bill?

48. *Rushing Yards* In 5 games a football player rushed for 87, 110, 123, 76, and 94 yards. What was the average number of yards per game?

Writing and Concept Extension

49. Explain how to determine the average of two numbers.

50. Explain why parentheses are not necessary in the expression $30 - (3 \cdot 4)$.

In Exercises 51 and 52, evaluate the expression.

51. $5 \cdot 10^3$

52. $6 \cdot 10^2$

In Exercises 53 and 54, insert parentheses to make the equation true.

53. $3 \cdot 5 + 2 = 21$

54. $2 + 3^2 = 25$

Exploring with Real-World Data: Collaborative Activities

Retail Sales of Motor Vehicles The table shows the retail sales of motor vehicles for selected years during the period 1990 to 2000. All numbers are in thousands; that is, 6,897 means 6,897 thousand or 6,897,000. (Source: Wards Autoinfobank.)

Copyright © Houghton Mifflin Company. All rights reserved.

	1990	1995	2000
Passenger Cars			
Domestic	6,897	7,129	6,830
Import	2,403	1,506	2,016
Trucks			
Domestic	4,215	6,064	8,092
Import	631	417	873

In Exercises 55–62, ignore remainders when you divide.

55. What were the average yearly sales of domestic cars?

56. What were the average yearly sales of imported trucks?

57. What were the average yearly sales of imported cars?

58. For which category did sales increase each given year?

59. For which given year were the total sales of passenger cars the highest?

60. What were the average yearly sales of passenger cars?

61. For which given year were the total sales of trucks the lowest?

62. Determine the average yearly sales of all vehicles.

Copyright © Houghton Mifflin Company. All rights reserved.

2.6 Expressions and Equations

A *Simplifying Products*
B *Evaluating Expressions*
C *Equations*

SUGGESTIONS FOR SUCCESS

We all develop mind cramps.

Sometimes, things seem clear and easy in class, but when you start the assignment, the very first exercise is a mystery. The worst thing you can do is throw up your hands and quit.

1. Check your excellent notes. There is a good chance that your instructor gave an example of the exercise.
2. Check the examples in the book. Exercises are keyed to the subsections so that you can easily find the examples.
3. If you have a *Student Solutions Manual,* look at the approach that was used to solve the problem. (But then do the rest of the exercises on your own.)
4. If you have a math lab with tutors, go there and ask. Be sure to have specific questions ready so that you don't just snarl at the tutors.

In short, be aggressive in getting the help you need.

A Simplifying Products

We *simplify* an expression by performing all the operations that we can. Simplifying also includes removing grouping symbols, if possible. Some products can be simplified with the Associative Property of Multiplication.

Example 1

Simplify the product $8(2x)$.

SOLUTION

$8(2x) = (8 \cdot 2) \cdot x$ — Associative Property of Multiplication

$= (16)x$ — Multiply $8 \cdot 2$.

$= 16x$ — The parentheses aren't needed.

Your Turn 1

Simplify $6(3y)$.

Answer: $18y$

Example 2

Simplify the product $(3y)(2z)$.

SOLUTION

$(3y)(2z) = 3 \cdot y \cdot 2 \cdot z$

$= 3 \cdot 2 \cdot y \cdot z$ — Commutative Property of Multiplication

$= (3 \cdot 2)(y \cdot z)$ — Associative Property of Multiplication

$= (6)(yz)$

$= 6yz$ — The parentheses aren't needed.

Your Turn 2

Simplify $(8a)(3b)$.

Answer: $24ab$

Copyright © Houghton Mifflin Company. All rights reserved.

We may be able to simplify a variable expression by applying the Distributive Property.

Example 3

Use the Distributive Property to write $7(x + 4)$ as a sum. Then simplify the result.

SOLUTION

$$7(x + 4) = 7(x) + 7(4) \quad \text{Distributive Property}$$
$$= 7x + 28 \quad \text{Multiply } 7 \cdot 4.$$

We can't add $7x$ and 28 because we don't know the value of $7x$.

Your Turn 3

Use the Distributive Property to write $9(y + 2)$ as a sum. Then simplify the result.

9y + 18

Answer: $9y + 18$

Simplifying may require the use of both the Distributive Property and the Associative Property of Multiplication.

Example 4

Use the Distributive Property to write $8(2x + 5)$ as a sum. Then simplify the result.

SOLUTION

$$8(2x + 5) = 8(2x) + 8(5) \quad \text{Distributive Property}$$
$$= (8 \cdot 2)x + 8(5) \quad \text{Associative Property of Multiplication}$$
$$= 16x + 40 \quad \text{Multiply.}$$

Your Turn 4

Use the Distributive Property to write $2(7a + 4)$ as a sum. Then simplify the result.

Answer: $14a + 8$

B Evaluating Expressions

Recall that the value of a variable expression cannot be found until we replace the variable(s) with specific values.

Example 5

Evaluate $4(8a)$ for $a = 1$.

SOLUTION

$$4(8a) \quad \text{The given variable expression}$$
$$\downarrow$$
$$4(8 \cdot 1) \quad \text{Replace } a \text{ with 1.}$$
$$= 4(8) \quad \text{Simplify: } 8 \cdot 1 = 8.$$
$$= 32$$

Here is another approach:

$$4(8a) = (4 \cdot 8)a \quad \text{Associative Property of Multiplication}$$
$$= 32a \quad \text{Simplify: } 4 \cdot 8 = 32.$$
$$\downarrow$$
$$= 32(1) = 32 \quad \text{Replace } a \text{ with 1.}$$

Your Turn 5

Evaluate $2(3x)$ for $x = 3$.

Answer: 18

Copyright © Houghton Mifflin Company. All rights reserved.

Example 6

Evaluate the following for $x = 4$ and $y = 5$:

(a) $3xy$ **(b)** $\frac{8y}{5x}$

SOLUTION

(a) $3xy = 3(4)(5)$ Replace x with 4 and y with 5.

$= 60$

(b) $\frac{8y}{5x} = \frac{8(5)}{5(4)}$ Replace x with 4 and y with 5.

$= \frac{40}{20} = 2$

Your Turn 6

Evaluate the following for $a = 3$ and $b = 2$.

(a) $6ab$ **(b)** $\frac{ab}{6}$

Answers: (a) 36; (b) 1

NOTE Make sure that you know the difference between $3xy$ and $3(x + y)$. The expression $3xy$ is simply the product of three numbers. The Distributive Property does *not* apply to $3xy$; that is, we do not multiply 3 times the x and then 3 times the y.

When a variable expression contains two or more operations, we use the Order of Operations to evaluate the expression for given values of the variables.

Example 7

Evaluate $a + bc$ for $a = 2$, $b = 5$, and $c = 3$.

SOLUTION

$a + b \cdot c$ The given variable expression

$\downarrow \quad \downarrow \quad \downarrow$

$2 + 5 \cdot 3$ Replace a with 2, b with 5, and c with 3.

$= 2 + 15$ Multiply first.

$= 17$ Then add.

Your Turn 7

Evaluate $x - yz$ for $x = 10$, $y = 4$, and $z = 1$.

Answer: 6

Example 8

Evaluate the following two expressions for $a = 5$ and $b = 3$:

(a) $a^2 - b^2$ **(b)** $(a + b)(a - b)$

SOLUTION

(a) $a^2 - b^2$ The given variable expression

$5^2 - 3^2$ Replace a with 5 and b with 3.

$= 25 - 9$ Exponents first

$= 16$ Then subtract.

(b) $(a + b)(a - b)$

$(5 + 3)(5 - 3)$ Replace a with 5 and b with 3.

$= (8)(2) = 16$ Parentheses first, then multiplication.

Your Turn 8

Evaluate the following two expressions for $a = 4$ and $b = 2$:

(a) $(a + b)^2$

(b) $a^2 + 2ab + b^2$

Answers: (a) 36; (b) 36

Copyright © Houghton Mifflin Company. All rights reserved.

NOTE In Example 8, did you observe that the two given expressions have the same value when $a = 5$ and $b = 3$? Later we will learn that $a^2 - b^2$ and $(a + b)(a - b)$ have the same value for *any* replacements of a and b.

C Equations

Recall that an equation is a statement that two expressions have the same value. We can see that the equation $3 \cdot 5 = 15$ is true and that the equation $7 \cdot 4 = 11$ is false. However, we don't know whether the equation $5x = 20$ is true or false until we replace the variable x with a specific value.

Solving an equation means finding the value(s) of the variable that make the equation true. Simple equations that involve products can be solved by inspection.

Example 9

Solve the equation $4x = 32$.

SOLUTION

We ask, "What number multiplied by 4 is 32?" From the basic multiplication facts, the solution is 8 because $4(8) = 32$.

Your Turn 9

Solve the equation $56 = 8y$.

Answer: 7

You may need to simplify as part of the equation-solving process.

Example 10

Solve the equation $2(3x) = 30$.

SOLUTION

$2(3x) = 30$	The given equation
$(2 \cdot 3)x = 30$	Associative Property of Multiplication
$6x = 30$	Simplify the left side.

By inspection, the solution is 5 because $6 \cdot 5 = 30$.

Your Turn 10

Solve the equation $4(2y) = 3 \cdot 8$.

Answer: 3

An equation that involves division may be simple enough that you can use the basic multiplication facts to solve the equation by inspection.

Example 11

Solve $\frac{x}{2} = 6$.

SOLUTION

The related multiplication is $6 \cdot 2 = x$. Therefore, $x = 12$.

Your Turn 11

Solve $\frac{x}{9} = 2$.

Answer: 18

Copyright © Houghton Mifflin Company. All rights reserved.

Example 12

Solve $\frac{20}{x} = 4$.

SOLUTION

The related multiplication is $4x = 20$. From the basic multiplication facts, we know that $x = 5$ because $4 \cdot 5 = 20$.

Your Turn 12

Solve $\frac{27}{x} = 3$.

Answer: 9

You also may be able to solve simple equations involving exponential expressions by inspection.

Example 13

Find the whole number for which the given equation is true.

(a) $s^2 = 49$ **(b)** $a^3 = 8$

SOLUTION

(a) For $s^2 = 49$, we ask, "For what whole number s does $s \cdot s = 49$?" The answer is 7 because $7 \cdot 7 = 49$.

(b) For $a^3 = 8$, we ask, "For what whole number a does $a \cdot a \cdot a = 8$?" The answer is 2 because $2 \cdot 2 \cdot 2 = 8$.

Your Turn 13

Find the whole number for which the given equation is true.

(a) $x^2 = 100$

(b) $z^3 = 27$

Answers: (a) 10; (b) 3

2.6 Quick Reference

A. Simplifying Products

1. We *simplify* an expression by performing all the operations that we can. Some products can be simplified with the Associative Property of Multiplication.

$$2(9x) = (2 \cdot 9)x = 18x$$

$$\begin{aligned}(5a)(2x) &= 5 \cdot a \cdot 2 \cdot x \\ &= 5 \cdot 2 \cdot a \cdot x \\ &= 10ax\end{aligned}$$

2. We may be able to simplify a variable expression by applying the Distributive Property.

$$\begin{aligned}4(x + 3) &= 4(x) + 4(3) \\ &= 4x + 12\end{aligned}$$

$$\begin{aligned}7(1 + 2y) &= 7(1) + 7(2y) \\ &= 7 + 14y\end{aligned}$$

Copyright © Houghton Mifflin Company. All rights reserved.

B. Evaluating Expressions

1. If a variable expression is a product or a quotient, we evaluate the expression by replacing each variable with a specific value and then performing the indicated operation.

If $a = 3$ and $y = 5$, then $2ay = 2(3)(5) = 30$.

If $a = 3$, $b = 4$, and $c = 6$, then

$$\frac{ab}{2c} = \frac{3 \cdot 4}{2 \cdot 6} = \frac{12}{12} = 1$$

2. When a variable expression contains two or more operations, we use the Order of Operations to evaluate the expression for given values of the variables.

Evaluate $ab + c^2$ for $a = 2$, $b = 3$, and $c = 5$.

$$a \cdot b + c^2$$
$$\downarrow \quad \downarrow \quad \downarrow$$
$$2 \cdot 3 + 5^2$$
$$= 2 \cdot 3 + 25$$
$$= 6 + 25$$
$$= 31$$

C. Equations

1. Simple equations that involve products can be solved by inspection.

The solution of $9x = 27$ is 3 because $9(3) = 27$.

2. You may need to simplify one or both sides of an equation before you can solve it by inspection.

$$3(2y) = 24$$
$$(3 \cdot 2)y = 24$$
$$6y = 24$$

The solution is 4 because $6(4) = 24$.

3. Related multiplications can be used to solve simple equations that involve division.

$\frac{x}{4} = 5$ means $5 \cdot 4 = x$.
Therefore, $x = 20$.

$\frac{63}{x} = 9$ means $9x = 63$.
Therefore, $x = 7$.

4. Simple equations involving exponential expressions can usually be solved by inspection.

The whole number for which $x^2 = 9$ is 3 because $3^2 = 9$.

Copyright © Houghton Mifflin Company. All rights reserved.

2.6 Exercises

A. Simplifying Products

1. Writing $(2x)(8y)$ as $16xy$ is called ______ the expression.

2. We use the ______ Property to write $3(z + 4)$ as $3z + 12$.

In Exercises 3–10, use the Associative Property of Multiplication to simplify the expression.

3. $5(3x)$

4. $8(7y)$

5. $2(4a)$

6. $9(6m)$

7. $(2a)(5x)$

8. $(7x)(3y)$

9. $(3x)(4y)$

10. $(5m)(8n)$

In Exercises 11–18, use the Distributive Property to write the expression as a sum. Then simplify the result.

11. $5(x + 6)$

12. $3(y + 4)$

13. $2(7 + a)$

14. $4(3 + c)$

15. $7(2x + 1)$

16. $6(5x + 9)$

17. $8(7a + 6)$

18. $2(6y + 5)$

B. Evaluating Expressions

19. To evaluate $a - b^2$ for $a = 6$ and $b = 1$, we must use the Order of ______

20. For the expression $5y$, replacing y with 3 and performing the indicated multiplication is called ______ the expression for $y = 3$.

Copyright © Houghton Mifflin Company. All rights reserved.

In Exercises 21–26, evaluate the expression for the given value(s) of the variable(s).

21. $8a$ for $a = 2$

22. $7x$ for $x = 1$

23. $(2a)b$ for $a = 4$ and $b = 3$

24. $c(3d)$ for $c = 4$ and $d = 2$

25. $3xy$ for $x = 5$ and $y = 10$

26. $4mn$ for $m = 5$ and $n = 7$

In Exercises 27–32, evaluate the expression for the given value(s) of the variable(s).

27. $\frac{y}{4}$ for $y = 24$

28. $\frac{36}{c}$ for $c = 12$

29. $\frac{mn}{10}$ for $m = 15$ and $n = 2$

30. $\frac{20}{xy}$ for $x = 2$ and $y = 5$

31. $\frac{5x}{6y}$ for $x = 12$ and $y = 10$

32. $\frac{2a}{3b}$ for $a = 18$ and $b = 4$

In Exercises 33–40, evaluate the expression for the given value(s) of the variable(s).

33. $2x - 3y$ for $x = 9$ and $y = 2$

34. $bc - a$ for $a = 12$, $b = 8$, and $c = 7$

35. $x^2 - 3x$ for $x = 5$

36. $b^2 + b - 6$ for $b = 3$

37. $x + 2(y - x)$ for $x = 3$ and $y = 10$

38. $7(b - c) - 3b$ for $b = 7$ and $c = 2$

39. $\frac{2b - c}{c - 2}$ for $b = 10$ and $c = 5$

40. $\frac{x + 3y}{x - 1}$ for $x = 8$ and $y = 2$

C. Equations

41. We call $7x$ a(n) expression, whereas we call $7x = 21$ a(n) equation

42. To solve the equation $4(2x) = 16$, first simplify the left side of the equation.

Copyright © Houghton Mifflin Company. All rights reserved.

In Exercises 43–50, solve the equation by inspection.

43. $7x = 56$

44. $5y = 30$

45. $4y = 4$

46. $10a = 50$

47. $72 = 9a$

48. $12 = 6x$

49. $4(2y) = 0$

50. $3(5x) = 15$

In Exercises 51–58, solve the equation by inspection.

51. $\frac{x}{9} = 9$

52. $\frac{a}{10} = 7$

53. $\frac{y}{5} = 0$

54. $\frac{x}{3} = 9$

55. $\frac{7}{n} = 1$

56. $\frac{12}{x} = 2$

57. $\frac{50}{y} = 5$

58. $\frac{48}{z} = 8$

In Exercises 59–62, find the whole number for which the equation is true.

59. $x^2 = 16$

60. $y^3 = 1$

61. $b^3 = 27$

62. $x^2 = 64$

Writing and Concept Extension

63. Identify the expression that uses the Distributive Property to remove the parentheses. Explain your answer.

(i) $5(3x)$

(ii) $5(3 + x)$

64. Write each of the following in symbols. Which is an equation and which is an expression?

(a) Three times x

(b) Three times x is 21.

65. Explain why any number is a solution of the equation $\frac{x}{1} = x$.

66. Explain why the equation $0 \cdot x = 7$ has no solution.

Copyright © Houghton Mifflin Company. All rights reserved.

67. Evaluate $x^0 + x^1 + x^2 + x^3$ for $x = 2$.

68. Simplify $(2x)(3y)(4z)$.

69. Simplify.

(a) $x(3 + 2y)$

(b) $2x(5a + 3b)$

Copyright © Houghton Mifflin Company. All rights reserved.

2.7 Area

SUGGESTIONS FOR SUCCESS

Remembering formulas follows the "use it or lose it" rule. If you use a particular formula often, you will remember it. Otherwise, you probably won't.

The formulas in this section are very basic, and expecting you to know them is reasonable. However, instructors have varying opinions about which formulas should be memorized. Asking about that is certainly a fair question.

The most important things to remember are that formulas exist and that you can look them up when you need them.

A Area of a Square

In Section 1.5, we described a *polygon* as a geometric figure whose sides are line segments that completely enclose the figure. We also defined the *perimeter* of a polygon as the sum of the lengths of the sides. In other words, the perimeter is the distance around the outside of the figure.

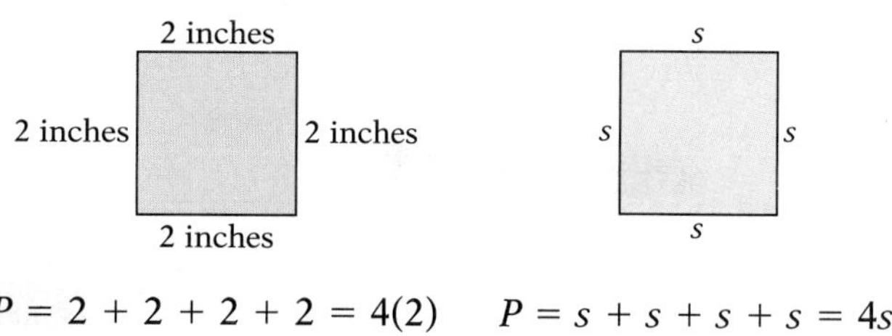

$P = 2 + 2 + 2 + 2 = 4(2)$ $\quad$ $P = s + s + s + s = 4s$

For a square whose sides are 2 inches long, the perimeter P is $4(2) = 8$ inches. We can more generally describe how to find the perimeter P of a square whose sides are s inches long with the *formula* $P = 4s$.

A **formula** is an equation with more than one variable. You can think of a formula as instructions for calculating certain quantities. For example, the formula $P = 4s$ tells us that we can calculate the perimeter P of a square by multiplying the length s of a side by 4.

In this section and the next, we consider formulas for three other features of geometric figures: area, volume, and surface area.

Area of a Square

Suppose that you decide to install ceramic tiles on the wall behind your kitchen counter. To determine the number of tiles that you will need, you will need to know how much wall surface you intend to cover.

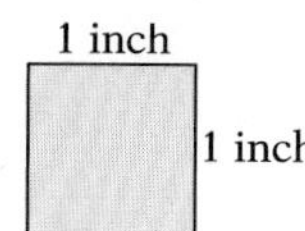

If each tile is a square whose sides are 1 inch long, then we say that the **area** of the square is 1 square inch. This measure of the area is a numerical way of describing the amount of surface that the tile has.

Copyright © Houghton Mifflin Company. All rights reserved.

Now consider the following two arrangements of tiles:

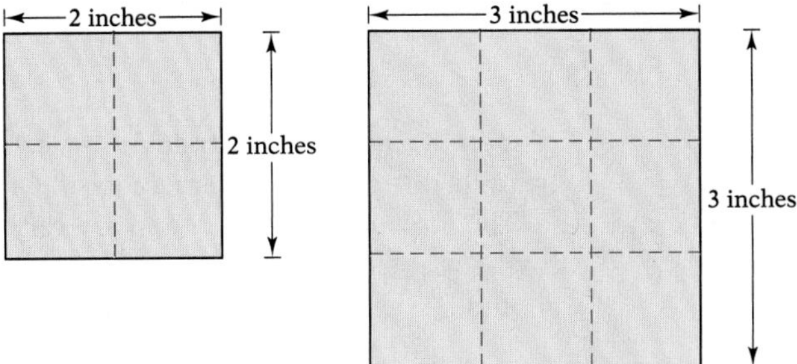

In each figure, the tiles are arranged to form a larger square. In the first arrangement, there are 4 tiles, each with an area of 1 square inch. Therefore, the total area is 4 square inches. Note that the sides of the overall square are 2 inches long and that the area, 4, can be found by $2 \cdot 2$, or 2^2.

Similarly, the second arrangement has 9 tiles, so the total area is 9 square inches. The sides of the overall square are 3 inches long, and the area, 9, can be found by $3 \cdot 3$, or 3^2.

We could continue to add more tiles to make larger squares. In general, if the side of the square is s units long, then the area of the square is s^2.

Area of a Square

The formula for the area A of a square whose sides are s units long is

$$A = s^2$$

NOTE We said that s^2 can be read as "s squared" rather than "s to the second power." This special name comes from the fact that s^2 is the area of a square whose sides are s units long.

A *unit square* is a square whose sides are 1 unit long and whose area is 1 square unit. The units used to describe the area of a square depend on the units used in the length of the sides. The following table shows some typical units, along with their common abbreviations.

Side of Square (s)	*Area of Square ($A = s^2$)*	*Abbreviation*
4 inches	16 square inches	16 sq in. or 16 in.^{2}
5 yards	25 square yards	25 sq yd or 25 yd^2
3 feet	9 square feet	9 sq ft or 9 ft^2
6 miles	36 square miles	36 sq mi or 36 mi^2

There are many other measures of length and area, including metric measures, which we will discuss in a later chapter.

Copyright © Houghton Mifflin Company. All rights reserved.

Example 1

What is the area of a square whose sides are 8 feet long?

SOLUTION

$A = s^2$ Formula for the area of a square

$A = (8)^2 = 64$ Replace s with 8.

The area is 64 square feet.

Your Turn 1

What is the area of a square whose sides are 9 inches long?

Answer: 81 square inches

Example 2

If the area of a square is 4 square yards, how long are the sides?

SOLUTION

$A = s^2$ Formula for the area of a square

$4 = s^2$ Replace A with 4.

$2 = s$ Ask, "What number squared equals 4?"

The sides are 2 yards long.

Your Turn 2

If the area of a square is 49 square inches, how long are the sides?

Answer: 7 inches

B Area of a Rectangle

We can think of the area of a rectangle as the number of unit squares needed to fill the rectangle.

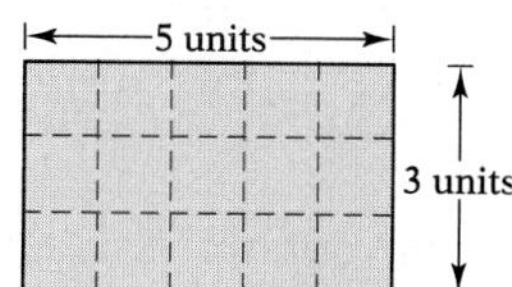

This rectangle contains 15 unit squares, which we can determine by multiplying the length, 5, by the width, 3. In general, the area of a rectangle is the product of the length and width.

Area of a Rectangle

If the length of a rectangle is L and the width is W, then the formula for the area A of the rectangle is

$$A = LW$$

Note that the perimeter of a rectangle is

$$P = L + W + L + W = 2L + 2W$$

Copyright © Houghton Mifflin Company. All rights reserved.

Example 3

A person bought a rectangular lot of land to build a home. The dimensions are 120 feet and 300 feet. How many square feet of land did the person buy?

SOLUTION

$A = LW$	Formula for the area of a rectangle
$A = (300)(120)$	Replace L with 300 and W with 120.
$= 36{,}000$	

The area of the land is 36,000 square feet.

Your Turn 3

A campaign poster is 30 inches wide and 42 inches high. What is the area of the poster?

Answer: 1,260 square inches

Example 4

The area of a greeting card is 40 square inches. How long is the card if the width is 5 inches?

SOLUTION

$A = LW$	Formula for the area of a rectangle
$40 = L \cdot 5$	Replace A with 40 and W with 5.
$8 = L$	Ask, "What number times 5 equals 40?"

The length of the card is 8 inches.

Your Turn 4

The area of a portrait photograph is 63 square inches. How wide is the picture if the length is 9 inches?

Answer: 7 inches

NOTE Formulas for areas of geometric figures can be used only when the dimensions all have the same units. For example, if the length of a rectangle is 2 feet and the width is 8 inches, we *cannot* find the area by multiplying $2 \cdot 8$. To use the formula $A = LW$, both the length and the width would need to be expressed in feet or in inches.

C Composite Figures

A *composite* figure is a combination of other geometric figures. In such cases, we usually can find the total area by dividing the figure into smaller parts whose areas can be found.

Typically, some dimensions of a composite figure are not given, but they can be calculated.

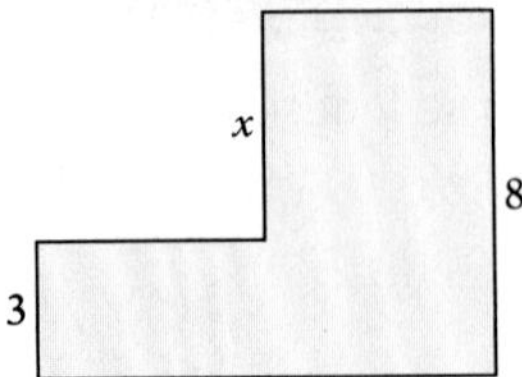

In the figure, the dimension labeled x is unknown. However, we see that $x + 3$ must be equal to 8. Therefore, solving the equation $x + 3 = 8$, we learn that the unknown dimension is 5.

Copyright © Houghton Mifflin Company. All rights reserved.

Example 5

The figure shows the design of a bank lobby.

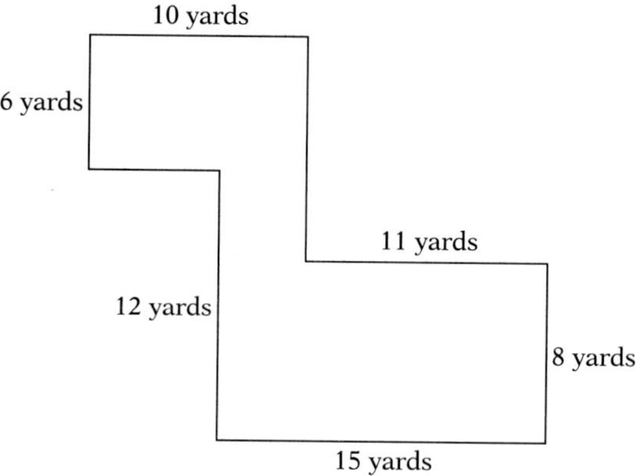

How many square yards of carpet are needed to cover the lobby floor?

SOLUTION

The dashed lines divide the lobby into three rectangular regions A, B, and C.

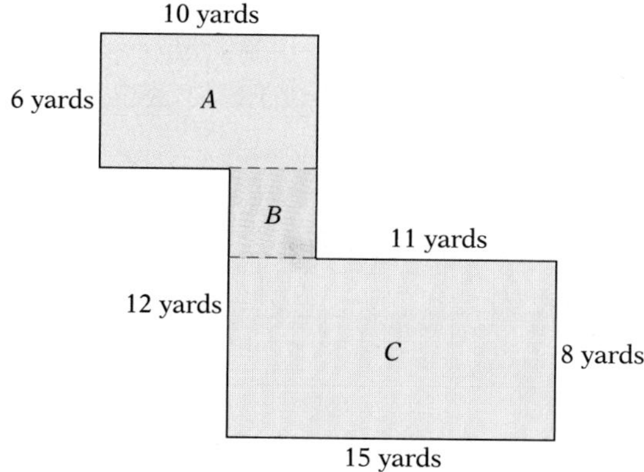

Using the given dimensions, we can find the missing dimensions of each rectangle.

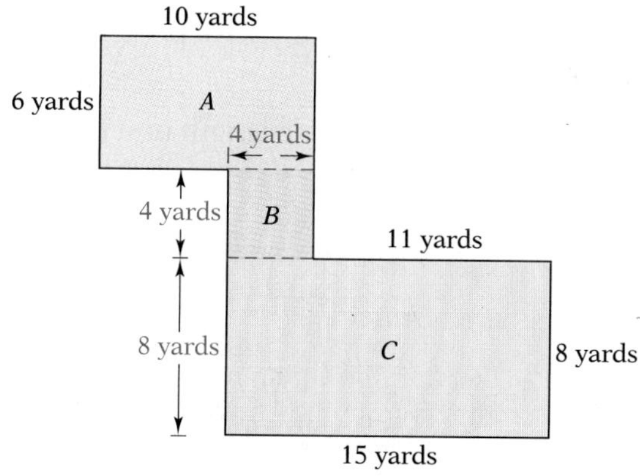

Your Turn 5

Find the total area of the figure.

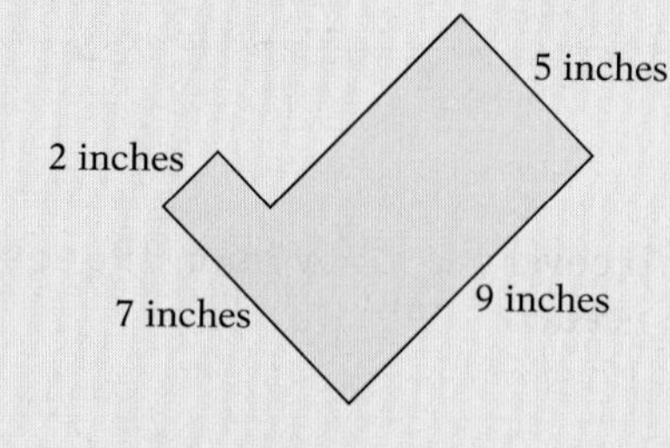

continued

Copyright © Houghton Mifflin Company. All rights reserved.

To find the total area, we add the areas of the three rectangles.

$$\begin{aligned}\text{Total area} &= \text{area } A + \text{area } B + \text{area } C\\ &= 10 \cdot 6 + 4 \cdot 4 + 15 \cdot 8\\ &= 60 + 16 + 120\\ &= 196\end{aligned}$$

To cover the lobby floor, 196 square yards of carpet are needed.

D Unit Conversion

When you describe the area of a geometric figure, you need to make a judgment about the most sensible units to use. For example, it makes little sense to write the area of a postage stamp in square yards or the area of an airport runway in square inches.

Generally the units used for the lengths of the sides of a figure will dictate the area units. However, we sometimes need to convert from one area unit to another.

Example 6

If the area of a square is 1 square yard, what is the area in square feet?

SOLUTION

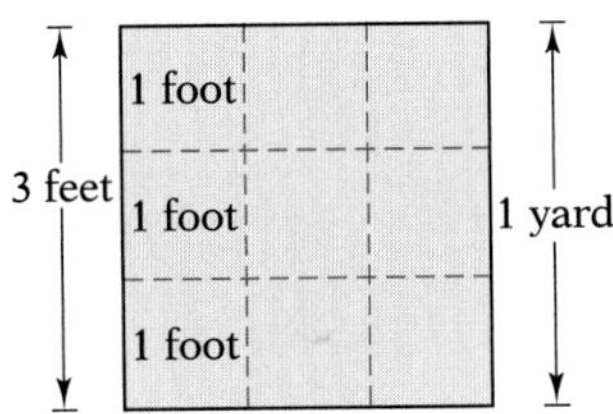

Because the area of the square is 1 square yard, the length of each side of the square is 1 yard, or 3 feet. Therefore, the area of the square in square feet is

$$A = s^2 = 3^2 = 9 \text{ square feet}$$

Your Turn 6

If the area of a square is 1 square foot, what is the area in square inches?

Answer: 144 square inches

The following tables can be used to convert from one length unit or one area unit to another.

LEARNING TIP

Converting from larger units to smaller units involves multiplication. Converting from smaller units to larger units involves division.

Length Units	*Area Units*
1 ft = 12 in.	1 sq ft = 144 sq in.
1 yd = 3 ft = 36 in.	1 sq yd = 9 sq ft = 1,296 sq in.
1 mi = 1,760 yd = 5,280 ft	1 acre = 43,560 sq ft
	1 sq mi = 640 acres

Copyright © Houghton Mifflin Company. All rights reserved.

Example 7

Suppose that a football field is laid out with a length of 100 yards and a width of 159 feet. What is the area of the field in (a) square feet and (b) square yards?

SOLUTION

(a) There are 3 feet in 1 yard, so

$$100 \text{ yards} = 3(100) = 300 \text{ feet}$$

Now we know that the length of the field is 300 feet and the width is 159 feet.

$$A = LW = 300 \cdot 159 = 47{,}700 \text{ square feet}$$

(b) There are 3 feet in 1 yard, so

$$159 \text{ feet} = \frac{159}{3} = 53 \text{ yards}$$

Now we know that the length of the field is 100 yards and the width is 53 yards.

$$A = LW = 100 \cdot 53 = 5{,}300 \text{ square yards}$$

Your Turn 7

A banner is 36 inches wide and 5 yards long. What is the area of the banner in square yards?

Answer: 5 square yards

2.7 Quick Reference

A. Area of a Square

1. A **formula** is an equation that gives us instructions for calculating certain quantities.

2. The **area** of a geometric figure is a measure of the amount of surface that the figure has.

3. A *unit square* is a square whose sides are 1 unit long and whose area is 1 square unit.

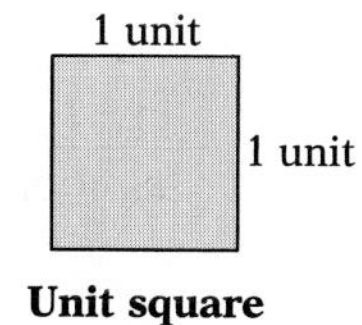

Unit square

4. The area A of a square whose sides are of length s is given by $A = s^2$.

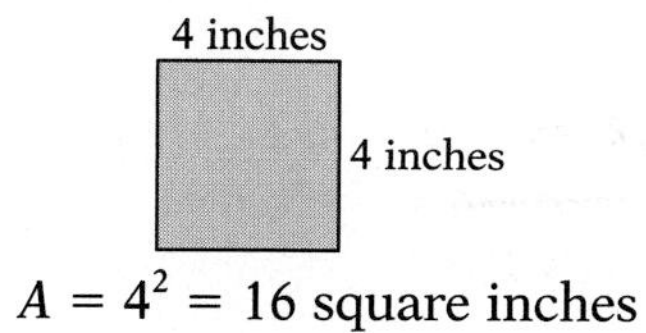

$A = 4^2 = 16$ square inches

Copyright © Houghton Mifflin Company. All rights reserved.

B. Area of a Rectangle

1. The area A of a rectangle whose length is L and whose width is W is $A = LW$.

6 feet

3 feet

$A = 6(3) = 18$ square feet

C. Composite Figures

1. If a geometric figure is a combination of other geometric figures (*composite* figure), try to divide it into smaller figures whose areas can be found with a formula.

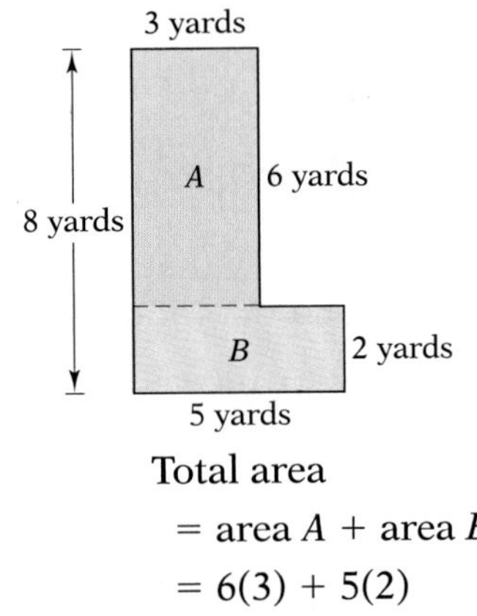

Total area

$= \text{area } A + \text{area } B$

$= 6(3) + 5(2)$

$= 18 + 10$

$= 28$ square yards

D. Unit Conversion

1. See Subsection D for tables that can be used to convert length or area units to other units.

Copyright © Houghton Mifflin Company. All rights reserved.

2.7 Exercises

A. Area of a Square

1. A(n) formula is an equation that gives instructions for calculating a specified quantity.

2. A square whose sides are 1 unit long is called a(n) unit square.

In Exercises 3–6, determine the area of the square whose sides have the given lengths.

3. 5 inches
4. 7 yards
5. 9 miles
6. 12 feet

In Exercises 7–10, determine the length of the sides of a square whose area is given.

7. 81 square miles
8. 36 square inches
9. 100 square yards
10. 9 square feet

B. Area of a Rectangle

Exercises 11–16 involve unknown areas of rectangles.

11. *Picture Frame* The glass for a picture frame is 8 inches long and 6 inches wide. What is the area of the glass?

12. *Parking Lot* A parking lot is 60 yards wide and 120 yards long. What is the area of the parking lot?

13. *Colorado* Assume that the state of Colorado is a rectangle that is approximately 383 miles long and 269 miles wide. What is the area of Colorado?

Copyright © Houghton Mifflin Company. All rights reserved.

14. *Baseball Infield* The infield of a baseball field is a square whose side is 90 feet long. What is the area of the infield?

15. *Coffee Table* The top of a coffee table is 3 feet long and 2 feet wide. Glass for the top of the table costs \$3 per square foot. What is the cost of enough glass to cover the top?

16. *Living Room Carpet* A living room is 6 yards wide and 9 yards long. If carpet costs \$18 per square yard, what is the cost of carpet to cover the floor?

Exercises 17–22 involve unknown dimensions of rectangles.

17. *Hallway Carpet* A hallway is 2 yards wide and requires 42 square yards of carpeting. What is the length of the hallway?

18. *Curtain Material* A curtain that is 8 feet long requires 72 square feet of material. What is the width of the curtain?

19. *Tiled Floor* A bathroom floor that is 9 feet long is covered with 63 square feet of tiles. What is the width of the tiled floor?

20. *Ranch Size* A ranch that is 5 miles long covers 20 square miles. What is the width of the ranch?

Copyright © Houghton Mifflin Company. All rights reserved.

21. *Walkway* A walkway that is 4 feet wide covers 120 square feet. What is the length of the walkway?

22. *Tablecloth* The area of a tablecloth is 4,800 square inches. If the length is 80 inches, what is the width of the tablecloth?

Exercises 23–26 involve areas of rectangles.

23. *Park Walkway* A park that is 130 feet wide and 180 feet long is surrounded by a brick walkway that is 4 feet wide. What is the area of the brick walkway?

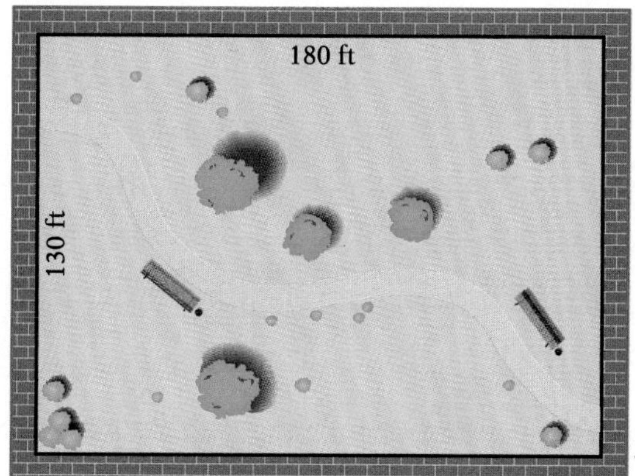

24. *Picture Frame* A picture frame that is 2 inches wide holds a picture that is 10 inches wide and 14 inches long. What is the area of the frame?

25. *Swimming Pool* A swimming pool that is 60 feet long and 25 feet wide is surrounded by a tiled deck that is 6 feet wide. How many 1-foot square tiles are in the deck?

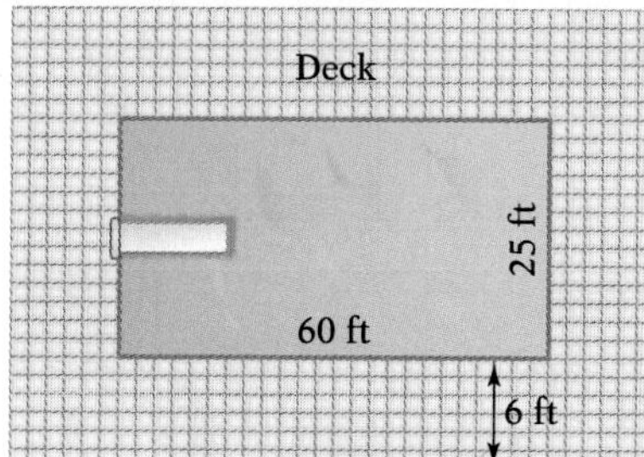

Copyright © Houghton Mifflin Company. All rights reserved.

26. *Bare Floor Area* A rug that is 8 feet wide and 10 feet long is placed on the floor of a room. This leaves a strip of uncarpeted floor 2 feet wide on all four sides of the rug. What is the area of the uncarpeted floor?

C. Composite Figures

In Exercises 27–34, determine the area.

27.

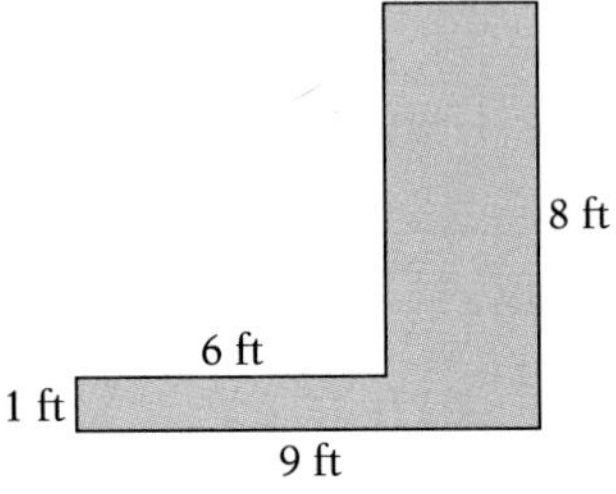

28.

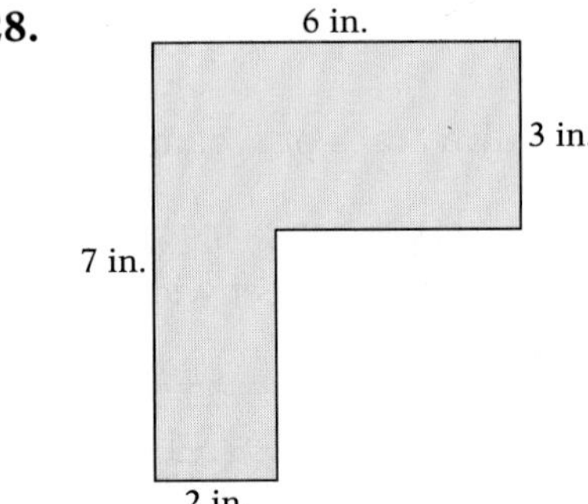

29.

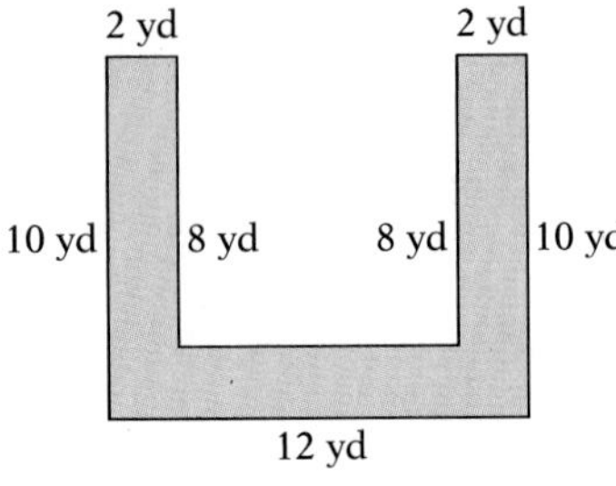

30.

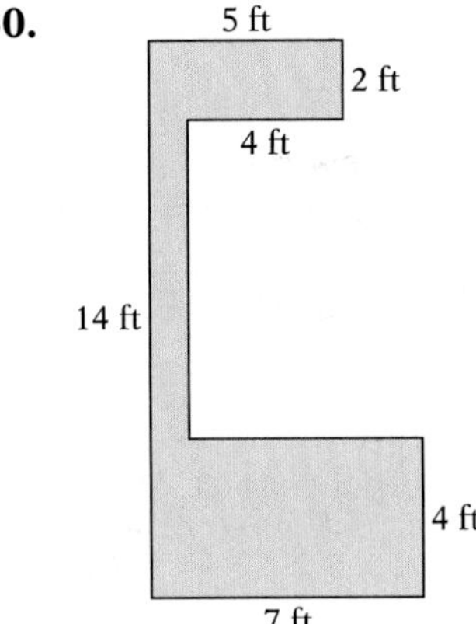

31.

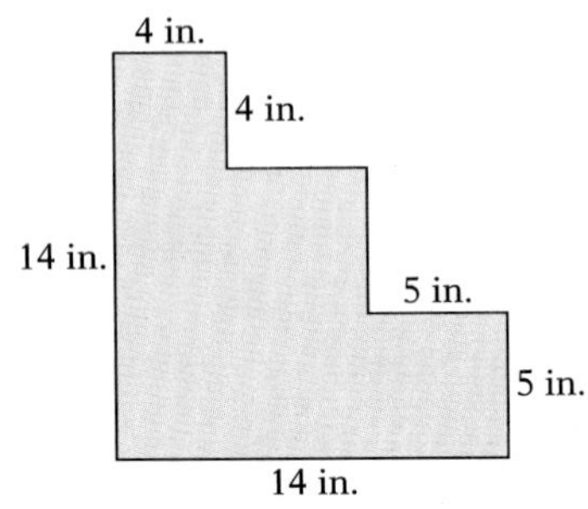

32.

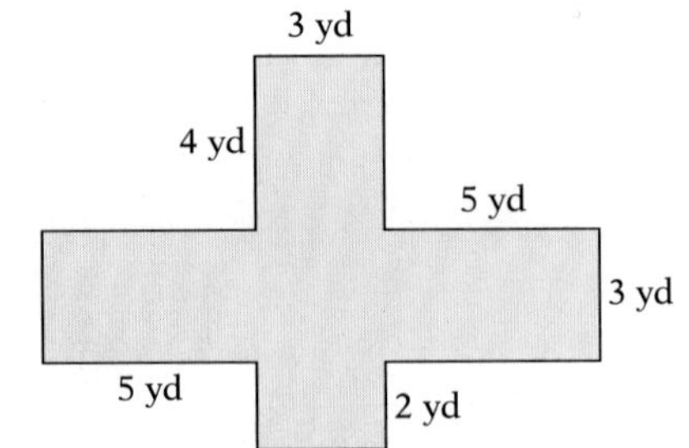

Copyright © Houghton Mifflin Company. All rights reserved.

33.

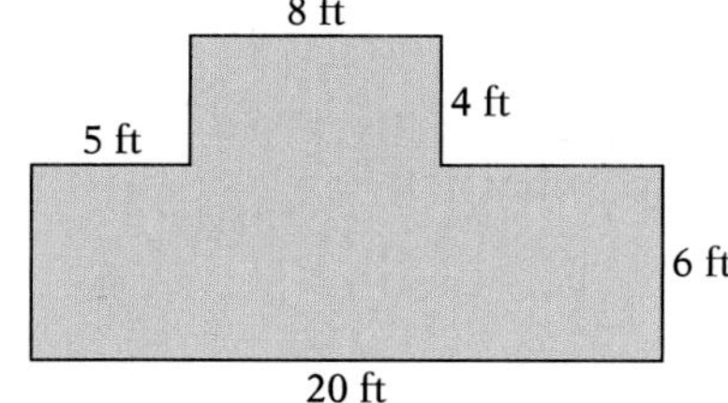

34.

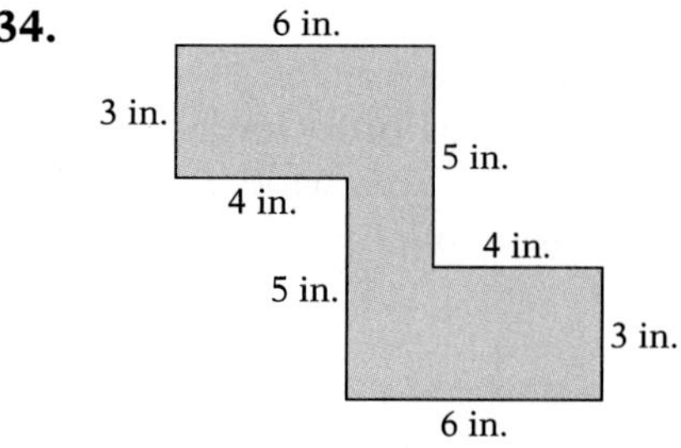

D. Unit Conversion

35. *Lawn Mower* A lawn mower cuts a strip 36 inches wide. If the length of a lawn is 70 feet, what is the area (in square feet) of the strip?

36. *Fireplace Mantle* A mantle is 6 feet long and 8 inches wide. What is the area (in square inches) of the top of the mantle?

37. *Road Easement* For a new road, a city acquired a strip of land that is 30 feet wide and 2 miles long. What is the area (in square feet) of the land?

38. *Molding* A strip of molding is 4 inches wide and 4 yards long. What is the area (in square inches) of the molding?

39. *Lawn Sod* Suppose that lawn sod costs \$3 per square yard. What is the cost of sod for a lawn that is 120 feet long and 60 feet wide?

40. *Scarf Sequins* A designer plans to cover a scarf that is 2 feet long and 6 inches wide with sequins. Sequins to cover 1 square inch cost \$2. What is the cost of enough sequins to cover the scarf?

Copyright © Houghton Mifflin Company. All rights reserved.

Calculator Exercises

41. *Library Carpet* A college library has 5 floors, each with a carpeted area that is 32 yards long and 26 yards wide. If carpet costs \$17 per square yard, what is the cost of recarpeting the entire library?

42. *Plastic Bags* Plastic trash bags are made from rectangular pieces of plastic 60 inches long and 36 inches wide. How many *square feet* of plastic are needed to manufacture 1,000 bags?

In Exercises 43 and 44, use the fact that 1 acre = 43,560 square feet.

43. *Lot Size* A 2-acre rectangular lot is 242 feet wide. What is the length of the lot?

44. *Development Acreage* For a new planned community, a developer bought a rectangular tract of land 7,920 feet long and 10,560 feet wide. How many acres did the developer purchase?

Writing and Concept Extension

45. If the length of a rectangle is given in inches and the width is given in feet, what must be done before you can compute the area of the rectangle?

46. Explain how to convert (a) feet to yards and (b) yards to inches.

47. The length of a side of a square is 3 inches. If the lengths of the sides are increased by 1 inch, what is the effect on (a) the perimeter and (b) the area?

Copyright © Houghton Mifflin Company. All rights reserved.

2.8 Volume and Surface Area

A *Volume*
B *Surface Area*

SUGGESTIONS FOR SUCCESS

When you are working problems that involve geometric figures, you should always sketch the figure and label the known and unknown dimensions. But how do you draw a three-dimensional object on a two-dimensional piece of paper? In early centuries, even artists didn't know how!

Look closely at the art in this section, and try drawing some of the figures yourself until you get the hang of it. Being able to draw these figures will be a big visual help in understanding the problem.

A Volume

Just as area is a measure of the amount of surface enclosed by a geometric figure, **volume** is a measure of the amount of space enclosed by a solid geometric figure.

We normally would not talk about the area of a geometric figure that is not fully enclosed. For example, if we draw just three sides of a rectangle, the missing boundary means that the surface extends indefinitely, and so the area cannot be determined.

In practice, we can treat volume somewhat differently. A box with a lid is fully enclosed, and we can measure the amount of space (volume) that is contained in the box. However, even if the lid is missing, which means that the box is not fully enclosed, it still makes sense to talk about the volume of the box, at least in terms of the capacity of the box to hold things.

As another example, a flower vase is not fully enclosed, but we can fill the vase with water and say that the volume of the vase is the same as the volume of the water.

DEVELOPING THE CONCEPT

Volume of a Rectangular Solid

Suppose that a company makes machine parts that are packed in boxes 1 foot wide, 1 foot long, and 1 foot high. These boxes are then packed in a large carton that is 2 feet wide, 4 feet long, and 2 feet high.

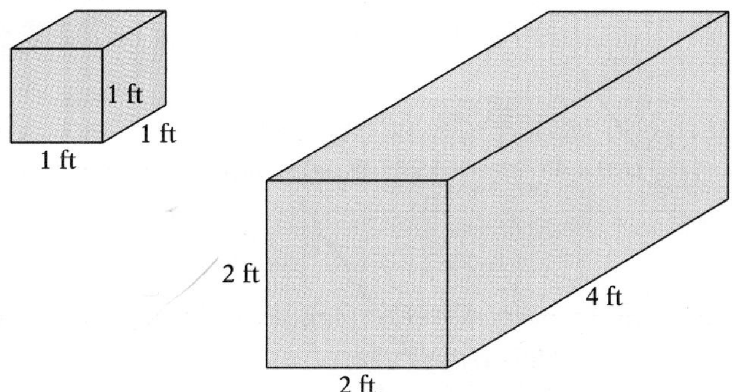

The small boxes and the large carton are both examples of *rectangular solids.* (Of course, the boxes are empty, not solid, but "rectangular solid" is the general name that we give to such geometric objects.)

The small box is called a cube because its length, width, and height are all the same. In fact, because all the dimensions are 1 foot, we call this cube a *unit cube,* and its volume is 1 cubic foot.

Copyright © Houghton Mifflin Company. All rights reserved.

We can think of the volume of the carton as the number of unit cubes that can be fit into it.

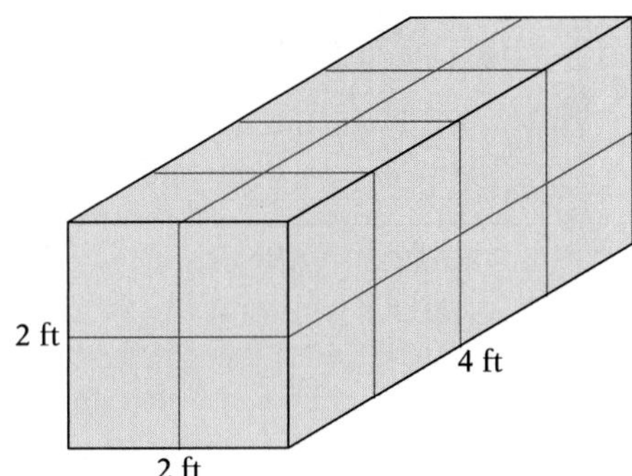

Each layer has $2 \cdot 4 = 8$ boxes. Because there are 2 layers, the box contains a total of $2 \cdot 8 = 16$ boxes. Therefore, we say that the volume of the carton is 16 cubic feet.

Note that the number of boxes, or unit cubes, in the carton can be found by multiplying the width, length, and height: $2 \cdot 4 \cdot 2 = 16$.

Volume of a Rectangular Solid

If a rectangular solid has a length L, a width W, and a height H, then the formula for the volume V is

$$V = LWH$$

A cube is a special kind of rectangular solid in which the length, width, and height are all the same.

Volume of a Cube

If s represents the length, width, and height of a cube, then the formula for the volume V is

$$V = s^3$$

NOTE Recall that s^3 can be read as "s cubed" rather than "s to the third power." This special name comes from the fact that s^3 is the volume of a cube whose sides are s units long.

As before, the units used for volume are often abbreviated. The most common abbreviations are cu in. or in.^{3} for cubic inches, and cu ft or ft^3 for cubic feet. The following table gives some common unit conversions for volume.

Unit Conversions for Volume
1 cubic foot = 1,728 cubic inches
1 cubic yard = 27 cubic feet

Copyright © Houghton Mifflin Company. All rights reserved.

Example 1

Find the volume of each solid.

(a)

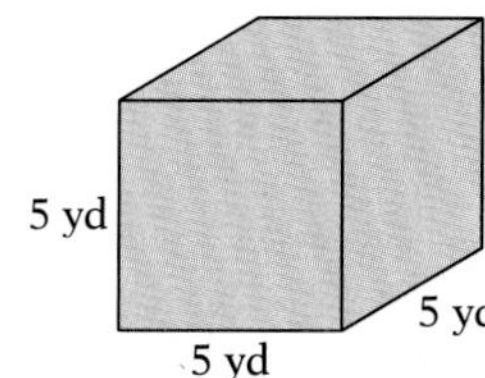

(b)

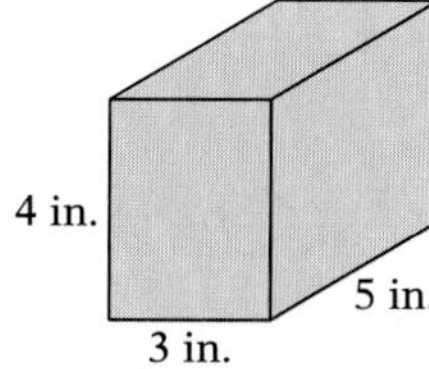

SOLUTION

(a) Because the dimensions are all the same, the solid is a cube.

$V = s^3$ Formula for the volume of a cube

$V = 5^3$ Replace s with 5.

$= 125$

The volume is 125 cubic yards.

(b) $V = LWH$ Formula for the volume of a rectangular solid

$= (5)(3)(4)$ Replace L with 5, W with 3, and H with 4.

$= 60$

The volume is 60 cubic inches.

Your Turn 1

Find the volume of the rectangular solid.

3 ft
8 ft
1 ft

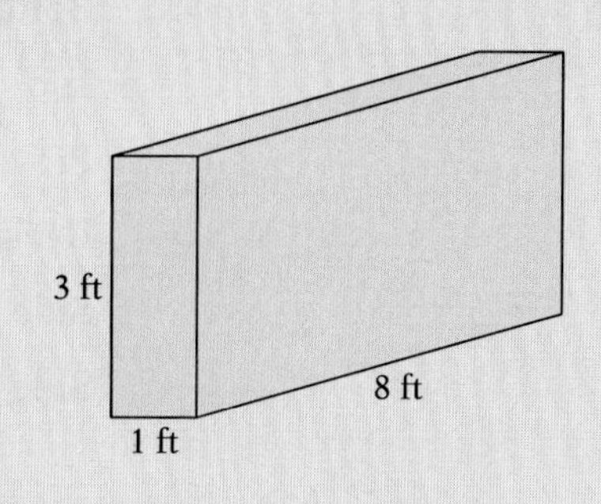

Answer: 24 cubic feet

B Surface Area

To construct a closed box, we would need a bottom and a top, a front and a back, and two sides.

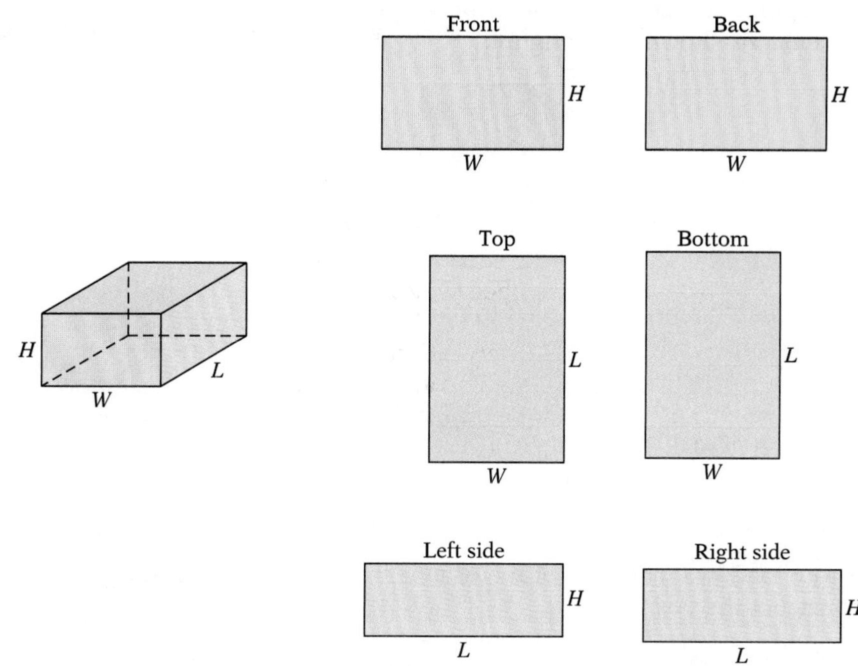

Copyright © Houghton Mifflin Company. All rights reserved.

By *surface area* of the box, we mean the total area of the six pieces.

Piece	*Area*
Bottom	LW
Top	LW
Front	WH
Back	WH
Side	LH
Side	LH

Total surface area $= S = 2LW + 2WH + 2LH$

Surface Area of a Rectangular Solid

If a rectangular solid has a length L, a width W, and a height H, then the formula for the total surface area S is

$$S = 2LW + 2WH + 2LH$$

If the rectangular solid is a cube, then the area of each of the six sides is s^2.

Surface Area of a Cube

If s represents the length, width, and height of a cube, then the formula for the total surface area S is

$$S = 6s^2$$

Example 2

Jet engines are packed in shipping crates with the dimensions shown in the figure.

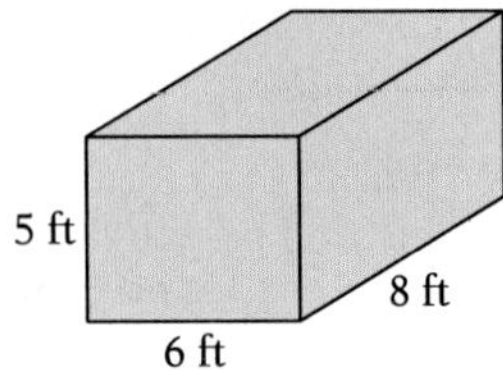

How much material is needed to construct the crate?

SOLUTION

$A = 2LW + 2WH + 2LH$ Formula for surface area

$= 2(8)(6) + 2(6)(5) + 2(8)(5)$

$= 96 + 60 + 80$

$= 236$

A total of 236 square feet of material is needed to construct the box.

Your Turn 2

Suppose that the top of the crate is installed later. What is the surface area of the open box with no top?

Answer: 188 square feet

Copyright © Houghton Mifflin Company. All rights reserved.

Example 3

A rectangular solid is 8 inches long, 3 inches wide, and 2 inches high. Compare its surface area with that of a cube whose sides are 4 inches.

SOLUTION

Rectangular solid: $L = 8$, $W = 3$, $H = 2$

$$\begin{aligned} S &= 2LW + 2WH + 2LH \\ &= 2(8)(3) + 2(3)(2) + 2(8)(2) \\ &= 48 + 12 + 32 \\ &= 92 \text{ square inches} \end{aligned}$$

Cube: $s = 4$

$$\begin{aligned} S &= 6s^2 \\ &= 6(4)^2 = 6 \cdot 16 = 96 \text{ square inches} \end{aligned}$$

The surface area of the rectangular solid is less than that of the cube.

Your Turn 3

By how much would the surface area of the cube in Example 3 be reduced if the sides were 1 inch shorter?

Answer: 42 square inches

2.8 Quick Reference

A. Volume

1. The **volume** of a geometric figure is a measure of the amount of space enclosed by the figure.

2. We can think of a **cube** as a box whose width, length, and height are all the same. All the dimensions of a *unit cube* are 1 unit, and the volume is 1 cubic unit.

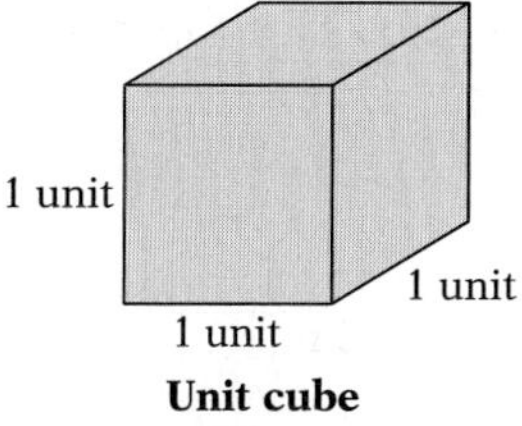

Unit cube

3. The volume V of a cube whose sides are s units long is given by $V = s^3$.

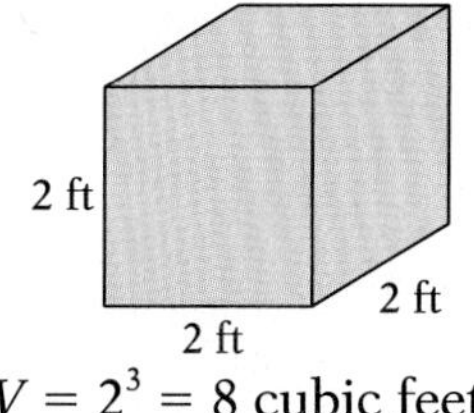

$V = 2^3 = 8$ cubic feet

Copyright © Houghton Mifflin Company. All rights reserved.

4. The volume V of a rectangular solid of length L, width W, and height H is given by $V = LWH$.

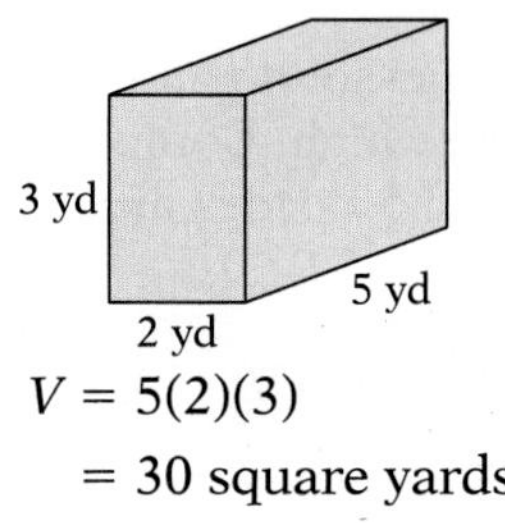

$$V = 5(2)(3)$$
$$= 30 \text{ square yards}$$

B. Surface Area

1. The *surface area* S of a rectangular solid is the total area of the six sides.

   (a) Cube: $S = 6s^2$

   (b) Rectangular solid: $S = 2LW + 2WH + 2LH$

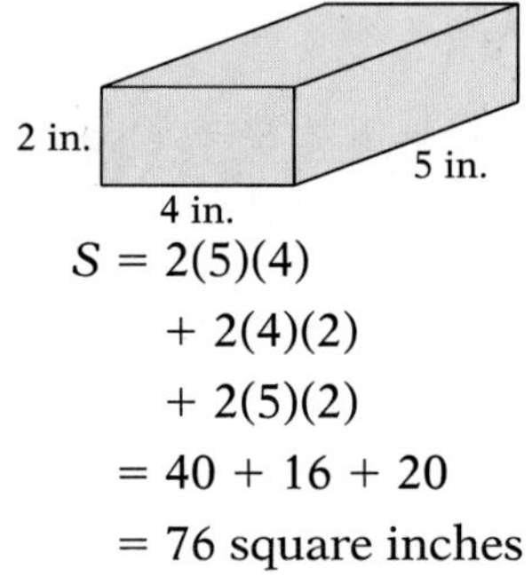

$$S = 2(5)(4) + 2(4)(2) + 2(5)(2)$$
$$= 40 + 16 + 20$$
$$= 76 \text{ square inches}$$

Copyright © Houghton Mifflin Company. All rights reserved.

2.8 Exercises

A. Volume

1. A measure of the amount of space enclosed by a solid geometric figure is called the volume of the geometric figure.

2. A rectangular solid whose length, width, and height are all the same is called a(n) cube .

In Exercises 3–10, determine the volume of the solid figure.

3.

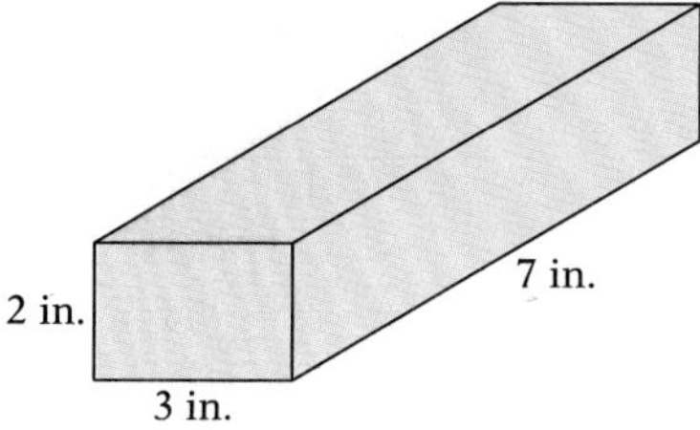

4.

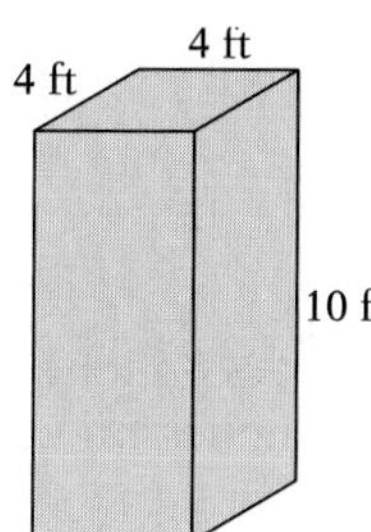

5.

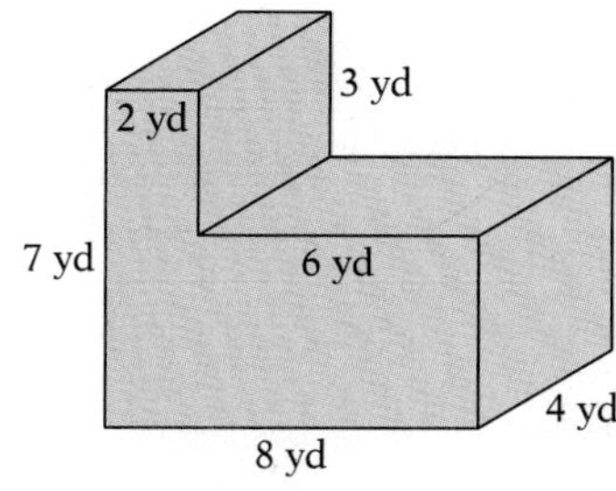

6.

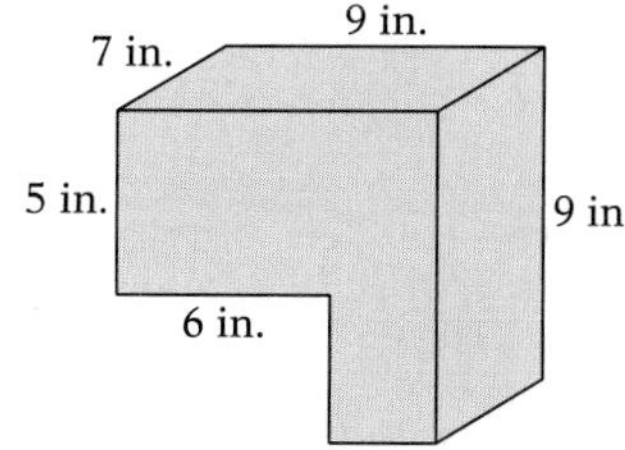

7.

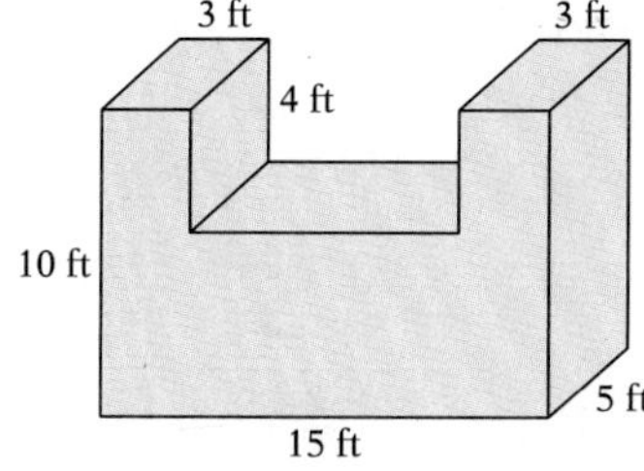

8.

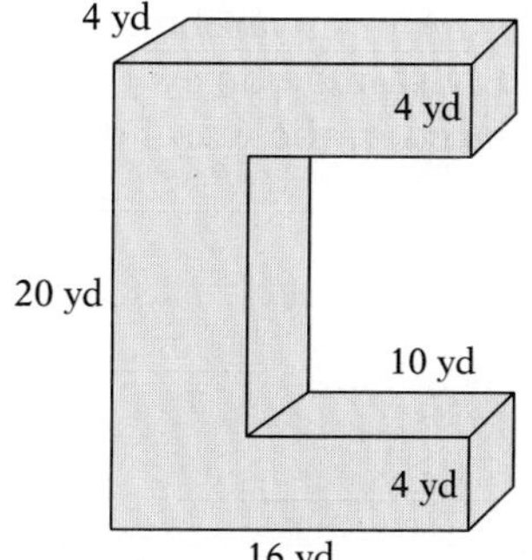

Copyright © Houghton Mifflin Company. All rights reserved.

9.

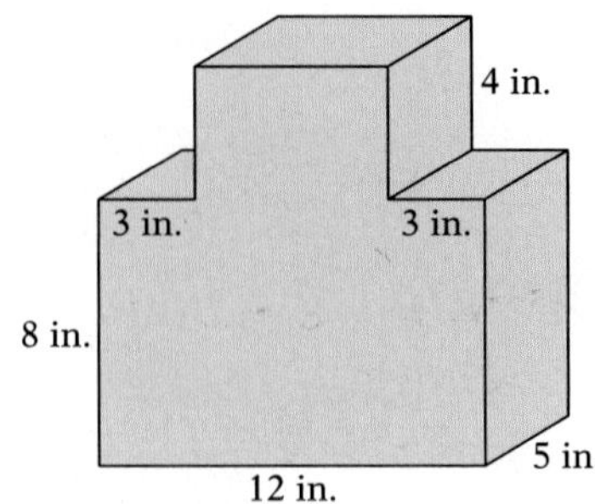

10.

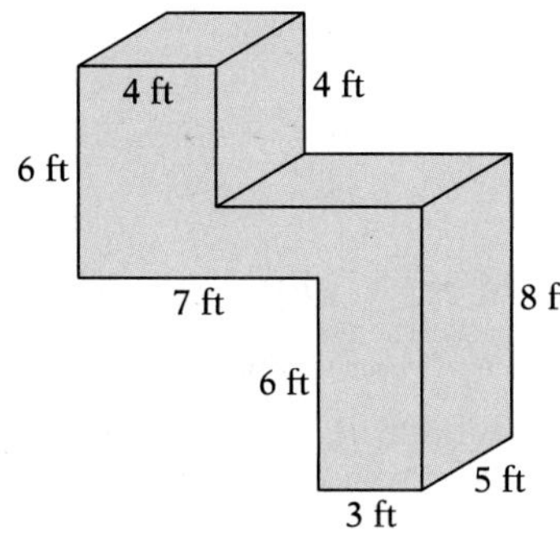

Exercises 11–14 involve volumes.

11. *Packaging* A cubic package that is 8 inches on each side is packed in a cubic shipping container that is 12 inches on each side. How many cubic inches of foam are needed to finish filling the shipping container?

12. *Pool Capacity* A swimming pool is 40 feet long, 25 feet wide, and 7 feet deep. How many cubic feet of water are needed to fill the pool to within 1 foot of the top?

13. *Insulation Cost* Suppose that a company charges $2 per cubic foot of material to insulate a home. What is the cost of installing 12 inches of insulation in an attic space that is 70 feet long and 40 feet wide?

14. *Miniwarehouse Storage* A miniwarehouse is 15 feet long, 12 feet wide, and 12 feet high. How many storage boxes that are 5 feet long, 4 feet wide, and 3 feet high can be placed in the warehouse?

Copyright © Houghton Mifflin Company. All rights reserved.

B. Surface Area

15. Volume is measured in cubic units, and area is measured in square units.

16. The total area of all six sides of a rectangular solid is called the surface area of the solid.

In Exercises 17 and 18, determine the surface area of the given geometric solid.

17. The solid in Exercise 3

18. The solid in Exercise 4

Exercises 19–22 involve surface area.

19. *Cardboard Box* An open cardboard box (no lid) is 24 inches long, 20 inches wide, and 14 inches deep. How many square inches of cardboard are needed to construct the box?

$24 \cdot 20 + 2 \cdot (24 \cdot 14) + 2 \cdot (20 \cdot 14) = 480 + 672 + 560 = 1712$ sq in.

Sum

20. *Wallpaper* A room is 10 feet wide and 12 feet long. The ceiling is 8 feet high. If a roll of wallpaper covers 44 square feet of surface, how many rolls of wallpaper are needed to cover the four walls?

21. *Paint* A small warehouse is 140 feet long and 80 feet wide. The ceiling is 12 feet high. If a gallon of paint covers 330 square feet of surface, how many gallons of paint are needed to paint the four walls of the warehouse?

22. *Roofing* One side of the roof of a house is 80 feet long and 40 feet wide. The other side is 80 feet long and 50 feet wide. If a bundle of shingles covers 100 square feet, how many bundles are needed to roof the house?

Copyright © Houghton Mifflin Company. All rights reserved.

Calculator Exercises

23. *Storage Area* The basement storage area for a large building is 235 feet long, 172 feet wide, and 9 feet high. How many cubic feet of storage space are in the building?

24. *Miniwarehouse Capacity* A miniwarehouse is divided into 15 sections. Each section is 12 feet wide, 20 feet deep, and 8 feet high. What is the total storage capacity of the miniwarehouse?

Writing and Concept Extension

25. Numerically, the area of one side of a unit cube is the same as its volume. How do the measures differ?

26. If the ends of a rectangular solid are squares whose sides are s, how might you rewrite the formula for the volume of the solid?

27. The length of a side of a cube is 4 inches. If the lengths of the sides are increased by 1 inch, what is the effect on (a) the volume and (b) the surface area?

Copyright © Houghton Mifflin Company. All rights reserved.

CHAPTER 2 REVIEW EXERCISES

Section 2.1

1. We call both $6 \cdot 9$ and the result 54 a(n) ______ and we call 6 and 9 ______.

2. Identify the factors in $7ab$.

In Exercises 3–6, identify the property that justifies the statement.

3. $0 \cdot x = 0$

4. $c = 1 \cdot c$

5. $6(3x) = (6 \cdot 3)x$

6. $yx = xy$

7. Explain how to determine the product of 5 and 3,000.

8. Multiply $68 \cdot 1{,}000$.

9. Multiply $6 \cdot 400$.

10. *Bicycling and Calories* Suppose that a person burns 300 calories by bicycling for 1 hour. How many calories does a person burn by cycling 1 hour per day for a week?

Section 2.2

11. By applying the ______ Property, we can write $5(20 + 7)$ as $5 \cdot 20 + 5 \cdot 7$.

12. Use the Distributive Property to write $2(400 + 7)$ as a sum.

13. Multiply 450 by 200.

In Exercises 14–17, determine the product.

14. $\begin{array}{r} 3{,}672 \\ \times \quad 8 \\ \hline \end{array}$

15. $\begin{array}{r} 604 \\ \times\ 38 \\ \hline \end{array}$

16. $\begin{array}{r} 4{,}284 \\ \times\ 380 \\ \hline \end{array}$

17. $\begin{array}{r} 593 \\ \times 245 \\ \hline \end{array}$

18. Estimate the product of 69,872 and 5,102.

19. *Nurse's Wages* A nurse makes $12 per hour for up to 40 hours work per week. For each hour above 40, the hourly pay increases to $18 per hour. How much does a nurse make for working 47 hours in 1 week?

Copyright © Houghton Mifflin Company. All rights reserved.

20. *Child's Food Costs* The weekly cost of food for a 10-year-old child is \$42. (Source: U.S. Department of Agriculture.) What is the cost of food for a 10-year-old child for 1 year (52 weeks)?

Section 2.3

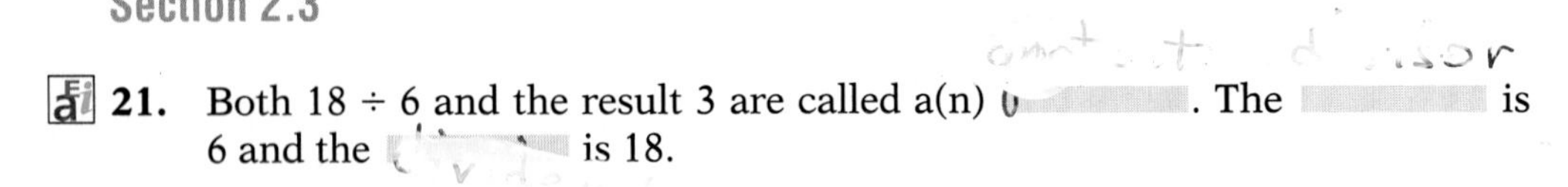

21. Both 18 ÷ 6 and the result 3 are called a(n) ________. The ________ is 6 and the ________ is 18.

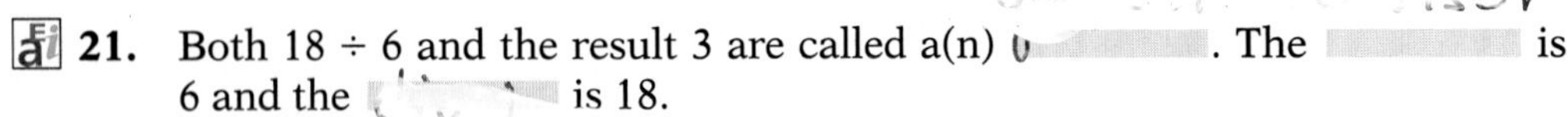

22. Explain how to check whether 32 ÷ 8 = 4 is true.

23. Determine the quotient, if possible.

(a) 0 ÷ 9 **(b)** 5 ÷ 0

24. Determine the quotient.

(a) 25 ÷ 25 **(b)** 17 ÷ 1

25. Determine the quotient of 1,196 and 23.

26. Divide: 15,120 ÷ 30.

In Exercises 27 and 28, determine the quotient and remainder.

27. $7\overline{)400}$

28. $172\overline{)15{,}022}$

29. *Truck Loading* A delivery company crew can load a truck with 1,092 packages in 7 hours. How many packages can the crew load in 1 hour?

30. *Reunion Catering* For a class reunion, the caterer estimates that 5 ounces of potato salad per person will be needed. If 112 people plan to attend the reunion, how many 64-ounce containers of potato salad are needed? How many ounces of potato salad will be left over?

Copyright © Houghton Mifflin Company. All rights reserved.

Section 2.4

31. In the ______ expression 6^2, 2 is called the exponent and 6 is called the Base.

32. Write $y \cdot y \cdot y \cdot y \cdot y$ in exponential form.

33. Write 6^4 in expanded form and in words.

34. Write "7 cubed" in exponential form.

In Exercises 35–39, evaluate the exponential expression.

35. 2^5

36. 1^{24}

37. 10^4

38. $9^0 = 1$

39. 15^2

40. Write "ten million" in exponential form.

Section 2.5

41. To evaluate an expression such as $4 - 7 \cdot 4^2$, we follow the ______.

42. Parentheses, brackets, and braces are examples of grouping symbols.

In Exercises 43–47, evaluate the expression.

43. $14 - 3 \cdot 2$

14 − 6
8

44. $5 + 2 \cdot 3^2$

45. $15 - 4(7 - 5)$

46. $21 \div 3 + 5[9 - 2(3 - 1)]$

47. $\dfrac{4 + 2 \cdot 3}{3 \cdot 4 - 7} =$

In Exercises 48 and 49, write a numerical expression corresponding to the given phrase. Then evaluate it.

48. Three less than the product of 6 and 4

49. The quotient of 20 and the difference of 9 and 5

50. *Basketball Scoring* In seven consecutive games a basketball player scored 15, 22, 16, 20, 12, 17, and 24 points. What was the average number of points per game?

Copyright © Houghton Mifflin Company. All rights reserved.

Section 2.6

51. To write $3(2a + 7)$ as a sum, we use the ________ Property.

52. Simplify $(5a)(3b)$.

53. Use the Distributive Property to write $4(3x + 7)$ as a sum. Then simplify the result.

In Exercises 54–57, evaluate the expression for the given values of the variable(s).

54. $4ab$ for $a = 7$ and $b = 10$

55. $\frac{5a}{3b}$ for $a = 6$ and $b = 2$

56. $n - 3(m - 6)$ for $n = 18$ and $m = 10$

57. $x^2 - 3x$ for $x = 4$

In Exercises 58–60, find the whole number for which the given equation is true.

58. $7(3n) = 0$

59. $\frac{y}{6} = 7$

60. $s^2 = 25$

Section 2.7

61. The ________ of a rectangle is the number of unit squares required to fill the rectangle.

62. An equation with more than one variable that gives instructions for calculating a specific quantity is called a(n) ________.

63. Explain how to convert inches to yards.

64. Explain how to convert feet to inches.

65. What is the area of a square whose sides are 7 yards long?

Copyright © Houghton Mifflin Company. All rights reserved.

66. Determine the length of the sides of a square whose area is 36 square feet.

67. *Carpet Cost* A rectangular recreation room is 24 feet wide and 11 yards long. If carpet costs $15 per square yard, what is the cost of carpet to cover the floor?

68. *Warehouse Size* A rectangular warehouse that is 260 feet long covers 45,500 square feet. What is the width of the building?

69. Determine the area of the figure.

3 mi
11 mi
8 mi
4 mi

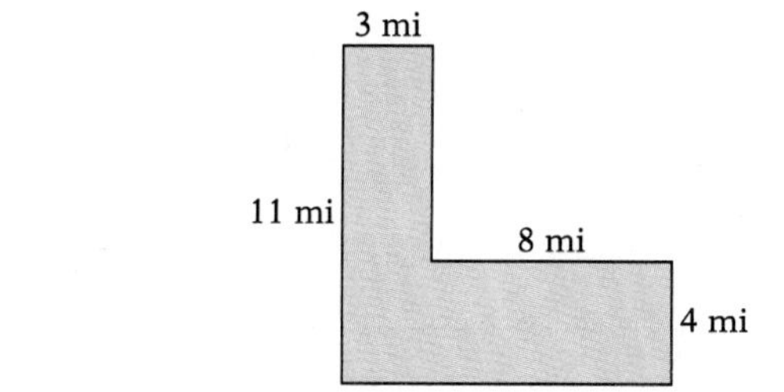

70. *Table Size* A rectangular table top is 30 inches wide and 4 feet long. What is the area of the table top in square feet?

Section 2.8

71. The amount of space enclosed by a solid geometric figure is called the ________.

72. A ________ solid whose length, width, and height are the same is called a cube.

Copyright © Houghton Mifflin Company. All rights reserved.

73. What is the formula for the volume of a cube?

74. Consider a cube whose sides are 5 inches long.
(a) What is its volume? **(b)** What is its surface area?

In Exercises 75 and 76, consider a rectangular solid whose dimensions are 3 feet, 2 feet, and 18 inches.

75. What is the volume (in cubic inches) of the solid?

76. What is the surface area (in square inches) of the solid?

77. *Aquarium* An aquarium is 32 inches long, 20 inches wide, and 25 inches deep. How many cubic inches of water are needed to fill the aquarium to within 2 inches of the top?

78. Determine the volume of the solid figure.

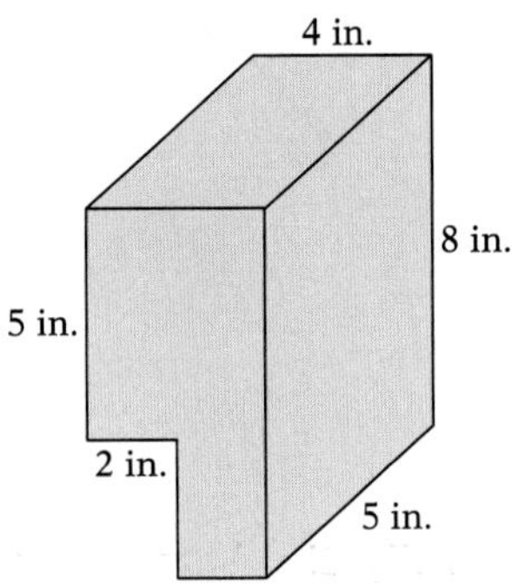

79. *Box Construction* How many square inches of cardboard are needed to construct a closed box that is 1 foot long, 10 inches wide, and 6 inches deep?

80. *Siding Cost* A small commercial building is 120 feet long and 75 feet wide. The ceiling is 12 feet high. If siding costs $4 per square foot, what is the cost of siding for the building?

Copyright © Houghton Mifflin Company. All rights reserved.

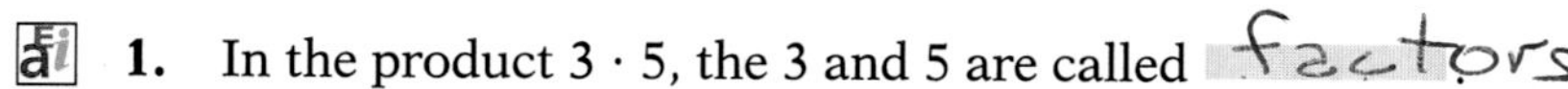

CHAPTER 2 TEST

1. In the product $3 \cdot 5$, the 3 and 5 are called ______

2. Identify the quotient that is not defined and explain your choice.

   (i) $8 \div 0$ (ii) $0 \div 8$

3. For parts (a) and (b), identify the property from the following list that justifies the given statement.

   (i) Associative Property of Multiplication

   (ii) Commutative Property of Multiplication

   (a) $6(3y) = (6 \cdot 3)y$ **(b)** $x \cdot 5 = 5x$

In Questions 4–8, perform the indicated operation.

4. $5 \cdot 700$

5. $\begin{array}{r} 3{,}024 \\ \times\ 400 \\ \hline \end{array}$

6. $\begin{array}{r} 6{,}307 \\ \times\ 926 \\ \hline \end{array}$

7. $30\overline{)15{,}120}$

8. $84\overline{)8{,}477}$

9. Write $7 \cdot 7 \cdot 7 \cdot 7$ in exponential form. Then identify the base and the exponent.

In Questions 10–13, evaluate the numerical expression.

10. 4^3

11. 15^0

12. $9 - 2(7 - 3)$

13. $21 - (16 - 4) \div 3$

14. Write a numerical expression corresponding to the phrase "twice the sum of 7 and 9." Then evaluate it.

15. Simplify $5(3a)$.

16. Use the Distributive Property to write $5(6x + 7)$ as a sum. Then simplify the result.

Copyright © Houghton Mifflin Company. All rights reserved.

17. Evaluate $x^2 + 3(x - y)$ for $x = 5$ and $y = 3$.

In Questions 18 and 19, solve the equation.

18. $\frac{45}{x} = 5$

19. $9y = 63$

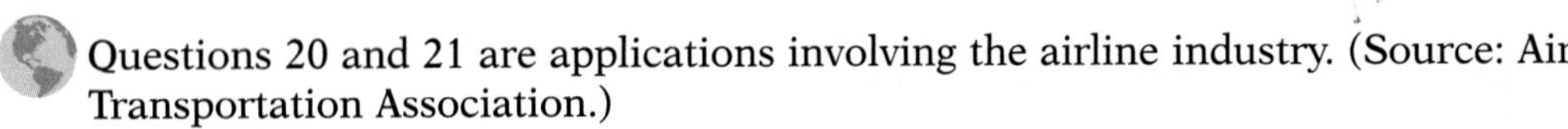

Questions 20 and 21 are applications involving the airline industry. (Source: Air Transportation Association.)

20. An L-1011 jetliner uses 2,400 gallons of fuel per hour. How many gallons of fuel does the plane use for a 4-hour flight?

21. A B-747 costs \$6,560 per hour to operate and holds 410 passengers. What is the hourly cost per passenger to operate the plane?

22. A student scored 85, 94, and 88 on three 100-point tests. What was the average score?

23. What is the area of a rectangular piece of plywood that is 5 feet wide and 8 feet long?

24. Determine the area of the geometric figure.

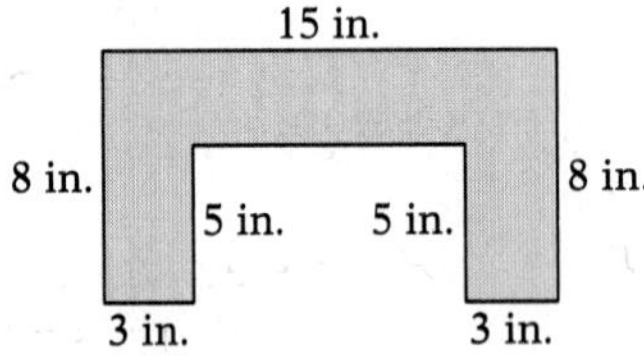

25. Determine (a) the volume and (b) the surface area of the rectangular solid.

5 ft
8 ft
5 ft

Copyright © Houghton Mifflin Company. All rights reserved.

CHAPTER 3

The Integers

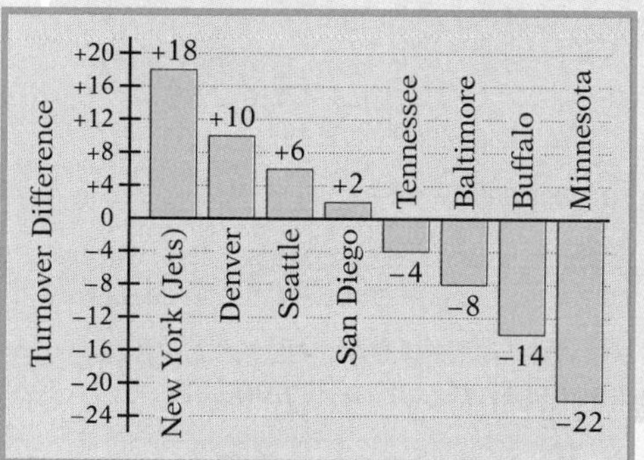

In football, if your team intercepts a pass or recovers an opposing team's fumble, your team is credited with a "take-away," and the opposing team is charged with a "give-away." A team's "turnover difference" is the total of its take-aways minus the total of its give-aways. Generally speaking, the greater the turnover difference, the more likely a team is to have a successful season.

The bar graph shows the turnover differences for 8 of the 30 teams in the National Football League during the 2001 season. (Source: NFL.com.)

Four of the teams had more give-aways than take-aways, so the turnover difference is given as a *negative* number. If we compare the turnover differences for New York and Minnesota, we would perform the subtraction $18 - (-22)$.

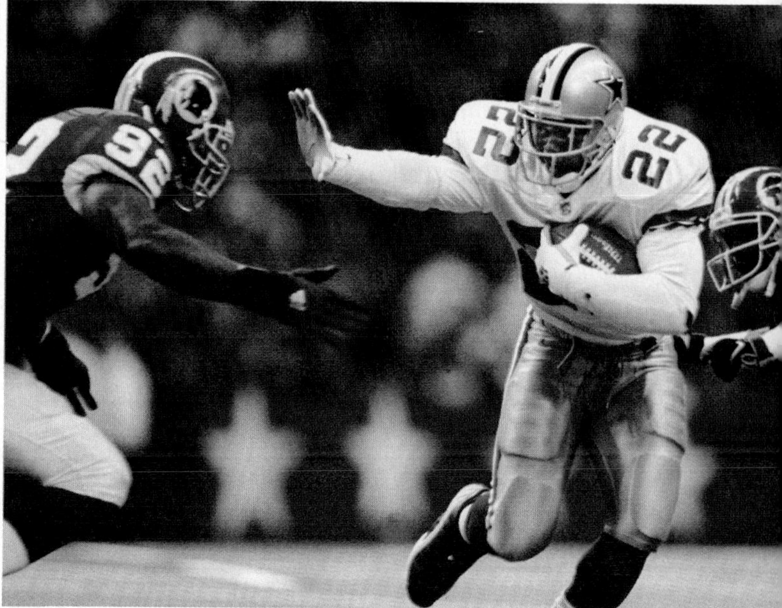

Chapter Snapshot

In this chapter, we focus on the rules for adding, subtracting, multiplying, and dividing with positive and negative numbers. We also extend our ability to evaluate and simplify expressions and to solve simple equations. The topic of positive and negative numbers is a big first step away from arithmetic and toward algebra.

Copyright © Houghton Mifflin Company. All rights reserved.

For online resources, visit the web site **math.college.hmco.com/students** and follow the links to Hubbard/Robinson, *Prealgebra.*

Some Friendly Advice . . .

You can't always rely on your intuition in mathematics. Some people explain their difficulty with math by saying, "It doesn't make sense." But math does make sense as long as you *play by the rules.*

In this section, we will define subtraction of integers. Think of a definition as a rule. Once we accept the definition, we must follow that rule to the letter.

The definition of subtraction will lead to some results that may surprise you. The key is to follow the rule and to avoid using your own intuition. The beauty of mathematics lies in its structure and logic. If you search for the beauty, a whole new way of thinking will be revealed to you.

WARM-UP SKILLS

The following questions review concepts and skills that you will need in Chapter 3.

1. Fill in the blank with $<$ or $>$ to make the statement true.

   (a) $0 < 7$

   (b) $12 > 8$

2. Simplify $(x + 8) + 6$.
3. Simplify $5(2a)$.
4. Evaluate $2x - y + 10$ for $x = 7$ and $y = 4$.
5. Write $3(2x + 7)$ as a sum.
6. Solve the equation $9 - x = 2$.
7. Evaluate 5^3.
8. Evaluate $\frac{3 \cdot 7 - 5}{8}$.
9. Determine the average of 88, 93, and 95.
10. What number is 5 units to the left of 7 on the number line?

Copyright © Houghton Mifflin Company. All rights reserved.

3.1 Introduction to the Integers

A Positive and Negative Numbers
B Opposites
C Absolute Value

SUGGESTIONS FOR SUCCESS

Now we are getting into some serious stuff. When you start learning about negative numbers, you are leaving basic arithmetic and laying the foundation for algebra.

Because we have a lot to do, you will notice that the exercise sets are a little longer and a little more varied. Don't let this scare you off. In fact, if you want to be Superstudent, do *all* the exercises. Even if your instructor assigns only the odd-numbered problems, try doing the even ones as well.

Another approach is to do the odd exercises one day and the even exercises a couple of days later. This is a great way to cycle back and review in an organized way.

A Positive and Negative Numbers

The scale on an outdoor thermometer looks much like a number line. Typically, temperature readings of 0°F (degrees Fahrenheit), 10°F, 20°F, and so on, are labeled, with other tick marks in between. However, because temperatures can fall below 0°F, the scale must also be labeled with other kinds of numbers.

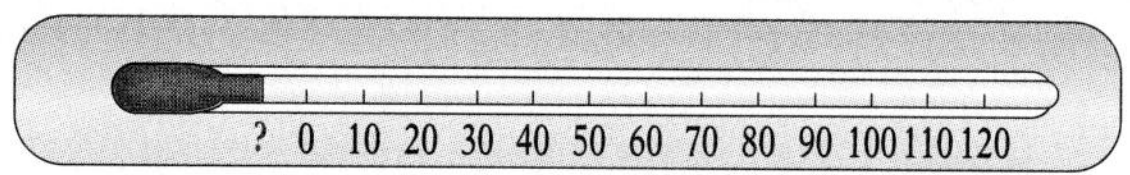

The number line for the whole numbers begins at 0 and extends forever to the right. We can extend the number line to the left simply by flipping the number line over and creating a mirror image.

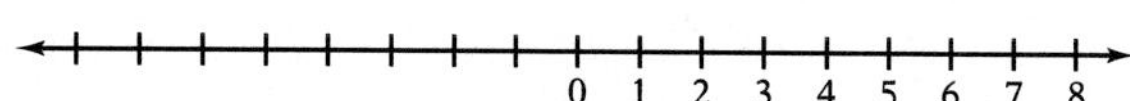

The point labeled with 0 is called the **origin.** We must have a way to label the tick marks so that the numbers to the left of 0 are different from the numbers to the right of 0. To do this, we use the *positive* symbol + in front of the numbers to the right of 0 and the *negative* symbol − in front of the numbers to the left of 0.

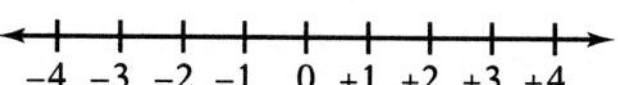

The numbers

$$\ldots, -4, -3, -2, -1, 0, +1, +2, +3, +4, \ldots$$

are called the **integers.** The three dots appear at both ends of the list because the number line extends forever to the left as well as to the right. We read a number such as +3 as "positive 3," and a number such as −5 as "negative 5."

NOTE We always indicate a negative number with the negative symbol −. However, we often omit the positive symbol + from positive numbers. For example, +4 and 4 both mean positive 4.

Copyright © Houghton Mifflin Company. All rights reserved.

On the number line, the numbers become larger as we move to the right and smaller as we move to the left. As before, we can use these facts to determine the order of the integers.

Order of the Integers

For any two numbers a and b on a number line, we say that:

1. $a < b$ if a is to the left of b.
2. $a > b$ if a is to the right of b.

Positive and Negative Integers

If n represents an integer, then:

1. n is a **positive integer** if $n > 0$.
2. n is a **negative integer** if $n < 0$.

Example 1

Insert < or > to make the statement true.

(a) -3 ___ 2 **(b)** -5 ___ -7

SOLUTION

(a)

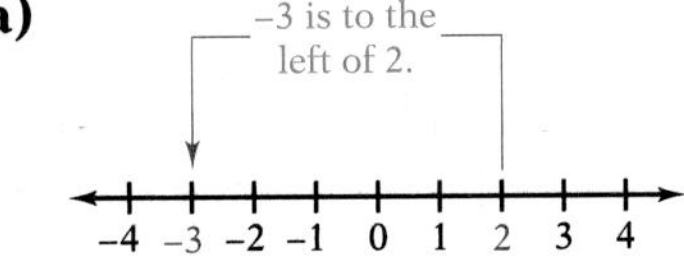

Therefore, $-3 < 2$.

(b)

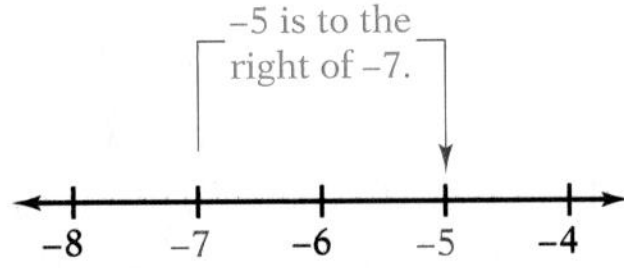

Therefore, $-5 > -7$.

Your Turn 1

Insert < or > to make the statement true.

(a) 0 ___ -1

(b) -100 ___ -99

Answers: (a) >; (b) <

Example 2

Write the numbers -4, 3, 0, 1, and -6 in order from smallest to largest.

SOLUTION

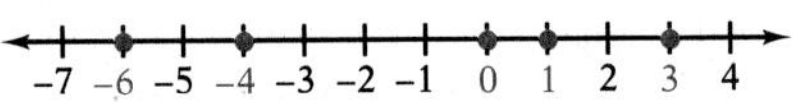

Starting with the leftmost number on the number line, we see that the order is

$$-6, -4, 0, 1, 3$$

Your Turn 2

Write the numbers 5, -7, 0, 2, and -1 in order from smallest to largest.

Answer: $-7, -1, 0, 2, 5$

Copyright © Houghton Mifflin Company. All rights reserved.

Example 3

What number is 5 units to the left of 2 on the number line?

SOLUTION

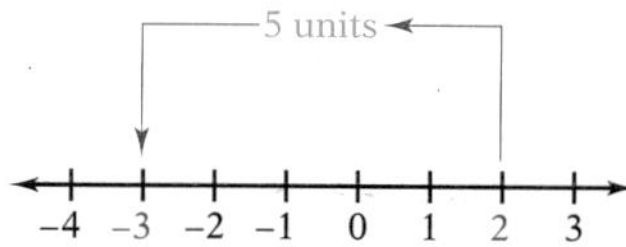

Counting 5 units to the left of 2, we see that the number is −3.

Your Turn 3

What number is 4 units to the right of −4?

Answer: 0

B Opposites

We extended the number line by making the distances between the tick marks to the left of 0 the same as the distances between the tick marks to the right of 0. This means that the distance from, say, 0 to +2 is the same as the distance from 0 to −2.

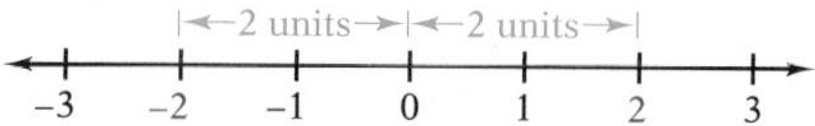

We call the numbers +2 and −2 *opposites* because they are on opposite sides of 0 and they are the same distance from 0.

Definition of Opposites

Two numbers are opposites if they are on opposite sides of 0 and are the same distance from 0. We will agree that the opposite of 0 is 0.

NOTE The more formal name for opposites is *additive inverses*. You will surely see that name in later courses. In this book, we will use *opposites*.

Example 4

What are the opposites of the given numbers?

(a) 7 **(b)** −5

SOLUTION

(a) The number 7 is 7 units to the right of 0. The number that is 7 units to the left of 0 is −7. Therefore, the opposite of 7 is −7.

(b) The number −5 is 5 units to the left of 0. The number that is 5 units to the right of 0 is 5. Therefore, the opposite of −5 is 5.

Your Turn 4

What are the opposites of the given numbers?

(a) −2 **(b)** 53

Answers: (a) 2; (b) −53

Copyright © Houghton Mifflin Company. All rights reserved.

In Example 4(a), we saw that -7 is the opposite of 7. This means that we can read the symbol -7 either as "negative 7" or as "the opposite of 7." For specific numbers, we usually just say "negative" rather than "the opposite of." However, if x represents some unknown number, how can we know whether $-x$ represents a positive number or a negative number?

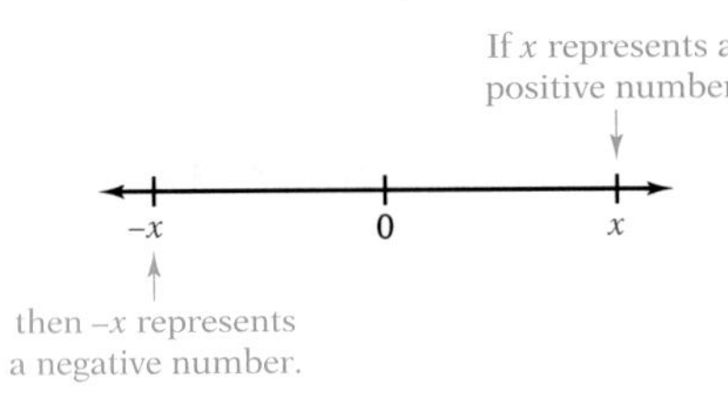

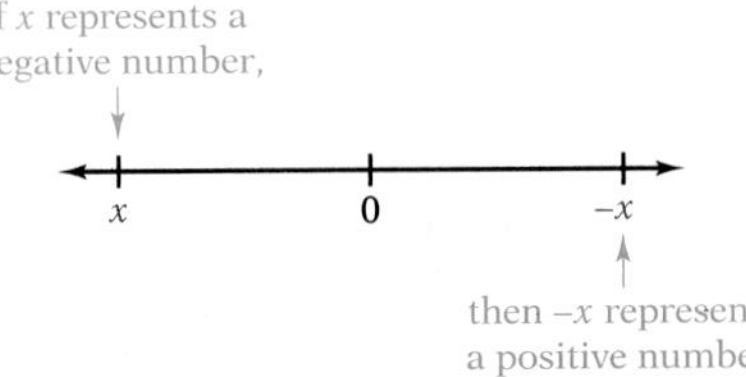

Because $-x$ can represent either a positive number or a negative number, depending on the value of x, we will always read $-x$ as "the opposite of x." In other words, $-x$ is the opposite of whatever number x represents.

> **LEARNING TIP**
>
> *Work hard at knowing when to say "minus," when to say "negative," and when to say "the opposite of." Yes, life would be easier if we had three different symbols, but you will get used to it.*

Unfortunately, the symbol $-$ is now being used in three different ways.

$5 - 3$ means "5 *minus* 3."
-4 means "*negative* 4."
$-x$ means "*the opposite of* x."

Also, the symbol $+$ is used in two different ways.

$6 + 7$ means "6 *plus* 7."
$+9$ means "*positive* 9."

Example 5

Write the given expression in words.

(a) $-8 - 2$ **(b)** $-y + 1$

(c) $3 - (-4)$ **(d)** $3 - (-x)$

SOLUTION

(a) *Negative* 8 *minus* 2

(b) *The opposite of y plus* 1

(c) *Three minus negative* 4

(d) *Three minus the opposite of x*

Your Turn 5

Write the given expression in words.

(a) $-x - 5$

(b) $-3 + a$

(c) $x + (-y)$

(d) $x - y$

Answers: **(a)** The opposite of x minus 5; **(b)** Negative 3 plus a; **(c)** x plus the opposite of y; **(d)** x minus y

Copyright © Houghton Mifflin Company. All rights reserved.

KEYS TO THE CALCULATOR

Your calculator may have a [+/−] key or a [(−)] key for entering a negative number. The following typical keystrokes can be used to display the number -17.

Scientific calculator: **17** [+/−] [=]

Graphing calculator: [(−)] **17** [ENTER]

Exercises

Display the given number with your calculator.

(a) -1 **(b)** -59 **(c)** -384 **(d)** $-12{,}573$

Finding the opposite of a number on the number line involves bouncing from one side of 0 to the other side. An expression such as $-(-3)$ involves two bounces. After the second bounce, we are right back where we started.

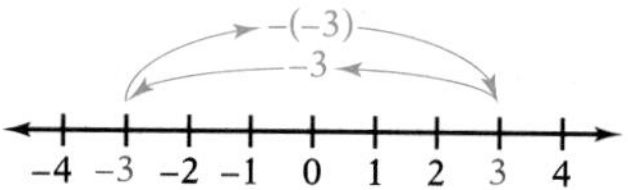

This means that $-(-3) = 3$. In words, the opposite of negative 3 is positive 3. We can state this as a general rule.

The Opposite of the Opposite of a Number

If x represents any number, then

$$-(-x) = x$$

In words, the opposite of the opposite of a number is the original number.

Example 6

Simplify the expression.

(a) $-(-23)$

(b) $-(-n)$

SOLUTION

(a) $-(-23) = 23$

(b) $-(-n) = n$

Your Turn 6

Simplify the expression.

(a) $-(-4)$

(b) $-(-a)$

Answers: (a) 4; (b) a

NOTE In Example 6(b), the expression involves a variable. Therefore, we read the expression as "the opposite of the opposite of n."

Copyright © Houghton Mifflin Company. All rights reserved.

C Absolute Value

We can find the distance between any two numbers on the number line simply by counting the number of units between them. For example, whether we count from 0 to 2 or from 2 to 0, the distance between 0 and 2 is 2 units. Similarly, the distance between 0 and −2 is 2 units.

The important thing to remember is that distance is not a negative number. One way to represent the distance between a number and 0 is as *absolute value.*

Definition of Absolute Value

The **absolute value** of a number is the distance between the number and 0. The symbol for the absolute value of a number n is $|n|$.

LEARNING TIP

You can think of finding the absolute value of a specific number as stripping away the + or − symbol from the number.

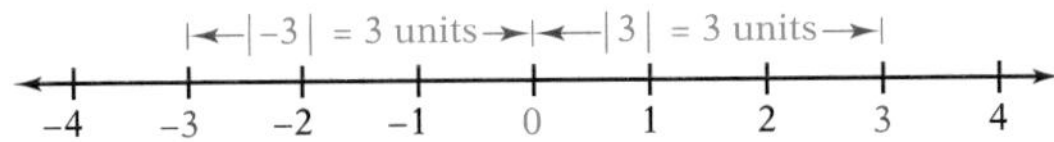

Example 7

Simplify the expression.

(a) $|4|$ **(b)** $|-5|$ **(c)** $|0|$

SOLUTION

(a) The expression $|4|$ means the distance between 4 and 0 on the number line. Because the distance is 4 units,

$$|4| = 4$$

(b) The expression $|-5|$ means the distance between −5 and 0 on the number line. Because the distance is 5 units,

$$|-5| = 5$$

(c) The distance between 0 and 0 is 0.

$$|0| = 0$$

Your Turn 7

Simplify the expression.

(a) $|-13|$

(b) $|48|$

Answers: (a) 13; (b) 48

Example 8

Insert =, <, or > to make the statement true.

(a) $|-7|$ ___ $|-5|$

(b) $|-9|$ ___ $|9|$

(c) $|4|$ ___ $|-6|$

Your Turn 8

Insert =, <, or > to make the statement true.

(a) $|0|$ ___ $|-1|$

(b) $|-3|$ ___ $|2|$

(c) $|8|$ ___ $|-8|$

Copyright © Houghton Mifflin Company. All rights reserved.

SOLUTION

(a) $|-7| > |-5|$ because $7 > 5$.

(b) $|-9| = |9|$ because $9 = 9$.

(c) $|4| < |-6|$ because $4 < 6$.

Answers: (a) <; (b) >; (c) =

Example 9

Write the following numbers in order from smallest to largest.

$$|-2|, -7, -(-6), 0, -|3|$$

SOLUTION

Be careful with $-|3|$. We are to find the absolute value of 3 and then the opposite of the result: $|3| = 3$, so $-|3| = -3$. Here is the required order.

Expression	*Value*
-7	-7
$-\|3\|$	-3
0	0
$\|-2\|$	2
$-(-6)$	6

The order is $-7, -|3|, 0, |-2|$, and $-(-6)$.

Your Turn 9

Write the following numbers in order from smallest to largest.

$$-(-3), |-4|, -1, -|4|$$

Answer: $-|4|, -1, -(-3), |-4|$

3.1 Quick Reference

A. Positive and Negative Numbers

1. We extended the number line to represent the integers.

$$\ldots, -3, -2, -1, 0, 1, 2, 3, \ldots$$

−4 −3 −2 −1 0 1 2 3 4

2. Integers to the right of 0 are called **positive integers;** integers to the left of 0 are called **negative integers.** The point labeled 0 is the **origin.**

3. For two numbers a and b, $a < b$ if a is to the left of b, and $a > b$ if a is to the right of b.

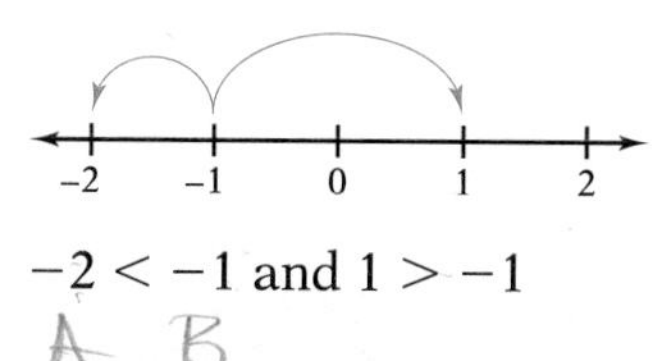

$-2 < -1$ and $1 > -1$

Copyright © Houghton Mifflin Company. All rights reserved.

B. Opposites

1. Two numbers are **opposites** if they are on opposite sides of 0 and are the same distance from 0.

   -3 is the opposite of 3.

   5 is the opposite of -5.

2. The symbol $-$ may be used to indicate subtraction, a negative number, or the opposite of a number.

   $9 - 4$: "9 *minus* 4"

   -7: "*negative* 7"

   $-x$: "*the opposite of x*"

3. If x represents any number, then

   $$-(-x) = x$$

   $-(-2) = 2$

   $-[-(-2)] = -[2] = -2$

C. Absolute Value

1. The **absolute value** of a number is the distance between the number and 0. The symbol for the absolute value of a number n is $|n|$.

   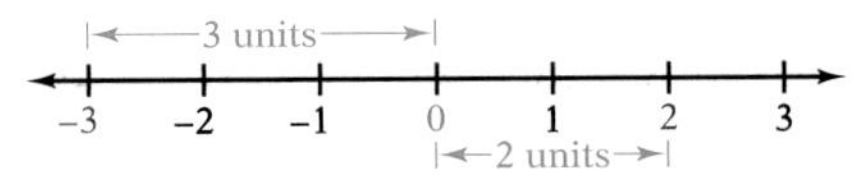

   $|-3| = 3$ $\quad |2| = 2$

Copyright © Houghton Mifflin Company. All rights reserved.

3.1 Exercises

A. Positive and Negative Numbers

1. The whole numbers and their opposites are called the ______.

2. On a number line, the point associated with 0 is called the ______.

In Exercises 3–10, insert < or > to make the statement true.

3. −6 ___ 4
4. 5 ___ −2
5. 0 ___ −3
6. −7 ___ 1
7. −8 ___ −6
8. −1 ___ −3
9. −24 ___ −37
10. −15 ___ −12

In Exercises 11–14, write the numbers in order from smallest to largest.

11. −1, −5, 1, 3, −3
12. −1, 0, −6, 2, 4
13. 0, 2, −4, −7, −3
14. 6, −4, −3, 1, −5

In Exercises 15–20, identify the number whose location on the number line is described. Drawing number lines will help.

15. 4 units to the right of −3

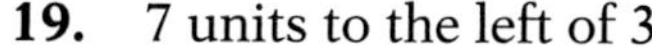

16. 5 units to the left of 2

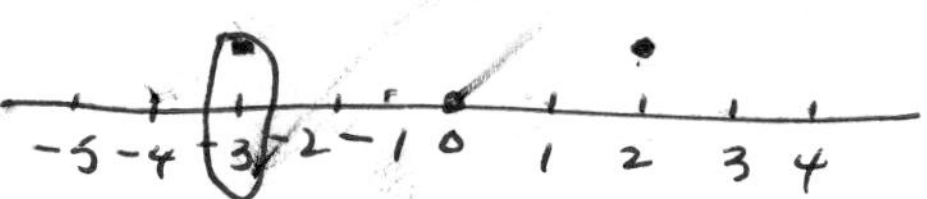

17. 2 units to the left of −4

18. 3 units to the right of −7

19. 7 units to the left of 3

20. 6 units to the right of −5

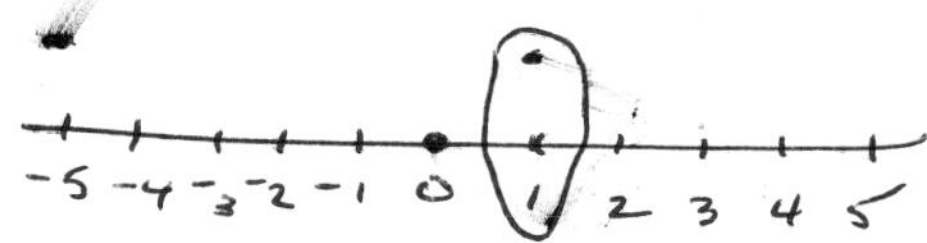

B. Opposites

21. The numbers −5 and 5 are the same distance from 0 and are called ______.

22. We read −8 as "______ 8," and we read $-a$ as "the ______ of a."

In Exercises 23–28, write the opposite of the number.

23. −3
24. 8
25. 22
26. −65
27. −7
28. 0

In Exercises 29–36, write the expression in words.

29. $-4 - 3$
30. $-a - 2$
31. $-6 + x$

Copyright © Houghton Mifflin Company. All rights reserved.

32. $-2 + 3$ **33.** $-3 - (-4)$ **34.** $5 - (-6)$

35. $-(-5)$ **36.** $-(-c)$

In Exercises 37–42, simplify the expression.

37. $-(-15)$ **38.** $-(-4)$ **39.** $-(-y)$

40. $-(-b)$ **41.** $-(8)$ **42.** $-(x)$

C. Absolute Value

43. The distance between a number and 0 is called the ____ of the number.

44. If two different numbers have the same absolute value, then the numbers are ____.

In Exercises 45–52, evaluate the expression.

45. $|7|$ **46.** $|-6|$ **47.** $|-12|$ **48.** $|23|$

49. $|-9|$ **50.** $|10|$ **51.** $-|5|$ **52.** $-|-3|$

In Exercises 53–58, insert <, >, or = to make the statement true.

53. $|7|$ ____ $|-7|$ **54.** $|-3|$ ____ $|2|$ **55.** $|8|$ ____ $|-12|$

56. $-|-4|$ ____ $-|4|$ **57.** $|6|$ ____ $|-4|$ **58.** $-|5|$ ____ $|-5|$

In Exercises 59–62, write the numbers in order from smallest to largest.

59. $|0|, -|2|, -(-4), |-5|, -1$ **60.** $-|-2|, |-2|, -3, -(-4), |3|$

61. $|-6|, -3, |2|, -|4|, -|-2|$ **62.** $-8, -|-3|, -(-5), -|0|, |-10|$

Applications

In Exercises 63–66, write a positive integer and a negative integer to describe the given information.

63. *Land Elevations* The highest point in Louisiana is Driskill Mountain, which is 535 feet above sea level. The lowest point is New Orleans, which is 8 feet below sea level.

Copyright © Houghton Mifflin Company. All rights reserved.

64. *Normal Temperatures* In Bismarck, North Dakota, the normal January low temperature is 2°F below zero and the normal January high temperature is 20°F above zero.

65. *Checking Account* A person deposits 350 dollars in a checking account. The next day the person writes a check for 128 dollars.

66. *Stock Prices* The price of a share of stock drops 5 dollars one week. The next week the price gains 3 dollars.

Writing and Concept Extension

67. Explain why $|a| = |-a|$ is always true.

68. What are two ways to read -7? Why should $-x$ be read only one way?

69. Explain why $-n$ does not necessarily represent a negative number.

70. Simplify the given expression.

(a) $-(-(-x))$ **(b)** $-(-(-(-x)))$

71. Determine all numbers that are 3 units from 0.

Exploring with Real-World Data: Collaborative Activities

Record Low Temperatures The table shows the lowest temperatures, in degrees Fahrenheit, recorded in Chicago for selected months.

Month	*Temperature*	*Month*	*Temperature*
January	−27° F	October	17° F
February	−17° F	November	1° F
July	40° F	December	−25° F

72. What are the highest and lowest temperatures given in the table?

73. Arrange the temperatures from smallest to largest.

Copyright © Houghton Mifflin Company. All rights reserved.

74. Construct a bar graph of the data in the table.

75. Write an inequality to compare the temperatures for February and December.

76. Write an inequality to compare the temperatures for January and February.

77. For which months are the temperatures opposites?

78. For which two months is the absolute value of the temperature largest?

79. For which month is the absolute value of the temperature the smallest?

Copyright © Houghton Mifflin Company. All rights reserved.

3.2 Addition of Integers

A *The Addition Rules for Integers*
B *Properties of Addition*

SUGGESTIONS FOR SUCCESS

Taking notes has been described as transferring information from the blackboard to a notebook without any of it passing through the note taker's brain. Here are some ways to avoid that trap.

1. Make listening and participating your highest priorities.
2. Don't bother making notes about things that you already know.
3. Write the date and topic so that you know what the notes are.
4. Review your notes as soon as possible after class, and flesh them out with information you didn't have time to include.
5. Get together with other students and share notes. You may be able to help others, and they may be able to help you.

A *The Addition Rules for Integers*

If your checking account statement is typical, each deposit is listed as a positive number, and each check that you wrote is listed as a negative number. Balancing your account involves adding these positive and negative numbers.

In Section 1.3, we used the number line to visualize the addition of whole numbers. To add $3 + 5$, we start at 3 and move 5 units to the right. Because we arrive at 8, we know that $3 + 5 = 8$.

The same method can be used for adding integers. However, when we add a negative number, we move to the left rather than to the right.

Example 1

Use the number line to add $2 + (-5)$.

SOLUTION

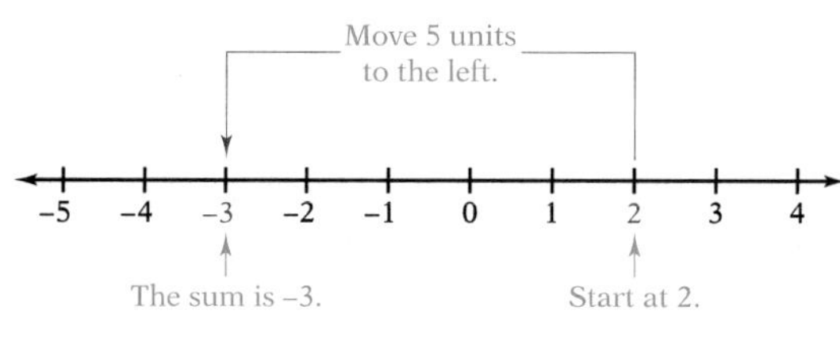

$$2 + (-5) = -3$$

Your Turn 1

Use the number line to add $1 + (-3)$.

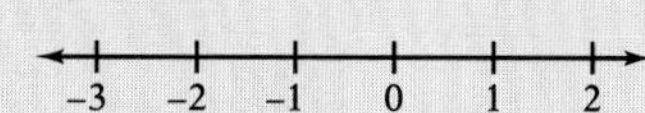

LEARNING TIP

Writing 2 + −5 would be confusing. We write 2 + (−5) to separate the plus symbol from the negative symbol. You will make fewer careless errors if you use the proper notation.

Answer: −2

Example 2

Use the number line to add $-4 + 6$.

SOLUTION

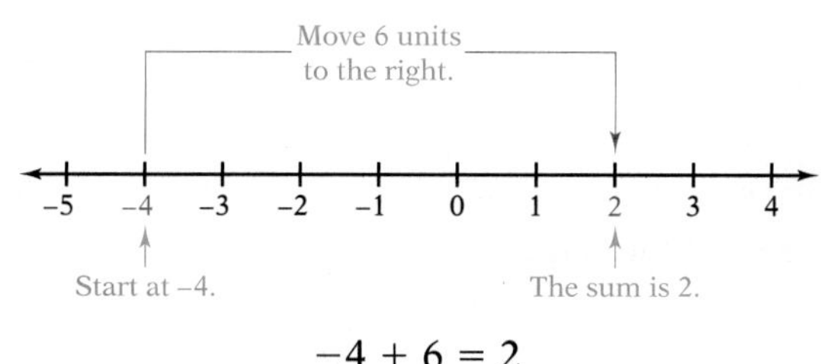

$$-4 + 6 = 2$$

Your Turn 2

Use the number line to add $-2 + 3$.

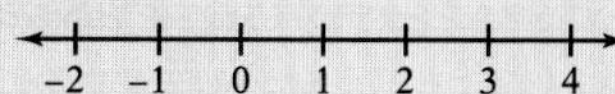

Answer: 1

Copyright © Houghton Mifflin Company. All rights reserved.

NOTE Some people like to enclose all negative numbers in parentheses. In Example 2, $-4 + 6$ can be written $(-4) + 6$. However, the parentheses aren't necessary for clarity because there is no symbol in front of the -4.

Example 3

Use the number line to add $-1 + (-3)$.

SOLUTION

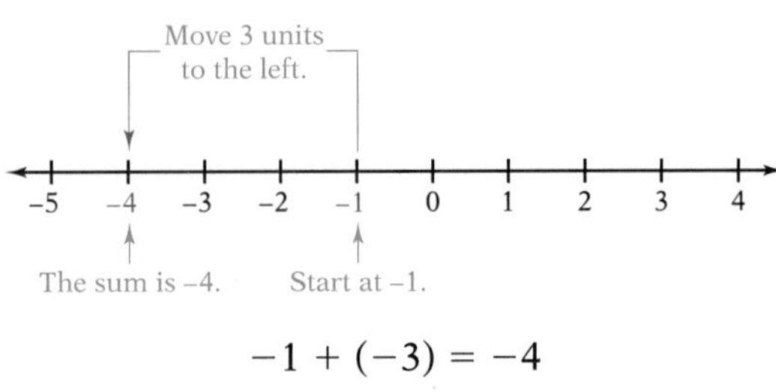

$$-1 + (-3) = -4$$

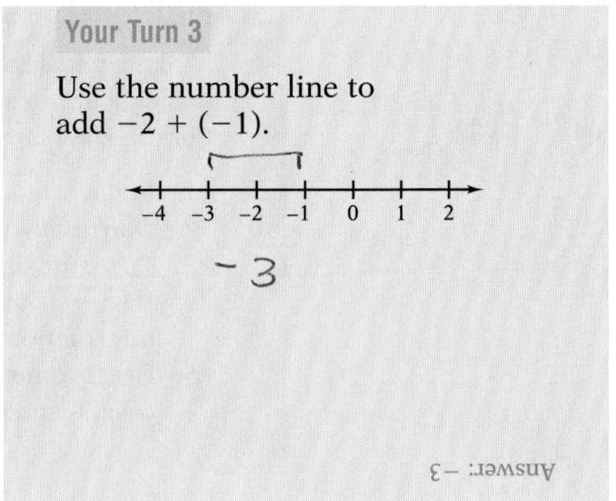

Your Turn 3

Use the number line to add $-2 + (-1)$.

Answer: –3

Now let's use the number line to add $287 + (-549)$.

Addition Rules

We were just kidding. No one wants to draw a number line so long that we could start at 287 and move 549 units to the left! Instead, we need to have rules that will allow us to add any integers without reference to the number line.

The results of our preceding examples (and Your Turns) will give us some patterns that should help us to discover what the rules should be. Here is a summary of what we have done so far.

Addends with the Same Signs	*Addends with Different Signs*
$3 + 5 = 8$	$2 + (-5) = -3$
$-1 + (-3) = -4$	$-4 + 6 = 2$
$-2 + (-1) = -3$	$1 + (-3) = -2$
	$-2 + 3 = 1$

In the first column, we see that adding the absolute values of the addends gives us the absolute value of the result. Also, the sign of the result is the same as the sign of the addends.

In the second column, subtracting the absolute values of the addends gives us the absolute value of the result. Also, the sign of the result is the same as the sign of the addend with the larger absolute value.

These observations lead to the following addition rules:

Rules for Adding Two Integers

1. If the addends have the same sign, then:
   (a) Add the absolute values of the addends.
   (b) Use the sign of the addends.
2. If the addends have different signs, then:
   (a) Subtract the absolute values of the addends.
   (b) Use the sign of the addend with the larger absolute value.

Copyright © Houghton Mifflin Company. All rights reserved.

Here is some other advice about adding integers.

1. Remember that we can think of the absolute value of a number as just the numerical part, with the sign stripped away.

$-18 + (-7)$ Adding the absolute values means adding $18 + 7$.

$9 + (-16)$ Subtracting the absolute values means subtracting $16 - 9$.

2. If both numbers are positive, then the sum is positive; if both numbers are negative, then the sum is negative.

Both numbers are positive. ↓ ↓
$9 + 12 = 21$ ← The sum is positive.

Both numbers are negative. ↓ ↓
$-7 + (-4) = -11$ ← The sum is negative.

3. If one number is positive and the other number is negative, then the sign of the sum is dictated by the number with the larger absolute value.

$-7 + 4 = -3$ ← The sign of the sum is the same as the sign of this number. (↑ -7)

$-5 + 9 = 4$ ← The sign of the sum is the same as the sign of this number. (↑ 9)

LEARNING TIP

Use flash cards, computer games, or any other means to learn the addition rules so well that you can add integers in your sleep. You should not rely on your calculator for these very basic skills.

Example 4

Add.

(a) $-10 + (-6)$ **(b)** $-8 + 5$ **(c)** $-4 + 11$

SOLUTION

(a) The addends have the same sign, so we add their absolute values.

$$10 + 6 = 16$$

The sign of the sum is the same as the sign of the addends.

$$-10 + (-6) = -16$$

(b) The addends have different signs, so we subtract their absolute values.

$$8 - 5 = 3$$

The addend with the larger absolute value is -8, which is negative. Therefore, the sum is negative.

$$-8 + 5 = -3$$

(c) The addends have different signs, so we subtract their absolute values.

$$11 - 4 = 7$$

The addend with the larger absolute value is 11, which is positive. Therefore, the sum is positive.

$$-4 + 11 = 7$$

Your Turn 4

Add.

(a) $7 + (-12)$

(b) $-9 + (-8)$

(c) $-5 + 13$

Answers: (a) -5; (b) -17; (c) 8

Copyright © Houghton Mifflin Company. All rights reserved.

Example 5

Add.

(a) $5 + (-8)$ **(b)** $-7 + (-6)$ **(c)** $-9 + 11$

(d) $-2 + (-15)$ **(e)** $12 + (-7)$ **(f)** $-8 + 3$

SOLUTION

(a) $5 + (-8) = -3$ **(b)** $-7 + (-6) = -13$

(c) $-9 + 11 = 2$ **(d)** $-2 + (-15) = -17$

(e) $12 + (-7) = 5$ **(f)** $-8 + 3 = -5$

Your Turn 5

Add.

(a) $-4 + 9$ **(b)** $-4 + (-9)$

(c) $4 + (-9)$ **(d)** $-5 + 3$

(e) $14 + (-8)$ **(f)** $-7 + (-4)$

Answers: (a) 5; (b) −13; (c) −5; (d) −2; (e) 6; (f) −11

KEYS TO THE CALCULATOR

To add positive and negative numbers with your calculator, you will need to remember to enter negative numbers either with the [+/−] key or with the [(−)] key. Here are two possible ways to add $-10 + (-6)$.

Scientific calculator: **10** [+/−] [+] **6** [+/−] [=]
Graphing calculator: [(−)] **10** [+] [(] [(−)] **6** [)] [ENTER]

If your calculator has a [(−)] key for negative numbers, then we recommend that you use parentheses just as you would if you were writing the expression.

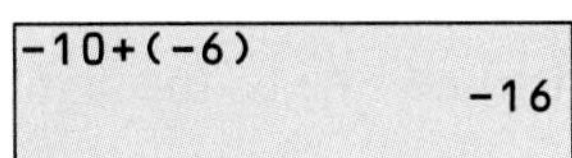

Exercises

(a) $-27 + 14$ **(b)** $-42 + (-98)$ **(c)** $576 + (-85)$ **(d)** $28 + (-173)$

B Properties of Addition

The properties of addition of whole numbers also hold for the addition of integers. The following are examples of those properties.

$-8 + 3 = 3 + (-8)$	Commutative Property of Addition
$(-2 + 7) + (-4) = -2 + [7 + (-4)]$	Associative Property of Addition
$-6 + 0 = -6$ and $0 + (-6) = -6$	Addition Property of 0

Another important property of addition involves the sum of two numbers that are opposites. According to the addition rules, all of the following statements are true:

$$-5 + 5 = 0 \qquad 8 + (-8) = 0 \qquad -23 + 23 = 0$$

Copyright © Houghton Mifflin Company. All rights reserved.

The Addition Property of Opposites

If a represents any number, then

$$a + (-a) = 0$$
and
$$-a + a = 0$$

In words, the sum of any number and its opposite is 0.

Example 6

Add.

(a) $-19 + 0$ **(b)** $-19 + 19$

SOLUTION

(a) $-19 + 0 = -19$ Addition Property of 0

(b) $-19 + 19 = 0$ Addition Property of Opposites

Your Turn 6

Add.

(a) $26 + (-26)$

(b) $0 + (-26)$

Answers: (a) 0; (b) −26

Many people find adding numbers with the same sign easier than adding numbers with different signs. The Commutative and Associative Properties are particularly helpful when we must add more than two positive and negative numbers.

Example 7

Add: $-3 + 8 + (-7) + 4$.

SOLUTION

Method 1

Following the Order of Operations, we add from left to right.

$$\begin{aligned} -3 + 8 + (-7) + 4 &= 5 + (-7) + 4 \\ &= -2 + 4 \\ &= 2 \end{aligned}$$

Method 2

We use the Commutative and Associative Properties of Addition to group the positive and negative numbers.

$$\begin{aligned} -3 + 8 + (-7) + 4 &= 8 + 4 + (-3) + (-7) \\ &= (8 + 4) + [-3 + (-7)] \\ &= 12 + (-10) \\ &= 2 \end{aligned}$$

Your Turn 7

Add.

$$9 + (-4) + (-5) + 2$$

LEARNING TIP

Method 2 is most effective when you can rearrange and regroup the numbers mentally. Make sure you are comfortable with the addition rules before you try that.

Answer: 2

Copyright © Houghton Mifflin Company. All rights reserved.

3.2 Quick Reference

A. The Addition Rules for Integers

1. To represent addition on the number line, we follow these steps.

   (a) Locate the first addend.

   (b) The second addend indicates the number of units to move. If the second addend is positive, move right; if the second addend is negative, move left.

$-3 + 5 = 2$

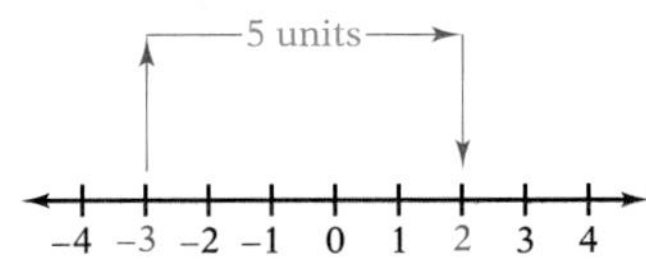

$2 + (-6) = -4$

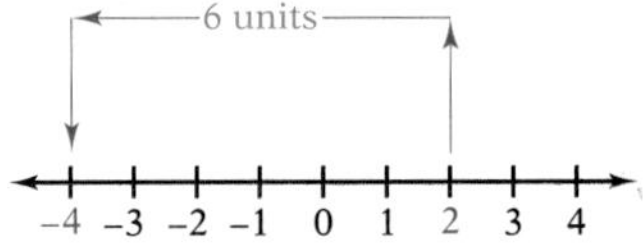

2. Rules for adding two integers

   (a) If the addends have the same sign, then add the absolute values of the addends and use the sign of the addends.

   $-7 + (-9) = -16$

   (b) If the addends have different signs, then subtract the absolute values of the addends and use the sign of the addend with the larger absolute value.

   $-8 + 3 = -5$

   $11 + (-4) = 7$

B. Properties of Addition

1. The following properties hold for the addition of integers:

   (a) Commutative Property of Addition — $5 + (-3) = -3 + 5$

   (b) Associative Property of Addition — $[2 + (-6)] + 4 = 2 + [-6 + 4]$

   (c) Addition Property of 0 — $-7 + 0 = 0 + (-7) = -7$

2. The Addition Property of Opposites

   If a represents any number, then

   $a + (-a) = 0$ and $-a + a = 0$

   $4 + (-4) = -4 + 4 = 0$

   In words, the sum of any two opposites is 0.

Copyright © Houghton Mifflin Company. All rights reserved.

3.2 Exercises

A. The Addition Rules for Integers

1. If two addends have the same sign, then we add the ______ of the addends.

2. If two addends have different signs, then we ______ the absolute values of the addends.

In Exercises 3–6, determine the sum.

3. $-7 + (-2)$ 4. $-4 + (-8)$ 5. $-10 + (-6)$ 6. $-9 + (-5)$

In Exercises 7–14, determine the sum.

7. $-10 + 3$ 8. $-4 + 10$ 9. $6 + (-2)$ 10. $5 + (-6)$

11. $-1 + 7$ 12. $-15 + 9$ 13. $5 + (-11)$ 14. $12 + (-8)$

In Exercises 15–44, determine the sum.

15. $-12 + (-8)$ 16. $-5 + (-10)$ 17. $17 + (-8)$ 18. $10 + (-3)$

19. $-12 + 8$ 20. $-15 + 7$ 21. $-9 + 9$ 22. $-15 + 15$

23. $-9 + 8$ 24. $-14 + 6$ 25. $-4 + 11$ 26. $-8 + 18$

27. $-5 + 0$ 28. $-8 + 0$ 29. $6 + (-13)$ 30. $9 + (-18)$

31. $8 + (-6)$ 32. $5 + (-1)$ 33. $-6 + (-9)$ 34. $-15 + (-5)$

35. $12 + (-12)$ 36. $7 + (-7)$ 37. $-20 + (-8)$ 38. $-12 + (-30)$

39. $12 + (-17)$ 40. $8 + (-14)$ 41. $-5 + 16$ 42. $-10 + 21$

43. $0 + (-12)$ 44. $0 + (-20)$

B. Properties of Addition

45. The sum of any two ______ is zero.

46. To add $-4 + 7 + (-3) + 9$, we can use the ______ and ______ Properties of Addition to group the positive numbers and the negative numbers.

In Exercises 47–50, use the indicated property to fill in the blank.

47. Addition Property of 0: ______ $+ 0 = -8$

48. Commutative Property of Addition: $-6 + 3 + (-8) = -6 +$ ______ $+ 3$

49. Addition Property of Opposites: ______ $+ 7 = 0$

50. Addition Property of Opposites: $0 = -3 +$ ______

Copyright © Houghton Mifflin Company. All rights reserved.

In Exercises 51–62, determine the sum.

51. $-4 + 5 + (-7)$

52. $8 + (-9) + 3$

53. $-9 + (-3) + (-8)$

54. $-2 + (-6) + 6$

55. $4 + (-6) + 10 + (-2)$

56. $-6 + 5 + (-1) + (-2)$

57. $-9 + 5 + 9 + (-2)$

58. $12 + (-8) + (-6) + 2$

59. $-6 + (-2) + 7 + 6 + (-1)$

60. $-2 + (-3) + (-6) + 9 + 3$

61. $7 + 3 + (-8) + (-1) + 5$

62. $-2 + 4 + (-3) + (-2) + (-6)$

≈ *Estimation*

In Exercises 63–66, select the number in column B that is the best estimate of the sum in column A.

	Column A	*Column B*
63.	$-307 + 793$	(A) -500
64.	$-78 + (-23)$	(B) -100
65.	$491 + (-989)$	(C) 100
66.	$811 + (-690)$	(D) 500

Calculator Exercises

In Exercises 67–70, use a calculator to determine the sum.

67. $-23 + 87$

68. $59 + (-128)$

69. $-472 + (-513)$

70. $-71 + 92$

Copyright © Houghton Mifflin Company. All rights reserved.

Applications

In Exercises 71–74, represent the given information as a sum of integers. Then answer the question.

71. *Checking Account Balance* A man had a balance of \$640 in a checking account. On Monday, he made a cash withdrawal of \$320 from an ATM, and on Wednesday he used his debit card to make a \$70 purchase. On Friday, he made a deposit of \$200. What was the account balance after Friday's deposit?

72. *Take-Home Pay* A student earned \$250 for one week, but she paid \$55 for taxes and \$20 for insurance. What was her take-home pay for the week?

73. *Diet Results* During a six-week diet, a person recorded the following weekly results (in pounds): lost 3, gained 1, lost 4, lost 1, gained 2, lost 3. At the end of the six weeks, what was the overall change in the person's weight?

74. *Profit and Loss* During the first year that a new business was in operation, the owner recorded the following quarterly results: lost \$3,500, made \$750, lost \$1,200, made \$2,300. What was the overall profit or loss for the year?

−3500 + 750 − 1200 + 2300

−4700
+3050
−1650 loss

Writing and Concept Extension

75. Explain how to use a number line to illustrate the sums $-2 + (-4)$ and $-2 + 4$.

76. To determine the sums $-5 + 3$ and $-5 + (-3)$, we use the absolute values of the addends: 5 and 3. Explain how we use the absolute values 5 and 3 in each sum.

77. The Addition Property of Opposites states that $a + (-a) = 0$ for *any* number a. For the property to hold for $a = 0$, what agreement must we make about the meaning of the symbol -0?

78. Explain how the symbol $-x$ can represent a positive number.

Copyright © Houghton Mifflin Company. All rights reserved.

In Exercises 79 and 80, determine the unknown number.

79. The sum of -3 and what number is 1?

$-3 + (-4) = 1$

80. What number is 5 more than -2? 3

81. Suppose that a represents a negative number and b represents a positive number. Determine the sign of each expression.

(a) $-a + b$ positive

(b) $a + (-b)$ negative

Copyright © Houghton Mifflin Company. All rights reserved.

3.3 Subtraction of Integers

A *Definition of Subtraction*
B *Combined Operations*
C *Applications*

SUGGESTIONS FOR SUCCESS

Your instructor may give you the opportunity to work in groups to complete certain assignments. Working with others can be an excellent way to help you learn the material.

To get the most benefit from group work, you should actively participate in the group. Arrange times and locations for the group to meet. Exchange phone numbers or e-mail addresses with others in the group so that you can stay in touch. When the group meets, stick to the project rather than give in to the urge to socialize. Explain to others how you would solve the problem and ask others to explain their strategies to you. Discuss the advantages and disadvantages of the methods.

A Definition of Subtraction

When we subtracted whole numbers, we were able to think of subtraction as "taking away." For example, if a cafeteria sets out 12 sandwiches and 9 of them are purchased, then $12 - 9$ represents taking 9 away from 12, and the result, 3, is the number of sandwiches that remain.

We can't use this interpretation very well when we subtract integers. An alternative is to use the method for checking a subtraction. For example, we know that $9 - 5 = 4$ because $4 + 5 = 9$. Here is how we might apply this approach to subtracting integers.

Subtracting Integers

Suppose that we want to subtract 3 from -7. We begin by writing the related addition.

$$-7 - 3 = \square \qquad \text{means that} \qquad \square + 3 = -7$$

Now we ask, "What number added to 3 gives us -7?" Possibly with the help of the number line, we can determine that the number is -10. Therefore,

$$-7 - 3 = -10$$

Although this method is not very efficient, it does at least remind us that there is a relationship between subtraction and addition. Consider the following lists of subtractions and additions:

Subtractions	*Additions*
$8 - 5 = 3$	$8 + (-5) = 3$
$9 - 4 = 5$	$9 + (-4) = 5$
$13 - 7 = 6$	$13 + (-7) = 6$
$11 - 2 = 9$	$11 + (-2) = 9$

Do you see that the answer to each subtraction is the same as the answer to the matching addition? This suggests that we can perform subtraction by writing the difference as a sum.

Copyright © Houghton Mifflin Company. All rights reserved.

According to the following rule, writing a difference as a sum involves two changes.

Definition of Subtraction

If a and b represent any numbers, then

$$a - b = a + (-b)$$

In words, to subtract $a - b$, we change the minus symbol to a plus symbol, and we change the number being subtracted to its opposite.

Now we can return to the subtraction $-7 - 3$ and perform the operation according to the definition.

Change the minus symbol to a plus symbol.

$$-7 - 3 = -7 + (-3) = -10$$

Change the number being subtracted to its opposite.

You will find that changing a subtraction to an addition will be easier if you can express the change in words. Go ahead and say the words aloud.

"Negative 7 *minus* 3 is equal to negative 7 *plus negative* 3."

Example 1

Subtract $9 - 6$.

SOLUTION

"9 *minus* 6 is equal to 9 *plus negative* 6."

$$9 - 6 = 9 + (-6) = 3$$

Your Turn 1

Subtract $11 - 8$.

Answer: 3

NOTE In Example 1, we already knew that $9 - 6 = 3$ without having to use the definition of subtraction. However, the example shows that the definition works, even for differences that we know how to find.

Example 2

Subtract $-12 - 7$.

SOLUTION

"Negative 12 *minus* 7 is equal to negative 12 *plus negative* 7."

$$-12 - 7 = -12 + (-7) = -19$$

Your Turn 2

Subtract $-8 - 6$.

Answer: −14

Copyright © Houghton Mifflin Company. All rights reserved.

Example 3

Subtract $5 - (-9)$.

SOLUTION

"5 *minus negative* 9 is equal to 5 *plus positive* 9."

$$5 - (-9) = 5 + (+9) = 14$$

We wrote the positive symbol in front of 9 just to emphasize the sign change. You can write

$$5 - (-9) = 5 + 9$$

Your Turn 3

Subtract $8 - (-2)$.

Answer: 10

Example 4

Subtract $-3 - (-7)$.

SOLUTION

"Negative 3 *minus negative* 7 is equal to negative 3 *plus positive* 7."

$$-3 - (-7) = -3 + 7 = 4$$

Your Turn 4

Subtract $-10 - (-4)$.

Answer: -6

As shown in the preceding examples, applying the rule in words is a good thing to do until the operation becomes second nature to you. In the rest of our examples, we won't write the words. But remember: When you see the symbols, say the words.

Example 5

Subtract.

(a) $-6 - 9$ **(b)** $5 - (-8)$

(b) $2 - 7$ **(d)** $-11 - (-4)$

SOLUTION

(a) $-6 - 9 = -6 + (-9) = -15$

(b) $5 - (-8) = 5 + 8 = 13$

(c) $2 - 7 = 2 + (-7) = -5$

(d) $-11 - (-4) = -11 + 4 = -7$

Your Turn 5

Subtract.

(a) $6 - (-3)$

(b) $-7 - 5$

(c) $-2 - (-8)$

(d) $4 - 12$

Answers: (a) 9; (b) -12; (c) 6; (d) -8

Before we show you how a calculator can be used to perform subtraction, we need to consider the most efficient way to think about a subtraction problem. When we use a calculator for $-73 - 16$, it is not necessary to use the definition of subtraction to write the problem as $-73 + (-16)$.

On the other hand, if the problem is $95 - (-28)$, then it makes sense to use the definition of subtraction to write the problem as $95 + 28$ because fewer keystrokes are needed. With practice, you will be able to make the conversion mentally and simply enter the sum. Of course, the calculator will return the correct result whichever form you use.

Copyright © Houghton Mifflin Company. All rights reserved.

EYS TO THE CALCULATOR

When you perform subtraction with a calculator, make sure that you know the difference between *minus* and *negative* and that you press the correct key for each word. Here are the typical keystrokes for Example 4.

Scientific calculator: **3** [+/−] [−] **7** [+/−] [=]
Graphing calculator: [(−)] **3** [−] [(] [(−)] **7** [)] [ENTER]

```
-3-(-7)
                4
```

Again, the parentheses around the −7 are optional but recommended.

Exercises

(a) $-23 - 19$ **(b)** $18 - (-34)$ **(c)** $-6 - (-51)$ **(d)** $-42 - (-25)$

B Combined Operations

Numerical expressions may involve more than one operation. As a general rule, each time you see a subtraction symbol, use the definition of subtraction to change the subtraction to a sum.

Example 6

Evaluate $-6 - (-4) + 2 - 8$.

SOLUTION

$$\begin{aligned} -6 - (-4) + 2 - 8 &= -6 + 4 + 2 + (-8) \\ &= (4 + 2) + [-6 + (-8)] \\ &= 6 + (-14) \\ &= -8 \end{aligned}$$

Your Turn 6

Evaluate $-5 - 3 - (-8) + 4$.

Answer: 4

Example 7

Evaluate $-9 - 4^2$.

SOLUTION

Remember that exponents rank above subtraction in the Order of Operations.

$$\begin{aligned} -9 - 4^2 &= -9 - 16 \\ &= -9 + (-16) \\ &= -25 \end{aligned}$$

Your Turn 7

Evaluate $4 - 3^2 - 5$.

Answer: −10

Copyright © Houghton Mifflin Company. All rights reserved.

C Applications

In many applications involving subtraction, we only have to find the difference of two positive numbers. For example, if an item costs \$73.28 at store A and \$61.57 at store B, then we can easily determine that the price at store A is \$73.28 − \$61.57 = \$11.71 higher. The definition of subtraction is not needed for such problems.

In other applications, such as accounting, temperatures, and measurements of distances, negative numbers can be involved. In these cases, using the definition of subtraction is an appropriate way to solve the problem.

Example 8

Suppose that your bank statement shows that you are overdrawn by \$23.50. However, your checkbook shows a balance of \$56. By how much do you and your bank differ with regard to your balance?

SOLUTION

We represent the bank's balance with the negative number -23.5 and your balance with the positive number 56. Then the difference is

$$56 - (-23.5) = 56 + 23.5 = 79.5$$

You and your bank differ by \$79.50.

Your Turn 8

Two nights ago, the low temperature was -5°F. Last night, the low was 26°F.

What was the difference in the low temperatures for the two nights?

Answer: 31°F

Assuming that your bank statement is correct in Example 8, you might be able to use the result in that example to track down the error. One possible explanation is that you wrote a check for \$79.50 and forgot to enter it in your checkbook ledger.

When you solve applications involving subtraction, the order in which you subtract usually doesn't matter very much, as long as you interpret the result in a meaningful way. In Example 8, we could just as well have subtracted $-23.5 - 56 = -79.50$ and then ignored the negative sign. We often care only about the absolute value of the difference.

3.3 Quick Reference

A. Definition of Subtraction

1. If a and b represent any numbers, then

$$a - b = a + (-b)$$

In words, to subtract $a - b$, we change the minus symbol to a plus symbol, and we change the number being subtracted to its opposite.

$$7 - 10 = 7 + (-10) = -3$$
$$7 - (-10) = 7 + 10 = 17$$
$$-7 - 10 = -7 + (-10) = -17$$
$$-7 - (-10) = -7 + 10 = 3$$

Copyright © Houghton Mifflin Company. All rights reserved.

B. Combined Operations

1. Numerical expressions may have more than one operation. Change each subtraction to a sum.

$$6 - (-3) + 2 - 9$$
$$= 6 + 3 + 2 + (-9)$$
$$= 11 + (-9)$$
$$= 2$$

2. When you evaluate a numerical expression, follow the Order of Operations.

$$3 - 2^3 - (-5)$$
$$= 3 - 8 - (-5)$$
$$= 3 + (-8) + 5$$
$$= 0$$

Copyright © Houghton Mifflin Company. All rights reserved.

3.3 Exercises

A. Definition of Subtraction

1. To subtract $3 - (-2)$, change the ______ symbol to a(n) ______ symbol and change -2 to its ______.

2. The expression $4 - (-7)$ is read as "4 ______ ______ 7."

In Exercises 3–40, determine the difference.

3. $8 - (-3)$
4. $7 - (-6)$
5. $9 - 7$
6. $8 - 3$
7. $-7 - 6$
8. $-2 - 9$
9. $-5 - (-10)$
10. $-8 - (-9)$
11. $4 - 7$
12. $2 - 3$
13. $-6 - (-4)$
14. $-9 - (-2)$
15. $-4 - 1$
16. $-5 - 6$
17. $9 - (-5)$
18. $3 - (-4)$
19. $12 - 5$
20. $17 - 9$
21. $-7 - (-14)$
22. $-4 - (-6)$
23. $-9 - 11$
24. $8 - 15$
25. $-12 - (-7)$
26. $-14 - (-4)$
27. $15 - (-6)$
28. $10 - (-6)$
29. $-10 - (-21)$
30. $12 - (-7)$
31. $0 - 5$
32. $0 - (-6)$
33. $3 - 18$
34. $9 - 20$
35. $-12 - (-12)$
36. $-23 - (-23)$
37. $-15 - (-8)$
38. $-16 - (-7)$
39. $-10 - 7$
40. $-3 - 12$

B. Combined Operations

41. To evaluate an expression with more than one operation, we follow the ______.

42. If a sum or difference of three or more numbers is written without grouping symbols, then we perform the operations from ______ to ______.

In Exercises 43–48, evaluate the expression.

43. $3 - 9 + 1 - (-4)$
44. $-5 - 2 + 6 - (-7)$
45. $-6 - 2 - (-6) + 3$
46. $6 - (-3) - 12 + 1$
47. $4 - 9 + 8 - (-8) - 15$
48. $8 - (-2) - 4 - (-10) + 12$

Copyright © Houghton Mifflin Company. All rights reserved.

In Exercises 49–54, perform the indicated operations.

49. $-(3-7)-(2-8)$

50. $4-[-3-(-7)]$

51. $7-5\cdot 3$

52. $-2-5\cdot 4$

53. $-7-3^2$

54. $-1-5(6-2)$

≈ *Estimation*

In Exercises 55–58, select the number in column B that is the best estimate of the difference in column A.

	Column A	*Column B*
55.	$-147-(-152)$	(A) 300
56.	$-510-192$	(B) 0
57.	$-496-(-210)$	(C) -300
58.	$-415-(-686)$	(D) -700

Calculator Exercises

In Exercises 59–62, use a calculator to determine the difference.

59. $-279-(-540)$

60. $386-1{,}255$

61. $-67-342$

62. $436-(-159)$

C. Applications

63. *Temperature Differences* A pilot flew from Chicago, where the temperature was -5°F, to San Antonio, where the temperature was 73°F. Write and evaluate a difference that gives the change in temperature.

64. *Diving Depths* A scuba diver descended from a depth of -30 feet to a depth of -43 feet. Write and evaluate a difference that gives the change in depth.

Copyright © Houghton Mifflin Company. All rights reserved.

65. *Checking Account Balance* The balance in a person's checking account was \$730. Write and evaluate a difference that gives the balance after the person wrote a check for \$875.

66. *Standardized Test Scores* Last year, students in a certain school system scored 18 points below the national average on a standardized test. This year, they scored 5 points above the national average. Write and evaluate a difference that gives the change in the scores.

Bringing It Together

In Exercises 67–76, perform the indicated operations.

67. $-3 - 12$

68. $-12 - (-18)$

69. $-7 + (-4)$

70. $5 + (-11)$

71. $16 - (-4)$

72. $6 - 12$

73. $-16 + 9 - 4$

74. $1 - (-8) + (-8)$

75. $2^3 - 3^2$

76. $3 - 4(7 - 1)$

Writing and Concept Extension

77. If a and b represent integers, what do you know about the values of $a - b$ and $b - a$?

78. Suppose that a represents a negative number and b represents a positive number. Determine the sign of each expression.

(a) $a - b$ **(b)** $b - a$

79. What number is 7 less than -3?

80. How much larger is 7 than -8?

81. Subtract -3 from the difference of 4 and -2.

82. Subtract the sum of -12 and 5 from 6.

Copyright © Houghton Mifflin Company. All rights reserved.

Exploring with Real-World Data: Collaborative Activities

Turnover Differences The following table summarizes the information about turnover differences that was described on the opening page of this chapter. (Source: NFL.com.)

Team	Turnover Difference	Team	Turnover Difference
New York Jets	+18	Tennessee	−4
Denver	+10	Baltimore	−8
Seattle	+6	Buffalo	−14
San Diego	+2	Minnesota	−22

83. What is the difference between the turnover differences for New York and Minnesota?

84. What is the difference between the turnover differences for Buffalo and Minnesota?

85. For which teams is the difference between the turnover differences 6?

86. For which teams is the difference between the turnover differences 10?

87. What is the combined turnover difference for the first three teams in the list?

88. What is the combined turnover difference for the last three teams in the list?

Copyright © Houghton Mifflin Company. All rights reserved.

3.4 Multiplication of Integers

A *The Multiplication Rules for Integers*
B *Products with More Than Two Factors*
C *Properties of Multiplication*
D *Numerical Expressions*

SUGGESTIONS FOR SUCCESS

Are you ready for the test?

If you don't have a test until next week, this might seem like a foolish question. The point is that you should be ready for a test each and every day. Preparing for tests should be an ongoing activity.

After you have completed a day's assignment, leaf back through the material in preceding sections. Review the Quick References. Make sure everything looks familiar and, if not, rework some examples and exercises. Make a note of anything you have forgotten and get some help with it.

Doing this just 15 minutes a day is worth more than hours of cramming the night before the test.

LEARNING TIP

We highly recommend that you go back and review addition and subtraction every day. Each operation has its own rules, and you need to be able to perform these operations quickly and accurately.

A The Multiplication Rules for Integers

You will be pleased to learn that multiplying positive and negative numbers is actually easier than adding them. For addition, we add absolute values when the addends have the same signs, and we subtract absolute values when the addends have different signs. As you will see, multiplication is much more straightforward.

We begin with the product of a positive number and a negative number.

Example 1

Multiply $5(-2)$.

SOLUTION

We can interpret the product $5(-2)$ as repeated addition.

$$5(-2) = (-2) + (-2) + (-2) + (-2) + (-2)$$
$$= -10$$

Your Turn 1

Multiply $3(-7)$.

Answer: -21

Example 2

Multiply $-8(4)$.

SOLUTION

As written, the expression would mean that 4 is to be added -8 times, but that doesn't make any sense. However, we can reverse the order of the factors.

$$4(-8) = (-8) + (-8) + (-8) + (-8) = -32$$

Your Turn 2

Multiply $-6(3)$.

Answer: -18

These examples suggest that the product of a positive number and a negative number is negative. Also, we can find the product without writing the repeated addition.

Copyright © Houghton Mifflin Company. All rights reserved.

The Product of a Positive Number and a Negative Number

To multiply a positive number and a negative number (in either order), multiply the absolute values. The product is *negative.*

Now we consider the product of two negative numbers.

The Product of Two Negative Numbers

We have seen that the product of a positive number and a negative number can be interpreted as repeated addition. For example, the product $5(-3) = (-3) + (-3) + (-3) + (-3) + (-3)$. However, we can't interpret the product of two negative numbers in the same way. For example, interpreting $-5(-3)$ as -3 added -5 times is meaningless. Instead, observe the pattern in the following list.

These factors are decreasing by 1. ↓ These products are increasing by 5. ↓

$$-5(3) = -15$$
$$-5(2) = -10$$
$$-5(1) = -5$$
$$-5(0) = 0$$
$$-5(-1) = \square$$
$$-5(-2) = \square$$
$$-5(-3) = \square$$

As the second factor decreases by 1, the product increases by 5. Continuing the pattern, we can write the last three lines of the list.

$$-5(-1) = 5$$
$$-5(-2) = 10$$
$$-5(-3) = 15$$

These results suggest the following rule for multiplying two negative numbers:

The Product of Two Negative Numbers

To multiply two negative numbers, multiply the absolute values. The product is *positive.*

Example 3

Multiply $-8(-3)$.

SOLUTION

$$-8(-3) = 24$$

Your Turn 3

Multiply $-7(-4)$.

Answer: 28

Copyright © Houghton Mifflin Company. All rights reserved.

If the two factors of a product are both positive or both negative, we say that the factors have *like signs.* If one factor is positive and the other factor is negative, we say that the factors have *unlike signs.* Here is a summary of the procedure for multiplying two integers.

The Product of Two Integers

To multiply two integers, multiply their absolute values.

1. If the factors have like signs, the product is positive.
2. If the factors have unlike signs, the product is negative.

Example 4

Multiply.

(a) $-6(9)$ **(b)** $4(3)$ **(c)** $-2(-8)$ **(d)** $5(-7)$

SOLUTION

(a) $-6(9) = -54$ Unlike signs, negative

(b) $4(3) = 12$ Like signs, positive

(c) $-2(-8) = 16$ Like signs, positive

(d) $5(-7) = -35$ Unlike signs, negative

Your Turn 4

Multiply.

(a) $3(-2)$ **(b)** $-3(-2)$

(c) $-3(2)$ **(d)** $3(2)$

Answers: (a) −6; (b) 6; (c) −6; (d) 6

KEYS TO THE CALCULATOR

The thorough way to use your calculator to multiply is to enter the signs of the factors. Here are the typical keystrokes for $-23(-17)$.

Scientific calculator: **23** [+/−] [×] **17** [+/−] [=]
Graphing calculator: [(−)] **23** [×] [(] [(−)] **17** [)] [ENTER]

```
-23*(-17)
                391
```

Exercises

(a) $42(-85)$ **(b)** $-6(128)$ **(c)** $-19(-54)$ **(d)** $256(-834)$

NOTE When we use a calculator to multiply, we can take advantage of the easy sign rules for multiplication to save keystrokes. For example, we know that the result of $18(-23)$ is negative and the result of $-18(-23)$ is positive. For such products, we can simply key in $18(23)$ and then supply the sign of the result ourselves.

Copyright © Houghton Mifflin Company. All rights reserved.

B Products with More Than Two Factors

When a product has more than two factors, we can follow the Order of Operations and multiply from left to right.

$$-3(2)(-1)(-4) = -6(-1)(-4) \qquad -3(2) = -6$$
$$= 6(-4) \qquad -6(-1) = 6$$
$$= -24 \qquad 6(-4) = -24$$

An easier approach is to use the following procedure.

Products with More Than Two Factors

To find the product of more than two factors, follow these steps.

1. Multiply the absolute values of all the factors.
2. Count the number of negative factors.
   (a) If the number of negative factors is even, then the product is positive.
   (b) If the number of negative factors is odd, then the product is negative.

Example 5

Multiply $-3(2)(-1)(-4)$.

SOLUTION

Multiply the absolute values of all the factors.

$$3 \cdot 2 \cdot 1 \cdot 4 = 24$$

Three of the factors are negative. Because the number of negative factors is odd, the product is negative.

$$-3(2)(-1)(-4) = -24$$

Your Turn 5

Multiply $-5(-3)(-1)(-2)$.

Answer: 30

Recall that 5^3 is an example of an *exponential expression* that means (5)(5)(5). The *base* is 5, and the *exponent* is 3. The base can also be a negative number, but the meaning is still the same.

Example 6

Evaluate $(-5)^3$.

SOLUTION

$$(-5)^3 = (-5)(-5)(-5) = -125$$

Your Turn 6

Evaluate $(-2)^4$.

Answer: 16

NOTE There is a subtle difference between $(-5)^2$ and $-(5^2)$. The expression $(-5)^2$ means $(-5)(-5) = 25$, whereas the expression $-(5^2)$ means that we square 5 and then take the opposite of the result: $-(5^2) = -(25) = -25$. The Order of Operations gives us different results because of the grouping symbols.

Copyright © Houghton Mifflin Company. All rights reserved.

C Properties of Multiplication

All of the properties of multiplication apply to the products of integers. Here is a list of those properties.

> ***Properties of Multiplication***
>
> Let a, b, and c represent any numbers.
>
> 1. $ab = ba$ — The Commutative Property of Multiplication
> 2. $(ab)c = a(bc)$ — The Associative Property of Multiplication
> 3. $a \cdot 0 = 0 \cdot a = 0$ — The Multiplication Property of 0
> 4. $a \cdot 1 = 1 \cdot a = a$ — The Multiplication Property of 1

Another important property of multiplication is suggested by products such as the following:

$$-1(4) = -4 \qquad -4 \text{ is the opposite of } 4.$$
$$-1(-6) = 6 \qquad 6 \text{ is the opposite of } -6.$$

Observe that multiplying a number by -1 changes the number to its opposite.

> ***The Multiplication Property of -1***
>
> If a represents any number, then $-1 \cdot a = -a$.

The Multiplication Property of -1 gives us another way to write the opposite of a number.

Example 7

Use the Multiplication Property of -1 to write the given expression in another way.

(a) $-x$ **(b)** $-1xy$

SOLUTION

(a) $-x = -1x$

(b) $-1xy = -xy$

Your Turn 7

Use the Multiplication Property of -1 to write the given expression in another way.

(a) $-1c$ **(b)** $-y$

Answers: (a) $-c$; (b) $-1y$

NOTE Mathematicians tend not to write understood 1s. We write x rather than $1x$, and we write $-x$ rather than $-1x$. However, there are times when writing the 1s improves clarity. The Multiplication Properties of 1 and -1 allow us to put in the 1s if we want to.

Copyright © Houghton Mifflin Company. All rights reserved.

D Numerical Expressions

As always, we use the Order of Operations to evaluate a numerical expression.

For your convenience, here is a review of the Order of Operations.

The Order of Operations

To evaluate a numerical expression, perform the indicated operations in the following order:

1. If the expression has grouping symbols, such as parentheses ( ), brackets [ ], or braces { }, then begin by performing all operations inside the grouping symbols.
2. Next, evaluate numbers that are in exponential form.
3. Working from left to right, perform all multiplications and divisions.
4. Working from left to right, perform all additions and subtractions.

Example 8

Evaluate each expression.

(a) $-5(-2) + 3(-6)$

(b) $5 - 2(-3)^2$

(c) $2^3(4 - 7)$

SOLUTION

(a) $-5(-2) + 3(-6)$

$= 10 + (-18)$ Multiplication before addition

$= -8$

(b) $5 - 2(-3)^2$

$= 5 - 2(9)$ Exponent first

$= 5 - 18$ Then multiplication

$= 5 + (-18)$ Definition of subtraction

$= -13$

(c) $2^3(4 - 7)$

$= 2^3[4 + (-7)]$ Definition of subtraction

$= 2^3(-3)$ Grouping symbols first

$= 8(-3)$ Then the exponent

$= -24$ Finally, the product

Your Turn 8

Evaluate each expression.

(a) $8 - (-2)(5)$

(b) $-3[5 + (-4)(2)]$

(c) $3(-2)^2 + 5(-4)$

Answers: (a) 18; (b) 9; (c) −8

Copyright © Houghton Mifflin Company. All rights reserved.

3.4 Quick Reference

A. The Multiplication Rules for Integers

1. To multiply two integers, multiply their absolute values.

(a) If the factors have like signs, then the product is positive.

$(7)(4) = 28$
$(-3)(-5) = 15$

(b) If the factors have unlike signs, then the product is negative.

$(-2)(8) = -16$
$(9)(-6) = -54$

B. Products with More Than Two Factors

1. To find the product of more than two factors, follow these steps.

(a) Multiply the absolute values of all the factors.

(b) Count the number of negative factors.

(i) If the number of negative factors is even, then the product is positive.

$-2(-3)5(-1)(-1) = 30$

(ii) If the number of negative factors is odd, then the product is negative.

$3(-4)(-1)(2)(-1) = -24$

2. The base of an exponential expression can be a negative number.

$(-2)^3 = -8$
$(-2)^4 = 16$

C. Properties of Multiplication

1. All of the properties of multiplication that were stated for whole numbers also apply to the multiplication of integers.

(a) Commutative Property of Multiplication — $2(-3) = -3(2)$

(b) Associative Property of Multiplication — $5[2(-4)] = [5 \cdot 2](-4)$

(c) Multiplication Property of 0 — $-6 \cdot 0 = 0 \cdot (-6) = 0$

(d) Multiplication Property of 1 — $-7 \cdot 1 = 1 \cdot (-7) = -7$

2. The Multiplication Property of -1

If a represents any number, then

$$-1 \cdot a = -a$$

In words, multiplying a number by -1 changes the number to its opposite.

$-1x = -x$
$-ac = -1ac$

Copyright © Houghton Mifflin Company. All rights reserved.

D. Numerical Expressions

1. We use the Order of Operations to evaluate a numerical expression.

$2(4 - 7)^2$
$= 2(-3)^2$
$= 2(9)$
$= 18$

Copyright © Houghton Mifflin Company. All rights reserved.

3.4 Exercises

A. The Multiplication Rules for Integers

1. If two numbers have like signs, then their product is ______.

2. The product of two numbers with ______ signs is negative.

In Exercises 3–28, multiply.

3. $6(-5)$

4. $9(-3)$

5. $(-9)(-7)$

6. $(-4)(-3)$

7. $8(5)$

8. $6(2)$

9. $-8(7)$

10. $-3(6)$

11. $-7(0)$

12. $0(-3)$

13. $(-8)(-6)$

14. $(-7)(-5)$

15. $100(-5)$

16. $19(-10)$

17. $12 \cdot 10$

18. $9 \cdot 100$

19. $-4(9)$

20. $-5(8)$

21. $(-10)(-7)$

22. $(-12)(-2)$

23. $0(-31)$

24. $-27(0)$

25. $2(-8)$

26. $7(-2)$

27. $-40(7)$

28. $-5(90)$

B. Products with More Than Two Factors

29. If the number of negative factors in a product is ______, then the product is positive.

30. If the number of negative factors in a product is odd, then the product is ______.

In Exercises 31–44, determine the product.

31. $-6(2)(-3)$

32. $(-4)(-2)(-5)$

33. $2(-4)(3)$

34. $-2(-3)(4)$

35. $(-1)(-4)(3)(-5)$

36. $(-2)(-6)(-3)(-4)$

37. $(-2)(-3)(-1)(-8)$

38. $2(-4)(-5)(-2)$

39. $(-2)^5$

40. $(-8)^2$

41. $(-3)^3$

42. $(-1)^8$

43. $(-1)^{12}$

44. $(-5)^3$

Copyright © Houghton Mifflin Company. All rights reserved.

C. Properties of Multiplication

45. The result of multiplying a number by -1 is the _______ of the number.

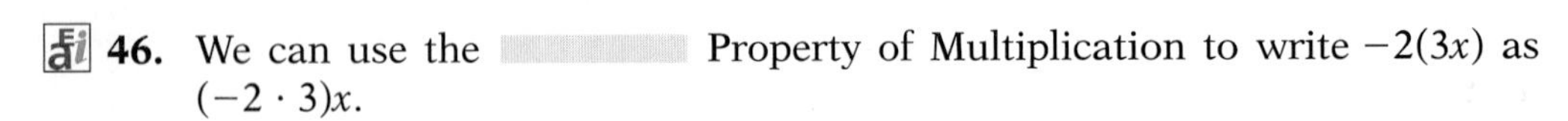

46. We can use the _______ Property of Multiplication to write $-2(3x)$ as $(-2 \cdot 3)x$.

In Exercises 47–50, use the Multiplication Property of -1 to write the given expression in another way.

47. $-x$

48. $-1y$

49. $-1a$

50. $-c$

D. Numerical Expressions

51. To evaluate a numerical expression, we begin by performing operations within _______ symbols.

52. To evaluate $-4 - 2(-3)$, the _______ tells us to perform the multiplication first.

In Exercises 53–64, use the Order of Operations to evaluate each expression.

53. $-5(9 - 15)$

54. $10(-8 - 6)$

55. $(5 - 7)(8 - 2)$

56. $6(1 - 8)(5 - 6)$

57. $-12 + 3(-6)$

58. $-6(-3) - 4(-1)$

59. $-6 - 3(4 - 9)$

60. $-2 + 5(-2 - 7)$

61. $-3(9 - 6) + 5(1 - 8)$

62. $3(-2)^3 + 12$

63. $2(7 - 5)^2 + 1$

64. $(4 - 3)^3 - (3 - 4)^2$

Copyright © Houghton Mifflin Company. All rights reserved.

≈ Estimation

In Exercises 65–68, select the number in column B that is the best estimate of the product in column A.

	Column A	*Column B*
65.	$-43(-88)$	(A) $-3{,}600$
66.	$357(-11)$	(B) $-2{,}000$
67.	$-98(-18)$	(C) $2{,}000$
68.	$-38(51)$	(D) $3{,}600$

Calculator Exercises

In Exercises 69–72, use a calculator to determine the product.

69. $(-799)(-64)$

70. $3{,}230(-59)$

71. $384(508 - 1{,}236)$

72. $-46(-87 - 42)$

Bringing It Together

In Exercises 73–82, perform the indicated operations.

73. $(-14)(-100)$

74. $-4(11)$

75. $(-4)^3$

76. $(-2)^6$

77. $7 - 2(-3)^2$

78. $4(8 - 2) - 6(-2 - 3)$

79. $-4 - 11$

80. $-6 + (-8)$

81. $15 + (-7)$

82. $10 - (-40)$

Writing and Concept Extension

83. Suppose that the product of two integers is 0. What do you know about one of the numbers?

84. Suppose that m represents a negative integer and n represents a positive integer. What is the sign of (a) mn and of (b) $(-m)n$?

Copyright © Houghton Mifflin Company. All rights reserved.

In Exercises 85–88, find a pair of numbers whose product is the first given number and whose sum is the second given number.

85. $-8, -2$

86. $10, -7$

87. $3, -4$

88. $-12, -4$

Copyright © Houghton Mifflin Company. All rights reserved.

3.5 Division of Integers

A *The Division Rules for Integers*
B *Numerical Expressions*
C *Applications*

SUGGESTIONS FOR SUCCESS

Are you a tree person or a forest person? A tree person looks at each of the four operations as different and separate topics. A forest person has a wider view and sees the relationships among the operations.

Just as every subtraction involving integers can be written as a related addition, we will see in this section that every division involving integers can be written as a related multiplication. Your knowledge of how the operations are related will greatly increase your overall understanding. Try to be a forest person.

A The Division Rules for Integers

Recall that in $8 \div 4 = 2$, for example, we call 8 the *dividend*, we call 4 the *divisor*, and we call 2 the *quotient*.

When we divide with whole numbers, we can check our answers with multiplication.

$$15 \div 5 = 3 \quad \text{because} \quad 3 \cdot 5 = 15$$
$$\frac{18}{2} = 9 \quad \text{because} \quad 9 \cdot 2 = 18$$

We can use this relationship between division and multiplication to discover the rules for dividing integers.

Dividing Integers

In each of the following experiments, we find the quotient of two integers by writing the related multiplication.

(a) $\frac{-12}{3} = \square$ means that $\square \cdot 3 = -12$.

Now we ask, "What number multiplied by 3 gives us -12?" The answer is -4. Therefore, $\frac{-12}{3} = -4$.

(b) $\frac{10}{-2} = \square$ means that $\square \cdot (-2) = 10$.

Now we ask, "What number multiplied by -2 gives us 10?" The answer is -5. Therefore, $\frac{10}{-2} = -5$.

(c) $\frac{-24}{-8} = \square$ means that $\square \cdot (-8) = -24$.

Now we ask, "What number multiplied by -8 gives us -24?" The answer is 3. Therefore, $\frac{-24}{-8} = 3$.

In all three parts, we see that the absolute values of the answers could have been found by dividing the absolute values of the dividend and the divisor.

In parts (a) and (b), the dividend and the divisor have unlike signs, and the quotient in each case is negative. In part (c), the dividend and the divisor have like signs and the quotient is positive.

In short, the sign rules for division are identical to the sign rules for multiplication.

Copyright © Houghton Mifflin Company. All rights reserved.

The Quotient of Two Integers

To divide two integers, divide their absolute values.

1. If the integers have like signs, the quotient is positive.
2. If the integers have unlike signs, the quotient is negative.

Example 1

Divide.

(a) $-15/3$

(b) $\frac{-18}{-6}$

(c) $21 \div (-7)$

SOLUTION

(a) $-15/3 = -5$ — Unlike signs, negative

(b) $\frac{-18}{-6} = 3$ — Like signs, positive

(c) $21 \div (-7) = -3$ — Unlike signs, negative

Your Turn 1

Divide.

(a) $\frac{27}{-3}$

(b) $\frac{-32}{4}$

(c) $-54 \div (-9)$

Answers: (a) −9; (b) −8; (c) 6

Keep in mind that we can't ever divide by 0. For example, the value of $\frac{0}{-4}$ is 0, but $-9 \div 0$ is undefined.

KEYS TO THE CALCULATOR

If your calculator uses the (−) key, enclosing negative numbers in parentheses is optional but recommended. Here are typical keystrokes for performing $\frac{-12}{-4}$.

Scientific calculator: **12** +/− ÷ **4** +/− =
Graphing calculator: ((−) **12**) ÷ ((−) **4**) ENTER

```
(-12)/(-4)
                 3
```

Exercises

(a) $-345 \div 23$ **(b)** $\frac{-988}{-52}$ **(c)** $\frac{0}{-100}$ **(d)** $\frac{-78}{0}$ **(e)** $\frac{144}{-12}$

Copyright © Houghton Mifflin Company. All rights reserved.

B Numerical Expressions

When we evaluate numerical expressions, the Order of Operations places grouping symbols and exponents ahead of division, which ranks equally with multiplication. A fraction bar indicates division, but it also serves as a grouping symbol.

$$\frac{16-2}{4+3} \quad \text{means} \quad \frac{(16-2)}{(4+3)}$$

Example 2

Evaluate $\frac{16-2}{4+3}$.

SOLUTION

$$\frac{16-2}{4+3} = \frac{14}{7} = 2$$

Your Turn 2

Evaluate $\frac{2(3)}{7-1}$.

Answer: 1

Example 3

Evaluate $\frac{3+4(-6)}{-2(-5)-3}$.

SOLUTION

We begin by evaluating the expressions (groups) above and below the fraction bar.

$\frac{3+4(-6)}{-2(-5)-3} = \frac{3+(-24)}{10-3}$	Multiplication first
$= \frac{-21}{7}$	Then addition and subtraction
$= -3$	Finally, the division

Your Turn 3

Evaluate $\frac{-3(-1)-4(2)}{-15 \div 3}$.

Answer: 1

Example 4

Evaluate $\frac{(-3)^2+7}{2^3-4}$.

SOLUTION

In the groups above and below the fraction bar, the exponents have the highest priority.

$$\frac{(-3)^2+7}{2^3-4} = \frac{9+7}{8-4} = \frac{16}{4} = 4$$

Your Turn 4

Evaluate $\frac{6-4^2}{2(-1)}$.

Answer: 5

C Applications

In Section 2.5, we said that the *average* of a list of numbers can be found by adding the numbers and then dividing the sum by the number of numbers in the list. Averages can be found in this way even when some or all of the numbers are negative.

Copyright © Houghton Mifflin Company. All rights reserved.

Example 5

During five particularly cold winter days in Boston, the low temperatures were as shown in the bar graph. (Note that the bars for the negative temperatures are below the horizontal axis.)

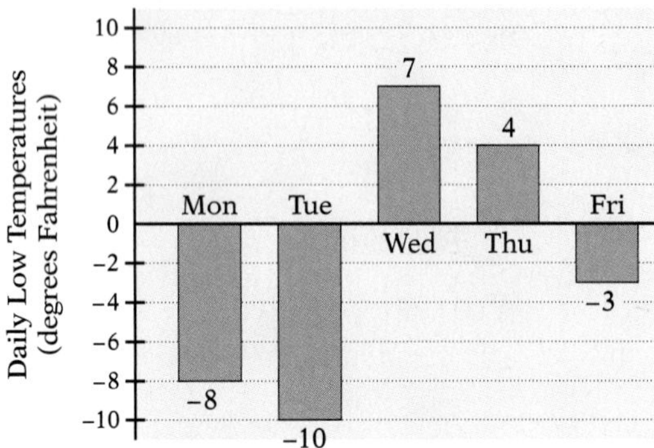

What was the average low temperature for this 5-day period?

SOLUTION

$$\frac{-8 + (-10) + 7 + 4 + (-3)}{5} = \frac{-21 + 11}{5}$$
$$= \frac{-10}{5}$$
$$= -2$$

The average low temperature was $-2°$F.

Your Turn 5

Suppose that each of the temperatures shown in the bar graph had been 1°F higher. What would the average low temperature have been?

Answer: $-1°$F

Example 6

An instructor grades quizzes by using a negative number to indicate the number of incorrect answers. Suppose that 8 students received the following grades.

$$-10, 0, -1, -2, -3, -2, -1, -5$$

What was the average score for these students?

SOLUTION

$$\frac{-10 + 0 + (-1) + (-2) + (-3) + (-2) + (-1) + (-5)}{8}$$
$$= \frac{-24}{8} = -3$$

The average score was -3.

Your Turn 6

What would the average score have been if the first student had not taken the quiz?

Answer: -2

NOTE Applications dealing with averages typically involve sums of many numbers. In Example 6, all the numbers (except 0) are negative, so the sum is easily found. However, in Example 5, we have a mix of positive and negative numbers. Remember that grouping the positive numbers and negative numbers, as we did in the example, helps in finding the sum.

Copyright © Houghton Mifflin Company. All rights reserved.

Sometimes we can most efficiently find averages by using multiplication as well as addition and division.

In golf, *par* refers to the number of strokes that a golfer should take on a given hole. A *birdie* is one stroke less than (under) par, and a *bogey* is one stroke more than (over) par. An *eagle* is two strokes less than (under) par, and a *double bogey* is two strokes more than (over) par.

Instead of recording the actual number of strokes on a given hole, golfers sometimes simply code their score in relation to par by using the following scheme:

Score	Code
Eagle	−2
Birdie	−1
Par	0
Bogey	+1
Double bogey	+2

Refer to this table for Example 7.

Example 7

The 40 golfers in a tournament had the following scores on the 18th hole.

Number of golfers	Score
2	Eagle
10	Birdie
18	Par
6	Bogey
4	Double bogey

Use coded scores to determine the average score for this hole.

SOLUTION

We multiply each coded score by the number of golfers who had that score.

Eagle: $2(-2) = -4$
Birdie: $10(-1) = -10$
Par: $18(0) = 0$
Bogey: $6(+1) = +6$
Double bogey: $4(+2) = +8$

To find the average score for the hole, we add these products and divide the sum by the total number of golfers.

$$\frac{[-4 + (-10) + 0 + 6 + 8]}{40}$$
$$= \frac{[-14 + 14]}{40}$$
$$= \frac{0}{40}$$
$$= 0$$

The average score for the hole was exactly par.

Your Turn 7

Rework Example 7 for 4 eagles, 6 birdies, 18 pars, 10 bogeys, and 2 double bogeys.

Answer: exactly par

Copyright © Houghton Mifflin Company. All rights reserved.

3.5 Quick Reference

A. The Division Rules for Integers

1. To divide two integers, divide their absolute values.

(a) If the integers have like signs, the quotient is positive.

$$-16 \div (-2) = 8$$

(b) If the integers have unlike signs, the quotient is negative.

$$\frac{-20}{5} = -4$$

$$\frac{27}{-3} = -9$$

B. Numerical Expressions

1. A fraction bar indicates division, but it is also a grouping symbol.

$$\frac{2(3-5)}{3^2-7} = \frac{2(-2)}{3^2-7}$$

$$= \frac{2(-2)}{9-7}$$

$$= \frac{-4}{9-7}$$

$$= \frac{-4}{2}$$

$$= -2$$

Copyright © Houghton Mifflin Company. All rights reserved.

3.5 Exercises

A. The Division Rules for Integers

1. The sign rules for division are the same as the sign rules for ______.

2. If b is any number except 0, then $b \div 0$ is ______, whereas $0 \div b$ is ______.

In Exercises 3–26, divide.

3. $-24 \div (-6)$
4. $-54 \div (-9)$
5. $\frac{15}{-3}$
6. $\frac{40}{-5}$
7. $\frac{-25}{5}$
8. $\frac{-56}{8}$
9. $24 \div (-2)$
10. $32 \div (-8)$
11. $-7 \div 0$
12. $0 \div (-3)$
13. $\frac{-30}{-10}$
14. $\frac{-21}{-3}$
15. $70 \div (-7)$
16. $18 \div (-9)$
17. $\frac{-12}{-12}$
18. $\frac{-4}{-4}$
19. $-48 \div 6$
20. $-14 \div 7$
21. $\frac{0}{-31}$
22. $\frac{-27}{0}$
23. $\frac{-28}{7}$
24. $\frac{-42}{6}$
25. $-35 \div (-5)$
26. $-24 \div (-3)$

B. Numerical Expressions

27. In the expression $\frac{2-7}{5}$, we regard the fraction bar as a ______ symbol.

28. In an expression involving only multiplication and division and no grouping symbols, the operations are performed from ______ to ______.

In Exercises 29–42, use the Order of Operations to evaluate each expression.

29. $\frac{4(-9)}{-36}$
30. $\frac{-5(-6)}{-10}$
31. $\frac{-3(-8)}{-2-4}$
32. $\frac{77-17}{6(-5)}$
33. $\frac{-2-10}{-7+3}$
34. $\frac{7-3}{9-11}$
35. $\frac{5+7(-2)}{-9}$
36. $\frac{3(-7)}{-4(-3)-5}$
37. $\frac{-2(5)+3(-4)}{12-1}$
38. $\frac{(-2)^2-6(8)}{2(7)+3(-6)}$
39. $-8 \div (4 \div 2)$
40. $-9 \cdot 8 \div (4 \cdot 9)$
41. $3^2 + 24 \div (-4)$
42. $\frac{7^2-1}{1-3^2}$

Copyright © Houghton Mifflin Company. All rights reserved.

≈ Estimation

In Exercises 43–46, select the number in column B that is the best estimate of the quotient in column A.

	Column A	*Column B*
43.	$-73 \div (-68)$	(A) -10
44.	$-1{,}490 \div (-152)$	(B) -1
45.	$-157 \div 160$	(C) 1
46.	$693 \div (-71)$	(D) 10

Calculator Exercises

In Exercises 47–50, use a calculator to determine the quotient.

47. $-55{,}120 \div (-848)$ **48.** $-24{,}120 \div 24$ **49.** $8{,}370 \div (-18)$ **50.** $-95{,}056 \div (-457)$

C. Applications

Record Low Temperatures The table gives the record low temperatures (in degrees Fahrenheit) for three cities for the three winter months. (Source: U.S. National Oceanic and Atmospheric Administration.) Refer to this table as you do Exercises 51–54.

	December	*January*	*February*
Atlanta	0	−8	5
Norfolk	−3	8	7
Raleigh	4	−9	5

51. What is the average record low winter temperature for Atlanta?

52. For which city is the average record low winter temperature a positive number? What is the average?

53. For which city is the average record low winter temperature 0°F?

54. What is the average January record low for the three cities?

Copyright © Houghton Mifflin Company. All rights reserved.

55. *Stock Market* During one week, the stock market recorded the following daily changes: -120, 90, 135, -98, and -147. What was the average daily change for the week?

56. *Weight Change* A weight-loss clinic recorded the following monthly weight changes for 8 clients: -5, -3, 4, -2, 6, -9, -8, and -7. For these 8 people, what was the average weight change for the month?

57. *Quiz Grades* An instructor grades 10-question quizzes by using a negative number for the number of incorrect answers. If a student's quiz grades were -3, -2, -1, -1, and -3, what was the average number of *correct* answers?

58. *Budget Accounting* An accountant uses negative numbers to indicate the amount by which expenses were over the budgeted amount. For the past 3 months, the accountant's report shows -56, -123, and -31 for a certain budget item. By what average amount were expenses over the budget for these 3 months?

Bringing It Together

In Exercises 59–68, perform the indicated operations.

59. $\dfrac{(-2)^3 + 5}{3^2 - 2(3)}$ **60.** $-7 + (-4)^3 \div (-8)$ **61.** $-18 + 6$ **62.** $-18 \div 6$

63. $(-7)(-3)$ **64.** $-7 - (-3)$ **65.** $\dfrac{-28}{-4}$ **66.** $8(-4)$

67. $14 - 20$ **68.** $-14 + (-20)$

Writing and Concept Extension

69. Explain how to check the division $24 \div 4 = 6$.

70. Suppose that the quotient $\frac{a}{b}$ is 0. What do you know about the two numbers?

71. Subtract -5 from the quotient of 28 and -4.

Copyright © Houghton Mifflin Company. All rights reserved.

72. What is 36 divided by the sum of -5 and 1?

73. Suppose that m represents a negative integer and n represents a positive integer. What is the sign of each expression?

(a) $\frac{m}{n}$ **(b)** $\frac{-m}{n}$

Exploring with Real-World Data: Collaborative Activities

Elevations Geographical elevations are usually given in feet above or below sea level. The table shows the highest and lowest elevations (in feet) for five continents. (Source: *World Almanac.*)

	Highest	*Lowest*
Africa	19,340	–512
Asia	29,028	–1312
Europe	18,510	–92
North America	20,320	–282
South America	22,834	–131

74. What is the largest difference between the highest and lowest elevations of a continent?

75. What is the smallest difference between the highest and lowest elevations of a continent?

76. What is the difference between the lowest elevations of South America and North America?

77. What is the difference between the lowest elevations of Europe and Africa?

78. What is the average of the highest and lowest elevations for Asia?

79. What is the average of the highest and lowest elevations for Europe?

80. To the nearest ten feet, what is the average of the lowest elevations for the five continents?

81. To the nearest thousand feet, what is the average of the highest elevations for the five continents?

Copyright © Houghton Mifflin Company. All rights reserved.

3.6 Expressions and Equations

A *Evaluating Expressions*
B *Equations*

SUGGESTIONS FOR SUCCESS

When asked what algebra is, a friend said, "I don't know, but it has something to do with x."

Unlike you, this person probably never understood that x, or any other variable, is simply a symbol that represents some number. In fact, if you say "some number" rather than "x," you may find that what you are doing seems less complicated.

For example, try reading the equation $3x = 12$ as "three times some number is equal to twelve." Say it aloud and think about what you have just said. You may find that you can understand the words better than the symbols.

Eventually, you will see that the symbols are nothing more than shorthand, and you will be comfortable with them.

A *Evaluating Expressions*

Recall that we *evaluate* (find the value of) a variable expression by replacing each occurrence of the variable with a specific value. In this section, we extend our ability to evaluate an expression by using integer values for the variables.

The addition and subtraction rules for integers may be needed to evaluate a variable expression.

Example 1

Evaluate $x + c + (-8)$ for $x = -2$ and $c = 7$.

SOLUTION

$$\begin{aligned} x + c + (-8) &= -2 + 7 + (-8) && x = -2 \text{ and } c = 7 \\ &= 5 + (-8) \\ &= -3 \end{aligned}$$

Your Turn 1

Evaluate $a + (-4) + b$ for $a = -3$ and $b = -10$.

Answer: −17

Example 2

Evaluate $-x - y + 6$ for $x = -2$ and $y = 3$.

SOLUTION

$$\begin{aligned} -x - y + 6 &= -(-2) - 3 + 6 && x = -2 \text{ and } y = 3 \\ &= 2 - 3 + 6 && -(-2) = 2 \\ &= 2 + (-3) + 6 \\ &= 8 + (-3) \\ &= 5 \end{aligned}$$

Your Turn 2

Evaluate $a - 1 - b$ for $a = -6$ and $b = 4$.

Answer: −11

Evaluating a variable expression may also involve multiplication and division.

Copyright © Houghton Mifflin Company. All rights reserved.

Example 3

Evaluate $-2xy$ for $x = 3$ and $y = -4$.

SOLUTION

$-2xy = -2(3)(-4)$ $x = 3$ and $y = -4$

$= 24$ Even number of negative factors

Your Turn 3

Evaluate $5ab$ for $a = -2$ and $b = -3$.

Answer: 30

Example 4

Evaluate $\frac{2x}{y}$ for $x = -9$ and $y = -3$.

SOLUTION

$\frac{2x}{y} = \frac{2(-9)}{-3}$ $x = -9$ and $y = -3$

$= \frac{-18}{-3}$

$= 6$

Your Turn 4

Evaluate $\frac{a}{3b}$ for $a = 12$ and $b = -1$.

Answer: −4

Recall that in the Order of Operations, exponents rank higher than multiplication.

Example 5

Evaluate $4x^2$ for $x = -3$.

SOLUTION

$4x^2 = 4(-3)^2$ Note that the base is in parentheses.

$= 4(9)$ Exponent first: $(-3)^2 = (-3)(-3) = 9$

$= 36$ Then multiplication

Your Turn 5

Evaluate $2a^4$ for $a = -3$.

Answer: 162

We use the Order of Operations to evaluate expressions that involve opposites and absolute values. For the purpose of evaluating, absolute value symbols are also grouping symbols.

Example 6

Evaluate the expressions for $x = -3$.

(a) $|x|$ **(b)** $|-x|$ **(c)** $-(-x)$

SOLUTION

(a) $|x| = |-3| = 3$

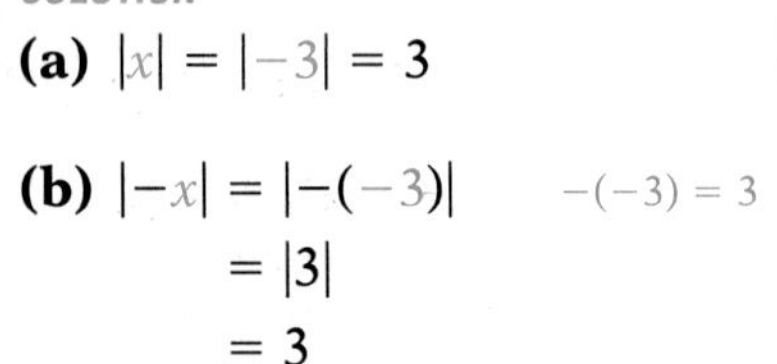

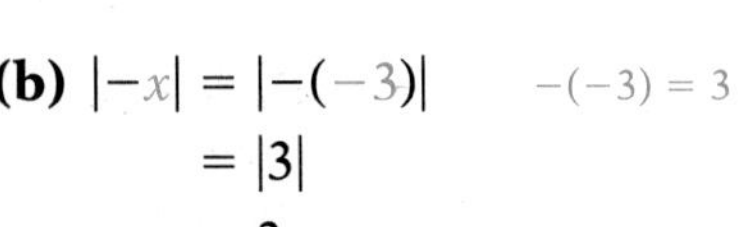

(b) $|-x| = |-(-3)|$ $-(-3) = 3$

$= |3|$

$= 3$

Your Turn 6

Evaluate the expressions for $c = -2$.

(a) $-|c|$

(b) $-(-c)$

(c) $-|-c|$

Copyright © Houghton Mifflin Company. All rights reserved.

(c) $-(-x) = -[-(-3)]$ Start inside the brackets.

$= -[3]$ $-(-3) = 3$

$= -3$

Answers: (a) −2; (b) −2; (c) −2

Example 7

Evaluate $-|-a + b|$ for $a = -2$ and $b = 5$.

SOLUTION

Because the absolute value symbols are grouping symbols, we begin by performing all operations inside the vertical bars.

$$\begin{aligned} -|-a + b| &= -|-(-2) + 5| && a = -2 \text{ and } b = 5 \\ &= -|2 + 5| && -(-2) = 2 \\ &= -|7| \\ &= -7 \end{aligned}$$

Your Turn 7

Evaluate $|-a| + |b|$ for $a = 4$ and $b = -1$.

Answer: 5

B *Equations*

Recall that solving an equation means finding a replacement for the variable that makes the equation true.

Example 8

Determine which of the given numbers is a solution of the equation $-2x + 5 = -11$.

(a) -3 **(b)** 8

SOLUTION

(a)

$-2x + 5$	-11
$-2(-3) + 5$	-11
$6 + 5$	-11
11	-11

When $x = -3$, the left and right sides of the equation do not have the same value, so -3 is not a solution.

(b)

$-2x + 5$	-11
$-2(8) + 5$	-11
$-16 + 5$	-11
-11	-11

When $x = 8$, the left and right sides of the equation do have the same value. Therefore, 8 is a solution.

Your Turn 8

Determine which of the given numbers is a solution of the equation $-5x + 3 = 13$.

(a) -2 **(b)** -4

Answers: (a) Yes; (b) No

Some equations involving opposites can be solved by simplifying. In other cases, such as $-x = 3$, you can ask this question: If the opposite of a number is 3, then what is the number?

Copyright © Houghton Mifflin Company. All rights reserved.

Example 9

Solve the given equation.

(a) $-(-x) = -2$ **(b)** $-x = 5$

SOLUTION

(a) Because $-(-x) = x$, we can write the equation as $x = -2$. Therefore, the solution is -2.

(b) If the opposite of x is 5, then x must be -5. The solution is -5.

Your Turn 9

Solve the given equation.

(a) $-y = -6$

(b) $-|3| = -(-a)$

Answers: (a) 6; (b) −3

Example 10

Solve $-3 + x = 4$.

SOLUTION

We ask, "What number added to -3 gives us 4?"

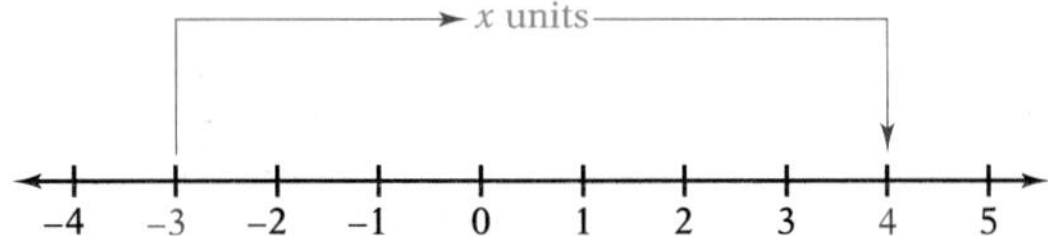

Starting at -3 on the number line, we see that we must move right 7 units to end at 4. We can check that 7 is the solution: $-3 + 7 = 4$.

Your Turn 10

Solve $x + (-2) = -5$.

Answer: −3

Certain simple equations involving subtraction can be solved by using the fact that if $a - b = c$, then $c + b = a$.

Example 11

Solve $x - 4 = -12$.

SOLUTION

$x - 4 = -12$ means that $-12 + 4 = x$. Therefore, $x = -8$.

We can check by replacing x with -8.

$$-8 - 4 = -12$$

Your Turn 11

Solve $x - (-3) = 7$.

Answer: 4

We use the multiplication rules to solve simple equations involving products.

Example 12

Solve $3x = -15$.

SOLUTION

We ask, "What number multiplied by 3 gives us -15?" The solution is -5 because $3(-5) = -15$.

Your Turn 12

Solve $-2x = -12$.

Answer: 6

Copyright © Houghton Mifflin Company. All rights reserved.

Simple equations involving division can be solved by using the related multiplication.

Example 13

Solve $\frac{x}{-2} = -5$.

SOLUTION

$\frac{x}{-2} = -5$ means that $-5(-2) = x$. Therefore, the solution is 10.

Your Turn 13

Solve $\frac{x}{-6} = 3$.

Answer: −18

Example 14

Solve $\frac{20}{x} = -4$.

SOLUTION

$\frac{20}{x} = -4$ means that $-4x = 20$. Because $-4(-5) = 20$, the solution is -5.

Your Turn 14

Solve $\frac{-32}{x} = -8$.

Answer: 4

3.6 Quick Reference

A. Evaluating Expressions

1. We evaluate variable expressions that involve addition in the usual way.

When $x = -5$, the expression

$9 + x = 9 + (-5) = 4$.

2. You may need to apply the definition of subtraction when you evaluate a variable expression.

If $x = -4$ and $y = 3$, then

$$\begin{aligned} 8 - x - y &= 8 - (-4) - 3 \\ &= 8 + 4 + (-3) \\ &= 12 + (-3) \\ &= 9 \end{aligned}$$

3. Evaluating a variable expression may involve multiplication.

When $x = -3$ and $y = 2$,

$-3xy = -3(-3)(2) = 18$.

4. Evaluating a variable expression may involve using the division rules.

Evaluate $\frac{-3x}{a}$ for $x = -4$ and $a = 6$.

$$\frac{-3x}{a} = \frac{-3(-4)}{6} = \frac{12}{6} = 2$$

Copyright © Houghton Mifflin Company. All rights reserved.

5. Expressions to be evaluated may contain opposites and absolute value symbols, which are grouping symbols.

If $x = -6$, then

$$|x| = |-6| = 6$$
$$|3 + (-x)| = |3 + [-(-6)]| = |3 + 6| = 9$$

B. Equations

1. *Solving* an equation means finding a replacement for the variable that makes the equation true.

We test whether -1 is a solution of $3x + 5 = 2$:

$$3x + 5 = 3(-1) + 5 = -3 + 5 = 2$$

Thus, -1 is a solution.

2. Equations that contain opposites can sometimes be solved by simplifying.

$$-(-x) + 4 = 6$$
$$x + 4 = 6$$
$$x = 2$$

3. For other equations, if we know the opposite of a number, then we can find the number.

$$-x = 8$$

If the opposite of x is 8, then x must be -8.

4. You can use the number line and your knowledge of the addition rules to solve equations that involve addition.

Solve $x + (-4) = 1$.

Starting at -4, we move 5 units to the right to reach 1. Therefore, the unknown number x is 5.

5. Some equations can be solved by using the fact that if $a - b = c$, then $c + b = a$.

If $x - 6 = -2$, then $-2 + 6 = x$, or $x = 4$. The solution of the equation $x - 6 = -2$ is 4.

6. The multiplication rules may be needed to solve equations.

We solve $-2x = 10$ by asking, "What number multiplied by -2 is 10?" The solution is -5.

7. Simple equations involving division can be solved by using the related multiplication.

$\frac{y}{-3} = 6$ means than $6(-3) = y$. Therefore, the solution is -18.

Copyright © Houghton Mifflin Company. All rights reserved.

3.6 Exercises

A. Evaluating Expressions

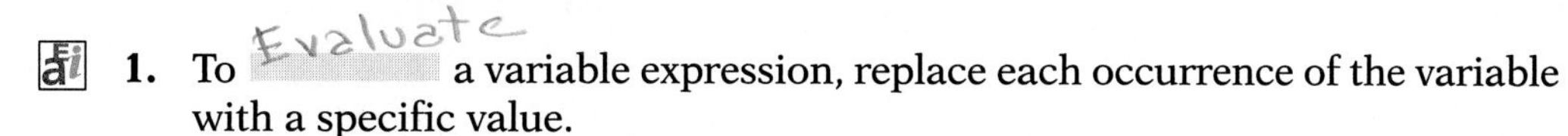

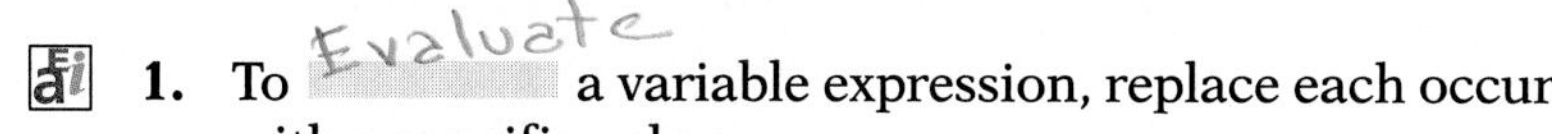

1. To ________ a variable expression, replace each occurrence of the variable with a specific value.

2. For the purpose of evaluating expressions, absolute value symbols are also ________ symbols.

In Exercises 3–10, evaluate the given expression for the indicated values of the variables.

3. $-5 + m + n$ for $m = -7$ and $n = -6$

4. $a + 3 + b$ for $a = 4$ and $b = -3$

5. $a - b - 7$ for $a = -2$ and $b = -3$

6. $-3 - m - n$ for $m = 4$ and $n = -6$

7. $y + (-7) + x$ for $x = -1$ and $y = 10$

8. $-2 + x + y$ for $x = -8$ and $y = 7$

9. $y - x - z$ for $x = -4$, $y = -9$, and $z = -1$

10. $-b - c - (-a)$ for $a = 1$, $b = -5$, and $c = 7$

In Exercises 11–20, evaluate the given expression for the indicated values of the variables.

11. $-9x$ for $x = 4$

12. $3y$ for $y = -6$

13. ab for $a = -4$ and $b = -7$

14. $-mn$ for $m = 3$ and $n = -7$

15. $-7xy$ for $x = -2$ and $y = -1$

16. $5ab$ for $a = -2$ and $b = -3$

17. $7a^3$ for $a = -2$

18. $-4y^2$ for $y = 5$

19. $-8x^2$ for $x = 3$

20. $3n^4$ for $n = -2$

Copyright © Houghton Mifflin Company. All rights reserved.

In Exercises 21–26, evaluate the given expression for the indicated values of the variables.

21. $\frac{a}{-3b}$ for $a = 30$ and $b = 2$

22. $\frac{-4x}{y}$ for $x = -5$ and $y = 10$

23. $\frac{xy}{x + y}$ for $x = 3$ and $y = -4$

24. $\frac{a + b}{a - b}$ for $a = -3$ and $b = -5$

25. $\frac{a + 7}{2b}$ for $a = -1$ and $b = -3$

26. $\frac{y - 5}{3x}$ for $x = 4$ and $y = -19$

In Exercises 27–30, for the indicated value of the variable, evaluate the given expressions.

27. $n = -5$: **(a)** $-(-n)$ **(b)** $|n|$

28. $y = -4$: **(a)** $|-y|$ **(b)** $-y$

29. $a = -1$: **(a)** $|-a|$ **(b)** $-|-a|$

30. $x = -8$: **(a)** $-|x|$ **(b)** $|x|$

In Exercises 31–34, evaluate the given expression for the indicated values of the variables.

31. $|-x - y|$ for $x = -7$ and $y = 2$

32. $-|m + n|$ for $m = -5$ and $n = 8$

33. $|a| - |b|$ for $a = -9$ and $b = -2$

34. $-x + |y|$ for $x = 3$ and $y = -7$

B. Equations

35. Finding a replacement for the variable that makes an equation true is called ________ the equation.

36. A statement that two expressions have the same value is called a(n) ________.

Copyright © Houghton Mifflin Company. All rights reserved.

In Exercises 37–40, determine which of the given numbers is a solution of the equation.

37. $-3n + 7 = 7$: $-4, 0$

38. $2x - 7 = -11$: $-2, 2$

39. $-13 = -3 + 2y$: $-5, 8$

40. $18 - 3y = 0$: $-6, 6$

In Exercises 41–46, solve the equation by inspection.

41. $-(-a) = 4$

42. $-6 = -(-b)$

43. $-x = 7$

44. $-n = -3$

45. $-12 = -c$

46. $15 = -y$

In Exercises 47–52, solve the equation by inspection.

47. $x + (-3) = 0$

48. $-2 + n = -6$

49. $-3 = 4 + y$

50. $0 = a + (-5)$

51. $-9 + c = -2$

52. $1 + x = -7$

In Exercises 53–58, solve the equation by inspection.

53. $x - 7 = -3$

54. $-5 = n - 4$

55. $5 = y - (-2)$

56. $x - (-4) = 9$

57. $b - 6 = -8$

58. $2 = a - 9$

In Exercises 59–66, solve the given equation.

59. $-2y = 8$

60. $4x = -20$

61. $-7n = -21$

62. $-5a = 30$

63. $7x = -7$

64. $-6m = -12$

65. $-27 = -3x$

66. $18 = -9a$

Copyright © Houghton Mifflin Company. All rights reserved.

In Exercises 67–72, solve the given equation.

67. $\frac{y}{4} = -9$

68. $\frac{x}{-6} = -5$

69. $\frac{-45}{b} = 5$

70. $-3 = \frac{6}{x}$

71. $\frac{42}{x} = -6$

72. $\frac{-32}{y} = -8$

Writing and Concept Extension

73. What question should you ask to help you solve the equation $-x = -2$?

74. What question should you ask to solve the equation $-4n = 12$?

75. Solve the equation $|x| = 7$.

76. If n is an integer, determine all values of n for which $|n| < 4$.

77. If n is any integer except 0, what is $n \div n$?

78. If any integer except 0 is divided by its opposite, what is the result?

79. What is true about a positive integer n if $(-7)^n$ is (a) positive and (b) negative?

Copyright © Houghton Mifflin Company. All rights reserved.

3.7 Simplifying Algebraic Expressions

A *Introduction to Simplifying Expressions*
B *The Distributive Property*
C *The Multiplication Properties of 1 and −1*
D *Combining Like Terms*

SUGGESTIONS FOR SUCCESS

There is an old saying: "When in doubt, simplify."

When things become too complicated and hectic, we try to simplify our lives. Mathematicians do the same thing with expressions. In this section, you will learn, for example, that $2x + 5x$ means the same thing as $7x$. Given the choice, we would rather work with the simpler expression $7x$.

Simplifying an expression often makes the expression easier to evaluate. We have also seen that equations are easier to solve if we can simplify one or both sides of the equation.

When you are asked to simplify an expression, ask yourself these questions: Can I remove the grouping symbols? Can I combine any of the terms? Are there any other operations that I can perform?

Simplifying expressions is a basic and important task in algebra. Getting the hang of it now will pay off in your future courses.

A Introduction to Simplifying Expressions

We have seen that an expression such as $2(7x)$ can be written as $(2 \cdot 7)x$ or $14x$. We say that $2(7x)$ and $14x$ are **equivalent expressions** because they have the same value for any given value of x.

Here are some other examples of equivalent expressions.

$x + 5 = 5 + x$	Commutative Property of Addition
$x \cdot 8 = 8x$	Commutative Property of Multiplication
$-1x = -x$	Multiplication Property of −1
$3(x + 4) = 3x + 12$	Distributive Property

When we write $2(7x) = 14x$ and $3(x + 4) = 3x + 12$, we are removing the parentheses from the original expression. Removing grouping symbols from an expression is one way in which we **simplify** an expression.

We begin our discussion with specific ways in which expressions can be simplified by removing parentheses.

The definition of subtraction allows us to write $x - 3$ as $x + (-3)$. The definition is also true in reverse.

$$x + (-3) = x - 3$$

We will usually think of $x - 3$ as being in the simpler form.

The Commutative and Associative Properties of Multiplication allow us to change the order of the factors and to group the factors in any way we wish. We can use these properties to simplify certain products.

Example 1

Simplify.

(a) $-6(2x)$ **(b)** $(-3y)(-5)$

Your Turn 1

Simplify.

(a) $5(-4x)$

(b) $(8a)(-2)$

continued

Copyright © Houghton Mifflin Company. All rights reserved.

SOLUTION

(a) $-6(2x)$

$= (-6 \cdot 2)x$ Associative Property of Multiplication

$= -12x$

(b) $(-3y)(-5)$

$= (-5)(-3y)$ Commutative Property of Multiplication

$= [-5(-3)]y$ Associative Property of Multiplication

$= 15y$

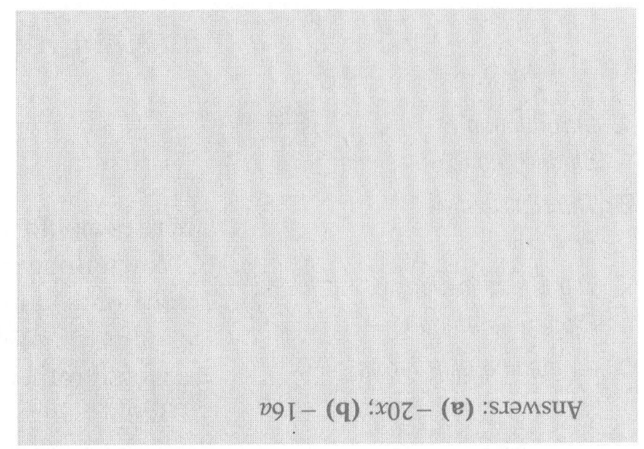

In Example 1, we indicated the properties that justify the steps in the solutions. Knowing why certain steps can be taken is important to becoming a good algebra student. However, once you are comfortable with these skills, you will find that you can often simplify mentally.

B The Distributive Property

As a reminder, the following is a restatement of the Distributive Property.

If a, b, and c represent any numbers, then $a(b + c) = ab + ac$.

The Distributive Property is frequently used to simplify an expression by removing parentheses.

Example 2

Simplify.

(a) $-7(x + 3)$

(b) $-6(3x + 2y)$

SOLUTION

(a) $-7(x + 3) = -7(x) + (-7)(3)$

$= -7x + (-21)$

$= -7x - 21$

(b) $-6(3x + 2y) = -6(3x) + (-6)(2y)$

$= -18x + (-12y)$

$= -18x - 12y$

Your Turn 2

Simplify.

(a) $-2(y + 9)$

(b) $-4(2a + 5b)$

Answers: (a) $-2y - 18$; (b) $-8a - 20b$

As stated, the Distributive Property involves multiplication and addition. The Distributive Property can also be stated with respect to multiplication and subtraction.

If a, b, and c represent any numbers, then $a(b - c) = ab - ac$.

This version of the Distributive Property can also be used to simplify expressions by removing parentheses.

Copyright © Houghton Mifflin Company. All rights reserved.

Example 3

Simplify.

(a) $5(a - 4)$ **(b)** $-2(y - 3)$

SOLUTION

(a) $5(a - 4) = 5(a) - 5(4) = 5a - 20$

(b) $-2(y - 3) = -2(y) - (-2)(3)$

$= -2y - (-6)$

$= -2y + 6$

Your Turn 3

Simplify.

(a) $8(x - 2)$

(b) $-4(a - 5)$

Answers: (a) $8x - 16$; (b) $-4a + 20$

The Distributive Property can be expanded to any number of addends inside the grouping symbols.

Example 4

Simplify $3(2x + y - 4)$.

SOLUTION

$$3(2x + y - 4) = 3(2x) + 3(y) - 3(4)$$
$$= 6x + 3y - 12$$

Your Turn 4

Simplify $2(x - y + 8)$.

2x−2y+16

Answer: $2x - 2y + 16$

C The Multiplication Properties of 1 and −1

The Multiplication Properties of 1 and -1 are written as $x = 1x$ and $-x = -1x$. However, in each case, x can also represent an expression. For example, $x + 5 = 1(x + 5)$ and $-(a - 2) = -1(a - 2)$.

Consider expressions such as $(x + 4)$ and $a + (b - c)$. How can we remove the parentheses? The answer lies in applying the Multiplication Property of 1 and the Distributive Property.

$$(x + 4) = 1(x + 4)$$
$$= 1(x) + 1(4)$$
$$= x + 4$$

$$a + (b - c) = a + 1(b - c)$$
$$= a + 1(b) - 1(c)$$
$$= a + b - c$$

Observe that the resulting expressions are the same as the original expressions but without the grouping symbols. In the first equation, there is no number initially written in front of the grouping symbol. In the second, the group is preceded by a + symbol. In such cases, the grouping symbols can simply be removed. No other steps need to be taken.

The expression $-(x + 3)$ is read as "the opposite of the quantity $x + 3$." (Note how the phrase *the quantity* signals that we are to take the opposite of the entire expression $x + 3$, not just the x.) To remove the parentheses from such expressions, we use the Multiplication Property of -1 and the Distributive Property.

Copyright © Houghton Mifflin Company. All rights reserved.

Example 5

Simplify.

(a) $-(x + 3)$ **(b)** $-(-2x + 5)$

SOLUTION

(a) $-(x + 3) = -1(x + 3)$ Multiplication Property of -1

$= -1(x) + (-1)(3)$ Distributive Property

$= -1x + (-3)$

$= -x - 3$

(b) $-(-2x + 5) = -1(-2x + 5)$ Multiplication Property of -1

$= -1(-2x) + (-1)(5)$

$= 2x + (-5)$

$= 2x - 5$

Your Turn 5

Simplify.

(a) $-(y + 5)$

(b) $-(-a + 3)$

Answers: (a) $-y - 5$; (b) $a - 3$

Example 6

Simplify $-(x - 4)$.

SOLUTION

$$-(x - 4) = -1(x - 4)$$
$$= -1(x) - (-1)(4)$$
$$= -1(x) - (-4)$$
$$= -x + 4$$

Your Turn 6

Simplify $-(2 - a)$.

Answer: $-2 + a$

The result in Example 6 can be written $4 + (-x)$ or $4 - x$.

$$-(x - 4) = 4 - x$$

We see that taking the opposite of $x - 4$ results in $4 - x$. In other words, the original order of the subtraction is reversed. This example illustrates the following important and useful property.

Property of the Opposite of a Difference

If a and b represent any numbers, then

$$-(a - b) = b - a$$

In words, taking the opposite of a difference reverses the difference.

Example 7

Simplify $-(3x - 5y)$.

SOLUTION

$$-(3x - 5y) = 5y - 3x$$

Your Turn 7

Simplify $-(2c - 9a)$.

Answer: $9a - 2c$

Copyright © Houghton Mifflin Company. All rights reserved.

D Combining Like Terms

In an algebraic expression, the addends are usually called **terms.**

$2x + y$ — The $2x$ and the y are called terms

$-5a + 2b + 6$ — The $-5a$, the $2b$, and the 6 are called terms.

Remember that terms are *addends.* Therefore, we must change any subtractions to addition in order to identify the terms.

$$4x - 3y - 7 = 4x + (-3y) + (-7)$$ The terms are $4x$, $-3y$, and -7.

Example 8

Identify the terms in the expressions.

(a) $y + 9$ **(b)** $-2a + 6b - 1$

SOLUTION

(a) The terms are y and 9.

(b) $-2a + 6b - 1 = -2a + 6b + (-1)$

The terms are $-2a$, $6b$, and -1.

Your Turn 8

Identify the terms in the following expressions.

(a) $9x - 5y$

(b) $-x + 8y - 2$

Answers: (a) $9x$, $-5y$; (b) $-x$, $8y$, -2

Like terms are terms that have the same variable parts. (We also regard numbers as like terms.)

Like Terms	*Not Like Terms*
$2x$, $5x$	$2x$, $5y$
$3y$, $-1y$	$3y$, -1
$1c$, $9c$	$1c$, 9
4, -7	4, $-7x$

In addition to removing grouping symbols, sometimes we can simplify an expression by *combining* like terms. To combine like terms, we use the following version of the Distributive Property.

If a, b, and c represent any numbers, then $ba + ca = (b + c)a$.

Example 9

If possible, simplify by combining like terms.

(a) $2x + 5x$ **(b)** $3y - y$

(c) $c + 9c$ **(d)** $4x + 8y$

SOLUTION

(a) $2x + 5x = (2 + 5)x = 7x$

(b) $3y - y = 3y + (-1)y$
$= [3 + (-1)]y$
$= 2y$

Your Turn 9

If possible, simplify by combining like terms.

(a) $7b + 3b$ = 10

(b) $5x + 5$ not possible

(c) $-a + 4a$ 3a

(d) $y - 10y$ −9

continued

Copyright © Houghton Mifflin Company. All rights reserved.

(c) $c + 9c = 1c + 9c = (1 + 9)c = 10c$

(d) Because $4x$ and $8y$ are not like terms, they cannot be combined.

LEARNING TIP
Have you noticed that like terms can be combined mentally just by adding the numerical parts?

Answers: **(a)** $10b$; **(b)** $5x + 5$; **(c)** $3a$; **(d)** $-9y$

If an expression has three or more terms, you may be able to combine some or all of them.

Example 10

Simplify.

(a) $2x + 3x + 4x$

(b) $5a + 2b + a - 6b + 4$

SOLUTION

(a) $2x + 3x + 4x = (2 + 3 + 4)x = 9x$

(b) $5a + 2b + a - 6b + 4$

$= (5a + 1a) + (2b - 6b) + 4$ Group like terms.

$= (5 + 1)a + (2 - 6)b + 4$ Distributive Property

$= 6a + (-4b) + 4$

$= 6a - 4b + 4$

Your Turn 10

Simplify.

(a) $y - 3y + 8y$

(b) $x + y - 3x + 2y$

Answers: **(a)** $6y$; **(b)** $-2x + 3y$

Simplifying an expression may involve removing parentheses *and* combining like terms.

Example 11

Simplify.

(a) $8 + (x - 3)$ **(b)** $(5 - 3x) - (-4)$

SOLUTION

(a) $8 + (x - 3) = 8 + x - 3$ Remove parentheses.

$= x + 5$ Combine like terms.

(b) $(5 - 3x) - (-4) = 5 - 3x - (-4)$

$= 5 - 3x + 4$

$= -3x + 9$

$= 9 - 3x$

Your Turn 11

Simplify.

(a) $-2 + (y - 4)$

(b) $(7 - 2x) - 3$

(a) = −2 + y − 4
= 6 + y
(b) = 7 − 2x − 3
= 4 − 2x

Answers: **(a)** $y - 6$; **(b)** $4 - 2x$

NOTE In Example 11(b), the expressions $-3x + 9$ and $9 - 3x$ are equivalent. Either form of the answer is correct.

Copyright © Houghton Mifflin Company. All rights reserved.

Example 12

Simplify.

(a) $-(2x - 8) + 5x$

(b) $2(x + 3) + 5(x - 7)$

SOLUTION

(a) $-(2x - 8) + 5x$

$= 8 - 2x + 5x$ Property of the Opposite of a Difference

$= 3x + 8$ Combine like terms.

(b) $2(x + 3) + 5(x - 7)$

$= 2x + 6 + 5x - 35$ Distributive Property

$= (2x + 5x) + (6 - 35)$ Group like terms.

$= 7x + (-29)$ Combine like terms.

$= 7x - 29$

Your Turn 12

Simplify.

(a) $-(3a + 4) + 2a - 1$

(b) $-3(x + 1) + 4(x - 2)$

Answers: (a) $-a - 5$; (b) $x - 11$

3.7 Quick Reference

A. Introduction to Simplifying Expressions

1. **Equivalent expressions** are expressions that have the same value for all values of the variable(s).

Equivalent expressions:

$3(5x)$,	$15x$
$-2x + 7$,	$7 - 2x$
$2(x + 6)$,	$2x + 12$

2. One way to **simplify** an expression is to remove grouping symbols.

$x + (-9) = x - 9$

$-3(6x) = [-3(6)]x = -18x$

B. The Distributive Property

1. The Distributive Property can be used to remove grouping symbols from expressions of the form $a(b + c)$.

$-2(x + 5) = -2x - 10$

$-3(7x + 3) = -21x - 9$

2. Another version of the Distributive Property is $a(b - c) = ab - ac$.

$6(x - 2) = 6x - 12$

$-3(a - 5) = -3a + 15$

3. The Distributive Property can be expanded to any number of addends inside the grouping symbols.

$4(x - 2y + 3)$
$= 4x - 8y + 12$

Copyright © Houghton Mifflin Company. All rights reserved.

C. The Multiplication Properties of 1 and −1

1. If no number is written in front of grouping symbols or if a group is preceded by a + symbol, then the parentheses can simply be removed.

$(x + 3) + y = x + 3 + y$

$a + (2 + b) = a + 2 + b$

2. Taking the opposite of a group is equivalent to multiplying the group by −1.

$$\begin{aligned} -(x + 2) &= -1(x + 2) \\ &= -1x + (-1)(2) \\ &= -1x + (-2) \\ &= -x - 2 \end{aligned}$$

3. If a and b represent any numbers, then

$$-(a - b) = b - a$$

In words, taking the opposite of a difference reverses the difference.

$-(x - 4) = 4 - x$

$-(2a - 3b) = 3b - 2a$

D. Combining Like Terms

1. The addends of an expression are called **terms.** To identify the terms of an expression, any subtractions must be changed to addition.

$7x + 4y + 3$
Terms: $7x$, $4y$, 3

$-2x - 9y = -2x + (-9y)$
Terms: $-2x$, $-9y$

2. **Like terms** are terms that have the same variable parts. Numbers are also regarded as like terms.

Like terms: $3x$, $8x$
Like terms: -4, 9

Unlike terms: $2x$, $2y$
Unlike terms: 7, $7a$

3. Another method for simplifying an expression is to *combine* like terms.

$3x + 8x = (3 + 8)x = 11x$

$-2y + 6y = (-2 + 6)y = 4y$

$5a - 7a = (5 - 7)a = -2a$

4. If an expression has three or more terms, you may be able to combine some or all of them.

$$\begin{aligned} &6x - 9x + 7y \\ &= (6 - 9)x + 7y \\ &= -3x + 7y \\ &= 7y - 3x \end{aligned}$$

5. Simplifying an expression may involve removing parentheses *and* combining like terms.

$$\begin{aligned} &5x + 2(3x - 4) \\ &= 5x + 2(3x) - 2(4) \\ &= 5x + 6x - 8 \\ &= 11x - 8 \end{aligned}$$

Copyright © Houghton Mifflin Company. All rights reserved.

3.7 Exercises

A. Introduction to Simplifying Expressions

1. Removing grouping symbols from an expression is one way to ________ an expression.

2. To simplify the expression $-2(3x)$, we use the ________ Property of Multiplication.

In Exercises 3–8, simplify the given expression.

3. $5(-2x)$

4. $-3(4b)$

5. $-6(-7a)$

6. $2(-8y)$

7. $(9x)(-4)$

8. $(-6c)(-4)$

B. The Distributive Property

9. The ________ Property allows us to remove parentheses in the expression $3(x - 7)$.

10. Because $2(x - 1)$ and $2x - 2$ have the same value for any number x, they are called ________ expressions.

In Exercises 11–28, use the Distributive Property to remove the parentheses.

11. $-5(x + 8)$

12. $-4(3 + y)$

13. $-7(2x + 7)$

14. $-3(x + 9)$

15. $8(x - 3y)$

16. $10(9 - 5a)$

17. $-9(a - 5)$

18. $-3(7 - b)$

19. $-2(5m - 6n)$

20. $-7(6a - 7b)$

21. $4(x + y + 3)$

22. $6(a - b + 1)$

23. $-3(m + n - 4)$

24. $-5(x - y - z)$

25. $7(-3a - 8b - c)$

26. $4(x - 8y - 1)$

27. $-3(2x - 5y + 7)$

28. $-6(m + 8n - 3)$

Copyright © Houghton Mifflin Company. All rights reserved.

C. The Multiplication Properties of 1 and −1

29. Taking the ______ of a difference reverses the order of subtraction.

30. We apply the ______ Property of 1 to write $(y - 5) = y - 5$.

In Exercises 31–38, remove the parentheses.

31. $-(5 + y)$

32. $-(y + 7)$

33. $-(-3x + 8)$

34. $-(-4y - z)$

35. $-(5 - 3b)$

36. $-(m - 6n)$

37. $-(2a - 5)$

38. $-(12 - 7y)$

D. Combining Like Terms

39. The ______ of an expression are separated by addition.

40. Terms with the same variable parts are called ______ terms.

In Exercises 41–50, if possible, simplify by combining like terms.

41. $3x + 9x$

42. $5b + 14b$

43. $8a - 9a$

44. $-5y + 6y$

45. $-8x - 12x$

46. $-9c - 10c$

47. $5a + 8b$

48. $-6 + 6y$

49. $n - 5n$

50. $12p - p$

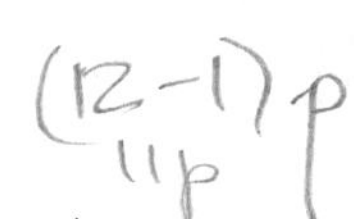

In Exercises 51–60, simplify by combining like terms.

51. $4y + 8y - 3y$

52. $8x - 10x + x$

53. $5x - 5 + 7x$

54. $2a - 9 - 5a$

55. $2x + 3 - 7 + 4x$

56. $-6 + 5b + 2 - 5b$

57. $8 + 3m + 5n - 6 - 3m$

58. $2x - 3y + 1 + 7y - 9x$

59. $x + 3 + 2y - 4x + 2y$

60. $-8a + 7 - b + 3b + 2$

Copyright © Houghton Mifflin Company. All rights reserved.

In Exercises 61–72, simplify.

61. $3 + (a - 7)$

62. $(3 - 2b) - 3$

63. $(-5 + 2n) - (-5)$

64. $-4y + (y + 10)$

65. $4(x - 5) - 7$

66. $2(3x + 1) - 5x$

67. $5(a - 2) + (4 - 3a)$

68. $(4a + 9) + 6(1 - a)$

69. $2(x + 3) + 5(x - 7)$

70. $-3(x - 2) + (3x + 4)$

71. $-(2x - 3) + 5x$

72. $-(4 - x) + 4$

Writing and Concept Extension

73. Explain how to identify the terms of an expression.

74. Name two ways to simplify an expression.

In Exercises 75–80, state whether the expressions are equivalent.

75. $1y, y$

76. $-a, -1a$

77. $x - 3, -3 + x$

78. $-4 + 2b, 2b - 4$

79. $n - 7, 7 - n$

80. $-(x - 2), -x - 2$

Copyright © Houghton Mifflin Company. All rights reserved.

CHAPTER 3 REVIEW EXERCISES

Section 3.1

1. The point of a number line that corresponds to 0 is called the ______.

2. Insert $<$ or $>$ to make the statement true.

   (a) -7 ___ 3 (b) -9 ___ -6

3. What number is 4 units to the left of 1 on the number line?

4. Write the opposite of the given number.

   (a) 9 (b) -12 (c) 0

5. Write the expression $-c - 5$ in words.

6. Simplify the given expression.

   (a) $-(-7)$ (b) $-(-x)$

7. Evaluate the given expression.

   (a) $|-9|$ (b) $-|-6|$

8. Write the numbers $-(-5)$, $|0|$, $|-7|$, $-|-3|$, -4 in order from smallest to largest.

9. Explain why the absolute values of -3 and 3 are the same.

10. A person wrote a check for \$97 and made a deposit of \$305. Write a positive integer and a negative integer to describe the given information.

Section 3.2

11. If m and n represent negative integers, the sign of the sum of m and n is ______.

In Exercises 12–15, add.

12. $-7 + (-9)$

13. $-10 + 3$

14. $-8 + 14$

15. $12 + (-17)$

Copyright © Houghton Mifflin Company. All rights reserved.

16. Use the indicated property to fill in the blank.

(a) Addition Property of Opposites: ____ $+ 5 = 0$

(b) Commutative Property of Addition:

$$3 + (-10) + 6 = 3 + 6 + \_\_\_\_$$

In Exercises 17 and 18, determine the sum.

17. $-3 + 5 + (-6)$

18. $12 + (-8) + (-9) + 5$

19. Use the Order of Operations to evaluate $(-2 + 9) + (-4 + 1)$.

20. A person earned $380. The employer withheld $60 for taxes, $18 for insurance, and $24 for the savings plan. Write and evaluate a sum to determine the take-home pay.

Section 3.3

21. Explain how to write $-5 - (-3)$ as a sum.

In Exercises 22–25, subtract.

22. $9 - 15$

23. $3 - (-6)$

24. $-7 - 5$

25. $-10 - (-2)$

In Exercises 26 and 27, evaluate the expression.

26. $-7 - (-9) - (-2) + 12$

27. $-8 - 3 + (-6) - 5$

In Exercises 28 and 29, use the Order of Operations to evaluate the given expression.

28. $-3 - (7 - 9)$

29. $6 - 3(9 - 4)$

30. In Australia, the highest elevation is 7,310 feet, and the lowest elevation is -52 feet. What is the difference between the highest and lowest elevations?

Copyright © Houghton Mifflin Company. All rights reserved.

Section 3.4

31. The product of two numbers with like signs is ________, whereas the product of two numbers with unlike signs is ________.

In Exercises 32–34, multiply.

32. $-5(-6)$ **33.** $-8(2)$ **34.** $7(-10)$

35. Evaluate $-3(-2)(4)(-1)(2)$.

36. Evaluate the given exponential expression.

(a) $(-1)^{10}$ **(b)** $(-2)^5$

37. Use the Multiplication Property of -1 to write the given expression in another way.

(a) $-c$ **(b)** $-1n$

In Exercises 38–40, use the Order of Operations to evaluate the given expression.

38. $-4 - 7(3 - 8)$ **39.** $-2(-3)^2 - (-21)$ **40.** $-9(4 - 7)$

Section 3.5

41. We say that the quotient $3 \div 0$ is ________.

In Exercises 42–45, divide.

42. $\frac{-28}{7}$ **43.** $\frac{0}{-8}$ **44.** $15 \div (-15)$ **45.** $-90 \div (-9)$

In Exercises 46–49, use the Order of Operations to evaluate the given expression.

46. $\frac{6(-8)}{-12}$ **47.** $\frac{-15 + 6}{4 - 7}$

48. $\frac{-5 - 3^2}{-15 - (-8)}$ **49.** $\frac{1 - 5^2}{3^2 - 1}$

50. During one week, the following low temperatures were recorded in Rapid City: -8°F, 10°F, -5°F, -12°F, 3°F, 0°F, and -9°F. What was the average low temperature for the week?

Copyright © Houghton Mifflin Company. All rights reserved.

Section 3.6

In Exercises 51–54, evaluate the given expression for the indicated values of the variables.

51. $-6 + a + b$ for $a = 10$ and $b = -2$

52. $a - c - b$ for $a = 2$, $b = 8$, and $c = -3$

53. $-4ab$ for $a = -2$ and $b = 0$

54. $\dfrac{ab}{a - b}$ for $a = -3$ and $b = -2$

55. Evaluate the given expression for $n = -6$.

(a) $-(-n)$ **(b)** $|n|$ **(c)** $|2n - 5|$

In Exercises 56–60, solve the given equation.

56. $-x = 10$

57. $-4 + y = 2$

58. $a - 5 = -1$

59. $-8n = -32$

60. $\dfrac{30}{x} = -3$

Section 3.7

61. Simplify $-3(-5x)$.

In Exercises 62–64, use the Distributive Property to remove the parentheses.

62. $-3(x + 4)$

63. $-4(2x - 8)$

64. $3(2a - 4b + 1)$

65. The expression $-(x - 5)$ is read "the ______ of the ______ $x - 5$."

66. Write $-(3m - 8n)$ without parentheses.

In Exercises 67–70, simplify.

67. $n - 5n$

68. $-7 - 3y + 2 + 3y$

69. $-(5a - 4) - 9a$

70. $4(x + 3) + 5(2 - 3x)$

Copyright © Houghton Mifflin Company. All rights reserved.

CHAPTER 3 TEST

1. Insert < or > to make the statement true.

   (a) $-3 > -10$ (b) $-|7|$ ______ $|-10|$

2. Arrange the numbers $-|0|, -2, |3|, -(-1)$, and $-|-7|$ from smallest to largest.

In Questions 3–6, perform the indicated operation.

3. (a) $-7 + (-8)$ (b) $-8 + 10$

4. (a) $-12 - 6$ (b) $-5 - (-8)$

5. (a) $-8(6)$ (b) $-7(-2)$

6. (a) $-16 \div (-4)$ (b) $-45 \div 5$

In Questions 7–9, perform the indicated operations.

7. $-3 + (-5) + 7$

8. $-3 - 10 - (-4) + 2$

9. $(-5)(-1)(2)(3)(-4)$

10. We can use the Multiplication Property of -1 to write $-m$ as ______ or to write $-1x$ as ______.

11. We say that $-7 \div 0$ is ______, whereas $0 \div (-7)$ is ______.

12. Evaluate $|-a - b|$ for $a = -9$ and $b = 4$.

13. Evaluate $a - b - c$ for $a = -3, b = -2$, and $c = 8$.

14. Suppose that n represents a positive integer. Explain how to determine the sign of $(-1)^n$.

Copyright © Houghton Mifflin Company. All rights reserved.

In Questions 15–17, use the Order of Operations to evaluate the given expression.

15. $-3 + 4(-1 - 5)$

16. $\dfrac{5 + 7(-3)}{2 - 10}$

17. $\dfrac{(-2)^3 - 7}{-3 - (-8)}$

18. Solve $n + (-4) = -7 + 3$.

19. Solve $\dfrac{x}{-5} = 10$.

20. On a certain day in International Falls, the high temperature was -7°F and the low temperature was -17°F. What was the average temperature for that day?

21. According to the __________ Property of -1, $-(a + 3) = -1(a + 3)$.

In Questions 22–25, simplify the given expression.

22. $4(x + 3y - 5)$

23. $-(-4b + 9)$

24. $-2(3x - 4) + 7x$

25. $4(b - 2a) + 7(a - 2b)$

Copyright © Houghton Mifflin Company. All rights reserved.

CHAPTERS 1–3 CUMULATIVE TEST

1. The distance between New York and Singapore is 9,534 miles.

(a) Write the number in words. **(b)** Round the number to the nearest thousand.

2. The distance between New Orleans and Pittsburgh is one thousand seventy miles. Write the number of miles in standard form.

3. Use the indicated property to fill in the blank.

(a) Addition Property of 0

$b + \_\_\_ = b$

(b) Associative Property of Multiplication

$(3 \cdot \_\_\_)x = 3(5x)$

(c) Multiplication Property of 1

$c = \_\_\_ \cdot c$

(d) Commutative Property of Addition

$3 + x + 7 = 3 + \_\_\_ + x$

In Questions 4–7, perform the indicated operation.

4. $\begin{array}{r} 532 \\ 987 \\ +\ 75 \\ \hline \end{array}$

5. $\begin{array}{r} 5{,}297 \\ -\ 3{,}748 \\ \hline \end{array}$

6. $\begin{array}{r} 7{,}029 \\ \times\ 76 \\ \hline \end{array}$

7. $54\overline{)14{,}256}$

8. Explain the difference between the quotients.

(i) $5 \div 0$ (ii) $0 \div 5$

9. The figure shows the floor plan for the lobby of a library. Determine the perimeter and the area of the lobby.

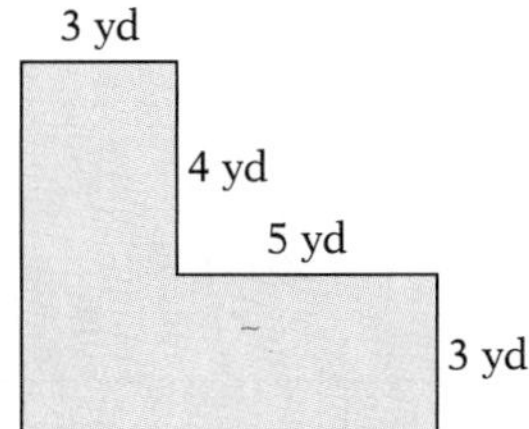

10. In the expression 2^4, we call 2 the ___ and 4 the ___.

11. The host of a Super Bowl party for 50 guests needs 6 ounces of Buffalo wings per guest.

(a) How many 64-ounce packages of wings should the host buy?

(b) How many ounces of wings will be left over?

Copyright © Houghton Mifflin Company. All rights reserved.

12. A closed metal box is 1 yard long, 2 feet wide, and 5 inches deep.

(a) How many square inches of metal are needed to construct the box?

(b) What is the volume of the box in cubic inches?

13. Arrange the numbers $-|-3|$, $-(-3)$, -7, $|-7|$, -0 in order from smallest to largest.

In Questions 14–18, perform the indicated operations.

14. **(a)** $-9 + (-6)$ **(b)** $-4 + 12$

15. **(a)** $-10 - 8$ **(b)** $-4 - (-6)$

16. **(a)** $9(-10)$ **(b)** $(-6)(-3)$

17. **(a)** $\frac{-27}{-9}$ **(b)** $\frac{-48}{6}$

18. $-3 + 10 - (-7) - 9$

19. Evaluate.

(a) $(-1)^5$ **(b)** $(-1)^8$

20. Evaluate $|2a - b|$ for $a = -5$ and $b = -3$.

In Questions 21 and 22, solve the equation.

21. $5x = 0$

22. $x - 4 = -1$

In Questions 23–25, evaluate the expression.

23. $2 \cdot 3^2 - 4(8 - 5)$

24. $-4 - 6(4 - 6)$

25. $\frac{10 - 3(-5)}{-3 - (-8)}$

In Questions 26 and 27, write the expression without parentheses.

26. $-(3x - y)$

27. $-5(-2x + 5)$

28. What are like terms?

29. Simplify $5x + (3 - 4x)$.

30. Simplify $3(x - 2y) + 6(y - 2x)$.

Copyright © Houghton Mifflin Company. All rights reserved.

CHAPTER 4

Equations

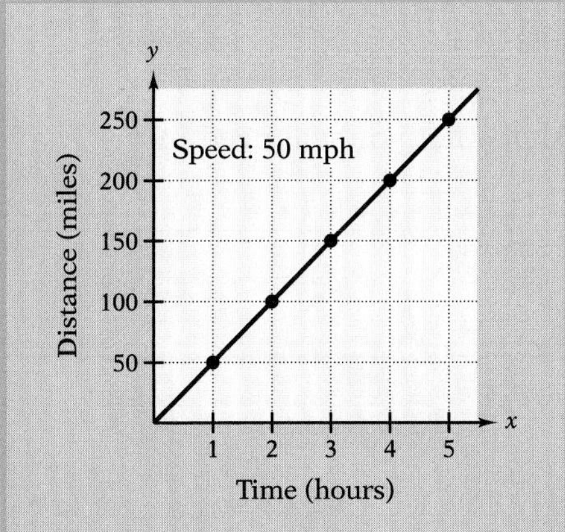

When you travel, your distance D, rate R (speed), and time T are related by the equation $D = RT$. If a truck's speed is 50 miles per hour, then the distance that it travels in 3 hours is $D = 50(3) = 150$ miles. In general, the truck's distance after T hours is given by the equation $D = 50T$.

A **graph** can be drawn to show the distance that the truck has traveled after any number of hours. From this graph, we can select a time and find the distance traveled during that time, or we can select a distance and find the time required to travel that distance.

Chapter Snapshot

We have seen that some equations are simple enough that we can solve them *by inspection.* In this chapter, we present some properties of equations, and we show how those properties can be used to solve a **linear equation in one variable.** Then we use your equation solving skills in a variety of application problems. We also discuss equations that have two variables, the meaning of solutions of such equations, and methods for visualizing those solutions with a graph.

Copyright © Houghton Mifflin Company. All rights reserved.

For online resources, visit the web site **math.college.hmco.com/students** and follow the links to Hubbard/Robinson, *Prealgebra.*

Some Friendly Advice . . .

Sometimes you can relate a mathematical concept or rule to things that you see about you in your everyday life.

In this chapter, we suggest that you think of an equation as being like a pan balance. If we add the same weight to both sides, the pans remain in balance. In the same way, if we add the same quantity to both sides of an equation, the equation remains in "balance." We also suggest the idea of a street map to think about moving around in a coordinate system.

Let your imagination run free. The more you can relate mathematical concepts to familiar things that you can visualize, the clearer the concepts will be to you.

WARM-UP SKILLS

The following questions review concepts and skills that you will need in Chapter 4.

1. Simplify.

(a) $x + 7 + (-7)$

(b) $y + 0$

In Exercises 2 and 3, simplify.

2. $5x + 3(x - 1)$

3. $5 - 3y + 4y$

In Exercises 4 and 5, evaluate the expression for the given value(s) of the variable(s).

4. $3x + 2$ for $x = 5$

5. $x - y$ for $x = 2$ and $y = -1$

In Exercises 6 and 7, solve the given equation by inspection.

6. $3a = 15$

7. $x + 5 = 12$

In Exercises 8–10, write and evaluate an expression.

8. 7 more than 3

9. 5 less than 2

10. The quotient of 24 and 6

Copyright © Houghton Mifflin Company. All rights reserved.

4.1 Properties of Equations

A *The Addition Property of Equations*
B *The Multiplication Property of Equations*

SUGGESTIONS FOR SUCCESS

Isolate: to set apart from others; place alone.

In this section you will learn that the goal in solving an equation is to isolate the variable—that is, to get the variable by itself on one side of the equation. Before you begin, ask yourself what steps are needed to accomplish this goal.

Keeping your eye on the goal is important in all of your study of mathematics. If you lose track of what you are trying to do, you will probably flounder. Identify the goal, plan how to meet that goal, and then work the plan.

A The Addition Property of Equations

In Section 1.7, we defined an **equation** as a statement that two expressions have the same value. The expressions may be numerical expressions or variable expressions, and we refer to them as the *left side* and *right side* of the equation.

An equation may be true for particular replacements for the variable and false for other replacements.

$$3x = 12 \qquad \text{True if } x = 4\text{; false if } x = 5$$

A **solution** of an equation is any replacement for the variable that makes the equation true. For the equation $3x = 12$, the only solution is 4. No other replacement for x is a solution.

We can determine whether a given number is a solution of an equation by replacing each occurrence of the variable with the number.

Example 1

Determine whether 5 is a solution of $3x + 2 = x + 12$.

SOLUTION

$3x + 2 = x + 12$	The equation
$3(5) + 2 = 5 + 12$	Replace x with 5.
$17 = 17$	True

Therefore, 5 is a solution.

Your Turn 1

Determine whether -2 is a solution of $x - 3 = 3x + 1$.

Answer: Yes

Example 2

Determine whether -1 is a solution of $-x + 1 = 2x$.

SOLUTION

$-x + 1 = 2x$	The equation
$-(-1) + 1 = 2(-1)$	Replace x with -1.
$1 + 1 = -2$	
$2 = -2$	False

Therefore, -1 is not a solution.

Your Turn 2

Determine whether 4 is a solution of $8 - x = 2x - 3$.

$8-4 = 2(4)-3$
$4 = 8-3$
$4 = 5$

Answer: No

Copyright © Houghton Mifflin Company. All rights reserved.

Examples 1 and 2 show how we can *test* a given number to determine whether it is a solution of a certain equation. *Finding* a solution of an equation is a different matter.

As we saw in earlier chapters, we can use our knowledge of basic arithmetic facts to solve simple equations by inspection. However, not all equations can be solved by inspection. For some equations, we must use an orderly method for arriving at the solution.

Adding to Both Sides of an Equation

Look at the pan balance shown in the figures.

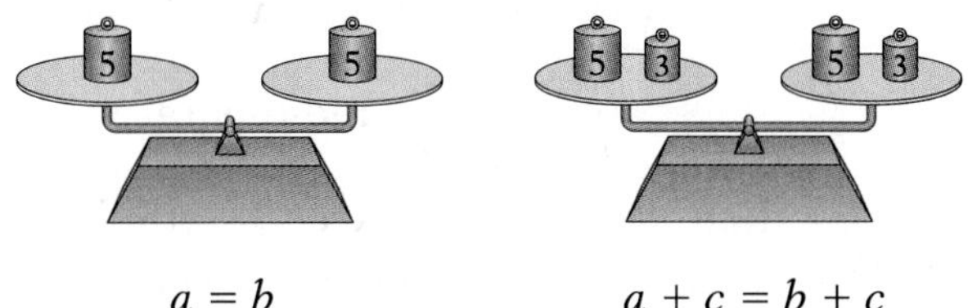

$a = b$ $\qquad$ $a + c = b + c$

Initially, each pan holds a 5-ounce weight, so the pans are perfectly in balance. This corresponds to $a = b$. If we now add a 3-ounce weight to each side, which corresponds to $a + c = b + c$, the pans remain in balance. In the same sense, an equation remains "balanced" when the same number is added to (or subtracted from) both sides.

The following property of equations is an important part of the equation solving process.

The Addition Property of Equations

Suppose that a, b, and c represent any numbers.

$$\text{If } a = b, \text{ then } a + c = b + c.$$
$$\text{If } a = b, \text{ then } a - c = b - c.$$

In words, any number can be added to or subtracted from both sides of an equation.

NOTE We can include subtraction in the Addition Property of Equations because subtraction is defined in terms of addition.

Now let's consider how the Addition Property of Equations can help us to solve equations.

Example 3

Solve $x + 7 = 15$.

SOLUTION

$x + 7 = 15$	The equation
$x + 7 + (-7) = 15 + (-7)$	Add -7 to both sides.
$x + 0 = 8$	$7 + (-7) = 0$
$x = 8$	$x + 0 = x$

The solution is 8.

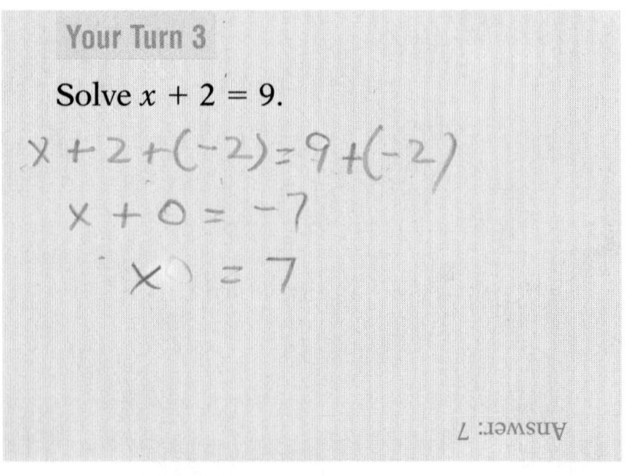

Your Turn 3

Solve $x + 2 = 9$.

Answer: 7

Copyright © Houghton Mifflin Company. All rights reserved.

In Example 3, how did we know to add -7 to both sides rather than 4 or -1 or some other number? Our goal is to write the equation so that the variable is by itself. In other words, we try to *isolate* the variable. To do that, we had to eliminate the 7. We chose to add -7 to both sides of the equation because we know that $7 + (-7) = 0$ and $x + 0 = x$.

We could have accomplished the same thing by subtracting 7 from both sides rather than adding -7.

> **LEARNING TIP**
> *Addition and subtraction undo each other. When you are trying to eliminate a number that is* added *to the variable,* subtract *that number from both sides. When you are trying to eliminate a number that is* subtracted *from the variable,* add *that number to both sides.*

$x + 7 = 15$	The equation
$x + 7 - 7 = 15 - 7$	Subtract 7 from both sides.
$x + 0 = 8$	$7 - 7 = 0$
$x = 8$	$x + 0 = x$

NOTE To check a solution, replace the variable with the solution. If the left and right sides then have the same value, your solution is correct. We will not always show the check of solutions. As you work through these examples, checking solutions on your own will be good practice.

In Example 4, we will solve the equation $x - 9 = 2$. Have you already guessed what we will do?

Example 4

Solve $x - 9 = 2$.

SOLUTION

$x - 9 = 2$	The equation
$x - 9 + 9 = 2 + 9$	Add 9 to both sides.
$x + 0 = 11$	$-9 + 9 = 0$
$x = 11$	

The solution is 11.

Your Turn 4

Solve $y - 6 = 8$.

y − 6 + 6 = 8 + 6
y + 0 = 14
y = 14

Answer: 14

The method is the same even when the variable appears on the right side.

Example 5

Solve $28 = x + 12$.

SOLUTION

$28 = x + 12$	
$28 - 12 = x + 12 - 12$	Subtract 12 from both sides.
$16 = x + 0$	
$16 = x$	

The equations $16 = x$ and $x = 16$ mean the same thing. The solution is 16.

Your Turn 5

Solve these equations.

(a) $15 = x + 9$

(b) $21 = c - 4$

A 15 − 9 = x + 9 − 9
6 = x + 0
6 = x

Answers: (a) 6; (b) 25

B 21 + 4 = c − 4 + 4
25 = c + 0
25 = c

Copyright © Houghton Mifflin Company. All rights reserved.

Sometimes we need to begin by simplifying one or both sides of the equation.

Example 6

Solve $5x + 3 - 4x = 8 - 7$.

SOLUTION

We begin by combining the like terms on the left side and performing the subtraction on the right side.

$$5x + 3 - 4x = 8 - 7$$

$$x + 3 = 1 \qquad 5x - 4x = 1x = x$$

$$x + 3 - 3 = 1 - 3 \qquad \text{Subtract 3 from both sides.}$$

$$x + 0 = -2$$

$$x = -2$$

The solution is -2.

Your Turn 6

Solve $-2y + 7 + 3y = 9 - 5$.

Answer: -3

B The Multiplication Property of Equations

Suppose we wish to solve the equation $2x = 10$. A common mistake is to try to isolate x by subtracting 2 from both sides.

$$2x = 10$$

$$2x - 2 = 10 - 2$$

However, we can't simplify $2x - 2$ because $2x$ and 2 are not like terms. We just can't use the Addition Property of Equations to solve this equation.

A second property of equations is needed to isolate the variable when it is being multiplied or divided by some number.

The Multiplication Property of Equations

Suppose that a, b, and c represent any numbers, where c is not 0.

$$\text{If } a = b, \text{ then } ac = bc.$$

$$\text{If } a = b, \text{ then } \tfrac{a}{c} = \tfrac{b}{c}.$$

In words, we can multiply or divide both sides of an equation by any number except 0.

When we use this property, we need to remember that division undoes multiplication. That is, if you multiply by a number and then divide by the same number, the result is the original number.

$$\frac{2 \cdot 5}{2} = \frac{10}{2} = 5 \qquad \frac{7 \cdot 3}{7} = \frac{21}{7} = 3 \qquad \frac{3 \cdot x}{3} = \frac{3x}{3} = x$$

Let's return to the equation $2x = 10$ and see how the property can be used.

Copyright © Houghton Mifflin Company. All rights reserved.

Example 7

Solve $2x = 10$.

SOLUTION

$$2x = 10$$

$$\frac{2x}{2} = \frac{10}{2} \quad \text{Divide both sides by 2.}$$

$$x = 5 \quad \frac{2x}{2} = x$$

The solution is 5.

Your Turn 7

Solve $6x = 18$.

Answer: 3

How did we know to divide both sides by 2? Because x is *multiplied* by 2, we *divide* by 2. When we divide $2x$ by 2, the x is isolated.

Example 8

Solve these equations.

(a) $-3x = 18$ **(b)** $-x = 5$

SOLUTION

(a) $-3x = 18$

$$\frac{-3x}{-3} = \frac{18}{-3} \quad \text{Divide both sides by } -3.$$

$$x = -6 \quad \frac{-3x}{-3} = x$$

The solution is -6.

(b) Because $-x = -1x$, we can write the equation as $-1x = 5$.

$$\frac{-1x}{-1} = \frac{5}{-1} \quad \text{Divide both sides by } -1.$$

$$x = -5 \quad \frac{-1x}{-1} = x$$

The solution is -5.

Your Turn 8

Solve these equations.

(a) $-2x = -14$

(b) $-4 = -y$

LEARNING TIP

When you use the Multiplication Property of Equations, the number that you multiply or divide by must have the same sign as the factor that you are trying to eliminate.

Answers: (a) 7; (b) 4

Sometimes we need to begin by simplifying one or both sides of the equation.

Example 9

Solve $5x - x = -3 - 9$.

SOLUTION

$$5x - x = -3 - 9$$

$$4x = -12 \quad \text{Combine like terms.}$$

$$\frac{4x}{4} = \frac{-12}{4} \quad \text{Divide both sides by 4.}$$

$$x = -3 \quad \frac{4x}{4} = x$$

The solution is -3.

Your Turn 9

Solve $4x + 3x = 2(3) + 8$.

Answer: 2

Copyright © Houghton Mifflin Company. All rights reserved.

The Addition and Multiplication Properties of Equations are the basis for the general equation solving procedure that we will present in the next section. Before you move ahead, make sure that you have mastered the use of these properties in solving the kinds of equations discussed in this section.

4.1 Quick Reference

A. The Addition Property of Equations

1. An **equation** is a statement that two expressions have the same value. An equation may be true or false.

   True: $1 \cdot x = x$
   False: $x + 3 = x + 5$

2. A **solution** of an equation is any replacement for the variable that makes the equation true.

   Equation: $x + 4 = 9$
   When we replace x with 5, $5 + 4 = 9$, which is true. Therefore, 5 is a solution.

3. Suppose that a, b, and c represent any numbers.

   If $a = b$, then $a + c = b + c$.

   If $a = b$, then $a - c = b - c$.

   In words, any number can be added to or subtracted from both sides of an equation.

$$\begin{aligned} x - 5 &= 11 \\ x - 5 + 5 &= 11 + 5 \\ x + 0 &= 16 \\ x &= 16 \end{aligned}$$

$$\begin{aligned} -4 &= x + 9 \\ -4 - 9 &= x + 9 - 9 \\ -13 &= x + 0 \\ -13 &= x \\ x &= -13 \end{aligned}$$

B. The Multiplication Property of Equations

1. Suppose that a, b, and c represent any numbers, where c is not 0.

   If $a = b$, then $ac = bc$.

   If $a = b$, then $\frac{a}{c} = \frac{b}{c}$.

   In words, we can multiply or divide both sides of an equation by any number except 0.

$$7x = 35$$
$$\frac{7x}{7} = \frac{35}{7}$$
$$x = 5$$

2. One or both sides of an equation may need to be simplified before the properties of equations are applied.

$$7x - 2x = 8(3) + 1$$
$$5x = 25$$
$$\frac{5x}{5} = \frac{25}{5}$$
$$x = 5$$

Copyright © Houghton Mifflin Company. All rights reserved.

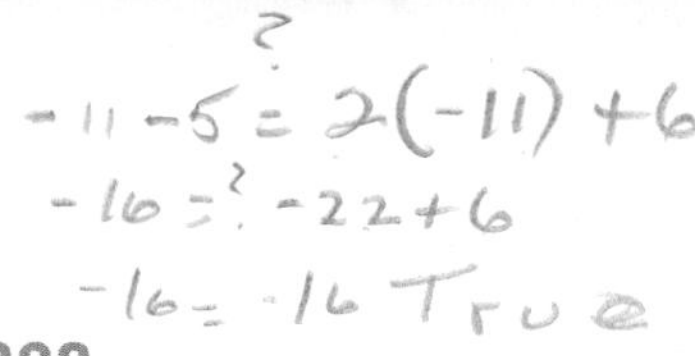

4.1 Exercises

A. The Addition Property of Equations

1. The _______ Property of Equations states that any number can be added to or subtracted from both sides of an equation.

2. When we write an equation so that the variable is by itself, we say that we _______ the variable.

In Exercises 3–8, determine whether the given number is a solution of the equation.

3. $x - 5 = 2x + 6;\quad -11$
4. $2x + 7 = x - 3;\quad 10$
5. $-3x + 7 = 1 - x;\quad 3$
6. $4 - x = 1 - 5x;\quad -1$
7. $12 - x = 2x;\quad -4$
8. $3x = -8 - x;\quad -2$

In Exercises 9–22, solve the given equation.

9. $x + 9 = 5$
10. $x + 7 = 12$
11. $y - 4 = 3$
12. $b - 8 = -2$
13. $2 = 9 + t$
14. $-5 = 3 + n$
15. $-5 = a - 10$
16. $-7 = c - 14$
17. $-3 + x = 1$
18. $-4 + y = -4$
19. $n + 7 = 0$
20. $0 = 3 + a$
21. $x - 9 = 5$
22. $z - 3 = -11$

In Exercises 23–30, simplify and solve the given equation.

23. $-3x - 7 + 4x = 2 - 7$
24. $5x - 2 - 4x = -3 - 5$
25. $-3 + 5 = 5 - 3y + 4y$
26. $4 - (-2) = 6a + 1 - 5a$

Copyright © Houghton Mifflin Company. All rights reserved.

27. $-4 + 2x - x = -5 - (-5)$

28. $3 - y - 4y + 6y = 7 - 2 - 5$

29. $-6 + 5 = 3n + 7 - 7n + 5n$

30. $-7 - (-5) = -6c - 6 + 7c$

B. The Multiplication Property of Equations

31. The ________ Property of Equations states that we can multiply or divide both sides of an equation by any number except zero.

32. To solve $-n = 8$, we can either ________ or ________ both sides of the equation by -1.

In Exercises 33–44, solve the given equation.

33. $8a = -40$

34. $-3y = 30$

35. $-6x = -12$

36. $4x = 28$

37. $-n = 3$

38. $-x = -6$

39. $45 = 5y$

40. $7 = -7c$

41. $0 = -3b$

42. $4x = 0$

43. $10x = 70$

44. $9z = -63$

In Exercises 45–50, simplify and solve.

45. $-2y - 3y = 3 - 6 - (-3)$

46. $n + 5n - 2n = -14 + 2$

47. $-2x + x = -3(-2)$

48. $5x - 6x = 1 - 10$

49. $-5x - x - 2x = -11 - 13$

50. $y + 8y + 3y = 4(-12)$

Copyright © Houghton Mifflin Company. All rights reserved.

Calculator Exercises

In Exercises 51–54, use a calculator to solve the given equation.

51. $y - 368 = -1{,}272$

52. $1{,}243 + x = 5{,}280$

53. $79x = -12{,}245$

54. $-128y = -6{,}912$

Bringing It Together

In Exercises 55–58, solve the given pairs of equations.

55. **(a)** $x + 7 = 4$ **(b)** $7x = -35$

56. **(a)** $-3n = -18$ **(b)** $n - 3 = 0$

57. **(a)** $-9y = -27$ **(b)** $6 = y - 9$

58. **(a)** $-3 = x + 4$ **(b)** $4x = -28$

Writing and Concept Extension

59. Explain why the Addition Property of Equations allows us to subtract a number from both sides of an equation.

60. Explain why we add 2 to both sides to solve $x - 2 = 6$, but we divide both sides by 2 to solve $2x = 6$.

61. To eliminate -3 in the equation $-3 + x = 15$, we add 3 to both sides. Why must the signs be different?

62. To eliminate -3 in the equation $-3x = 15$, we divide by -3. Why must we divide by -3 rather than 3?

63. Solve the equation $1 \cdot n = n$.

64. Solve the equation $0 \cdot n = 0$.

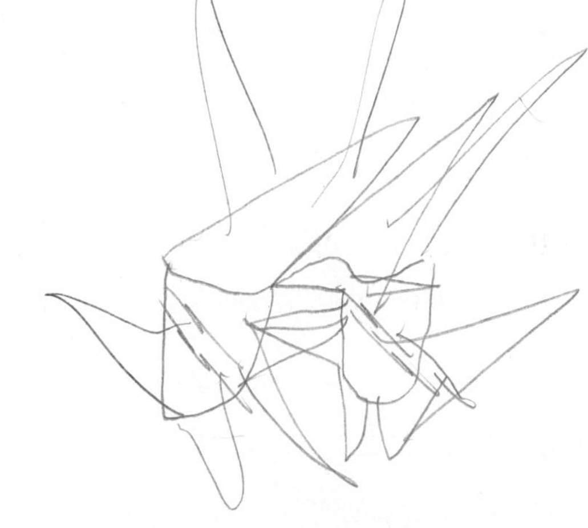

Copyright © Houghton Mifflin Company. All rights reserved.

In Exercises 65–68, simplify and solve.

65. $3(x - 2) + 6 = 15$

66. $-24 = -4 + 2(2 - 3x)$

67. $2(1 - x) + 3x = 7$

68. $0 = 7x - 2(3x + 5)$

Copyright © Houghton Mifflin Company. All rights reserved.

4.2 Linear Equations in One Variable

SUGGESTIONS FOR SUCCESS

Take one step at a time.

A *procedure* is a list of steps or instructions that are to be carried out in a certain order: Do A, then B, then C. If you do B before A or if you try to do A and B together, the wheels are likely to come off.

In this section, we present a procedure for solving a certain type of equation. Try writing the procedure on a card and keeping it beside you as you work the exercises. Refer to it, step by step. Eventually, the procedure will become second nature to you, and you will be a successful equation solver.

A Applying Both Properties of Equations

In Section 4.1, all the equations that we solved are called *linear equations in one variable.* As the name implies, such equations contain only one variable. Also, the exponent on the variable is 1, and the variable cannot appear in the denominator of a fraction.

These Are Linear Equations in One Variable.	*These Are Not Linear Equations in One Variable.*	
$3x + 2 = 7$	$2x + 3y = 8$	More than one variable
$x - 4 = 5 - 6x$	$x^2 + 1 = 4$	Exponent on x is not 1.
$2(x + 1) = 9$	$\frac{3}{x} = -2$	Variable in the denominator

In this section, we will continue to focus on methods for solving linear equations in one variable.

So far, we have been able to solve equations by applying either the Addition Property of Equations or the Multiplication Property of Equations. We often need to apply both properties.

Applying Both Properties

When we solve an equation such as $x + 5 = 9$, our goal is to isolate the variable term, x. This means that we must eliminate the 5.

$$x + 5 = 9$$
$$x + 5 - 5 = 9 - 5$$
$$x + 0 = 4$$
$$x = 4$$

For an equation such as $2x + 5 = 9$, our first goal is the same: Isolate the variable term, $2x$. Again, we must eliminate the 5.

$$2x + 5 = 9$$
$$2x + 5 - 5 = 9 - 5$$
$$2x + 0 = 4$$
$$2x = 4$$

Copyright © Houghton Mifflin Company. All rights reserved.

Now the variable term, $2x$, is isolated.

To complete the problem, we isolate x itself by dividing both sides by 2.

$$\frac{2x}{2} = \frac{4}{2}$$

$$x = 2$$

In summary, we begin by isolating the variable term, and then we isolate the variable itself.

The following is the beginning of a procedure for solving linear equations in one variable. We have left the first step blank. We will fill in that step later.

Equation-Solving Procedure

1.
2. Use the Addition Property of Equations to isolate the variable term.
3. Use the Multiplication Property of Equations to isolate the variable.
4. Check the solution.

Example 1

Solve $5x - 2 = -17$.

SOLUTION

$$5x - 2 = -17$$

$$5x - 2 + 2 = -17 + 2$$ Add 2 to both sides to isolate the variable term.

$$5x = -15$$

$$\frac{5x}{5} = \frac{-15}{5}$$ Divide both sides by 5 to isolate the variable.

$$x = -3$$

Check

$5x - 2$	-17
$5(-3) - 2$	-17
$-15 - 2$	-17
-17	-17

The solution is -3.

Your Turn 1

Solve $-4x + 9 = -7$.

$-4x+9-9=-7-9$

$-4x=-16$

$x=4$

Answer: 4

Step 2 of the equation-solving procedure is to isolate the variable term by using the Addition Property of Equations. However, sometimes we can isolate the variable term simply by combining like terms.

$$8x + 2x - 7x = 9$$ Begin by combining like terms.

$$3x = 9$$ Now the variable term is isolated.

In fact, combining like terms may isolate the variable term and the variable itself without using either of the properties of equations.

Copyright © Houghton Mifflin Company. All rights reserved.

Example 2

Solve $16 = 10 - 3x$.

SOLUTION

$$16 = 10 - 3x$$

$$-10 + 16 = -10 + 10 - 3x$$ Add −10 to both sides to isolate the variable term.

$$6 = -3x$$

$$\frac{6}{-3} = \frac{-3x}{-3}$$ Divide both sides by −3 to isolate the variable.

$$-2 = x$$

Check

16	$10 - 3x$
16	$10 - 3(-2)$
16	$10 - (-6)$
16	16

The solution is −2.

Your Turn 2

Solve $-19 = 5x - 4$.

Answer: −3

NOTE Although we will not always show the checking step, you will benefit from checking solutions as you work through the remaining examples. Of course, you should check your solutions in all the exercises that you do.

B Variable Terms on Both Sides

If variable terms appear on both sides of an equation, we can use the Addition Property of Equations to eliminate the variable term from one of the sides.

Example 3

Solve $8x + 7 = 3x - 13$.

SOLUTION

We begin by subtracting $3x$ from both sides. This will eliminate the variable term from the right side.

$$8x + 7 = 3x - 13$$

$$8x - 3x + 7 = 3x - 3x - 13$$ Subtract $3x$ from both sides.

$$5x + 7 = -13$$ Combine like terms.

$$5x + 7 - 7 = -13 - 7$$ Subtract 7 from both sides to isolate the variable term.

$$5x = -20$$

$$\frac{5x}{5} = \frac{-20}{5}$$ Divide both sides by 5 to isolate the variable.

$$x = -4$$

The solution is −4.

Your Turn 3

Solve $6y - 1 = 2y + 11$.

Answer: 3

Copyright © Houghton Mifflin Company. All rights reserved.

When variable terms appear on both sides of an equation, you can choose the one that you want to eliminate. In Example 3, we chose to eliminate $3x$ from the right side, but we could have eliminated $8x$ from the left side.

$$8x + 7 = 3x - 13$$

$$8x - 8x + 7 = 3x - 8x - 13 \qquad \text{Subtract } 8x \text{ from both sides.}$$

$$7 = -5x - 13 \qquad \text{Combine like terms.}$$

$$7 + 13 = -5x - 13 + 13 \qquad \text{Add 13 to both sides to isolate the variable term.}$$

$$20 = -5x$$

$$\frac{20}{-5} = \frac{-5x}{-5} \qquad \text{Divide both sides by } -5 \text{ to isolate the variable.}$$

$$-4 = x$$

As you gain more experience, you will be able to select the method that requires the fewest steps.

C Simplifying Both Sides

Before we use the Addition and Multiplication Properties of Equations, we might need to simplify one or both sides of an equation. Recall that simplifying an expression involves removing grouping symbols and combining like terms. Let's add this step to our equation-solving procedure.

Equation-Solving Procedure

1. If necessary, simplify the two sides of the equation.
   (a) Remove grouping symbols.
   (b) Combine like terms.
2. Use the Addition Property of Equations to isolate the variable term.
3. Use the Multiplication Property of Equations to isolate the variable.
4. Check the solution.

Example 4

Solve $2(x + 1) = 8$.

SOLUTION

$$2(x + 1) = 8$$

$$2 \cdot x + 2 \cdot 1 = 8 \qquad \text{Remove the parentheses with the Distributive Property.}$$

$$2x + 2 = 8$$

$$2x + 2 - 2 = 8 - 2 \qquad \text{Subtract 2 from both sides to isolate the variable term.}$$

$$2x = 6$$

$$\frac{2x}{2} = \frac{6}{2} \qquad \text{Divide both sides by 2 to isolate the variable.}$$

$$x = 3$$

The solution is 3.

Your Turn 4

Solve.

$3(x - 5) = 6$

Answer: 7

Copyright © Houghton Mifflin Company. All rights reserved.

Example 5

Solve $5x + 3(x - 1) = 13$.

SOLUTION

$$5x + 3(x - 1) = 13$$

$$5x + 3 \cdot x - 3 \cdot 1 = 13$$ Remove the parentheses with the Distributive Property.

$$5x + 3x - 3 = 13$$

$$8x - 3 = 13$$ Combine like terms.

$$8x - 3 + 3 = 13 + 3$$ Add 3 to both sides to isolate the variable term.

$$8x = 16$$

$$\frac{8x}{8} = \frac{16}{8}$$ Divide both sides by 8 to isolate the variable.

$$x = 2$$

The solution is 2.

Your Turn 5

Solve.

$-2x + 4(x + 3) = 4$

Answer: −4

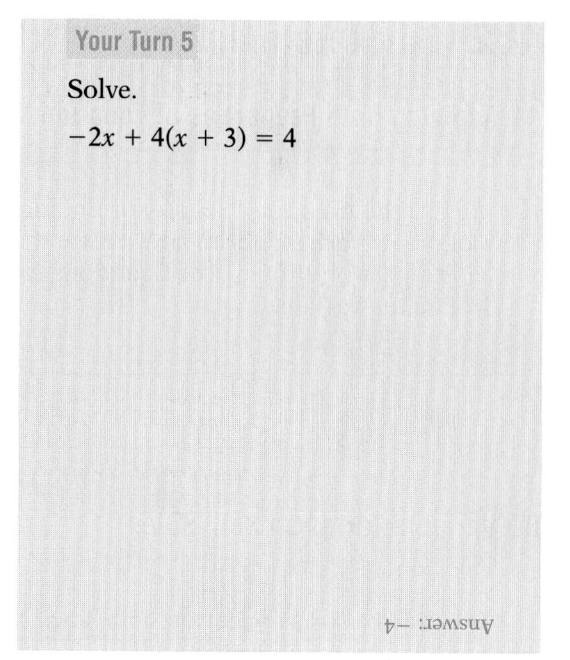

Recall that $-a = -1a$. When you solve an equation that involves the opposite of a quantity, replacing the opposite symbol with -1 can help you to see what to do.

Example 6

Solve $-(x + 3) + 2x = 0$.

SOLUTION

$$-(x + 3) + 2x = 0$$

$$-1(x + 3) + 2x = 0$$ $-a = -1a$

$$-1 \cdot x + (-1)(3) + 2x = 0$$ Distributive Property

$$-1x + (-3) + 2x = 0$$

$$x + (-3) = 0$$ Combine like terms.

$$x - 3 = 0$$ $a + (-b) = a - b$

$$x - 3 + 3 = 0 + 3$$ Add 3 to both sides.

$$x = 3$$

The solution is 3.

Your Turn 6

Solve $-(x - 2) = 5$.

Answer: −3

Parentheses can be preceded by an opposite symbol (as in Example 6) or by a minus symbol. The procedure for removing the parentheses is essentially the same in both cases. Consider the following equation:

$$1 - (2x + 1) = 6$$

$$1 + [-(2x + 1)] = 6$$ Definition of subtraction

$$1 + [-1(2x + 1)] = 6$$ $-a = -1a$

$$1 + (-2x) + (-1) = 6$$ Distributive Property

Now the parentheses around $2x + 1$ are removed, and we can perform the remaining steps in the usual manner.

Copyright © Houghton Mifflin Company. All rights reserved.

4.2 Quick Reference

A. Applying Both Properties of Equations

1. We use the Addition Property of Equations to isolate the variable term, and we use the Multiplication Property of Equations to isolate the variable.

$$7x - 4 = 10$$
$$7x - 4 + 4 = 10 + 4$$
$$7x = 14$$
$$\frac{7x}{7} = \frac{14}{7}$$
$$x = 2$$

B. Variable Terms on Both Sides

1. If variable terms appear on both sides of an equation, we can use the Addition Property of Equations to eliminate the variable term from one of the sides.

$$4x + 10 = 9x$$
$$4x - 4x + 10 = 9x - 4x$$
$$10 = 5x$$
$$\frac{10}{5} = \frac{5x}{5}$$
$$2 = x$$

C. Simplifying Both Sides

1. We might need to simplify one or both sides of an equation by removing grouping symbols and combining like terms.

$$3(x - 1) + 2x = 12$$
$$3x - 3 + 2x = 12$$
$$5x - 3 = 12$$
$$5x - 3 + 3 = 12 + 3$$
$$5x = 15$$
$$x = 3$$

2. When the opposite of a quantity appears in an equation, we can replace the opposite symbol with -1 and then apply the Distributive Property.

$$-(x - 4) = 13$$
$$-1(x - 4) = 13$$
$$-1x + 4 = 13$$
$$-1x + 4 - 4 = 13 - 4$$
$$-1x = 9$$
$$\frac{-1x}{-1} = \frac{9}{-1}$$
$$x = -9$$

Copyright © Houghton Mifflin Company. All rights reserved.

4.2 Exercises

A. Applying Both Properties of Equations

1. The equation $2x - 7 = 1$ is an example of a Linear equation in one variable

2. When we solve an equation, we use the ________ Property of Equations to isolate the variable term.

In Exercises 3–18, solve the given equation.

3. $3x + 8 = -13$

4. $2a - 11 = 9$

5. $4n - 10 = -30$

6. $3x + 1 = 19$

7. $8 = -3a - 7$

8. $2 = -b - 10$

9. $-x + 3 = 1$

10. $-3t + 5 = -7$

11. $9 - 8y = 17$

12. $-7 + 2b = 7$

13. $0 = 6n - 24$

14. $8 - p = 3$

15. $21 = 5 + 8x$

16. $-3 = 15 - 6y$

17. $3a - 2 = -2$

18. $18 = -8x - 6$

Copyright © Houghton Mifflin Company. All rights reserved.

B. Variable Terms on Both Sides

19. For the equation $3x - 2 = 3 - 2x$, we can ________ $2x$ to both sides to eliminate the variable term on the ________ side of the equation.

20. For the equation $3x - 2 = 3 - 2x$, we can ________ $3x$ from both sides to eliminate the variable term on the ________ side of the equation.

In Exercises 21–32, solve the given equation.

21. $7x - 6 = 4x + 12$

22. $5x + 2 = x - 6$

23. $x + 4 = 3x - 2$

24. $6y - 5 = 10y + 3$

25. $5n - 14 = 6 - 5n$

26. $5 + b = 7 - b$

27. $3x - 20 = 7x$

28. $x = -4x$

29. $4a + 7 = 7 - 8a$

30. $4 - 4y = -2 - y$

31. $-5x = 10 - 3x$

32. $3w - 16 = -w$

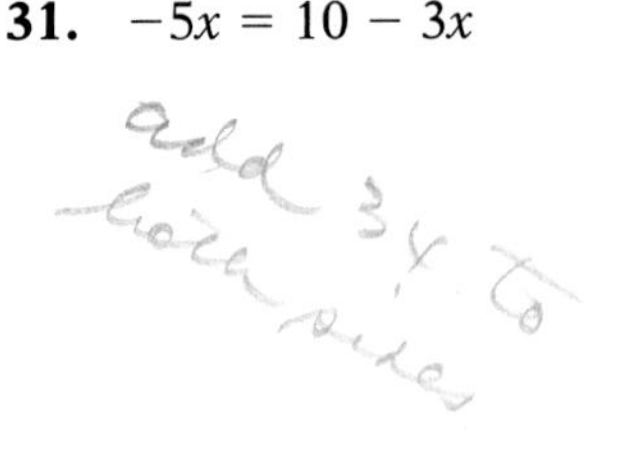

C. Simplifying Both Sides

33. We remove parentheses and combine like terms to ________ an expression.

34. If possible, we usually ________ both sides of an equation before applying the Addition or Multiplication Property of Equations.

In Exercises 35–48, solve the given equation.

35. $7 - 3x + 8x = 10 + x - 3$

36. $7x + 8 - 6x = 2 - 5x - 12$

37. $3(x + 2) = 12$

Copyright © Houghton Mifflin Company. All rights reserved.

38. $4(1 - x) = 24$

39. $17 = -(4y - 5)$

40. $0 = 11 - (-1 - 3z)$

41. $6 + 7x - 4x = 2(3 - x)$

42. $5(a - 2) = 9a + 4a + 6$

43. $-5 = -2(2x - 5) - 3$

44. $6x + 2(7 - x) = -2$

45. $-(2y - 5) + y = 0$

46. $12 = -(3 - 2b) + 5$

47. $3(x - 7) = 6x - (1 - 2x)$

48. $-(2y + 1) - 31 = 2(3y - 4)$

Bringing It Together

In Exercises 49–62, solve the given equation.

49. $7y + 5 = 11 + 4y$

50. $-5 - 2y = y + 1$

51. $9 = 15 + 3c$

52. $10 - 6x = -2$

53. $-7x = 70$

54. $-3 = x - 9$

55. $7 - w = 12$

56. $-7 = 2 - 3n$

57. $2(1 - 3n) = 8$

58. $20 = 5(3c - 2)$

59. $3b + 7 = 4b$

60. $3z = 8z - 25$

Copyright © Houghton Mifflin Company. All rights reserved.

61. $-2 = 4 - 3(2 - x)$

62. $(2x + 1) - (4 - x) = x - 9$

Calculator Exercises

In Exercises 63–66, use a calculator to solve the given equation.

63. $47x - 840 = 194$

64. $1{,}350x + 690 = 16{,}890$

65. $88x + 4{,}806 = 51x + 1{,}513$

66. $1{,}199 - 28x = 38x + 2{,}651$

Writing and Concept Extension

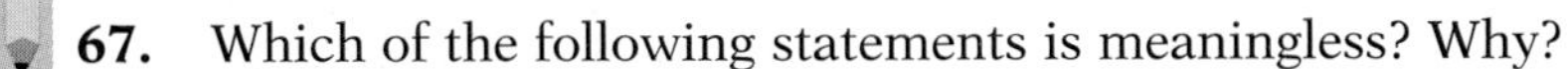

67. Which of the following statements is meaningless? Why?

(i) Solve $2(x - 1) + 10$.

(ii) Solve $2(x - 1) = 10$.

68. Which of the following statements is meaningless? Why?

(i) Simplify $3 - 5x = -8$.

(ii) Simplify $3 - 5x - 8$.

An equation has no solution when no replacement for the variable makes the equation true. We call such equations **contradictions.** There are also special cases in which *every* replacement for the variable makes the equation true. Such equations are called **identities.**

In Exercises 69 and 70, use the preceding information to solve the equation.

69. $x + 7 = 2 + x$

70. $3(x - 2) = 3x - 6$

Copyright © Houghton Mifflin Company. All rights reserved.

Do this (7/7)

4.3 Applications

A *Translations into Equations*
B *General Strategy*

SUGGESTIONS FOR SUCCESS

"I can do a story problem once it's set up."

If you are like most students, the hardest part of application problems is getting started. Assuming that you are now a good equation solver, we urge you to pay special attention to the first five steps of the general procedure that we will present.

In particular, make sure that you read the problem carefully enough to know what your goal is. What unknown quantity are you trying to find and how is it related to the other given information? Try to state the problem in your own words as though you were explaining it to someone else.

Above all, don't let story problems scare you. Trust the general procedure and all will be well.

A Translations into Equations

The basic practical use of mathematics is for the solution of a problem in which some unknown quantity is to be found. Typically, we are given certain known information about the problem. The information usually consists of specific known values and of the relationships between those values and the unknown quantity.

LEARNING TIP

Take the time to write the specific meaning of the variable. Writing c = car is not meaningful if you mean c = the cost of the car. Remember that a variable represents a number, not a thing.

To solve such problems, we need to be able to *translate* the given information into mathematical expressions and equations. The process begins with our assigning a variable to the unknown quantity. Here are examples of the way most application problems begin.

Let x = the number.
Let n = the number of people at the concert.
Let w = the boy's weight.
Let L = the length of the rectangle.
Let c = the cost of the computer.

Of course, you can choose any variable that you want to use to represent the unknown quantity.

In English, a sentence contains a verb, whereas a phrase does not.

Phrase	*Sentence*
4 more than a number	4 more than a number is 12. (↑ verb)

In mathematics, we translate a *phrase* into an *expression*; we translate a *sentence* into an *equation*. The main difference is that an equation contains the symbol =, whereas an expression does not.

Some Words That Mean =

Equals
Is
Results in

Copyright © Houghton Mifflin Company. All rights reserved.

Example 1

Translate the given information into an equation.

(a) 6 more than a number is 13.

(b) 3 less than 4 times a number is −2.

(c) 4 times the difference of a number and 5 is 9.

SOLUTION

(a)

6	6
$+ 6$	more than (added to)
$x + 6$	a number (unknown)
$x + 6 =$	is
$x + 6 = 13$	13

The equation is $x + 6 = 13$.

(b)

3	3
$- 3$	less than (subtracted from)
$4x - 3$	4 times a number (unknown)
$4x - 3 =$	is
$4x - 3 = -2$	−2

The equation is $4x - 3 = -2$.

(c)

4	4
$4 \cdot$	times
$4 \cdot (\ \ -\ \ )$	the difference of
$4 \cdot (x -\ \ )$	a number (unknown)
$4 \cdot (x - 5)$	and 5
$4 \cdot (x - 5) =$	is
$4 \cdot (x - 5) = 9$	9

The equation is $4(x - 5) = 9$.

Your Turn 1

Translate the given information into an equation.

(a) The difference between a number and 7 is −4.

(b) Adding 6 to twice a number results in 8.

(c) Twice the sum of a number and 3 is −5.

Answers: (a) $x - 7 = -4$; (b) $2x + 6 = 8$; (c) $2(x + 3) = -5$

B General Strategy

Translating information into an equation is only part of the process of solving an application problem. The following general strategy for solving application problems is a useful guide in helping you to think about the problem, to organize information, and to work toward a solution in a systematic way.

General Strategy for Solving Application Problems

1. Read the problem carefully to get a sense of what the problem is about. Make sure you understand what information is given and what the problem is asking you to determine.
2. If appropriate, draw a figure or a diagram or make a chart or table to organize the information.

Copyright © Houghton Mifflin Company. All rights reserved.

3. Assign a variable to the unknown quantity.
4. If more than one unknown quantity is involved, represent each quantity with an expression in the same variable.
5. Translate the given information into an equation.
6. Solve the equation.
7. Answer the question. (The solution in step 6 is not necessarily the answer to the question.)
8. Check the answer to see whether it meets the conditions stated in the problem.

Application problems come in many flavors, and no single procedure can be used to solve every type. Nevertheless, the preceding general strategy can be applied to a wide variety of problems.

We don't claim that the following examples of applications are very practical. Their main purpose is to illustrate the use of the general strategy and to give you practice with the problem-solving process. The circled numbers refer to the steps in the general procedure. (Not all steps are used in every example.)

Example 2

If decreasing a number by 3 results in 4 more than twice the number, what is the number?

SOLUTION

① *What is the goal?*

Find a certain number.

③ *Assign a variable.*

Let x = the unknown number.

⑤ *Translate into an equation.*

$(\ ) - (\ )$	decreasing (subtracting)
$x - (\ )$	a number (unknown)
$x - 3$	by 3
$x - 3 =$	results in (equals)
$x - 3 = (\ ) + 4$	4 more than (added to)
$x - 3 = 2x + 4$	twice the number

⑥ *Solve the equation.*

$x - 3 = 2x + 4$	
$x - x - 3 = 2x - x + 4$	Subtract x from both sides so that the variable is on one side.
$-3 = x + 4$	Combine like terms.
$-3 - 4 = x + 4 - 4$	Subtract 4 from both sides.
$-7 = x$	

⑦ *Answer the question.*

The unknown number is -7.

Your Turn 2

If 5 more than a number is 1 less than 3 times the number, what is the number?

continued

Copyright © Houghton Mifflin Company. All rights reserved.

⑧ *Check the answer.*

Decreasing -7 by 3 means

$$-7 - 3 = -10$$

4 more than twice the number means

$$2(-7) + 4 = -14 + 4 = -10$$

Because the results are the same, the answer is verified.

Observe that we check the answer by going back into the wording of the problem. We don't just substitute the value of x into the equation because we might not have written the equation correctly.

NOTE Although checking is an essential step, in our remaining examples we will leave this part of the problem-solving procedure to you.

Example 3

The length of a rectangle is 3 times its width. If the perimeter is 64 inches, what is the length of the rectangle?

SOLUTION

① *What is the goal?*

Find the length of a rectangle.

② *Draw a figure.*

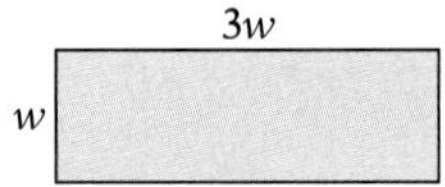

③ *Assign a variable*

Note in the figure that we have chosen to let $w =$ the width of the rectangle.

④ *Represent other unknown quantities.*

The length is 3 times the width.

$$\text{Length} = 3w$$

The perimeter is the sum of the lengths of the sides.

$$\text{Perimeter} = w + 3w + w + 3w$$

⑤ *Translate into an equation.*

We know that the perimeter is 64 inches.

$$w + 3w + w + 3w = 64$$

Your Turn 3

One side of a triangle is twice as long as another side. The third side is 6 feet long. If the perimeter is 21 feet, how long is the shortest side of the triangle?

Copyright © Houghton Mifflin Company. All rights reserved.

⑥ *Solve the equation.*

$$w + 3w + w + 3w = 64$$
$$8w = 64 \quad \text{Combine like terms.}$$
$$w = 8 \quad \text{Divide both sides by 8.}$$

⑦ *Answer the question.*

Now we know that the width w is 8 inches, but the problem asks for the length, $3w$.

$$3w = 3(8) = 24$$

The length of the rectangle is 24 inches.

LEARNING TIP

When there are two unknown quantities, such as width and length in Example 3, we often let the variable represent the smaller quantity.

Answer: 5 feet

Example 4

An electrician charges \$50 to make a house call and \$30 for each hour that he works. How many hours did the electrician work if his total bill was \$260?

SOLUTION

① *What is the goal?*

Find the number of hours that the electrician worked.

③ *Assign a variable.*

Let h = the number of hours worked.

④ *Represent other unknown quantities.*

The electrician charges \$30 per hour.

$$\text{Charge for } h \text{ hours} = 30h$$

⑤ *Translate into an equation.*

The total bill includes the charge for h hours plus \$50 for the house call.

$$30h + 50 = 260 \quad \text{The total charge was \$260.}$$

⑥ *Solve the equation.*

$$30h + 50 = 260$$
$$30h + 50 - 50 = 260 - 50 \quad \text{Subtract 50 from both sides.}$$
$$30h = 210$$
$$h = 7 \quad \text{Divide both sides by 30.}$$

⑦ *Answer the question.*

The electrician worked 7 hours.

Your Turn 4

A car rental agency charges a flat rate of \$25 plus \$30 for each day that the car is rented. For how many days was a car rented if the final bill was \$115?

Answer: 3

Remember that when a problem involves two unknown quantities, we need to represent both quantities with expressions in the same variable.

Copyright © Houghton Mifflin Company. All rights reserved.

Example 5

A 30-foot rope is cut so that one piece is 6 feet longer than the other piece. How long are the two pieces?

SOLUTION

① *What is the goal?*

Find the lengths of the two pieces.

② *Draw a figure.*

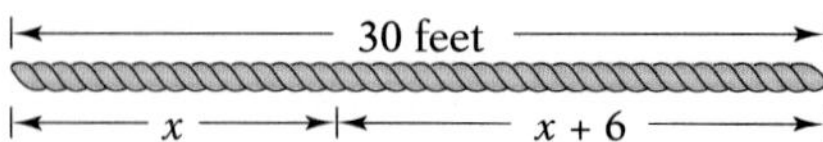

③ *Assign a variable.*

Note in the figure that we have chosen to let $x =$ the length of the shorter piece.

④ *Represent other unknown quantities.*

The other piece is 6 feet longer.

Length of the longer piece $= x + 6$

⑤ *Translate into an equation.*

$x + (x + 6) = 30$ — The sum of the lengths of the two pieces is 30 feet.

⑥ *Solve the equation.*

$x + (x + 6) = 30$

$x + x + 6 = 30$ — Remove the parentheses.

$2x + 6 = 30$ — Combine like terms.

$2x + 6 - 6 = 30 - 6$ — Subtract 6 from both sides.

$2x = 24$

$x = 12$ — Divide both sides by 2.

⑦ *Answer the question.*

The length x of the shorter piece is 12 feet. The length $x + 6$ of the longer piece is $12 + 6 = 18$ feet.

Your Turn 5

A steam fitter has a 12-foot pipe that he needs to cut into two pieces. If one piece is to be twice as long as the other, how long is the shorter piece?

Answer: 4 feet

Copyright © Houghton Mifflin Company. All rights reserved.

4.3 Exercises

A. Translations into Equations

In Exercises 1–8, translate the given phrase into a variable expression.

1. A number n increased by 7

2. 4 more than a number x

3. The sum of -3 and y

4. The difference between 8 and a number y

5. Five less than a number b

6. -9 subtracted from x

7. Twice a number z

8. The quotient of a number c and 7

In Exercises 9–14, let n represent the unknown number and translate the given phrase into a variable expression.

9. 3 more than twice a number

10. 5 less than four times a number

11. 8 less than half of a number

12. Three more than the quotient of a number and 6

13. Three times the sum of a number and 7

14. Twice the difference of 8 and a number

In Exercises 15–20, let n represent the unknown number and translate the given information into an equation.

15. Five less than a number is 8.

16. When 5 times a number is increased by 8, the result is -17.

17. Three times the sum of a number and 3 is -21.

Copyright © Houghton Mifflin Company. All rights reserved.

18. Twice the difference of a number and −4 is −12.

19. Six less than the product of a number and 5 is −1.

20. The product of 7 and a number is twice the number.

B. General Strategy

In Exercises 21–26, write and solve an equation to determine the unknown number. (Let x represent the unknown number.)

21. When a number is increased by 8, the result is −4.

22. The difference of a number and −7 is 8.

23. Seven less than twice a number is −9.

24. When the product of a number and 4 is decreased by 9, the result is 11.

25. Twice the difference of a number and −6 is 20.

26. Five subtracted from twice the difference of a number and 2 is 5.

Copyright © Houghton Mifflin Company. All rights reserved.

In Exercises 27–30, write and solve an equation to determine the unknown number. (Let x represent the unknown number.)

27. The difference of twice a number and 3 is five times the number.

28. Four times a number is 12 less than the number.

29. One less than 5 times a number is the difference of 7 and three times the number.

30. Three more than twice a number is the sum of 10 and three times the number.

In Exercises 31–34, write and solve an equation to determine the unknown dimensions.

31. The length of a rectangle is 3 feet more than its width. The perimeter is 26 feet. Find the length and width of the rectangle.

32. The width of a rectangle is 6 yards less than the length. The perimeter is 32 yards. Find the length and width of the rectangle.

33. One side of a triangle is 3 inches shorter than a second side. The longest side is 13 inches, and the perimeter of the triangle is 26 inches. Find the length of the shortest side.

Copyright © Houghton Mifflin Company. All rights reserved.

34. Two sides of a triangle are the same length and the third side is 3 times the length of the two equal sides. The perimeter is 125 feet. What are the lengths of the sides of the triangle?

In Exercises 35–40, use the general strategy that was presented in this section to solve the problem.

35. *Planned Community* The number of apartments in a planned community is 5 less than 3 times the number of single family homes. If there are 151 apartments, how many single family homes are in the community?

36. *Gender Mix* The number of women in an economics class is 3 less than twice the number of men. If 21 women are in the class, how many men are in the class?

37. *Playground Enclosure* The residents of a housing development use 360 feet of fencing to enclose a rectangular playground. The length of the playground is twice the width. What are the dimensions of the playground?

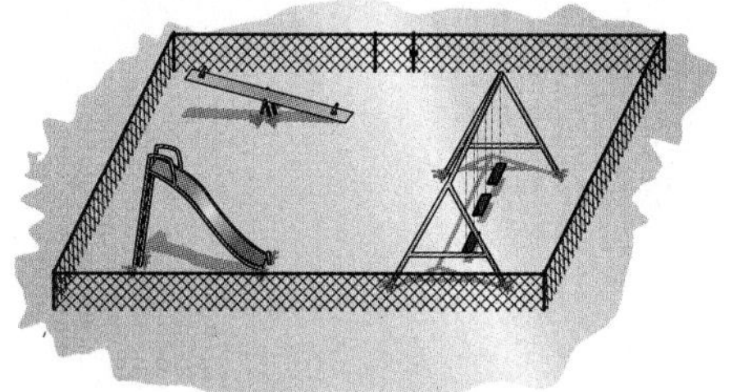

38. *Wire Frame* A wire, 40 inches long, is bent to form a rectangular frame. The width of the frame is 4 inches less than the length. What are the dimensions of the frame?

40 in.
W L W L → W L

39. *Plumbing Job* A plumber charged \$40 plus \$18 per hour. If the total bill was \$130, how many hours did the plumber work?

Copyright © Houghton Mifflin Company. All rights reserved.

40. *Car Depreciation* Suppose that the value of a new car decreases by \$2,600 each year. If the price of a new car is \$18,000, after how many years will the value of the car be \$5,000?

In Exercises 41–46, two quantities are unknown. Be sure to represent both quantities with variable expressions in the same variable before you write the equation.

41. One number is 3 more than another. If the sum of the numbers is 27, what are the two numbers?

42. One number is 5 less than another number. If the difference of the larger number and twice the smaller number is 8, what are the two numbers?

43. *Board Lengths* A board 10 feet long is cut into 2 pieces. If the second piece is 2 feet longer than the first, determine the length of each piece.

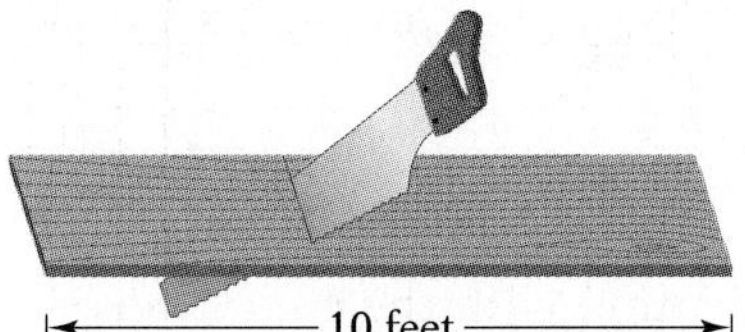

44. *Cloth Lengths* A bolt of cloth 50 yards long is cut into two pieces. If the length of one piece is 5 yards more than twice the length of the other piece, what is the length of each piece?

45. *Books and Tuition* A student spent \$1,000 on tuition and books. If the cost of tuition was \$100 more than 3 times the cost of books, what was the cost of tuition?

Copyright © Houghton Mifflin Company. All rights reserved.

46. *Ornamental Trees* A garden center advertises Bradford pear trees and Yoshino cherry trees. If the center has a total of 85 trees available and the number of pear trees is 15 less than 3 times the number of cherry trees, how many pear trees does the center have in stock?

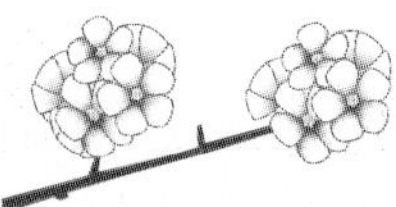

White House Pets While occupying the White House, 23 presidents have had a dog as a pet. Presidents also have had a variety of other pets. (Source: *USA Today.*) In Exercises 47 and 48, solve an equation to determine the number of presidents who had other pets.

47. The number of presidents who have had dogs is 7 less than 3 times the number of those who have had cats. How many presidents had cats?

48. Three times the number of presidents who have had horses is 10 more than the number of those who have had dogs. How many presidents had horses?

Writing and Concept Extension

49. How do you know whether to translate information into an expression or an equation?

50. Why should you check the answer to an application with the conditions of the problem rather than just substituting the value of the variable into the equation?

51. *Masonry Job* A brick mason charges $20 per hour and a helper charges $14 per hour. On a certain job, the brick mason worked twice the number of hours as the helper. How many hours did each work if the total charge was $378?

52. *Land Parcels* One parcel of land is 3 times as large as the other. If the size of each parcel were increased by 50 acres, one would be twice as large as the other. What is the size of each parcel of land?

Copyright © Houghton Mifflin Company. All rights reserved.

4.4 Equations in Two Variables

A *Translations*
B *Solutions of Equations in Two Variables*
C *Determining Solutions*

SUGGESTIONS FOR SUCCESS

Try to anticipate.

When we solve an equation with one variable, we find a value of that variable that makes the equation true. Suppose an equation has two variables. What do you suppose we need to find? If you said values for both variables that make the equation true, you are right!

As you study mathematics, don't be afraid to ask "What if?" questions. Predicting how something is probably going to be done will make the process more meaningful to you, and you are likely to understand and remember it better.

A Translations

The general strategy for solving an application problem includes assigning a variable to the unknown quantity and representing any other unknown quantities in that same variable. Then the equation that we write is an equation that has just one variable.

For some application problems, an alternative is to use two different variables to represent the unknown quantities. For example, if the sum of two numbers is 6, then we might let x represent one number and let y represent the other number. Then the equation would be $x + y = 6$. This time, the equation has two variables.

Example 1

Suppose that the difference between two numbers is 3. Use two variables to write an equation that describes this information.

SOLUTION

Let x = the larger number.
Let y = the smaller number.
Then $x - y = 3$.

Your Turn 1

Suppose that one number is 4 more than another number. Use two variables to write an equation that describes this information.

Answer: x = smaller number, y = larger number, $y = x + 4$

NOTE When you use two variables, you need to be especially specific about what they represent. In Example 1, if we had just let x and y represent numbers, then we wouldn't know whether to write the equation as $x - y = 3$ or $y - x = 3$. That's why we let x represent the *larger* number.

If we had used just one variable in Example 1, then we might have represented the two numbers this way:

Let x = the smaller number.
Let $x + 3$ = the larger number.

Observe, however, that no other information is given that would allow us to write an equation.

B Solutions of Equations in Two Variables

For an equation in *one* variable, a solution is any replacement for the variable that makes the equation true. What is a solution of an equation in *two* variables?

Copyright © Houghton Mifflin Company. All rights reserved.

DEVELOPING THE CONCEPT

Solutions of Equations in Two Variables

Consider the equation $x + y = 10$.

To solve this equation, we must find a replacement for x and a replacement for y that make the equation true. In other words, we must find a *pair* of replacements for the variables such that their sum is 10.

One such pair is $x = 3$ and $y = 7$.

$$
\begin{aligned}
x + y &= 10 \\
\downarrow \quad \downarrow & \\
3 + 7 &= 10 \\
10 &= 10 \quad \text{True}
\end{aligned}
$$

Here are some other pairs of numbers that make the equation true.

$x = 9, y = 1$	$x = 10, y = 0$	$x = -2, y = 12$
$x + y = 10$	$x + y = 10$	$x + y = 10$
$\downarrow \quad \downarrow$	$\downarrow \quad \downarrow$	$\downarrow \quad \downarrow$
$9 + 1 = 10$	$10 + 0 = 10$	$-2 + 12 = 10$
$10 = 10$	$10 = 10$	$10 = 10$

Each of these pairs of numbers is called a *solution* of the equation. Are you beginning to realize that there is no end to the number of solutions of $x + y = 10$?

We say that the equation $x + y = 10$ has *infinitely many* solutions, which means that the list of solutions would go on forever.

Rather than write $x = 4$ and $y = 6$ as a solution of the equation $x + y = 10$, we use a shorter, more convenient notation: (4, 6). The 4 is understood to be the x-value, and the 6 is understood to be the y-value. Note that the numbers are separated by a comma, and they are enclosed in parentheses. We call (4, 6) an *ordered pair.*

Example 2

Determine whether the given ordered pair is a solution of $x - y = 3$.

(a) (8, 5)

(b) (5, 8)

(c) (2, −1)

SOLUTION

(a)
$$
\begin{aligned}
x - y &= 3 \quad (8, 5) \text{ means } x = 8, y = 5. \\
8 - 5 &= 3 \\
3 &= 3 \quad \text{True}
\end{aligned}
$$

Therefore, (8, 5) is a solution.

(b)
$$
\begin{aligned}
x - y &= 3 \quad (5, 8) \text{ means } x = 5, y = 8. \\
5 - 8 &= 3 \\
-3 &= 3 \quad \text{False}
\end{aligned}
$$

Therefore, (5, 8) is *not* a solution.

Your Turn 2

Determine whether the given ordered pair is a solution of $y = x + 2$.

(a) (1, 3)

(b) (−1, 1)

(c) (−2, 2)

Copyright © Houghton Mifflin Company. All rights reserved.

(c)

$$\begin{aligned} x - y &= 3 \quad && (2, -1) \text{ means } x = 2, y = -1. \\ 2 - (-1) &= 3 \\ 2 + 1 &= 3 \\ 3 &= 3 && \text{True} \end{aligned}$$

Therefore, $(2, -1)$ is a solution.

Answers: (a) Yes; (b) Yes; (c) No

LEARNING TIP

Keep in mind that ordered pairs have the form (x-value, y-value). This should help you to remember that if you swap the numbers, you will have a different ordered pair.

In Example 2, we found that $(8, 5)$ is a solution of $x - y = 3$, but $(5, 8)$ is not a solution. This is the reason that we call $(8, 5)$ and $(5, 8)$ *ordered* pairs; the order of the two numbers is important! The ordered pairs $(8, 5)$ and $(5, 8)$ are entirely different.

C Determining Solutions

As illustrated in Example 2, testing a given ordered pair to determine whether it is a solution of an equation in two variables is easy. We just replace the x and the y in the equation with the x- and y-values of the ordered pair. If the resulting equation is true, then the given ordered pair is a solution.

Determining a solution involves other methods. We replace one variable with a given value (or a value of our choice), and then we solve the resulting equation for the corresponding value of the other variable.

Example 3

Complete the following ordered pairs so that they are solutions of $2x + 3y = 12$.

(a) (3, ___)

(b) (___ , 0)

(c) (−6, ___)

SOLUTION

(a)

$$\begin{aligned} 2x + 3y &= 12 \\ 2(3) + 3y &= 12 \quad && \text{The given } x\text{-value is 3.} \\ 6 + 3y &= 12 \\ 3y &= 6 && \text{Subtract 6 from both sides.} \\ y &= 2 && \text{Divide both sides by 3.} \end{aligned}$$

When $x = 3$, the corresponding y-value is 2. The solution is $(3, 2)$.

(b)

$$\begin{aligned} 2x + 3y &= 12 \\ 2x + 3(0) &= 12 \quad && \text{The given } y\text{-value is 0.} \\ 2x + 0 &= 12 \\ 2x &= 12 && \text{Divide both sides by 2.} \\ x &= 6 \end{aligned}$$

When $y = 0$, the corresponding x-value is 6. The solution is $(6, 0)$.

Your Turn 3

Complete the following ordered pairs so that they are solutions of $x - 2y = 5$.

(a) (___ , 0)

(b) (3, ___)

(c) (___ , −2)

continued

Copyright © Houghton Mifflin Company. All rights reserved.

(c)

$$\begin{aligned} 2x + 3y &= 12 \\ 2(-6) + 3y &= 12 \quad \text{The given } x\text{-value is } -6. \\ -12 + 3y &= 12 \\ 3y &= 24 \quad \text{Add 12 to both sides.} \\ y &= 8 \quad \text{Divide both sides by 3.} \end{aligned}$$

When $x = -6$, the corresponding y-value is 8. The solution is $(-6, 8)$.

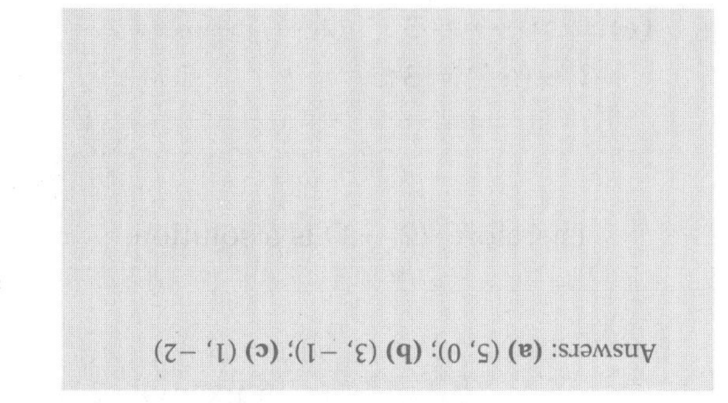

Example 3 includes only three of the infinitely many solutions of the equation $2x + 3y = 12$. A convenient way to organize solutions is with a table of values. The following table shows the solutions in Example 3.

x	y	(x, y)
3	2	(3, 2)
6	0	(6, 0)
−6	8	(−6, 8)

Example 4

Complete the following table of values for the equation $x - 3y = 15$.

x	y	(x, y)
3		(,)
	1	(,)
	−2	(,)
0		(,)

SOLUTION

For $x = 3$

$$\begin{aligned} x - 3y &= 15 \\ 3 - 3y &= 15 \\ -3y &= 12 \\ y &= -4 \end{aligned}$$

For $y = 1$

$$\begin{aligned} x - 3y &= 15 \\ x - 3(1) &= 15 \\ x - 3 &= 15 \\ x &= 18 \end{aligned}$$

For $y = -2$

$$\begin{aligned} x - 3y &= 15 \\ x - 3(-2) &= 15 \\ x - (-6) &= 15 \\ x + 6 &= 15 \\ x &= 9 \end{aligned}$$

For $x = 0$

$$\begin{aligned} x - 3y &= 15 \\ 0 - 3y &= 15 \\ -3y &= 15 \\ y &= -5 \end{aligned}$$

Your Turn 4

Complete the following table of values for the equation $2x + y = 7$.

x	y	(x, y)
−1		(,)
0		(,)
	5	(,)
	−1	(,)

Copyright © Houghton Mifflin Company. All rights reserved.

Now we can complete the table.

x	y	(x, y)
3	-4	$(3, -4)$
18	1	$(18, 1)$
9	-2	$(9, -2)$
0	-5	$(0, -5)$

Answer:

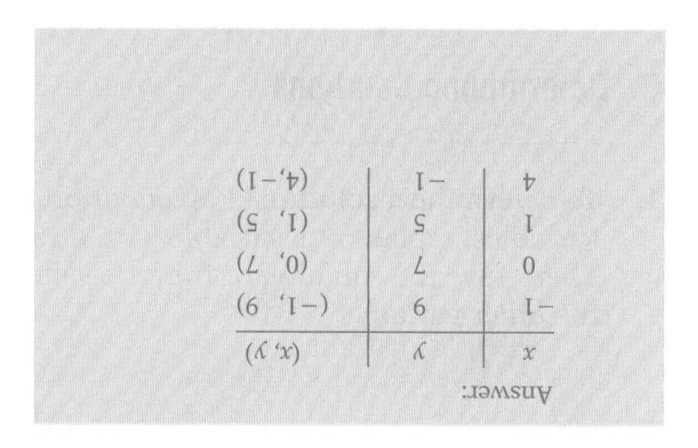

x	y	(x, y)
-1	9	$(-1, 9)$
0	7	$(0, 7)$
1	5	$(1, 5)$
4	-1	$(4, -1)$

Suppose that one number is 4 more than another number. We now have two ways in which that information can be represented.

One Variable	*Two Variables*
Let x = the smaller number.	Let x = the smaller number.
Let $x + 4$ = the larger number.	Let y = the larger number.

As we observed earlier in this section, we don't have enough information to write an equation for the quantities in one variable.

For the quantities in two variables, we can use the fact that the larger number is 4 more than the smaller number to write an equation $y = x + 4$. However, as we have seen, this equation has infinitely many solutions. Therefore, neither method gives us a definite solution.

4.4 Quick Reference

A. Translations

1. When an application problem involves two unknown quantities, we can assign two different variables to the quantities. Then the equation that we write has two variables.

Suppose that one number is 6 more than another number.

Let x = the smaller number.
Let y = the larger number.

Then $y = x + 6$.

B. Solutions of Equations in Two Variables

1. A *solution* of an equation in two variables x and y is an *ordered pair* of the form (x, y). Equations in two variables have infinitely many solutions.

$$y = x + 6$$
$$\downarrow \quad \downarrow$$
$$9 = 3 + 6$$
$$9 = 9 \qquad \text{True}$$

Therefore, (3, 9) is a solution of the equation.

2. We call (3, 9) an *ordered* pair because the order in which the numbers are written is important.

The ordered pair (9, 3) is different from the ordered pair (3, 9).

Copyright © Houghton Mifflin Company. All rights reserved.

C. Determining Solutions

1. To determine a solution of an equation in two variables, replace one variable with a value. Then solve the equation to find the value of the other variable.

Equation: $3x - y = 5$.
If $y = 1$, then

$$\begin{aligned} 3x - 1 &= 5 \\ 3x - 1 + 1 &= 5 + 1 \\ 3x &= 6 \\ x &= 2 \end{aligned}$$

A solution of the equation is (2, 1).

2. A table of values is a convenient way to list solutions of an equation in two variables.

$$3x - y = 5$$

x	y	(x, y)
0	−5	(0, −5)
1	−2	(1, −2)
2	1	(2, 1)

Copyright © Houghton Mifflin Company. All rights reserved.

4.4 Exercises

A. Translations

1. To write an equation describing two numbers whose sum is 7, we often use an equation with two ________.

2. When we write $x - y = 2$, x represents the ________ of two numbers.

In Exercises 3–8, translate the information into an equation in two variables. (Let x represent the larger number and y represent the smaller number.)

3. The sum of two numbers is 7.

4. The difference of two numbers is 6.

5. One number is twice the other number.

6. One number is three times the other number.

7. One number is 4 less than the other number.

8. One number is 2 more than the other number.

B. Solutions of Equations in Two Variables

9. A solution of an equation in two variables is a(n) ________ of numbers.

10. An equation such as $x + y = 12$ has ________ many solutions.

In Exercises 11–18, determine which of the given ordered pairs are solutions of the equation.

11. $x + y = -2$; $(2, 4), (-6, 4), (1, -3)$

12. $x - y = 4$; $(9, 5), (3, 7), (3, -1)$

13. $2x - 5y = 15$; $(-3, 0), (5, -1), (0, -3)$

14. $4x + 3y = 20$; $(5, 0), (0, 5), (2, 4)$

15. $y = 3 - x$; $(-2, 1), (3, 0), (-1, 4)$

16. $y = 2x + 1$; $(-1, -1), (0, 1), (-1, 1)$

17. $y = 3x$; $(0, 0), (-1, -3), (-2, 6)$

18. $y = -2x$; $(-2, 1), (3, -6), (0, 0)$

Copyright © Houghton Mifflin Company. All rights reserved.

C. Determining Solutions

19. To determine a(n) ______ of an equation in two variables, replace one variable with any number and solve for the other variable.

20. We use a(n) ______ of values to organize solutions of an equation such as $x + y = 10$.

In Exercises 21–28, complete the ordered pairs so that they are solutions of the given equation.

21. $x + y = 7$; (____, 8), (3, ____), (____, −2)

22. $2x - y = 6$; (____, 0), (4, ____), (0, ____)

23. $3x + 2y = 18$; (____, 12), (0, ____), (6, ____)

24. $x - 3y = 15$; (____, −3), (9, ____), (____, 1)

25. $y = 2x - 1$; (____, −3), (____, −1), (3, ____)

26. $y = -x + 3$; (2, ____), (____, 4), (5, ____)

27. $y = -x$; (4, ____), (____, 2), (____, 5)

28. $y = 3x$; (−2, ____), (____, 0), (1, ____)

In Exercises 29–34, complete the table of values for the given equation.

29. $x + 3y = 6$

x	y	(x, y)
	4	(____, ____)
0		(____, ____)
	1	(____, ____)
6		(____, ____)

30. $3x - 2y = 4$

x	y	(x, y)
2		(____, ____)
	−5	(____, ____)
4		(____, ____)
	7	(____, ____)

31. $2x - 4y = 4$

x	y	(x, y)
	−3	(____, ____)
	0	(____, ____)
0		(____, ____)
6		(____, ____)

32. $3x + 2y = 4$

x	y	(x, y)
	5	(____, ____)
0		(____, ____)
2		(____, ____)
	−4	(____, ____)

Copyright © Houghton Mifflin Company. All rights reserved.

33. $y = 2x$

x	y	(x, y)
0		(,)
	−2	(,)
	2	(,)
6		(,)

34. $y - x = 0$

x	y	(x, y)
−4		(,)
0		(,)
	1	(,)
	2	(,)

In Exercises 35–38, complete the table of values for the given equation.

35. $y = x + 1$

x	y	(x, y)
−2		(,)
4		(,)
6		(,)
−4		(,)

36. $y = -x - 4$

x	y	(x, y)
6		(,)
3		(,)
−3		(,)
0		(,)

37. $y = -2x - 1$

x	y	(x, y)
−2		(,)
2		(,)
1		(,)
0		(,)

38. $y = 3x + 2$

x	y	(x, y)
1		(,)
−1		(,)
−3		(,)
0		(,)

In Exercises 39–44, complete the table of values for the given equation.

39. $2x + y = 2$

x	y	(x, y)
−2		(,)
0		(,)
2		(,)
3		(,)

40. $x + y = -2$

x	y	(x, y)
−5		(,)
−2		(,)
1		(,)
3		(,)

41. $x = 3 + y$

x	y	(x, y)
−2		(,)
−1		(,)
2		(,)
3		(,)

42. $2x = y - 4$

x	y	(x, y)
−4		(,)
−2		(,)
0		(,)
1		(,)

Copyright © Houghton Mifflin Company. All rights reserved.

43. $x = 1 - y$

x	y	(x, y)
-4		(,)
-1		(,)
0		(,)
4		(,)

44. $2x = 1 - y$

x	y	(x, y)
-2		(,)
1		(,)
2		(,)
3		(,)

Calculator Exercises

In Exercises 45 and 46, determine which of the ordered pairs is a solution of the given equation.

45. $37x + 24y = -588$; $(12, -43)$, $(-12, 43)$

46. $12x - 56y = -1{,}332$; $(-13, -21)$, $(-13, 21)$

In Exercises 47 and 48, complete the ordered pairs so that they are solutions of the given equation.

47. $55x - 32y = 400$; $(16, \quad)$, $(\quad, 70)$

48. $12x + 25y = 1{,}000$; $(\quad, 28)$, $(-50, \quad)$

Writing and Concept Extension

49. If we say that (1, 2) is a solution of an equation in two variables, how do you know which is the value of x and which is the value of y?

50. Suppose that to determine a solution of an equation in two variables, we replace y with -2. Solving the resulting equation reveals that x has a value of 3. Is the solution of the original equation $(-2, 3)$ or $(3, -2)$? Why?

51. Complete the ordered pairs $(-3, \quad)$, $(3, \quad)$, $(-2, \quad)$, and $(2, \quad)$ so that they are solutions of the equation $y = x^2$. On the basis of your observations, determine all pairs $(\quad, 25)$ that are also solutions.

52. Complete the ordered pairs $(1, \quad)$, $(-1, \quad)$, $(5, \quad)$, and $(-5, \quad)$ so that they are solutions of the equation $y = |x|$. On the basis of your observations, determine all pairs $(\quad, 8)$ that are also solutions.

Copyright © Houghton Mifflin Company. All rights reserved.

4.5 The Coordinate Plane

A *The Rectangular Coordinate System*
B *Applications*

SUGGESTIONS FOR SUCCESS

How do I get to City Hall?

Walk 3 blocks east and 5 blocks north.

As you study the material in this section, imagine yourself using a map to travel from a starting point to some destination point. Given the directions, you will move a certain distance east or west, and then some distance north or south. If you know how to do that, then you already understand the basic concepts in this section.

Whenever you can relate mathematics to your own experience, you will have a much better sense of what you are doing.

A The Rectangular Coordinate System

The solution of the equation $2x + 5 = 11$ is 3. We can visualize this solution by highlighting the point of the number line that corresponds to 3. This is called *plotting* the point, and the result is called the *graph* of 3.

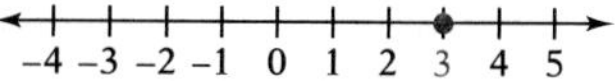

The graph of 3 is not very interesting, but at least we can see where the solution is located relative to all the other numbers of the number line.

As we saw in Section 4.4, the solution of an equation in two variables is an ordered pair of numbers of the form (x, y). How might we visualize the solutions of such equations with a graph?

The Rectangular Coordinate System

Because the ordered pair (3, 5) consists of two numbers, we need two number lines, one for the x-value and one for the y-value. We draw these two number lines (called **axes**) at right angles. The horizontal number line is called the ***x*-axis,** and the vertical number line is called the ***y*-axis.** The point at which the axes intersect is the **origin.**

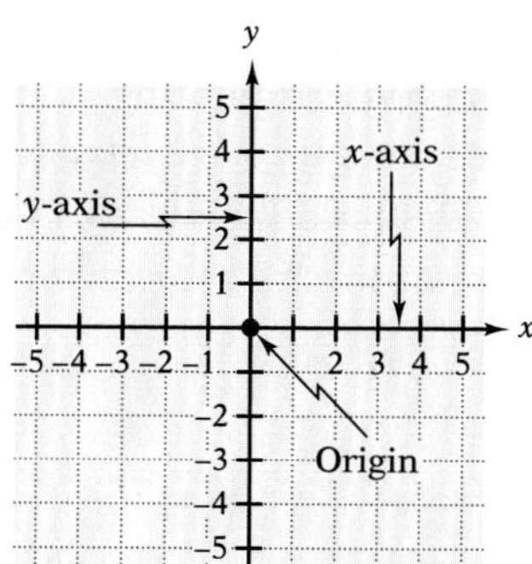

On the x-axis, numbers to the right of the origin are positive, and numbers to the left of the origin are negative. On the y-axis, numbers above the origin are positive, and numbers below the origin are negative.

Copyright © Houghton Mifflin Company. All rights reserved.

This arrangement is called the **rectangular coordinate system,** and the surface on which we draw it (such as this page or your paper) is called the **coordinate plane.**

To plot the point corresponding to the ordered pair (3, 5), we begin at the origin and move 3 units to the right. From there, we move 5 units upward. Our destination is the point that corresponds to (3, 5).

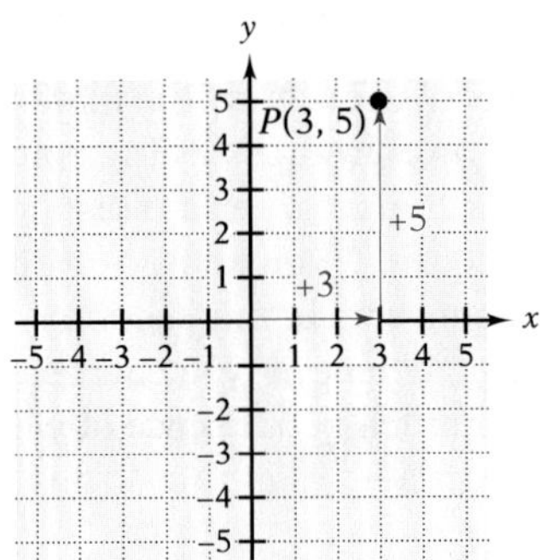

NOTE A graph consists of points that correspond to ordered pairs. However, to keep our language simple, we will regard plotting points and plotting ordered pairs as meaning the same thing.

We often label a plotted point so that we can refer to it. In the preceding figure, we labeled the point $P(3, 5)$. The ***x*-coordinate** of point P is 3, and the ***y*-coordinate** is 5.

Here is a summary of the method for plotting points in the rectangular coordinate system.

Plotting Points in the Rectangular Coordinate System

1. Always begin at the origin.
2. If the x-coordinate is positive, move that many units to the right. If the x-coordinate is negative, move that many units to the left.
3. If the y-coordinate is positive, move that many units upward. If the y-coordinate is negative, move that many units downward.
4. The destination is the point (x, y).

When you are plotting points, it is essential to remember that any point $P(x, y)$ represents an *ordered* pair. The first number of the pair (x-coordinate) indicates movement right and left; the second number (y-coordinate) indicates movement up and down. The points representing (3, 5) and (5, 3) are completely different points.

Example 1

Plot the following points.

(a) $A(-2, 3)$ **(b)** $B(-3, -4)$ **(c)** $C(4, -2)$

Your Turn 1

Plot the following points.

(a) $E(2, 4)$ **(b)** $F(2, -4)$

(c) $G(-2, 4)$ **(d)** $H(-2, -4)$

Copyright © Houghton Mifflin Company. All rights reserved.

SOLUTION

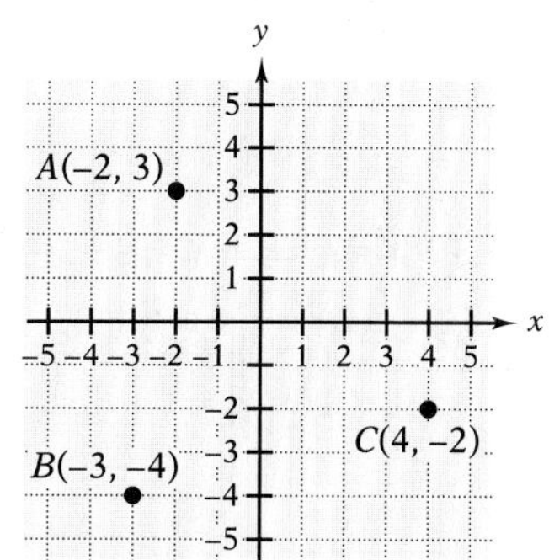

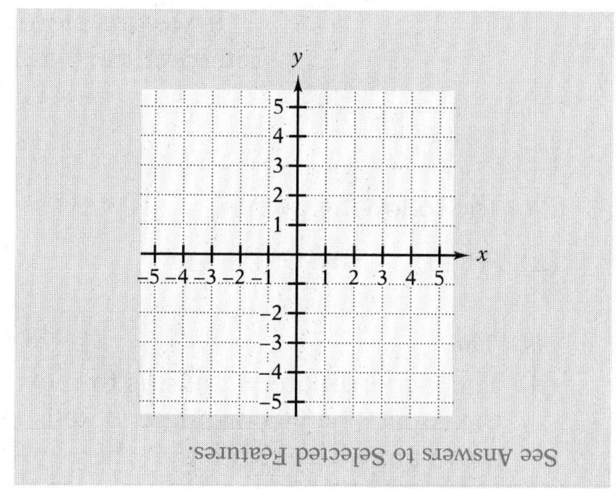

See Answers to Selected Features.

The x- and y-axes divide up the coordinate plane into four regions, called **quadrants.** The following figure shows how the quadrants are numbered.

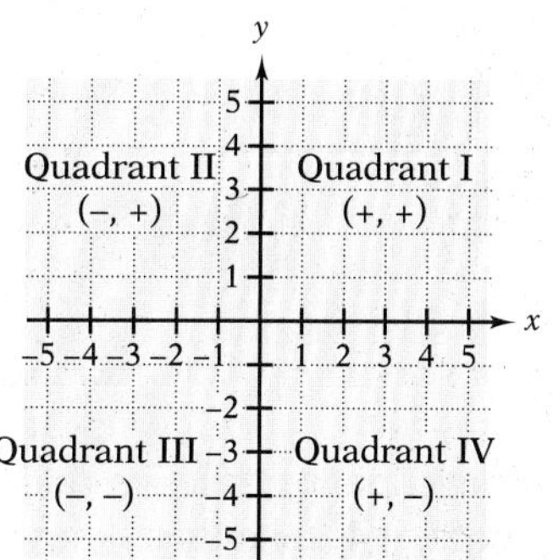

LEARNING TIP

There is no need to memorize these sign patterns. If you know how to plot points, then you can figure out the sign patterns when you need to.

Earlier, we saw that $P(3, 5)$ is in Quadrant I. In Example 1, $A(-2, 3)$ is in Quadrant II, $B(-3, -4)$ is in Quadrant III, and $C(4, -2)$ is in Quadrant IV. The figure shows the general sign patterns for the coordinates of points in the four quadrants.

Example 2

Determine the quadrant in which the given point is located.

(a) $P(-4, 9)$ **(b)** $Q(53, -1)$ **(c)** $R(-28, -17)$

SOLUTION

We don't have to plot the points to answer the questions. Just look at the signs of the coordinates.

(a) Because the x-coordinate is negative and the y-coordinate is positive, point P lies in Quadrant II.

(b) Because the x-coordinate is positive and the y-coordinate is negative, point Q lies in Quadrant IV.

(c) Because both coordinates are negative, point R lies in Quadrant III.

Your Turn 2

Determine the quadrant in which the given point is located.

(a) $A(-3, -9)$

(b) $B(15, -43)$

(c) $C(-24, 6)$

Answers: (a) III; (b) IV; (c) II

Copyright © Houghton Mifflin Company. All rights reserved.

If we are plotting an ordered pair in which one of the coordinates is 0, we interpret the 0 as meaning "don't move anywhere."

Example 3

Plot the following points.

(a) $P(5, 0)$ **(b)** $Q(0, -4)$

SOLUTION

(a) The x-coordinate, 5, tells us to move 5 units to the right. The y-coordinate, 0, tells us not to move anywhere. Note that the plotted point P is a point of the x-axis.

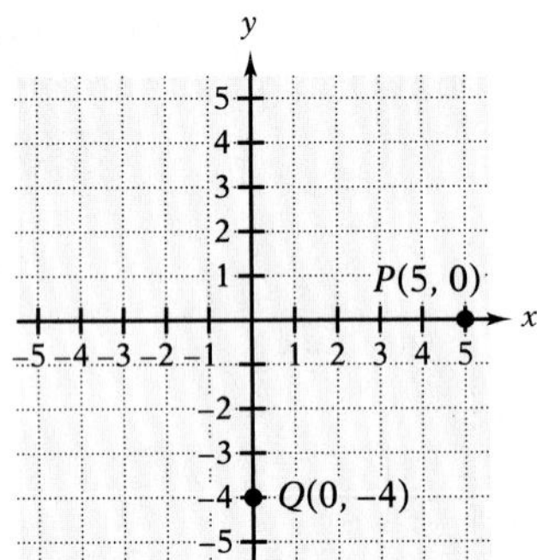

(b) The x-coordinate, 0, tells us not to leave the origin. The y-coordinate, -4, tells us to move 4 units downward. Note that the plotted point Q is a point of the y-axis.

Your Turn 3

Plot the following points.

(a) $A(0, 3)$

(b) $B(-2, 0)$

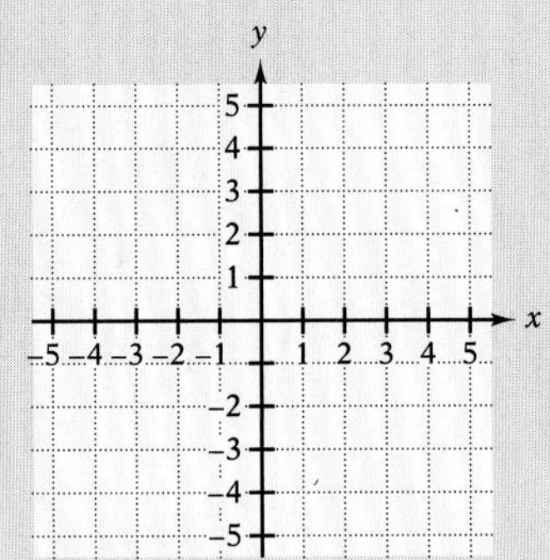

See Answers to Selected Features.

NOTE If at least one coordinate of an ordered pair is 0, then the point is a point of one of the axes. Such a point is not considered to lie in a quadrant.

Example 4

Determine the coordinates of the plotted points.

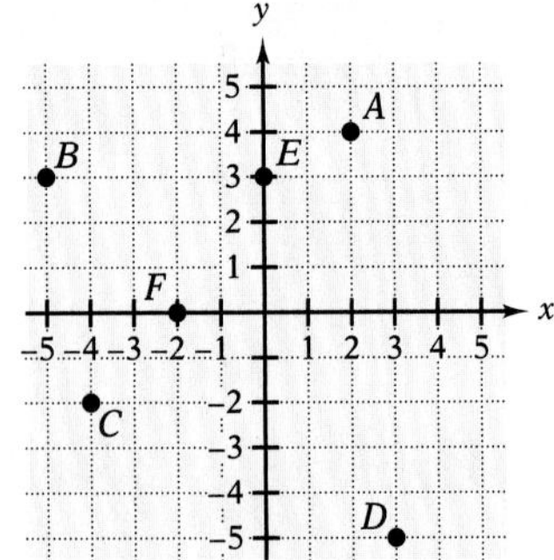

SOLUTION

$A(2, 4)$ $B(-5, 3)$ $C(-4, -2)$
$D(3, -5)$ $E(0, 3)$ $F(-2, 0)$

Your Turn 4

Determine the coordinates of the plotted points.

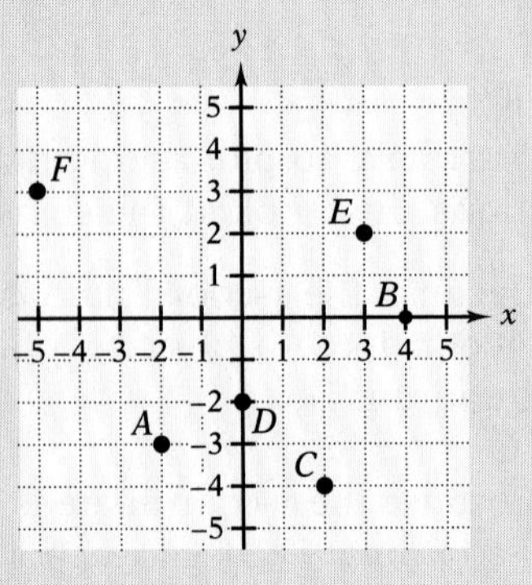

Answer: $A(-2, -3)$, $B(4, 0)$, $C(2, -4)$, $D(0, -2)$, $E(3, 2)$, $F(-5, 3)$

Copyright © Houghton Mifflin Company. All rights reserved.

Again, keep in mind that the order of the coordinates is important. The ordered pairs $(3, -5)$ and $(-5, 3)$ are different. As shown in Example 4, those pairs correspond to different points in the rectangular coordinate system.

B Applications

Paired data can be represented by points in a coordinate system. We call such a graph a **scatterplot.**

Because scatterplots don't always stand out visually, we often connect the points with line segments. The resulting picture is called a **line graph.** Line graphs are useful in seeing trends in the data.

Example 5

The table shows the distances required for a car traveling at selected speeds to stop. (Source: AAA.)

Speed (mph)	*Distance (feet)*
55	273
65	355
75	447

(a) Write the data as ordered pairs in the form (speed, distance).

(b) Plot the points corresponding to the data and connect the points to form a line graph.

(c) From the graph, what can you tell about braking distances as speed increases?

SOLUTION

(a) (55, 273), (65, 355), (75, 447)

(b)

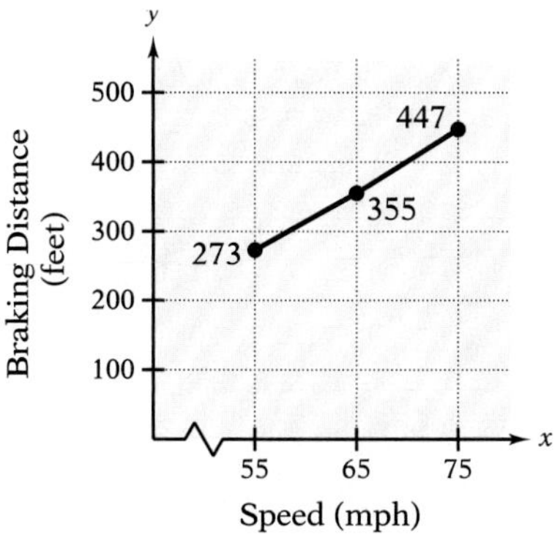

(c) The graph shows that as speed increases, braking distances also increase.

Your Turn 5

Use the graph in Example 5 to estimate the braking distance for a car traveling 60 miles per hour.

Answer: 315 feet

Copyright © Houghton Mifflin Company. All rights reserved.

4.5 Quick Reference

A. The Rectangular Coordinate System

1. The **rectangular coordinate system** consists of two number lines, called the ***x*-axis** and the ***y*-axis,** that are drawn at right angles to each other. The axes intersect at the **origin** and form four regions called **quadrants.**

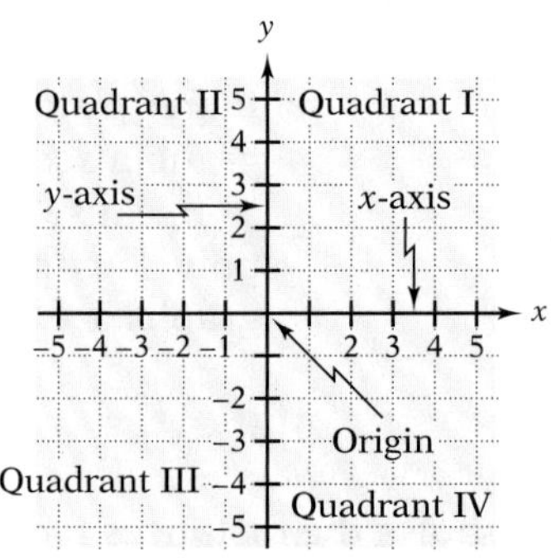

2. Points that correspond to ordered pairs can be plotted in a coordinate system. The figure shows the point *P*(3, 2), where 3 is called the ***x*-coordinate,** and 2 is called the ***y*-coordinate.**

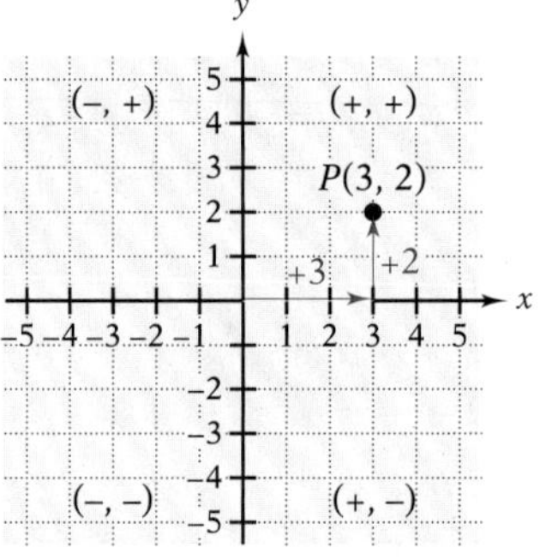

3. Points with coordinates of the form $(a, 0)$ are points of the x-axis; points with coordinates of the form $(0, b)$ are points of the y-axis. Such points are not in any quadrant.

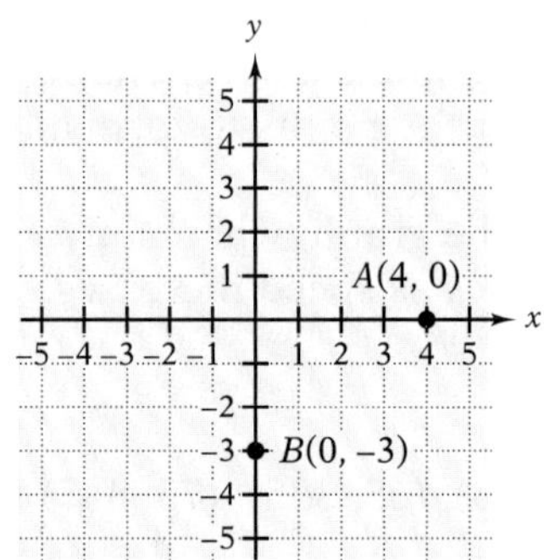

Copyright © Houghton Mifflin Company. All rights reserved.

4.5 Exercises

A. The Rectangular Coordinate System

1. The system formed by two perpendicular number lines is called the ______.

2. In a rectangular coordinate system, the vertical number line is called the ______, and the horizontal number line is called the ______.

In Exercises 3–6, plot the given ordered pairs.

3. (−2, 1), (3, −2), (4, 5), (0, 4), (−4, −3)

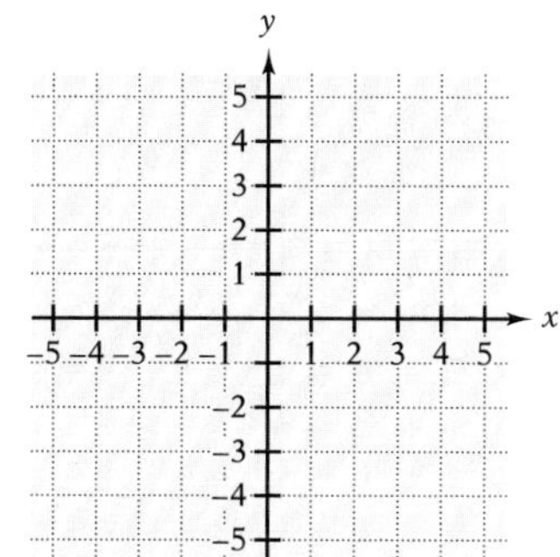

4. (2, −3), (3, 0), (−4, −1), (4, 1), (−3, −2)

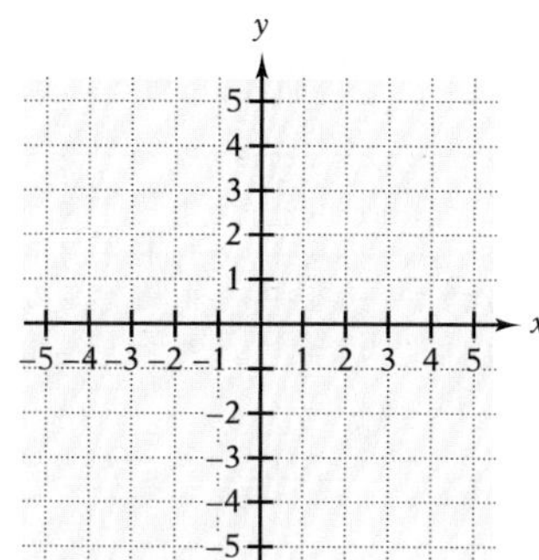

5. (−5, 0), (1, −3), (−2, 4), (4, 3), (−1, −5)

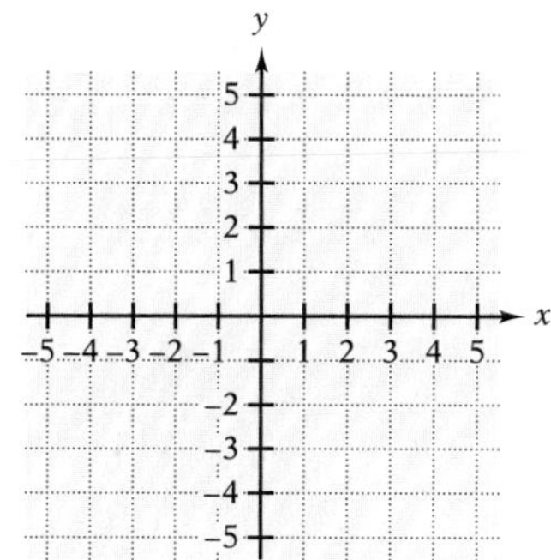

6. (0, −1), (−3, 0), (3, 3), (−4, 5), (4, −4)

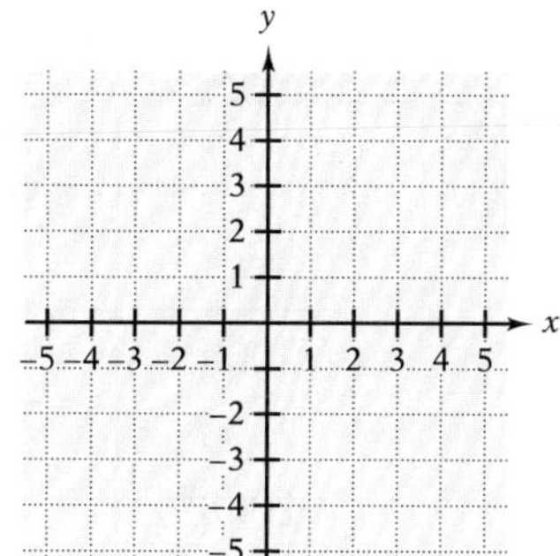

In Exercises 7–18, name the quadrant or axis in which the point lies.

7. (−15, 23)

8. (16, −17)

9. (0, 12)

10. (32, 0)

11. (−22, −9)

12. (46, 20)

13. (5, −8)

14. (−16, 9)

15. (17, 0)

16. (0, −4)

17. (19, 13)

18. (−1, −1)

Copyright © Houghton Mifflin Company. All rights reserved.

In Exercises 19–22, determine the coordinates of the plotted points.

19.

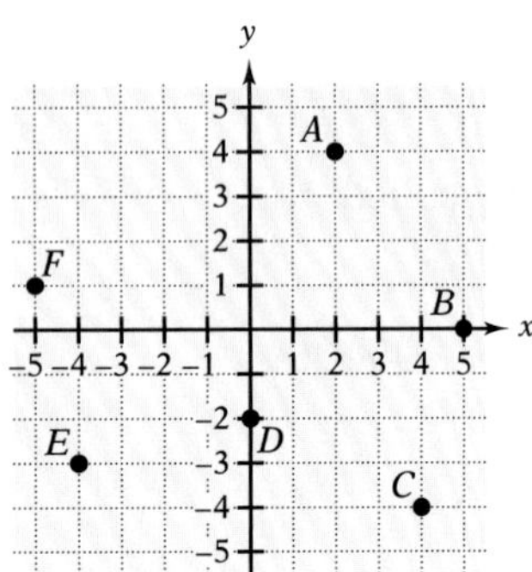

20.

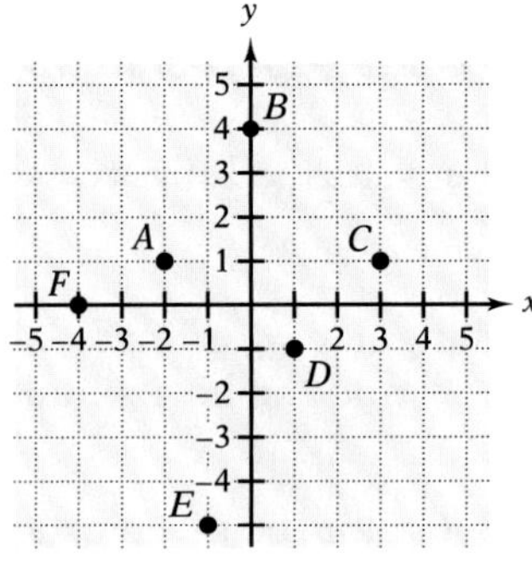

21.

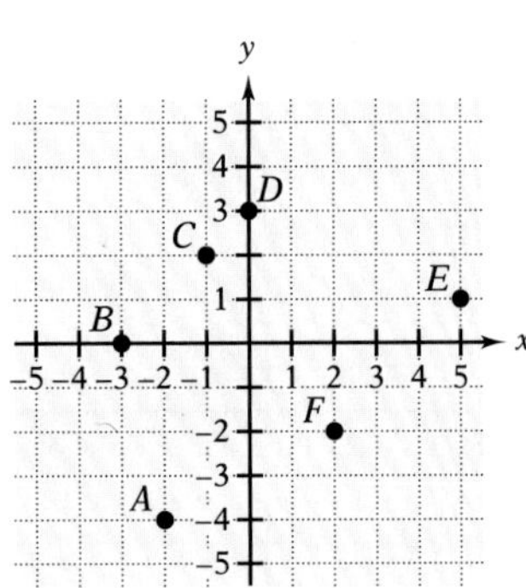

22.

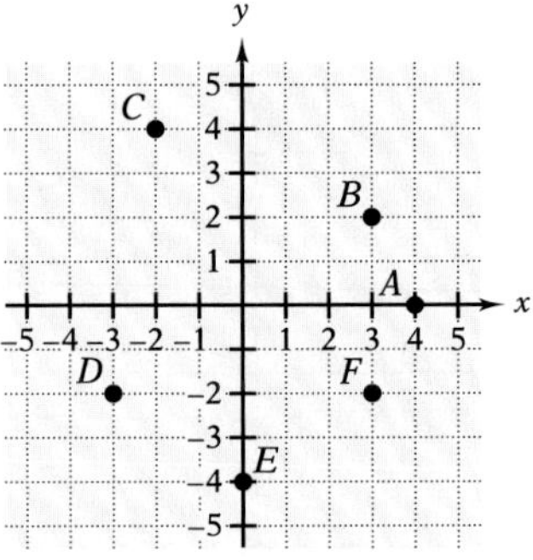

23. ***Births in the United States*** The table shows the actual and projected numbers of births (in thousands) in the United States during selected years in the period 2000 to 2013. (Source: National Center for Health Statistics.)

Year	*Number of Births (in thousands)*
2000	3,900
2003	4,100
2008	4,300
2013	4,600

(a) Write the data as ordered pairs (year, number of births).

(b) Plot the points corresponding to the data, and connect the points to form a line graph.

Copyright © Houghton Mifflin Company. All rights reserved.

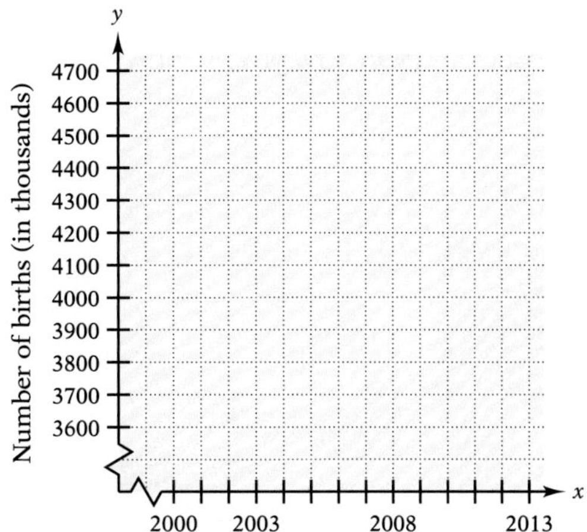

(c) From the graph, what is the projected trend in the number of births during the period 2000 to 2013?

24. ***Digital Camera Sales*** The table shows the actual and projected sales (in millions) of digital cameras during the period 2000 to 2005. (Source: InfoTrends Research Group.)

Year	*Digital Camera Sales (in millions)*
2000	7
2001	15
2002	21
2003	29
2004	38
2005	42

(a) Write the data as ordered pairs (year, sales).

(b) Plot the points corresponding to the data, and connect the points to form a line graph.

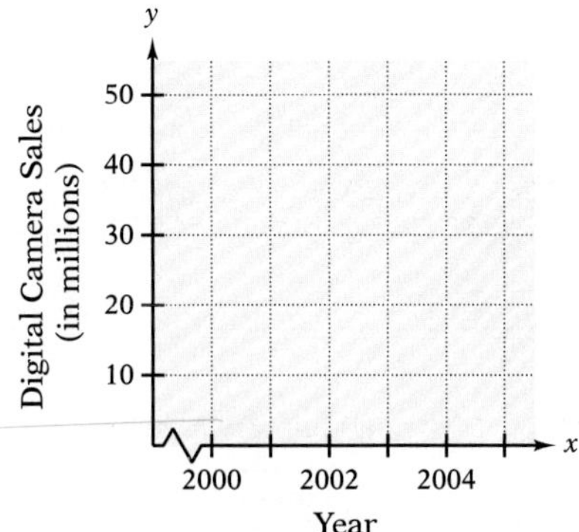

(c) From the graph, what is the projected trend in digital camera sales during the period 2000 to 2005?

Copyright © Houghton Mifflin Company. All rights reserved.

25. *Air Bags* The table shows the number (in millions) of vehicles equipped with air bags during selected years in the period 1996 to 2001. (Source: National Safety Council.)

Year	*Number of Vehicles (in millions)*
1996	20
1998	45
2000	70
2001	107

(a) Write the data as ordered pairs (year, number of vehicles).

(b) Plot the points corresponding to the data, and connect the points to form a line graph.

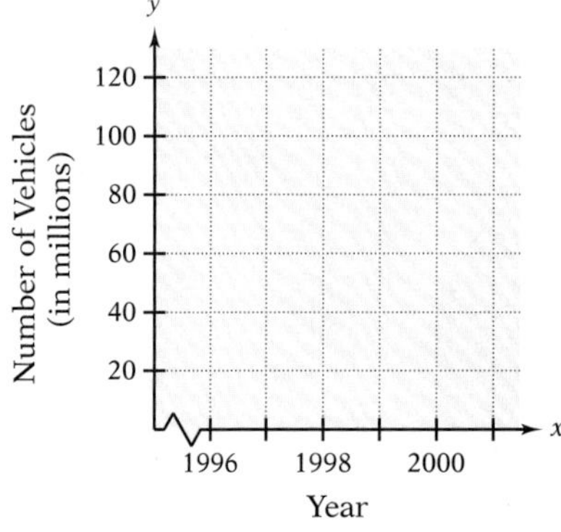

(c) From the graph, what was the trend in the number of vehicles that were equipped with air bags during the period 1996 to 2001?

26. *Air Quality* The table shows the number of days per year that New Jersey failed to meet acceptable air quality standards for selected years during the period 1988 to 2001. (Source: New Jersey Department of Environmental Protection.)

Year	*Number of Days*
1988	59
1991	37
1994	16
1997	12
2001	11

Copyright © Houghton Mifflin Company. All rights reserved.

(a) Write the data as ordered pairs (year, number of days)

(b) Plot the points corresponding to the data, and connect the points to form a line graph.

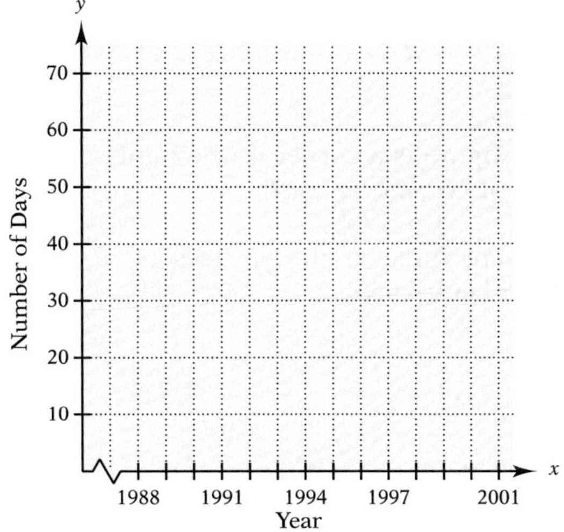

(c) From the graph, what was the trend in the number of days with unsatisfactory air quality during the period 1988 to 2001?

Writing and Concept Extension

27. What is true about any point of the x-axis?

28. What is true about any point of the y-axis?

29. Why are (1, 2) and (2, 1) considered to be different pairs?

30. Explain the meanings of (2)(4) and (2, 4).

31. How can $P(a, b)$ and $Q(b, a)$ be the same point?

32. Consider the points $A(6, 7)$ and $B(-13, -7)$. Which point is farther from the x-axis?

Copyright © Houghton Mifflin Company. All rights reserved.

In Exercises 33–36, in which possible quadrants could the point $P(a, b)$ lie? (Assume that a and b are not 0.)

33. a is positive

34. b is positive

35. b is negative

36. a is negative

For Exercises 37 and 38, a *vertex* of a geometric figure is a corner of the figure or the point at which two sides meet. The plural of "vertex" is "vertices."

37. The points $A(3, 5)$, $B(7, 5)$, and $C(7, 1)$ are three of the vertices of a rectangle. What are the coordinates of the other vertex?

38. Determine the perimeter of the rectangle whose vertices are $A(2, 4)$, $B(7, 4)$, $C(7, -3)$, and $D(2, -3)$.

Copyright © Houghton Mifflin Company. All rights reserved.

4.6 Graphing Equations in Two Variables

SUGGESTIONS FOR SUCCESS

"That was then, this is now." A person who studies mathematics in that way fails to see how everything is tied together.

In Section 4.4, you learned how to find solutions of equations in two variables. In this section, we will discuss how to draw a "picture" of those solutions. These are not different and separate topics. In fact, they are the *same* topic.

Each day, you should cycle back and review previous material. When you do, ask yourself how that material relates to what you are doing now. When you do this, you will see how "that was then" is connected to "this is now." When you understand how topics flow together, you are on your way to knowledge and success.

A Solutions and Graphs

Plotting Points

As we saw in Section 4.5, an equation in two variables, such as $x + y = 5$, has infinitely many ordered pair solutions. We know how to find as many of those solutions as we want to. The following table lists eight of the solutions of $x + y = 5$:

x	y	(x, y)
-4	9	$(-4, 9)$
-2	7	$(-2, 7)$
0	5	$(0, 5)$
1	4	$(1, 4)$
3	2	$(3, 2)$
5	0	$(5, 0)$
6	-1	$(6, -1)$
9	-4	$(9, -4)$

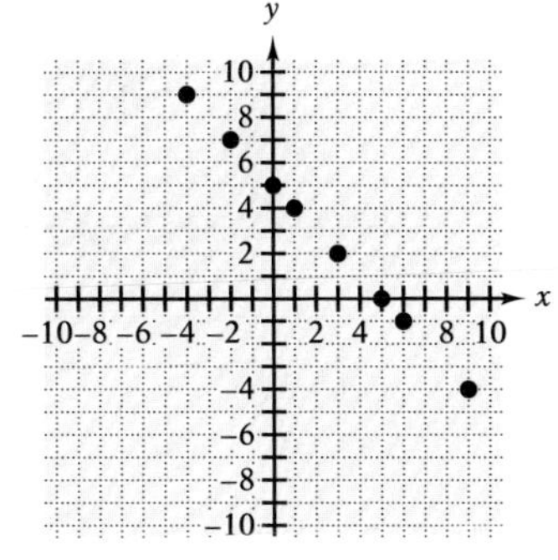

The figure beside the table shows these eight ordered pairs plotted in a rectangular coordinate system.

Do you notice that the eight points appear to be lined up? We could draw a straight line that would contain every one of the points. In fact, if we were to add 10 or 100 or 1,000 more solutions to the table and then plot those points, they would continue to be points of that straight line.

Let's be clear on the language that mathematicians like to use. The graph of the equation $x + y = 5$ is a "picture" of all the solutions of that equation. Because the points are all points of a line, we often refer to the graph as "the line $x + y = 5$."

Copyright © Houghton Mifflin Company. All rights reserved.

Here is a summary of the procedure for graphing an equation whose graph is a straight line.

LEARNING TIP

Only two points are needed to draw a straight line, but we recommend three. If the three points appear to form a triangle, you will know that you have made an error in at least one of them.

Graphing an Equation Whose Graph is a Straight Line

1. By selecting values for one variable and determining the corresponding values of the other variable, find three solutions of the equation.
2. Plot these three ordered pairs.
3. Draw a line through the three points.

Example 1

Complete the following solutions of $y = x - 4$ and use the solutions to draw the line that is the graph of the equation.

(9, ▭) (4, ▭) (−4, ▭)

SOLUTION

For $x = 9$	$y = 9 - 4 = 5$	(9, 5)
For $x = 4$	$y = 4 - 4 = 0$	(4, 0)
For $x = -4$	$y = -4 - 4 = -8$	(−4, −8)

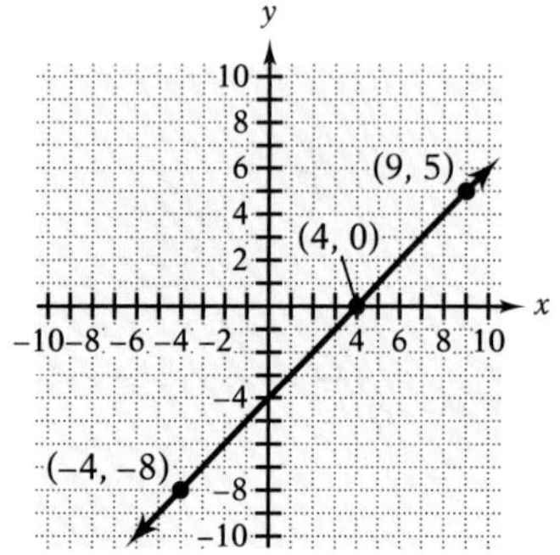

Your Turn 1

Complete the following solutions of $y = x + 3$ and use the solutions to draw the line that is the graph of the equation.

(−5, ▭)

(1, ▭)

(2, ▭)

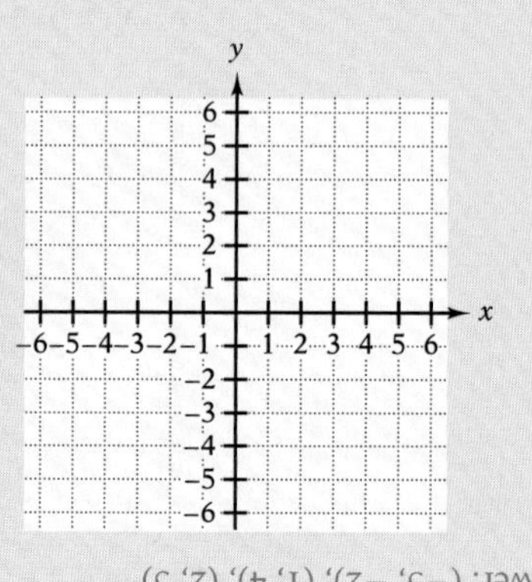

Answer: (−5, −2), (1, 4), (2, 5)

In Example 1, we gave x-values to you in order to show how the graphs can be drawn. When you study algebra, you will usually be asked to select your own values. Two values that are often easy to use are $x = 0$ and $y = 0$.

Example 2

Draw the line $2x + y = 6$.

Your Turn 2

Draw the line $x + 2y = 4$. Use $x = 0$, $y = 0$, and choose another value of y.

Copyright © Houghton Mifflin Company. All rights reserved.

SOLUTION

We need three solutions to draw the line.

For one solution, we choose $x = 0$.

$$2x + y = 6$$
$$2(0) + y = 6$$
$$y = 6$$

Because $y = 6$ when $x = 0$, one solution is $(0, 6)$.

For another solution, we choose $y = 0$.

$$2x + y = 6$$
$$2x + 0 = 6$$
$$2x = 6$$
$$x = 3$$

Because $x = 3$ when $y = 0$, another solution is $(3, 0)$.

We leave it to you to select a value of x to find a third solution and to verify that it is a point of the line.

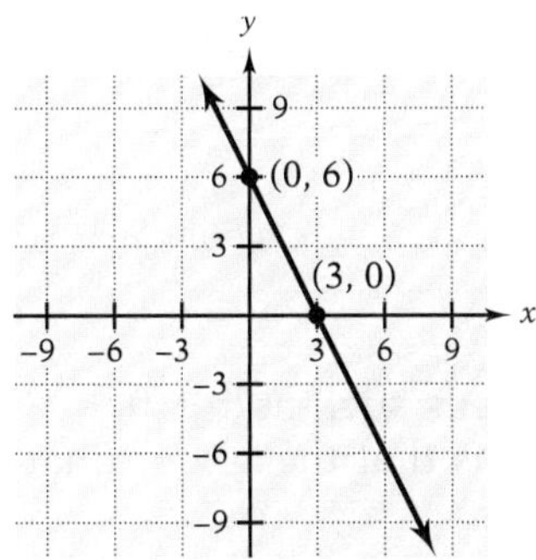

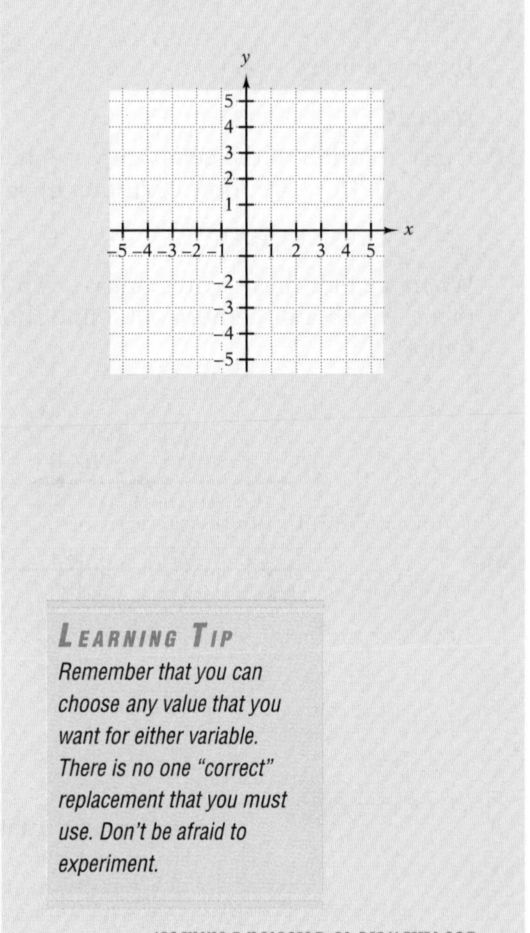

LEARNING TIP

Remember that you can choose any value that you want for either variable. There is no one "correct" replacement that you must use. Don't be afraid to experiment.

See Answers to Selected Features.

In Section 4.5, we noted that we are sometimes a little loose with our language. For example, we often say, "Plot the point (1, 2)." Of course, (1, 2) is an ordered pair of numbers, not a point. Technically speaking, we should say, "Plot the point corresponding to the ordered pair (1, 2)." Saying "Plot the point (1, 2)" is just easier.

A similar situation arises when we graph equations in two variables. You may have noted in Example 2 that we said, "Draw the line $2x + y = 6$." However, $2x + y = 6$ is an equation, not a line. The technically correct wording would be, "Draw the line whose points represent solutions of the equation $2x + y = 6$." That's quite a mouthful, and our version is easier.

Nevertheless, even when the technical wording is not used, it is important to keep in mind the relationship between an equation and its graph.

1. Every point of the graph represents a solution of the equation.
2. Every solution of the equation is represented by a point of the graph.

B Special Cases

Officially, the equation $0x + y = 4$ is an equation in two variables. However, the simpler form is $y = 4$ because the term $0x$ is just 0.

The equation $y = 4$ tells us that the y-coordinate of every solution is 4, no matter what the x-coordinate is.

Copyright © Houghton Mifflin Company. All rights reserved.

Example 3

Draw the line $y = 4$.

SOLUTION

Every solution of the equation $y = 4$ has the form (, 4). Here are three examples of solutions.

$$(-3, 4) \qquad (1, 4) \qquad (5, 4)$$

When we plot these points and draw a line through them, we obtain the horizontal line shown in the figure.

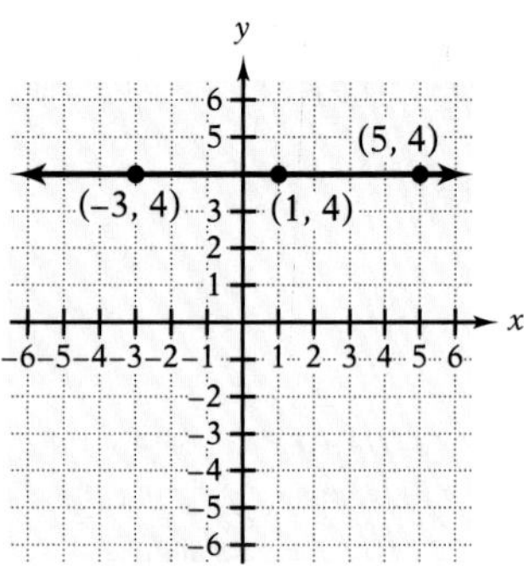

Your Turn 3

Draw the line $y = -3$.

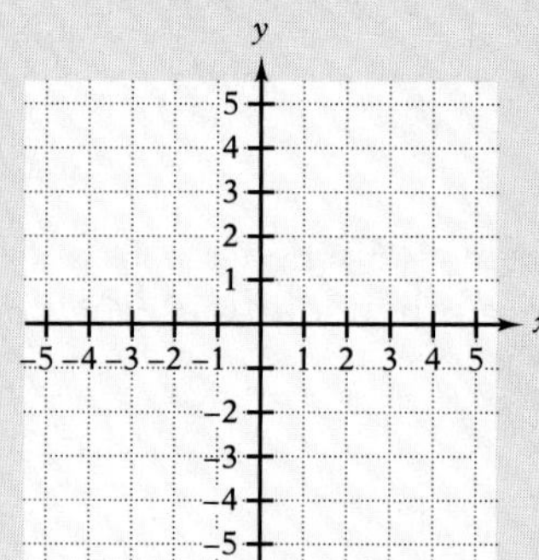

See Answers to Selected Features.

A similar situation arises for equations such as $x + 0y = 3$, which in simpler form is just $x = 3$. This equation tells us that the x-coordinate of every solution is 3, no matter what the y-coordinate is.

Example 4

Draw the line $x = 3$.

SOLUTION

Every solution of the equation $x = 3$ has the form (3,). Some examples of solutions are

$$(3, -5) \qquad (3, 0) \qquad (3, 4)$$

The figure shows that the line that contains these three points is a vertical line.

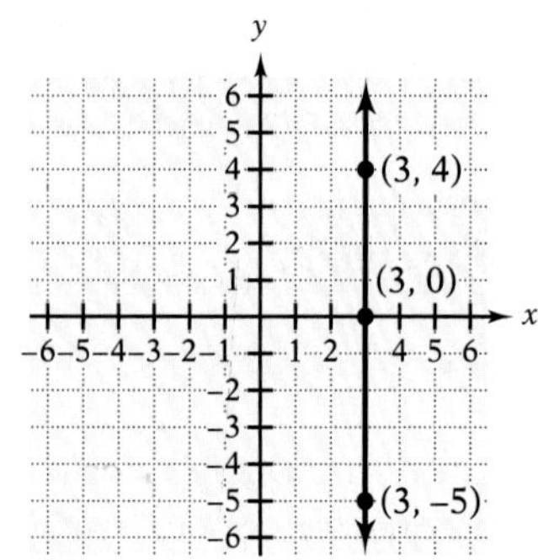

Your Turn 4

Draw the line $x = -2$.

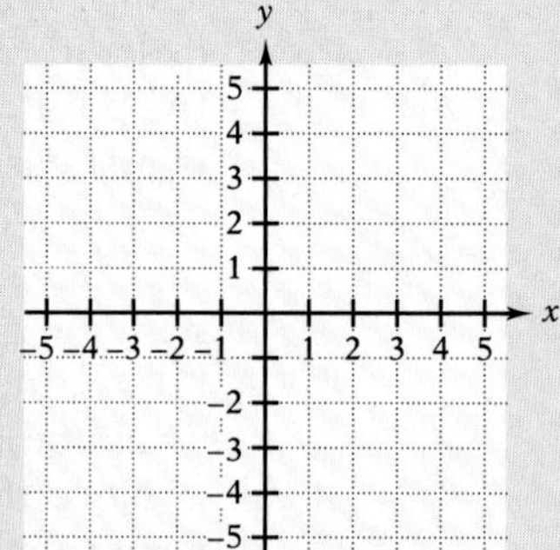

See Answers to Selected Features.

Copyright © Houghton Mifflin Company. All rights reserved.

Example 5

Use a graph to determine the ordered pair that is a solution of *both* of the following equations:

$$x = 2 \quad \textit{and} \quad y = -3$$

SOLUTION

Every solution of $x = 2$ has a first coordinate 2. Every solution of $y = -3$ has a second coordinate -3. Here are a few solutions for each equation.

$x = 2$:	$(2, 3)$	$(2, -3)$	$(2, -5)$
$y = -3$:	$(5, -3)$	$(2, -3)$	$(-4, -3)$

Observe that $(2, -3)$ is among the solutions of both of the equations, so we already know the answer to the question.

Nevertheless, we will graph the two equations to learn the significance of the solution.

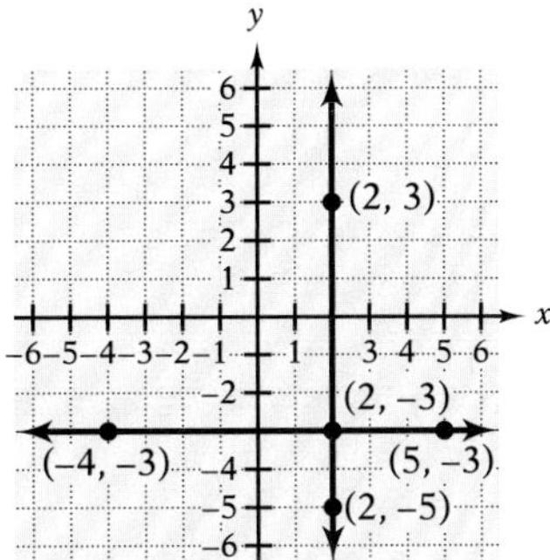

The vertical line is the graph of $x = 2$, and the horizontal line is the graph of $y = -3$. The two lines intersect at the point $(2, -3)$. In other words, the ordered pair that is a solution of both equations is the point of intersection of the graphs of the equations.

Your Turn 5

Use a graph to determine the ordered pair that is a solution of *both* of the following equations:

$$x = -3 \quad \textit{and} \quad y = -2$$

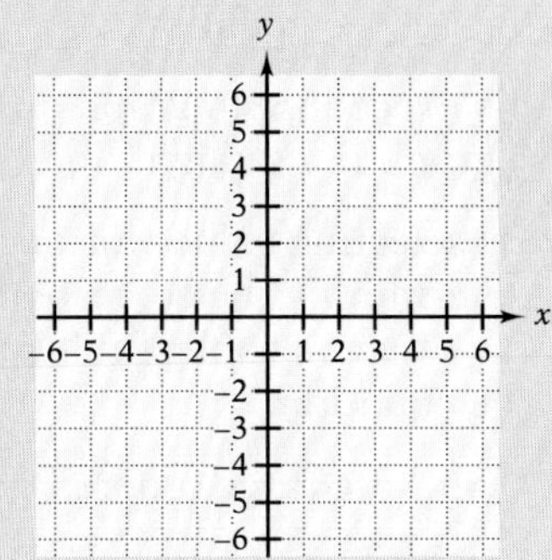

Answer: $(-3, -2)$

In Example 5, we were given two equations, $x = 2$ and $y = -3$, and we found an ordered pair that is a solution of both equations at the same time. We didn't really need to graph the two equations to arrive at the solution $(2, -3)$, but the graphs help us to see that the solution is represented by the point of intersection of the two lines.

This example is a preview of an important topic that you will study in algebra. Suppose, for example, that we are given the following two equations:

$$x + y = 6 \quad \text{and} \quad x - y = 4$$

Both equations have infinitely many solutions, but you can easily verify that $(5, 1)$ is a solution of both equations at the same time. If you were to draw the graphs of these equations, what do you suppose would be the coordinates of the point of intersection? If you said $(5, 1)$, then you are already on your way to understanding a very important concept.

Copyright © Houghton Mifflin Company. All rights reserved.

4.6 Quick Reference

A. Solutions and Graphs

1. The graph of an equation is a picture of the solutions of the equation.

   Use this procedure to graph an equation whose graph is a straight line.

   (a) By selecting values for one variable and determining the corresponding values of the other variable, find three solutions of the equation.

   (b) Plot these three ordered pairs.

   (c) Draw a line through the three points.

Equation: $y = x + 2$

x	y	(x, y)
-2	0	$(-2, 0)$
0	2	$(0, 2)$
2	4	$(2, 4)$

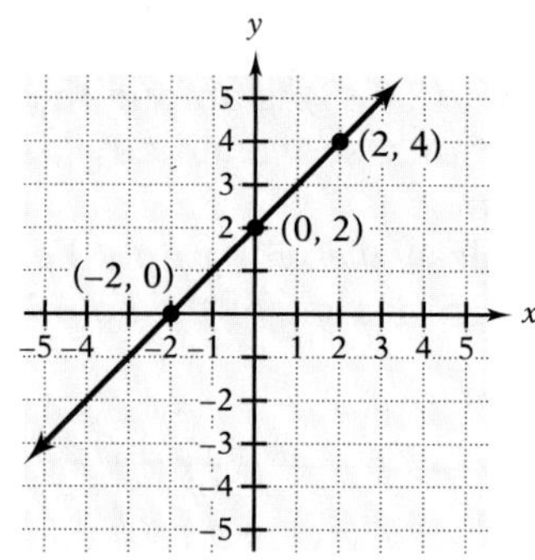

B. Special Cases

1. In simpler form, the equation $0x + y = 2$ is $y = 2$. Its graph is a horizontal line whose points all have a y-coordinate of 2.

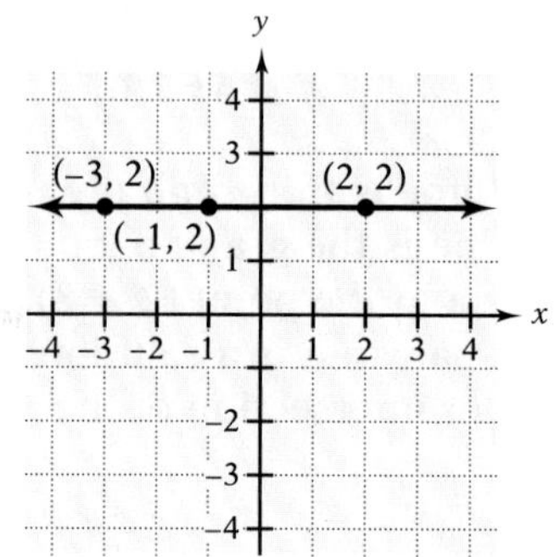

2. In simpler form, the equation $x + 0y = 1$ is $x = 1$. Its graph is a vertical line whose points all have an x-coordinate of 1.

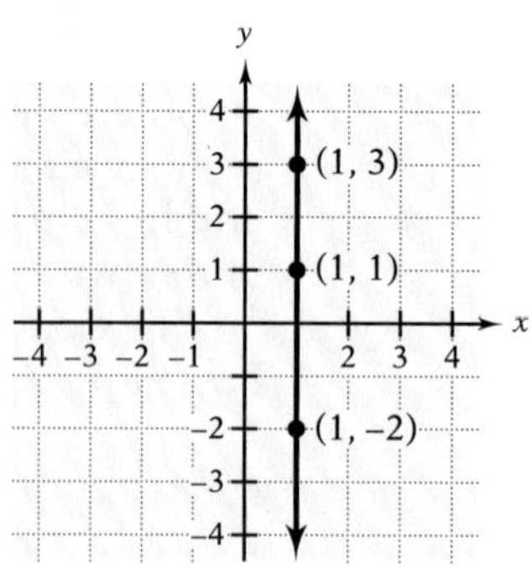

Copyright © Houghton Mifflin Company. All rights reserved.

4.6 Exercises

A. Solutions and Graphs

1. To graph an equation such as $y = x - 5$, we can begin by determining three ____ of the equation.

2. The graph of an equation such as $x + y = 3$ is a ____.

In Exercises 3–6, complete the solutions of the given equation and use the solutions to draw the graph of the equation.

3. $y - x = 3$; $(0, \_\_\_)$, $(\_\_\_, 0)$, $(1, \_\_\_)$

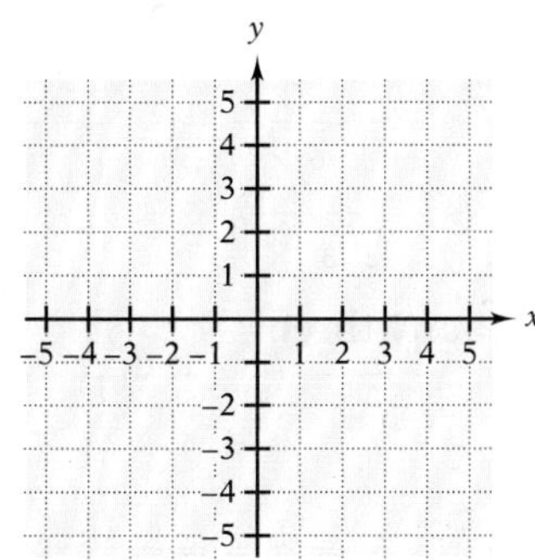

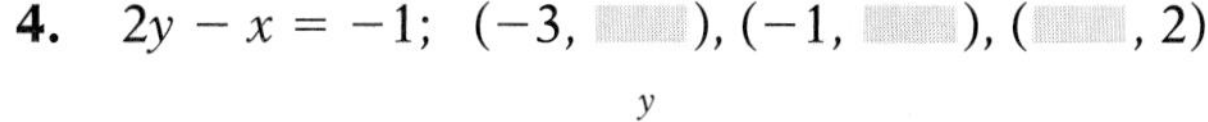

4. $2y - x = -1$; $(-3, \_\_\_)$, $(-1, \_\_\_)$, $(\_\_\_, 2)$

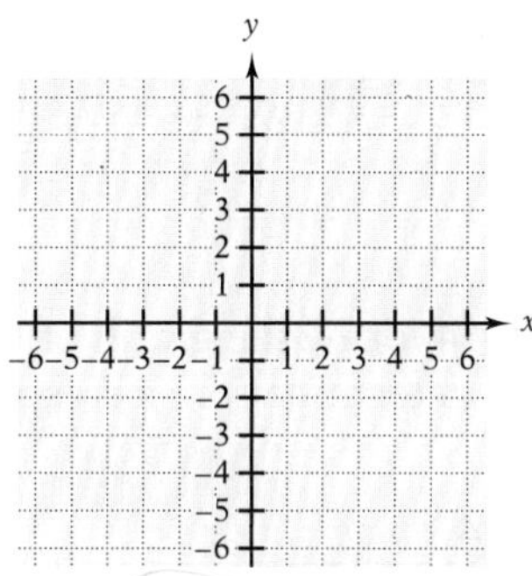

5. $y = 2 - 2x$; $(-1, \_\_\_)$, $(0, \_\_\_)$, $(2, \_\_\_)$

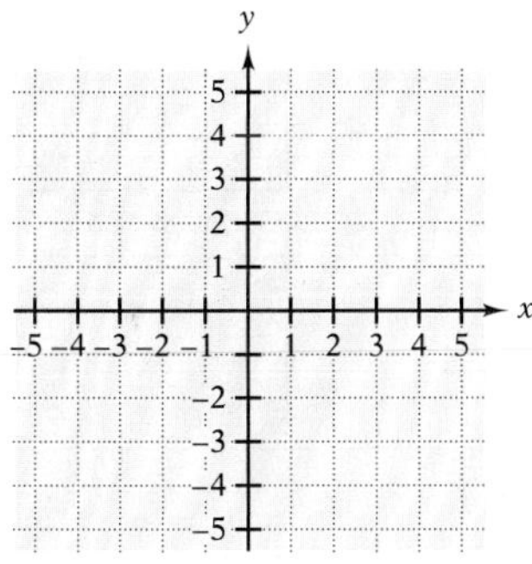

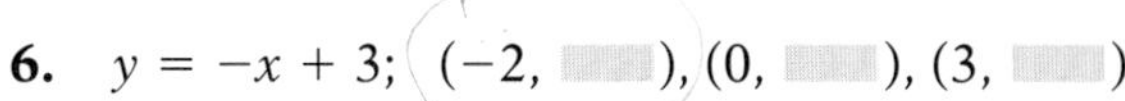

6. $y = -x + 3$; $(-2, \_\_\_)$, $(0, \_\_\_)$, $(3, \_\_\_)$

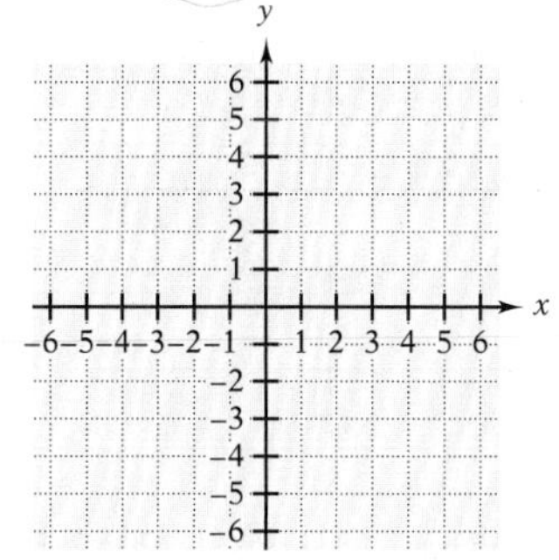

In Exercises 7–10, complete the table of values for the given equation. Then use the solutions to draw the graph of the equation.

7. $x + 2y = 0$

x	y	(x, y)
	1	(,)
	0	(,)
4		(,)

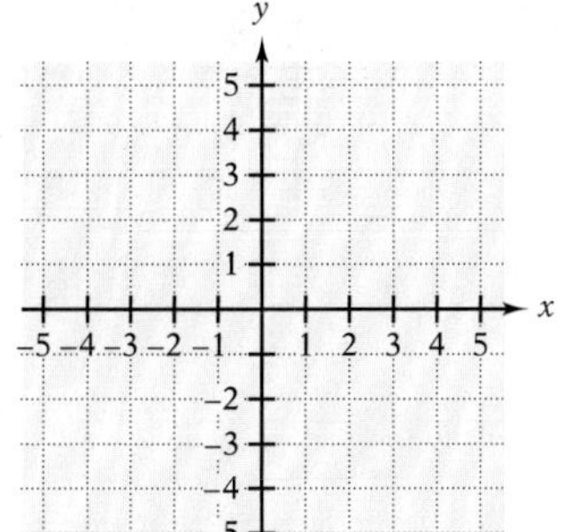

8. $3x - y = 1$

x	y	(x, y)
−1		(,)
	2	(,)
	5	(,)

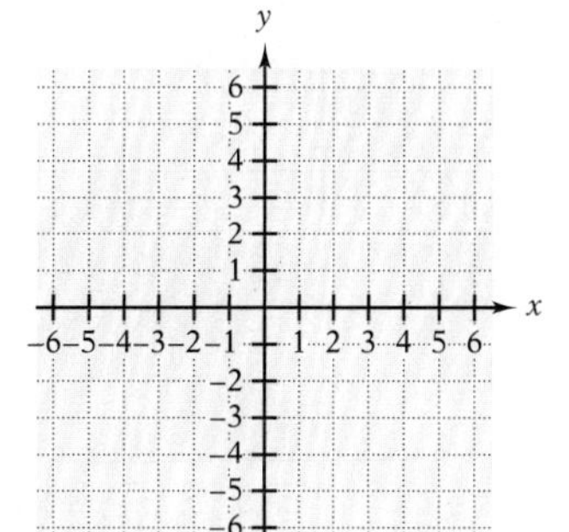

Copyright © Houghton Mifflin Company. All rights reserved.

9. $y = 2x - 1$

x	y	(x, y)
-2		(,)
0		(,)
3		(,)

10. $y = -3x + 2$

x	y	(x, y)
-1		(,)
1		(,)
2		(,)

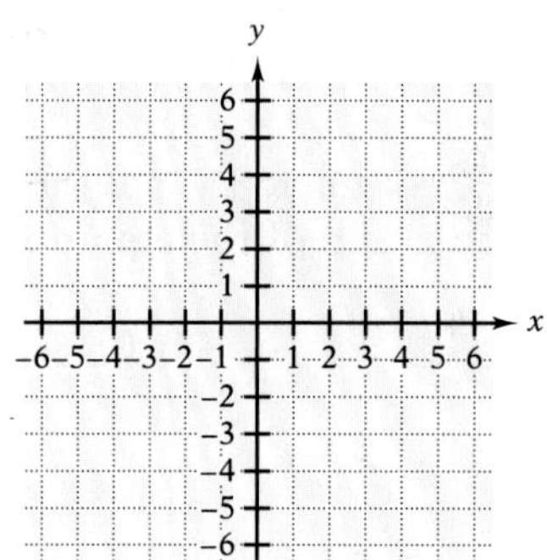

In Exercises 11–14, draw the graph of the given equation. (*Hint:* Select any three values for x and determine the corresponding values of y.)

11. $y = -2x + 4$

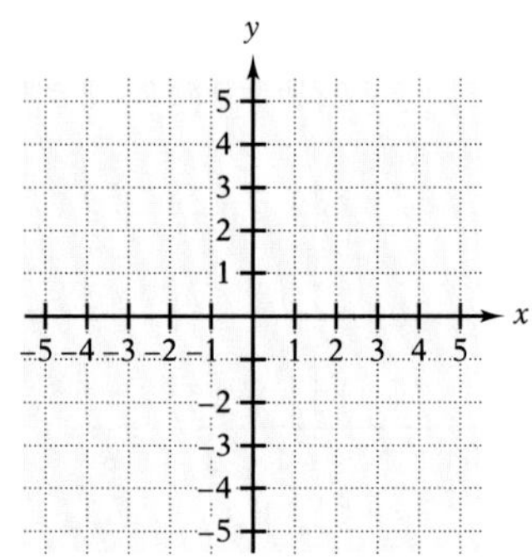

12. $y = 3x - 5$

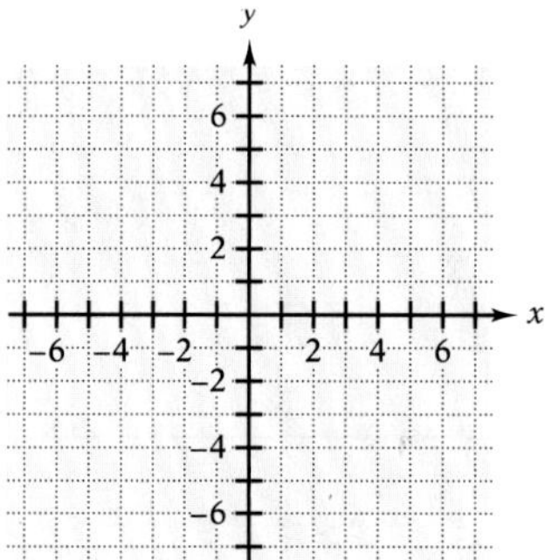

13. $3x + y = 1$

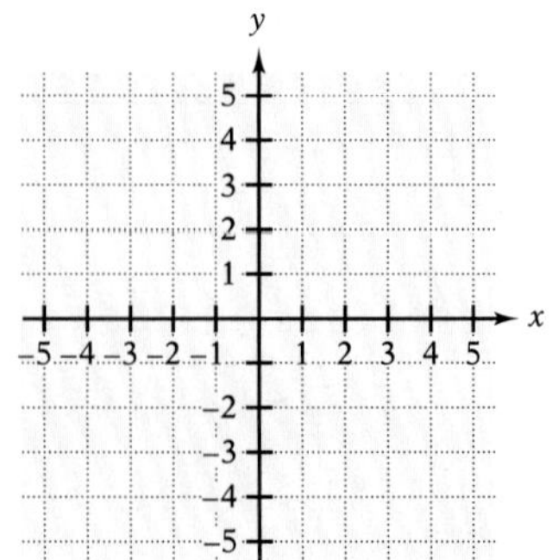

14. $2x + y = 3$

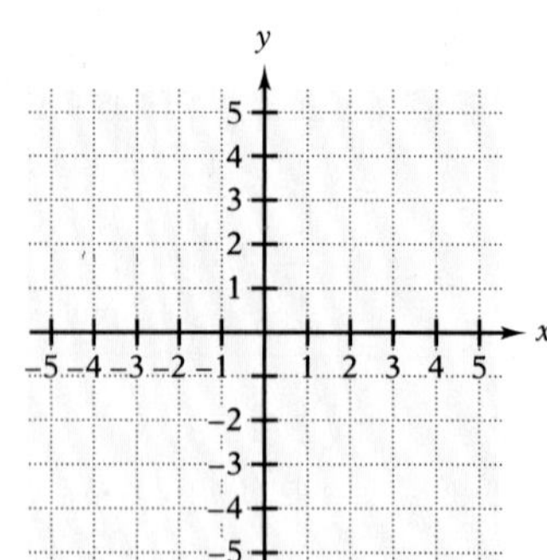

Copyright © Houghton Mifflin Company. All rights reserved.

In Exercises 15–18, draw the graph of the given equation. (*Hint:* Select any three values for y and determine the corresponding values of x.)

15. $x = 3 - 2y$

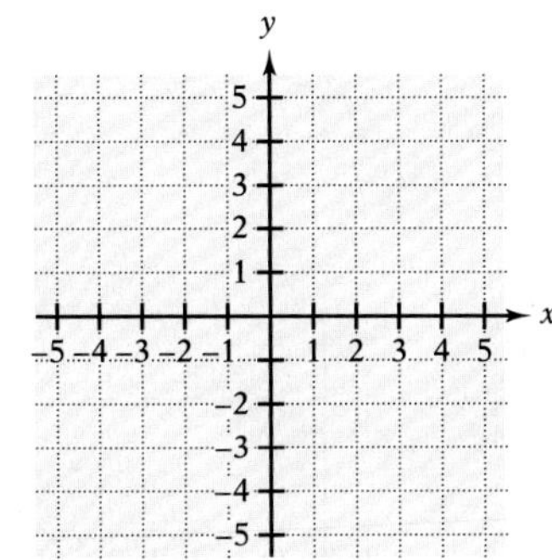

16. $x = 3y + 1$

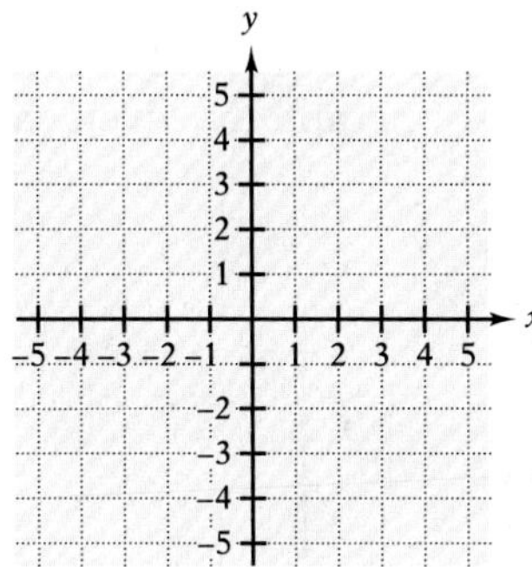

17. $x - 2y = 1$

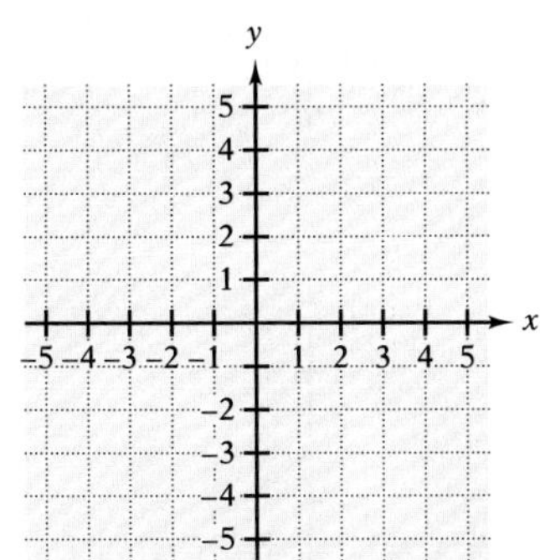

18. $x + 2y = 8$

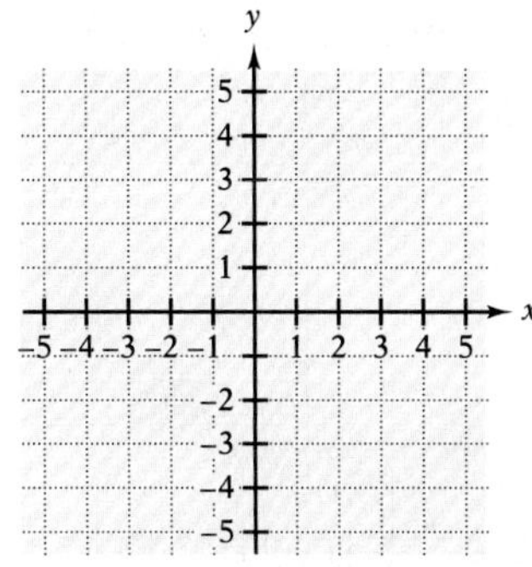

In Exercises 19–22, draw the graph of the given equation.

19. $x = y$

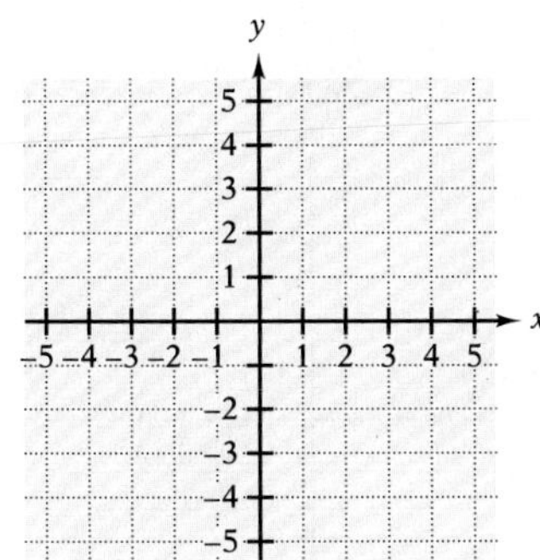

20. $y = -x$

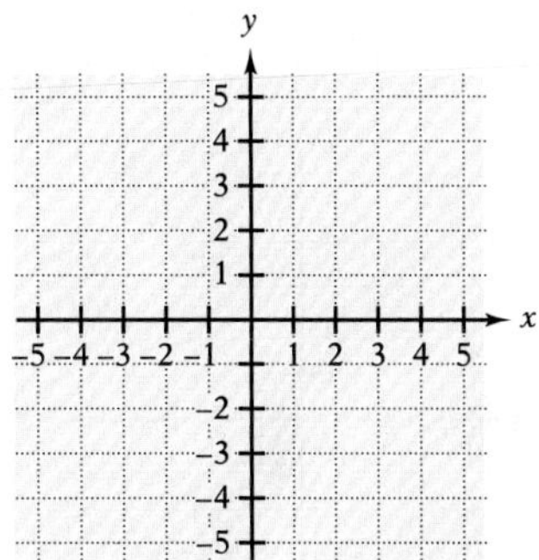

21. $x + y = 5$

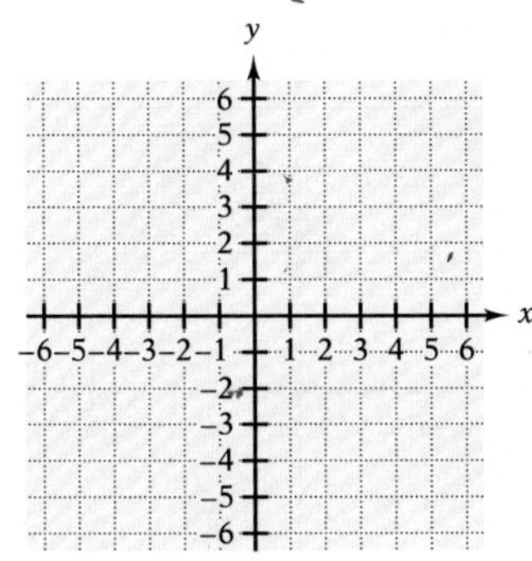

22. $x - y = 2$

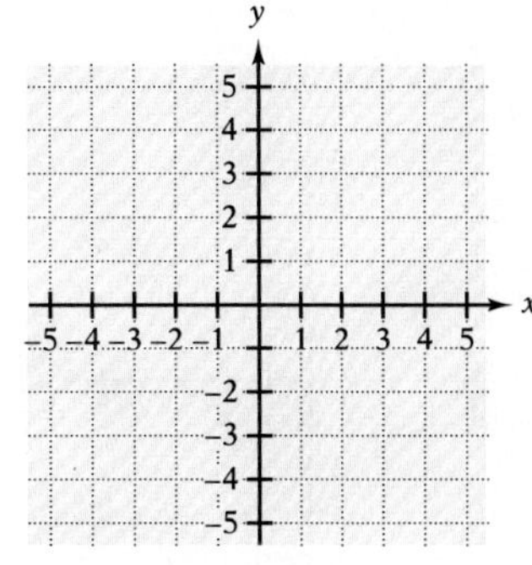

Copyright © Houghton Mifflin Company. All rights reserved.

B. Special Cases

23. The equation $y = 2$ means that the ________ of every solution is 2.

24. The graph of $x = -1$ is a(n) ________ line.

In Exercises 25–32, graph the given equation.

25. $x = 2$

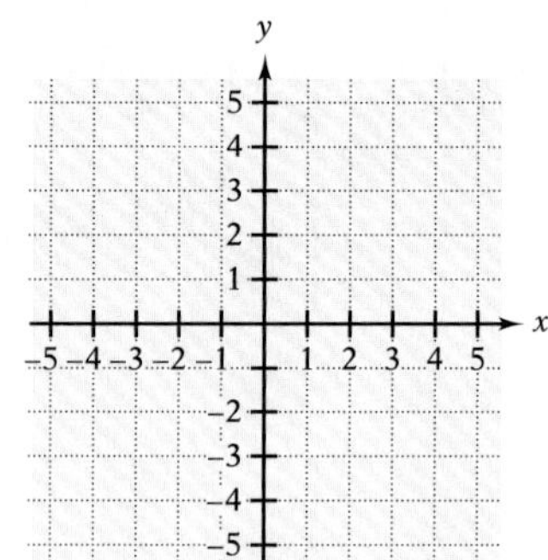

26. $x = 5$

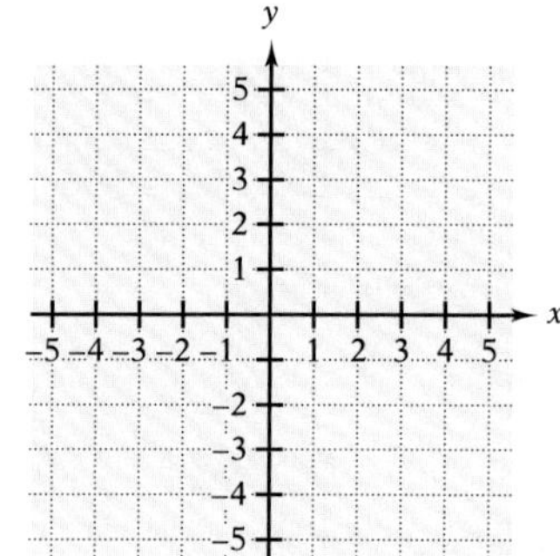

27. $x = -1$

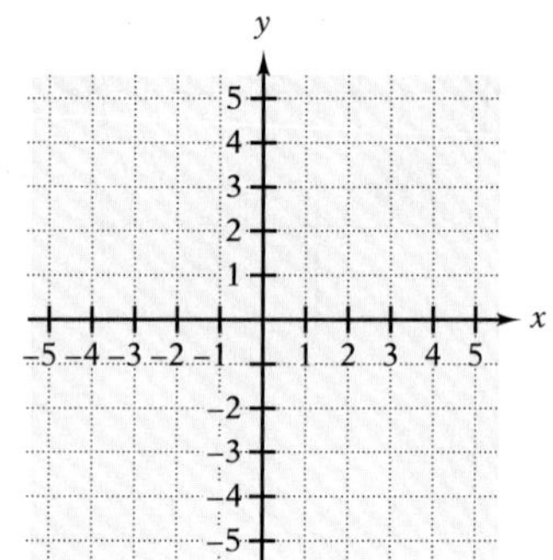

28. $x = -3$

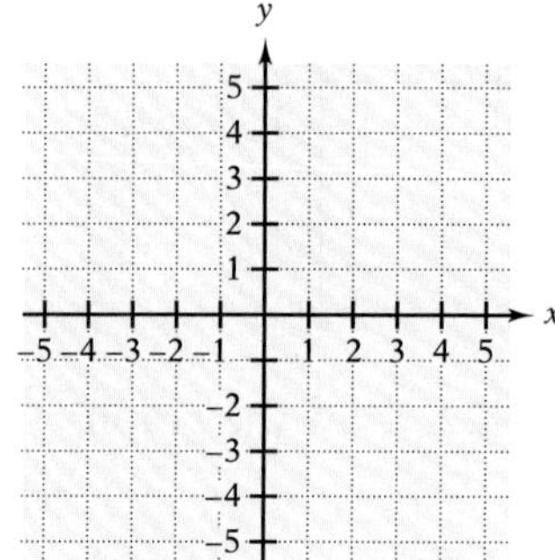

29. $y = 3$

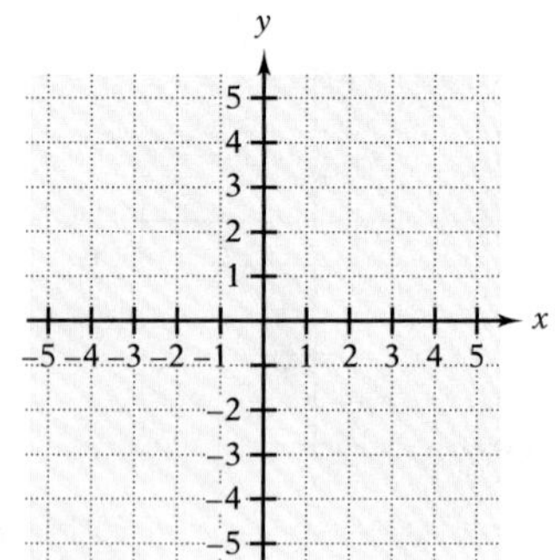

30. $y = 1$

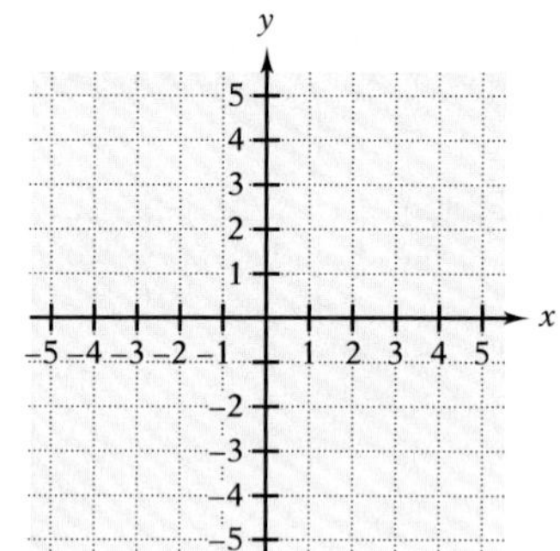

31. $y = -2$

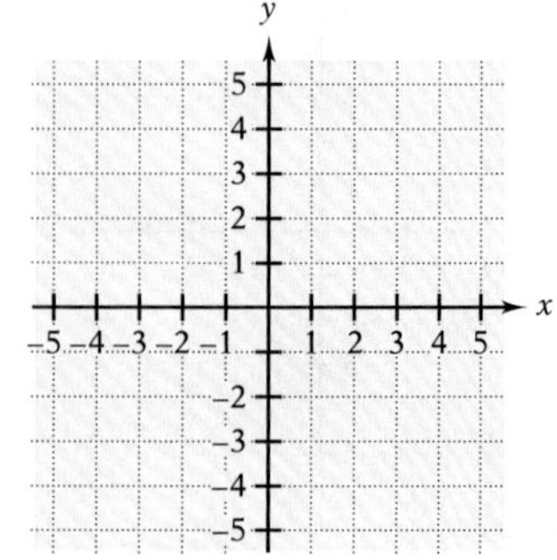

32. $y = -4$

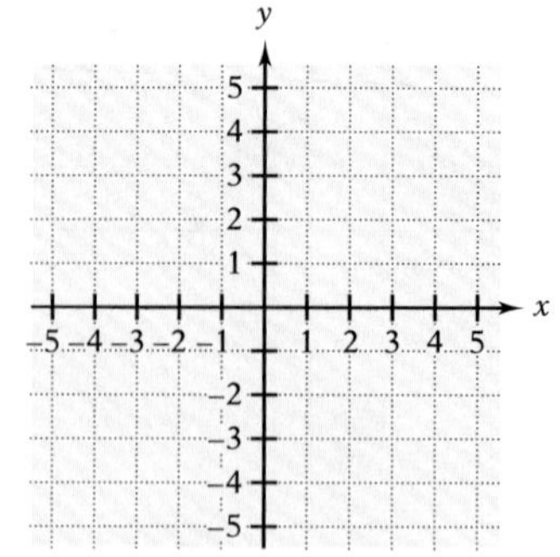

Copyright © Houghton Mifflin Company. All rights reserved.

Applications

33. The formula $D = 60T$ relates time T and distance D for a car traveling at 60 miles per hour.

(a) Complete the table of values for the equation.

T	D	(T, D)
1		(,)
2		(,)
3		(,)
4		(,)
5		(,)

(b) Use the table in part (a) to draw a first-quadrant graph.

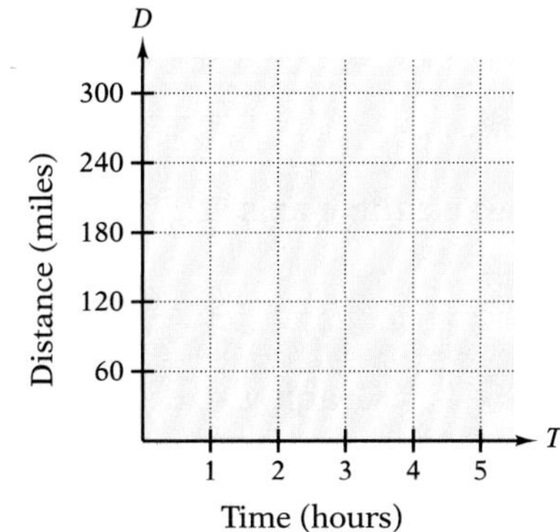

(c) Using only the graph, find the distance that the car travels after 4 hours.

(d) Using only the graph, find the time required to travel 120 miles.

(e) What is the significance of the point (0, 0) in the graph in part (b)?

(f) Why do we restrict the graph in part (b) to the first quadrant?

34. The equation $5F - 9C = 160$ relates Celsius and Fahrenheit temperatures. Complete the table of values for the equation. Then use the table to draw the graph.

C	F	(C, F)
−5		(,)
	32	(,)
5		(,)
10		(,)
	59	(,)

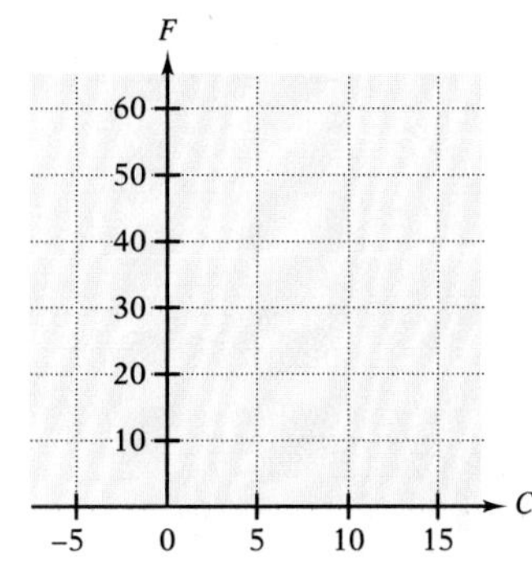

Copyright © Houghton Mifflin Company. All rights reserved.

Writing and Concept Extension

35. The graph of $y = x - 5$ is a straight line. Although only two solutions are needed to draw the line, explain why finding three solutions is a good idea.

36. Explain how to determine a solution of $x + y = -2$.

37. Why do you suppose that an equation such as $x + 2y = 8$ is called a *linear* equation in two variables?

38. Suppose that you draw the graphs of $x = 2$ and $y = 3$ on the same coordinate system. What are the coordinates of the point where the two lines intersect?

39. Where does the graph of $3x - 2y = 6$ cross the x-axis? the y-axis?

40. Graph $y = |x|$.

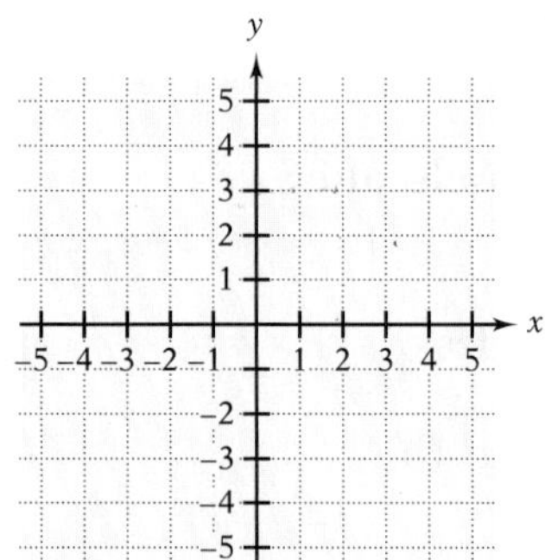

41. Graph $y = x^2$.

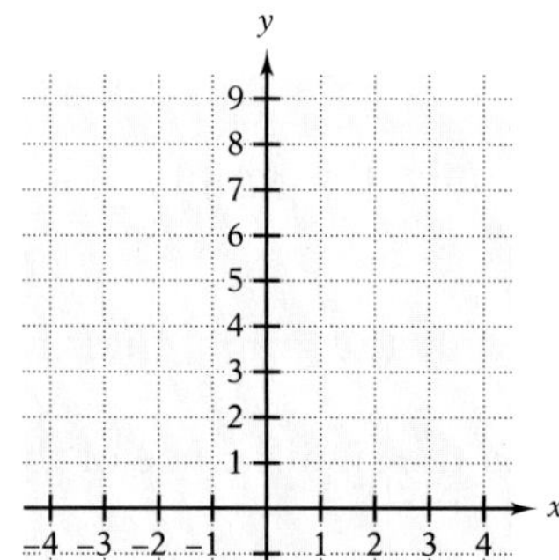

Copyright © Houghton Mifflin Company. All rights reserved.

CHAPTER 4 REVIEW EXERCISES

Section 4.1

1. Show that -1 is a solution of $5 - 2x = x + 8$.

2. To solve $x + 5 = 4$, we use the ______ Property of Equations.

3. Determine whether 1 is a solution of $2x - 1 = 6 - 5x$.

In Exercises 4–6, solve the given equation.

4. $5 + x = -3$

5. $-1 = n - 5$

6. $10 - 3 = 5x - x + 4 - 3x$

7. To solve $4x = 8$, we use the ______ Property of Equations.

In Exercises 8–10, solve the given equation.

8. $6x = -30$

9. $0 = -9y$

10. $2y - 3y = -1 + 8$

Section 4.2

11. Explain the difference between an expression and an equation.

12. If an equation contains parentheses, we often begin by ______ both sides of the equation.

In Exercises 13–20, solve the given equation.

13. $3x - 7 = -25$

14. $2 = 3x - 13$

15. $3x = 8x - 20$

16. $x + 7 = 3x - 11$

17. $-7x + 9 + 2x = 5 + 3x - 20$

18. $3(x - 7) = 15$

19. $x - 2(1 - 3x) = 12$

20. $-(4 - 5x) - 6x = 4$

Copyright © Houghton Mifflin Company. All rights reserved.

Section 4.3

In Exercises 21 and 22, translate the given information into an algebraic expression.

21. The sum of -5 and a number

22. Three times the difference of 5 and a number

23. When a number is decreased by 5, the result is -14. Translate this information into an equation.

24. The difference of 7 and twice a number is 3. What is the number?

25. Twice the sum of a number and 5 is 6 more than the number. What is the number?

26. The length of a rectangle is 3 less than twice its width. If the perimeter is 36 inches, determine the length and width of the rectangle.

27. An officer used 260 feet of police tape to enclose a rectangular area. If the length of the area was 80 feet, what was the width?

28. An auto repair service charges $45 plus $25 per hour. If the total bill was $170, for how many hours was the customer billed?

29. One number is 8 more than another number. The sum of the smaller number and 3 times the larger number is 0. What are the two numbers?

30. A 25-foot cable is cut into two pieces. If the second piece is 1 more than twice the length of the first piece, determine the length of each piece.

Copyright © Houghton Mifflin Company. All rights reserved.

Section 4.4

31. Translate the following information into an equation in two variables: The difference of two numbers is 7. (Let x represent the larger number and y represent the smaller number.)

32. We call $(3, -6)$ a(n) ______.

In Exercises 33–35, determine whether the given pair is a solution of $2x - 4y = 8$.

33. $(-6, -5)$ **34.** $(2, -3)$ **35.** $(6, 1)$

In Exercises 36–38, complete the ordered pairs so that they are solutions of the given equation.

36. $3x + 5y = 30$; $(\ \_\_\ , 6), (5, \_\_\ )$

37. $y = -x + 6$; $(-4, \_\_\ ), (\_\_\ , 4)$

38. $y = 7$; $(3, \_\_\ ), (-5, \_\_\ )$

In Exercises 39 and 40, complete the table of values for the given equation.

39. $x + 3y = 3$

x	y	(x, y)
	2	(,)
	1	(,)
3		(,)
6		(,)

40. $y = 2x$

x	y	(x, y)
−2		(,)
0		(,)
1		(,)
3		(,)

Section 4.5

41. Highlighting the point associated with an ordered pair is called ______ the point.

In Exercises 42–44, what name do we give to the given part of the rectangular coordinate system?

42. The vertical number line **43.** The horizontal number line

44. The point whose coordinates are $(0, 0)$.

Copyright © Houghton Mifflin Company. All rights reserved.

45. Plot the following ordered pairs.

$(2, 4), (3, -1), (0, 1), (-3, 2), (-1, 0), (-4, -3)$

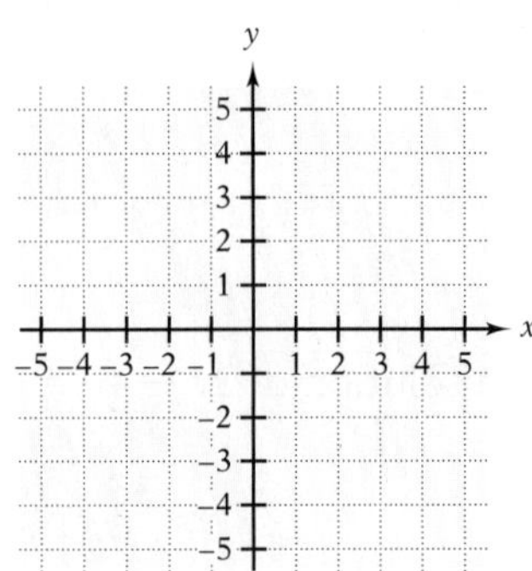

In Exercises 46–49, determine the quadrant or the axis in which the point lies.

46. $(-3, -6)$ **47.** $(4, -1)$ **48.** $(0, -1)$ **49.** $(5, 0)$

50. Determine the coordinates of the plotted points.

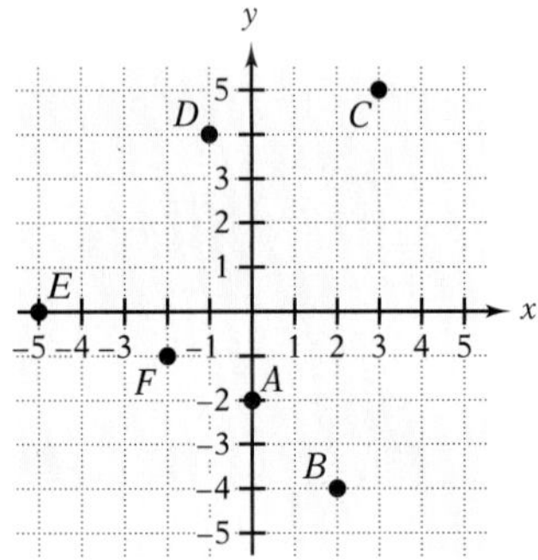

Section 4.6

51. To determine a solution of $2x - 5y = 20$, you might begin by choosing a value for x. What number should you use?

52. For the equation $2x + 6y = 6$, complete the following solutions and use the solutions to draw the graph of the equation. $(-3, \quad)$, $(0, \quad)$, $(\quad, -1)$

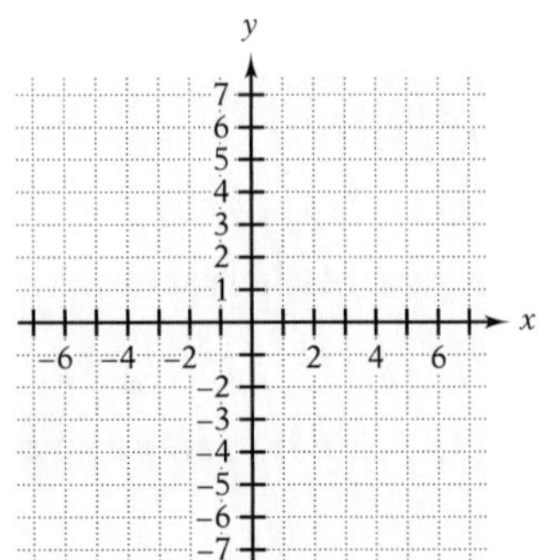

Copyright © Houghton Mifflin Company. All rights reserved.

53. Complete the table of values for the equation $4x - 5y = 20$. Then use the solutions to draw the graph of the equation.

x	y	(x, y)
	-4	(,)
5		(,)
-5		(,)

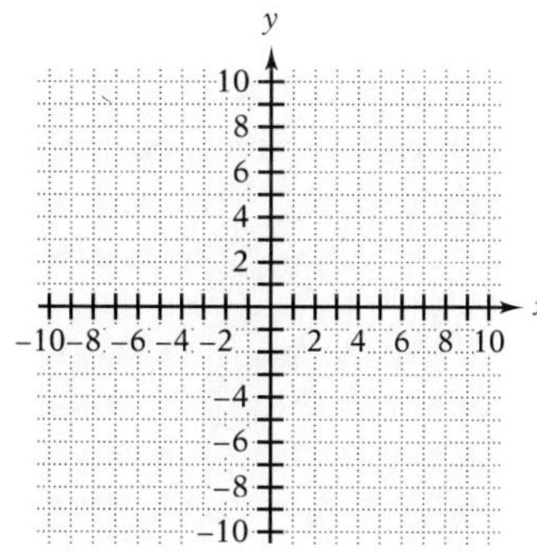

54. By choosing values of x and determining the corresponding values of y, graph $y = -2x$.

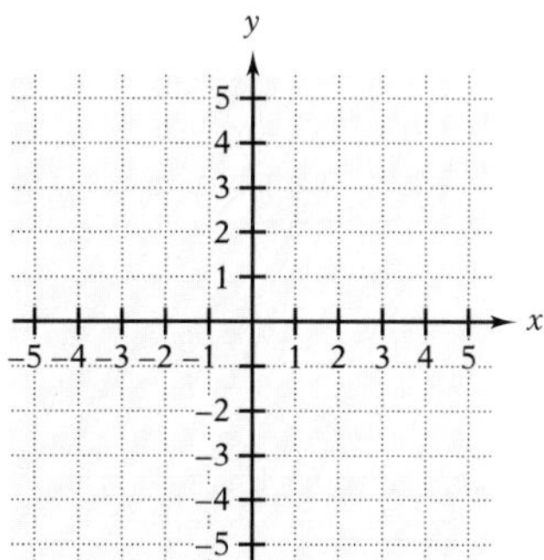

55. By choosing values of y and determining the corresponding values of x, graph $x - 2y = 4$.

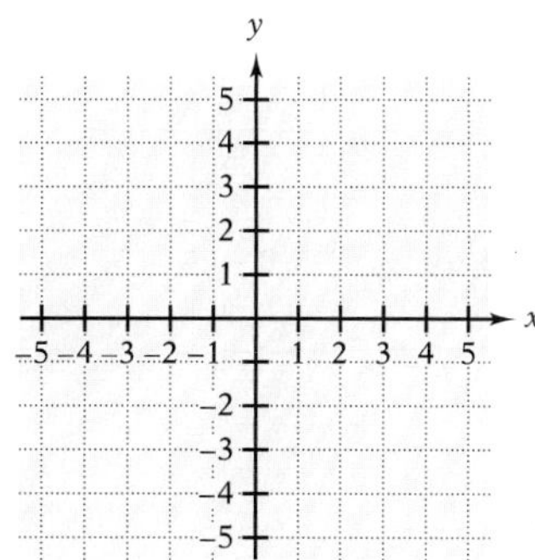

56. Graph $y - x = -2$.

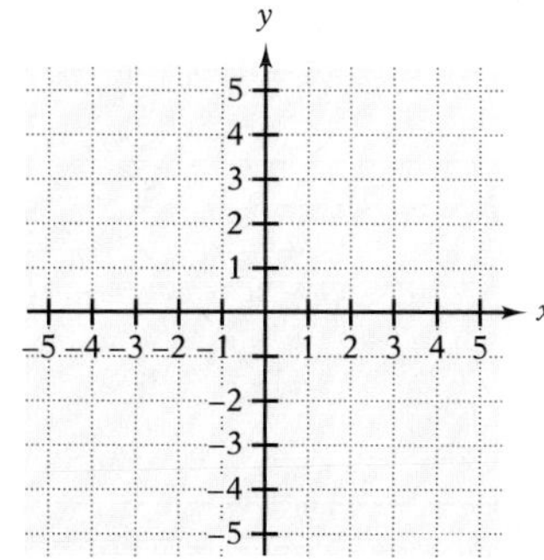

Copyright © Houghton Mifflin Company. All rights reserved.

57. The equation $x = -1$ means that the ________ of every solution is -1.

58. The equation $y = 4$ means that the ________ of every solution is 4.

In Exercises 59 and 60, draw the graph of the given equation.

59. $x = 4$

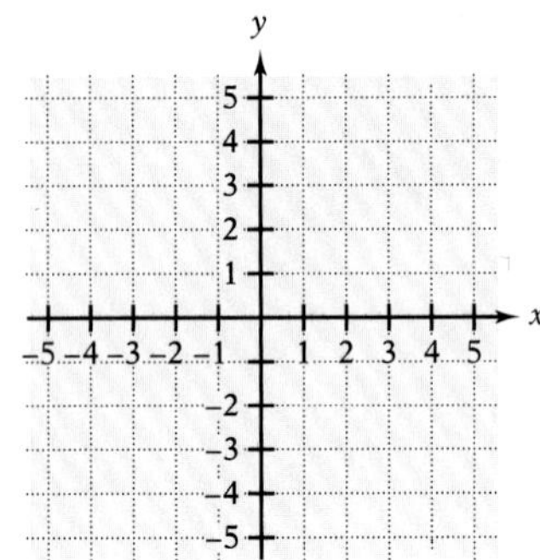

60. $y = -2$

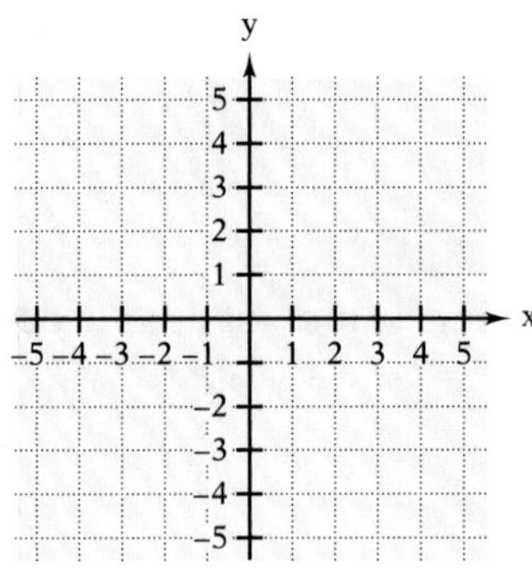

Copyright © Houghton Mifflin Company. All rights reserved.

CHAPTER 4 TEST

1. Which of the following is an equation and which is an expression?

   (i) $3x + 2(x + 3)$ (ii) $3x = 2(x + 3)$

2. To solve $-x = 6$, we can ________ or ________ both sides by -1.

3. To solve $-4x = 8$, we use the ________ Property of Equations, whereas to solve $x - 4 = 8$, we use the ________ Property of Equations.

In Questions 4–9 solve the given equation.

4. $5y + 8 - 4y = 5 - 7$

5. $x - 8x + 3x = -6 - 2$

6. $-35 = 7a$

7. $11 - 2n = 3$

8. $4b + 7 = 7 - 5b$

9. $7 + 4x = 6x - 9$

10. A basketball team won 1 game less than 3 times the number of games they lost. If the team won 20 games, how many games did they lose?

11. While in the White House, 23 presidents have owned dogs. Six times the number who owned goats is 1 more than the number who owned dogs. How many presidents owned goats?

12. A 15-foot board is cut into two pieces. The first piece is 3 feet shorter than the second piece. What is the length of each piece?

13. The coordinate plane is divided into four ________.

In Questions 14–17, name the quadrant or the axis in which the point lies.

14. $(-12, 0)$

15. $(-21, 15)$

16. $(43, -54)$

17. $(0, -46)$

18. Determine the coordinates of the plotted points.

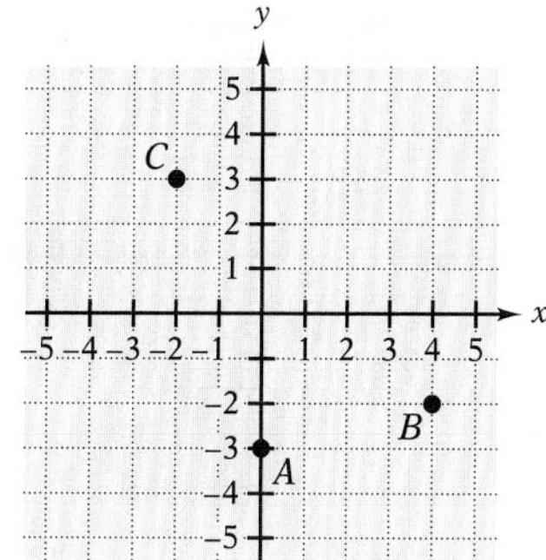

Copyright © Houghton Mifflin Company. All rights reserved.

19. The graph of $y = 2$ is a(n) ______ line, whereas the graph of $x = 2$ is a(n) ______ line.

20. Of the three ordered pairs (6, 0), (0, −8), (−3, 12), which are solutions of $4x - 3y = 24$?

21. Consider the equation $4x + 3y = 9$. If y is −5, what is the value of x?

22. Complete the pairs (___ , 3), (0, ___), and (−3, ___) so that they are solutions of $y - 3x = 6$. Then draw the graph of the equation.

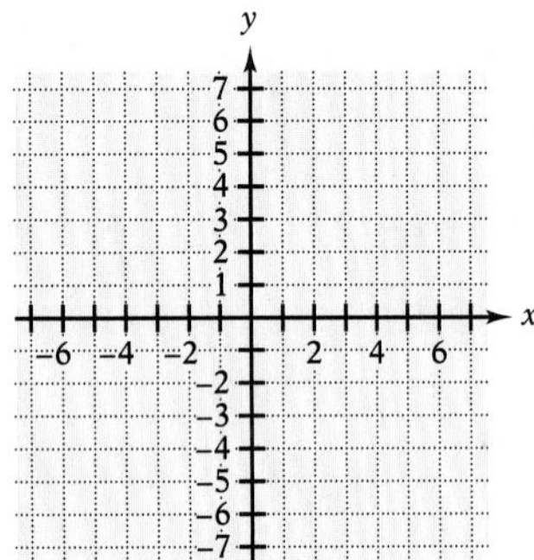

23. Complete the table of values for $2x - 5y = 10$. Then draw the graph of the equation.

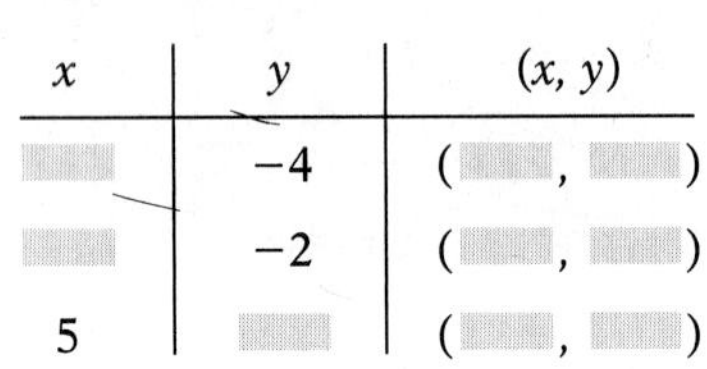

x	y	(x, y)
___	−4	(___ , ___)
___	−2	(___ , ___)
5	___	(___ , ___)

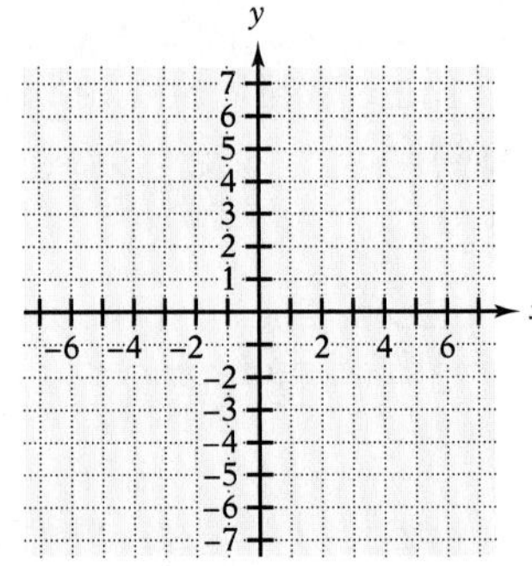

24. Draw the graph of $y = 2x + 3$. (*Hint:* Select any three values for x and determine the corresponding values of y.)

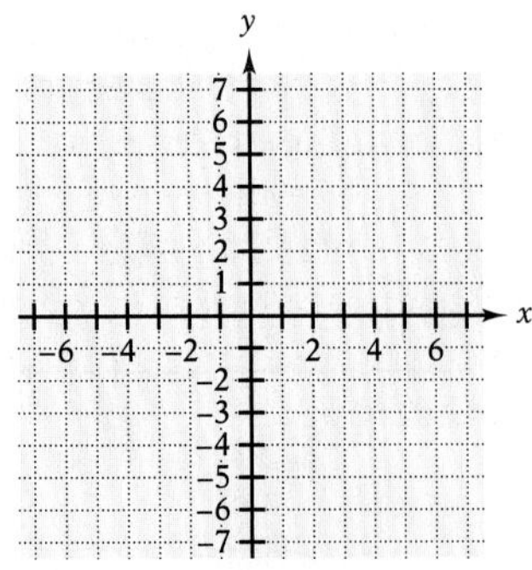

Copyright © Houghton Mifflin Company. All rights reserved.

CHAPTER 5

Fractions

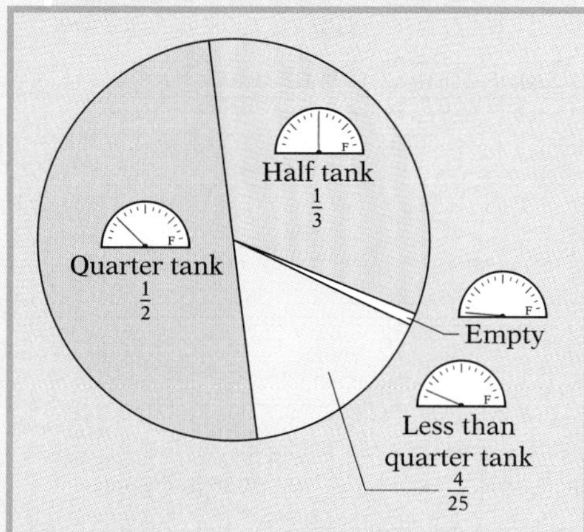

When do you stop to put gas in your car?

The accompanying pie graph shows the fraction of vacation drivers who stop for gas when the fuel gauge reaches a certain level.

To determine the fraction of drivers who stop for gas when their tanks are at least one-quarter full, we would need to add the fractions $\frac{1}{2} + \frac{1}{3}$. The method for performing that addition is not what you might think!

Chapter Snapshot

In this chapter, we begin by describing what a fraction is and how it can be interpreted. Then you will learn about prime numbers and how they can be used to simplify a fraction. Most of the chapter deals with the four basic operations with fractions: multiplication and division, addition and subtraction. We also show how those operations can be performed with mixed numbers. We conclude with methods for solving equations that contain fractions and formulas from geometry, business, and science.

Because fractions play an important role in algebra, your mastery of the topics in this chapter will give you a good start in your further studies of mathematics.

Copyright © Houghton Mifflin Company. All rights reserved.

For online resources, visit the web site **math.college.hmco.com/students** and follow the links to Hubbard/Robinson, *Prealgebra.*

Some Friendly Advice . . .

By now, you have probably made some friends in your class. Some of them may share your goals and are willing to work toward them as hard as you do. If so, consider joining forces.

Many students benefit from small study groups that they organize on their own. A group of no more than three seems to work best. They meet at each other's homes or somewhere on campus to do homework together and to help each other prepare for tests. They share notes and practice teaching each other the material.

This is not only an outstanding way to stay focused and improve your learning, but it also serves as a support group that can be counted on during the good and bad times. Some students make lifelong friends through study groups. Try it!

WARM-UP SKILLS

The following questions review concepts and skills that you will need in Chapter 5.

In Exercises 1 and 2, divide.

1. $\frac{-10}{-5}$ **2.** $\frac{0}{7}$

In Exercises 3–5, perform the indicated operations.

3. $-5(-6)$ **4.** $2 - (-8)$ **5.** $2^3 \cdot 3 \cdot 5$

6. Divide $23 \div 4$.

7. Simplify $3x + 5x$.

8. Evaluate $\frac{bh}{2}$ for $b = 5$ and $h = 8$.

9. Solve $4x + 30 = 2$.

10. Determine the perimeter and the area of a rectangle that is 8 inches wide and 12 inches long.

Copyright © Houghton Mifflin Company. All rights reserved.

5.1 Introduction to Fractions

A Basic Concepts
B Multiplication of Fractions
C Applications

SUGGESTIONS FOR SUCCESS

Here is the best way to make fractions unpleasant and hard to understand: Keep telling yourself that fractions are unpleasant and hard to understand.

As with any subject that you study, not just mathematics, much of your success will depend on your attitude about it. Bad past experiences don't matter because you are older and wiser now.

Forget what other people may tell you. Fractions are *not* hard. If you are determined to have a positive attitude and if you truly believe that you can succeed, then you will; we guarantee it.

A Basic Concepts

We have seen that expressions of the form $\frac{a}{b}$ represent division and are read as "a divided by b." We can also regard $\frac{a}{b}$ as a number, which we call a **fraction.**

Definition of a Fraction

A **fraction** is a number that is written in the form $\frac{a}{b}$, where b is not 0. The number a is called the **numerator,** and the number b is called the **denominator.**

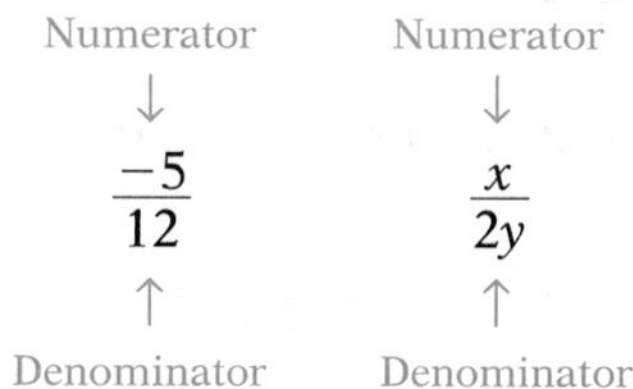

$$\frac{-5}{12} \qquad \frac{x}{2y}$$

NOTE The numerator and the denominator of a fraction can be any kind of number. In this chapter, we will limit the numerator and the denominator to integers. Such fractions are called *rational numbers.*

One way in which fractions can be used is to indicate the number of equal parts of a whole.

DEVELOPING THE CONCEPT

Interpretations of Fractions

Suppose that we divide a rectangle into 2, 4, and 8 equal parts.

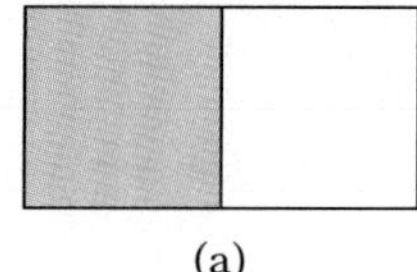

(a)

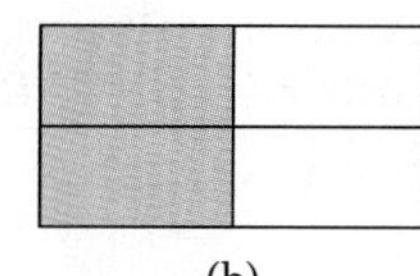

(b)

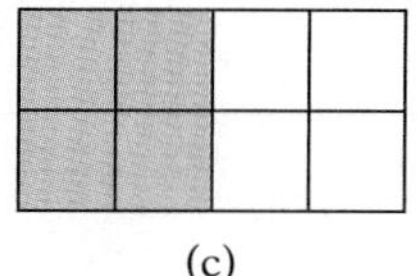

(c)

Copyright © Houghton Mifflin Company. All rights reserved.

Figure	*Shaded Area*	*Represented as a Fraction*
(a)	1 of the 2 equal areas	$\frac{1}{2}$
(b)	2 of the 4 equal areas	$\frac{2}{4}$
(c)	4 of the 8 equal areas	$\frac{4}{8}$

If you compare the shaded areas in the three rectangles, you will see that they are all the same. Therefore, the fractions that represent those shaded areas must also be the same: $\frac{1}{2} = \frac{2}{4} = \frac{4}{8}$.

The fractions $\frac{1}{2}$, $\frac{2}{4}$, and $\frac{4}{8}$ have different "names," but we have seen that they have the same value. We call such fractions **equivalent fractions.**

We can also use the number line to represent fractions and to show equivalent fractions.

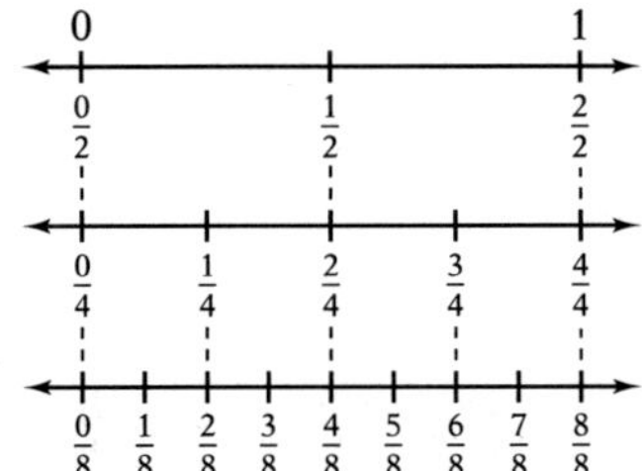

Again, we see that fractions with different names are associated with the same point of the number line. For example, $\frac{3}{4}$ and $\frac{6}{8}$ have the same value because they are both associated with the same point. Therefore, $\frac{3}{4}$ and $\frac{6}{8}$ are equivalent fractions.

Example 1

In the preceding figure, find a fraction that is equivalent to $\frac{2}{8}$.

SOLUTION

The fraction $\frac{1}{4}$ is equivalent to $\frac{2}{8}$ because both fractions are associated with the same point.

Your Turn 1

In the preceding figure, find two fractions that are equivalent to $\frac{4}{8}$.

Answer: $\frac{2}{4}$, $\frac{1}{2}$

In the preceding figure, did you observe that the fractions $\frac{0}{2}$, $\frac{0}{4}$, and $\frac{0}{8}$ are all associated with 0? Similarly, the fractions $\frac{2}{2}$, $\frac{4}{4}$, and $\frac{8}{8}$ are all associated with 1. These results are consistent with previously stated rules for division.

Special Fractions

If n is any number except 0, then:

1. $\frac{0}{n} = 0$
2. $\frac{n}{n} = 1$

Copyright © Houghton Mifflin Company. All rights reserved.

Example 2

Write the given fraction as an integer.

(a) $\frac{12}{12}$ **(b)** $\frac{0}{-7}$

SOLUTION

(a) $\frac{12}{12} = 1$ **(b)** $\frac{0}{-7} = 0$

Your Turn 2

Write the given fraction as an integer.

(a) $\frac{0}{31}$ **(b)** $\frac{-8}{-8}$

Answers: (a) 0; (b) 1

Another division rule that carries over to fractions is $\frac{n}{1} = n$. This rule allows us to write any number as a fraction.

Example 3

Write each number as a fraction.

(a) 5 **(b)** −2

SOLUTION

(a) $5 = \frac{5}{1}$ **(b)** $-2 = \frac{-2}{1}$

Your Turn 3

Write each number as a fraction.

(a) 16 **(b)** x

Answers: (a) $\frac{16}{1}$; (b) $\frac{x}{1}$

LEARNING TIP

There is nothing "improper" about an improper fraction. It's just the name that we give to fractions such as $\frac{9}{5}$. Sometimes the language of mathematics can seem misleading.

Special names are given to certain kinds of fractions.

Proper and Improper Fractions

A **proper fraction** is a fraction in which the numerator is less than the denominator. An **improper fraction** is a fraction in which the numerator is greater than or equal to the denominator.

Example 4

Determine whether these fractions are proper or improper fractions.

(a) $\frac{11}{7}$ **(b)** $\frac{2}{5}$

SOLUTION

(a) Because the numerator is greater than the denominator, $\frac{11}{7}$ is an improper fraction.

(b) Because the numerator is less than the denominator, $\frac{2}{5}$ is a proper fraction.

Your Turn 4

Determine whether these fractions are proper or improper fractions.

(a) $\frac{8}{9}$ **(b)** $\frac{3}{2}$

Answers: (a) proper; (b) improper

Copyright © Houghton Mifflin Company. All rights reserved.

KEYS TO THE CALCULATOR

If your calculator has a fraction key, such as [a b/c], you can enter fractions and perform operations with them. Here are typical keystrokes for entering the fraction $\frac{5}{8}$.

Scientific calculator: **5** [a b/c] **8** Display: [5 ⌋ 8]

Note that the symbol ⌋ separates the numerator and the denominator. Make sure that you clear the display before you enter another fraction.

To enter a negative fraction, use the change of sign key, possibly [+/−] or [+○−]. Here are typical keystrokes for entering $-\frac{3}{5}$.

3 [a b/c] **5** [+○−] Display: [−3 ⌋ 5]

A graphing calculator will display a fraction in fraction form only if you select the *Frac* option. Here are keystrokes for entering $\frac{5}{8}$ and $-\frac{3}{5}$.

Graphing calculator: **5** [÷] **8** [MATH] [Frac] [ENTER]

[(−)] **3** [÷] **5** [MATH] [Frac] [ENTER]

```
5/8▶Frac
                5/8
-3/5▶Frac
               -3/5
```

Exercises

Enter the following fractions in your calculator:

(a) $\frac{2}{3}$ **(b)** $-\frac{4}{7}$ **(c)** $-\frac{3}{10}$ **(d)** $\frac{8}{15}$

B Multiplication of Fractions

Because fractions are numbers, we want to be able to perform all the basic operations with them. The easiest operation to perform with fractions is multiplication.

Multiplication of Fractions

If a, b, c, and d are any numbers, where b and d are not 0, then

$$\frac{a}{b} \cdot \frac{c}{d} = \frac{a \cdot c}{b \cdot d}$$

In words, to multiply fractions, we multiply the numerators and multiply the denominators.

Copyright © Houghton Mifflin Company. All rights reserved.

Example 5

Multiply.

(a) $\frac{2}{3} \cdot \frac{7}{5}$

(b) $\frac{8}{9} \cdot \frac{x}{y}$

SOLUTION

(a) $\frac{2}{3} \cdot \frac{7}{5} = \frac{2 \cdot 7}{3 \cdot 5} = \frac{14}{15}$

(b) $\frac{8}{9} \cdot \frac{x}{y} = \frac{8 \cdot x}{9 \cdot y} = \frac{8x}{9y}$

Your Turn 5

Multiply.

(a) $\frac{7}{2} \cdot \frac{5}{6}$

(b) $\frac{4}{x} \cdot \frac{y}{3}$

Answers: (a) $\frac{35}{12}$; (b) $\frac{4y}{3x}$

When we multiply a fraction and an integer, writing the integer as a fraction helps us to apply the multiplication rule.

Example 6

Multiply $\frac{3}{8} \cdot 5$.

SOLUTION

$$\begin{aligned} \frac{3}{8} \cdot 5 &= \frac{3}{8} \cdot \frac{5}{1} && \text{Write 5 as } \frac{5}{1}. \\ &= \frac{3 \cdot 5}{8 \cdot 1} \\ &= \frac{15}{8} \end{aligned}$$

Your Turn 6

Multiply $4 \cdot \frac{7}{3}$.

Answer: $\frac{28}{3}$

NOTE In mathematics, the word *of* means "times." In Example 6, the expression $\frac{3}{8} \cdot 5$ is sometimes read as "$\frac{3}{8}$ of 5."

Example 7

Multiply $-6 \cdot \frac{5}{7}$.

SOLUTION

$$\begin{aligned} -6 \cdot \frac{5}{7} &= \frac{-6}{1} \cdot \frac{5}{7} && \text{Write } -6 \text{ as } \frac{-6}{1}. \\ &= \frac{-6 \cdot 5}{1 \cdot 7} \\ &= \frac{-30}{7} \end{aligned}$$

Your Turn 7

Multiply $\frac{2}{3} \cdot (-8)$.

Answer: $\frac{-16}{3}$

Consider the following division problems:

$$\frac{-12}{4} = -3 \qquad \frac{12}{-4} = -3 \qquad -\frac{12}{4} = -3$$

Note that $\frac{-12}{4}$, $\frac{12}{-4}$, and $-\frac{12}{4}$ all have the same value. We can summarize this observation with the following rule.

Copyright © Houghton Mifflin Company. All rights reserved.

Signs in Fractions

If a and b are any numbers, where b is not 0, then

$$\frac{-a}{b} = \frac{a}{-b} = -\frac{a}{b}$$

In words, an opposite symbol in a fraction can be placed in the numerator, in the denominator, or in front of the fraction.

Example 8

Write the given fractions in two other ways.

(a) $\frac{5}{-11}$ **(b)** $\frac{-3}{8}$ **(c)** $-\frac{1}{2}$

SOLUTION

(a) $\frac{5}{-11} = \frac{-5}{11} = -\frac{5}{11}$

(b) $\frac{-3}{8} = \frac{3}{-8} = -\frac{3}{8}$

(c) $-\frac{1}{2} = \frac{-1}{2} = \frac{1}{-2}$

Your Turn 8

Write the given fractions in two other ways.

(a) $\frac{-2}{3}$ **(b)** $-\frac{6}{5}$ **(c)** $\frac{1}{-8}$

Answers: (a) $\frac{-2}{3} = -\frac{2}{3} = \frac{2}{-3}$; (b) $-\frac{6}{5} = \frac{-6}{5} = \frac{6}{-5}$; (c) $\frac{1}{-8} = \frac{-1}{8} = -\frac{1}{8}$

Example 9

Multiply $\frac{-2}{7} \cdot \frac{8}{3}$ and write the result in three different ways.

SOLUTION

$$\frac{-2}{7} \cdot \frac{8}{3} = \frac{-2 \cdot 8}{7 \cdot 3} = \frac{-16}{21}$$
$$= \frac{16}{-21}$$
$$= -\frac{16}{21}$$

Your Turn 9

Multiply $\frac{5}{2} \cdot \frac{3}{-2}$ and write the result in three different ways.

Answer: $\frac{15}{-4} = -\frac{15}{4} = \frac{-15}{4}$

The rule $\frac{-a}{b} = \frac{a}{-b} = -\frac{a}{b}$ gives us a lot of flexibility when we multiply fractions. Example 10 shows three different ways to multiply $\frac{-2}{3} \cdot \frac{5}{-7}$.

Example 10

Multiply $\frac{-2}{3} \cdot \frac{5}{-7}$.

SOLUTION

Method 1

$$\frac{-2}{3} \cdot \frac{5}{-7} = \frac{-2}{3} \cdot \frac{-5}{7} = \frac{(-2)(-5)}{(3)(7)} = \frac{10}{21}$$

Your Turn 10

Multiply $-\frac{1}{6} \cdot \frac{5}{-3}$.

Copyright © Houghton Mifflin Company. All rights reserved.

Method 2

$$\frac{-2}{3} \cdot \frac{5}{-7} = \frac{2}{-3} \cdot \frac{5}{-7} = \frac{(2)(5)}{(-3)(-7)} = \frac{10}{21}$$

Method 3

$$\frac{-2}{3} \cdot \frac{5}{-7} = \left(-\frac{2}{3}\right)\left(-\frac{5}{7}\right)$$

$$= \frac{10}{21}$$ The product of two negative numbers is positive.

Look ahead at the expressions in parts (a) and (b) of Example 11. These look somewhat complicated. Fortunately, the Associative Property of Multiplication gives us the flexibility to move parentheses in a product and, by doing so, we can sometimes simplify such expressions.

Example 11

Simplify.

(a) $3\left(\frac{1}{3}x\right)$

(b) $-\frac{5}{2}\left(-\frac{2}{5}y\right)$

SOLUTION

(a) $3\left(\frac{1}{3}x\right) = \left(3 \cdot \frac{1}{3}\right)x$ Associative Property of Multiplication

$= \left(\frac{3}{1} \cdot \frac{1}{3}\right)x$

$= \frac{3}{3}x$

$= 1x$

$= x$

(b) $-\frac{5}{2}\left(-\frac{2}{5}y\right) = \left[-\frac{5}{2}\left(-\frac{2}{5}\right)\right]y$

$= \frac{10}{10}y$

$= 1y$

$= y$

Your Turn 11

Simplify.

(a) $-\frac{1}{2}(-2x)$

(b) $\frac{3}{4}\left(\frac{4}{3}a\right)$

Answers: (a) x; (b) a

C Applications

A *pie graph* is sometimes used to show quantities relative to each other. Each "slice" of the pie is associated with a particular quantity, and the size of the slice indicates the quantity's fractional part of the total.

Sometimes a pie graph shows actual data. Other times, a pie graph simply shows the fractional part of the whole that each slice represents.

Copyright © Houghton Mifflin Company. All rights reserved.

Example 12

The following pie graph shows the number of students in a math class who study the indicated number of hours per week.

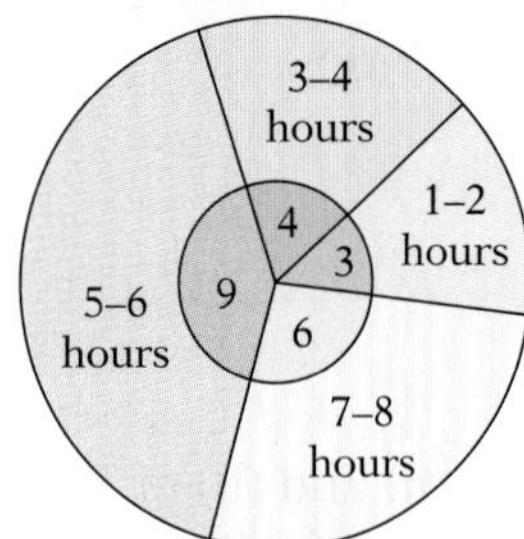

(a) What fraction of the students study 5–6 hours?

(b) What fraction of the students study less than 5 hours?

SOLUTION

First, we add the number of students in each slice to find the total number of students:

$$3 + 4 + 9 + 6 = 22$$

(a) In the slice for 5–6 hours, we see that 9 of the 22 students fall into this category: $\frac{9}{22}$.

(b) There are 3 students who study 1–2 hours, and there are 4 students who study 3–4 hours. Therefore, 7 of the 22 students study less than 5 hours: $\frac{7}{22}$.

Your Turn 12

(a) What fraction of the students study 1–2 hours?

(b) What fraction of the students study more than 2 hours?

Answers: (a) $\frac{3}{22}$; (b) $\frac{19}{22}$

This introductory section on fractions contains quite a lot of vocabulary. Words such as *numerator, denominator, equivalent fractions,* and so on are needed in order to describe and discuss fractions intelligently. The sooner you can build these words into your own vocabulary, the better your understanding of fractions will be.

5.1 Quick Reference

A. Basic Concepts

1. A **fraction** is a number that is written in the form $\frac{a}{b}$, where b is not 0. The number a is called the **numerator,** and the number b is called the **denominator.**

$\frac{3}{5}$ — 3: Numerator; 5: Denominator

Copyright © Houghton Mifflin Company. All rights reserved.

2. A fraction can be used to indicate the number of equal parts of a whole.

The shaded area is $\frac{2}{3}$ of the total area.

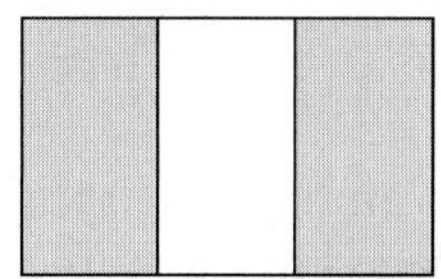

3. Fractions with different names but with the same value are called **equivalent fractions.**

$$\frac{1}{2} = \frac{2}{4} = \frac{3}{6} = \frac{4}{8} = \dots$$

$$\frac{1}{3} = \frac{2}{6} = \frac{3}{9} = \frac{4}{12} = \dots$$

4. Special fractions: If n is any number except 0, then:

(a) $\frac{0}{n} = 0$

$\frac{0}{6} = 0 \qquad \frac{0}{-9} = 0$

(b) $\frac{n}{n} = 1$

$\frac{4}{4} = 1 \qquad \frac{-3}{-3} = 1$

5. The rule $\frac{n}{1} = n$ allows us to write any number as a fraction.

$$7 = \frac{7}{1}$$

$$2x = \frac{2x}{1}$$

6. A **proper fraction** is a fraction in which the numerator is less than the denominator. An **improper fraction** is a fraction in which the numerator is greater than or equal to the denominator.

Proper fraction: $\frac{7}{12}$

Improper fraction: $\frac{9}{8}$

B. Multiplication of Fractions

1. If a, b, c, and d are any numbers, where b and d are not 0, then

$$\frac{a}{b} \cdot \frac{c}{d} = \frac{a \cdot c}{b \cdot d}$$

In words, to multiply fractions, we multiply the numerators and multiply the denominators.

$$\frac{5}{3} \cdot \frac{2}{9} = \frac{5 \cdot 2}{3 \cdot 9} = \frac{10}{27}$$

2. Writing integers as fractions helps us to apply the multiplication rule.

$$8\left(\frac{2}{3}\right) = \frac{8}{1} \cdot \frac{2}{3} = \frac{8 \cdot 2}{1 \cdot 3} = \frac{16}{3}$$

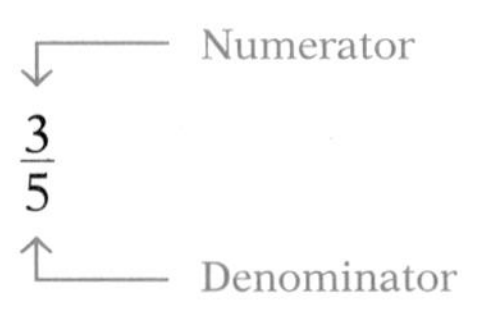

Copyright © Houghton Mifflin Company. All rights reserved.

3. If a and b are any numbers, where b is not 0, then

$$\frac{-a}{b} = \frac{a}{-b} = -\frac{a}{b}$$

In words, an opposite symbol in a fraction can be placed in the numerator, in the denominator, or in front of the fraction.

$$\frac{-2}{3} = \frac{2}{-3} = -\frac{2}{3}$$

$$\frac{-2x}{y} = \frac{2x}{-y} = -\frac{2x}{y}$$

4. The multiplication rule for fractions is used to simplify certain expressions.

$$\frac{5}{3}\left(\frac{3}{5}x\right) = \left(\frac{5 \cdot 3}{3 \cdot 5}\right)x = \frac{15}{15}x = x$$

Copyright © Houghton Mifflin Company. All rights reserved.

5.1 Exercises

A. Basic Concepts

1. In the ______ $\frac{3}{7}$, 7 is called the ______, and 3 is called the ______.

2. Fractions with the same value but with different names are called ______ fractions.

In Exercises 3–8, write a fraction that describes the area that is shaded.

3. 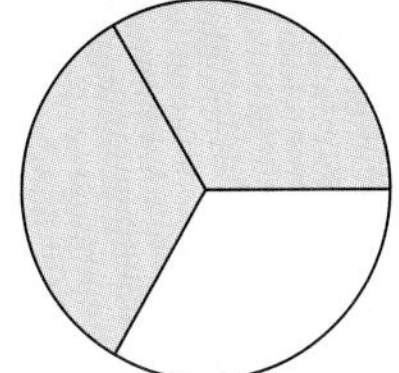

4. 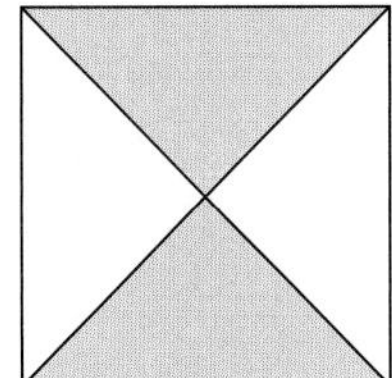

5. 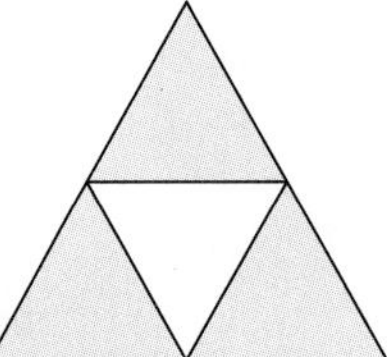

6. 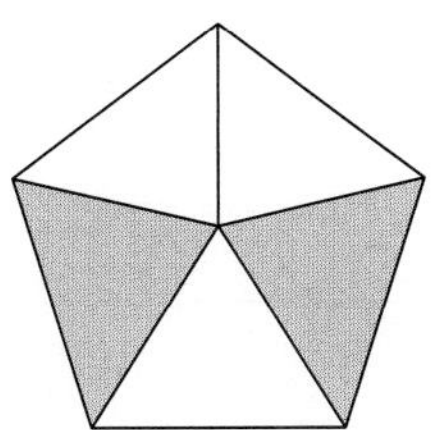

7. 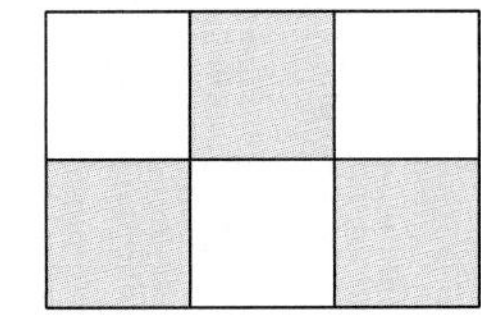

8. 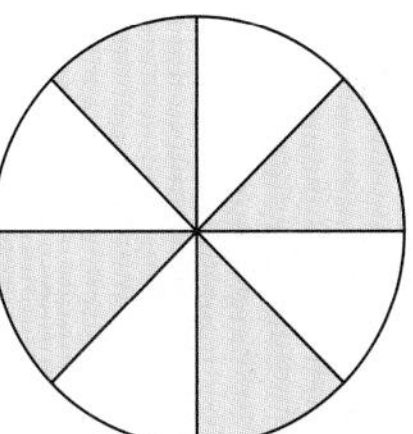

In Exercises 9–12, use the following number lines to determine a fraction that is equivalent to the given fraction:

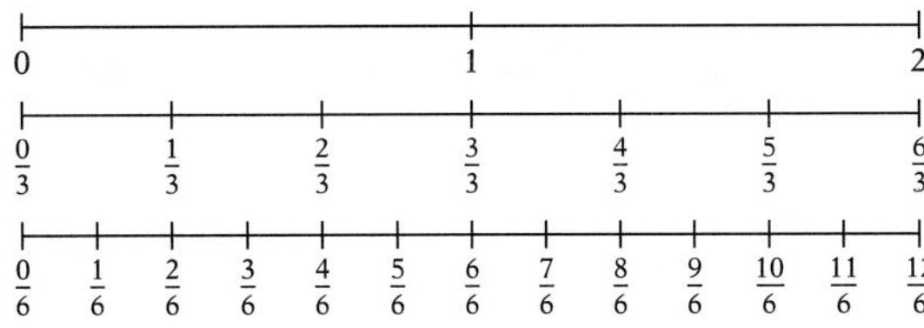

9. $\frac{4}{6}$

10. $\frac{1}{3}$

11. $\frac{4}{3}$

12. $\frac{10}{6}$

In Exercises 13–16, write an integer that is equivalent to the given fraction.

13. $\frac{-8}{8}$

14. $\frac{10}{10}$

15. $\frac{0}{5}$

16. $\frac{0}{-3}$

In Exercises 17–20, write the given integer as a fraction.

17. 4

18. -8

19. -12

20. 15

In Exercises 21–24, from the given list of fractions, identify (a) the proper fractions and (b) the improper fractions.

21. $\frac{5}{9}, \frac{12}{7}, \frac{14}{11}, \frac{12}{19}$

22. $\frac{10}{3}, \frac{21}{10}, \frac{5}{12}, \frac{6}{13}$

23. $\frac{11}{4}, \frac{8}{4}, \frac{9}{20}, \frac{6}{17}$

24. $\frac{0}{6}, \frac{2}{9}, \frac{2}{1}, \frac{6}{3}$

Copyright © Houghton Mifflin Company. All rights reserved.

B. Multiplication of Fractions

25. To multiply fractions, we multiply the ________ and multiply the ________.

26. To multiply an integer and a fraction, we can write the integer as a ________ and then multiply.

In Exercises 27–30, write the given fraction in two other ways.

27. $\frac{-5}{12}$ 28. $\frac{1}{-4}$ 29. $\frac{x}{-3}$ 30. $-\frac{6}{n}$

In Exercises 31–44, multiply.

31. $\frac{3}{5} \cdot \frac{3}{2}$ 32. $\frac{8}{3} \cdot \frac{2}{5}$ 33. $\frac{1}{4} \cdot \frac{3}{2}$ 34. $\frac{9}{2} \cdot \frac{1}{5}$

35. $-\frac{2}{3} \cdot \frac{5}{7}$ 36. $\frac{8}{5}\left(\frac{-1}{3}\right)$ 37. $\left(\frac{-6}{11}\right)\left(\frac{-2}{7}\right)$ 38. $-\frac{7}{10}\left(\frac{-9}{17}\right)$

39. $\frac{9}{10}\left(-\frac{10}{9}\right)$ 40. $-\frac{5}{8}\left(-\frac{8}{5}\right)$ 41. $\frac{3}{4} \cdot \frac{x}{y}$ 42. $-\frac{7}{2} \cdot \frac{x}{y}$

43. $\frac{3}{2b} \cdot \frac{-5a}{7}$ 44. $\frac{-4y}{5} \cdot \frac{-2}{9x}$

In Exercises 45–52, multiply.

45. $5 \cdot \frac{3}{4}$ 46. $8 \cdot \frac{5}{7}$ 47. $\frac{5}{12}(-7)$ 48. $-\frac{4}{15} \cdot 2$

49. $8\left(-\frac{2}{13}\right)$ 50. $-6\left(-\frac{5}{11}\right)$ 51. $4 \cdot \frac{n}{15}$ 52. $-5 \cdot \frac{2}{m}$

In Exercises 53–58, simplify.

53. $\frac{1}{4}(4x)$ 54. $-\frac{1}{5}(5y)$ 55. $-6\left(\frac{1}{6}n\right)$

56. $-10\left(-\frac{1}{10}b\right)$ 57. $\frac{3}{4}\left(\frac{4}{3}a\right)$ 58. $-\frac{8}{5}\left(\frac{5}{8}x\right)$

C. Applications

59. *Used Cars* A used car lot has 18 cars, 10 trucks, and 3 vans. What fraction of the vehicles are cars?

60. *Cable TV Technicians* A cable TV company employees 23 men and 12 women as service technicians. What fraction of the technicians are men?

Copyright © Houghton Mifflin Company. All rights reserved.

61. *Pets* Of the 129 million pets in the United States, cats outnumber dogs by 59 million to 53 million. (Source: American Veterinary Medical Association.) What fraction of the total number of pets are cats?

62. *Super Bowl* In Super Bowl XXXIII, Denver defeated Atlanta by the score 34 to 19. What fraction of the total points did Denver score?

63. *Cinnamon Rolls* A recipe for cinnamon rolls calls for $\frac{3}{4}$ cup of sugar. How much sugar is needed for half the recipe?

64. *Meat Loaf* A recipe for meat loaf calls for 4 tablespoons of catsup. How much catsup is needed for two-thirds the recipe?

65. *Photo Negative* What is the area of a rectangular negative that is $\frac{5}{8}$ inch wide and $\frac{3}{4}$ inch long?

66. *Doormat* What is the area of a rectangular doormat whose length is $\frac{4}{5}$ yard and whose width is $\frac{2}{3}$ yard?

Writing and Concept Extension

67. What is the difference between a proper fraction and an improper fraction?

68. Explain why $\frac{5}{5}$, $\frac{6}{6}$, and $\frac{-7}{-7}$ are equivalent fractions.

In Exercises 69–72, determine the product.

69. $\frac{1}{3}\left(\frac{5}{2} \cdot \frac{7}{6}\right)$ **70.** $(7 + 5)\left(\frac{-1}{4} \cdot \frac{-1}{3}\right)$ **71.** $5\left(\frac{3}{4}\right)^2$ **72.** $\left(-\frac{4}{3}\right)^2$

73. Show how to write $\frac{1}{4}x$ as $\frac{x}{4}$.

74. Show that $\frac{2x}{5}$ and $\frac{2}{5}x$ are equivalent.

75. What number is halfway between 0 and $\frac{7}{8}$ on a number line?

Copyright © Houghton Mifflin Company. All rights reserved.

Exploring with Real-World Data: Collaborative Activities

Casual Office Dress The following pie graph shows the fraction of office workers, ages 25–29, who dress casually at the office. (Source: *USA Today.*)

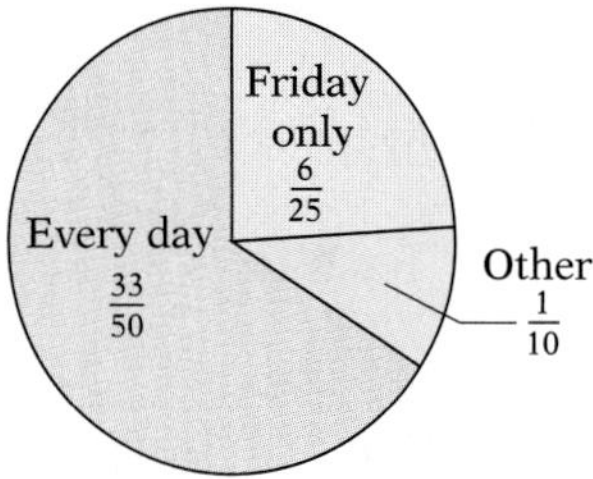

Use the pie graph for Exercises 76–81.

76. Which fraction is the best approximation of the number of workers who dress casually on Friday only?

(i) $\frac{1}{4}$ (ii) $\frac{1}{2}$ (iii) $\frac{2}{3}$

77. Which fraction is the best approximation of the number of workers who dress casually every day?

(i) $\frac{1}{4}$ (ii) $\frac{1}{2}$ (iii) $\frac{2}{3}$

78. Use an inequality symbol to compare the fraction of workers who dress casually on Friday only and the fraction who dress casually every day.

79. Use an inequality symbol to compare the fraction of workers who dress casually on Friday only and the fraction who are in the "other" category.

In the remaining exercises, assume that an office has 300 employees in the age range 25–29.

80. How many dress casually every day?

81. How many dress casually on Friday only?

Copyright © Houghton Mifflin Company. All rights reserved.

5.2 Simplifying Fractions

A *Prime Numbers and Factors*
B *Factoring and Simplifying*

SUGGESTIONS FOR SUCCESS

Which fraction do you like better: 117/156 or 3/4?

Using methods presented in this section, we can simplify 117/156 and show that it has the same value as 3/4. If 117/156 and 3/4 are equivalent, wouldn't we rather work with 3/4? Simplifying fractions is good because it makes them friendlier.

Learning how to simplify fractions is one more step along the road to success. A good way to boost your confidence is to sit back and admire your work occasionally and remind yourself that you are succeeding. Taking the time to congratulate yourself once in awhile may sound goofy, but it will increase your energy and enthusiasm.

A Prime Numbers and Factors

Writing a fraction with the smallest possible numerator and denominator is called *simplifying* or *reducing* the fraction. To get started with simplifying fractions, we need to understand some new words.

Recall that a number x is *divisible* by a number y if x can be divided by y with a remainder of 0. For example, 6 is divisible by 2 because $\frac{6}{2} = 3$ R 0. On the other hand, 9 is not divisible by 7 because $\frac{9}{7} = 1$ R 2. The word *divisible* is important in the following two definitions.

Definition of a Prime Number

A **prime number** is any whole number greater than 1 that is divisible only by itself and 1.

Here is a list of the first ten prime numbers.

$$2, 3, 5, 7, 11, 13, 17, 19, 23, 29$$

We say that 5, for example, is a prime number because 5 is divisible only by 5 (itself) and 1.

NOTE Although the list of prime numbers continues with larger and larger prime numbers, for our purposes we will need only the smaller numbers in the list.

Definition of a Composite Number

A **composite number** is any whole number greater than 1 that is not a prime number.

For example, the number 8 is divisible by itself and 1, but it is also divisible by 2 and 4. Therefore, 8 is a composite number.

Copyright © Houghton Mifflin Company. All rights reserved.

Example 1

Indicate whether the given number is a prime number or a composite number.

(a) 10 **(b)** 7

SOLUTION

(a) Because 10 is divisible by 2 and 5, it is a composite number.

(b) Because 7 is divisible only by itself and 1, it is a prime number.

Your Turn 1

Indicate whether the given number is a prime number or a composite number.

(a) 13 **(b)** 18

Answers: (a) prime; (b) composite

When we say that the composite number 6 is divisible by 2 and 3, we mean that 6 can be written as the product $2 \cdot 3$ and that 2 and 3 are *factors* of 6. In fact, we call 2 and 3 *prime factors* of 6 because 2 and 3 are prime numbers.

We can write any composite number as a product of prime factors.

$$20 = 2 \cdot 10$$ 10 is still a composite number.

$$= 2 \cdot 2 \cdot 5$$ Write 10 as $2 \cdot 5$. Now all the factors are prime numbers.

There is no one correct way to prime-factor a composite number. For example, we could have prime-factored 20 in the following way:

$$20 = 4 \cdot 5$$ 4 is still a composite number.

$$= 2 \cdot 2 \cdot 5$$ Write 4 as $2 \cdot 2$. Now all the factors are prime numbers.

Example 2

Prime-factor these numbers.

(a) 30 **(b)** 27

SOLUTION

(a) $30 = 2 \cdot 15$

$= 2 \cdot 3 \cdot 5$ $\quad 15 = 3 \cdot 5$

(b) $27 = 3 \cdot 9$

$= 3 \cdot 3 \cdot 3$ $\quad 9 = 3 \cdot 3$

Your Turn 2

Prime-factor these numbers.

(a) 14 **(b)** 24

Answers: (a) $2 \cdot 7$; (b) $2 \cdot 2 \cdot 2 \cdot 3$

NOTE Sometimes we want to write a prime factorization with exponents. For example, $20 = 2 \cdot 2 \cdot 5 = 2^2 \cdot 5$ and $27 = 3 \cdot 3 \cdot 3 = 3^3$.

Copyright © Houghton Mifflin Company. All rights reserved.

Knowing where to begin is usually the hardest part of prime factoring larger numbers. Here are some divisibility rules that should help you.

1. If a number ends in 0, then the number is divisible by 10.

   Examples: 90, 1,730

2. If a number is an even number—that is, if the number ends with 0, 2, 4, 6, or 8—then the number is divisible by 2.

   Examples: 90, 132, 394, 276, 518

3. If a number ends with 5, then the number is divisible by 5.

   Examples: 135, 865

4. If the sum of the digits of a number is divisible by 3, then the number is divisible by 3.

   Example: The digits of 279 are 2, 7, and 9, and $2 + 7 + 9 = 18$. Because 18 is divisible by 3, we know that 279 is divisible by 3.

Example 3

Prime-factor these numbers.

(a) 120

(b) 1,323

SOLUTION

(a) Because 120 ends with 0, we know 120 is divisible by 10.

$$\begin{aligned} 120 &= 10 \cdot 12 \\ &= 5 \cdot 2 \cdot 2 \cdot 6 \\ &= 5 \cdot 2 \cdot 2 \cdot 2 \cdot 3 \end{aligned}$$

Answers are usually written from smallest factors to largest factors or with exponents.

$$2 \cdot 2 \cdot 2 \cdot 3 \cdot 5 = 2^3 \cdot 3 \cdot 5$$

(b) By adding the digits, we can use the divisibility by 3 rule three times.

$$\begin{aligned} 1{,}323 &= 3 \cdot 441 \\ &= 3 \cdot 3 \cdot 147 \\ &= 3 \cdot 3 \cdot 3 \cdot 49 \\ &= 3 \cdot 3 \cdot 3 \cdot 7 \cdot 7 \\ &= 3^3 \cdot 7^2 \end{aligned}$$

Your Turn 3

Prime-factor these numbers.

(a) 180 **(b)** 735

LEARNING TIP

You can use a calculator to help you prime-factor numbers. For example, to prime-factor 165, enter $165 \div 5 = 33$. This means that $165 = 5 \cdot 33$. Now you can break 33 down into prime factors.

Answers: **(a)** $2^2 \cdot 3^2 \cdot 5$; **(b)** $3 \cdot 5 \cdot 7^2$

Copyright © Houghton Mifflin Company. All rights reserved.

B Factoring and Simplifying

Earlier, we described a simplified fraction as one in which the numerator and denominator are as small as possible. Here is a more formal definition of a simplified fraction.

Definition of a Simplified Fraction

A fraction is **simplified** (or **reduced**) if the numerator and the denominator have no factors in common other than 1 or −1.

By "no factors in common," we mean that there are no factors that are the same. For example, the fraction $\frac{10}{21} = \frac{2 \cdot 5}{3 \cdot 7}$ has 2 and 5 as prime factors in the numerator, and it has 3 and 7 as prime factors in the denominator. Because the numerator and the denominator have no factors in common, we consider $\frac{10}{21}$ to be simplified.

On the other hand, the fraction $\frac{6}{8} = \frac{2 \cdot 3}{2 \cdot 2 \cdot 2}$ has 2 as a factor in both the numerator and the denominator. Therefore, $\frac{6}{8}$ is not simplified.

We can use the rule for multiplying fractions and the methods of prime-factoring numbers to simplify a fraction.

DEVELOPING THE CONCEPT

Simplifying Fractions

We can reverse the rule for multiplying fractions and write it as

$$\frac{a \cdot c}{b \cdot d} = \frac{a}{b} \cdot \frac{c}{d}$$

Now we can apply this version of the rule to a fraction such as $\frac{10}{15}$.

$$\frac{10}{15} = \frac{2 \cdot 5}{3 \cdot 5} \qquad \text{Prime-factor the numerator and the denominator.}$$

$$= \frac{2}{3} \cdot \frac{5}{5} \qquad \text{Multiplication rule for fractions}$$

$$= \frac{2}{3} \cdot 1 \qquad \frac{5}{5} = 1$$

$$= \frac{2}{3} \qquad \text{Multiplication Property of 1}$$

We have simplified $\frac{10}{15}$ to $\frac{2}{3}$. Although the two fractions are equivalent, we regard $\frac{2}{3}$ as the simplified (or reduced) form.

In the preceding steps, we divided the common factor 5 by itself to obtain 1. A shortcut to those steps is to prime-factor the numerator and the denominator and then "divide out" the common factor(s) mentally.

$$\frac{10}{15} = \frac{2 \cdot \cancel{5}}{3 \cdot \cancel{5}} = \frac{2}{3}$$

Crossing out the common factor of 5 in the numerator and the denominator is sometimes called "canceling" the fives. However, we prefer to say that we are "dividing out" the fives.

We can summarize the shortcut method for simplifying a fraction.

Copyright © Houghton Mifflin Company. All rights reserved.

Simplifying a Fraction

1. Prime-factor the numerator and the denominator.
2. Divide out any common factors in the numerator and the denominator.

Example 4

Simplify $\frac{9}{12}$.

SOLUTION

$$\frac{9}{12} = \frac{3 \cdot 3}{2 \cdot 2 \cdot 3}$$ Prime-factor the numerator and the denominator.

$$= \frac{3 \cdot \cancel{3}}{2 \cdot 2 \cdot \cancel{3}}$$ Divide out the common factor 3.

$$= \frac{3}{2 \cdot 2}$$

$$= \frac{3}{4}$$

Your Turn 4

Simplify $\frac{8}{10}$.

Answer: $\frac{4}{5}$

Sometimes more than one common factor can be divided out.

Example 5

Simplify.

(a) $\frac{30}{42}$ **(b)** $\frac{16}{20}$

SOLUTION

(a) $\frac{30}{42} = \frac{2 \cdot 3 \cdot 5}{2 \cdot 3 \cdot 7}$ Prime-factor the numerator and the denominator.

$$= \frac{\cancel{2} \cdot \cancel{3} \cdot 5}{\cancel{2} \cdot \cancel{3} \cdot 7}$$ Divide out the common factors 2 and 3.

$$= \frac{5}{7}$$

(b) $\frac{16}{20} = \frac{2 \cdot 2 \cdot 2 \cdot 2}{2 \cdot 2 \cdot 5}$ Prime-factor the numerator and the denominator.

$$= \frac{\cancel{2} \cdot \cancel{2} \cdot 2 \cdot 2}{\cancel{2} \cdot \cancel{2} \cdot 5}$$ Divide out the common factor 2 two times.

$$= \frac{2 \cdot 2}{5}$$

$$= \frac{4}{5}$$

Your Turn 5

Simplify.

(a) $\frac{12}{18}$ **(b)** $\frac{18}{45}$

Answers: (a) $\frac{2}{3}$; (b) $\frac{2}{5}$

Remember that 1 is a factor of every number. Therefore, if you divide out all the factors in the numerator or denominator, a factor of 1 is understood to remain.

Copyright © Houghton Mifflin Company. All rights reserved.

Example 6

Simplify.

(a) $\frac{15}{75}$ **(b)** $\frac{24}{12}$

SOLUTION

(a) $\frac{15}{75} = \frac{3 \cdot 5}{3 \cdot 5 \cdot 5}$

$= \frac{1 \cdot \cancel{3} \cdot \cancel{5}}{\cancel{3} \cdot \cancel{5} \cdot 5}$ Write the understood 1 in the numerator.

$= \frac{1}{5}$

(b) $\frac{24}{12} = \frac{2 \cdot 2 \cdot 2 \cdot 3}{2 \cdot 2 \cdot 3}$

$= \frac{\cancel{2} \cdot \cancel{2} \cdot 2 \cdot \cancel{3}}{1 \cdot \cancel{2} \cdot \cancel{2} \cdot \cancel{3}}$ Write the understood 1 in the denominator.

$= \frac{2}{1}$

$= 2$

Note that we would obtain the same result simply by dividing 24 by 12.

Your Turn 6

Simplify.

(a) $\frac{6}{90}$ **(b)** $\frac{27}{9}$

Answers: (a) $\frac{1}{15}$; (b) 3

KEYS TO THE CALCULATOR

You can use a calculator to display a fraction in simplified form. The following are typical scientific calculator keystrokes for simplifying the proper fraction $\frac{12}{18}$ and the improper fraction $\frac{24}{15}$.

12 [a b/c] **18** [=] Display: [2 ⌋ 3]

24 [a b/c] **15** [=] [2nd] [d/c] Display: [8 ⌋ 5]

To simplify $\frac{24}{15}$ with a graphing calculator,

24 [÷] **15** [MATH] [Frac] [ENTER]

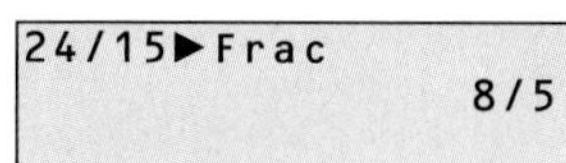

Exercises

Use your calculator to simplify the given fraction.

(a) $\frac{117}{156}$ **(b)** $\frac{126}{105}$ **(c)** $\frac{63}{36}$ **(d)** $\frac{34}{153}$

Copyright © Houghton Mifflin Company. All rights reserved.

When you simplify a fraction that has an opposite symbol in the numerator or in the denominator, you can use the rule $\frac{-a}{b} = \frac{a}{-b} = -\frac{a}{b}$ to move the opposite symbol to the front of the fraction.

Example 7

Simplify $\frac{-9}{21}$.

SOLUTION

$\frac{-9}{21} = -\frac{9}{21}$ Place the opposite symbol in front of the fraction.

$= -\frac{\cancel{3} \cdot 3}{\cancel{3} \cdot 7}$ Factor the numerator and the denominator and divide out the common factor.

$= -\frac{3}{7}$

Your Turn 7

Simplify $\frac{8}{-20}$.

Answer: $-\frac{2}{5}$

According to the sign rules for division,

$$\frac{-12}{-3} = 4 \qquad \text{and} \qquad \frac{12}{3} = 4$$

In other words,

$$\frac{-12}{-3} = \frac{12}{3}$$

We generalize this with the following rule.

Signs in Quotients

If a and b represent any numbers, where b is not 0, then

$$\frac{-a}{-b} = \frac{a}{b}$$

In words, if opposite symbols appear in both the numerator and the denominator, then both symbols can be removed.

Example 8

Simplify $\frac{-12}{-30}$.

SOLUTION

$\frac{-12}{-30} = \frac{12}{30}$ $\frac{-a}{-b} = \frac{a}{b}$

$= \frac{\cancel{2} \cdot 2 \cdot \cancel{3}}{\cancel{2} \cdot \cancel{3} \cdot 5}$ Divide out common factors.

$= \frac{2}{5}$

Your Turn 8

Simplify $\frac{-9}{-6}$.

Answer: $\frac{3}{2}$

Copyright © Houghton Mifflin Company. All rights reserved.

The numerator or the denominator of a fraction can contain factors that are variables. However, the method for simplifying such fractions is the same.

Example 9

Simplify $\frac{5xy}{10x}$.

SOLUTION

$$\frac{5xy}{10x} = \frac{\cancel{5} \cdot \cancel{x} \cdot y}{2 \cdot \cancel{5} \cdot \cancel{x}} \quad \text{Divide out the common factors.}$$

$$= \frac{y}{2}$$

Your Turn 9

Simplify $\frac{12a}{3abc}$.

Answer: $\frac{4}{bc}$

Repeated variable factors are usually indicated with exponents. One approach to simplifying such fractions is to write the exponential expressions in expanded form.

Example 10

Simplify $\frac{30a^2b}{45ab^3}$.

SOLUTION

$$\frac{30a^2b}{45ab^3} = \frac{2 \cdot \cancel{3} \cdot \cancel{5} \cdot \cancel{a} \cdot a \cdot \cancel{b}}{3 \cdot \cancel{3} \cdot \cancel{5} \cdot \cancel{a} \cdot \cancel{b} \cdot b \cdot b}$$

$$= \frac{2 \cdot a}{3 \cdot b \cdot b}$$

$$= \frac{2a}{3b^2}$$

Your Turn 10

Simplify $\frac{20x^3y}{6xy^2z}$.

Answer: $\frac{10x^2}{3yz}$

Sometimes factoring and simplifying fractions comes into play in application problems that involve data.

Example 11

A theater chain is interested in knowing which age groups are most likely to attend weekday matinees. On a certain afternoon, the management conducted a survey and obtained the following information.

Age Group	*Number of Customers*
16 and under	5
17–50	45
over 50	10

What fraction of the total movie attendance does the number of customers in each age group represent?

Your Turn 11

In Example 11, what fraction of the total attendance would the number of customers in the 17–50 age group have represented if 30 more customers in that age group had attended?

Copyright © Houghton Mifflin Company. All rights reserved.

SOLUTION

First, we determine the total number of customers.

$$5 + 45 + 10 = 60$$

$$\text{16 and under: } \frac{5}{60} = \frac{\cancel{5} \cdot 1}{\cancel{5} \cdot 12} = \frac{1}{12}$$

$$\text{17–50: } \frac{45}{60} = \frac{\cancel{15} \cdot 3}{\cancel{15} \cdot 4} = \frac{3}{4}$$

$$\text{over 50: } \frac{10}{60} = \frac{\cancel{10} \cdot 1}{\cancel{10} \cdot 6} = \frac{1}{6}$$

5.2 Quick Reference

A. Prime Numbers and Factors

1. A **prime number** is any whole number greater than 1 that is divisible only by itself and 1.

   Because 7 is divisible only by 7 (itself) and 1, 7 is a prime number.

2. A **composite number** is any whole number greater than 1 that is not a prime number.

   Because 8 is divisible by 2 and 4, it is a composite number.

3. If a number a is divisible by a number b, then b is a *factor* of a.

   Because 10 is divisible by 5, we say that 5 is a factor of 10: $10 = 5 \cdot 2$.

4. A composite number can be written as a product of prime factors.

   $$18 = 2 \cdot 9 = 2 \cdot 3 \cdot 3 = 2 \cdot 3^2$$

5. Certain divisibility rules can be helpful in prime factoring large numbers.

   (a) Numbers that end in 0 are divisible by 10. $540 = 10 \cdot 54$

   (b) Even numbers are divisible by 2. $74 = 2 \cdot 37$

   (c) Numbers that end in 5 are divisible by 5. $65 = 5 \cdot 13$

   (d) If the sum of the digits of a number is divisible by 3, then the number is divisible by 3.

   For 87, $8 + 7 = 15$, which is divisible by 3.

   $87 = 3 \cdot 29$

Copyright © Houghton Mifflin Company. All rights reserved.

B. Factoring and Simplifying

1. A fraction is **simplified** (or **reduced**) if the numerator and the denominator have no factors in common other than 1 or −1.

$$\frac{6}{5} = \frac{2 \cdot 3}{5}$$

Because the same factor does not appear in the numerator and the denominator, $\frac{6}{5}$ is considered simplified.

2. To simplify a fraction, we follow these steps.

(a) Prime-factor the numerator and the denominator.

(b) Divide out any common factors in the numerator and the denominator.

$$\frac{15}{21} = \frac{\cancel{3} \cdot 5}{\cancel{3} \cdot 7} = \frac{5}{7}$$

$$\frac{9}{27} = \frac{1 \cdot \cancel{3} \cdot \cancel{3}}{3 \cdot \cancel{3} \cdot \cancel{3}} = \frac{1}{3}$$

$$\frac{-6}{14} = -\frac{\cancel{2} \cdot 3}{\cancel{2} \cdot 7} = -\frac{3}{7}$$

$$\frac{4xy^2}{6y} = \frac{2 \cdot \cancel{2} \cdot x \cdot \cancel{y} \cdot y}{\cancel{2} \cdot 3 \cdot \cancel{y}}$$

$$= \frac{2xy}{3}$$

Copyright © Houghton Mifflin Company. All rights reserved.

5.2 Exercises

A. Prime Numbers and Factors

1. When 12 is divided by 3 the remainder is 0. Thus we say that 12 is ________ by 3.

2. A number that is divisible by a number other than 1 and itself is called a(n) ________ number.

In Exercises 3 and 4, identify (a) the prime numbers and (b) the composite numbers.

3. 9, 11, 23, 27
4. 2, 10, 19, 29

In Exercises 5–10, prime-factor the given number.

5. 42
6. 60
7. 48
8. 56
9. 72
10. 54

In Exercises 11–14, prime-factor the given number.

11. 360
12. 441
13. 297
14. 1,575

B. Factoring and Simplifying

15. If the numerator and the denominator of a fraction have no factors in common other than -1 or 1, we say that the fraction is ________ or ________.

16. Crossing out the common factor of 3 in the fraction $\frac{7 \cdot 3}{4 \cdot 3}$ is called ________ the threes.

In Exercises 17–20, simplify.

17. $\frac{15}{25}$
18. $\frac{18}{21}$
19. $\frac{28}{49}$
20. $\frac{12}{22}$

In Exercises 21–24, simplify.

21. $\frac{40}{24}$
22. $\frac{42}{18}$
23. $\frac{27}{63}$
24. $\frac{36}{44}$

In Exercises 25–28, simplify.

25. $\frac{15}{30}$
26. $\frac{12}{36}$
27. $\frac{15}{5}$
28. $\frac{21}{3}$

Copyright © Houghton Mifflin Company. All rights reserved.

In Exercises 29–32, simplify.

29. $\frac{-33}{15}$ **30.** $\frac{-18}{15}$ **31.** $\frac{-54}{-12}$ **32.** $\frac{-35}{-42}$

In Exercises 33–44, simplify.

33. $\frac{21}{12}$ **34.** $\frac{63}{35}$ **35.** $\frac{-20}{36}$ **36.** $\frac{42}{-70}$

37. $\frac{36}{96}$ **38.** $\frac{28}{70}$ **39.** $\frac{-24}{-48}$ **40.** $\frac{-8}{-56}$

41. $\frac{54}{36}$ **42.** $\frac{48}{36}$ **43.** $\frac{10}{50}$ **44.** $\frac{8}{48}$

In Exercises 45–50, simplify.

45. $\frac{20a}{12ab}$ **46.** $\frac{3xy}{15y}$ **47.** $\frac{8mn}{6m}$

48. $\frac{24xy}{8x}$ **49.** $\frac{6a}{18ab}$ **50.** $\frac{9m}{36mn}$

In Exercises 51–56, simplify.

51. $\frac{15x^2}{25x}$ **52.** $\frac{14n^3}{21n}$ **53.** $\frac{20ab^3}{24a^2b}$

54. $\frac{24x^3y^2}{10x^2y}$ **55.** $\frac{36xy}{16x^2y^3}$ **56.** $\frac{30ab}{20ab^2}$

Calculator Exercises

In Exercises 57 and 58, use a calculator to factor the number.

57. 2,025 **58.** 28,224

Copyright © Houghton Mifflin Company. All rights reserved.

In Exercises 59 and 60, use a calculator to help factor the numerator and the denominator of the given fraction. Then simplify the fraction.

59. $\frac{252}{270}$

60. $\frac{504}{945}$

Applications

61. What part of a foot is 8 inches?

62. What part of a pound is 12 ounces?

63. What part of an hour is 24 minutes?

64. What part of a day is 16 hours?

65. *Exam Scores* Of the 20 questions on a history exam, a student answered 18 correctly. What fraction of the questions were answered *incorrectly?*

66. *Passing Statistics* A quarterback completed 14 of 24 passes that he attempted. What fraction of the passes were *incomplete*?

67. *Ohio State Football* As of 2001, Ohio State University's all-time football record was 724 wins, 287 losses, and 53 ties. (Source: *World Almanac.*) What fraction of its games did Ohio State win?

68. *Olympic Medals* In the 2002 Winter Olympic Games, Russia won 6 gold medals, 6 silver medals, and 4 bronze medals. (Source: *World Almanac.*) What fraction of the medals were (a) gold, and (b) bronze?

Family Income and Expenses The pie graph shows how a certain family's monthly income was spent.

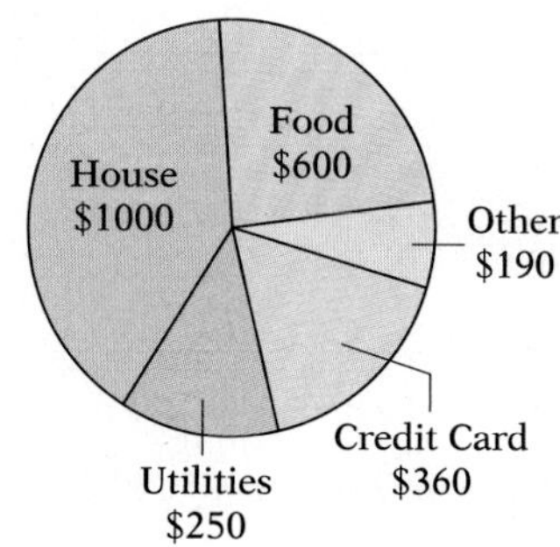

Copyright © Houghton Mifflin Company. All rights reserved.

Use the pie graph on the preceding page for Exercises 69–72.

69. What fraction of the income was spent on credit card payments?

70. What fraction of the income was spent on food?

71. What fraction of the income was spent on the house and utilities?

72. What fraction of the income was spent on food and credit card payments?

Writing and Concept Extension

73. What do we mean when we say that a number is *prime factored?*

74. Explain how to determine whether a number is divisible by 3.

75. How do you know that 760 and 995 are divisible by 5 but that 5,552 is not?

76. Explain why 1 is neither a prime number nor a composite number.

77. Identify the fraction that is *not* equivalent to the other fractions

$$-\frac{-3}{-4}, \quad -\frac{-3}{4}, \quad \frac{-3}{-4}, \quad -\frac{3}{-4}$$

78. What are all the positive factors of 36? What are the prime factors of 36?

Copyright © Houghton Mifflin Company. All rights reserved.

5.3 Multiplication and Division of Fractions

A Multiplying and Simplifying
B Division of Fractions
C Applications

SUGGESTIONS FOR SUCCESS

At this point you are probably about halfway through your course. Your instructor may give a midterm exam that covers all the material that you have studied so far. Even if you don't have a midterm exam, now is still a good time to review everything. Revisit the chapter tests and any tests or quizzes that your instructor has given. Frequent reviewing is an excellent way to maintain your mathematical skills.

A Multiplying and Simplifying

In the first two sections of this chapter, we have used the multiplication rule for fractions in two different ways.

Multiplying Fractions	*Simplifying Fractions*
$\frac{a}{b} \cdot \frac{c}{d} = \frac{ac}{bd}$	$\frac{ac}{bc} = \frac{a}{b}$
$\frac{3}{2} \cdot \frac{1}{7} = \frac{3 \cdot 1}{2 \cdot 7} = \frac{3}{14}$	$\frac{15}{21} = \frac{5 \cdot \cancel{3}}{7 \cdot \cancel{3}} = \frac{5}{7}$

Before we multiply fractions, it is worth taking a look at them to see whether they can be simplified. Consider this product:

$$\frac{26}{39} \cdot \frac{55}{77} = \frac{26 \cdot 55}{39 \cdot 77} = \frac{1{,}430}{3{,}003}$$

Multiplying right away results in a cumbersome fraction. If we had first simplified the fractions to $\frac{2}{3}$ and $\frac{5}{7}$, respectively, the arithmetic and the result would have been more manageable.

Example 1 shows that fractions might be simplified, but their product might not be.

Example 1

Multiply $\frac{8}{3} \cdot \frac{3}{10}$ and simplify the result.

SOLUTION

$$\frac{8}{3} \cdot \frac{3}{10} = \frac{8 \cdot 3}{3 \cdot 10} = \frac{24}{30} \quad \text{Multiply the fractions.}$$

$$\frac{24}{30} = \frac{\cancel{2} \cdot 2 \cdot 2 \cdot \cancel{3}}{\cancel{2} \cdot \cancel{3} \cdot 5} \quad \text{Simplify the result.}$$

$$= \frac{2 \cdot 2}{5}$$

$$= \frac{4}{5}$$

Therefore, $\frac{8}{3} \cdot \frac{3}{10} = \frac{4}{5}$.

Your Turn 1

Multiply $\frac{3}{2} \cdot \frac{6}{7}$ and simplify the result.

Answer: $\frac{9}{7}$

A more efficient way to multiply fractions involves simplifying *before* we carry out the multiplication.

Copyright © Houghton Mifflin Company. All rights reserved.

DEVELOPING THE CONCEPT

Simplifying and Multiplying

Here is another way in which the fractions in Example 1 can be multiplied.

$$\frac{8}{3}\cdot\frac{3}{10}=\frac{2\cdot 2\cdot 2}{3}\cdot\frac{3}{2\cdot 5}$$ Prime-factor the numerators and the denominators.

$$=\frac{\cancel{2}\cdot 2\cdot 2\cdot \cancel{3}}{\cancel{2}\cdot \cancel{3}\cdot 5}$$ Multiply the numerators and the denominators.

$$=\frac{2\cdot 2}{5}$$ Divide out the common factors.

$$=\frac{4}{5}$$ Multiply the remaining factors.

Look closely at the step in which we multiply the numerators and the denominators. All the prime factors of the two numerators now appear in one numerator; and all the prime factors of the two denominators now appear in one denominator. Therefore, we could have simply divided out the common factors *before* we performed the multiplication.

$$\frac{8}{3}\cdot\frac{3}{10}=\frac{\cancel{2}\cdot 2\cdot 2}{\cancel{3}}\cdot\frac{\cancel{3}}{\cancel{2}\cdot 5}$$ Prime-factor the numerators and the denominators. Divide out the common factors.

$$=\frac{2\cdot 2}{1}\cdot\frac{1}{5}$$ Remember the understood factors of 1.

$$=\frac{4}{5}$$ Now multiply the fractions.

Here is a summary of the more efficient procedure for multiplying fractions.

Multiplying Fractions

To multiply two or more fractions, follow these steps.

1. Prime-factor the numerators and the denominators of the fractions.
2. Divide out any common factors that appear in the numerator of any fraction and the denominator of any fraction.
3. Multiply the remaining factors in the numerators and the denominators to obtain the simplified result.

Example 2

Multiply $\frac{9}{10}\cdot\frac{5}{21}$.

SOLUTION

$$\frac{9}{10}\cdot\frac{5}{21}=\frac{\cancel{3}\cdot 3}{2\cdot \cancel{5}}\cdot\frac{\cancel{5}}{\cancel{3}\cdot 7}$$ Prime-factor the numerators and the denominators and divide out common factors.

$$=\frac{3}{14}$$ Multiply the remaining factors in the numerator and the denominator.

Your Turn 2

Multiply $\frac{14}{15}\cdot\frac{10}{7}$.

Answer: $\frac{4}{3}$

Copyright © Houghton Mifflin Company. All rights reserved.

The same procedure can be used to multiply fractions that include variables.

Example 3

Multiply $\frac{5x}{2} \cdot \frac{4}{x^2}$.

SOLUTION

$$\frac{5x}{2} \cdot \frac{4}{x^2} = \frac{5 \cdot \not{x}}{\not{2}} \cdot \frac{\not{2} \cdot 2}{\not{x} \cdot x}$$ Prime-factor the numerators and the denominators and divide out common factors.

$$= \frac{10}{x}$$ Multiply the remaining factors in the numerator and the denominator.

Your Turn 3

Multiply $\frac{3}{6y} \cdot \frac{4xy}{9}$.

Answer: $\frac{2x}{9}$

We can extend this procedure to products with any number of factors.

Example 4

Multiply $\frac{3}{5} \cdot \frac{10}{7} \cdot \frac{14}{9}$.

SOLUTION

$$\frac{3}{5} \cdot \frac{10}{7} \cdot \frac{14}{9} = \frac{\not{3}}{\not{5}} \cdot \frac{2 \cdot \not{5}}{\not{7}} \cdot \frac{2 \cdot \not{7}}{\not{3} \cdot 3}$$

$$= \frac{4}{3}$$

Your Turn 4

Multiply $\frac{5}{6} \cdot \frac{3}{4} \cdot \frac{2}{3}$.

Answer: $\frac{5}{12}$

Remember that an opposite sign in a fraction can be placed in front of the fraction, and then the usual sign rules for multiplication can be used. Also, when one of the factors is an integer, writing it as a fraction is a good idea.

Example 5

Multiply.

(a) $\frac{4}{-5} \cdot \frac{15}{16}$ **(b)** $\frac{5}{6} \cdot 18$

SOLUTION

(a) $\frac{4}{-5} \cdot \frac{15}{16} = -\frac{4}{5} \cdot \frac{15}{16}$

$$= -\frac{\not{2} \cdot \not{2}}{\not{5}} \cdot \frac{3 \cdot \not{5}}{\not{2} \cdot \not{2} \cdot 2 \cdot 2}$$

$$= -\frac{3}{4}$$

Your Turn 5

Multiply.

(a) $\frac{2}{-3}\left(-\frac{3}{4}\right)$

(b) $20 \cdot \frac{3}{5}$

continued

Copyright © Houghton Mifflin Company. All rights reserved.

(b) $\frac{5}{6} \cdot 18 = \frac{5}{6} \cdot \frac{18}{1}$

$= \frac{5}{\cancel{2} \cdot \cancel{3}} \cdot \frac{\cancel{2} \cdot \cancel{3} \cdot 3}{1}$

$= \frac{15}{1}$

$= 15$

Answers: (a) $\frac{1}{2}$; (b) 12

As you become more skilled and confident, you will discover your own shortcuts. For instance, in Example 5(b), rather than prime-factor the 6 and the 18, you may have noticed that the expression could have been written

$$\frac{5}{6} \cdot \frac{18}{1} = \frac{5}{\cancel{6}} \cdot \frac{\cancel{6} \cdot 3}{1} = \frac{5 \cdot 3}{1} = 15$$

With experience, you will find that fractions can be multiplied quickly and easily.

B Division of Fractions

To perform division with fractions, you need to understand what we mean by the reciprocal of a number.

Property of Reciprocals

Every number n except 0 has a **reciprocal** $\frac{1}{n}$ such that $n \cdot \frac{1}{n} = 1$.

NOTE The number 0 does not have a reciprocal because $\frac{1}{0}$ is undefined.

Because the product of a number and its reciprocal is 1, we can form the reciprocal of a number in the following ways:

1. If the number is an integer, the reciprocal is 1 over the integer.

   Number: 8 Reciprocal: $\frac{1}{8}$

2. If the number is a fraction, the reciprocal is the fraction turned upside down.

   Number: $\frac{3}{5}$ Reciprocal: $\frac{5}{3}$

   Number: $\frac{1}{7}$ Reciprocal: $\frac{7}{1} = 7$

3. If the number is negative, then the reciprocal is also negative.

   Number: -2 Reciprocal: $-\frac{1}{2}$

   Number: $-\frac{11}{5}$ Reciprocal: $-\frac{5}{11}$

Copyright © Houghton Mifflin Company. All rights reserved.

Example 6

Write the reciprocal of the given number.

(a) $\frac{5}{8}$ **(b)** $\frac{3}{-7}$ **(c)** 10 **(d)** $\frac{1}{9}$

SOLUTION

	Number	*Reciprocal*
(a)	$\frac{5}{8}$	$\frac{8}{5}$
(b)	$\frac{3}{-7}$	$\frac{-7}{3} = -\frac{7}{3}$
(c)	10	$\frac{1}{10}$
(d)	$\frac{1}{9}$	$\frac{9}{1} = 9$

Your Turn 6

Write the reciprocal of the given number.

(a) $\frac{9}{2}$ **(b)** -5

(c) $\frac{-1}{8}$ **(d)** 1

Answers: (a) $\frac{2}{9}$; (b) $\frac{1}{-5}$; (c) -8; (d) 1

Just as every subtraction problem can be written as a related addition, every division problem can be written as a related multiplication.

Division	*Related Multiplication*
$10 \div 5 = 2$	$10 \cdot \frac{1}{5} = 2$
$18 \div 6 = 3$	$18 \cdot \frac{1}{6} = 3$
$32 \div 4 = 8$	$32 \cdot \frac{1}{4} = 8$

These examples show that dividing by a number is the same as multiplying by the reciprocal of that number.

Definition of Division

If a and b are any numbers, where b is not 0, then

$$a \div b = a \cdot \frac{1}{b}$$

In words, dividing a number by b is the same as multiplying the number by the reciprocal of b.

To divide $32 \div 8$, there is no advantage in writing $32 \cdot \frac{1}{8}$. However, when one or both of the numbers in a division problem are fractions, we use the related multiplication.

Copyright © Houghton Mifflin Company. All rights reserved.

Example 7

Divide.

(a) $6 \div \frac{7}{5}$

(b) $\frac{2}{3} \div 9$

(c) $\frac{3}{5} \div \frac{2}{7}$

SOLUTION

(a) $6 \div \frac{7}{5} = \frac{6}{1} \cdot \frac{5}{7}$ Multiply by the reciprocal of $\frac{7}{5}$.

$= \frac{30}{7}$

(b) $\frac{2}{3} \div 9 = \frac{2}{3} \cdot \frac{1}{9}$ Multiply by the reciprocal of 9.

$= \frac{2}{27}$

(c) $\frac{3}{5} \div \frac{2}{7} = \frac{3}{5} \cdot \frac{7}{2}$ Multiply by the reciprocal of $\frac{2}{7}$.

$= \frac{21}{10}$

Your Turn 7

Divide.

(a) $\frac{6}{7} \div 5$

(b) $\frac{8}{3} \div \frac{1}{4}$

(c) $3 \div \frac{2}{3}$

Answers: (a) $\frac{6}{35}$; (b) $\frac{32}{3}$; (c) $\frac{9}{2}$

Example 8

Divide.

(a) $-\frac{2}{9} \div 5$

(b) $-6 \div \left(-\frac{5}{3}\right)$

(c) $\frac{3}{5} \div \left(-\frac{2}{3}\right)$

SOLUTION

(a) $-\frac{2}{9} \div 5 = -\frac{2}{9} \cdot \frac{1}{5} = -\frac{2}{45}$

(b) $-6 \div \left(-\frac{5}{3}\right) = -\frac{6}{1}\left(-\frac{3}{5}\right) = \frac{18}{5}$

(c) $\frac{3}{5} \div \left(-\frac{2}{3}\right) = \frac{3}{5}\left(-\frac{3}{2}\right) = -\frac{9}{10}$

Your Turn 8

Divide.

(a) $-\frac{1}{3} \div (-2)$

(b) $8 \div \left(-\frac{3}{2}\right)$

(c) $-\frac{1}{9} \div \left(-\frac{3}{2}\right)$

Answers: (a) $\frac{1}{6}$; (b) $-\frac{16}{3}$; (c) $\frac{2}{27}$

After you convert a division problem into the related multiplication problem, you may find that you can simplify, as we discussed at the beginning of this section.

Copyright © Houghton Mifflin Company. All rights reserved.

Example 9

Divide and simplify.

(a) $\frac{6}{7} \div \frac{3}{14}$ **(b)** $-8 \div \frac{10}{7}$

(c) $\frac{5}{6} \div \left(-\frac{15}{2}\right)$ **(d)** $\frac{3}{8} \div 9$

SOLUTION

(a) $\frac{6}{7} \div \frac{3}{14} = \frac{6}{7} \cdot \frac{14}{3}$ Multiply by the reciprocal of $\frac{3}{14}$.

$= \frac{2 \cdot \cancel{3}}{\cancel{7}} \cdot \frac{2 \cdot \cancel{7}}{\cancel{3}}$ Prime-factor and divide out common factors.

$= \frac{2 \cdot 2}{1}$

$= 4$

(b) $-8 \div \frac{10}{7} = -\frac{8}{1} \cdot \frac{7}{10}$ Multiply by the reciprocal of $\frac{10}{7}$.

$= -\frac{\cancel{2} \cdot 2 \cdot 2}{1} \cdot \frac{7}{\cancel{2} \cdot 5}$ Prime-factor and divide out common factors.

$= -\frac{2 \cdot 2 \cdot 7}{5}$

$= -\frac{28}{5}$

(c) $\frac{5}{6} \div \left(-\frac{15}{2}\right) = \frac{5}{6} \cdot \left(-\frac{2}{15}\right)$ Multiply by the reciprocal of $-\frac{15}{2}$.

$= \frac{\cancel{5}}{\cancel{2} \cdot 3} \cdot \left(-\frac{\cancel{2}}{3 \cdot \cancel{5}}\right)$ Prime-factor and divide out common factors.

$= -\frac{1}{3 \cdot 3}$

$= -\frac{1}{9}$

(d) $\frac{3}{8} \div 9 = \frac{3}{8} \cdot \frac{1}{9}$ Multiply by the reciprocal of 9.

$= \frac{1 \cdot \cancel{3}}{8} \cdot \frac{1}{\cancel{3} \cdot 3}$ Prime-factor and divide out common factors.

$= \frac{1}{8} \cdot \frac{1}{3}$

$= \frac{1}{24}$

Your Turn 9

Divide and simplify.

(a) $-\frac{3}{4} \div \frac{9}{2}$

(b) $16 \div \left(-\frac{12}{5}\right)$

(c) $-\frac{1}{2} \div \left(-\frac{5}{6}\right)$

(d) $\frac{12}{5} \div 6$

Answers: (a) $-\frac{1}{6}$; (b) $-\frac{20}{3}$; (c) $\frac{3}{5}$; (d) $\frac{2}{5}$

You can use a calculator to perform multiplication and division of fractions without having to convert quotients into products or to simplify results. Your calculator will do those things automatically. On the other hand, quite a few keystrokes are needed. Eventually, you might find that you can perform these operations more quickly by hand than with a calculator.

Copyright © Houghton Mifflin Company. All rights reserved.

KEYS TO THE CALCULATOR

Here are the typical keystrokes for multiplying $\frac{2}{3}$ times $\frac{3}{4}$.

Scientific calculator:

2 [a b/c] 3 [x] 3 [a b/c] 4 [=]   Display: [1 ⌋ 2]

Graphing calculator (see Figure A):

[(] 2 [÷] 3 [)] [(] 3 [÷] 4 [)] [MATH] [Frac] [ENTER]

```
(2/3)(3/4)▶Frac
            1/2
```

(A)

```
(4/9)/(2/15)▶Frac
             10/3
```

(B)

To perform $\frac{4}{9} \div \frac{2}{15}$, try these keystrokes.

Scientific calculator:

4 [a b/c] 9 [÷] 2 [a b/c] 15 [=] [2nd] [d/c]   Display: [10 ⌋ 3]

Graphing calculator (see Figure B):

[(] 4 [÷] 9 [)] [÷] [(] 2 [÷] 15 [)] [MATH] [Frac] [ENTER]

Note that all the results are displayed as simplified fractions.

Exercises

(a) $\frac{7}{8} \cdot \frac{4}{3}$ **(b)** $6 \div \frac{3}{5}$ **(c)** $\frac{1}{4} \div \frac{7}{12}$ **(d)** $\frac{5}{6} \cdot \frac{3}{20}$

C Applications

Products and quotients of fractions can occur in a variety of application problems. In the context of fraction arithmetic, we need to be aware that the word *of* means *times*.

At a baseball game, $\frac{1}{10}$ *of* the 30,000 fans were from out of state.

The number of out-of-state fans was $\frac{1}{10}$ *times* 30,000.

In many application problems, you will usually need to know (or be able to determine) some total number.

A charitable organization spent $\frac{1}{20}$ of its receipts for administrative costs.

In this instance, you can't calculate how much was spent for administrative costs because you don't know what the total receipts were.

Copyright © Houghton Mifflin Company. All rights reserved.

Example 10

A college seminar was attended by 120 students. Of those students, two-thirds were women, and three-fourths of those women were seniors. How many female seniors attended the seminar?

SOLUTION

The number of women who attended was

$$\frac{2}{3} \cdot 120 = \frac{2}{3} \cdot \frac{120}{1} = \frac{240}{3} = 80$$

Of those 80 women, three-fourths were seniors.

$$\frac{3}{4} \cdot 80 = \frac{3}{4} \cdot \frac{80}{1} = \frac{240}{4} = 60$$

Therefore, 60 female seniors attended the seminar. The problem can also be done in one step.

$$\frac{2}{3} \cdot \frac{3}{4} \cdot 120 = \frac{\cancel{2}}{\cancel{3}} \cdot \frac{\cancel{3}}{\cancel{2} \cdot 2} \cdot \frac{120}{1} = \frac{120}{2} = 60$$

Your Turn 10

In Example 10, suppose that half of those attending had been women, rather than two-thirds. How many female seniors attended the seminar?

Answer: 45

5.3 Quick Reference

A. Multiplying and Simplifying

1. To multiply two or more fractions, follow these steps.

(a) Prime-factor the numerators and the denominators of the fractions.

(b) Divide out any common factors that appear in the numerator of any fraction and in the denominator of any fraction.

(c) Multiply the remaining factors in the numerators and the denominators to obtain the simplified result.

$$\frac{6}{5} \cdot \frac{15}{4}$$
$$= \frac{2 \cdot 3}{5} \cdot \frac{3 \cdot 5}{2 \cdot 2}$$
$$= \frac{\cancel{2} \cdot 3}{1 \cdot \cancel{5}} \cdot \frac{3 \cdot \cancel{5}}{\cancel{2} \cdot 2}$$
$$= \frac{3}{1} \cdot \frac{3}{2}$$
$$= \frac{3 \cdot 3}{1 \cdot 2}$$
$$= \frac{9}{2}$$

B. Division of Fractions

1. If b is not 0, then the number $\frac{1}{b}$ is called the **reciprocal** of b.

The reciprocal of 5 is $\frac{1}{5}$.

Copyright © Houghton Mifflin Company. All rights reserved.

2. Every number a except 0 has a reciprocal $\frac{1}{a}$ such that $a \cdot \frac{1}{a} = 1$.

Number	*Reciprocal*
$\frac{1}{2}$	2
$\frac{3}{5}$	$\frac{5}{3}$

3. A number and its reciprocal have the same sign.

Number	*Reciprocal*
$-\frac{8}{5}$	$-\frac{5}{8}$

4. If a and b are any numbers, where b is not 0, then

$$a \div b = a \cdot \frac{1}{b}$$

In words, dividing a number by b is the same as multiplying the number by the reciprocal of b.

$$18 \div 6 = 18 \cdot \frac{1}{6}$$

5. When one or both of the numbers in a division problem are fractions, we use the related multiplication.

$$4 \div \frac{5}{3} = 4 \cdot \frac{3}{5} = \frac{12}{5}$$

$$\frac{2}{7} \div 3 = \frac{2}{7} \cdot \frac{1}{3} = \frac{2}{21}$$

$$\frac{3}{5} \div \frac{2}{9} = \frac{3}{5} \cdot \frac{9}{2} = \frac{27}{10}$$

6. After you write a division as the related multiplication, you may be able to simplify before multiplying.

$$\frac{4}{9} \div \frac{1}{3} = \frac{4}{9} \cdot \frac{3}{1} = \frac{2 \cdot 2}{\cancel{3} \cdot 3} \cdot \frac{\cancel{3}}{1} = \frac{4}{3}$$

Copyright © Houghton Mifflin Company. All rights reserved.

5.3 Exercises

A. Multiplying and Simplifying

1. The multiplication rule for fractions can be used to ______ fractions as well as to multiply fractions.

2. In the product of two or more fractions, we can ______ any common factors that appear in the numerator of any fraction and in the denominator of any fraction.

In Exercises 3–20, multiply and simplify.

3. $\frac{21}{8} \cdot \frac{2}{15}$

4. $\frac{12}{35} \cdot \frac{14}{9}$

5. $-\frac{15}{12} \cdot \frac{2}{10}$

6. $\frac{10}{21} \cdot \frac{-49}{15}$

7. $-10\left(-\frac{4}{5}\right)$

8. $\frac{3}{4}(-32)$

9. $\frac{28}{-15} \cdot \frac{9}{35}$

10. $\frac{6}{-10} \cdot \frac{-5}{18}$

11. $-\frac{25}{4} \cdot \frac{4}{5}$

12. $-\frac{5}{3} \cdot \frac{3}{-5}$

13. $\frac{5}{6} \cdot \frac{3}{2} \cdot \frac{4}{5}$

14. $\frac{7}{4} \cdot \frac{4}{5} \cdot \frac{10}{4}$

15. $\frac{3}{4} \cdot \frac{10}{9} \cdot \frac{7}{5}$

16. $\frac{15}{8} \cdot \frac{3}{10} \cdot \frac{28}{9}$

17. $\frac{6}{10ab} \cdot \frac{5a}{9}$

18. $\frac{18mn}{3n} \cdot \frac{1}{12m}$

19. $\frac{15}{a} \cdot \frac{a^2}{35}$

20. $\frac{9b^2}{10b} \cdot \frac{5}{6b}$

B. Division of Fractions

21. To divide a number by $\frac{9}{2}$, multiply the number by the ______ of $\frac{9}{2}$.

22. If a number is not 0, then the product of that number and its ______ is always 1.

In Exercises 23–28, write the reciprocal of the given number.

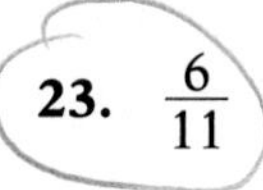

23. $\frac{6}{11}$

24. $\frac{-5}{9}$

25. -10

26. $\frac{1}{-5}$

27. $\frac{a}{3b}$

28. $\frac{2}{x}$

Copyright © Houghton Mifflin Company. All rights reserved.

In Exercises 29–34, divide.

29. $\frac{1}{3} \div \frac{2}{5}$

30. $\frac{3}{2} \div \frac{4}{7}$

31. $\frac{3}{4} \div \left(-\frac{2}{3}\right)$

32. $-\frac{5}{4} \div \frac{4}{5}$

33. $\frac{x}{5} \div \frac{y}{2}$

34. $\frac{a}{b} \div \frac{5}{9}$

In Exercises 35–52, divide and simplify.

35. $\frac{2}{3} \div \frac{4}{9}$

36. $\frac{3}{10} \div \frac{8}{5}$

37. $-\frac{10}{24} \div \left(-\frac{25}{16}\right)$

38. $-\frac{12}{7} \div \frac{8}{21}$

39. $\frac{9}{4} \div \frac{3}{8}$

40. $\frac{6}{7} \div \frac{24}{14}$

41. $\frac{15}{16} \div \left(\frac{-9}{4}\right)$

42. $-\frac{15}{14} \div \left(-\frac{6}{7}\right)$

43. $\frac{10}{42} \div \frac{5}{32}$

44. $\frac{14}{33} \div \frac{28}{15}$

45. $\frac{-15}{7} \div 5$

46. $\frac{5}{12} \div 10$

47. $-9 \div \left(-\frac{3}{2}\right)$

48. $5 \div \frac{3}{10}$

49. $\frac{x}{3} \div \frac{y}{9}$

50. $-\frac{6}{a} \div \frac{8}{b}$

51. $\frac{ab}{10} \div \left(-\frac{b}{15}\right)$

52. $\frac{8}{xy} \div \frac{28}{x}$

C. Applications

53. *Stack of Boards* How high is a stack of 20 boards if each board is $\frac{3}{4}$ inch thick?

54. *Sour Cream Container* A small container of sour cream contains $\frac{1}{2}$ pint. How many containers are needed to fill a 6-pint container?

55. *Wall Tiles* Tiles that are $\frac{3}{8}$ foot wide are to be placed side by side along a wall that is 15 feet long. How many tiles are needed?

56. *Vest Material* To make a child's vest, a manufacturer needs $\frac{3}{4}$ yard of cloth. How many vests can be made from 9 yards of cloth?

Copyright © Houghton Mifflin Company. All rights reserved.

57. *Sandwich Meat* A deli uses $\frac{3}{8}$ pound of meat in a sandwich. How many sandwiches can be made from 9 pounds of meat?

58. *Ice Cream Carton* How many $\frac{2}{3}$-cup servings of ice cream are in a carton containing 8 cups?

59. *Home-Based Business* A survey shows that $\frac{3}{5}$ of all people who have a home-based business as well as a full-time job work less than 40 hours per week at their own businesses. (Source: *USA Today.*) For a sample of 80 home-based businesses, how many owners would we expect to work less than 40 hours per week?

60. *Local First-Year Students* At a community college, $\frac{3}{5}$ of all first-year students attended high school within 50 miles of the campus. If the college enrolled 240 first-year students, how many first-year students attended high school within 50 miles of the campus?

61. *Registered Voters* In a community of 3,200 adults, $\frac{7}{8}$ are registered voters. If $\frac{2}{5}$ of the registered voters actually vote in a school board election, how many people will vote?

62. *Recall Petition* For a recall petition, a citizens group collected 2,400 signatures of which $\frac{31}{32}$ were determined to be valid signatures. If $\frac{3}{5}$ of the signatures were from women, how many women's signatures are predicted to be valid?

63. *Working College Students* Two-thirds of college students work while attending college. Of those who work, $\frac{2}{5}$ make over $400 per month. (Source: *Campus Concepts.*) In a college with an enrollment of 1,350 students, how many are predicted to make over $400 per month?

64. *Extraterrestrial Life.* A survey found that of all registered voters, $\frac{1}{3}$ of the women and $\frac{11}{20}$ of the men believe that intelligent life exists on other planets. (Source: Fox News.) Of the approximately 100 million men who are voting age, $\frac{3}{5}$ are registered to vote. How many men who are registered voters believe that intelligent life exists on other planets?

Copyright © Houghton Mifflin Company. All rights reserved.

Writing and Concept Extension

65. Explain why 10 divided by one-half is not 5.

66. Explain why multiplying a number by 2 is the same as dividing the number by $\frac{1}{2}$.

67. Which of the following is greater?

(i) 12 divided by 3 **(ii)** 12 divided by $\frac{1}{3}$

68. What positive number is the same as its reciprocal?

69. On a map, $\frac{1}{4}$ inch represents 20 miles. How many miles apart are two cities that are separated by $\frac{15}{16}$ inch on the map?

Copyright © Houghton Mifflin Company. All rights reserved.

5.4 Least Common Multiples

A *Renaming Fractions*
B *Finding Least Common Multiples*
C *Order of Fractions*

SUGGESTIONS FOR SUCCESS

Sometimes it may seem as though we are taking a detour.

Although we can multiply and divide fractions just as they are, you will see in the next section that we can add and subtract fractions only if they have the same denominator. This section prepares you for the topic of adding and subtracting of fractions.

You can become confused and frustrated if you feel that topics seem to be in random order. Just as looking back can help you to tie material together, looking forward to see what's coming can give you a greater sense of purpose. Don't be afraid to leaf ahead a little. Knowing where you are going can be comforting.

A *Renaming Fractions*

When we simplify a fraction, we divide out common factors in the numerator and the denominator. The purpose is to reduce the numerator and the denominator to smaller numbers so that we can work with the fraction more easily.

Sometimes we need to reverse the simplifying process. Instead of reducing the numerator and the denominator, we need to build them up to larger numbers. Usually, we do this when we want to choose a new denominator for a fraction.

Renaming a Fraction

We have seen that a fraction such as $\frac{1}{2}$ has many different names.

$$\frac{1}{2} = \frac{2}{4} = \frac{3}{6} = \frac{4}{8} = \ldots$$

Suppose that we want to rename $\frac{1}{2}$ so that its denominator is 12. Here is how we can do that.

$$\begin{aligned} \frac{1}{2} &= \frac{1}{2} \cdot 1 && \text{Multiplication Property of 1} \\ &= \frac{1}{2} \cdot \frac{6}{6} && 1 = \frac{6}{6} \\ &= \frac{1 \cdot 6}{2 \cdot 6} && \text{Multiplication rule for fractions} \\ &= \frac{6}{12} \end{aligned}$$

We wrote 1 as $\frac{6}{6}$ because we knew that multiplying the original denominator, 2, by 6 would result in the desired denominator, 12. In effect, multiplying $\frac{1}{2}$ by $\frac{6}{6}$ is the same as simply multiplying the numerator and the denominator of $\frac{1}{2}$ by 6. In fact, we can multiply the numerator and the denominator of a fraction by *any* number we choose (except 0), and the new fraction will be equivalent to the original fraction.

NOTE In theory, we can rename a fraction so that it has any denominator we choose. It was easy to rename $\frac{1}{2}$ so that its denominator was 12 because we

Copyright © Houghton Mifflin Company. All rights reserved.

know that $6 \cdot 2 = 12$. However, suppose we want the denominator to be 17. There is no whole number that we can multiply times 2 to obtain 17. Fortunately, we don't have to do something like that very often.

The following rule summarizes our results:

Renaming a Fraction

If a, b, and c are any numbers, where b and c are not 0, then

$$\frac{a}{b} = \frac{ac}{bc}$$

In words, we can multiply the numerator and the denominator of a fraction by any number except 0.

Example 1

Rename $\frac{2}{3}$ so that its denominator is 15.

SOLUTION

We ask, "By what number must we multiply the denominator, 3, to obtain the new denominator, 15?" The answer is 5.

$$\frac{2}{3} = \frac{2 \cdot 5}{3 \cdot 5} = \frac{10}{15}$$

Your Turn 1

Rename $\frac{4}{7}$ so that its denominator is 14.

Answer: $\frac{8}{14}$

We build up a fraction by multiplying the numerator and the denominator by a certain number. Usually we can figure out what number to use just from the basic multiplication facts. Another method is to divide the desired denominator by the original denominator.

Example 2

Determine the unknown numerator.

$$\frac{3}{4} = \frac{\square}{60}$$

SOLUTION

Because $60 \div 4 = 15$, we multiply the numerator and the denominator of $\frac{3}{4}$ by 15.

$$\frac{3}{4} = \frac{3 \cdot 15}{4 \cdot 15} = \frac{45}{60}$$

The unknown numerator is 45.

Your Turn 2

Determine the unknown numerator.

$$\frac{2}{5} = \frac{\square}{50}$$

Answer: 20

To rename a fraction that has a negative sign in the denominator, the easiest approach is to move the symbol to the numerator or to the front of the fraction.

Copyright © Houghton Mifflin Company. All rights reserved.

Example 3

Determine the unknown numerator.

$$\frac{5}{-6} = \frac{\square}{18}$$

SOLUTION

$$\frac{5}{-6} = \frac{-5}{6} = \frac{-5 \cdot 3}{6 \cdot 3} = \frac{-15}{18}$$

The unknown numerator is -15.

Your Turn 3

Determine the unknown numerator.

$$\frac{2}{-9} = \frac{\square}{36}$$

Answer: −8

The same methods can be used on fractions that involve variables.

Example 4

Determine the unknown numerator.

$$\frac{3}{x} = \frac{\square}{xy}$$

SOLUTION

$$\frac{3}{x} = \frac{3 \cdot y}{x \cdot y} = \frac{3y}{xy}$$

The unknown numerator is $3y$.

Your Turn 4

Determine the unknown numerator.

$$\frac{a}{2} = \frac{\square}{2z}$$

Answer: az

B Finding Least Common Multiples

For reasons that will be more apparent later, we want to be able to rename two or more fractions so that they have the same denominator. Before we can discuss how to do that, you need to understand the topic of multiples of numbers.

A **multiple** of a natural number is the product of that number and some natural number. The following lists show some multiples of 3 and 5.

Multiples of 3	*Multiples of 5*
$1 \cdot 3 = 3$	$1 \cdot 5 = 5$
$2 \cdot 3 = 6$	$2 \cdot 5 = 10$
$3 \cdot 3 = 9$	$3 \cdot 5 = 15$
$4 \cdot 3 = 12$	$4 \cdot 5 = 20$
$5 \cdot 3 = 15$	$5 \cdot 5 = 25$
$6 \cdot 3 = 18$	$6 \cdot 5 = 30$
$7 \cdot 3 = 21$	$7 \cdot 5 = 35$

Note that 15 is a multiple of both 3 and 5. In fact, 15 is the smallest such number. Therefore, we call 15 the *least common multiple* of 3 and 5.

Definition of Least Common Multiple

The **least common multiple (LCM)** of two natural numbers a and b is the smallest natural number that is a multiple of both a and b.

Copyright © Houghton Mifflin Company. All rights reserved.

LEARNING TIP
Another way of saying that "12 is a multiple of 3" is "12 is divisible by 3."

Finding the LCM of two numbers is easy if one number is a multiple of the other. For example, if the given numbers are 3 and 12, then 12 is the LCM because 12 is a multiple of 3.

If the larger of the given numbers is *not* a multiple of the smaller number, we can keep listing the multiples of the larger number until we find one that *is* a multiple of the smaller number.

Example 5

Find the LCM of 6 and 10.

SOLUTION

We see that 10 is not a multiple of 6, so we start listing multiples of 10.

$2 \cdot 10 = 20$ 20 is not a multiple of 6.

$3 \cdot 10 = 30$ 30 *is* a multiple of 6.

The LCM of 6 and 10 is 30.

Your Turn 5

Find the LCM of 4 and 9.

Answer: 36

The method shown in Example 5 works well for small numbers. However, the method can be tedious and time-consuming when you are trying to find the LCM of larger numbers. Also, the method of Example 5 does not apply to fractions that have variables in the denominator. A more systematic method involves prime factoring.

Finding an LCM by Prime Factoring

The LCM of 6 and 10 is a number that is divisible by both 6 and 10. This means that the LCM must contain all the factors of both 6 and 10. Therefore, to "build" the LCM, we begin by prime factoring.

$$6 = 2 \cdot 3 \qquad 10 = 2 \cdot 5$$

The LCM must contain the factors of 6.

$$\text{LCM (so far): } 2 \cdot 3$$

The LCM must also contain the factors of 10, which are 2 and 5. However, the LCM (so far) already contains the factor 2, so we need to build in only the 5.

Factors of 6

$$\text{LCM: } 2 \cdot 3 \cdot 5 = 30$$

Factors of 10

If we had simply multiplied 6 and 10, then the result, 60, is certainly a common multiple of the two numbers. However, the method just discussed gives us the *least* common multiple, which is usually what we want.

Here is a summary of the procedure.

Copyright © Houghton Mifflin Company. All rights reserved.

Finding the LCM of Two Numbers by Factoring

1. Prime-factor both numbers.
2. The LCM (so far) contains all the factors of one of the numbers.
3. Build in all factors of the other number that do not already appear in the LCM (so far).
4. The resulting product is the LCM of the two numbers.

Example 6

Find the LCM of 12 and 15.

SOLUTION

$12 = 2 \cdot 2 \cdot 3$ — Prime-factor both numbers.

$15 = 3 \cdot 5$

LCM (so far): $2 \cdot 2 \cdot 3$ — The LCM must contain all the factors of 12.

For 15, the factor 3 already appears in the LCM (so far), so we build in only the 5.

$$\text{LCM: } 2 \cdot 2 \cdot 3 \cdot 5 = 60$$

Your Turn 6

Find the LCM of 10 and 42.

LEARNING TIP

You can start with either number to write the LCM (so far). Choosing the number with the most factors is usually easier.

Answer: 210

Example 7

Find the LCM of 20 and 24.

SOLUTION

$$20 = 2 \cdot 2 \cdot 5$$
$$24 = 2 \cdot 2 \cdot 2 \cdot 3$$

We start with the prime factors of 20.

$$\text{LCM (so far): } 2 \cdot 2 \cdot 5$$

We see that 24 has three factors of 2, but the LCM (so far) already has two factors of 2. We need to build in one more factor of 2 and a factor of 3.

$$\text{LCM: } 2 \cdot 2 \cdot 2 \cdot 3 \cdot 5 = 120$$

Your Turn 7

Find the LCM of 8 and 12.

Answer: 24

The procedure for finding the LCM of two numbers can be applied to three or more numbers.

Example 8

Find the LCM of 12, 20, and 42.

SOLUTION

$12 = 2 \cdot 2 \cdot 3$

$20 = 2 \cdot 2 \cdot 5$

$42 = 2 \cdot 3 \cdot 7$

LCM (so far): $2 \cdot 2 \cdot 3$ — The factors of 12

Your Turn 8

Find the LCM of 6, 9, and 60.

continued

Copyright © Houghton Mifflin Company. All rights reserved.

For 20, the two factors of 2 already appear in the LCM (so far), so we build in the 5.

$$\text{LCM (so far): } 2 \cdot 2 \cdot 3 \cdot 5$$

For 42, the factors 2 and 3 already appear in the LCM (so far), so we build in the 7.

$$\text{LCM: } 2 \cdot 2 \cdot 3 \cdot 5 \cdot 7 = 420$$

The procedure for finding the LCM of variable expressions is exactly the same.

Example 9

Find the LCM of x^2y and xy^2.

SOLUTION

$$x^2y = x \cdot x \cdot y$$
$$xy^2 = x \cdot y \cdot y$$

We begin with x^2y.

$$\text{LCM (so far): } x \cdot x \cdot y$$

For xy^2, the factor x and one factor of y already appear in the LCM (so far). We need to build in one more factor of y.

$$\text{LCM: } x \cdot x \cdot y \cdot y = x^2y^2$$

Your Turn 9

Find the LCM of $2x^2$ and $3xy$.

Answer: $6x^2y$

C Order of Fractions

If two fractions have the same denominator, we can determine their order on the number line by comparing the numerators.

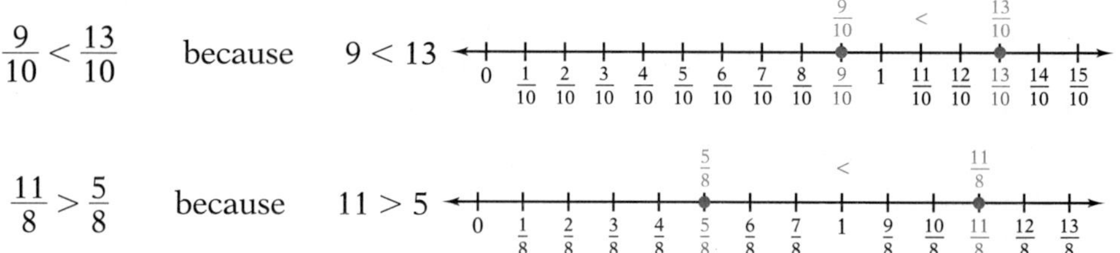

To compare fractions with unlike denominators, we need to rename the fractions so that they have the same (common) denominator. For convenience, we would like the fractions to have the *least* common denominator.

Definition of Least Common Denominator

The **least common denominator (LCD)** of two or more fractions is the least common multiple (LCM) of the denominators of the fractions.

Copyright © Houghton Mifflin Company. All rights reserved.

NOTE Practically speaking, the terms *least common denominator* (LCD) and *least common multiple* (LCM) mean the same thing. We use LCM when we are referring to numbers in general, whereas we use LCD when we are referring specifically to denominators of fractions.

Example 10

What is the LCD for $\frac{3}{4}$ and $\frac{7}{10}$?

SOLUTION

The LCD is the least common multiple (LCM) of 4 and 10.

$$4 = 2 \cdot 2$$
$$10 = 2 \cdot 5$$
$$\text{LCD: } 2 \cdot 2 \cdot 5 = 20$$

Your Turn 10

What is the LCD for $\frac{5}{9}$ and $\frac{7}{12}$?

Answer: 36

Example 11

What is the order of $\frac{3}{4}$ and $\frac{7}{10}$?

SOLUTION

From Example 10, we know that the LCD is 20. We rename both fractions so that their denominators are 20.

$$\frac{3}{4} = \frac{3 \cdot 5}{4 \cdot 5} = \frac{15}{20}$$
$$\frac{7}{10} = \frac{7 \cdot 2}{10 \cdot 2} = \frac{14}{20}$$

Now we can write the order of the fractions by comparing their numerators.
Because $15 > 14$, $\frac{15}{20} > \frac{14}{20}$.
Therefore, $\frac{3}{4} > \frac{7}{10}$.

Your Turn 11

What is the order of $\frac{5}{9}$ and $\frac{7}{12}$?

Answer: $\frac{5}{9} < \frac{7}{12}$

5.4 Quick Reference

A. Renaming Fractions

1. We use the following rule to build up the numerator and the denominator of a fraction to larger numbers.

   If a, b, and c are any numbers, where b and c are not 0, then

   $$\frac{a}{b} = \frac{ac}{bc}$$

   In words, we can multiply the numerator and the denominator of a fraction by any number except 0.

$$\frac{3}{5} = \frac{3 \cdot 4}{5 \cdot 4} = \frac{12}{20}$$
$$\frac{-2}{7} = \frac{-2 \cdot 3}{7 \cdot 3} = \frac{-6}{21}$$
$$\frac{5}{2a} = \frac{5 \cdot b}{2a \cdot b} = \frac{5b}{2ab}$$

Copyright © Houghton Mifflin Company. All rights reserved.

B. Finding Least Common Multiples

1. The **least common multiple (LCM)** of two natural numbers a and b is the smallest natural number that is a multiple of both a and b.

Common multiples of 2 and 3 are

$$6, 12, 18, 24, 30, \ldots$$

The LCM is 6.

2. We use the following steps to find the LCM of two numbers:

(a) Prime-factor both numbers.

(b) The LCM (so far) contains all the factors of one of the numbers.

(c) Build in all factors of the other number that do not already appear in the LCM (so far).

(d) The resulting product is the LCM of the two numbers.

Find the LCM of 9 and 12.

$$9 = 3 \cdot 3$$
$$12 = 2 \cdot 2 \cdot 3$$

LCM (so far): $2 \cdot 2 \cdot 3$

For 9, the LCM (so far) already contains one factor of 3, so we build in one more factor of 3.

LCM: $2 \cdot 2 \cdot 3 \cdot 3 = 36$

C. Order of Fractions

1. If two fractions have the same denominator, we can determine their order by comparing the numerators.

Because $8 < 10$, $\frac{8}{11} < \frac{10}{11}$.

2. The **least common denominator (LCD)** of two or more fractions is the least common multiple (LCM) of the denominators of the fractions.

For $\frac{1}{6}$ and $\frac{1}{14}$,

$$6 = 2 \cdot 3$$
$$14 = 2 \cdot 7$$

LCD: $2 \cdot 3 \cdot 7 = 42$

3. To compare fractions with unlike denominators, we rename the fractions so that they have an LCD.

For $\frac{8}{9}$ and $\frac{13}{15}$, the LCD is 45.

$$\frac{8}{9} = \frac{8 \cdot 5}{9 \cdot 5} = \frac{40}{45}$$

$$\frac{13}{15} = \frac{13 \cdot 3}{15 \cdot 3} = \frac{39}{45}$$

Therefore, $\frac{8}{9} > \frac{13}{15}$.

Copyright © Houghton Mifflin Company. All rights reserved.

5.4 Exercises

A. Renaming Fractions

1. Writing a fraction with a different denominator is called ______ the fraction.

2. To rename $\frac{3}{5}$ so that the denominator is 10, we multiply both the ______ and the ______ by 2.

In Exercises 3–10, rename the fraction.

3. $\frac{3}{4} = \frac{\square}{12}$

4. $\frac{3}{5} = \frac{\square}{20}$

5. $\frac{1}{4} = \frac{\square}{28}$

6. $\frac{1}{9} = \frac{\square}{54}$

7. $\frac{2}{7} = \frac{\square}{21}$

8. $\frac{3}{10} = \frac{\square}{70}$

9. $\frac{7}{12} = \frac{\square}{60}$

10. $\frac{8}{15} = \frac{\square}{45}$

In Exercises 11–14, rename the fraction.

11. $\frac{5}{-6} = \frac{\square}{24}$

12. $\frac{-4}{7} = \frac{\square}{35}$

13. $\frac{-7}{10} = \frac{\square}{30}$

14. $-\frac{2}{9} = \frac{\square}{63}$

In Exercises 15–20, rename the fraction.

15. $\frac{4}{5} = \frac{\square}{5a}$

16. $\frac{3}{7} = \frac{\square}{7x}$

17. $\frac{7}{b} = \frac{\square}{bc}$

18. $\frac{y}{z} = \frac{\square}{xz}$

19. $\frac{x}{y} = \frac{\square}{y^2}$

20. $\frac{y}{2} = \frac{\square}{6x}$

B. Finding Least Common Multiples

21. Because 15 is divisible by 5, we say that 15 is a ______ of 5.

22. The smallest natural number that is a multiple of both 6 and 8 is called the ______ of 6 and 8.

In Exercises 23–28, determine the LCM of the given numbers.

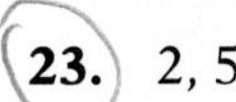

23. 2, 5

24. 16, 8

25. 4, 10

26. 9, 15

27. 45, 30

28. 24, 30

Copyright © Houghton Mifflin Company. All rights reserved.

In Exercises 29–34, determine the LCM of the given numbers.

29. 15, 21, 8 **30.** 12, 18, 15 **31.** 30, 21, 14

32. 10, 20, 14 **33.** 45, 12, 25 **34.** 20, 30, 40

In Exercises 35–40, determine the LCM of the given expressions.

35. $3a, 4b$ **36.** $5m, 2n$ **37.** $6x, 2xy$

38. $10b, 5ab$ **39.** $5a^2b, 2ab$ **40.** $6x^2, 3xy$

C. Order of Fractions

41. To determine the order of two fractions with the same denominator, we compare the ______.

42. To determine the ______ of two fractions, determine the least common multiple of their denominators.

In Exercises 43–46, determine the LCD for the given fractions.

43. $\frac{3}{7}, \frac{1}{2}$ **44.** $\frac{2}{5}, \frac{2}{3}$ **45.** $\frac{5}{8}, \frac{1}{12}$ **46.** $\frac{6}{35}, \frac{3}{40}$

In Exercises 47–52, insert < or > to make the statement true.

47. $\frac{4}{7}$ ______ $\frac{2}{3}$ **48.** $\frac{7}{9}$ ______ $\frac{3}{4}$ **49.** $\frac{3}{5}$ ______ $\frac{3}{4}$

50. $\frac{4}{7}$ ______ $\frac{5}{8}$ **51.** $\frac{5}{12}$ ______ $\frac{7}{18}$ **52.** $\frac{8}{21}$ ______ $\frac{4}{9}$

In Exercises 53–56, write the fractions in order from smallest to largest.

53. $\frac{1}{2}, \frac{3}{4}, \frac{2}{5}$ **54.** $\frac{5}{6}, \frac{2}{9}, \frac{5}{7}$ **55.** $\frac{2}{3}, \frac{4}{9}, \frac{5}{6}$ **56.** $\frac{4}{5}, \frac{1}{2}, \frac{3}{10}$

Copyright © Houghton Mifflin Company. All rights reserved.

≈ Estimation

In Exercises 57–60, match the fraction to its approximate value.

57. $\frac{7}{16}$ **(A)** $\frac{1}{4}$

58. $\frac{11}{15}$ **(B)** $\frac{1}{2}$

59. $\frac{14}{25}$ **(C)** $\frac{2}{3}$

60. $\frac{3}{16}$ **(D)** $\frac{3}{5}$

Applications

61. *Shuttle Buses* Three WDW shuttle buses leave a central parking lot at 7:00 A.M. to transport visitors to three theme parks. The three roundtrip routes require 15, 20, and 30 minutes, respectively. After how many minutes will the three buses be back at the lot at the same time?

62. *Truck Routes* Three truckers drive roundtrip routes that require 6, 10, and 15 days, respectively. If the three drivers leave the terminal on the same day, after how many days will all three arrive back at the terminal?

63. *Long Distance Walking* Two friends begin walking on a 1-mile track at the same time. They walk the mile in 15 and 18 minutes, respectively. After how many minutes will both be back at the starting point at the same time?

64. *Copier Times* Three copiers produce copies of employee benefits packages in 4, 6, and 9 seconds, respectively. If the copiers are started at the same time, after how many seconds will all three complete a manual at the same time?

Copyright © Houghton Mifflin Company. All rights reserved.

Writing and Concept Extension

65. Suppose that m and n are two different prime numbers. Explain how to determine the LCM of m and n.

66. Suppose that m is divisible by n. What is the LCM of m and n?

In Exercises 67 and 68, determine the LCM of the given expressions.

67. $3xy, 2y^2, 4x^2$

68. $9a^2, 3ab^2, 2b$

In Exercises 69 and 70, rename the fraction.

69. $\frac{3}{7} = \frac{12}{\square}$

70. $\frac{7}{8} = \frac{14}{\square}$

Exploring with Real-World Data: Collaborative Activities

Sunscreen Users The pie graphs show the fractions of females and males age 12–19 who use sunscreen. (Source: *USA Today*.)

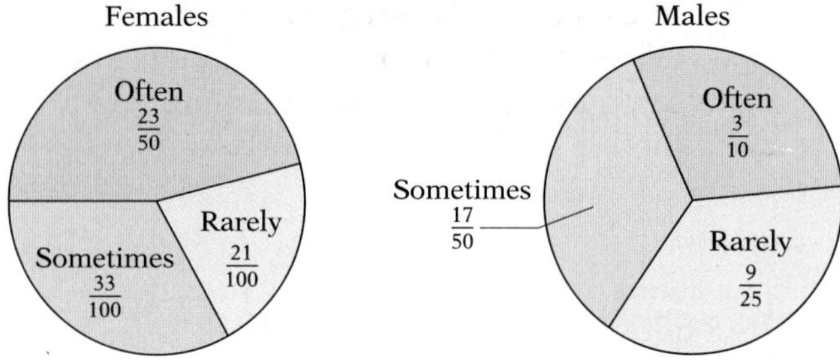

71. For which category or categories of males is $\frac{1}{3}$ a good approximation of the fraction?

72. For which category or categories of females is $\frac{1}{3}$ a good approximation of the fraction?

73. For females, write the fractions with a denominator of 100. Then arrange the fractions in order from smallest to largest.

74. For males, write the fractions with a denominator of 100. Then arrange the fractions from smallest to largest.

75. Use an inequality symbol to compare the fraction of females in the categories "often" and "sometimes."

76. Use an inequality symbol to compare the fraction of males in the categories "often" and "rarely."

77. In which category(s) is the fraction of females greater than the fraction of males?

78. In which category(s) is the fraction of males greater than the fraction of females?

Copyright © Houghton Mifflin Company. All rights reserved.

5.5 Addition and Subtraction of Fractions

SUGGESTIONS FOR SUCCESS

How do you add 5 nails plus 9 bolts? We want to say 14, but 14 what? One approach is to give "nails" and "bolts" the same name. If we call them "fasteners," for example, then the sum is 14 fasteners.

You might think about adding fractions in the same way. As you will see, we can't add fractions unless they have the same denominator, and this means that we may need to rename the fractions.

As always, if you can relate a mathematical concept to something that makes sense in your everyday life, then you will have a better understanding of what you are doing and why.

A Like Denominators

At a family picnic, $\frac{3}{5}$ of the family members were between the ages of 20 and 50, and $\frac{1}{6}$ of the family members were over age 50. To determine the fraction of the family that was over age 20, we need to add $\frac{3}{5} + \frac{1}{6}$.

Because the product of two fractions is found by multiplying the numerators and multiplying the denominators, it might be tempting to imagine that the sum of two fractions is found by adding the numerators and adding the denominators. However, as you will see, this is not the case. In fact, it is not even possible to add two fractions unless they have the same denominators.

We begin our discussion of adding and subtracting fractions with fractions that have the same (like) denominators.

Adding Fractions with Like Denominators

Consider the expression $\frac{1}{7}(2 + 4)$. We have two ways in which we can remove the parentheses.

Order of Operations

$$\frac{1}{7}(2 + 4) = \frac{1}{7}(6) = \frac{1}{7} \cdot \frac{6}{1} = \frac{6}{7}$$

Distributive Property

$$\frac{1}{7}(2 + 4) = \frac{1}{7} \cdot 2 + \frac{1}{7} \cdot 4 = \frac{1}{7} \cdot \frac{2}{1} + \frac{1}{7} \cdot \frac{4}{1} = \frac{2}{7} + \frac{4}{7}$$

Because both methods are valid, the two results must be equal.

$$\frac{2}{7} + \frac{4}{7} = \frac{6}{7}$$

Note that $\frac{2}{7}$ and $\frac{4}{7}$ have the same (like) denominators.

The left side of this equation is the sum of two fractions that have the same denominator. This experiment suggests that we can add two fractions that have like denominators by adding the numerators $(2 + 4 = 6)$ and keeping the same denominator.

The same principle applies to subtraction of fractions. We can state these results as a general rule.

Copyright © Houghton Mifflin Company. All rights reserved.

Adding and Subtracting Fractions

If a, b, and c are any numbers, where c is not 0, then

$$\frac{a}{c} + \frac{b}{c} = \frac{a+b}{c} \quad \text{and} \quad \frac{a}{c} - \frac{b}{c} = \frac{a-b}{c}$$

In words, if two fractions have like denominators, then we add (subtract) the fractions by adding (subtracting) the numerators and keeping the denominator.

Example 1

Add $\frac{3}{10} + \frac{6}{10}$.

SOLUTION

$$\frac{3}{10} + \frac{6}{10} = \frac{3+6}{10} \qquad \text{Add the numerators.}$$

$$= \frac{9}{10} \qquad \text{Keep the denominator.}$$

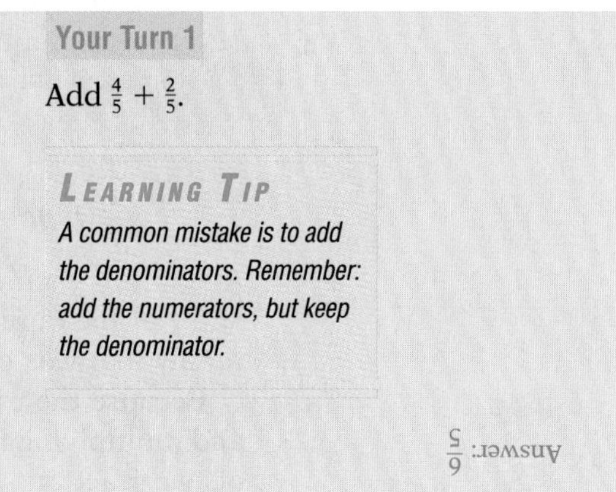

Your Turn 1

Add $\frac{4}{5} + \frac{2}{5}$.

LEARNING TIP

A common mistake is to add the denominators. Remember: add the numerators, but keep the denominator.

Answer: $\frac{6}{5}$

The same rule applies even if the denominator is a variable or a variable expression. Remember, though, that the denominators must be the same.

Example 2

Add $\frac{2}{x} + \frac{9}{x}$.

SOLUTION

$$\frac{2}{x} + \frac{9}{x} = \frac{2+9}{x} \qquad \text{Add the numerators.}$$

$$= \frac{11}{x} \qquad \text{Keep the denominator.}$$

Example 3

Subtract $\frac{9}{16} - \frac{5}{16}$.

SOLUTION

$$\frac{9}{16} - \frac{5}{16} = \frac{9-5}{16} \qquad \text{Subtract the numerators.}$$

$$= \frac{4}{16} \qquad \text{Keep the denominator.}$$

$$= \frac{1}{4} \qquad \text{Simplify the result.}$$

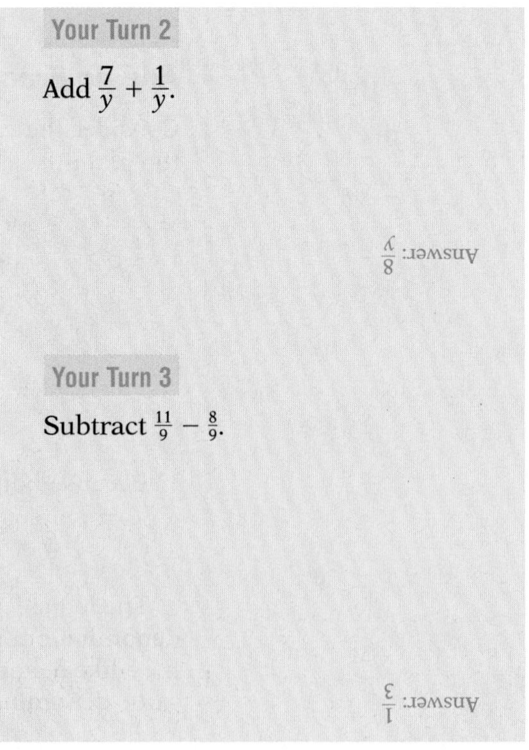

Your Turn 2

Add $\frac{7}{y} + \frac{1}{y}$.

Answer: $\frac{8}{y}$

Your Turn 3

Subtract $\frac{11}{9} - \frac{8}{9}$.

Answer: $\frac{1}{3}$

Copyright © Houghton Mifflin Company. All rights reserved.

When you subtract the numerators, you may need to use the definition of subtraction: $a - b = a + (-b)$.

Example 4

Subtract $\frac{2}{5} - \frac{9}{5}$.

SOLUTION

$$\frac{2}{5} - \frac{9}{5} = \frac{2-9}{5} \quad \text{Subtract the numerators.}$$

$$= \frac{2+(-9)}{5} \quad \text{Definition of subtraction}$$

$$= \frac{-7}{5} = -\frac{7}{5}$$

Your Turn 4

Subtract $\frac{4}{7} - \frac{9}{7}$.

Answer: $-\frac{5}{7}$

If the fraction that you are subtracting is negative, then applying the definition of subtraction first makes the operation easier to perform.

Example 5

Subtract $\frac{2}{3} - (-\frac{10}{3})$.

SOLUTION

$$\frac{2}{3} - \left(-\frac{10}{3}\right) = \frac{2}{3} + \frac{10}{3} \quad \text{Definition of subtraction}$$

$$= \frac{2+10}{3} \quad \text{Add the numerators.}$$

$$= \frac{12}{3} \quad \text{Keep the denominator.}$$

$$= 4 \quad \text{Simplify the result.}$$

Your Turn 5

Subtract $\frac{1}{5} - (-\frac{9}{5})$.

Answer: 2

Before we leave the topic of sums of fractions with like denominators, let's look deeper into why the addition rule applies. In the following, note how the Distributive Property allows us to determine the sum.

$$\frac{3}{7} + \frac{2}{7} = \frac{1 \cdot 3}{7 \cdot 1} + \frac{1 \cdot 2}{7 \cdot 1} = \frac{1}{7} \cdot 3 + \frac{1}{7} \cdot 2 = \frac{1}{7}(2+3) \quad \text{Distributive Property}$$

$$= \frac{1}{7} \cdot 5 = \frac{1}{7} \cdot \frac{5}{1} = \frac{5}{7}$$

Had the denominators not been the same, there would have been no common factor at the end of the first line, and the Distributive Property could not have been applied to finish the steps.

LEARNING TIP

Like denominators are needed for adding or subtracting fractions. However, we do not need like denominators for multiplication or division. Think about the operation you are performing before you begin.

B Unlike Denominators

We can't apply our addition rule for fractions to the sum $\frac{1}{3} + \frac{2}{5}$ because the fractions don't have like denominators. In such instances, we must rename one or both of the fractions so that the denominators are the same. Only then can we perform the addition.

For convenience, we rename the fractions so that they have a least common denominator (LCD).

Copyright © Houghton Mifflin Company. All rights reserved.

Example 6

Add $\frac{3}{4} + \frac{1}{8}$.

SOLUTION

Note that 8 is a multiple of 4, so the LCD is 8. This means that we need to rename only the first fraction.

$$\frac{3}{4} + \frac{1}{8} = \frac{3 \cdot 2}{4 \cdot 2} + \frac{1}{8}$$
$$= \frac{6}{8} + \frac{1}{8}$$
$$= \frac{7}{8}$$

Your Turn 6

Add $\frac{4}{9} + \frac{2}{3}$.

Answer: $\frac{10}{9}$

In Section 5.4, we saw that factoring numbers is helpful in finding the LCM of those numbers. We can use the same method for finding an LCD.

Example 7

Add $\frac{5}{6} + \frac{1}{9}$.

SOLUTION

First, we factor the denominators.

$$\frac{5}{6} + \frac{1}{9} = \frac{5}{2 \cdot 3} + \frac{1}{3 \cdot 3}$$

By using the methods of Section 5.4, we find that the LCD is $2 \cdot 3 \cdot 3$. Now we can rename the fractions by comparing the original denominators to the LCD.

$$\frac{5}{6} = \frac{5}{2 \cdot 3} = \frac{5 \cdot 3}{2 \cdot 3 \cdot 3} = \frac{15}{18}$$
$$\frac{1}{9} = \frac{1}{3 \cdot 3} = \frac{2 \cdot 1}{2 \cdot 3 \cdot 3} = \frac{2}{18}$$

Finally, we can add the fractions.

$$\frac{5}{6} + \frac{1}{9} = \frac{15}{18} + \frac{2}{18} = \frac{15 + 2}{18} = \frac{17}{18}$$

Your Turn 7

Add $\frac{3}{2} + \frac{2}{7}$.

Answer: $\frac{25}{14}$

The following is a summary of the procedure for adding and subtracting fractions that have unlike denominators.

Adding and Subtracting Fractions with Unlike Denominators

1. Prime-factor the denominators.
2. Build the least common denominator (LCD).
3. For each fraction, compare the original denominator to the LCD. Multiply the numerator and the denominator by any factors in the LCD that are missing in the original denominator.
4. Add (subtract) the renamed fractions by adding (subtracting) the numerators and keeping the denominator.
5. Simplify the result, if possible.

Copyright © Houghton Mifflin Company. All rights reserved.

Note that the last step of the procedure is to simplify the result, if possible. Because simplifying means dividing out common factors, leaving the LCD in factored form until the last step is a good idea.

Example 8

Add $\frac{5}{6} + \frac{3}{10}$.

SOLUTION

$$\frac{5}{6} + \frac{3}{10} = \frac{5}{2 \cdot 3} + \frac{3}{2 \cdot 5}$$ Prime-factor the denominators.

The LCD is $2 \cdot 3 \cdot 5$.

Now we multiply the numerator and the denominator of each fraction by factors in the LCD that are not factors of the original denominator.

$$\frac{5}{2 \cdot 3} + \frac{3}{2 \cdot 5} = \frac{5 \cdot 5}{2 \cdot 3 \cdot 5} + \frac{3 \cdot 3}{2 \cdot 3 \cdot 5}$$

$$= \frac{25}{2 \cdot 3 \cdot 5} + \frac{9}{2 \cdot 3 \cdot 5}$$

$$= \frac{34}{2 \cdot 3 \cdot 5}$$ Add the fractions.

$$= \frac{\cancel{2} \cdot 17}{\cancel{2} \cdot 3 \cdot 5}$$ Factor the numerator and divide out the common factor, 2.

$$= \frac{17}{15}$$ The simplified result

Your Turn 8

Add $\frac{2}{15} + \frac{1}{5}$.

Answer: $\frac{1}{3}$

KEYS TO THE CALCULATOR

The following are typical keystrokes for the sum $\frac{5}{8} + \frac{3}{24}$. Note that the results are displayed in simplified form.

Scientific calculator:

5 [a b/c] **8** [+] **3** [a b/c] **24** [=]   Display: [3 ⌋ 4]

Graphing calculator:

5 [÷] **8** [+] **3** [÷] **24** [MATH] [Frac] [ENTER]

```
5/8+3/24▶Frac
            3/4
```

To perform subtraction, substitute the [−] key for the [+] key.

Exercises

(a) $\frac{7}{15} + \frac{1}{3}$ **(b)** $\frac{3}{4} - \frac{5}{16}$ **(c)** $2 - \frac{11}{8}$ **(d)** $\frac{5}{12} + \frac{3}{8}$

Copyright © Houghton Mifflin Company. All rights reserved.

When a sum or difference includes an integer, we write the integer as a fraction.

Example 9

Subtract $\frac{4}{7} - 2$.

SOLUTION

$$\frac{4}{7} - 2 = \frac{4}{7} - \frac{2}{1} \quad \text{Write 2 as a fraction.}$$

$$= \frac{4}{7} - \frac{2 \cdot 7}{1 \cdot 7} \quad \text{The LCD is 7.}$$

$$= \frac{4}{7} - \frac{14}{7}$$

$$= \frac{4 - 14}{7} \quad \text{Subtract the numerators.}$$

$$= \frac{4 + (-14)}{7} \quad \text{Definition of subtraction}$$

$$= \frac{-10}{7} = -\frac{10}{7}$$

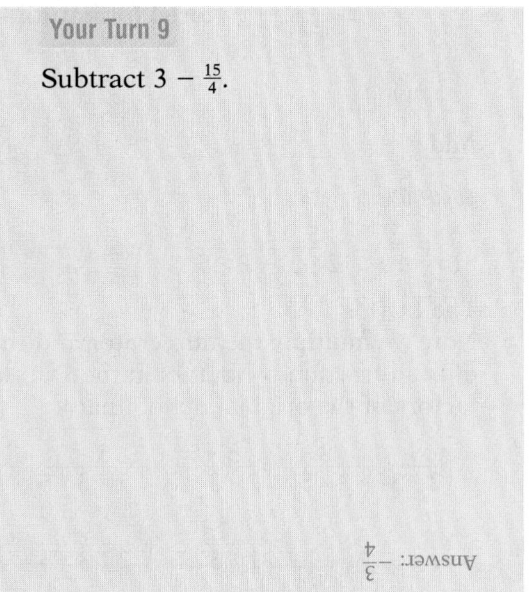

Your Turn 9

Subtract $3 - \frac{15}{4}$.

Answer: $-\frac{3}{4}$

Numerical expressions can contain more than two fractions and a combination of operations.

Example 10

Combine $\frac{1}{2} + \frac{1}{3} - \frac{1}{5}$.

SOLUTION

$$\frac{1}{2} + \frac{1}{3} - \frac{1}{5}$$

$$= \frac{1 \cdot 3 \cdot 5}{2 \cdot 3 \cdot 5} + \frac{1 \cdot 2 \cdot 5}{2 \cdot 3 \cdot 5} - \frac{1 \cdot 2 \cdot 3}{2 \cdot 3 \cdot 5}$$

$$= \frac{15}{30} + \frac{10}{30} - \frac{6}{30}$$

$$= \frac{15 + 10 - 6}{30}$$

$$= \frac{19}{30}$$

Your Turn 10

Combine $\frac{2}{3} - \frac{1}{6} + \frac{1}{7}$.

LEARNING TIP

After you have built up fractions, you will be tempted to divide out common factors. This would ruin all your good work! Wait until the last step to try to simplify.

Answer: $\frac{9}{14}$

In algebra, we often need to add or subtract fractions that involve variables in the numerator or the denominator or both. This raises some interesting questions. Suppose we need to perform the following addition:

$$\frac{5}{x} + \frac{8}{3}$$

First, if we don't know the value of x, then how do we know whether the denominators are the same? Simply put, we don't. We assume that these are unlike denominators and follow the usual procedure of renaming the fractions with an LCD.

Copyright © Houghton Mifflin Company. All rights reserved.

Second, how do we know that x doesn't represent 0? Again, we don't, but we must insist that x is not 0. Most algebra textbooks emphasize this by including $x \neq 0$ in the problem statement. In this text, we assume that no denominator is 0.

Example 11

(a) Add $\frac{2}{x} + \frac{3}{y}$.

(b) Subtract $6 - \frac{5}{x}$.

SOLUTION

(a) $\frac{2}{x} + \frac{3}{y} = \frac{2 \cdot y}{x \cdot y} + \frac{3 \cdot x}{x \cdot y}$ The LCD is xy.

$= \frac{2y}{xy} + \frac{3x}{xy}$

$= \frac{3x + 2y}{xy}$ Add the numerators. Keep the denominator.

(b) $6 - \frac{5}{x} = \frac{6}{1} - \frac{5}{x}$ Write 6 as a fraction.

$= \frac{6 \cdot x}{1 \cdot x} - \frac{5}{x}$ The LCD is x.

$= \frac{6x}{x} - \frac{5}{x}$

$= \frac{6x - 5}{x}$ Subtract the numerators. Keep the denominator.

Your Turn 11

(a) Add $\frac{1}{a} + \frac{3}{5}$.

(b) Subtract $\frac{5a}{b} - 2$.

Answers: (a) $\frac{5 + 3a}{5a}$; (b) $\frac{5a - 2b}{b}$

5.5 Quick Reference

A. Like Denominators

1. The following are the general rules for adding and subtracting fractions that have like denominators.

   If a, b, and c are any numbers, where c is not 0, then

   $$\frac{a}{c} + \frac{b}{c} = \frac{a + b}{c} \qquad \frac{a}{c} - \frac{b}{c} = \frac{a - b}{c}$$

   In words, if two fractions have like denominators, then we add (subtract) the fractions by adding (subtracting) the numerators and keeping the denominator.

$$\frac{2}{7} + \frac{3}{7} = \frac{2 + 3}{7} = \frac{5}{7}$$

$$\frac{5}{9} - \frac{1}{9} = \frac{5 - 1}{9} = \frac{4}{9}$$

$$\frac{3}{x} + \frac{8}{x} = \frac{3 + 8}{x} = \frac{11}{x}$$

$$\frac{7}{3y} - \frac{5}{3y} = \frac{7 - 5}{3y} = \frac{2}{3y}$$

$$\frac{2}{9} - \frac{7}{9} = \frac{2 - 7}{9} = \frac{-5}{9} = -\frac{5}{9}$$

$$\frac{1}{5} - \left(-\frac{3}{5}\right) = \frac{1}{5} + \frac{3}{5} = \frac{4}{5}$$

Copyright © Houghton Mifflin Company. All rights reserved.

B. Unlike Denominators

1. We can add fractions that have unlike denominators by following these steps.

To add $\frac{7}{12} + \frac{3}{20}$:

(a) Prime-factor the denominators.

$$12 = 2 \cdot 2 \cdot 3$$
$$20 = 2 \cdot 2 \cdot 5$$

(b) Build the least common denominator (LCD).

$$\text{LCD: } 2 \cdot 2 \cdot 3 \cdot 5$$

(c) For each fraction, compare the original denominator to the LCD. Multiply the numerator and the denominator by any factors in the LCD that are missing in the original denominator.

$$\frac{7}{12} = \frac{7}{2 \cdot 2 \cdot 3} = \frac{7 \cdot 5}{2 \cdot 2 \cdot 3 \cdot 5}$$
$$\frac{3}{20} = \frac{3}{2 \cdot 2 \cdot 5} = \frac{3 \cdot 3}{2 \cdot 2 \cdot 3 \cdot 5}$$

(d) Add (subtract) the renamed fractions by adding (subtracting) the numerators and keeping the denominator.

$$\frac{7}{12} + \frac{3}{20} = \frac{35}{2 \cdot 2 \cdot 3 \cdot 5} + \frac{9}{2 \cdot 2 \cdot 3 \cdot 5}$$
$$= \frac{44}{2 \cdot 2 \cdot 3 \cdot 5}$$
$$= \frac{\cancel{2} \cdot \cancel{2} \cdot 11}{\cancel{2} \cdot \cancel{2} \cdot 3 \cdot 5}$$

(e) Simplify the result if possible.

$$= \frac{11}{3 \cdot 5} = \frac{11}{15}$$

Copyright © Houghton Mifflin Company. All rights reserved.

5.5 Exercises

A. Like Denominators

1. To add two fractions with like denominators, we add the ________ and keep the ________.

2. To add or subtract two fractions, the ________ must be the same.

In Exercises 3–8, perform the indicated operation.

3. $\frac{2}{7} + \frac{4}{7}$

4. $\frac{5}{6} + \frac{8}{6}$

5. $\frac{2}{3} - \frac{1}{3}$

6. $\frac{5}{17} - \frac{8}{17}$

7. $\frac{6}{y} - \frac{10}{y}$

8. $\frac{3}{a} + \frac{7}{a}$

In Exercises 9–12, add and simplify.

9. $\frac{5}{12} + \frac{1}{12}$

10. $\frac{3}{8} + \frac{5}{8}$

11. $\frac{-3}{10} + \left(-\frac{5}{10}\right)$

12. $\frac{2}{15} + \frac{-8}{15}$

In Exercises 13–16, subtract and simplify.

13. $\frac{13}{16} - \frac{1}{16}$

14. $-\frac{1}{6} - \frac{1}{6}$

15. $-\frac{4}{9} - \frac{2}{9}$

16. $\frac{1}{4} - \left(-\frac{3}{4}\right)$

B. Unlike Denominators

17. To add $\frac{1}{2}$ and $\frac{2}{3}$, we must first determine the ________.

18. One way to determine the LCD for two fractions is to begin by ________ the denominators.

In Exercises 19–30, perform the indicated operation.

19. $\frac{1}{4} + \frac{2}{3}$

20. $\frac{5}{8} + \frac{1}{3}$

21. $\frac{5}{3} - \frac{1}{4}$

22. $\frac{3}{10} - \frac{1}{7}$

23. $\frac{3}{4} + \frac{5}{12}$

24. $\frac{7}{18} - \frac{2}{9}$

25. $\frac{7}{15} - \frac{2}{5}$

26. $\frac{2}{7} + \frac{3}{14}$

27. $3 + \frac{2}{5}$

28. $\frac{1}{2} + 2$

29. $\frac{4}{5} - 1$

30. $4 - \frac{5}{3}$

In Exercises 31–38, perform the indicated operation.

31. $\frac{-3}{10} + \frac{5}{6}$

32. $\frac{5}{12} + \frac{9}{20}$

33. $\frac{1}{12} - \frac{7}{30}$

34. $\frac{9}{20} - \frac{1}{30}$

Copyright © Houghton Mifflin Company. All rights reserved.

35. $\frac{7}{15} + \frac{9}{20}$

36. $\frac{5}{12} - \frac{3}{20}$

37. $\frac{3}{10} - \frac{7}{15}$

38. $\frac{2}{15} - \frac{5}{6}$

In Exercises 39–42, perform the indicated operations.

39. $\frac{5}{6} - \frac{2}{3} + \frac{1}{2}$

40. $\frac{3}{5} - \frac{1}{2} - \frac{7}{10}$

41. $-\frac{3}{8} + \frac{3}{4} - \frac{2}{3}$

42. $\frac{1}{2} - \left(-\frac{1}{3}\right) - \frac{1}{4}$

In Exercises 43–48, perform the indicated operation.

43. $\frac{3}{4} + \frac{1}{x}$

44. $\frac{3}{a} - \frac{5}{b}$

45. $\frac{4}{b} - \frac{3}{c}$

46. $\frac{x}{y} - \frac{2}{3}$

47. $4 - \frac{1}{a}$

48. $\frac{x}{5} + 7$

Bringing It Together

In Exercises 49–58, perform the indicated operation.

49. $\frac{3}{7} + \frac{1}{2}$

50. $\frac{5}{8} + \frac{7}{24}$

51. $-6\left(\frac{-5}{9}\right)$

52. $-\frac{15}{28} \cdot \frac{21}{20}$

53. $\frac{1}{12} - \frac{1}{20}$

54. $\frac{3}{14} - \frac{8}{21}$

55. $\frac{-26}{30} \div \frac{39}{25}$

56. $\frac{2}{5} \div (-6)$

57. $\frac{5}{9} + \frac{1}{10} - \frac{4}{5}$

58. $\frac{2}{3} - \frac{3}{4} - \left(\frac{-4}{5}\right)$

Applications

59. *Roofing Job* In one day an experienced roofer can complete $\frac{3}{8}$ of a job. If an inexperienced helper can complete half as much of the job in a day, what fractional part of the job can the two complete in one day?

Copyright © Houghton Mifflin Company. All rights reserved.

60. *Window Washing* A window washer cleaned $\frac{3}{8}$ of the windows of a building at the same time a helper cleaned $\frac{1}{6}$ of the windows. What portion of the windows were cleaned?

61. *Swimming* Two swimmers began swimming across a lake that is $\frac{3}{4}$ mile wide. At the time that the first swimmer completed the trip, the slower swimmer had completed $\frac{3}{5}$ mile of the trip. What fraction of a mile remained for the slower swimmer to complete the trip?

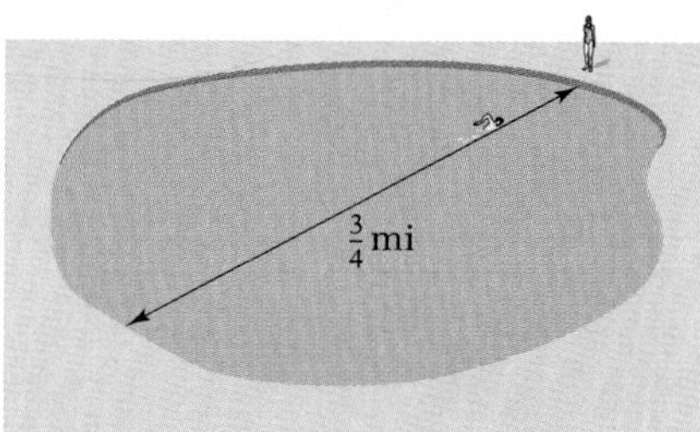

62. *Water Tank* In one hour one pipe can fill $\frac{1}{4}$ of a tank and another pipe can drain $\frac{1}{5}$ of a tank. If both pipes are open, what fractional part of a tank can be filled in one hour?

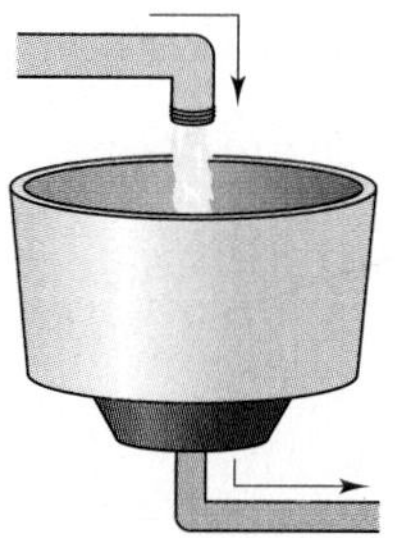

63. *Business Partners* Three partners own an auto body shop. One partner has a $\frac{1}{3}$ interest and another partner has a $\frac{3}{8}$ interest. What fraction of the business does the other partner own?

64. *Estate Division* A grandmother willed her entire estate to her four grandchildren. The oldest received $\frac{2}{5}$ of the estate and the next two received $\frac{1}{6}$ and $\frac{1}{8}$, respectively. What fraction of the estate did the youngest child receive?

Copyright © Houghton Mifflin Company. All rights reserved.

65. *Class Grades* Of 40 students in a history class, $\frac{3}{8}$ received A's and $\frac{2}{5}$ received B's. What fraction of the class received grades other than A's or B's? How many students received A's or B's?

66. *Stock Prices* The price of a share of McDonald's stock rose $\frac{3}{4}$ of a point one day, declined $\frac{3}{16}$ the next day, and rose $\frac{5}{8}$ the third day. What was the change in price for the three-day period?

Household Expenses The pie graph shows the fraction of a couple's $2,100-per-month income that is spent for selected items.

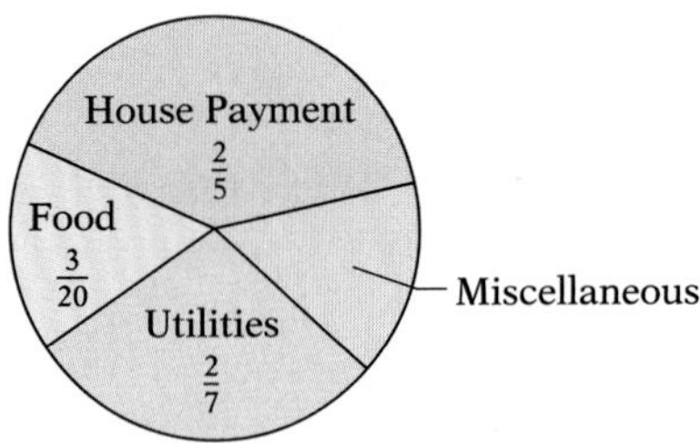

67. What fraction of the income is spent for the house and utilities?

68. What fraction of the income is used for miscellaneous expenses?

69. What is the dollar expenditure for the house and utilities?

Copyright © Houghton Mifflin Company. All rights reserved.

70. What is the dollar expenditure for food?

Writing and Concept Extension

71. Describe the LCD of two fractions if their denominators are different numbers with no factors in common.

72. When you add or subtract fractions, why is keeping the LCD in factored form until the last step a good idea?

73. Identify the operations with fractions for which a common denominator is necessary.

74. Identify the operations with fractions for which a common denominator is *not* necessary.

In Exercises 75 and 76, perform the indicated operation and simplify.

75. $\frac{4}{5a} + \frac{1}{5a}$

76. $\frac{7a}{10} - \frac{3a}{10}$

Exploring with Real-World Data: Collaborative Activities

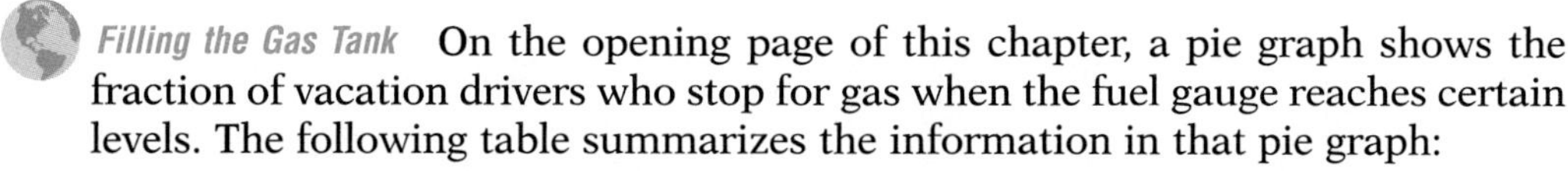

Filling the Gas Tank On the opening page of this chapter, a pie graph shows the fraction of vacation drivers who stop for gas when the fuel gauge reaches certain levels. The following table summarizes the information in that pie graph:

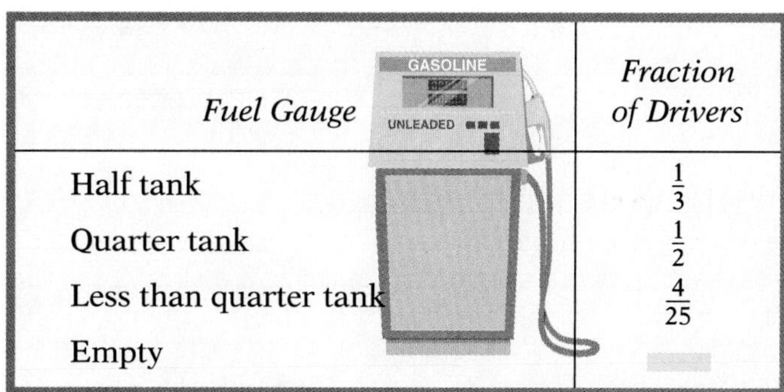

Fuel Gauge	*Fraction of Drivers*
Half tank	$\frac{1}{3}$
Quarter tank	$\frac{1}{2}$
Less than quarter tank	$\frac{4}{25}$
Empty	

(*Source:* Exxon Corporation.)

Copyright © Houghton Mifflin Company. All rights reserved.

77. What fraction of drivers stop when the tank is a quarter or more full?

78. What fraction of drivers stop when the tank is less than half full?

79. What fraction of drivers wait until the gauge reaches empty?

80. What fraction of drivers wait until the gauge is empty or less than a quarter?

Safe Drivers The pie graph shows the fraction of men's and women's responses to the question "Who are the safest drivers?" (Source: *USA Today.*)

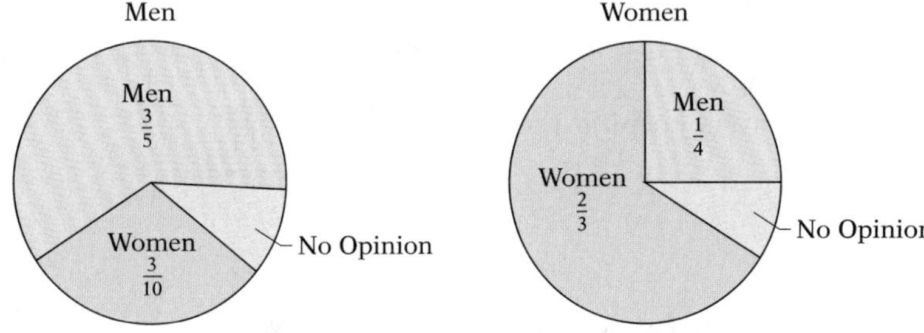

81. What is the difference between the fraction of men and women who responded "men"?

82. What is the difference between the fraction of men and women who responded "women"?

83. What fraction of women had no opinion?

84. What fraction of men had no opinion?

85. What fraction of men had an opinion?

86. What fraction of women had an opinion?

Copyright © Houghton Mifflin Company. All rights reserved.

5.6 Operations with Mixed Numbers

A *Introduction to Mixed Numbers*
B *Multiplication and Division*
C *Addition and Subtraction*

SUGGESTIONS FOR SUCCESS

Believe it or not, there was a time when there were no calculators. Operations with fractions had to be performed the old-fashioned way.

The good news about calculators is that they make our lives easier. The bad news is that blindly relying on them might mean that you are getting results without knowing why. Do your best to avoid using a calculator as a crutch. Make sure that you fully understand *how* an operation is performed before you routinely reach for the machine.

Your instructor may enforce this by not letting you use a calculator on tests. If so, don't despair. Instead, take pride in knowing.

A Introduction to Mixed Numbers

A certain recipe calls for $1\frac{1}{2}$ cups of flour. The finished width of a baseboard molding might be $3\frac{5}{16}$ inches. A quilt block is $17\frac{3}{8}$ inches long. The numbers $1\frac{1}{2}$, $3\frac{5}{16}$, and $17\frac{3}{8}$ are all examples of **mixed numbers.**

Definition of Mixed Number

A **mixed number** is the sum of a whole number and a proper fraction.

In the definition, the word *sum* is important. To use the recipe that calls for $1\frac{1}{2}$ cups of flour, you would probably use measuring cups.

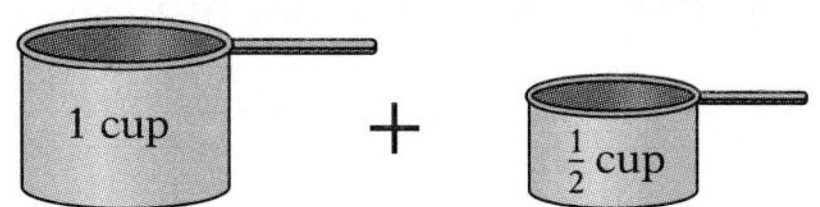

The total amount of flour that you would add is 1 cup *plus* $\frac{1}{2}$ cup. The number $1\frac{1}{2}$ is usually read as "one and a half," but the more precise meaning is $1 + \frac{1}{2}$.

NOTE A mixed number represents a *sum,* not a product. The mixed number $3\frac{5}{16}$ does *not* mean 3 times $\frac{5}{16}$, which we would write as $3(\frac{5}{16})$ or $3 \cdot \frac{5}{16}$. Whenever you see a mixed number, mentally insert a plus symbol between the whole number and the fraction.

The use of mixed numbers is far more common in everyday life than in mathematics. In fact, you will rarely see mixed numbers at all in algebra. If you tell a clerk in a fabric store that you want $\frac{5}{2}$ yards of material, you might be surprised at what you get. If, instead, you say that you want $2\frac{1}{2}$ yards of material, your clerk is more likely to know what you mean.

Although arithmetic can be performed with mixed numbers just as they are, the procedures can be tedious and messy. Generally speaking, writing mixed numbers as improper fractions is a more convenient way to perform operations. However, when we discuss addition and subtraction of mixed numbers, we will illustrate both methods.

Because a mixed number represents a sum, we can actually perform the addition to change the mixed number to an improper fraction.

Copyright © Houghton Mifflin Company. All rights reserved.

Example 1

Write $5\frac{2}{3}$ as an improper fraction.

SOLUTION

$$5\frac{2}{3} = 5 + \frac{2}{3}$$ Write the mixed number as a sum.

$$= \frac{5}{1} + \frac{2}{3}$$ Write 5 as $\frac{5}{1}$.

$$= \frac{5 \cdot 3}{1 \cdot 3} + \frac{2}{3}$$ The LCD is 3.

$$= \frac{15}{3} + \frac{2}{3}$$

$$= \frac{17}{3}$$

Your Turn 1

Write $4\frac{3}{8}$ as an improper fraction.

Answer: $\frac{35}{8}$

NOTE You will rarely see mixed numbers in algebra. For example, we do *not* write the expression $3 + \frac{1}{x}$ as $3\frac{1}{x}$.

The following pattern gives us a shortcut for converting a mixed number into an improper fraction.

②15 + 2 = 17

①3 · 5 = 15

$$5\frac{2}{3} = \frac{17}{3}$$

① Multiply the denominator of the fraction times the whole number.
② Add the result to the numerator.

This shortcut gives us the *numerator* of the improper fraction. The denominator is the same as the denominator of the fraction.

Example 2

Use the shortcut to write $6\frac{1}{2}$ as an improper fraction.

SOLUTION

$6\frac{1}{2}$ Multiply the 2 in the denominator by the whole number 6.

2 · 6 = 12

12 + 1 = 13

$6\frac{1}{2}$ Now add 12 to the numerator.

The result, 13, is the numerator of the improper fraction.

$$6\frac{1}{2} = \frac{13}{2}$$

Your Turn 2

Use the shortcut to write $2\frac{5}{8}$ as an improper fraction.

Answer: $\frac{21}{8}$

Because a fraction can be interpreted as division, one way to write an improper fraction as a mixed number is to perform the division.

Copyright © Houghton Mifflin Company. All rights reserved.

Example 3

Write $\frac{17}{5}$ as a mixed number.

SOLUTION

$$5\overline{)17}\quad \text{quotient } 3,\quad 15,\quad \text{remainder } 2$$

$\frac{17}{5}$ means $17 \div 5$.

In this division, 3 is the whole number of times that 5 can be divided into 17. However, the remainder, 2, still must be divided by 5. Therefore, we can write the result as $3 + \frac{2}{5}$ or the mixed number $3\frac{2}{5}$.

Your Turn 3

Write $\frac{20}{9}$ as a mixed number.

Answer: $2\frac{2}{9}$

B Multiplication and Division

The easiest way to multiply and divide with mixed numbers is to begin by changing the mixed numbers to improper fractions. After you have performed the operation, you can change the result back to a mixed number.

Example 4

Multiply $4\frac{1}{2} \cdot 6\frac{2}{3}$.

SOLUTION

$$4\frac{1}{2} \cdot 6\frac{2}{3} = \frac{9}{2} \cdot \frac{20}{3}$$ Write the mixed numbers as improper fractions.

$$= \frac{3 \cdot \cancel{3}}{\cancel{2}} \cdot \frac{\cancel{2} \cdot 2 \cdot 5}{\cancel{3}}$$ Divide out common factors.

$$= \frac{3}{1} \cdot \frac{2 \cdot 5}{1}$$

$$= \frac{30}{1} = 30$$

Your Turn 4

Multiply $2\frac{2}{5} \cdot 1\frac{7}{8}$.

Answer: $4\frac{1}{2}$

Example 5

Divide $2\frac{1}{5} \div \frac{3}{10}$.

SOLUTION

$$2\frac{1}{5} \div \frac{3}{10} = \frac{11}{5} \div \frac{3}{10}$$ Write the mixed number as an improper fraction.

$$= \frac{11}{5} \cdot \frac{10}{3}$$ Multiply by the reciprocal.

$$= \frac{11}{\cancel{5}} \cdot \frac{2 \cdot \cancel{5}}{3}$$ Divide out common factors.

$$= \frac{22}{3} = 7\frac{1}{3}$$

Your Turn 5

Divide $1\frac{2}{7} \div \frac{3}{14}$.

Answer: 6

Copyright © Houghton Mifflin Company. All rights reserved.

KEYS TO THE CALCULATOR

The following keystrokes are typical for finding the product $2\frac{1}{3} \cdot 1\frac{3}{7}$.

Scientific calculator:

2 [a b/c] 1 [a b/c] 3 [×] 1 [a b/c] 3 [a b/c] 7 [=] Display: [3_1⌋3]

In the display, the underline separates the whole number 3 from the fraction. The result can also be displayed as an improper fraction.

[2nd] [d/c] Display: [10⌋3]

Graphing calculator:

[(] 2 [+] 1 [÷] 3 [)] [(] 1 [+] 3 [÷] 7 [)] [MATH] [Frac] [ENTER]

```
(2+1/3)(1+3/7)▶
Frac
              10/3
```

Note that a mixed number usually can be entered directly in a scientific calculator, whereas in a graphing calculator a mixed number is entered as a sum enclosed in parentheses.

To divide, substitute the [÷] key for the [×] key.

Exercises

(a) $3\frac{3}{5} \cdot 2\frac{1}{3}$ **(b)** $1\frac{5}{6} \div 4\frac{1}{8}$ **(c)** $5 \div 3\frac{1}{3}$ **(d)** $2\frac{3}{10} \cdot \frac{2}{3}$

C Addition and Subtraction

One approach to adding and subtracting mixed numbers is to begin by changing the mixed numbers to improper fractions. Then we can perform the operations in the usual way.

Example 6

Perform the indicated operation.

(a) $2\frac{1}{3} + 5\frac{1}{2}$ **(b)** $3\frac{1}{8} - 1\frac{1}{5}$

SOLUTION

(a) $2\frac{1}{3} + 5\frac{1}{2} = \frac{7}{3} + \frac{11}{2}$ Write the mixed numbers as improper fractions.

$= \frac{7 \cdot 2}{3 \cdot 2} + \frac{11 \cdot 3}{2 \cdot 3}$ The LCD is 6.

$= \frac{14}{6} + \frac{33}{6}$

Your Turn 6

Perform the indicated operation.

(a) $1\frac{3}{8} + 4\frac{1}{4}$

(b) $5\frac{1}{2} - 1\frac{3}{4}$

Copyright © Houghton Mifflin Company. All rights reserved.

$= \frac{47}{6}$ Add the numerators. Keep the denominator.

$= 7\frac{5}{6}$ Write the result as a mixed number.

(b) $3\frac{1}{8} - 1\frac{1}{5} = \frac{25}{8} - \frac{6}{5}$ Write the mixed numbers as improper fractions.

$= \frac{25 \cdot 5}{8 \cdot 5} - \frac{6 \cdot 8}{5 \cdot 8}$ The LCD is 40.

$= \frac{125}{40} - \frac{48}{40}$

$= \frac{77}{40}$ Subtract the numerators. Keep the denominator.

$= 1\frac{37}{40}$ Write the result as a mixed number.

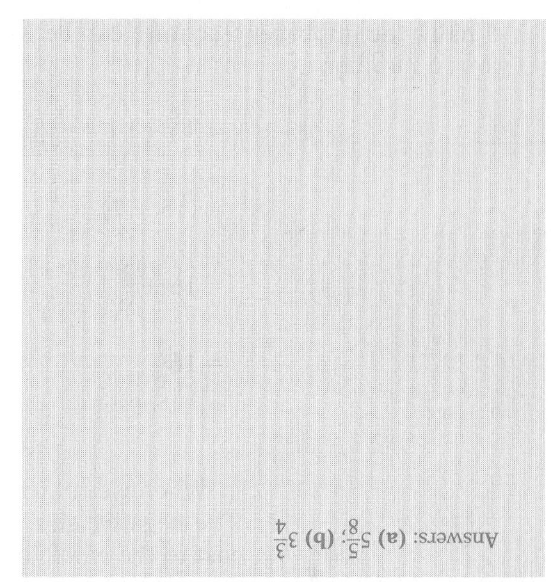

Another approach to adding and subtracting mixed numbers is to arrange the numbers in columns.

$$2\frac{1}{3} = 2 + \frac{1}{3} = 2 + \frac{1 \cdot 2}{3 \cdot 2} = 2 + \frac{2}{6} \quad \text{The LCD is 6.}$$

$$5\frac{1}{2} = 5 + \frac{1}{2} = 5 + \frac{1 \cdot 3}{2 \cdot 3} = 5 + \frac{3}{6}$$

$$7 + \frac{5}{6} = 7\frac{5}{6}$$

Therefore, $2\frac{1}{3} + 5\frac{1}{2} = 7\frac{5}{6}$, the same result that we obtained in Example 6(a).

When we use this column method, sometimes the result includes an improper fraction, and we might need to simplify the fraction. Example 7 illustrates these possibilities.

Example 7

Use the column method to add $6\frac{3}{4} + 9\frac{5}{12}$.

SOLUTION

The LCD of the fractions is 12.

$$6\frac{3}{4} = 6 + \frac{3}{4} = 6 + \frac{3 \cdot 3}{4 \cdot 3} = 6 + \frac{9}{12}$$

$$9\frac{5}{12} = 9 + \frac{5}{12} = 9 + \frac{5}{12} = 9 + \frac{5}{12}$$

$$15 + \frac{14}{12}$$

The fraction $\frac{14}{12} = \frac{2 \cdot 7}{2 \cdot 6} = \frac{7}{6}$. Now the result can be written as $15 + \frac{7}{6}$.

Your Turn 7

Use the column method to add $2\frac{5}{18} + 7\frac{8}{9}$.

continued

Copyright © Houghton Mifflin Company. All rights reserved.

Finally, the improper fraction $\frac{7}{6}$ can be written as a mixed number.

$$15 + \frac{7}{6} = 15 + \left(1 + \frac{1}{6}\right)$$
$$= (15 + 1) + \frac{1}{6}$$
$$= 16 + \frac{1}{6}$$
$$= 16\frac{1}{6}$$

When we subtract mixed numbers, we need to be aware of an important fact. The negative sign in front of a mixed number applies to the entire number, not just to the whole number.

$$-2\frac{3}{4} = -\left(2 + \frac{3}{4}\right) = -2 - \frac{3}{4} \qquad \text{Note that } -2\frac{3}{4} \neq -2 + \frac{3}{4}.$$

The column method can be used to subtract mixed numbers.

Example 8

Subtract $5\frac{7}{10} - 2\frac{3}{10}$.

SOLUTION

$$5\frac{7}{10} = 5 + \frac{7}{10}$$
$$-2\frac{3}{10} = -2 - \frac{3}{10} \qquad \text{Note that } -2\frac{3}{10} = -2 - \frac{3}{10}.$$
$$3 + \frac{4}{10} \qquad \text{Subtract the whole numbers and the fractions.}$$

After simplifying the fraction $\frac{4}{10}$ to $\frac{2}{5}$, we write the result: $3\frac{2}{5}$.

Your Turn 8

Subtract $4\frac{5}{8} - 1\frac{3}{8}$.

Answer: $3\frac{1}{4}$

What do we do when the fraction that we are subtracting is greater than the fraction above it? Recall that we sometimes had a similar situation when we subtracted whole numbers. Just as we did in those instances, we solve the problem by borrowing.

Example 9

Subtract $9\frac{1}{6} - 5\frac{5}{6}$.

SOLUTION

Here is the problem in column form.

$$\begin{array}{r} 9\frac{1}{6} \\ -5\frac{5}{6} \\ \hline \end{array}$$

Your Turn 9

Subtract $8\frac{3}{10} - 5\frac{7}{10}$.

Copyright © Houghton Mifflin Company. All rights reserved.

Observe that $\frac{5}{6}$ is greater than the $\frac{1}{6}$ above it. To fix this problem, we borrow 1 from the 9.

$$9 + \frac{1}{6} = 8 + 1 + \frac{1}{6} \qquad 9 = 8 + 1$$

$$= 8 + \frac{6}{6} + \frac{1}{6} \qquad 1 = \frac{6}{6}$$

$$= 8 + \frac{7}{6} \qquad \text{Add } \frac{6}{6} + \frac{1}{6} = \frac{7}{6}$$

$$= 8\frac{7}{6}$$

Now we can rewrite the problem.

$$\begin{aligned} 8\frac{7}{6} &= 8 + \frac{7}{6} \\ -5\frac{5}{6} &= -5 - \frac{5}{6} \\ \hline & 3 + \frac{2}{6} \end{aligned}$$

After simplifying $\frac{2}{6}$ to $\frac{1}{3}$, we can write the result: $3\frac{1}{3}$.

Answer: $2\frac{3}{5}$

A subtraction problem may involve finding an LCD as well as borrowing.

Example 10

Subtract $6\frac{1}{8} - 2\frac{5}{6}$.

SOLUTION

First, we write the fractions with their LCD, which is 24.

$$6\frac{1}{8} = 6 + \frac{1 \cdot 3}{8 \cdot 3} = 6 + \frac{3}{24}$$

$$-2\frac{5}{6} = -2 - \frac{5 \cdot 4}{6 \cdot 4} = -2 - \frac{20}{24}$$

To subtract the fractions, we must borrow 1 from the 6.

$$6 + \frac{3}{24} = 5 + 1 + \frac{3}{24} = 5 + \frac{24}{24} + \frac{3}{24} = 5 + \frac{27}{24}$$

Now we rewrite the problem as follows:

$$\begin{aligned} & 5 + \frac{27}{24} \\ & -2 - \frac{20}{24} \\ \hline & 3 + \frac{7}{24} = 3\frac{7}{24} \end{aligned}$$

Your Turn 10

Subtract $9\frac{2}{3} - 6\frac{4}{5}$.

Answer: $2\frac{13}{15}$

Copyright © Houghton Mifflin Company. All rights reserved.

KEYS TO THE CALCULATOR

A calculator can be used to add and subtract mixed numbers. The following typical keystrokes are for the sum $6\frac{3}{4} + 9\frac{5}{12}$ in Example 7:

Scientific calculator:

6 [a b/c] **3** [a b/c] **4** [+] **9** [a b/c] **5** [a b/c] **12** [=]   Display: [16_1⌋6]

Graphing calculator:

6 [+] **3** [÷] **4** [+] **9** [+] **5** [÷] **12** [MATH] [Frac] [ENTER]

```
6+3/4+9+5/12►Frac
             97/6
```

Note that the scientific calculator displays the result as a mixed number, whereas the graphing calculator displays the improper fraction.

Exercises

(a) $2\frac{3}{20} + 3\frac{1}{8}$ **(b)** $5\frac{3}{16} - 2\frac{5}{8}$ **(c)** $6 - 4\frac{5}{7}$ **(d)** $5\frac{1}{2} + 4\frac{1}{3}$

5.6 Quick Reference

A. Introduction to Mixed Numbers

1. A **mixed number** is the sum of a whole number and a proper fraction.

$$4\frac{2}{3} = 4 + \frac{2}{3}$$

2. To change a mixed number to an improper fraction, add the whole number and the fraction.

$$4\frac{3}{5} = 4 + \frac{3}{5} = \frac{4}{1} + \frac{3}{5}$$
$$= \frac{4 \cdot 5}{1 \cdot 5} + \frac{3}{5}$$
$$= \frac{20}{5} + \frac{3}{5} = \frac{23}{5}$$

3. We can use a shortcut for changing a mixed number to an improper fraction: Multiply the denominator of the fraction by the whole number and add the numerator of the fraction. The result is the numerator of the improper fraction.

② $24 + 2 = 26$

$$8\frac{2}{3} = \frac{26}{3}$$

① $3 \cdot 8 = 24$

Copyright © Houghton Mifflin Company. All rights reserved.

4. One way to write an improper fraction as a mixed number is to perform division.

$\frac{5}{2}$ means $5 \div 2$, which is $2\frac{1}{2}$.

B. Multiplication and Division

1. To multiply and divide with mixed numbers, begin by changing the mixed numbers to improper fractions.

$$2\frac{1}{3} \cdot 4\frac{1}{2} = \frac{7}{3} \cdot \frac{9}{2}$$
$$= \frac{7}{\cancel{3}} \cdot \frac{\cancel{3} \cdot 3}{2}$$
$$= \frac{21}{2} = 10\frac{1}{2}$$

2. When you divide by a fraction, multiply by the reciprocal of the fraction.

$$1\frac{1}{10} \div \frac{2}{3} = \frac{11}{10} \div \frac{2}{3}$$
$$= \frac{11}{10} \cdot \frac{3}{2}$$
$$= \frac{33}{20}$$

C. Addition and Subtraction

1. Mixed numbers can be added or subtracted by changing them to improper fractions.

$$1\frac{1}{2} + 3\frac{1}{4} = \frac{3}{2} + \frac{13}{4}$$
$$= \frac{6}{4} + \frac{13}{4}$$
$$= \frac{19}{4} = 4\frac{3}{4}$$

2. A convenient way to add or subtract mixed numbers is to arrange the numbers in columns.

$$1\frac{1}{2} = 1 + \frac{1}{2} = 1 + \frac{2}{4}$$
$$3\frac{1}{4} = 3 + \frac{1}{4} = 3 + \frac{1}{4}$$
$$4 + \frac{3}{4} = 4\frac{3}{4}$$

3. For a negative mixed number, the negative sign applies to the entire number, not just to the whole number.

$$-5\frac{2}{3} = -5 - \frac{2}{3}$$

Copyright © Houghton Mifflin Company. All rights reserved.

4. The column method can be used to subtract mixed numbers. However, if the fraction that you are subtracting is greater than the fraction above it, borrowing is needed before the subtraction is performed.

$$\begin{array}{rcrcr} 5\frac{2}{7} & = & 5 + \frac{2}{7} & = & 4 + \frac{9}{7} \\ -1\frac{5}{7} & = & -1 - \frac{5}{7} & = & -1 - \frac{5}{7} \\ \hline & & & & 3 + \frac{4}{7} = 3\frac{4}{7} \end{array}$$

Observe that we wrote $5 + \frac{2}{7}$ as

$$4 + 1 + \frac{2}{7} = 4 + \frac{7}{7} + \frac{2}{7} = 4 + \frac{9}{7}$$

Copyright © Houghton Mifflin Company. All rights reserved.

5.6 Exercises

A. Introduction to Mixed Numbers

1. The sum of a whole number and a proper fraction is called a(n) ________ number.

2. A mixed number can be written as a(n) ________ fraction.

In Exercises 3–6, write the expression as a mixed number.

3. $5 + \frac{4}{9}$
4. $2 + \frac{5}{7}$
5. $-3 - \frac{2}{3}$
6. $-1 - \frac{3}{4}$

In Exercises 7–10, write the mixed number as a sum or difference.

7. $3\frac{5}{8}$
8. $1\frac{3}{7}$
9. $-2\frac{3}{4}$
10. $-5\frac{1}{3}$

In Exercises 11–14, write the mixed number as an improper fraction.

11. $2\frac{3}{4}$
12. $5\frac{7}{8}$
13. $4\frac{1}{6}$
14. $10\frac{3}{7}$

In Exercises 15–18, write the improper fraction as a mixed number.

15. $\frac{15}{8}$
16. $\frac{32}{5}$
17. $\frac{17}{4}$
18. $\frac{55}{6}$

B. Multiplication and Division

19. To multiply or divide mixed numbers, we usually write the numbers as ________ fractions.

20. If the product of two mixed numbers is $\frac{20}{9}$, we usually write the result as a ________ number.

In Exercises 21–26, multiply. Express the result as a mixed number.

21. $1\frac{5}{7} \cdot 4\frac{2}{3}$
22. $1\frac{7}{8} \cdot 1\frac{5}{6}$
23. $5\frac{1}{4} \cdot 3\frac{1}{3}$
24. $4\frac{4}{5} \cdot 3\frac{5}{8}$
25. $5 \cdot 1\frac{3}{7}$
26. $1\frac{3}{10} \cdot \frac{5}{9}$

Copyright © Houghton Mifflin Company. All rights reserved.

In Exercises 27–32, divide. Express the result as a mixed number.

27. $6\frac{2}{5} \div 2\frac{3}{10}$

28. $4\frac{2}{3} \div 2\frac{2}{9}$

29. $7\frac{1}{2} \div 5\frac{5}{6}$

30. $2\frac{1}{10} \div 1\frac{7}{8}$

31. $2\frac{1}{4} \div \frac{3}{10}$

32. $7\frac{2}{3} \div 5$

C. Addition and Subtraction

33. One way to add mixed numbers is to add the ______ numbers and add the ______.

34. One approach to adding mixed numbers is to begin by writing the mixed numbers as ______ fractions.

In Exercises 35–44, perform the indicated operation.

35. $2\frac{2}{5} + 4\frac{1}{5}$

36. $3\frac{4}{7} + 6\frac{1}{7}$

37. $6\frac{5}{8} - 4\frac{1}{8}$

38. $4\frac{7}{10} - 1\frac{3}{10}$

39. $3\frac{2}{5} - 2$

40. $5 - \frac{2}{3}$

41. $2\frac{3}{4} + \frac{1}{2}$

42. $\frac{7}{8} + 4\frac{3}{5}$

43. $1\frac{5}{9} - \frac{2}{3}$

44. $5\frac{5}{12} - \frac{3}{4}$

Copyright © Houghton Mifflin Company. All rights reserved.

In Exercises 45–52, perform the indicated operation.

45. $9\frac{1}{2} + 5\frac{2}{3}$

46. $1\frac{1}{4} + 5\frac{5}{6}$

47. $8\frac{3}{4} - 4\frac{4}{5}$

48. $6\frac{2}{3} - 3\frac{4}{5}$

49. $3\frac{7}{8} + 7\frac{5}{12}$

50. $8\frac{7}{10} + 3\frac{5}{6}$

51. $2\frac{3}{4} - 1\frac{9}{10}$

52. $3\frac{5}{9} - 2\frac{7}{12}$

Applications

In Exercises 53–66, express your answers as mixed numbers.

53. *Cookie Recipe* A recipe for cookies uses $4\frac{1}{2}$ cups of flour. How many cups of flour are needed for two-thirds of the recipe?

54. *Gas Mileage* A car averages $22\frac{3}{4}$ miles per gallon of gasoline. How many miles can the car travel on 15 gallons of gasoline?

55. *Carpeting* How many square yards of carpet are needed for a room that is $7\frac{1}{3}$ yards long and $5\frac{3}{4}$ yards wide?

56. *Fuel Consumption* A family drove $313\frac{9}{10}$ miles to the beach. If they used $17\frac{1}{5}$ gallons of gasoline for the trip, how many miles per gallon did the car average?

Copyright © Houghton Mifflin Company. All rights reserved.

57. *Laundry Detergent* A giant size of laundry detergent contains $5\frac{1}{2}$ pounds of detergent. If you use $\frac{3}{8}$ pound for each load of laundry, how many loads can you wash?

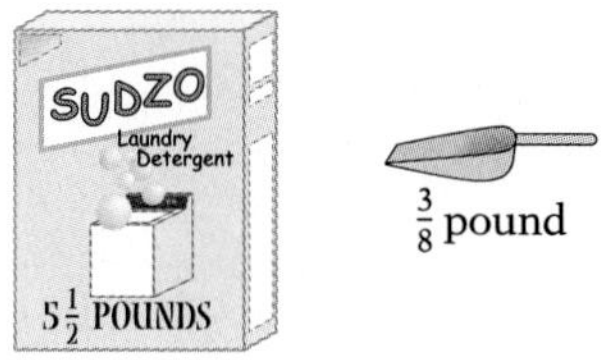

58. *Meat Loaf* A package of ground beef weighs 6 pounds. How many meat loaves, each requiring $1\frac{1}{3}$ pounds of meat, can be made?

59. *Gingerbread* A gingerbread recipe calls for $\frac{1}{4}$ cup oil, $1\frac{2}{3}$ cups milk, and $\frac{3}{8}$ cup molasses. How many cups of liquid are used?

60. *Candy Sampler* A large box of candy contains $1\frac{1}{6}$ pounds of chocolate nut pieces, $2\frac{1}{3}$ pounds of mints, and $4\frac{3}{8}$ pounds of cream filled pieces. How many pounds of candy are in the box?

61. *Base Molding* How many feet of base molding are needed for a room that is $15\frac{1}{2}$ feet long and $12\frac{3}{4}$ feet wide?

62. *Cereal Box Sizes* A large box of cereal contains $23\frac{3}{4}$ ounces and a small box contains $15\frac{1}{2}$ ounces. How many more ounces are in a large box than a small box?

Copyright © Houghton Mifflin Company. All rights reserved.

63. *Volunteer Work* As part of the membership requirement in a service club, each member must do 40 hours of volunteer work in the community. Suppose that you worked $3\frac{1}{2}$ hours one week and $5\frac{3}{4}$ hours another week. How many additional hours must you work?

64. *Waist Size* The waist size of a pair of slacks is $36\frac{1}{2}$ inches. After washing they shrink $1\frac{3}{4}$ inches. What is the new waist size?

65. *Subdivision* A subdivision has 70 houses built on $\frac{3}{4}$-acre lots and 100 houses on $\frac{2}{3}$-acre lots. What is the total number of acres in the subdivision?

66. *Electric Wire* An electrician has 5 rolls of wire with $10\frac{3}{4}$ feet per roll and 8 rolls of wire with $17\frac{2}{3}$ feet per roll. How many feet of additional wire does he need for a job that requires 220 feet of wire?

Writing and Concept Extension

67. Explain the difference between $5\frac{1}{2}$ and $5 \cdot \frac{1}{2}$.

68. Which of the following has the same meaning as $-2\frac{1}{3}$? Why?

(i) $-2 - \dfrac{1}{3}$ (ii) $-2 + \dfrac{1}{3}$

69. Why is $2\frac{4}{3}$ not a mixed number?

Copyright © Houghton Mifflin Company. All rights reserved.

70. Show how to write $2 + \frac{4}{3}$ as a mixed number.

In Exercises 71 and 72, evaluate the given expression.

71. $\left(2\frac{3}{4}\right)^2$

72. $\left(3 + 4\frac{2}{5}\right) \div \left(1 + 2\frac{1}{10}\right)$

A fraction such as $\frac{\frac{3}{4}}{\frac{5}{8}}$ is called a **complex fraction.** To simplify the fraction we write the fraction as $\frac{3}{4} \div \frac{5}{8}$. In Exercises 73 and 74, use the preceding information to simplify the complex fraction.

73. $\dfrac{\frac{3}{4}}{\frac{5}{8}}$

74. $\dfrac{\frac{6}{7}}{\frac{9}{5}}$

Exploring with Real-World Data: Collaborative Activities

Online Travel Bookings The bar graph shows the number (in millions) of online travel bookings (tickets purchased on the Internet) for selected years. (Source: Jupiter Communications.)

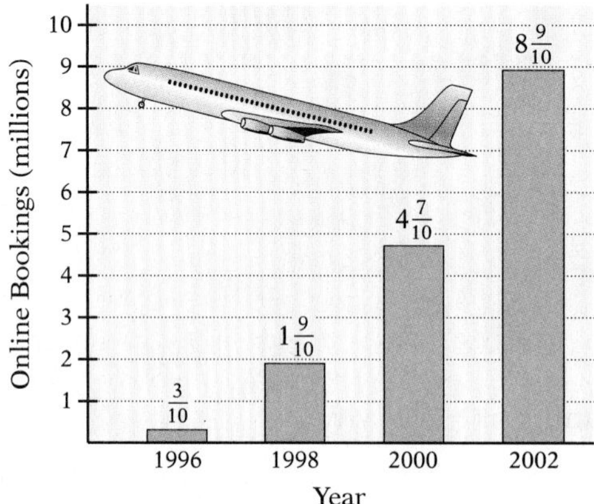

75. How many more bookings were there in 2002 than in 1998?

76. What were the total bookings for 1996 and 1998?

77. What were the total bookings for 1996, 1998, and 2000?

78. How many more bookings were there in 2002 than in the other three years combined?

Copyright © Houghton Mifflin Company. All rights reserved.

5.7 Equations with Fractions

A *Basic Operations*
B *Clearing Fractions*

SUGGESTIONS FOR SUCCESS

If you turn to the left and then turn to the right, you are facing the same direction that you were originally. If you take one step forward and then one step backward, you are back to your starting point.

In a similar way, if you add 3 to a number and then subtract 3 from the result, you are left with the original number. If you divide a number by 5 and then multiply the result by 5, you are again left with the original number.

Think of addition and subtraction as undoing each other, and the same for multiplication and division. When you are solving an equation, ask yourself what operation you want to undo. Thinking in this way will help to guide you in knowing what operation you need to perform.

A Basic Operations

Recall that the goal of equation solving is to isolate the variable. As Example 1 shows, the solution of an equation is not necessarily an integer.

Example 1

Solve $5x = 8$.

SOLUTION

$5x = 8$ We see that x is *multiplied* by 5.

$\frac{\cancel{5} \cdot x}{\cancel{5}} = \frac{8}{5}$ To undo this, we *divide* both sides by 5.

$x = \frac{8}{5}$ The solution is a fraction.

The solution is $\frac{8}{5}$.

Your Turn 1

Solve $7y = 3$.

Answer: $\frac{3}{7}$

NOTE To check a solution, we replace the variable with the solution and verify that the left and right sides of the equation have the same value. Although we do not show checking in these examples, remember that checking solutions is an important step in the equation solving process.

Example 2

Solve $\frac{x}{3} = 7$.

SOLUTION

$\frac{x}{3} = 7$ We see that x is *divided* by 3, so we

$\frac{x}{\cancel{3}} \cdot \frac{\cancel{3}}{1} = 7 \cdot 3$ multiply both sides by 3. Write 3 as $\frac{3}{1}$.

$x = 21$

The solution is 21.

Your Turn 2

Solve $\frac{y}{5} = 4$.

Answer: 20

Copyright © Houghton Mifflin Company. All rights reserved.

Example 3

Solve $-\frac{x}{2} = 7$.

SOLUTION

Method 1

$$-\frac{x}{2} = 7$$ Place the negative sign in the denominator.

$$\frac{x}{-2} = 7$$ Because x is *divided* by -2,

$$\frac{x}{\cancel{-2}} \cdot \frac{\cancel{-2}}{1} = 7(-2)$$ we *multiply* both sides by -2.

$$x = -14$$

Method 2

$$-\frac{x}{2} = 7$$

$$-\frac{x}{2} \cdot (-1) = 7(-1)$$ Multiply both sides by -1.

$$\frac{x}{2} = -7$$

$$\frac{x}{\cancel{2}} \cdot \frac{\cancel{2}}{1} = -7(2)$$ Multiply both sides by 2.

$$x = -14$$

The solution is -14.

Your Turn 3

Solve $-6 = -\frac{x}{3}$.

Answer: 18

Consider the equation $\frac{1}{3}x = 5$. Because the variable is *multiplied* by $\frac{1}{3}$, we would normally *divide* both sides by $\frac{1}{3}$ to isolate the x. However, dividing by $\frac{1}{3}$ is the same as multiplying by the reciprocal of $\frac{1}{3}$, which is 3. Therefore, we can save a step with the following rule of thumb:

> If the variable is multiplied by a fraction, multiply both sides of the equation by the reciprocal of the fraction.

Example 4

Solve $\frac{1}{3}x = 5$.

SOLUTION

$$\frac{1}{3}x = 5$$

$$3 \cdot \left(\frac{1}{3}x\right) = 3(5)$$ Multiply both sides by 3, which is the reciprocal of $\frac{1}{3}$.

$$\left(3 \cdot \frac{1}{3}\right)x = 15$$ Associative Property of Multiplication

$$1x = 15$$ $3 \cdot \frac{1}{3} = 1$

$$x = 15$$

The solution is 15.

Your Turn 4

Solve $\frac{1}{5}x = 2$.

Answer: 10

Copyright © Houghton Mifflin Company. All rights reserved.

Example 5

Solve $\frac{2}{5}y = \frac{3}{10}$.

SOLUTION

We want to eliminate the $\frac{2}{5}$, so we multiply both sides by the reciprocal, $\frac{5}{2}$.

$$\frac{2}{5}y = \frac{3}{10}$$

$$\frac{5}{2}\left(\frac{2}{5}y\right) = \frac{5}{2}\left(\frac{3}{10}\right)$$

$$\left(\frac{5}{2}\cdot\frac{2}{5}\right)y = \frac{\cancel{5}}{2}\cdot\frac{3}{2\cdot\cancel{5}} \qquad \text{Associative Property of Multiplication}$$

$$1y = \frac{1}{2}\cdot\frac{3}{2} \qquad \text{Divide out common factors.}$$

$$y = \frac{3}{4}$$

The solution is $\frac{3}{4}$.

Your Turn 5

$\frac{7}{3}x = \frac{5}{6}$.

Answer: $\frac{5}{14}$

In Example 5, we could have begun by multiplying both sides by 10 to obtain $4y = 3$. Then we would have divided both sides by 4 to obtain the solution, $\frac{3}{4}$. By multiplying by the reciprocal, we combine those two steps into one.

Remember that we use the Addition Property of Equations to isolate the variable term. Then we use the Multiplication Property of Equations to isolate the variable itself.

Example 6

Solve $\frac{2}{3}x + 1 = 5$.

SOLUTION

$$\frac{2}{3}x + 1 = 5$$

$$\frac{2}{3}x + 1 - 1 = 5 - 1 \qquad \text{Subtract 1 from both sides to isolate the variable term.}$$

$$\frac{2}{3}x = 4$$

$$\frac{3}{2}\left(\frac{2}{3}x\right) = \frac{3}{2}\cdot\frac{4}{1} \qquad \text{Multiply both sides by } \frac{3}{2} \text{ to isolate the variable.}$$

$$\left(\frac{3}{2}\cdot\frac{2}{3}\right)x = \frac{3}{\cancel{2}}\cdot\frac{\cancel{2}\cdot 2}{1}$$

$$1x = 6$$

$$x = 6$$

The solution is 6.

Your Turn 6

Solve $\frac{3}{5}x - 2 = 7$.

Answer: 15

Copyright © Houghton Mifflin Company. All rights reserved.

B Clearing Fractions

Example 7 shows one method for solving an equation that contains more than one fraction.

Example 7

Solve $x + \frac{1}{3} = \frac{5}{6}$.

SOLUTION

$$x + \frac{1}{3} = \frac{5}{6}$$

$$x + \frac{1}{3} - \frac{1}{3} = \frac{5}{6} - \frac{1}{3}$$ To isolate x, subtract $\frac{1}{3}$ from both sides.

$$x + 0 = \frac{5}{6} - \frac{2}{6}$$ The LCD is 6.

$$x = \frac{3}{6} = \frac{1}{2}$$ Simplify the fraction.

The solution is $\frac{1}{2}$.

Your Turn 7

Solve $y + \frac{1}{2} = \frac{5}{4}$.

Answer: $\frac{3}{4}$

An alternative to the method shown in Example 7 is to begin by eliminating the fractions from the equation. We call this step *clearing* the fractions. To clear the fractions, we multiply both sides of the equation by the LCD of all the fractions. In Example 8, we repeat Example 7 with this new approach.

Example 8

Solve $x + \frac{1}{3} = \frac{5}{6}$ by clearing fractions.

SOLUTION

$$x + \frac{1}{3} = \frac{5}{6}$$ The LCD of the two fractions is 6.

$$6\left(x + \frac{1}{3}\right) = 6\left(\frac{5}{6}\right)$$ Multiply both sides by 6 to clear the fractions.

$$6x + \frac{6}{1} \cdot \frac{1}{3} = \frac{\cancel{6}}{1} \cdot \frac{5}{\cancel{6}}$$ Distributive Property

$$6x + 2 = 5$$

$$6x + 2 - 2 = 5 - 2$$ Subtract 2 from both sides to isolate the variable term.

$$6x = 3$$

$$\frac{\cancel{6} \cdot x}{\cancel{6}} = \frac{3}{6}$$ Divide both sides by 6 to isolate the variable.

$$x = \frac{1}{2}$$

The solution is $\frac{1}{2}$.

Your Turn 8

Use the method of clearing fractions to solve $y + \frac{1}{2} = \frac{5}{4}$.

Answer: $\frac{3}{4}$

Copyright © Houghton Mifflin Company. All rights reserved.

As you can see, clearing fractions adds a few more steps to the solving process. However, when you become really good at it, you will find that you can clear fractions mentally and shorten the procedure. For now, writing all the steps is the safest way to go.

Example 9

Solve $\frac{2}{5}x + 3 = \frac{1}{10}$.

SOLUTION

$$\frac{2}{5}x + 3 = \frac{1}{10}$$ The LCD of the two fractions is 10.

$$10\left(\frac{2}{5}x + 3\right) = 10\left(\frac{1}{10}\right)$$ Multiply both sides by 10 to clear the fractions.

$$10\left(\frac{2}{5}x\right) + 10(3) = 10\left(\frac{1}{10}\right)$$ Distributive Property

$$\left(\frac{2 \cdot \cancel{5}}{1} \cdot \frac{2}{\cancel{5}}\right)x + 30 = \frac{\cancel{10}}{1} \cdot \frac{1}{\cancel{10}}$$ Write 10 as $\frac{2 \cdot 5}{1}$.

$$4x + 30 = 1$$

$$4x + 30 - 30 = 1 - 30$$ Subtract 30 from both sides to isolate the variable term.

$$4x = -29$$

$$\frac{\cancel{4} \cdot x}{\cancel{4}} = \frac{-29}{4}$$ Divide both sides by 4 to isolate the variable.

$$x = -\frac{29}{4}$$

The solution is $-\frac{29}{4}$.

Your Turn 9

Solve $2 + \frac{2}{3}x = \frac{5}{6}$.

Answer: $-\frac{7}{4}$

NOTE In Example 9, multiplying both sides by 10 involved the Distributive Property on the left side. This means that *every term* is multiplied by 10, not just the fractions. Be sure to see that the term 3 is multiplied by 10, even though it is not a fraction.

In Example 10, we don't show the step that involves the Distributive Property. We simply multiply every term by the LCD.

Example 10

Solve $\frac{1}{3}x + \frac{1}{6} = \frac{1}{9}x$.

SOLUTION

$$\frac{1}{3}x + \frac{1}{6} = \frac{1}{9}x$$ The LCD is 18.

$$18\left(\frac{1}{3}x\right) + 18\left(\frac{1}{6}\right) = 18\left(\frac{1}{9}x\right)$$ Multiply every term by 18.

$$\left(\frac{18}{1} \cdot \frac{1}{3}\right)x + \frac{18}{1} \cdot \frac{1}{6} = \left(\frac{18}{1} \cdot \frac{1}{9}\right)x$$

Your Turn 10

Solve $\frac{1}{2} + \frac{1}{3}x = \frac{1}{4}x$.

continued

Copyright © Houghton Mifflin Company. All rights reserved.

$$6x + 3 = 2x$$ There are variable terms on both sides.

$$6x - 2x + 3 = 2x - 2x$$ Subtract $2x$ from both sides.

$$4x + 3 = 0$$

$$4x + 3 - 3 = 0 - 3$$ Subtract 3 from both sides to isolate the variable term.

$$4x = -3$$

$$\frac{\not{4} \cdot x}{\not{4}} = \frac{-3}{4}$$ Divide both sides by 4 to isolate the variable.

$$x = -\frac{3}{4}$$

The solution is $-\frac{3}{4}$.

Answer: −1

Example 11

The Forestry Service needed to clear trees from a fire break. On the first, second, and third days of the project, $\frac{1}{4}$, $\frac{1}{3}$, and $\frac{1}{5}$ of the total number of trees were cleared. If 188 trees were removed during those three days, what was the original total number of trees?

SOLUTION

Let x = the original total number of trees.

Number of trees cleared on day 1: $\frac{1}{4}x$

Number of trees cleared on day 2: $\frac{1}{3}x$

Number of trees cleared on day 3: $\frac{1}{5}x$

To write an equation we add the number of trees cleared during the 3 days.

$$\frac{1}{4}x + \frac{1}{3}x + \frac{1}{5}x = 188$$

$$60 \cdot \frac{1}{4}x + 60 \cdot \frac{1}{3}x + 60 \cdot \frac{1}{5}x = 60 \cdot 188$$ The LCD is 60.

$$\frac{60}{1} \cdot \frac{1}{4}x + \frac{60}{1} \cdot \frac{1}{3}x + \frac{60}{1} \cdot \frac{1}{5}x = 11{,}280$$

$$15x + 20x + 12x = 11{,}280$$

$$47x = 11{,}280$$

$$x = 240$$ Divide both sides by 47.

Originally, there were 240 trees.

Your Turn 11

A student wrote $\frac{1}{2}$ of an assigned term paper in one week, and during the second week, she wrote another $\frac{3}{7}$ of the paper. If she wrote 26 pages during these two weeks, how long was the completed paper?

Answer: 28 pages

The decision about whether to clear fractions is yours. It is usually the best approach, and it should be your first step. You will find a summary of the equation solving procedure in the Quick Reference.

Copyright © Houghton Mifflin Company. All rights reserved.

5.7 Quick Reference

A. Basic Operations

1. To isolate the variable in simple equations, we use the fact that multiplication and division undo each other.

   (a) If the variable is multiplied by a number, we divide both sides by that number.

$$7x = 3$$
$$\frac{7 \cdot x}{7} = \frac{3}{7}$$
$$x = \frac{3}{7}$$

   (b) If the variable is divided by a number, we multiply both sides by that number.

$$\frac{x}{5} = -3$$
$$\frac{5}{1} \cdot \frac{x}{5} = 5(-3)$$
$$x = -15$$

2. If the variable is multiplied by a fraction, multiply both sides of the equation by the reciprocal of the fraction.

$$\frac{3}{5}x = \frac{9}{10}$$
$$\frac{5}{3}\left(\frac{3}{5}x\right) = \frac{5}{3}\left(\frac{9}{10}\right)$$
$$\left(\frac{5}{3} \cdot \frac{3}{5}\right)x = \frac{5}{3} \cdot \frac{3 \cdot 3}{2 \cdot 5}$$
$$1x = \frac{3}{2}$$
$$x = \frac{3}{2}$$

B. Clearing Fractions

1. We *clear* fractions from an equation by multiplying both sides by the LCD of all the fractions. In effect, we multiply *every term* by the LCD.

$$\frac{2}{3}x + 2 = \frac{1}{5}$$
$$15\left(\frac{2}{3}x\right) + 15(2) = 15\left(\frac{1}{5}\right)$$
$$\left(15 \cdot \frac{2}{3}\right)x + 30 = 3$$
$$10x + 30 = 3$$
$$10x + 30 - 30 = 3 - 30$$
$$10x = -27$$
$$x = -\frac{27}{10}$$

Copyright © Houghton Mifflin Company. All rights reserved.

2. The following is a summary of the equation solving procedure:

   (a) Clear any fractions from the equation by multiplying both sides by the LCD of all the fractions.

   (b) If necessary, simplify the two sides of the equation.

      (i) Remove grouping symbols.

      (ii) Combine like terms.

   (c) Use the Addition Property of Equations to isolate the variable term.

   (d) Use the Multiplication Property of Equations to isolate the variable.

   (e) Check the solution.

Copyright © Houghton Mifflin Company. All rights reserved.

5.7 Exercises

A. Basic Operations

1. To solve $\frac{2}{3}n = 5$, we can ________ by $\frac{2}{3}$ or we can multiply by the ________ of $\frac{2}{3}$.

2. A statement that two expressions have the same value is called a(n) ________.

In Exercises 3–6, solve the given equation.

3. $5x = 12$
4. $3y = -7$
5. $4n = -2$
6. $-6x = 8$

In Exercises 7–24, solve the given equation.

7. $\frac{x}{5} = -7$
8. $\frac{x}{8} = 2$
9. $-\frac{a}{3} = 4$
10. $\frac{n}{-7} = 1$
11. $\frac{x}{4} = \frac{2}{5}$
12. $\frac{y}{6} = -\frac{4}{3}$
13. $-\frac{n}{6} = \frac{4}{3}$
14. $\frac{-c}{8} = -\frac{5}{6}$
15. $\frac{3}{4}x = -6$
16. $\frac{1}{8}y = 0$
17. $-\frac{1}{4}x = \frac{1}{3}$
18. $\frac{6}{7}y = \frac{3}{2}$
19. $-\frac{2}{5}x = 10$
20. $0 = -\frac{4}{5}x$
21. $\frac{5}{6}x = -10$
22. $\frac{3}{2}x = -1$
23. $-\frac{5}{2}y = -\frac{10}{9}$
24. $\frac{4}{3}x = -3$

In Exercises 25–28, solve the given equation.

25. $\frac{3}{4}x - 5 = 1$
26. $\frac{1}{2}x + 9 = -1$
27. $7 - \frac{4}{3}y = -1$
28. $4 - \frac{3}{8}z = 10$

B. Clearing Fractions

29. When we eliminate fractions in an equation, we say that we ________ the fractions.

30. To clear the fractions in an equation, multiply both sides by the ________.

Copyright © Houghton Mifflin Company. All rights reserved.

In Exercises 31–36, solve the given equation.

31. $n + \frac{5}{8} = \frac{1}{8}$

32. $-\frac{5}{9} + x = \frac{7}{9}$

33. $-\frac{1}{6} = x - \frac{2}{3}$

34. $a - \frac{2}{5} = -\frac{3}{10}$

35. $a - \frac{3}{8} = \frac{-7}{12}$

36. $\frac{4}{9} = c + \frac{5}{6}$

In Exercises 37–50, solve the given equation.

37. $\frac{x}{2} = \frac{5}{2} - \frac{x}{3}$

38. $\frac{1}{2} - \frac{y}{6} = \frac{1}{3}$

39. $\frac{2}{3}x - 4 = \frac{1}{3}$

40. $\frac{1}{4}x + 1 = \frac{1}{4}$

41. $2 - \frac{3}{10}n = \frac{2}{5}$

42. $\frac{3}{4}b - 2 = \frac{1}{8}$

43. $\frac{1}{2} = -\frac{3}{8} + \frac{a}{8}$

44. $2y = \frac{5}{9} - \frac{y}{6}$

45. $1 + \frac{1}{2}y = \frac{1}{4}$

46. $\frac{1}{2} + \frac{7a}{4} = 4$

47. $\frac{1}{9}x + 1 = \frac{1}{3}x$

48. $\frac{7}{4}c - 2 = \frac{5}{4}c$

49. $\frac{3}{4}x - \frac{3}{8} = \frac{1}{2}x$

50. $2x + \frac{1}{2} = \frac{3}{4}x$

Copyright © Houghton Mifflin Company. All rights reserved.

Applications

In Exercises 51–56, write and solve an equation to determine the unknown number. (Let n represent the unknown number.)

51. Two-thirds of a number is 24.

52. Five-fourths of a number is 10.

53. When half of a number is decreased by 7, the result is -4.

54. Eight less than the quotient of a number and 3 is 1.

55. The product of 3 and a number is half the number.

56. One more than one-third of a number is half the number.

In Exercises 57–60, write an equation that describes the given information. Then use the equation to solve the problem.

57. *Driving Distance* After driving 225 miles, a motorist had completed three-fourths of the trip from Atlanta to Mobile. What is the distance from Atlanta to Mobile?

Copyright © Houghton Mifflin Company. All rights reserved.

58. *Homework Assignment* After solving 28 problems, a student had completed four-fifths of a homework assignment. How many problems were included in the homework assignment?

59. *Soccer Tickets* A community organization purchased a block of tickets to a soccer game. After the organization sold 18 tickets, the number of tickets remaining was 6 more than half the number purchased. How many tickets did the organization purchase?

60. *Recreation Area* After 7 acres were added to a recreation area, the size of the area was four-thirds of its original size. What was the original size of the area?

Writing and Concept Extension

61. Describe two ways to solve $\frac{1}{4}x = 3$.

62. Explain why the Multiplication Property of Equations allows us to divide both sides of an equation by any number except zero.

In Exercises 63–66, solve the given equation.

63. $\frac{1}{3}x - \frac{1}{2} = \frac{5}{6}x + \frac{3}{2}$

64. $\frac{9}{10}y - 1 = \frac{1}{2}y - \frac{3}{5}$

65. $-n + \frac{1}{2} = -\frac{n}{6} + \frac{4}{3}$

66. $\frac{3t}{2} - \frac{1}{2} = \frac{3}{2} + t$

In Exercises 67 and 68, solve the given equation.

67. $\frac{3}{x} = \frac{1}{4}$

68. $2 + \frac{1}{x} = \frac{7}{x}$

Copyright © Houghton Mifflin Company. All rights reserved.

5.8 Formulas

A *Geometry Formulas*
B *Business Formulas*
C *Science Formulas*

SUGGESTIONS FOR SUCCESS

From time to time we have given you a procedure, which is a list of steps to take in order to perform a certain task, such as solving an equation. You can think of a formula as a procedure, except that a formula is written in symbols rather than in words.

When you use a formula, try stating the formula's instructions in your own words. For example, if the sides of a triangle are a, b, and c, then the perimeter P is given by the formula $P = a + b + c$. You might say, "To find the perimeter of a triangle, add the lengths of the three sides."

Formulas will be easier to use if you understand that they are simply instructions that you can translate into words.

A Geometry Formulas

In Section 2.7, we described a **formula** as an equation with more than one variable. In preceding sections, we have used geometric formulas for calculating perimeters, areas, volumes, and surface areas of various geometric figures.

A *parallelogram* is a four-sided figure whose opposite sides are parallel. A *trapezoid* is a four-sided figure in which one pair of opposite sides are parallel. In the following figures, the part labeled b is called the *base*, and the part labeled h is called the *height*. Note that the parallel sides of a trapezoid are both called bases, and they are labeled b_1 (read as "b sub 1") and b_2 (read as "b sub 2").

The formulas appearing below the figures are for the area A of the figure.

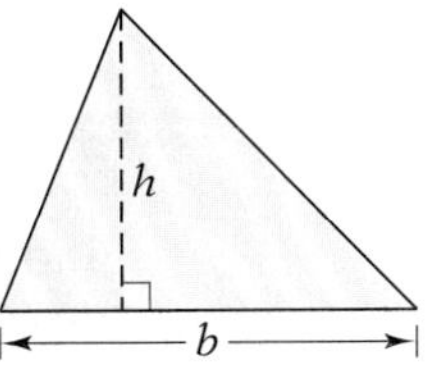

Triangle

$A = \frac{1}{2}bh$

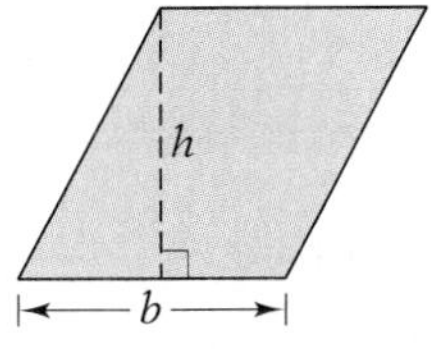

Parallelogram

$A = bh$

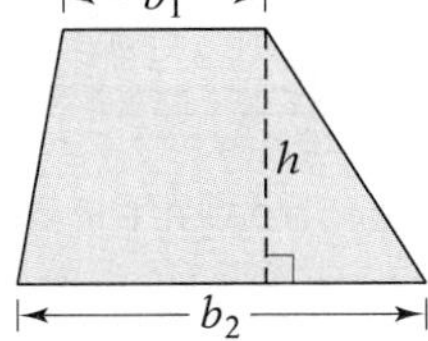

Trapezoid

$A = \frac{1}{2}h(b_1 + b_2)$

Example 1

Find the area of the given triangle.

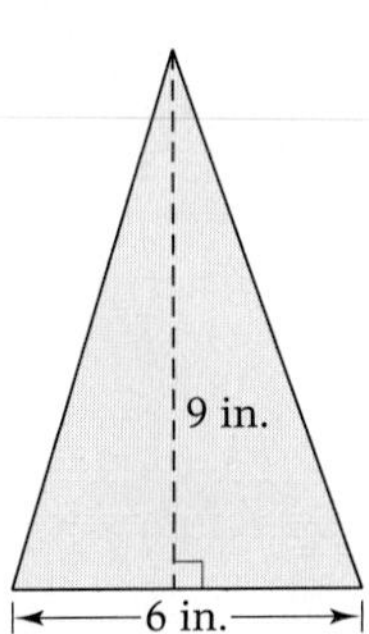

Your Turn 1

Find the area of the given triangle.

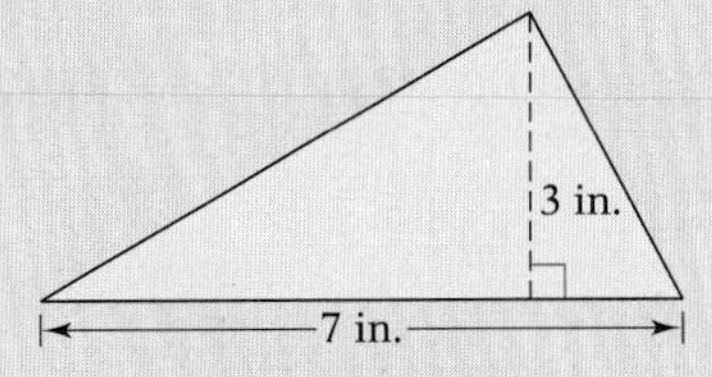

continued

Copyright © Houghton Mifflin Company. All rights reserved.

SOLUTION

$$A = \frac{1}{2}bh = \frac{1}{2} \cdot 6 \cdot 9$$
$$= \frac{1}{\cancel{2}} \cdot \frac{\cancel{2} \cdot 3}{1} \cdot \frac{9}{1}$$
$$= \frac{1}{1} \cdot \frac{3}{1} \cdot \frac{9}{1}$$
$$= 27 \text{ square inches}$$

Answer: $10\frac{1}{2}$ square inches

Example 2

Find the area of the given parallelogram.

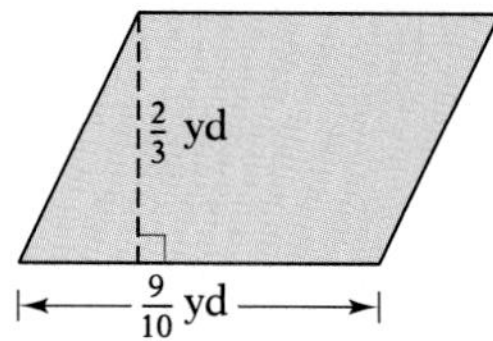

SOLUTION

$$A = bh$$
$$= \frac{9}{10} \cdot \frac{2}{3}$$
$$= \frac{\cancel{3} \cdot 3}{\cancel{2} \cdot 5} \cdot \frac{\cancel{2}}{\cancel{3}}$$
$$= \frac{3}{5} \text{ square yard}$$

Your Turn 2

Find the area of the given parallelogram.

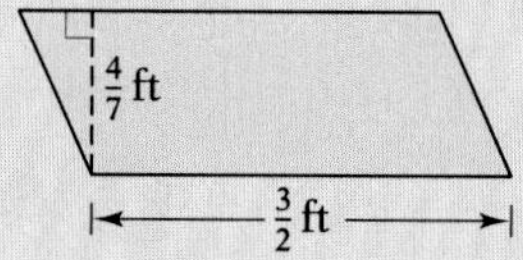

Answer: $\frac{6}{7}$ square foot

Example 3

Find the area of the given trapezoid.

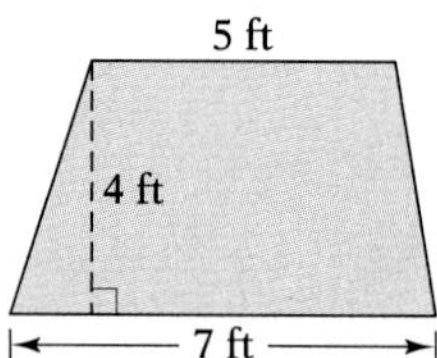

SOLUTION

$$A = \frac{1}{2}h(b_1 + b_2)$$
$$= \frac{1}{2} \cdot 4 \cdot (5 + 7) \quad \text{Start with the grouping symbols.}$$
$$= \frac{1}{2} \cdot 4 \cdot 12$$
$$= \frac{1}{2} \cdot 48$$
$$= \frac{48}{2}$$
$$= 24 \text{ square feet}$$

Your Turn 3

Find the area of the given trapezoid.

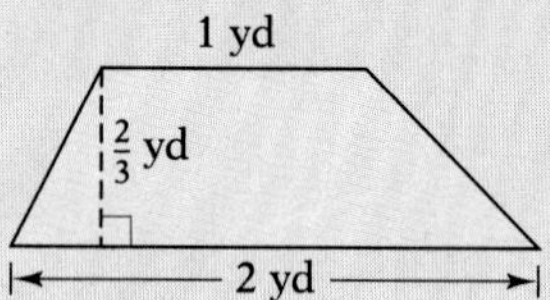

Answer: 1 square yard

Copyright © Houghton Mifflin Company. All rights reserved.

Now that we know how to solve certain kinds of equations, we can also find the value of any one of the variables of a formula, provided that we know the values of the other variables.

Example 4

If the area of the given triangle is $\frac{1}{2}$ square yard, what is the height?

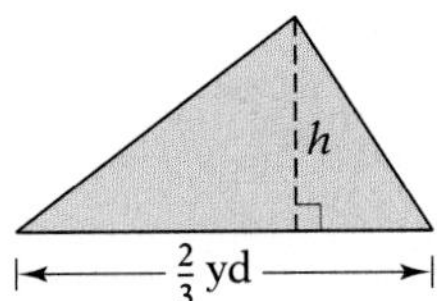

SOLUTION

$$A = \frac{1}{2}bh \quad \text{Formula for the area}$$

$$\frac{1}{2} = \frac{1}{2} \cdot \frac{2}{3} \cdot h \quad \text{Let } A = \frac{1}{2} \text{ and } b = \frac{2}{3}.$$

$$\frac{1}{2} = \frac{1}{\not{2}} \cdot \frac{\not{2}}{3} \cdot h \quad \text{Divide out common factors.}$$

$$\frac{1}{2} = \frac{1}{3}h$$

$$\frac{6}{1} \cdot \frac{1}{2} = \frac{6}{1} \cdot \frac{1}{3}h \quad \text{The LCD is 6; clear the fractions.}$$

$$3 = 2h$$

$$\frac{3}{2} = h \quad \text{Divide both sides by 2.}$$

The height is $\frac{3}{2}$ or $1\frac{1}{2}$ yards.

Your Turn 4

If the area of the given triangle is $\frac{1}{3}$ square foot, what is the base?

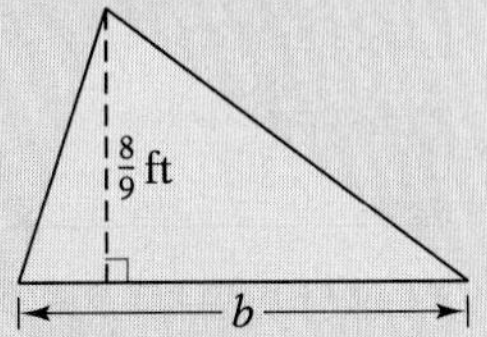

Answer: $\frac{3}{4}$ foot

B Business Formulas

As the owner of a business, your goal is to make a *profit*. Roughly speaking, profit is the difference between the *retail* (selling) *price* of an item and the *wholesale price* (the amount that it cost you to buy the item).

Letting P represent profit, R represent retail price, and W represent wholesale price, the relationship among the three quantities is given by the formula $P = R - W$. Because the formula $P = R - W$ has three variables, we will need to know (or be able to determine) the values of two of those variables in order to solve for the third variable.

As it is written, the formula $P = R - W$ is best suited for finding profit P if we know the retail price R and the wholesale price W. There are actually three ways in which the formula can be written.

For finding the profit: $P = R - W$

For finding the retail price: $R = W + P$

For finding the wholesale price: $W = R - P$

However, memorizing all three of these formulas is not useful. The best approach is to choose one and use it for all such applications, regardless of the quantity that you are trying to determine.

Copyright © Houghton Mifflin Company. All rights reserved.

Example 5

As a hardware owner, you purchased 20 electric drills for $8 each. After you sold all 20 drills, your total profit was $260. What was the retail price of these drills?

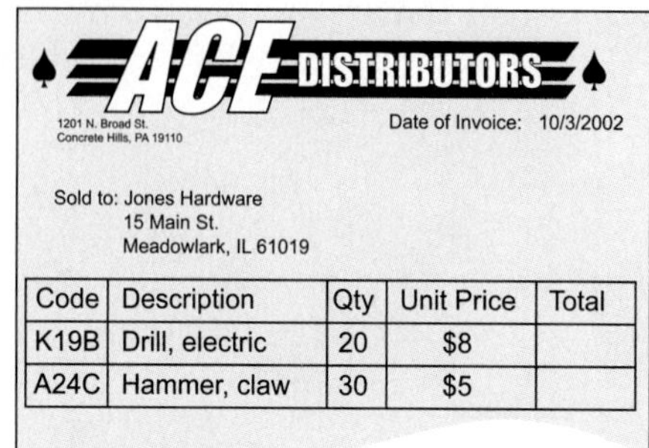

ACE DISTRIBUTORS

1201 N. Broad St.
Concrete Hills, PA 19110

Date of Invoice: 10/3/2002

Sold to: Jones Hardware
15 Main St.
Meadowlark, IL 61019

Code	Description	Qty	Unit Price	Total
K19B	Drill, electric	20	$8	
A24C	Hammer, claw	30	$5	

SOLUTION

First, we need to know the profit on each drill.

$$P = \frac{\textit{total profit}}{\textit{number of drills sold}} = \frac{260}{20} = 13$$

The profit on each drill was $13.

$P = R - W$ — The profit formula

$13 = R - 8$ — Profit was $13; wholesale price was $8.

$13 + 8 = R - 8 + 8$ — Add 8 to both sides.

$21 = R$

The retail price was $21.

Your Turn 5

Suppose that you bought an antique vase and later made a profit of $48 by selling the vase for $200. How much did you pay for the vase?

Answer: $152

When you buy something that is on sale, the price that you pay is the original price decreased by a *discount*. If S is the sale price, P is the original price, and D is the discount, then $S = P - D$.

Example 6

Last week, a blouse cost $40. This week, the blouse is on sale for $28. What is the discount?

Your Turn 6

Computers are on sale at a discount of $150. If the sale price is $1,420, what was the price before the sale?

Copyright © Houghton Mifflin Company. All rights reserved.

SOLUTION

$S = P - D$	The sale price formula
$28 = 40 - D$	The sale price S is \$28; the original price P was \$40.
$28 + D = 40 - D + D$	Add D to both sides.
$28 + D = 40$	
$28 - 28 + D = 40 - 28$	Subtract 28 from both sides.
$D = 12$	

The discount is \$12.

C Science Formulas

In the United States the Fahrenheit (F) temperature scale is most often used for weather reports. In other parts of the world and in science, the Celsius (C) temperature scale is more often used. To give you a rough idea of how the two scales compare, the freezing temperature of water is 32°F and 0°C. The temperature at which water boils is 212°F and 100°C.

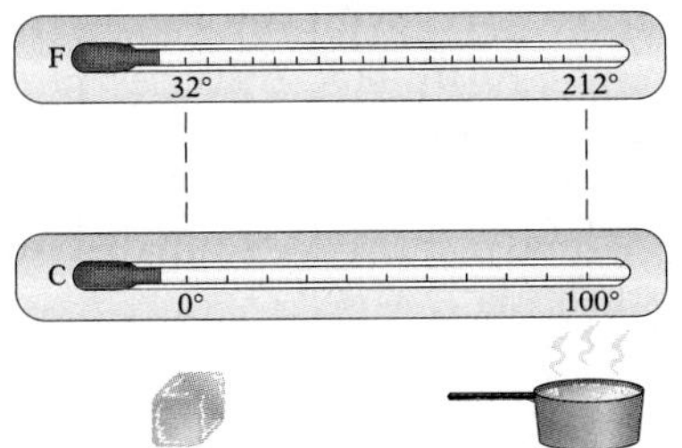

To convert a Celsius temperature to a Fahrenheit temperature, we use the formula $F = \frac{9}{5}C + 32$. We can use this formula to verify the freezing and boiling temperatures just given.

$$\text{If } C = 0, \text{ then } F = \frac{9}{5}(0) + 32 = 0 + 32 = 32.$$

$$\text{If } C = 100, \text{ then } F = \frac{9}{5}(100) + 32 = 180 + 32 = 212.$$

Example 7

Suppose that you check your thermostat and find that the room temperature is 68°F. What would the temperature be on the Celsius scale?

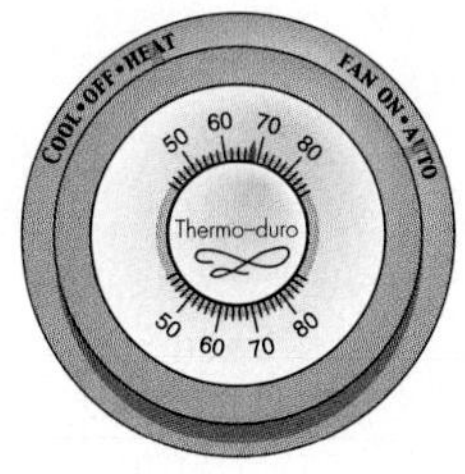

Your Turn 7

On a very hot day, the temperature was 95°F. What was the temperature on the Celsius scale?

continued

Copyright © Houghton Mifflin Company. All rights reserved.

SOLUTION

$$F = \frac{9}{5}C + 32$$ Conversion formula

$$68 = \frac{9}{5}C + 32$$ The Fahrenheit temperature is 68°F.

$$68 - 32 = \frac{9}{5}C + 32 - 32$$ Subtract 32 from both sides.

$$36 = \frac{9}{5}C$$

$$\frac{5}{9}(36) = \frac{5}{9}\left(\frac{9}{5}C\right)$$ Multiply both sides by $\frac{5}{9}$.

$$\frac{5}{\cancel{9}} \cdot \frac{4 \cdot \cancel{9}}{1} = \left(\frac{5}{9} \cdot \frac{9}{5}\right)C$$

$$20 = 1C$$

$$20 = C$$

The room temperature is 20°C.

Answer: 35°C

NOTE Although Fahrenheit temperatures can be converted to Celsius temperatures with the method shown in Example 7, an easier method is to use the formula $C = \frac{5}{9}(F - 32)$. In your later studies, you will learn how to convert a formula for one variable to a formula for another variable.

When you travel, your distance D, rate R (speed), and time T are related by the formula $D = RT$.

Example 8

Suppose that you drove 275 miles in 5 hours. What was your average speed?

SOLUTION

$$D = RT$$ The distance-rate-time formula

$$275 = R \cdot 5$$ $D = 275$ miles; $T = 5$ hours.

$$\frac{275}{5} = \frac{R \cdot \cancel{5}}{\cancel{5}}$$ Divide both sides by 5.

$$55 = R$$

Your average speed was 55 miles per hour.

Your Turn 8

How long would driving 135 miles take if your average speed were 45 miles per hour?

Answer: 3 hours

Copyright © Houghton Mifflin Company. All rights reserved.

5.8 Quick Reference

A. Geometry Formulas

1. Some additional area formulas were introduced in this section.

Triangle: $A = \frac{1}{2}bh$

Parallelogram: $A = bh$

Trapezoid: $A = \frac{1}{2}h(b_1 + b_2)$

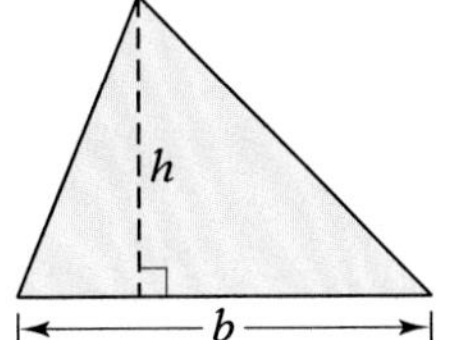

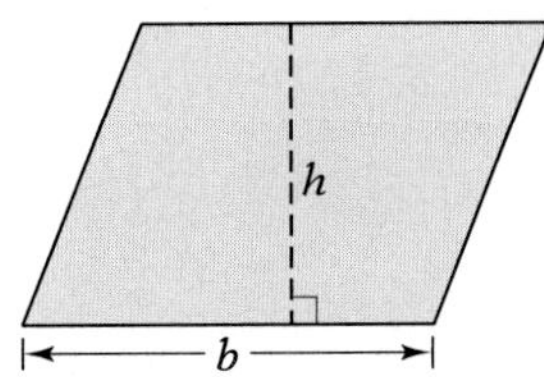

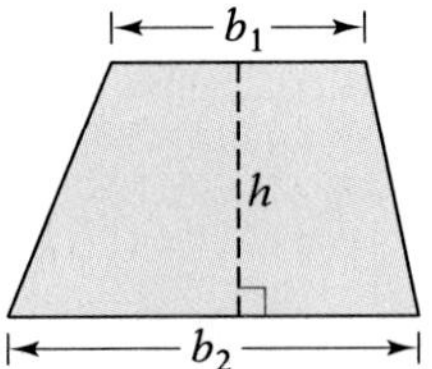

2. We can find the value of any one of the variables of a formula provided that we know the values of the other variables.

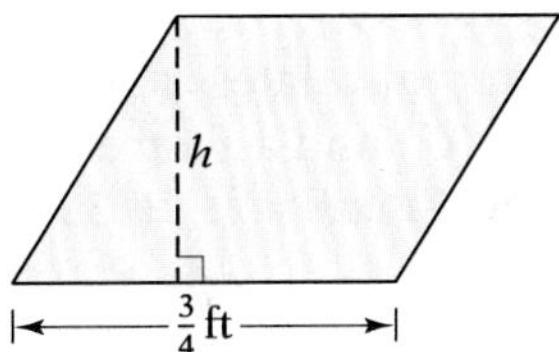

If the area of the given parallelogram is $\frac{1}{2}$ square foot, what is the height?

$$A = bh$$

$$\frac{1}{2} = \frac{3}{4} \cdot h$$

$$\frac{4}{3} \cdot \frac{1}{2} = \frac{4}{3} \cdot \frac{3}{4} \cdot h$$

$$\frac{\not{2} \cdot 2}{3} \cdot \frac{1}{\not{2}} = \frac{\not{4}}{\not{3}} \cdot \frac{\not{3}}{\not{4}} \cdot h$$

$$\frac{2}{3} = h$$

The height is $\frac{2}{3}$ foot.

B. Business Formulas

1. The *retail price* (R) of an item is its selling price. The *wholesale price* (W) is the amount that a retailer pays for an item. *Profit* (P) is given by $P = R - W$.

By selling an item for \$18, a retailer makes a profit of \$4. What was the wholesale price?

$$P = R - W$$

$$4 = 18 - W$$

$$4 + W = 18 - W + W$$

$$4 + W = 18$$

$$4 + W - 4 = 18 - 4$$

$$W = 14$$

The wholesale price was \$14.

Copyright © Houghton Mifflin Company. All rights reserved.

2. A *discount* (D) is the amount by which the *original price* (P) is reduced. The *sale price* S is given by $S = P - D$.

A printer is on sale for \$318. If the discount is \$46, what was the original price of the printer?

$$S = P - D$$
$$318 = P - 46$$
$$318 + 46 = P - 46 + 46$$
$$364 = P$$

The original price was \$364.

C. Science Formulas

1. The formula that relates a Celsius temperature (C) and a Fahrenheit temperature (F) is $F = \frac{9}{5}C + 32$.

If the temperature is 50° on the Fahrenheit scale, what is the temperature on the Celsius scale?

$$F = \frac{9}{5}C + 32$$
$$50 = \frac{9}{5}C + 32$$
$$50 - 32 = \frac{9}{5}C + 32 - 32$$
$$18 = \frac{9}{5}C$$
$$\frac{5}{9} \cdot 18 = \frac{5}{9} \cdot \frac{9}{5} \cdot C$$
$$\frac{5}{\cancel{9}} \cdot \frac{\cancel{9} \cdot 2}{1} = \frac{\cancel{5}}{\cancel{9}} \cdot \frac{\cancel{9}}{\cancel{5}} \cdot C$$
$$10 = C$$

The Celsius temperature is 10°C.

2. Distance (D), rate (R), and time (T) are related by the formula $D = RT$.

If a 126-mile trip took 3 hours, what was the average rate (speed)?

$$D = RT$$
$$126 = R \cdot 3$$
$$\frac{126}{3} = \frac{R \cdot \cancel{3}}{\cancel{3}}$$
$$42 = R$$

The average rate of speed was 42 miles per hour.

Copyright © Houghton Mifflin Company. All rights reserved.

5.8 Exercises

A. Geometry Formulas

In Exercises 1–4, determine the area of the given figure.

1.

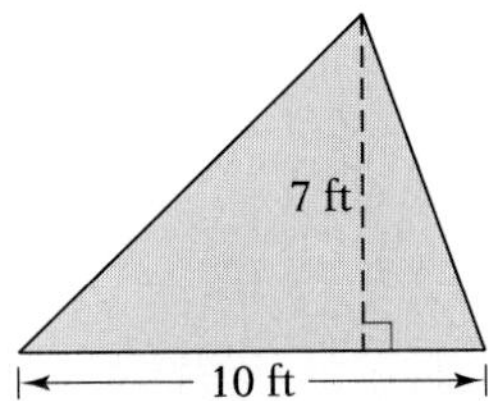

2.

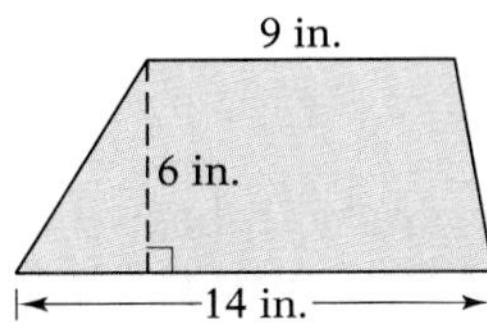

3.

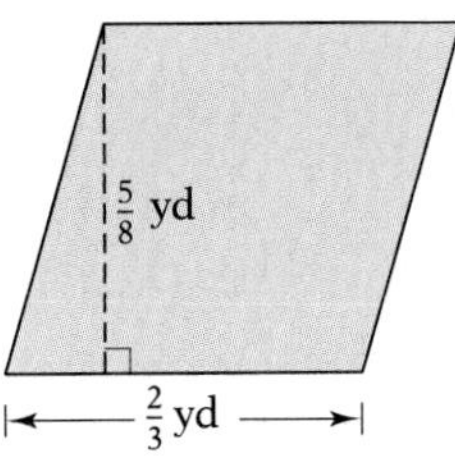

4.

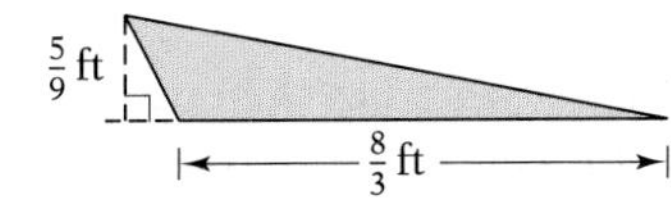

The area A of a triangle with base b and height h is $A = \frac{1}{2}bh$. In Exercises 5–10, use this formula to determine the indicated dimension.

5. $b = 3$ inches, $h = 5$ inches, $A = ?$

6. $b = 8$ feet, $h = \frac{7}{10}$ foot, $A = ?$

7. $A = 21$ square miles, $h = 7$ miles, $b = ?$

8. $A = 3\frac{3}{8}$ square yards, $b = 9$ yards, $h = ?$

9. *Deck Carpet* A homeowner needs 54 square yards of outdoor carpet to cover a deck that is triangular. What is the dimension labeled h?

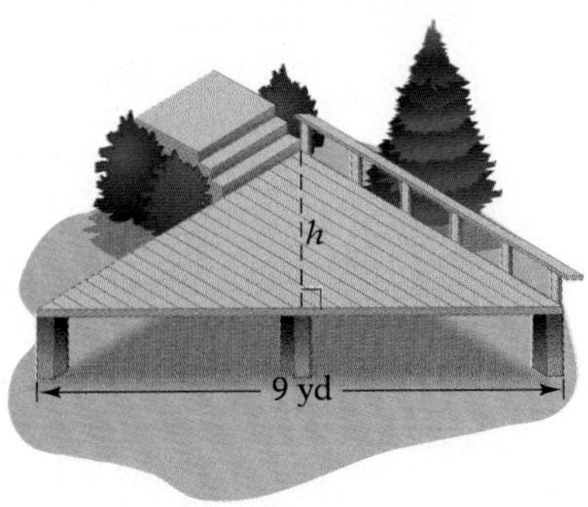

Copyright © Houghton Mifflin Company. All rights reserved.

10. *Plywood Platform* To construct the floor of a triangular platform, a builder used 225 square feet of plywood. What is the dimension labeled b?

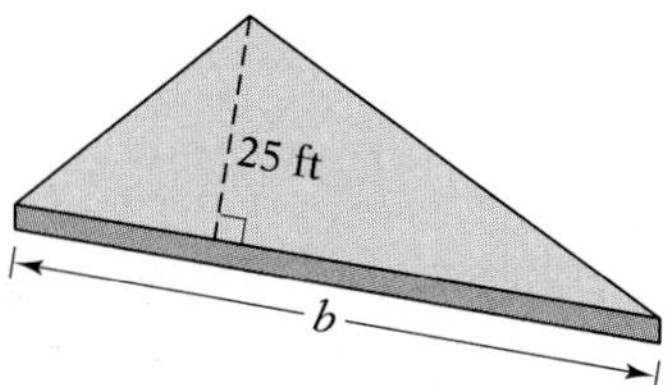

The area A of a parallelogram with base b and height h is $A = bh$. In Exercises 11–16, use this formula to determine the indicated dimension.

11. $b = 8$ inches, $h = 3\frac{1}{2}$ inches, $A = ?$

12. $b = \frac{3}{4}$ mile, $b = \frac{2}{3}$ mile, $A = ?$

13. $A = 9$ square feet, $b = 4$ feet, $h = ?$

14. $A = 3\frac{1}{8}$ square inches, $h = 1\frac{1}{4}$ inches, $b = ?$

15. *Apartment Complex* An apartment complex is on $3\frac{1}{2}$ square miles of land that is in the shape of a parallelogram. What is the dimension labeled h?

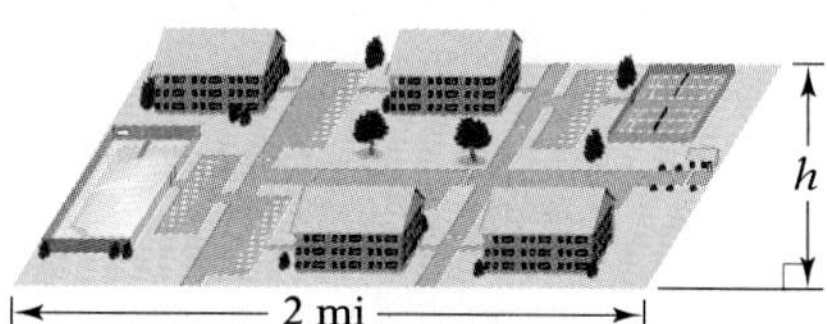

16. *Electric Component* The shape of a small electric component is a parallelogram, and it has an area of $\frac{3}{8}$ square inch. What is the dimension labeled b?

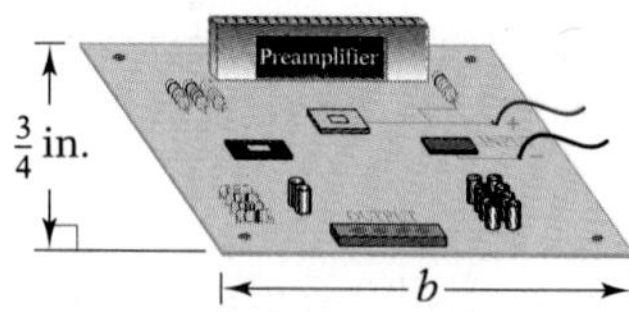

The area A of a trapezoid with height h and bases b_1 and b_2 is $A = \frac{1}{2}h(b_1 + b_2)$. In Exercises 17–22, use this formula to determine the indicated quantity.

17. $h = 3$ inches, $b_1 = 6$ inches, $b_2 = 10$ inches, $A = ?$

18. $A = 2\frac{1}{4}$ square inches, $b_1 = 4$ inches, $b_2 = 2$ inches, $h = ?$

Copyright © Houghton Mifflin Company. All rights reserved.

19. $A = 3$ square feet, $h = \frac{3}{4}$ foot, $b_1 = 5$ feet, $b_2 = ?$

20. $A = 34$ square yards, $h = 4$ yards, $b_2 = 7$ yards, $b_1 = ?$

21. *Flower Bed* A flower bed is in the shape of a trapezoid and occupies $6\frac{1}{8}$ square yards. What is the dimension labeled b_1?

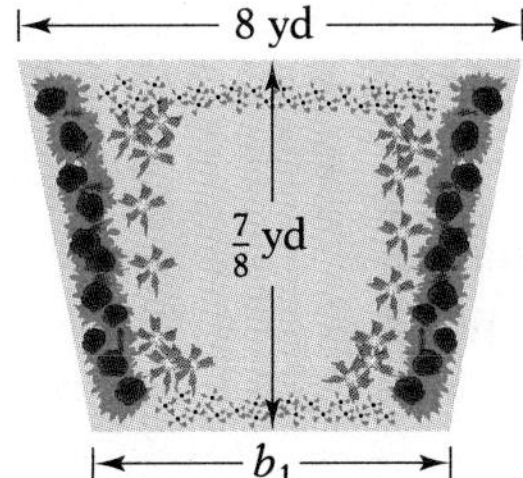

22. *Wall Tile* A construction worker used 40 square feet of tile to cover a portion of a wall that is in the shape of a trapezoid. What is the dimension labeled h?

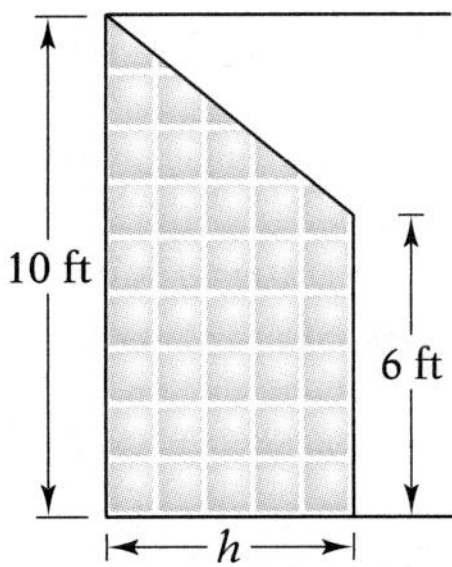

B. Business Formulas

Profit P can be found with the formula $P = R - W$, where R is the retail price and W is the wholesale price. In Exercises 23 and 24, use this formula to determine the indicated price.

23. *Coffee Mugs* A case of coffee mugs contains 36 mugs. If all mugs are sold at \$5 each, the total profit would be \$108. What is the wholesale price of a case of mugs?

24. *Tool Sets* A discount store purchased 350 tool sets at \$32 each. If the total profit from the sale of all sets was \$2,450, what was the retail price of each tool set?

Copyright © Houghton Mifflin Company. All rights reserved.

For items whose prices have been reduced, the sale price S is $S = P - D$, where P is the original price and D is the discount. In Exercises 25 and 26, use this formula to determine the indicated quantity.

25. *Sofas on Sale* A furniture store advertises that all sofas are discounted by \$70. If the sale price of a sofa is \$469, what was the original price?

26. *Ticket Discount* Students are given a \$4 discount on amusement park tickets. If a student pays \$19 for a ticket, what is the regular price?

When you borrow money, as in a loan or a credit card charge, the amount A that you must pay back is $A = P + I$, where P is the amount that you borrowed and I is the interest. In Exercises 27 and 28, use this formula to determine the indicated amount.

27. *Credit Card* Suppose that your credit card balance is \$2,450, which includes \$320 in interest. How much did you charge on the card?

28. *Loan Interest* Suppose that you borrowed \$1,230 and you paid off the loan in one year. However, the total amount that you paid was \$1,810. How much interest did you pay for the year?

C. Science Formulas

The formula $F = \frac{9}{5}C + 32$ relates temperature in degrees Celsius (C) and in degrees Fahrenheit (F). In Exercises 29–32, convert the temperature given in one scale to the temperature in the other scale.

29. The normal July high temperature in Norfolk, Virginia is 86°F.

30. The normal January low temperature in Barrow, Alaska, is −22°F.

Copyright © Houghton Mifflin Company. All rights reserved.

31. The lowest temperature recorded in Memphis, Tennessee, is $-25°C$.

32. The highest temperature recorded in Phoenix, Arizona, is 50°C.

The formula $D = RT$ relates distance D, rate R, and time T. In Exercises 33–36, use this formula to determine the indicated quantity.

33. If a trucker drove 93 miles at 62 miles per hour, how many hours did the trip take?

34. A motorist drove 42 miles at 56 miles per hour. How many hours did the trip take?

35. A cyclist rode 54 miles in $2\frac{1}{4}$ hours. What was the cyclist's average speed?

36. A hiker walked 13 miles in $4\frac{1}{3}$ hours. At what average speed did the hiker walk?

Writing and Concept Extension

37. What is a formula?

38. Which length is possible to determine? Why?

(i) The length of a rectangle with area 36

(ii) The length of a side of a square with area 36

Copyright © Houghton Mifflin Company. All rights reserved.

In Exercises 39 and 40, determine the area of the given figure.

39.

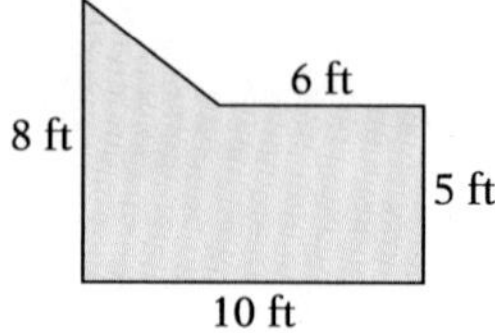

40.

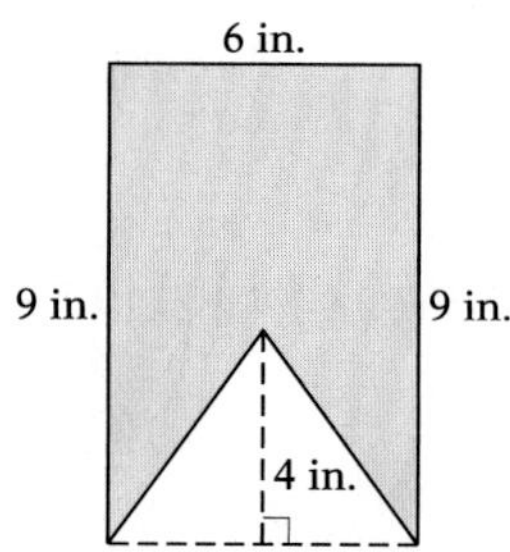

The average A of three tests and a final exam, which counts twice as much as a test, is given by $A = \frac{1}{5}(a + b + c + 2d)$ where a, b, and c are the grades on the tests and d is the grade on the final exam. In Exercises 41–44, use this formula to determine the indicated grade.

41. $A = 85, a = 74, b = 82, d = 89, c = ?$

42. $A = 74, a = 65, b = 82, c = 73, d = ?$

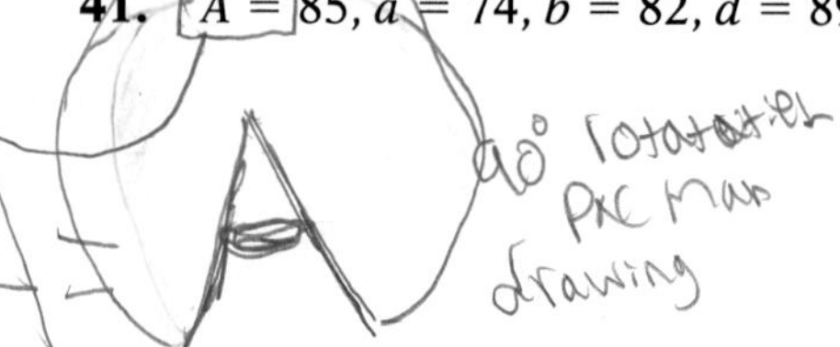

43. *Test Grades* Suppose that your grades on three tests are 71, 94, and 87. What grade do you need on the final exam to have an average of 80 for the course?

44. *Grade Average* Suppose that your grades on two tests are 94 and 98. Because you cut class, you received a 0 on the third test. What grade must you make on the final exam to have an average of 70 for the course?

45. Determine the surface area of the solid figure.

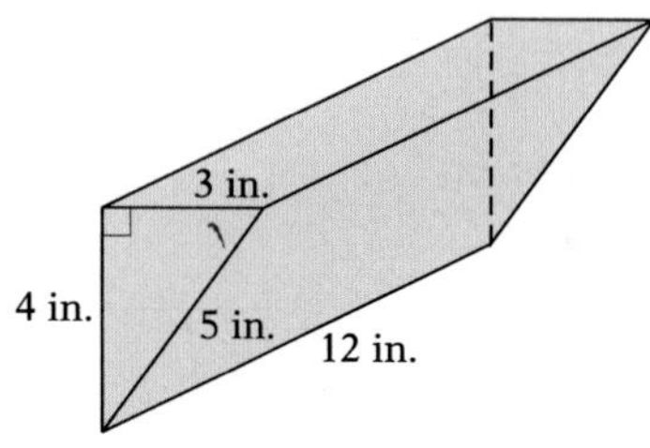

Copyright © Houghton Mifflin Company. All rights reserved.

CHAPTER 5 REVIEW EXERCISES

Section 5.1

 **1.** The ________ of $\frac{3}{4}$ is 3 and the ________ of $\frac{5}{7}$ is 7.

2. Explain why $\frac{3}{3}$ and $\frac{-8}{-8}$ are equivalent fractions.

3. Simplify the given fraction.

(a) $\frac{0}{7}$ **(b)** $\frac{2x}{1}$

 **4.** We call $\frac{15}{7}$ a(n) ________ fraction, whereas we call $\frac{5}{18}$ a(n) ________ fraction.

5. Write $\frac{-3}{8}$ in two other ways.

In Exercises 6–8, multiply.

6. $-\frac{5}{6} \cdot \frac{7}{2}$

7. $8\left(-\frac{4}{3}\right)$

8. $\frac{5}{9} \cdot \frac{a}{b}$

9. Simplify $\frac{3}{5}\left(\frac{5}{3}y\right)$.

10. *Grade Distribution* On an English quiz, 9 students received A's, 6 students received B's, and the remaining 11 students received C's. What fraction of the students received A's or B's?

Section 5.2

 **11.** A number that is divisible only by itself and 1 is called a(n) ________ number.

In Exercises 12 and 13, prime-factor the number.

12. 80

13. 2,475

In Exercises 14–18, simplify.

14. $\frac{24}{15}$

15. $\frac{42}{54}$

16. $\frac{18}{36}$

Copyright © Houghton Mifflin Company. All rights reserved.

17. $\frac{-16}{-40}$

18. $\frac{6ab}{8b}$

19. What part of an hour is 36 minutes?

20. *College Enrollment* A small college enrolled 1,500 men and 900 women. What fraction of the total enrollment are men?

Section 5.3

In Exercises 21–23, multiply and simplify.

21. $-15\left(\frac{9}{20}\right)$

22. $\frac{14}{36} \cdot \frac{30}{35}$

23. $\frac{10xy}{21} \cdot \frac{14}{5x}$

24. To divide a number by a fraction, we ________ the number by the ________ of the fraction.

25. Write the reciprocal of the given number.

(a) $-\frac{15}{7}$

(b) 9

In Exercises 26–28, divide and simplify.

26. $\frac{-15}{22} \div \frac{35}{6}$

27. $\frac{21}{16} \div 14$

28. $\frac{4a^2}{9} \div \frac{a}{3}$

29. *Bread Recipe* A recipe for banana nut bread uses $\frac{2}{3}$ cup of nuts. How many cups of nuts are needed for $\frac{3}{4}$ of the recipe?

30. *Stack of Bricks* If bricks $\frac{1}{3}$ foot thick are laid one on the other, how many bricks are needed to create a stack that is 4 feet high?

Section 5.4

31. Writing $\frac{3}{4}$ as $\frac{9}{12}$ is called ________ the fraction.

In Exercises 32–34, rename the fraction.

32. $\frac{5}{8} = \frac{}{48}$

33. $\frac{3}{-4} = \frac{}{20}$

34. $\frac{b}{3} = \frac{}{3a}$

Copyright © Houghton Mifflin Company. All rights reserved.

35. The numbers 3, 6, 9, and 12 are ______ of 3.

In Exercises 36 and 37, determine the LCM of the given numbers.

36. 12, 20

37. 10, 15, 21

38. Determine the LCM of $3x^2$ and $2xy$.

39. Insert < or > to make the statement true.

$$\frac{9}{16} \quad \_\_\_ \quad \frac{13}{24}$$

40. *Candy Boxes* Three people fill boxes of Valentine's candy in 8, 10, and 15 seconds, respectively. If they begin filling boxes at the same time, after how many seconds will they simultaneously complete filling a box?

Section 5.5

In Exercises 41 and 42, perform the indicated operation.

41. $\frac{-5}{14} + \frac{12}{14}$

42. $\frac{3}{10} - \frac{7}{10}$

43. To add two fractions, we first write the fractions with the same ______, then we add the ______.

In Exercises 44–49, perform the indicated operation.

44. $\frac{4}{5} - \frac{3}{4}$

45. $\frac{5}{9} + \frac{1}{3}$

46. $\frac{8}{3} - 2$

47. $\frac{1}{6} + \frac{7}{9}$

48. $\frac{3x}{7} + \frac{x}{7}$

49. $\frac{m}{2} - 6$

50. *Emergency Room Patients* Of 150 patients visiting an emergency medical facility, one-sixth had minor accident injuries and two-fifths had flu symptoms.

(a) What fraction of the patients visited for other reasons?

(b) How many patients had accident injuries or flu?

Copyright © Houghton Mifflin Company. All rights reserved.

Section 5.6

51. A number such as $3\frac{5}{8}$ is called a(n) ________ number.

52. A year has $52\frac{1}{7}$ weeks. Write the number as an improper fraction.

53. 145 seconds is $\frac{145}{60}$ minutes. Write the number of minutes as a mixed number.

In Exercises 54–58, perform the indicated operation.

54. $2\frac{2}{5} \cdot 1\frac{1}{9}$

55. $1\frac{5}{9} \div 5\frac{5}{6}$

56. $4\frac{2}{3} + 1\frac{5}{6}$

57. $3\frac{1}{4} - 1\frac{3}{4}$

58. $7\frac{2}{9} + 3\frac{5}{12}$

59. *Lawn Weed Killer* The instructions for a lawn weed killer call for $2\frac{1}{4}$ tablespoons of chemical for each 200 square feet of lawn. How much weed killer is needed for a lawn whose area is 500 square feet?

60. *Plywood Sheets* A building supply store stacked 50 sheets of $\frac{3}{8}$-inch plywood on top of 75 sheets of $\frac{1}{2}$-inch plywood. What is the height (in inches) of the stack?

Section 5.7

61. Explain how to eliminate fractions in an equation.

Copyright © Houghton Mifflin Company. All rights reserved.

In Exercises 62–68, solve the given equation.

62. $4y = 10$

63. $\frac{5}{4}c = -30$

64. $-\frac{x}{6} = \frac{3}{4}$

65. $6 - \frac{1}{3}a = 2$

66. $-\frac{1}{2} = b - \frac{5}{8}$

67. $\frac{3}{2}w - \frac{4}{3} = \frac{w}{2}$

68. $\frac{6}{5} = 1 - \frac{x}{2}$

In Exercises 69 and 70, write an equation to describe the given conditions. Then use the equation to solve the problem.

69. Four less than one-third of a number is -2. What is the number?

70. *Video Tape* After 54 minutes, three-fifths of a video tape had been played. How many minutes are required to play the entire tape?

Section 5.8

(For a summary of the formulas presented in this section, see the Quick Reference for Section 5.8.)

71. A(n) ________, such as $A = \frac{1}{2}bh$, is a special kind of equation.

72. Determine the area of the triangle.

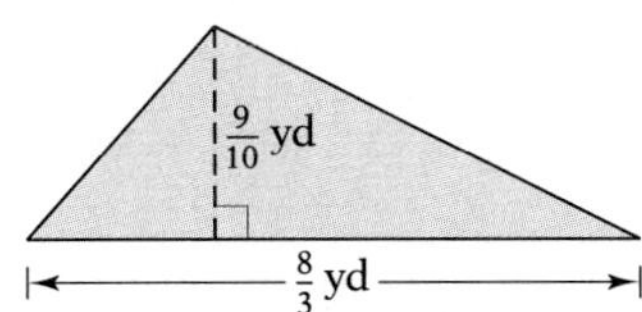

73. If the area of a trapezoid is 42 square inches and the lengths of the bases are 5 inches and 7 inches, what is the height of the trapezoid?

Copyright © Houghton Mifflin Company. All rights reserved.

74. *Cloth Banner* To make a banner that is in the shape of a parallelogram, $8\frac{3}{4}$ square feet of cloth are needed. If the base of the parallelogram is $4\frac{3}{8}$ feet, what is the height?

75. *Motor Oil Price* A case of motor oil contains 24 cans. If all cans are sold at a retail price R of \$5 each, the total profit P will be \$72. What is the wholesale price W of a case of motor oil?

76. *Used Cars* A used car dealer advertises that all cars are sold at a discount D of \$1,400. If the sale price S of a car is \$7,900, what was the original price P?

77. *Low Temperature* The normal January low temperature in Montreal, Canada, is $-15°C$. What is the corresponding Fahrenheit temperature?

78. *High Temperature* The normal July high temperature in Bombay, India, is 86°F. What is the corresponding Celsius temperature?

79. *Driving Time* A motorist drove 126 miles at 54 miles per hour. How many hours did the trip take?

80. *Driving Speed* Suppose that you drove $265\frac{5}{8}$ miles in $4\frac{1}{4}$ hours. What was your average speed in miles per hour?

Copyright © Houghton Mifflin Company. All rights reserved.

CHAPTER 5 TEST

 **1.** For the fraction $\frac{5}{3}$, we call 3 the ________, 5 the ________, and $\frac{3}{5}$ the ________.

2. Simplify.

(a) $\frac{0}{8}$ **(b)** $\frac{3y}{1}$

3. Prime-factor 90.

4. Simplify the fraction.

(a) $\frac{20}{35}$ **(b)** $\frac{72}{60}$ **(c)** $\frac{8x}{6xy}$

5. Rename each fraction.

(a) $\frac{3}{7} = \frac{\quad}{28}$ **(b)** $\frac{3}{x} = \frac{\quad}{5x}$

6. What is the LCM of 10, 14, and 15?

7. Write **(a)** $2\frac{3}{8}$ as an improper fraction and **(b)** $\frac{15}{2}$ as a mixed number.

8. Insert < or > to make the statement true.

$$\frac{5}{9} \quad \square \quad \frac{7}{12}$$

In Questions 9–19, perform the indicated operation.

9. $\frac{7}{12} \cdot \frac{8}{21}$ **10.** $-6\left(\frac{-3}{4}\right)$ **11.** $-\frac{9}{10} \div 2$ **12.** $\frac{3}{4} \div \frac{9}{4}$

13. $\frac{17}{18} - \frac{5}{18}$ **14.** $\frac{3}{8} - \frac{1}{16}$ **15.** $5 + \frac{2}{x}$ **16.** $\frac{7}{15} + \frac{5}{6}$

Copyright © Houghton Mifflin Company. All rights reserved.

17. $2\frac{6}{7} \div 1\frac{1}{14}$

18. $1\frac{3}{8} + 2\frac{3}{4}$

19. $5\frac{1}{3} - 2\frac{1}{2}$

In Questions 20 and 21, solve the given equation.

20. $\frac{7}{9} = -\frac{y}{6}$

21. $\frac{4}{3} = \frac{1}{3}x - \frac{3}{2}$

22. Psychology Exam A 20-question psychology exam has 15 multiple-choice questions. What fraction of the questions are multiple choice?

23. Cycling Time Three cyclists began riding around a circular track at the same time and took 10, 15, and 18 seconds, respectively, to complete a lap. After how many seconds were all three back at the starting point at the same time?

24. Welding In one day an experienced welder can complete $\frac{5}{8}$ of a job. An apprentice can complete $\frac{2}{5}$ as much of the job in a day as the experienced welder. If both welders work on the job together, what fractional part of the job will remain unfinished at the end of the day?

25. Determine the area of a triangle whose base is 5 feet and height is 3 feet.

26. Baseball Tickets Children receive a $5 discount on tickets to a baseball game. If a father pays $12 for his son's ticket, what was the price of the father's ticket? ($S = P - D$)

Copyright © Houghton Mifflin Company. All rights reserved.

CHAPTERS 4–5 CUMULATIVE TEST

1. We call $\frac{5}{3}$ the _____ of $\frac{3}{5}$.

2. The number $1\frac{2}{5}$ is called a(n) _____ number whereas $\frac{7}{5}$ is called a(n) _____ fraction.

3. Simplify $\frac{36}{48}$.

In Questions 4–12, perform the indicated operation and simplify the result if possible.

4. $\frac{12}{15} \cdot \frac{35}{6}$

5. $\frac{5}{2} \div 10$

6. $-\frac{7}{8} \div \frac{21}{4}$

7. $\frac{7}{12} + \frac{1}{12}$

8. $\frac{5}{6} - \frac{1}{3}$

9. $\frac{7}{10} - \frac{4}{15}$

10. $1\frac{3}{7} \div 1\frac{1}{14}$

11. $4 - 1\frac{2}{3}$

12. $5\frac{1}{2} + 2\frac{2}{3}$

In Questions 13–17, solve the given equation.

13. $2x + 7 - 3x = 10$

14. $-\frac{x}{7} = 5$

15. $3 - 2x = 3x - 7$

16. $3(2 - x) = x - 6$

17. $\frac{x}{2} - \frac{1}{3} = 1 - \frac{2}{3}x$

Copyright © Houghton Mifflin Company. All rights reserved.

18. Write the numbers $\frac{5}{9}$, $\frac{5}{6}$, and $\frac{4}{7}$ in order from smallest to largest.

19. *Burger Meat* A restaurant uses $\frac{5}{8}$ pound of ground meat for a burger. How many burgers can be made from 10 pounds of meat?

20. *Fitness Center* Three partners own a fitness center. One owns a $\frac{1}{3}$ interest, and another owns a $\frac{2}{5}$ interest. What portion of the business does the third partner own?

21. Determine the area of a trapezoid whose height is $\frac{3}{4}$ yard and whose bases are 3 yards and 5 yards. $\left[A = \frac{1}{2}h(b_1 + b_2)\right]$

22. *Truck Speed* A trucker traveled 128 miles in $2\frac{2}{3}$ hours. What was the trucker's average speed? $(D = RT)$

23. Seven less than 3 times a number is the product of the number and 2. What is the number?

24. *Biology Class* A biology class has 6 more women than men. If 36 students are in the class, how many women are in the class?

25. For the ordered pair (2, 4), we call 2 the ________ and 4 the ________.

26. Name the quadrant or axis in which the point lies.

(a) (14, 0) **(b)** (−30, −50)

Copyright © Houghton Mifflin Company. All rights reserved.

27. Determine the coordinates of the plotted points.

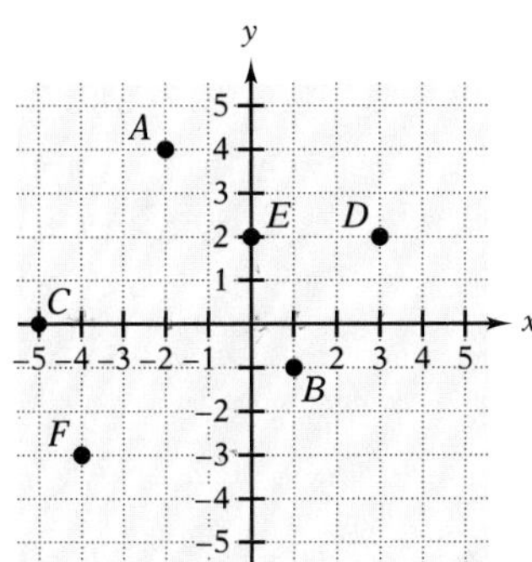

28. Complete the pairs (___ , 3), (0, ___), and (−2, ___) so that they are solutions of $3x - 2y = 0$. Then draw the graph of the equation.

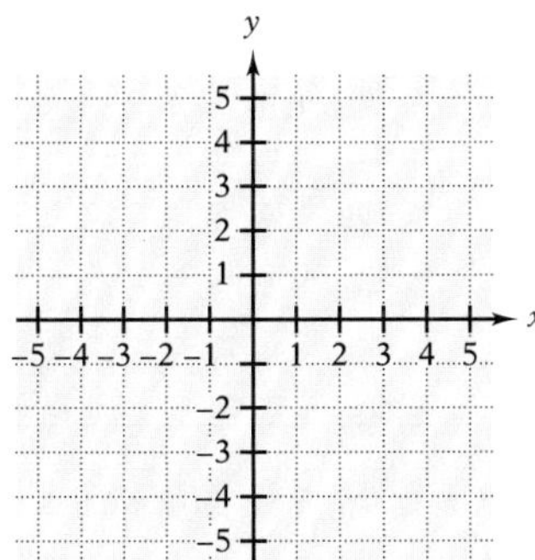

29. Complete the table of values for $y = 1 - \frac{2}{3}x$. Then draw the graph of the equation.

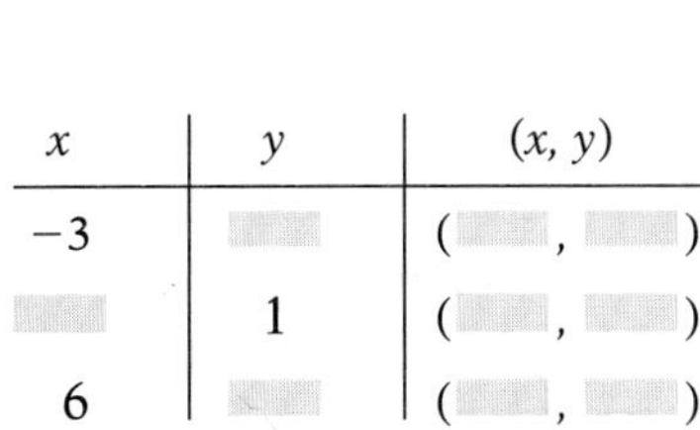

x	y	(x, y)
−3		(,)
	1	(,)
6		(,)

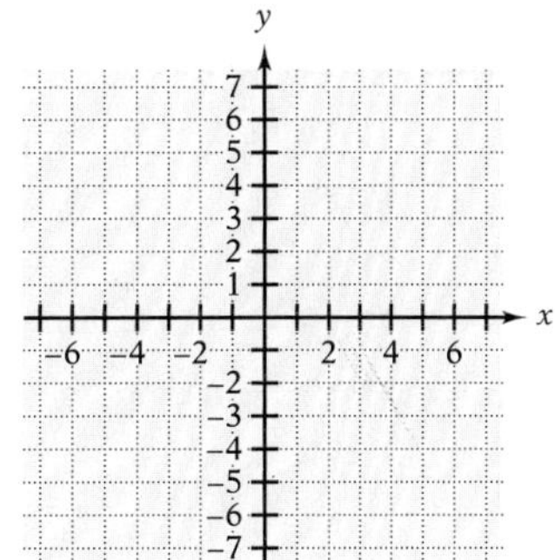

30. Graph $x = 4$.

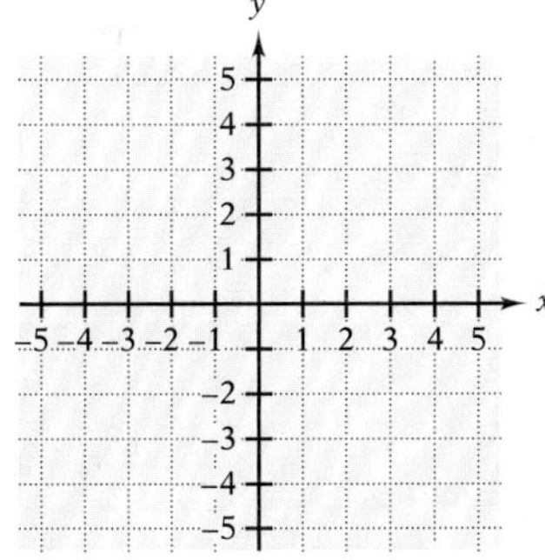

Copyright © Houghton Mifflin Company. All rights reserved.

CHAPTER 6

Decimals

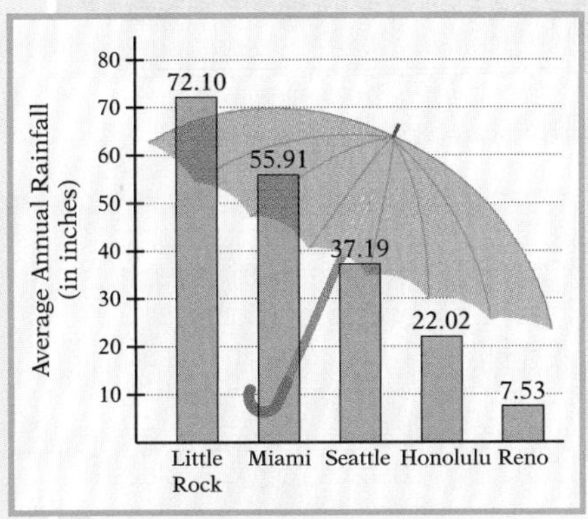

The bar graph shows that the average annual rainfall in Little Rock, Arkansas, is nearly 10 times that in Reno, Nevada. The rainfall amounts are written as **decimal numbers.** To find the total rainfall for a year and to find the average annual rainfall, we need to know how to add and divide decimals.

Chapter Snapshot

In this chapter, we extend our number system by introducing decimal numbers and the basic operations that we perform with them. We will see that the methods for evaluating expressions and solving equations are the same, even when decimal numbers are involved. We learn how the decimal representations of rational and irrational numbers differ, with a particular focus on square roots of numbers. Finally, we see how the Pythagorean Theorem can be used with right triangles, and we introduce the number π, which plays a prominent role in many geometry formulas.

Measurements of such quantities as distance, time, area, and volume are rarely whole numbers. Therefore, your understanding of the decimal number system is essential to your ability to solve real-life problems.

For online resources, visit the web site **math.college.hmco.com/students** and follow the links to Hubbard/Robinson, *Prealgebra.*

Copyright © Houghton Mifflin Company. All rights reserved.

Some Friendly Advice . . .

Whenever possible, you should draw a figure to help you visualize the problem you are working. Highlighting important parts of the figure with color is an excellent idea. Some students even use color very effectively in their class and reading notes. Adding color takes a little time, but it's worth it.

Sidebar: One of the authors advised a student to use a magic marker to highlight important passages in the text. The student bought a black magic marker and—you know the rest. Choose and use your colors wisely.

WARM-UP SKILLS

The following questions review concepts and skills that you will need in Chapter 6.

Perform the indicated operations.

1. $23 \cdot 1{,}000$

2. $(-4)^2$

3. $\frac{3}{10} + \frac{7}{100}$

4. $2\frac{3}{10} + 1\frac{1}{10}$

5. $\begin{array}{r} 1{,}627 \\ -\ 353 \\ \hline \end{array}$

6. $\begin{array}{r} 1{,}286 \\ +\ 490 \\ \hline \end{array}$

7. $\begin{array}{r} 847 \\ \times\ 21 \\ \hline \end{array}$

8. Solve $\frac{3}{10}x - 2 = \frac{1}{10}$.

9. Evaluate $a^2 + b^2$ for $a = 3$ and $b = 4$.

10. Round 1,462 to the nearest hundred.

Copyright © Houghton Mifflin Company. All rights reserved.

6.1 Introduction to Decimals

SUGGESTIONS FOR SUCCESS

Why do you suppose our place value system is based on powers of 10? One theory is because we have ten fingers. The word *decimal* comes from a Latin word that means *ten.*

You will probably think that this chapter is something new and different, but it really isn't. We will extend the place value system that we first introduced in Chapter 1, and we will see that decimal numbers are simply another way to write certain special fractions.

Once again, we urge you to step back and look at the big picture. The material in this chapter is based on concepts that you already know.

A Decimal Numbers

When you write a check, you indicate the amount of the check in two ways.

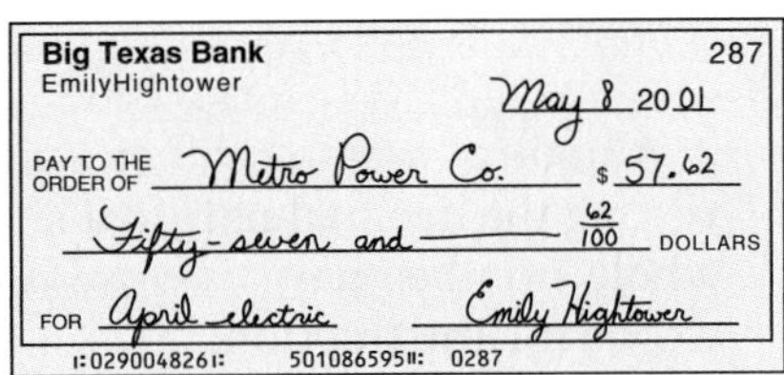

The dollar amount is written as the **decimal number** 57.62. The amount is also written in words as "Fifty-seven and $\frac{62}{100}$ dollars." In the decimal number, the 57 corresponds to the words *fifty-seven,* and the .62 corresponds to the fraction $\frac{62}{100}$.

The Decimal Number System

The decimal number system is an extension of the system for writing whole numbers. In Chapter 1, we showed how whole numbers can be written with the *digits* 0, 1, 2, 3, 4, 5, 6, 7, 8, and 9, where each digit has a *place value.* For example, the digits of the whole number 28,573 have the following place values.

Digit	*Place Value*	
2	10,000	
8	1,000	$1{,}000 = \frac{1}{10} \cdot 10{,}000$
5	100	$100 = \frac{1}{10} \cdot 1{,}000$
7	10	$10 = \frac{1}{10} \cdot 100$
3	1	$1 = \frac{1}{10} \cdot 10$

Copyright © Houghton Mifflin Company. All rights reserved.

Note that each place value is $\frac{1}{10}$ of the previous place value. The decimal number system is formed by continuing this pattern. The next place value is $\frac{1}{10} \cdot 1 = \frac{1}{10}$. Then the place value after that is $\frac{1}{10} \cdot \frac{1}{10} = \frac{1}{100}$. We can continue in this way, with each place value being $\frac{1}{10}$ of the place value to the left.

The following extended place value chart shows the number 38.257. Note that we separate the whole number place values from the fractional place values with a **decimal point.**

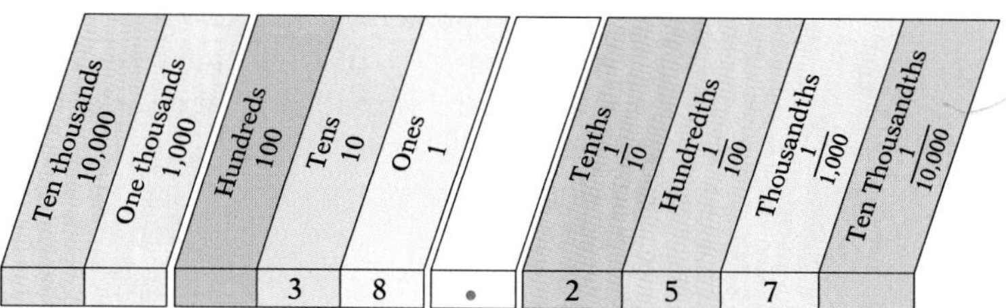

In expanded form, the number 38.257 means

$$3 \cdot 10 + 8 \cdot 1 + 2 \cdot \frac{1}{10} + 5 \cdot \frac{1}{100} + 7 \cdot \frac{1}{1,000}$$

A natural way to read or write a decimal number in words is to begin by converting the decimal part to a single fraction. Then we use the usual words for the whole number part, we use *and* for the decimal point, and we use the usual words for the fraction.

Example 1

Write 38.257 in words.

SOLUTION

The decimal part is 0.257, which can be written as follows:

$$0.257 = \frac{2}{10} + \frac{5}{100} + \frac{7}{1,000} \quad \text{The LCD is 1,000.}$$

$$= \frac{200}{1,000} + \frac{50}{1,000} + \frac{7}{1,000}$$

$$= \frac{257}{1,000}$$

We read this fraction as "two hundred fifty-seven thousandths."

Now the number 38.257 is written as "thirty-eight *and* two hundred fifty-seven thousandths."

Your Turn 1

Write 9.61 in words.

LEARNING TIP

If the whole number part of a decimal number is 0, we usually write the 0. For example, we write 0.257 rather than .257. This reminds us that the decimal point is there and helps us to avoid losing it.

Answer: Nine and sixty-one hundredths

When we are working with whole numbers and decimal numbers, the letters *th* make a big difference in the meanings of words. Compare the word names for the following numbers:

10	ten	0.1	one ten*th*
100	one hundred	0.01	one hundred*th*

Copyright © Houghton Mifflin Company. All rights reserved.

LEARNING TIP

*The letters th make a big difference in the meaning of a word. Ten means 10, whereas one-ten**th** means $\frac{1}{10}$ or 0.1.*

To write a decimal number as a fraction or mixed number, we can use a rule of thumb that is easier than the preceding method. Because 0.257 has three digits in the decimal part, we write the numerator as 257 and the denominator with three zeros.

$$\underbrace{0.257}_{\text{3 decimal places}} = \frac{257}{\underbrace{1{,}000}_{\text{3 zeros}}}$$

Example 2

Write the given decimal number in words.

(a) 27.4 **(b)** 2.91 **(c)** 8.365

SOLUTION

(a) $27.4 = 27 + \frac{4}{10}$

"Twenty-seven and four tenths"

(b) $2.91 = 2 + \frac{91}{100}$

"Two and ninety-one hundredths"

(c) $8.365 = 8 + \frac{365}{1{,}000}$

"Eight and three hundred sixty-five thousandths"

Your Turn 2

Write the given decimal number in words.

(a) 273.6 **(b)** 19.43 **(c)** 5.842

Answers: (a) Two hundred seventy-three and six tenths; (b) Nineteen and forty-three hundredths; (c) Five and eight hundred forty-two thousandths

A decimal number such as nine hundredths means $\frac{9}{100}$. Be sure to include the necessary zero in the tenths place: 0.09. If you omit the 0, you would have 0.9, which is nine tenths, not nine hundredths.

NOTE Word names for decimal numbers can be quite long. For example, the number 36.8719 is "thirty-six and eight thousand seven hundred nineteen ten thousandths." In practice, we don't often use words for such long decimals. A more common way to say this number is "thirty-six point eight seven one nine."

Example 3

Write the given number as a decimal number.

(a) Sixteen and three tenths

(b) Five and four hundred sixty-one thousandths

SOLUTION

(a) "Sixteen and three tenths" means $16 + \frac{3}{10} = 16.3$.

(b) "Five and four hundred sixty-one thousandths" means $5 + \frac{461}{1{,}000} = 5.461$.

Your Turn 3

Write the given number as a decimal number.

(a) Five and one tenth

(b) Twenty-one and eighty-two hundredths

Answers: (a) 5.1; (b) 21.82

Copyright © Houghton Mifflin Company. All rights reserved.

When you write a decimal number as a fraction, you might be able to simplify the fraction.

Example 4

Write 0.32 as a simplified fraction.

SOLUTION

$$0.32 = \frac{32}{100} = \frac{\not{4} \cdot 8}{\not{4} \cdot 25} = \frac{8}{25}$$

Your Turn 4

Write 0.14 as a simplified fraction.

Answer: $\frac{7}{50}$

Unless the denominator of a fraction is a power of 10—that is, 10, 100, 1,000, and so on—you can't automatically write the fraction as a decimal. However, sometimes you can rename the fraction so that the denominator is a power of 10.

Example 5

Write $\frac{6}{25}$ as a decimal number.

SOLUTION

$$\frac{6}{25} = \frac{6 \cdot 4}{25 \cdot 4} = \frac{24}{100} = 0.24$$

Your Turn 5

Write $\frac{3}{50}$ as a decimal number.

Answer: 0.06

KEYS TO THE CALCULATOR

By interpreting a fraction as division, you can use a calculator to convert a fraction to a decimal. The keystrokes are the same for both scientific and graphing calculators. Here is the conversion in Example 5.

6 [÷] **25** [=] (or [ENTER]) Display: [0.24]

You can also convert from a decimal to a fraction.

Scientific calculator:

.24 [2nd] [F↔D] Display: [6 ⌋ 25]

Graphing Calculator:

.24 [MATH] [Frac] [ENTER]

```
.24►Frac
                6/25
```

Exercises

Verify the results in the following table by displaying the fractions as decimals and the decimals as fractions.

Copyright © Houghton Mifflin Company. All rights reserved.

Some decimals and fractions are so frequently used that knowing their equivalents can be very useful in your day-to-day life. The following table gives you some common equivalents:

Common Fraction and Decimal Equivalents

Half	*Fourths*	*Fifths*	*Eighths*
$\frac{1}{2} = 0.5$	$\frac{1}{4} = 0.25$	$\frac{1}{5} = 0.2$	$\frac{1}{8} = 0.125$
	$\frac{3}{4} = 0.75$	$\frac{2}{5} = 0.4$	$\frac{3}{8} = 0.375$
		$\frac{3}{5} = 0.6$	$\frac{5}{8} = 0.625$
		$\frac{4}{5} = 0.8$	$\frac{7}{8} = 0.875$

B Rounding Decimal Numbers

Decimal numbers are often used for relatively precise measurements. If a person times a 100-meter freestyler with a wristwatch, the most accurate time might be 49 seconds. With a stopwatch, the time might be 48.7 seconds. Newer technologies that are used at the Olympics, for example, can report times to thousandths of a second.

When we measure such quantities as time, distance, length, speed, and so on, we need to be reasonable about the precision that is required. We also need to take into account the precision that our measuring instruments can give us. For example, we cannot measure the width of a table with a yardstick and expect to report the width as 27.6238 inches.

Usually we round off decimal numbers to a number of decimal places that makes sense. The rounding procedure for decimal numbers is similar to that for rounding whole numbers. If the digit to the right of the roundoff column is 5 or greater, we add 1 to the digit in the roundoff column. Otherwise, we leave the digit in the roundoff column as it is. However, there are a few other details that we must note.

Example 6

Round 17.2385 to the nearest hundredth.

SOLUTION

This number is greater than 5, so we round up.
↓
$$17.2385 \approx 17.24$$
↑
We are rounding to this decimal place.

Note that all digits to the right of the hundredths place are discarded.

Your Turn 6

Round 17.2385 to the nearest thousandth.

Answer: 17.239

Copyright © Houghton Mifflin Company. All rights reserved.

Example 7

Round 6.1398 to the nearest thousandth.

SOLUTION

This number is greater than 5, so we round up.

↓

6.1398

↑

We are rounding to this decimal place.

Rounding the 9 up makes it 10. As we did with whole numbers, we write the 0 and carry the 1 to the next column, which means that the 3 becomes 4.

$$6.1398 \approx 6.140$$

Your Turn 7

Round 28.099 to the nearest hundredth.

Answer: 28.10

NOTE Although 6.140 and 6.14 are the same number, we leave the 0 on the answer to Example 7 because we want to indicate the number of decimal places to which the number was rounded.

Keep in mind that when we round a whole number, we fill in the columns to the right of the roundoff column with zeros. However, when we round a decimal number, we discard all digits to the right of the roundoff column.

Example 8

Round 26,416.935 to the nearest:

(a) Hundred **(b)** One

(c) Tenth **(d)** Hundredth

SOLUTION

$1 < 5$

↓

(a) $26{,}416.935 \approx 26{,}400$

↑

Roundoff column

$9 > 5$

↓

(b) $26{,}416.935 \approx 26{,}417$

↑

Roundoff column

$3 < 5$

↓

(c) $26{,}416.935 \approx 26{,}416.9$

↑

Roundoff column

Your Turn 8

Round 8,153.689 to the nearest:

(a) Ten (not tenth)

(b) One

(c) Tenth

(d) Hundredth

(a) 8153.689 → 8150

(b) 8153.689 → 8154

(c) 8153.689 → 8153.7

(d) 8153.689 → 8153.69

Copyright © Houghton Mifflin Company. All rights reserved.

5 = 5
↓
(d) 26,416.935 ≈ 26,416.94
↑
Roundoff column

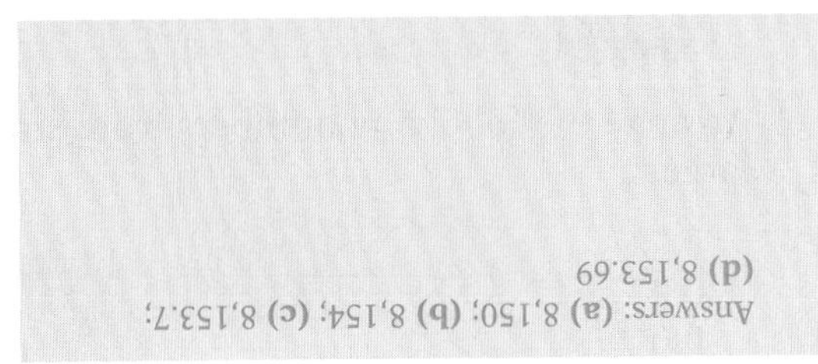

C Order of Decimal Numbers

Recall that numbers can be represented by points of a number line. If the point that corresponds to the number a is to the left of the point that corresponds to the number b, then the *order* of a and b is described by $a < b$.

Decimal numbers can also be plotted on a number line, but only roughly. No one could hope to find the exact point that corresponds to the number 2.7963.

The number line is not a reliable way to determine the order of two decimal numbers if they are close in value. Instead, starting from the left, we compare the numbers digit by digit until one digit is different from the other.

Example 9

What is the order of 36.578 and 36.562?

SOLUTION

36.578
↕↕ ↕
36.562

Note that the first three digits of the two numbers are the same. When we reach the fourth digit, we see that $7 > 6$. This means that the number on top is larger than the number on the bottom.

$36.578 > 36.562$ or $36.562 < 36.578$

Your Turn 9

What is the order of 1.2435 and 1.2453?

1.2453 > 1.2435
1.2435 < 1.2453

Answer: 1.2435 < 1.2453

Sometimes you will find it helpful to add zeros to one of the given numbers in order to write the numbers with the same number of decimal places.

Example 10

What is the order of 7.098 and 7.0986?

SOLUTION

7.0980 Write both numbers with four decimal places.
↕ ↕↕↕
7.0986

We see that all the digits are the same until we reach the last digit. Because the 0 in the upper number is less than the 6 in the lower number, the number on top is less than the number on the bottom.

$7.098 < 7.0986$ or $7.0986 > 7.098$

Your Turn 10

What is the order of 9.1234 and 9.12?

Answer: 9.1234 > 9.12

Copyright © Houghton Mifflin Company. All rights reserved.

Example 11

Arrange the following numbers from smallest to largest:

0.02
0.0216
0.019
0.0203

SOLUTION

There is no difference in the digits until we reach the hundredths column. Because $1 < 2$, the smallest of the numbers is 0.019.

For the remaining numbers, we move over to the thousandths column. Again, we add some 0s to make comparisons easier.

0.0200
0.0216
0.0203

Because $1 > 0$, we see that 0.0216 is the largest of the numbers.

For the remaining two numbers, we move to the last column.

0.0200
0.0203

Because $0 < 3$, we see that $0.0200 < 0.0203$.

Therefore, from smallest to largest, the arrangement is

0.019, 0.02, 0.0203, 0.0216

Your Turn 11

Arrange the following numbers from largest to smallest.

0.3901
0.298
0.391
0.389

Answer: 0.391, 0.3901, 0.389, 0.298

6.1 Quick Reference

A. Decimal Numbers

1. The decimal number system is an extension of the system for writing whole numbers. A decimal point separates the whole number part from the decimal part.

Hundreds	Tens	Ones		Tenths	Hundredths	Thousandths	Ten Thousandths
4	2	8	.	7	1	6	3

Copyright © Houghton Mifflin Company. All rights reserved.

2. To write the decimal part of a number as a fraction, write the decimal part (without the decimal point) in the numerator. The denominator is 1, followed by the same number of zeros as there are digits in the decimal part.

The number 0.92 has two decimal places, so the denominator has two zeros.

$$0.92 = \frac{92}{100}$$

3. To read or write a decimal number in words, begin by converting the decimal part to a single fraction. Then use the usual words for the whole number part, use *and* for the decimal point, and use the usual words for the fraction.

$5.47 = 5 + \frac{47}{100}$

"Five and forty-seven hundredths"

$16.239 = 16 + \frac{239}{1,000}$

"Sixteen and two hundred thirty-nine thousandths"

4. To change from words to decimal notation, translate the words to a fraction or mixed number and use that to write the decimal.

"Fifty-three and nine hundredths" means $53 + \frac{9}{100}$, which is 53.09.

5. A decimal number might be written as a simplified fraction. Also, you might be able to rename a fraction so that its denominator is a power of 10, which allows you to write the fraction as a decimal.

$$0.54 = \frac{54}{100} = \frac{2 \cdot 27}{2 \cdot 50} = \frac{27}{50}$$

$$\frac{9}{20} = \frac{9 \cdot 5}{20 \cdot 5} = \frac{45}{100} = 0.45$$

B. Rounding Decimal Numbers

1. The rounding procedure for decimal numbers is similar to that for rounding whole numbers. However, digits to the right of the roundoff column are discarded.

Round 6.7931 to the nearest tenth.

9 > 5, so round up.

↓

$6.7931 \approx 6.8$

↑

Roundoff column

2. When rounding involves carrying, we usually leave a 0 in the roundoff column to indicate the number of decimal places to which the number has been rounded.

Round 2.198 to the nearest hundredth.

The 8 in the thousandths column means that we round up the 9 in the hundredths column to 10. Write 0 and carry 1 to the tenths column.

$2.198 \approx 2.20$

Copyright © Houghton Mifflin Company. All rights reserved.

C. Order of Decimal Numbers

1. To determine the order of two decimal numbers, starting from the left, compare the numbers digit by digit until one digit is different from the other. If necessary, write additional zeros so that the numbers have the same number of decimal places.

Compare 59.24 and 59.2415.

59.2400
↕↕ ↕↕
59.2415

The first four digits are the same. For the fifth digits, $1 > 0$. Therefore, $59.2415 > 59.24$.

Copyright © Houghton Mifflin Company. All rights reserved.

6.1 Exercises

A. Decimal Numbers

1. To separate whole number place values from fractional place values, we use a ________.

2. The number 500 is read "five ________," whereas the number $\frac{5}{100}$ is read "five ________."

In Exercises 3–6, identify the place value of the 7 in the given number.

3. 3.1765
4. 3.7165
5. 3.1657
6. 3.1675

In Exercises 7–10, write the given decimal number as a proper fraction or mixed number. (Do not simplify the fraction.)

7. 0.43
8. 0.067
9. 15.009
10. 40.08

In Exercises 11–14, write the fraction as a decimal.

11. $\frac{7}{10}$
12. $\frac{34}{1,000}$
13. $18\frac{3}{1,000}$
14. $7\frac{39}{100}$

In Exercises 15–18, write the decimal number in words.

15. 0.45
16. 21.6
17. 6.045
18. 40.06

In Exercises 19–22, write the decimal number in each sentence in words.

19. The winning time in the Olympic men's 100-meter butterfly was 52.27 seconds. (Source: *World Almanac.*)

20. One cubic foot is 7.481 gallons.

21. In Alaska, 2.283 hours of each workday is needed to pay taxes. (Source: *USA Today.*)

22. The average commute in Pittsburgh is 18.3 minutes. (Source: Scarborough Research.)

In Exercises 23–26, write the number in decimal form.

23. Twenty-four hundredths
24. Thirty-five thousandths
25. Eleven and six tenths
26. Three hundred and four thousandths

Copyright © Houghton Mifflin Company. All rights reserved.

In Exercises 27–30, write the number in each sentence in decimal form.

27. One cubic inch is five hundred fifty-four thousandths fluid ounce.

28. In New York, three and fifteen hundredths hours of each workday is needed to pay taxes. (Source: *USA Today*.)

29. Vatican City, the world's smallest nation, occupies seventeen hundredths square mile.

30. The winning time in the Olympic women's 50-meter free style was twenty-four and eighty-seven hundredths seconds. (Source: *World Almanac*.)

In Exercises 31–34, write the decimal as a simplified fraction.

31. 0.8 **32.** 0.45 **33.** 0.06 **34.** 0.025

In Exercises 35–38, write the fraction as a decimal.

35. $\frac{7}{20}$ **36.** $\frac{9}{50}$ **37.** $\frac{19}{500}$ **38.** $\frac{21}{200}$

B. Rounding Decimal Numbers

39. When we round decimal numbers, we discard all ________ to the right of the roundoff column.

40. In a decimal number, the first column to the right of the decimal point is the ________ column, and the third column is the ________ column.

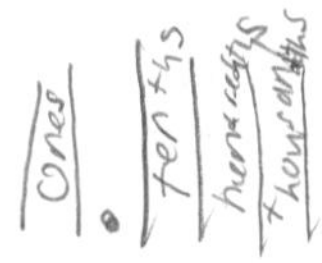

In Exercises 41–46, round the number to the indicated place value.

41.	5.386	Tenth	**42.**	5.39	Whole number
43.	6.398	Hundredth	**44.**	49.04962	Thousandth
45.	35.997	Hundredth	**46.**	39.976	Tenth

In Exercises 47–50, round the number to the indicated place values.

47.	34.746	Ten, tenth	**48.**	562.643	Hundred, hundredth
49.	3,456.4606	Thousand, thousandth	**50.**	12,638.3994	Thousand, thousandth

Copyright © Houghton Mifflin Company. All rights reserved.

In Exercises 51–54, round the number to the nearest (a) whole number, (b) tenth, (c) hundredth, and (d) thousandth.

51. The winning time in the Olympic women's marathon was 2.4347 hours. (Source: *World Almanac*.)

52. The winning time in the Olympic men's Giant Slalom was 2.64183 minutes. (Source: *World Almanac*.)

53. A quart of water weighs approximately 2.0825 pounds.

54. An approximate value of a number whose symbol is π is 3.14159.

C. Order of Decimal Numbers

55. When we determine the ______ of two decimal numbers, we compare the numbers digit by digit until one digit is different.

56. In a decimal number, the second column to the right of the decimal point is the ______ column, and the fourth column is the ______ column.

In Exercises 57–60, insert $<$ or $>$ to make the statement true.

57. 0.23 ______ 0.4 **58.** 2.06 ______ 2.059 **59.** 0.05 ______ c **60.** 1.0095 ______ 1.036

In Exercises 61–64, arrange the numbers in the indicated order.

61. 0.071, 0.069, 0.07, 0.0701; smallest to largest

62. 0.925, 0.92, 0.918, 0.9247; largest to smallest

63. 1.02, 1.0201, 1.009, 1.019; largest to smallest

64. 3.43, 3.407, 3.429, 3.4301; smallest to largest

Copyright © Houghton Mifflin Company. All rights reserved.

Writing and Concept Extension

65. We use a comma and a period to separate numbers. Explain the role of each.

66. Round 3.999 to the nearest (a) whole number, (b) tenth, and (c) hundredth.

In Exercises 67 and 68, insert $<$ or $>$ to make the statement true.

67. -0.0021 ____ -0.002

68. -3.001 ____ -3.01

Exploring with Real-World Data: Collaborative Activities

Chemical Elements in the Human Body The table shows the number of pounds of certain chemical elements that are found in the body of a person who weighs 155 pounds.

Chemical Element	*Weight (in pounds)*
Carbon	27.9
Calcium	2.79
Hydrogen	15.5
Nitrogen	4.805
Oxygen	100.44

69. Arrange the numbers in the Weight column from largest to smallest.

70. A 155-pound person has 20 or more pounds of which element(s)?

71. Use the symbol $<$ to compare the weight of nitrogen to the weight of calcium.

72. Use the symbol $>$ to compare the weight of hydrogen to the weight of carbon.

73. Round each of the weights to the nearest whole number.

Use the rounded weights in Exercise 73 to answer the remaining questions.

74. The weight of oxygen is approximately 20 times the weight of what other element?

75. The weight of calcium is approximately $\frac{1}{10}$ the weight of what other element?

76. Oxygen accounts for approximately ____ of the body weight of a person who weighs 155 pounds. Select your answer from the following choices:

(i) $\frac{1}{2}$ (ii) $\frac{3}{4}$ (iii) $\frac{2}{3}$ (iv) $\frac{3}{5}$

Copyright © Houghton Mifflin Company. All rights reserved.

6.2 Addition and Subtraction of Decimals

A Finding Sums and Differences
B Estimating and Simplifying
C Applications

SUGGESTIONS FOR SUCCESS

You can think of a decimal point as a wall that separates the whole number digits from the decimal digits.

When you add or subtract decimal numbers, the digits of the numbers must be in their proper columns. For this to happen, the "walls" (decimal points) must be in line. One way to remind yourself of this is to begin by writing just the decimal points in a column. Then fill in the digits on either side of the walls. By setting up the problem in this way, you can be sure that all the digits are correctly arranged.

A Finding Sums and Differences

Suppose that you make a quick stop at the grocery store where you buy two items that cost \$5.23 and \$7.14. The total cost is the *sum* of the decimal numbers: \$5.23 + \$7.14 = \$12.37. If you hand the clerk a \$20 bill, your change is the *difference* of the decimal numbers: \$20.00 − \$12.37 = \$7.63. In this section, we will discuss the procedures for adding and subtracting decimal numbers.

DEVELOPING THE CONCEPT

Sums of Decimal Numbers

One approach to adding 5.23 + 7.14 is to write the decimal numbers in expanded form and then add the fractions that have the same denominator.

$$\begin{aligned} 5.23 &= 5 + \frac{2}{10} + \frac{3}{100} \\ +\ 7.14 &= 7 + \frac{1}{10} + \frac{4}{100} \\ \hline &= 12 + \frac{3}{10} + \frac{7}{100} \\ &= 12.37 \end{aligned}$$

Note how we line up the whole numbers, the tenths, and the hundredths.

We can accomplish the same thing by lining up the decimal numbers with the whole numbers, the tenths, and the hundredths in columns. To do this, we must make sure that the decimal points are in line.

Line up the decimal points. ↓

$$\begin{array}{r} 5.23 \\ +\ 7.14 \\ \hline 12.37 \end{array}$$

Lining up the decimal points also lines up the tenths and hundredths columns, which allows us to add by columns.

↑ Make sure that this decimal point is in line with the others.

As long as we keep the decimal points in line, we can use our usual methods for adding. Recall that addition sometimes requires carrying.

Copyright © Houghton Mifflin Company. All rights reserved.

If the decimal numbers that you are adding don't have the same number of decimal places, you can insert zeros in unfilled columns.

Example 1

Add.

(a) 13.9 + 5.427 **(b)** 128.63 + 49

SOLUTION

(a)
```
  13.900
+  5.427
```
Write 13.9 as 13.900 so that both numbers have the same number of decimal places.

```
   1
  13.900
+  5.427
  19.327
```
Note that we had to carry in the tenths column.

(b)
```
   1
  128.63
+  49.00
  177.63
```
The integer 49 can be written as 49.00. Note the carry in the tens column.

Your Turn 1

Add.

(a) 6.23 + 14.8

(b) 24 + 9.531

Answers: **(a)** 21.03; **(b)** 33.531

Our usual methods for subtraction (possibly with borrowing) apply to decimal numbers. Again, we must line up the decimal points, and we might need to fill in columns with zeros.

Example 2

Subtract 16.27 − 3.53.

SOLUTION

```
   5 12
  16.27
-  3.53
  12.74
```
We must borrow in the tenths column.

Your Turn 2

Subtract 84.16 − 19.03.

Answer: 65.13

Because decimal numbers can be negative as well as positive, we apply the usual addition and subtraction rules for signed numbers. Here is a brief review of the addition rules.

1. For addends with like signs, we add their absolute values; the sign of the sum is the same as the sign of the addends.
2. For addends with unlike signs, we subtract their absolute values; the sign of the sum is the sign of the addend with the larger absolute value.

For subtraction, using the Definition of Subtraction to write the difference as a sum is often helpful.

$$a - b = a + (-b)$$

Copyright © Houghton Mifflin Company. All rights reserved.

Example 3

Perform the indicated operations.

(a) $-14.21 + 8.5$

(b) $-26.3 - (-37.14)$

SOLUTION

(a) The numbers have unlike signs, so we find the difference of the numbers.

$$\begin{array}{r} 14.21 \\ -\ 8.50 \\ \hline 5.71 \end{array}$$

Because the negative number -14.21 has the larger absolute value, the result is negative.

$$-14.21 + 8.5 = -5.71$$

(b) First, we write the subtraction as an addition.

$$-26.3 - (-37.14) = -26.3 + 37.14$$

In the sum, the numbers have unlike signs, so we find the difference of the numbers.

$$\begin{array}{r} 37.14 \\ -\ 26.30 \\ \hline 10.84 \end{array}$$

Because the positive number 37.14 has the larger absolute value, the result is positive.

$$-26.3 - (-37.14) = 10.84$$

Your Turn 3

Perform the indicated operations.

(a) $1.6 + (-9.72)$

(b) $8 - (-11.3)$

Answers: (a) −8.12; (b) 19.3

KEYS TO THE CALCULATOR

You can use your calculator to add and subtract positive and negative decimal numbers. The one additional key that you need to find is the decimal point key [.]. Here are the typical keystrokes for the sum in Example 3(a).

Scientific calculator:

14.21 [+/−] [+] **8.5** [=] Display: [−5.71]

Graphing calculator:

[(−)] **14.21** [+] **8.5** [ENTER]

Exercises

(a) $-153.48 - 27.6$

(b) $18.3 + (-21.92)$

(c) $26.4 - 31.92 + 5$

(d) $-9.9 - (-4.31) + 6.524$

Copyright © Houghton Mifflin Company. All rights reserved.

B Estimating and Simplifying

Just as we did with whole numbers, we estimate a sum or difference by rounding off the numbers before adding or subtracting. Estimating gives you a rough check of your answer. Also, an estimate can sometimes be used when precision isn't needed.

Although there are no fixed rules for estimating, we usually round the numbers to their highest place value.

≈ **Example 4**

Estimate the sum of 19.43 and 6.25.

SOLUTION

To round to the highest place values, we round 19.43 to the nearest ten and 6.25 to the nearest one.

$$\begin{array}{r} 19.43 \\ +\ 6.25 \\ \hline \end{array} \quad \rightarrow \quad \begin{array}{r} 20 \\ +\ 6 \\ \hline 26 \end{array}$$

You can verify that the exact sum is 25.68.

Your Turn 4

Estimate the difference of 50.25 and 29.4.

Answer: 20

≈ **Example 5**

Estimate the difference of 4.7 and 0.532.

SOLUTION

To round to the highest place values, we round 4.7 to the nearest one and 0.532 to the nearest tenth.

$$\begin{array}{r} 4.700 \\ -\ 0.532 \\ \hline \end{array} \quad \rightarrow \quad \begin{array}{r} 5.0 \\ -\ 0.5 \\ \hline 4.5 \end{array}$$

The exact difference is 4.168.

Your Turn 5

Estimate the sum of 10.18 and 0.49.

Answer: 10.5

NOTE Estimating requires some judgment about rounding. In Example 5, you might prefer to round both numbers to the nearest tenth. Then the approximate difference would be $4.7 - 0.5 = 4.2$, which is a better estimate than the one we obtained in the example.

One of the steps in equation solving is to combine any like terms, which are terms with the same variable parts. We use the Distributive Property to simplify an expression by combining like terms.

Example 6

Simplify $3.2x + 5.7x$.

SOLUTION

$$\begin{aligned} 3.2x + 5.7x &= (3.2 + 5.7)x && \text{Distributive Property} \\ &= 8.9x \end{aligned}$$

Your Turn 6

Simplify $8.4y - 2.3y$.

Answer: $6.1y$

Copyright © Houghton Mifflin Company. All rights reserved.

Example 7

Simplify $5a + 6.1b - 2.3a + 4b$.

SOLUTION

$$\begin{aligned} &5a + 6.1b - 2.3a + 4b \\ &= (5 - 2.3)a + (6.1 + 4)b \qquad \text{Distributive Property} \\ &= 2.7a + 10.1b \end{aligned}$$

Your Turn 7

Simplify.

$$2.4x + 5y - 3.4y + 8x.$$

Answer: $10.4x + 1.6y$

C Applications

Real-life problems often involve decimal numbers.

Example 8

Two stores offer the same spreadsheet software, but their prices differ by \$45.92. If the lower price is \$453.25, what is the higher price?

SOLUTION

The higher price is \$45.92 *more than* \$453.25.

$$\begin{array}{r} 453.25 \\ +\ 45.92 \\ \hline 499.17 \end{array}$$

The lower price of \$453.25 increased by \$45.92 gives us the higher price.

The higher price is \$499.17.

Your Turn 8

With a rebate of \$7.65, your cost for a box of computer disks is \$19.23. What is the selling price of the box of disks?

Answer: \$26.88

Example 9

In an Olympics track and field event, four runners ran a 400-meter relay in 37.40 seconds. If the times of three of the runners were 9.32 seconds, 9.36 seconds, and 9.37 seconds, what was the time of the fourth runner?

SOLUTION

We begin by finding the total time for the first three runners.

$$\begin{array}{r} 9.32 \\ 9.36 \\ +\ 9.37 \\ \hline 28.05 \end{array}$$

The time of the fourth runner is the difference between the total time of 37.40 seconds and the total time of the first three runners.

$$\begin{array}{r} 37.40 \\ -\ 28.05 \\ \hline 9.35 \end{array}$$

The total time for the relay minus the time for the first three runners gives us the time for the fourth runner.

The fourth runner's time was 9.35 seconds.

Your Turn 9

The lengths of two sides of a triangle are 18.3 inches and 12.9 inches. If the perimeter is 40.8 inches, how long is the third side?

Answer: 9.6 inches

Copyright © Houghton Mifflin Company. All rights reserved.

6.2 Quick Reference

A. Finding Sums and Differences

1. To add or subtract decimal numbers, we line up the decimal points and add columns in the usual way. Addition might require carrying, and subtraction might require borrowing. If necessary, add zeros so that the numbers have the same number of decimal places.

$$\begin{array}{r} \scriptstyle 1 \\ 16.273 \\ +\ 2.154 \\ \hline 18.427 \end{array} \qquad \begin{array}{r} \scriptstyle 6\ 11 \\ 27.\not{1}8 \\ -\ 4.80 \\ \hline 22.38 \end{array}$$

2. We apply the usual addition and subtraction rules for signed decimal numbers. We write a difference $a - b$ as $a + (-b)$.

$-4.9 + 2.3 = -2.6$

$$\begin{aligned} &6.74 - (-5.2) \\ &= 6.74 + 5.20 \\ &= 11.94 \end{aligned}$$

B. Estimating and Simplifying

1. When we estimate sums and differences of decimal numbers, we usually round the numbers to their highest place value.

$$\begin{array}{r} 11.63 \\ +\ 5.85 \\ \hline \end{array} \quad \begin{array}{c} \rightarrow \\ \rightarrow \\ \end{array} \quad \begin{array}{r} 12 \\ +\ 6 \\ \hline 18 \end{array}$$

2. We use the Distributive Property to simplify an expression by combining like terms.

$$\begin{aligned} &12.4x + 7.5x \\ &= (12.4 + 7.5)x \\ &= 19.9x \end{aligned}$$

Copyright © Houghton Mifflin Company. All rights reserved.

6.2 Exercises

A. Finding Sums and Differences

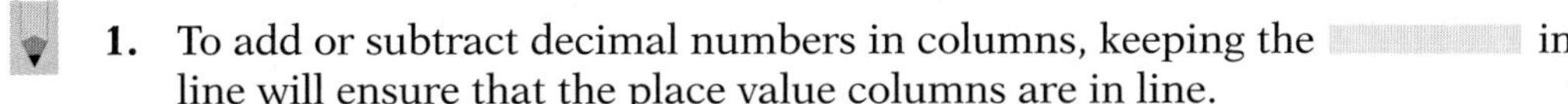

1. To add or subtract decimal numbers in columns, keeping the ________ in line will ensure that the place value columns are in line.

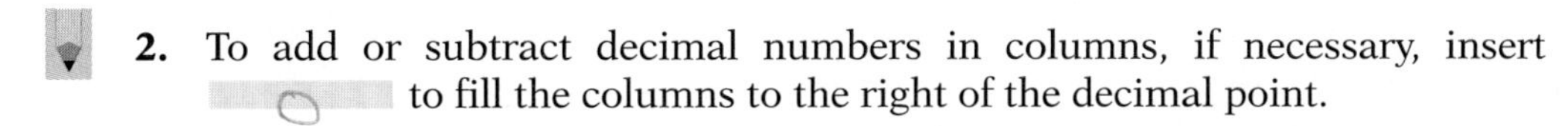

2. To add or subtract decimal numbers in columns, if necessary, insert ________ to fill the columns to the right of the decimal point.

In Exercises 3–14, add.

3. $\begin{array}{r} 12.5 \\ +\ 3.24 \\ \hline \end{array}$

4. $\begin{array}{r} 10.1 \\ +\ 9.23 \\ \hline \end{array}$

5. $\begin{array}{r} 94.215 \\ +\ 3.476 \\ \hline \end{array}$

6. $\begin{array}{r} 21.3649 \\ +\ 3.1493 \\ \hline \end{array}$

7. $\begin{array}{r} 5.37 \\ +\ 2.369 \\ \hline \end{array}$

8. $\begin{array}{r} 21.36 \\ +\ 0.495 \\ \hline \end{array}$

9. 27.142 + 8.935

10. 487.36 + 210.92

11. 0.32 + 0.05 + 0.96

12. 1.19 + 3.02 + 0.45

13. 4.6 + 12.67 + 2.391

14. 14.36 + 213.4 + 5.697

In Exercises 15–26, subtract.

15. $\begin{array}{r} 0.97 \\ -\ 0.5 \\ \hline \end{array}$

16. $\begin{array}{r} 5.4 \\ -\ 1.97 \\ \hline \end{array}$

17. $\begin{array}{r} 4.63 \\ -\ 1.59 \\ \hline \end{array}$

18. $\begin{array}{r} 12.397 \\ -\ 8.462 \\ \hline \end{array}$

19. $\begin{array}{r} 6.03 \\ -\ 5.1 \\ \hline \end{array}$

20. $\begin{array}{r} 12.62 \\ -\ 9.7 \\ \hline \end{array}$

21. 32.1 − 9.607

22. 8.3 − 2.79

23. 6 − 2.349

24. 1 − 0.999

25. 425.69 − 391.004

26. 23.598 − 19.03

In Exercises 27–38, perform the indicated operations.

27. 6.24 + (−3.5)

28. 1.23 + (−9.3)

29. −12.5 − 3.62

30. −4.1 − 7.64

31. −23.71 + 15.6

32. −14.6 + 19.37

Copyright © Houghton Mifflin Company. All rights reserved.

33. $-46.05 + 63.1$

34. $-100.9 + 67.002$

35. $4.67 - 9.003$

36. $70.3 - 102.01$

37. $-7.006 - (-3.9)$

38. $-12.9 - (-0.02)$

B. Estimating and Simplifying

39. To estimate the sum of 12.2 and 5.9, we might ______ the numbers to 12 and 6 and then add.

40. To combine like terms, we apply the ______ Property.

In Exercises 41–44, estimate the sum or difference.

41. $0.31 + 0.72$

42. $4.93 + 0.871$

43. $0.469 - 0.163$

44. $25.69 - 19.26$

In Exercises 45–52, simplify.

45. $4.2x + 1.3x$

46. $9.36x + 1.4x$

47. $7.3c - 5c$

48. $y + 0.03y$

49. $4.3y - 2.1y + 5.4y$

50. $6.2n - 1.3n - 0.4n$

51. $3a - 5.2b - 6.2a + 1.5a$

52. $2.5a - b + 3.1a + 1.3b$

C. Applications

53. *Fenced Enclosure* A landscaper used 33.45 feet of decorative edging to enclose a triangular area. If two of the sides were 12.75 feet and 6.2 feet, what was the length of the other side?

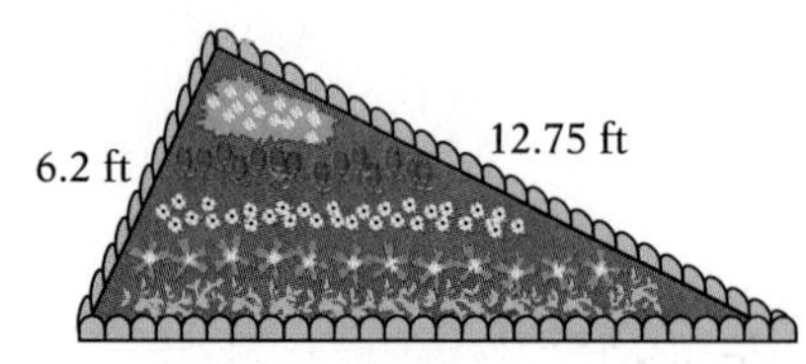

Copyright © Houghton Mifflin Company. All rights reserved.

54. *400-Meter Relay* A swim team recorded a time of 4.083 minutes for a 400-meter relay. If the times for three of the relay swimmers were 1.023, 1.021, and 1.02 minutes, what was the time for the fourth swimmer?

55. *Price Difference* Two stores sell the same drill, but the prices differ by $6.47. If the more expensive drill is $45.98, what is the price of the drill at the store with the cheaper drill?

56. *Rebate* After a rebate of $4.50, the cost of a coffeemaker was $27.39. What was the original price?

57. *Checking Account Balance* Suppose that your checking account balance was $509.06. If you wrote checks for $98.50 for groceries and $129.47 for a car payment, and you deposited a tax refund check of $130.72, what was your new balance?

58. *Checking Account Balance* Suppose that your checking account balance was $1,987.32. If you wrote checks of $253.30 for books and $847.45 for tuition, and you deposited a scholarship check for $500, what was your new balance?

59. *Take-Home Pay* Suppose that a clerk earned $346.15 for one week's work. After deductions of $52.03 for federal tax, $13.82 for social security, and $9.30 for state tax, what was the clerk's take-home pay?

60. *Pretax Earnings* Suppose that after deductions of $74.82 for federal tax, $14.90 for social security, and $29.91 for state tax, a nurse's aide's weekly take-home pay was $378.93. What were the aide's pretax earnings for the week?

61. *Change* A shopper purchases a paperback book for $4.92, shower cleaner for $5.79, shredded cheese for $8.39, and apples for $4.49. If the shopper gives the clerk two $20 bills, how much change should the shopper receive?

Copyright © Houghton Mifflin Company. All rights reserved.

62. *Restaurant Tip* At a restaurant, you and a friend ordered onion rings for \$4.70, a seafood platter for \$15.49, a steak for \$19.69, and dessert for \$5.95. If you used a \$50 gift certificate to pay the bill and told the server to keep the change for a tip, how much was the tip?

Calculator Exercises

In Exercises 63–66, use a calculator to evaluate the expression.

63. $3{,}976.902 + 456.004 + 9.8976$

64. $2{,}367.0127 - 1{,}999.15748$

65. $3.0069 - (1.097002 - 0.009099)$

66. $10.9004 - (3.89037 - 1.100489)$

Writing and Concept Extension

67. In which of the following can the zero be omitted?

(i) 1.50 (ii) 0.5 (iii) 1.05

68. Show how to use the Distributive Property to simplify $x - 0.08x$.

69. Suppose that you wrote a check for \$57.85. However, the bank made an error by transposing the digits 5 and 7. The bank deducted \$75.85 and reported your balance as \$970.41. What was the correct balance?

70. Suppose that you made a payment of \$391.92 on your credit card balance. The credit card company incorrectly recorded the payment as another charge and is now reporting that you owe \$1,570.12. What is the actual correct amount that you owe?

71. The lengths of the sides of a triangle are 6.89 inches, 9.23 inches, and 12.01 inches.

(a) If all three sides are increased by 0.5 inches, by how much would the perimeter be increased?

(b) If all three sides are increased by x inches, by how much would the perimeter be increased?

Copyright © Houghton Mifflin Company. All rights reserved.

72. A rectangle is 10.45 yards long and 6.25 yards wide. By how much would either the length or the width need to be increased in order for the rectangle to have a perimeter of 36 yards?

Exploring with Real-World Data: Collaborative Activities

Average Annual Rainfall The bar graph shows the average annual rainfall (in inches) in selected cities. (Source: National Climatic Data Center.)

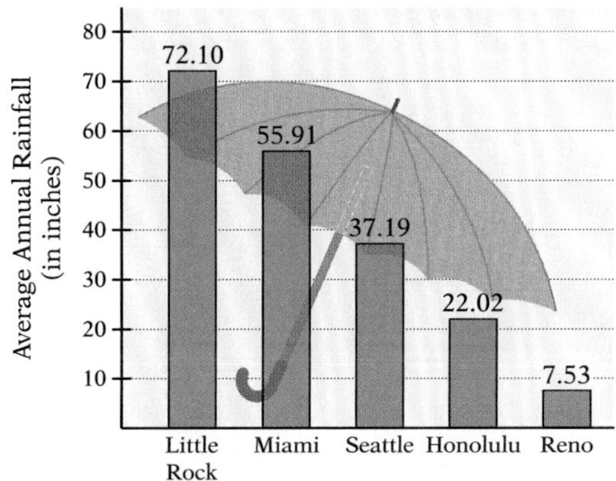

73. By how many inches does the rainfall in Miami exceed the rainfall in Honolulu?

74. What is the difference between the rainfall in Little Rock and Reno?

75. By how many inches does the rainfall in Little Rock exceed the combined rainfall for Reno, Honolulu, and Seattle?

76. How many more inches of rain falls in Seattle than in Reno and Honolulu combined?

Dog Bite Victims The bar graph shows the number of dog bite victims per 10,000 population. (Source: Allegheny University of the Health Sciences.)

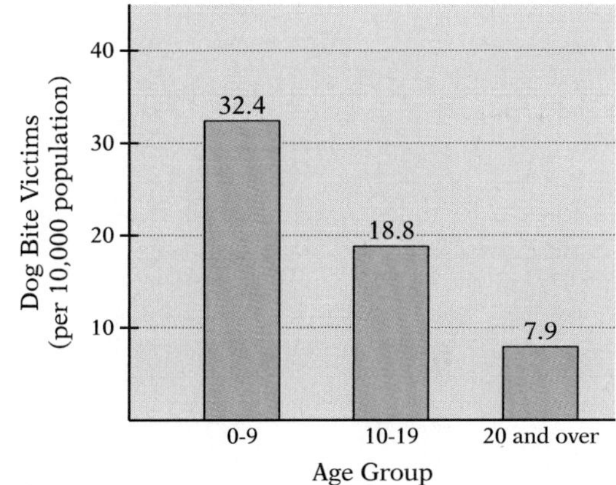

Copyright © Houghton Mifflin Company. All rights reserved.

77. What is the total number of victims per 10,000 population under 20 years old?

78. What is the total number of victims per 10,000 population?

79. How many more victims per 10,000 population are in the 0–9 age group than in the 10–19 age group?

80. How many more victims per 10,000 population are under 20 than in the 20-and-over age group?

Copyright © Houghton Mifflin Company. All rights reserved.

6.3 Multiplication of Decimals

SUGGESTIONS FOR SUCCESS

Decimal points can wander or get lost completely.

When you check your answer in the Answers section of this book, you may find that your answer is not the same as the answer given. The first thing you should do is look to see whether you copied the problem correctly. For example, you might have written 236.7 when the number should have been 23.67.

Incorrectly writing down the problem is a common error, particularly with problems involving decimal numbers. After all, decimal points are pretty tiny. You will save time and frustration if you take a minute to make sure you are working the right problem.

A Multiplying Decimal Numbers

Suppose that a person earns $9.26 per hour and, on a given day, works 7.5 hours. To determine the person's earnings for that day, we must be able to find the product 7.5×9.26.

NOTE In products of decimal numbers, we often use the $\times$ symbol rather than the multiplication dot, which could be confused with a decimal point.

Multiplying Decimal Numbers

One approach to multiplying decimal numbers is to write the numbers as fractions. Here are three examples.

This factor has 1 decimal place. This factor has 1 decimal place. The answer has 1 + 1 = 2 decimal places.

$$0.4 \times 0.9 = \frac{4}{10} \cdot \frac{9}{10} = \frac{36}{100} = 0.36$$

Observations: Ignoring the decimal points, $4 \times 9 = 36$.
The result 0.36 has 2 decimal places, which is the sum of the number of decimal places in the two factors.

This factor has 1 decimal place. This factor has 2 decimal places. The answer has 1 + 2 = 3 decimal places.

$$0.5 \times 0.03 = \frac{5}{10} \cdot \frac{3}{100} = \frac{15}{1{,}000} = 0.015$$

Observations: Ignoring the decimal points, $5 \times 3 = 15$.
The result 0.015 has 3 decimal places, which is the sum of the number of decimal places in the two factors.

This factor has 2 decimal places. This factor has 2 decimal places. The answer has 2 + 2 = 4 decimal places.

$$0.62 \times 0.01 = \frac{62}{100} \cdot \frac{1}{100} = \frac{62}{10{,}000} = 0.0062$$

Observations: Ignoring the decimal points, $62 \times 1 = 62$.
The result 0.0062 has 4 decimal places, which is the sum of the number of decimal places in the two factors.

Copyright © Houghton Mifflin Company. All rights reserved.

These examples illustrate the fact that we can learn what the digits of the answer are by pretending that the factors are whole numbers rather than decimal numbers.

Also, we can see that the number of decimal places in the answer is equal to the sum of the number of decimal places in the factors. This tells us where to place the decimal point in the answer.

Multiplying Decimal Numbers

1. Ignore the decimal points in the factors and multiply the factors as though they were whole numbers.
2. Place the decimal point in the answer so that the number of decimal places in the answer is equal to the sum of the number of decimal places in the factors.

Example 1

Multiply 3.21 × 5.847.

SOLUTION

```
     5.847    ← This factor has 3 decimal places.
   × 3.21     ← This factor has 2 decimal places.
      5847
    11694
   17541
  18.76887    ← The answer has 3 + 2 = 5 decimal places.
```

Your Turn 1

Multiply 2.87 × 41.5.

Answer: 119.105

NOTE In Example 1, we ignored the decimal points in the factors as we performed the multiplication. Then we placed the decimal point in the answer after we were done multiplying.

Example 2

Multiply 2.3 × 0.02.

SOLUTION

We can begin by mentally multiplying $23 \cdot 2 = 46$. However, we need to be careful about placing the decimal point.

The factor 2.3 has 1 decimal place, and the factor 0.02 has 2 decimal places. Therefore, the answer must have $1 + 2 = 3$ decimal places.

$$2.3 \times 0.02 = 0.046$$

↑ Note that we have to insert this 0 in order for the answer to have 3 decimal places.

Your Turn 2

Multiply 8 × 0.003.

Answer: 0.024

Copyright © Houghton Mifflin Company. All rights reserved.

KEYS TO THE CALCULATOR

Few people will multiply decimal numbers by hand if a calculator is available. Your calculator will display the answer with the decimal point in its proper position. Here are typical keystrokes for the product in Example 1. (The keystrokes are the same for both scientific and graphing calculators.)

3.21 [×] **5.847** [=] (or [ENTER]) Display: [18.76887]

Exercises

(a) 19.26×53.84 **(b)** 0.008×6.71
(c) 0.12×0.012 **(d)** 587.1×48.63

LEARNING TIP
To multiply signed numbers, multiply the numbers as though they were both positive. When you are done, affix the proper sign on the answer.

The usual rules for multiplying positive and negative numbers apply to products of decimal numbers. If the two factors have like signs, then the product is positive; if the two factors have unlike signs, then the product is negative.

Example 3

Multiply.

(a) -8.4×3.5 **(b)** $(-1.9)(-5.7)$

SOLUTION

(a) $-8.4 \times 3.5 = -29.40$

Because the last 0 is not needed, we can discard it and write the answer as -29.4. (A calculator will not display unneeded zeros.)

(b) $(-1.9)(-5.7) = 10.83$

Your Turn 3

Multiply.

(a) $-6.4(-9.1)$

(b) $1.7 \times (-2.8)$

Answers: (a) 58.24; (b) −4.76

B Powers of 10 and Estimating

Consider the following products in which one factor is a power of 10:

$3.5817 \times 10 = 35.817$ There is 1 zero in 10, and the decimal point moved 1 place to the right.

$3.5817 \times 100 = 358.17$ There are 2 zeros in 100, and the decimal point moved 2 places to the right.

$3.5817 \times 1{,}000 = 3{,}581.7$ There are 3 zeros in 1,000, and the decimal point moved 3 places to the right.

Such products can be found mentally simply by counting the number of zeros in the power of 10 and then shifting the decimal point that many places to the right.

Sometimes the power of 10 is written with an exponent, which tells us how many places to the right to shift the decimal point.

$27.6135 \times 10^2 = 2{,}761.35$ The exponent is 2, and the decimal point moved 2 places to the right.

$27.6135 \times 10^3 = 27{,}613.5$ The exponent is 3, and the decimal point moved 3 places to the right.

$27.6135 \times 10^5 = 2{,}761{,}350$ The exponent is 5, and the decimal point moved 5 places to the right.

Copyright © Houghton Mifflin Company. All rights reserved.

Note that when we multiplied by 10^5, shifting the decimal point 5 places to the right meant that we had to add an extra 0 at the end of the answer.

Example 4

Multiply.

(a) $1.32 \times 1{,}000$

(b) 41.9578×10^2

SOLUTION

(a) There are 3 zeros in 1,000, so we shift the decimal point 3 places to the right. (To do this, we need to add an extra 0.)

$$1.320. = 1{,}320$$

(b) The exponent on 10 is 2, so we shift the decimal point 2 places to the right.

$$41.95.78 = 4{,}195.78$$

Your Turn 4

Multiply.

(a) 5.47×10^4

(b) 1.32867×100

Answers: (a) 54,700; (b) 132.867

If an expression has decimal numbers in it, multiplying the expression by 10, 100, or some other power of 10 might result in an expression with no decimal numbers.

Example 5

Simplify.

(a) $10(4.5x - 3)$

(b) $100(3.2x + 0.15)$

SOLUTION

(a) $10(4.5x - 3)$

$= 10(4.5x) - 10(3)$ Distributive Property

$= 45x - 30$ Multiplying by 10 moves the decimal point 1 place to the right.

(b) $100(3.2x + 0.15)$

$= 100(3.2x) + 100(0.15)$ Distributive Property

$= 320x + 15$ Multiplying by 100 moves the decimal point 2 places to the right.

Your Turn 5

Simplify.

(a) $10(0.1 + 2.7x)$

(b) $1{,}000(5.86x - 0.027)$

Answers: (a) $1 + 27x$; (b) $5{,}860x - 27$

Except for those arithmetic mistakes that we all make, the most common error in multiplying decimal numbers is in the placement of the decimal point in the answer. Estimating products can help us to check whether our answers are reasonable. As always, we round off numbers in a way that makes them easy to multiply.

Copyright © Houghton Mifflin Company. All rights reserved.

Example 6

Estimate the product 4.81×19.7.

SOLUTION

$$4.81 \approx 5$$
$$19.7 \approx 20$$

The estimated answer is $5 \cdot 20 = 100$.
Verify that the actual product is 94.757.

Your Turn 6

Estimate the product 9.1×5.92.

Answer: 54

C Applications

News media and business reports often include large numbers that are written as a combination of numbers and words. For example, the population of the United States might be written as 287 million. The actual number is 287,000,000, which we obtain by interpreting 287 million as $287 \times 1{,}000{,}000$.

Large amounts of money are also often expressed in numbers and words.

Example 7

Suppose that the federal government's budget surplus is \$80 billion (read as "80 billion dollars") and that 0.18 of that amount is to be used for new programs. What is the dollar amount that is budgeted for new programs?

SOLUTION

$$(0.18 \times 80) \text{ billion} = \$14.4 \text{ billion}$$

We can obtain the actual numerical amount by multiplying 14.4 by 1 billion.

$$14.4 \times 1{,}000{,}000{,}000 = \$14{,}400{,}000{,}000$$

Your Turn 7

If the population of the United States is 287 million, of whom 0.1 are Hispanic, what is the Hispanic population?

Answer: 28.7 million

Example 8

Suppose that an auto worker earns \$16.78 per hour plus a time-and-a-half hourly wage of \$25.17 for every hour over 40 hours per week. How much does the worker earn for a 46-hour work week?

SOLUTION

Regular pay	40(16.78) =	671.20
Overtime pay	6(25.17) =	151.02
Total pay		822.22

For a 46-hour week, the worker earns \$822.22.

Your Turn 8

"Time-and-a-half" means 1.5 times the regular hourly wage. Verify that the worker's overtime pay is \$25.17 per hour.

Answer: $1.5 \times \$16.78 = \$25.17$

Copyright © Houghton Mifflin Company. All rights reserved.

6.3 Quick Reference

A. Multiplying Decimal Numbers

1. We use the following steps to multiply decimal numbers:

(a) Ignore the decimal points in the factors and multiply the factors as though they were whole numbers.

(b) Place the decimal point in the answer so that the number of decimal places in the answer is equal to the sum of the number of decimal places in the factors.

$$\begin{array}{r} 6.23 \\ \times \quad 5.8 \\ \hline 4984 \\ 3115 \\ \hline 36.134 \end{array}$$

6.23 — 2 decimal places

5.8 — 1 decimal place

The answer has 1 + 2 = 3 decimal places.

2. The usual sign rules apply to products of positive and negative decimal numbers.

$-2.6(3.9) = -10.14$

$-2.6(-3.9) = 10.14$

B. Powers of 10 and Estimating

1. When we multiply a decimal number by a power of 10, we look at the number of zeros (or at the exponent) in the power of 10. Then we shift the decimal point that many places to the right.

$17.8435 \times 100 = 1{,}784.35$

$17.8435 \times 10^3 = 17{,}843.5$

2. To estimate a product of decimal numbers, we round off the numbers in a way that makes them easy to multiply.

$$\begin{array}{lcr} 9.7 & \rightarrow & 10 \\ 6.25 & \rightarrow & \times \ 6 \\ \hline & & 60 \end{array}$$

$9.7 \times 6.25 \approx 60$

Copyright © Houghton Mifflin Company. All rights reserved.

6.3 Exercises

A. Multiplying Decimal Numbers

1. To determine the digits of the product of two decimal numbers, we multiply the factors as though they were ________ numbers.

2. The number of decimal places in the product of two decimal numbers is the ________ of the number of decimal places in the two factors.

In Exercises 3–16, multiply.

3. (0.3)(0.6)
4. 9(0.4)
5. (0.2)(0.07)
6. (0.09)(0.8)
7. 0.01(9.4)
8. 60(0.005)
9. (−0.2)(−0.08)
10. (−0.9)(−0.007)
11. (3.7)(−0.1)
12. (−0.32)(0.1)
13. −400(−0.09)
14. −60(0.08)
15. 5,000(−0.0003)
16. −900(−0.002)

In Exercises 17–30, multiply.

17. $\begin{array}{r} 4.3 \\ \times\, 0.4 \\ \hline \end{array}$

18. $\begin{array}{r} 2.6 \\ \times\, 0.03 \\ \hline \end{array}$

19. $\begin{array}{r} 7.002 \\ \times\, 0.04 \\ \hline \end{array}$

20. $\begin{array}{r} 5.06 \\ \times\, 0.002 \\ \hline \end{array}$

21. $\begin{array}{r} 3.375 \\ \times\quad 16 \\ \hline \end{array}$

22. $\begin{array}{r} 3.125 \\ \times\quad 3.2 \\ \hline \end{array}$

23. −8.4(−3.7)
24. 9.3(−2.1)
25. −3.6(7.5)
26. −2.5(2.8)
27. 8.9(−2.5)
28. −3.4(−3.3)
29. −7.2(−4.8)
30. 1.9(−9.9)

Copyright © Houghton Mifflin Company. All rights reserved.

In Exercises 31–34, multiply.

31. $\begin{array}{r} 40.39 \\ \times\ 0.24 \\ \hline \end{array}$

32. $\begin{array}{r} 210.17 \\ \times\ \ 0.32 \\ \hline \end{array}$

33. $\begin{array}{r} 11.2 \\ \times\ 9.5 \\ \hline \end{array}$

34. $\begin{array}{r} 1.56 \\ \times\ 0.74 \\ \hline \end{array}$

B. Powers of 10 and Estimating

35. In the product 2.8753×10^3, the ______ indicates the number of places to shift the decimal point.

36. To estimate a product, ______ the factors so they are easy to multiply.

In Exercises 37–44, multiply.

37. 14.306×100

38. $56.003 \times 1{,}000$

39. $8.583 \times 10{,}000$

40. $3.57 \times 100{,}000$

41. 3.78×10^2

42. 7.5879×10^3

43. 6.783×10^5

44. 1.63×10^4

In Exercises 45–48 simplify the given expression.

45. $100(4.56x + 3.2)$

46. $10(3x + 4.7)$

47. $1{,}000(3.247n + 0.035)$

48. $100(6.24 + 4.01a)$

In Exercises 49–52, estimate the product.

49. 7.79×28.67

50. 19.03×301.39

51. $(0.079)(51.039)$

52. $(0.214)(0.497)$

C. Applications

53. *Watermelons* Each year, Americans eat an average of 17.4 pounds of watermelon. (Source: U.S. Department of Agriculture.) Assuming a population of 0.27 billion, how many pounds of watermelon do Americans eat in a year?

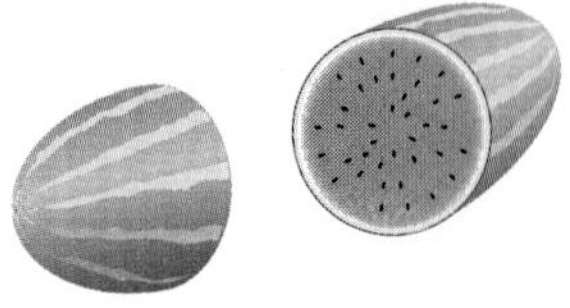

Copyright © Houghton Mifflin Company. All rights reserved.

54. *Overweight Adults* According to a Harris poll, 0.76 of the 203 million adults in the United States say that they exceed their suggested weight. How many adults say they are overweight?

55. *Commuting Time* In Washington, D.C., the average commuting time is 0.47 hours per day. (Source: *USA Today*.) In a 5-day workweek, how many hours does the average person spend commuting?

56. *United Nations Budget* The budget for the United Nations is $2,500 million. The United States pays 0.25 of that amount. (Source: United Nations.) How much does the United States pay?

57. *Cost of Gasoline* What is the cost of 12.4 gallons of gasoline at $1.45 per gallon?

58. *Car Rental* A rental company rents a luxury car for $19.89 per day plus $.32 per mile. What is the cost of renting the car for a 3-day trip of 624 miles?

59. *Fast-Food Employee* A fast-food employee earns $6.20 per hour for the first 40 hours of a week and $9.30 per hour for hours over 40. How much is the person paid for working 48.5 hours in one week?

Copyright © Houghton Mifflin Company. All rights reserved.

60. *Part-Time Wages* Suppose that a student makes $5.60 at a part-time job at the library and $6.40 at a part-time job at a restaurant. How much did the student make from working 14.5 hours at the library and 22.25 hours at the restaurant?

61. *Furniture Financing* A furniture store sells a set of living room furniture for $1,565.20. Rather than pay immediately, you decide to pay the down payment of $120 and $147.85 per month for a year. How much more do you pay for the furniture than if you had paid in full at the time of the purchase?

62. *Withholding Tax* Each month your employer withholds $407.45 for federal tax. However when you file your tax return you receive a $314.36 refund. How much tax did you pay?

Calculator Exercises

In Exercises 63–66, use a calculator to evaluate the expression.

63. $56.478(372.4 - 46.329)$

64. $-4.395(46.34 - 120.347)$

65. $(8.4)^2 - 5.67(7.53)$

66. $(2.1)^2 + (1.7)^2 + (5.6)^2$

Bringing It Together

In Exercises 67–72, perform the indicated operation.

67. $14.8 - 8.06$

68. $307.6 - 30.76$

69. $0.0001(24.6)$

Copyright © Houghton Mifflin Company. All rights reserved.

70. 200(0.03)

71. 45.9 + 24.02

72. 6.92 + 4.03

Writing and Concept Extension

73. Explain how to multiply 0.2 by 0.04 mentally.

74. What is an advantage of estimating the product of two decimal numbers?

75. Write 573 as the product of a decimal number and 100.

76. Suppose that you pay $1.07 for each $1,000 of insurance coverage. What is the premium for $54,700 of coverage?

Exploring with Real-World Data: Collaborative Activities

Recycling The bar graph shows the number of pounds of certain types of recyclable wastes that each person in the United States generates each day. (Source: U.S. Environmental Protection Agency.)

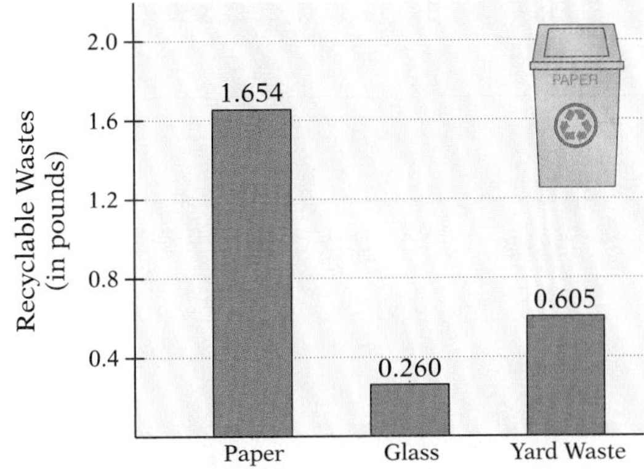

77. What is the difference between the number of pounds of yard waste and of glass that each person generates each day?

78. For the three given categories, what is the total number of pounds of recyclables per person per day?

79. For the three given categories, what is the total number of pounds of recyclables per person per *week*?

Copyright © Houghton Mifflin Company. All rights reserved.

80. For each person, by how much does the number of pounds of recyclable paper exceed the combined total number of pounds of glass and yard waste?

In Exercises 81–84, assume that the population of the United States is 287 million and that a year is 365 days.

81. How many pounds of waste paper are generated each day in the United States?

82. How many pounds of yard waste are generated each day in the United States?

83. How many pounds of glass waste does each person in the United States generate in one year?

84. How many pounds of waste paper does each person in the United States generate in one year?

Copyright © Houghton Mifflin Company. All rights reserved.

6.4 Division of Decimals

A *Dividing by a Whole Number*
B *Dividing by a Decimal*
C *Powers of 10 and Estimating*
D *Applications*

SUGGESTIONS FOR SUCCESS

Get rid of the scratch-out-and-scribble method.

In this section, you often will be asked to shift decimal points. Doing this by scratching out the old decimal point and scribbling in the new one is inviting disaster. Even if you carefully erase the decimal points, you are likely to forget where they were.

Try this: Write the original problem and use arrows to show you where the decimal points are to be moved. Then rewrite the problem with the decimal points in their new locations. If you are patient enough to do that, you will avoid many careless errors.

A Dividing by a Whole Number

In Section 6.1, we listed certain fraction and decimal equivalents: $\frac{1}{4} = 0.25$, $\frac{4}{5} = 0.8$, $\frac{3}{8} = 0.375$, and so on. To find such decimal names for fractions, we must realize that a fraction also indicates division.

Dividing by a Whole Number

If we want to write $\frac{7}{4}$ as a decimal, we perform the division $7 \div 4$. Using the long division format, we write 7 as 7.00, and we place the decimal point of the answer directly above the decimal point in 7.00.

The decimal point in the answer is directly above the decimal point in 7.00.

$$\begin{array}{r} . \\ 4\overline{)7.00} \end{array}$$ ← Write 7 as 7.00.

Divide in the usual way, but continue dividing past the decimal point.

$$\begin{array}{r} 1. \\ 4\overline{)7.00} \\ \underline{4} \\ 3\,0 \end{array}$$ Bring down the first 0.

$$\begin{array}{r} 1.7 \\ 4\overline{)7.00} \\ \underline{4} \\ 3\,0 \\ \underline{2\,8} \\ 20 \end{array}$$ Bring down the second 0.

$$\begin{array}{r} 1.75 \\ 4\overline{)7.00} \\ \underline{4} \\ 3\,0 \\ \underline{2\,8} \\ 20 \\ \underline{20} \\ 0 \end{array}$$ We stop dividing because the remainder is 0.

Dividing 7 by 4 results in $\frac{7}{4} = 1.75$.

Copyright © Houghton Mifflin Company. All rights reserved.

This method can be used to divide any number, including a decimal number, by a whole number. Recall that a *dividend* is the number you are dividing, and a *divisor* is the number you are dividing by. We use these words in the following summary of the procedure:

Dividing by a Whole Number

1. Write the division problem in the long division format.
2. If the dividend is a whole number, place a decimal point after the last digit.
3. Place the decimal point of the quotient directly above the decimal point in the dividend.
4. Carry out the long division until the remainder is 0 or to a specified number of decimal places.

NOTE To carry out the division to a specified number of decimal places, you might need to write extra zeros after the last digit of the dividend.

Example 1

Divide 73 ÷ 16 and round your answer to the nearest hundredth.

SOLUTION

Because we are to round the answer to the nearest hundredth, we will carry out the division to three decimal places. Therefore, we write 73 as 73.000.

```
     .
16)73.000
```

Place the decimal point of the quotient directly above the decimal point in 73.000.

```
    4.562
16)73.000
   64
    9 0
    8 0
    1 00
      96
       40
       32
        8
```

Rounded to the nearest hundredth, $73 \div 16 \approx 4.56$.

Your Turn 1

Divide 183 ÷ 51 and round your answer to the nearest hundredth.

Answer: 3.59

Example 2 illustrates division in which the dividend is a decimal number.

Copyright © Houghton Mifflin Company. All rights reserved.

Example 2

Divide 2.05 by 7 and round your answer to the nearest thousandth.

SOLUTION

Because we are to round the answer to the nearest thousandth, we write extra zeros so that the dividend has 4 decimal places.

```
      .2928
   7)2.0500
     1 4
       65
       63
        20
        14
         60
         56
          4
```

Rounded to the nearest thousandth, $2.05 \div 7 \approx 0.293$.

Your Turn 2

Divide $13.15 \div 4$ and round your answer to the nearest thousandth.

Answer: 3.288

B Dividing by a Decimal

In the division problem $2.4 \div 0.19$, the divisor 0.19 is a decimal number. We know how to divide by a whole number, but how do we divide by a decimal? First, we can write the problem in fraction form. Then, observe how we can change the original fraction into an equivalent fraction.

$$\frac{2.4}{0.19} = \frac{2.4 \times 100}{0.19 \times 100} \quad \text{Multiply the numerator and the denominator by 100.}$$

$$= \frac{240}{19}$$

Now the divisor is a whole number, and we can carry out the division in the usual way.

How did we know to multiply the numerator and the denominator by 100? Our goal was to move the decimal point 2 places to the right so that the decimal number 0.19 would become the whole number 19. Multiplying by 100 accomplished that goal.

Remember, though, that if you multiply the divisor (denominator) by some power of 10, then you must also multiply the dividend (numerator) by that same power of 10. In effect, we move the decimal point to the right the same number of places in both the divisor and the dividend.

Example 3

Divide $3.92 \div 1.8$.

SOLUTION

Initially, the problem looks like this.

```
        .
   1.8)3.92
```

Your Turn 3

Divide $2.735 \div 1.62$ and round your answer to the nearest hundredth.

continued

Copyright © Houghton Mifflin Company. All rights reserved.

To make the divisor 1.8 a whole number, we move the decimal point 1 place to the right. But then we must also move the decimal point in the dividend 1 place to the right.

$$1.8.\overline{)3.9.2} \quad \rightarrow \quad 18\overline{)39.2}$$

Now the problem should look familiar. We write some extra zeros in the dividend, and we divide as before.

```
     2.177
 18)39.200
    36
     3 2
     1 8
     1 40
     1 26
        140
        126
         14
```

If we were to continue dividing, the remainder would be 14 over and over again. Therefore, we will stop dividing and just round the answer (quotient) to the nearest hundredth.

$$3.92 \div 1.8 \approx 2.18$$

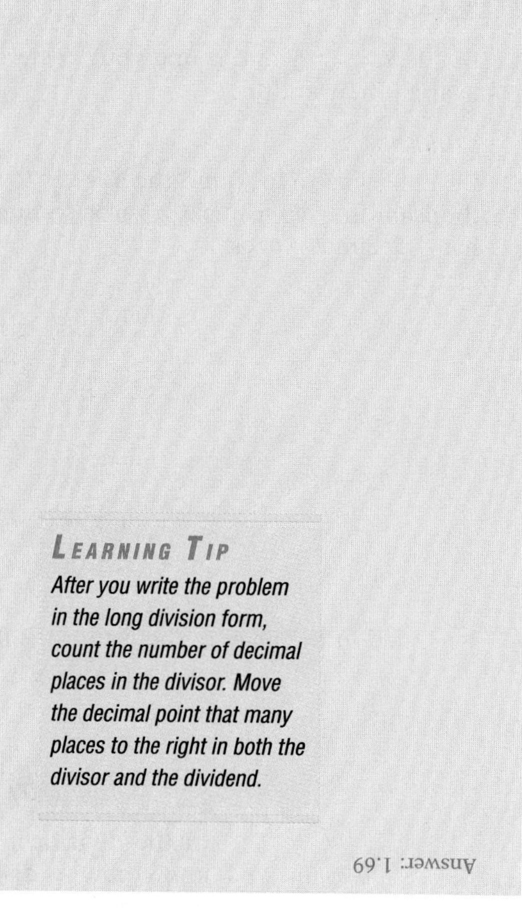

LEARNING TIP

After you write the problem in the long division form, count the number of decimal places in the divisor. Move the decimal point that many places to the right in both the divisor and the dividend.

Answer: 1.69

NOTE If you aren't told how to round off a quotient, you can use the following guidelines. Stop dividing whenever:

1. The remainder is 0.
2. The remainder starts repeating.
3. Additional precision of your answer is not meaningful.

Shifting the decimal point in the dividend may mean that you have to write extra zeros in the whole number part of the dividend.

Example 4

Divide $0.2 \div 0.54$ and round your answer to the nearest hundredth.

SOLUTION

$$0.54.\overline{)0.20.}$$

For 0.2, moving the decimal point 2 places to the right gives us 20 so we must write an extra 0.

Now we write some extra zeros after the decimal point so that we can round the answer to the nearest hundredth.

Your Turn 4

Divide $0.8 \div 0.39$ and round your answer to the nearest hundredth.

Copyright © Houghton Mifflin Company. All rights reserved.

```
      .370      Remember to line up the decimal points.
 54)20.000
    16 2
     3 80
     3 78
       20
```

To the nearest hundredth, $0.2 \div 0.54 \approx 0.37$.

NOTE The sign rules remain the same for division of decimals. If the numbers have like signs, the quotient is positive; if the numbers have unlike signs, the quotient is negative.

KEYS TO THE CALCULATOR

Most calculators allow you to set the number of decimal places to which your answers are to be displayed. These typical keystrokes round your answers to the nearest hundredth (2 decimal places).

Scientific calculator: 2nd FIX 2

Graphing calculator: MODE (Select 2 in the second line.)

One advantage of using a calculator to divide decimals is that you don't have to shift decimal points. Just enter the problem as it is given. The following are the typical keystrokes for Example 4 (with the result rounded to the nearest hundredth). The keystrokes are the same for both scientific and graphing calculators.

.2 ÷ **.54** = (or ENTER) Display: .37

Exercises

Round all answers to the given number of decimal places.

(a) $265.78 \div 94$; 2 **(b)** $-18.7 \div 5.732$; 3

(c) $4.29 \div 0.007$; 4 **(d)** $-654.1 \div (-1{,}846.53)$; 1

C *Powers of 10 and Estimating*

In Section 6.3, we found that multiplying a decimal by some power of 10 moves the decimal point to the right. Now we consider what happens when we divide a decimal by a power of 10.

$$\frac{28.63}{10} = 2.863$$ There is 1 zero in 10, and the decimal point moved 1 place to the left.

$$\frac{28.63}{100} = 0.2863$$ There are 2 zeros in 100, and the decimal point moved 2 places to the left.

$$\frac{28.63}{1{,}000} = 0.02863$$ There are 3 zeros in 1,000, and the decimal point moved 3 places to the left. Note that we had to write an extra 0 after the decimal point.

We see that dividing a decimal by a power of 10 has the effect of moving the decimal point to the left. If the power of 10 is written in exponential form, then the exponent tells us how many places to the left to move the decimal point. In either case, we can perform such divisions mentally.

Copyright © Houghton Mifflin Company. All rights reserved.

Example 5

Divide.

(a) $\frac{7.96}{1{,}000}$ **(b)** $\frac{250{,}000}{10^5}$

SOLUTION

(a) 0.007.96 — 1,000 has 3 zeros, so move the decimal point 3 places to the left. To do this, we had to write 2 extra zeros.

$\frac{7.96}{1{,}000} = 0.00796$

(b) 2.50000. — The exponent is 5, so move the decimal point 5 places to the left.

$\frac{250{,}000}{10^5} = 2.5$ — The zeros after the 5 can be discarded.

Your Turn 5

Divide.

(a) $\frac{8.1}{10^2}$ **(b)** $\frac{643.9}{10{,}000}$

Answers: (a) 0.081; (b) 0.06439

As is true for multiplication, incorrectly placing the decimal point in a quotient is the most common arithmetic error. Estimating the quotient in advance can help you avoid such a mistake.

Example 6

Estimate the quotient 17.918 ÷ 2.16.

SOLUTION

$$17.918 \approx 18$$
$$2.16 \approx 2$$

The estimated quotient is 18 ÷ 2 = 9. Verify that the actual quotient is approximately 8.3.

Your Turn 6

Estimate the quotient 25.04 ÷ 4.83.

Answer: 5

D Applications

To calculate a *grade-point average* (*GPA*), points are usually assigned to letter grades. In Example 7, we use A = 4 points, B = 3 points, and C = 2 points. For each course, the letter grade points are multiplied by the number of credit hours earned. To determine the GPA, we divide the total of these products by the total number of credit hours.

Example 7

A student received the following grades at the end of the first semester.

Course	*Grade*	*Hours*
English	A	3
Prealgebra	B	3
First-year seminar	A	2
History	C	3

Your Turn 7

In Example 7, what would the GPA have been if the student had earned an A in prealgebra?

Copyright © Houghton Mifflin Company. All rights reserved.

What is the student's GPA?

SOLUTION

Course	*Grade · Hours*
English	$4 \cdot 3 = 12$
Prealgebra	$3 \cdot 3 = 9$
First-year seminar	$4 \cdot 2 = 8$
History	$2 \cdot 3 = 6$
Total	35

The total number of credit hours was 11. Rounded to the nearest hundredth, the GPA is

$$\frac{35}{11} = 3.18$$

At a grocery, the *unit cost* of an item is found by dividing the cost of the item by the weight or volume.

Example 8

An 8-ounce jar of decaffeinated coffee sells for \$4.98. A 10-ounce jar costs \$5.99. Which size is the better buy in terms of unit cost?

SOLUTION

Size	*Cost*	*Unit Cost*
8 ounce	\$4.98	$\frac{\$4.98}{8} = \$0.6225$
10 ounce	\$5.99	$\frac{\$5.99}{10} = \$0.5990$

The unit cost of the 10-ounce jar is slightly less than that of the 8-ounce jar.

Your Turn 8

How could you have found the unit cost of the 10-ounce jar mentally?

Answer: Move the decimal point one place to the left.

6.4 Quick Reference

A. Dividing by a Whole Number

1. To divide any number by a whole number:

(a) Write the division problem in the long division format.

(b) If the dividend is a whole number, place a decimal point after the last digit.

(c) Place the decimal point of the quotient directly above the decimal point in the dividend.

(d) Carry out the long division until the remainder is 0 or to a specified number of decimal places.

```
      2.483
 11)27.313
    22
     5 3
     4 4
       91
       88
        33
        33
         0
```

Copyright © Houghton Mifflin Company. All rights reserved.

2. When you are to round off your answer, divide 1 decimal place beyond the roundoff column. To do this, you may need to write extra zeros.

Find 3.1 ÷ 9 to the nearest hundredth.

$$\begin{array}{r} .344 \\ 9\overline{)3.100} \\ \underline{2\,7} \\ 40 \\ \underline{36} \\ 40 \\ \underline{36} \\ 4 \end{array}$$

$3.1 \div 9 \approx 0.34$

B. Dividing by a Decimal

1. When the divisor is a decimal, we change it to a whole number by moving the decimal to after the last digit. Then we move the decimal point in the dividend the same number of places to the right.

$0.23\overline{)0.5} \rightarrow 0.23.\overline{)0.50.}$

Add zeros as needed.

$$\begin{array}{r} 2.17 \\ 23\overline{)50.00} \\ \underline{46} \\ 4\,0 \\ \underline{2\,3} \\ 1\,70 \\ \underline{1\,61} \\ 9 \end{array}$$

$0.5 \div 0.23 \approx 2.2$

C. Powers of 10 and Estimating

1. Dividing a decimal by a power of 10 moves the decimal point to the left. The number of places is given by the number of zeros in the power of 10 or by the exponent on 10.

$\frac{9.43}{100} = .09.43 = 0.0943$

$\frac{2{,}687.1}{10^3} = 2.687.1 = 2.6871$

2. Estimating a quotient helps us to avoid errors in the placement of the decimal point.

$$\begin{array}{lcl} 54.13 & \div & 5.82 \\ \downarrow & & \downarrow \\ 54 & \div & 6 = 9 \end{array} \quad \text{Estimate}$$

$54.13 \div 5.82 \approx 9$

Copyright © Houghton Mifflin Company. All rights reserved.

6.4 Exercises

A. Dividing by a Whole Number

1. In the division 4.8 ÷ 15, we call 15 the ________ and 4.8 the ________.

2. When we use long division to divide 4.8 by 15, we align the ________ in the quotient and the dividend.

In Exercises 3–6, divide until the remainder is 0.

3. 37 ÷ 5 **4.** 78 ÷ 30 **5.** 314 ÷ 25 **6.** 57 ÷ 24

In Exercises 7–10, divide until the remainder is 0.

7. 7.86 ÷ 6 **8.** 97.92 ÷ 8 **9.** 0.23 ÷ 46 **10.** 0.99 ÷ 22

In Exercises 11–14, divide. Round the result as indicated.

11. 20 ÷ 11, tenth **12.** 4.2 ÷ 9, tenth **13.** 44.3 ÷ 18, hundredth **14.** 3.23 ÷ 9, thousandth

B. Dividing by a Decimal

15. When you use long division to divide 3.15 by 4.2, move the decimal point ________ place(s) to the ________ in both the dividend and the divisor.

16. When you use long division to divide 8.1 by 2.78, you must insert extra ________ in the dividend.

In Exercises 17–28, divide to find the exact quotient.

17. 1.2 ÷ 0.03 **18.** 0.24 ÷ 0.6 **19.** 0.36 ÷ 0.0009 **20.** 15 ÷ 0.003

Copyright © Houghton Mifflin Company. All rights reserved.

21. 6.39 ÷ 0.01

22. 0.379 ÷ 0.001

23. $\frac{600}{1.2}$

24. $\frac{240}{0.12}$

25. $\frac{0.3}{1.5}$

26. $\frac{6}{2.4}$

27. $\frac{0.035}{0.07}$

28. $\frac{0.032}{0.8}$

In Exercises 29–32, divide to find the exact quotient.

29. 0.051 ÷ 0.03

30. 24.067 ÷ 4.1

31. 7.852 ÷ 1.3

32. 322.94 ÷ 0.67

In Exercises 33–36, divide. Round the result as indicated.

33. 17.3 ÷ 0.035, tenths

34. 0.0359 ÷ 0.054, thousandths

35. 9.1 ÷ 0.6, hundredths

36. 8.2 ÷ 0.03, hundredths

C. Powers of 10 and Estimating

37. To estimate a quotient, ________ the dividend and the divisor so that they are easy to divide.

38. In $\frac{6.4}{10^3}$, the ________ indicates the number of places to shift the decimal point to the left.

In Exercises 39–46, divide.

39. 43.62 ÷ 100

40. 3.72 ÷ 10

41. 398.6 ÷ 1,000

42. 57.1 ÷ 10,000

43. $0.4 \div 10^3$

44. $65.8 \div 10^4$

45. $5.698 \div 10^2$

46. $7{,}849.6 \div 10^5$

Copyright © Houghton Mifflin Company. All rights reserved.

≈ In Exercises 47–50, match the quotient in column A with the best estimate in column B.

	Column A	*Column B*
47.	0.007162 ÷ 0.07	(A) 0.01
48.	7.162 ÷ 0.7	(B) 0.1
49.	7.162 ÷ 0.07	(C) 10
50.	0.007162 ÷ 0.7	(D) 100

D. Applications

51. *Unit Cost* A 16-ounce box of cheese crackers costs $3.59 and a 10-ounce box costs $2.39. Which size is the better value in terms of unit cost?

52. *Unit Cost* A 5.6-pound box of laundry detergent costs $4.89, and an 8.5-pound box costs $6.95. Which size is the better value in terms of unit cost?

53. *Housing Density* Suppose that developers plan to construct 185 houses on 60.4 acres of land. To the nearest tenth, how many houses per acre will be built?

54. *Weekly Salary* Suppose that your annual salary is $35,583.60. For a 52-week work year, what is your weekly salary?

Copyright © Houghton Mifflin Company. All rights reserved.

55. *Unit Cost* Suppose that a case of light bulbs costs $80.64. If a case contains 72 bulbs, what is the cost of 1 bulb?

56. *Miles per Gallon* At the beginning of a trip, a motorist noted that the gas tank was full and the odometer reading was 45,879.4. At the end of the trip, the motorist filled the tank and noted that the odometer reading was 46,704.3. If the gas purchases were for 12.5, 14.3, and 9.8 gallons, to the nearest tenth, what was the gas mileage in miles per gallon?

57. *Appliance Financing* Suppose that a refrigerator, priced at $680.50, is advertised for $175 down, with the balance to be paid in 6 equal payments. How much is each payment?

58. *Garbage Generation* People in the United States generate 208.1 million tons of garbage each year. (Source: U.S. Environmental Protection Agency.) To the nearest tenth, how many tons per person are generated annually? (Assume that the population of the United States is 287 million.)

Grade-Point Average In Exercises 59 and 60, calculate the grade-point average (GPA) for the given semester grades.

59.

	Grade	*Hours*
Economics	A	3
English	C	5
History	B	3
Mathematics	A	4

Copyright © Houghton Mifflin Company. All rights reserved.

60.

	Grade	*Hours*
Algebra	A	5
Drama	C	3
Art	B	4
Biology	A	5

Calculator Exercises

In Exercises 61–64, use a calculator to evaluate the given expression.

61. $6{,}018.962 \div 105.8$ **62.** $156.1208 \div 24.86$ **63.** $(28.35 - 4.3^2) \div 2.4$ **64.** $24.6^2 \div [(9.7)(2.3)]$

Bringing It Together

In Exercises 65–72, perform the indicated operation.

65. $1.187 - 0.84$ **66.** $45.7 - 10.004$ **67.** $2.242 \div 0.59$ **68.** $256.04 \div 7.4$

69. $350.5 + 48.92$ **70.** $4.5 + 8.076$ **71.** $1.4(0.3)$ **72.** $6.03(2.1)$

Writing and Concept Extension

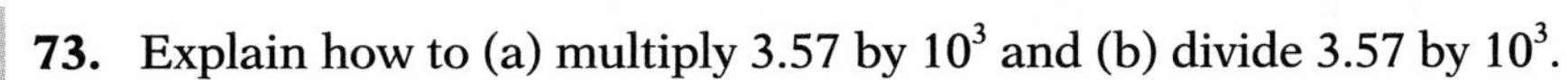

73. Explain how to (a) multiply 3.57 by 10^3 and (b) divide 3.57 by 10^3.

74. To how many decimal places should you carry a division in order to round to the indicated place value?

(a) Whole number **(b)** Tenth **(c)** Hundredth

75. Write 0.06 as the quotient of a whole number and 100.

Copyright © Houghton Mifflin Company. All rights reserved.

76. Refer to Exercise 51. Suppose that you use a $.50 coupon for a box of cheese crackers. Which size is the better value in terms of unit cost?

77. Refer to Exercise 52. Suppose that you use a $1 coupon for a box of detergent. Which size is the better value in terms of unit cost?

Exploring with Real-World Data: Collaborative Activities

NFL Rushing Statistics The table shows rushing statistics for four National Football League players. (Source: NFL.)

	Years	*Attempts*	*Yards*
Walter Payton	13	3,838	16,726
Eric Dickerson	11	2,996	13,259
Tony Dorsett	12	2,936	12,739
Jim Brown	9	2,359	12,312

Round answers to the nearest hundredth.

78. What was the average number of years that the 4 players were in the NFL?

79. What was the average number of attempts for the 4 players?

80. Determine the average yards per season for Jim Brown.

81. Determine the average yards per season for Tony Dorsett.

82. Determine the average yards per attempt for Walter Payton.

83. Determine the average yards per attempt for Eric Dickerson.

84. What is the difference between the average yards per attempt for Brown and Payton?

85. What is the difference between the average number of yards per season for Brown and Payton?

Copyright © Houghton Mifflin Company. All rights reserved.

6.5 Expressions and Equations

A *Evaluating Algebraic Expressions*
B *Equations with Decimals*
C *Applications*

SUGGESTIONS FOR SUCCESS

The good news: There is little that is new in this section. The bad news: None.

You already know how to evaluate an expression and how to solve a certain kind of equation. Even though the expressions and equations in this section involve decimal numbers, the methods are the same.

This would be a good time to revisit the Order of Operations in Section 2.5 and the procedure for solving equations introduced in Sections 4.1 and 4.2. A thorough review of that material will help you hit the ground running in this section.

A Evaluating Algebraic Expressions

Recall that we evaluate an algebraic expression by replacing each variable with a given value and then performing the indicated operations. We use this same procedure when the replacements are decimal numbers.

Example 1

Evaluate $a - b - c$ for $a = 12.4$, $b = -9.51$, and $c = 7.365$.

SOLUTION

$$a - b - c = 12.4 - (-9.51) - 7.365$$
$$= 12.4 + 9.51 - 7.365$$

$$\begin{array}{r} 12.40 \\ +\ 9.51 \\ \hline 21.910 \\ -\ 7.365 \\ \hline 14.545 \end{array}$$

Add 12.4 and 9.51.

Now subtract 7.365.

For the given values of the variables, the value of the expression is 14.545.

Your Turn 1

Evaluate $x + y - z$ for $x = 1.52$, $y = -2.7$, and $z = -3.846$.

Answer: 2.666

In Example 1, we showed how the operations are performed "by hand." However, performing operations with decimals is an appropriate use of a calculator. In the remaining examples, will we show only the results of operations.

An algebraic expression might include more than one operation. When we evaluate such an expression, we follow the Order of Operations.

Example 2

Evaluate $x(3.1 + y)$ for $x = 1.6$ and $y = 8.4$.

SOLUTION

$x(3.1 + y) = 1.6(3.1 + 8.4)$ Replace x with 1.6 and y with 8.4.

$= 1.6(11.5)$ Add inside the parentheses.

$= 18.4$ Then multiply.

Your Turn 2

Evaluate $2.4(x - y)$ for $x = 9.21$ and $y = 3.76$.

Answer: 13.08

Copyright © Houghton Mifflin Company. All rights reserved.

NOTE In Example 2, we could have used the Distributive Property to write the given expression as $3.1x + xy$. Evaluating this expression for the given values of x and y would produce the same result.

Example 3

Evaluate $a + b^2 \div c$ for $a = 2.8$, $b = -3.1$, and $c = 5.4$.

SOLUTION

$a + b^2 \div c$

$= 2.8 + (-3.1)^2 \div 5.4$ Replace a with 2.8, b with -3.1, and c with 5.4.

$= 2.8 + 9.61 \div 5.4$ The exponent has the highest priority.

$\approx 2.8 + 1.78$ Next, we divide and round to the nearest hundredth.

≈ 4.58 Finally, we add.

Note that rounding in the division step means that the result is only approximate.

Your Turn 3

Evaluate $\frac{x}{y} + 3y$ for $x = 25.2$ and $y = 8.4$.

Answer: 28.2

B Equations with Decimals

The Addition Property of Equations (Example 4) and the Multiplication Property of Equations (Example 5) can be used to solve equations that contain decimal numbers.

Example 4

Solve $x - 4.72 = 11.23$.

SOLUTION

$x - 4.72 = 11.23$

$x - 4.72 + 4.72 = 11.23 + 4.72$ Add 4.72 to both sides.

$x = 15.95$

The solution is 15.95.

Your Turn 4

Solve $x + 6.3 = 10.8$.

Answer: 4.5

Example 5

Solve $\frac{x}{5.3} = 6.21$.

SOLUTION

$\frac{x}{5.3} = 6.21$

$\left(\frac{5.3}{1}\right)\left(\frac{x}{5.3}\right) = (5.3)(6.21)$ Multiply both sides by 5.3.

$x = 32.913$

The solution is 32.913.

Your Turn 5

Solve $\frac{x}{0.06} = 9.1$.

Answer: 0.546

Copyright © Houghton Mifflin Company. All rights reserved.

Just as we have the option of clearing fractions from an equation, *clearing the decimals* might make an equation easier to solve. The method involves multiplying both sides of the equation by some power of 10.

Example 6

Solve $0.13x = 2.6$.

SOLUTION

We can solve the equation as it is written by dividing both sides by 0.13.

$$x = \frac{2.6}{0.13} = 20$$

However, we can change the decimal numbers to whole numbers by multiplying both sides by 100.

$$100(0.13x) = 100(2.6)$$
$$13x = 260$$
$$x = \frac{260}{13} = 20$$

The solution is 20.

Your Turn 6

Solve $0.03x = 0.12$.

LEARNING TIP

Count the number of decimal places in all the decimals in the equation and note the highest number. That number tells you the number of zeros in the power of 10 by which you multiply both sides of the equation in order to clear the decimals.

Answer: 4

Example 7

Solve $5.12x - 13.416 = 2x$.

SOLUTION

The 3 decimal places in 13.416 is the highest number of decimal places in all the numbers. Therefore, to clear the decimals, we multiply both sides by $10^3 = 1{,}000$. This moves the decimal point of each number 3 places to the right.

$$1{,}000(5.12x) - 1{,}000(13.416) = 1{,}000(2x)$$
$$5{,}120x - 13{,}416 = 2{,}000x$$
$$3{,}120x - 13{,}416 = 0 \quad \text{Subtract } 2{,}000x \text{ from both sides.}$$
$$\text{Add 13,416 to both sides.} \quad 3{,}120x = 13{,}416$$
$$\text{Divide both sides by 3,120.} \quad x = \frac{13{,}416}{3{,}120} = 4.3$$

The solution is 4.3.

Your Turn 7

Solve $1.7x + 17.214 = 6.23x$.

LEARNING TIP

Remember that multiplying both sides of an equation by a number means that every term is multiplied by that number.

Answer: 3.8

NOTE We clear decimals from *equations,* not *expressions.* Multiplying the terms of an expression by some power of 10 would change the value of the expression.

As with clearing fractions, the decision to clear decimals is yours. If you are able to use a calculator to perform the necessary arithmetic, there is not much advantage in clearing decimals. If you really prefer to work with whole numbers, then the extra steps in clearing fractions may be worth it to you.

Copyright © Houghton Mifflin Company. All rights reserved.

C Applications

Application problems, especially those that involve money, often involve decimal numbers.

Example 8

For children under 12, the cost of a movie ticket is \$5.50. Adult tickets cost \$8.25. For a group of people attending a movie, the total cost was \$57.75. If there were 3 more children than adults in the group, how many tickets of each kind were purchased?

SOLUTION

Let x = the number of adults.
Then $x + 3$ = the number of children.
$8.25x$ = cost of adult tickets.
$5.50(x + 3)$ = cost of children's tickets.

$$8.25x + 5.50(x + 3) = 57.75 \quad \text{The total cost was \$57.75.}$$
$$8.25x + 5.50x + 16.50 = 57.75 \quad \text{Distributive Property}$$
$$13.75x + 16.50 = 57.75 \quad \text{Combine like terms.}$$
$$13.75x = 41.25 \quad \text{Subtract 16.50 from both sides.}$$
$$x = \frac{41.25}{13.75} \quad \text{Divide both sides by 13.75.}$$
$$x = 3$$

Adult tickets (x): 3

Children's tickets ($x + 3$): 6

Your Turn 8

A ratchet wrench was on sale for \$2.65 off the original price. If a dozen wrenches cost \$111, what was the original price?

Answer: \$11.90

Example 9

A person is paid \$9.85 per hour for up to 40 hours per week and \$15.50 per hour for overtime. If the worker's pay for one week was \$487.00, how many overtime hours were worked?

SOLUTION

Let h = the number of overtime hours.
Regular pay: 40(\$9.85) = \$394.00
Overtime pay: \$15.50h
Total pay: \$487.00

$$394.00 + 15.50h = 487.00$$
$$15.50h = 93.00 \quad \text{Subtract 394.00 from both sides.}$$
$$h = \frac{93.00}{15.50} \quad \text{Divide both sides by 15.50.}$$
$$h = 6$$

The person worked 6 overtime hours.

Your Turn 9

Suppose a person earns \$11.33 per hour for up to 40 hours and double that for overtime hours. If the person earns \$521.18 for one week, how many overtime hours were worked?

Answer: 3

Copyright © Houghton Mifflin Company. All rights reserved.

Example 10

A veterinarian treats a dog with 30 milligrams of an antibiotic and prescribes one 12.5-milligram pill each day thereafter. After how many days has the dog received a total of 92.5 milligrams of the antibiotic?

SOLUTION

Let $d =$ the number of days of treatment after the first day. Then $12.5d =$ the number of milligrams taken after the first day. Number of milligrams taken on the first day $= 30$.

$$30 + 12.5d = 92.5$$

$$12.5d = 62.5 \quad \text{Subtract 30 from both sides.}$$

$$d = \frac{62.5}{12.5} \quad \text{Divide both sides by 12.5.}$$

$$d = 5$$

The dog received 92.5 milligrams over a 6-day period.

Your Turn 10

A cyclist travels 9.6 miles in the first hour of a ride, but then averages only 7.2 miles each hour after the first. What is the total number of hours that the cyclist rode if the total distance was 31.2 miles?

Answer: 4

6.5 Quick Reference

A. Evaluating Algebraic Expressions

1. We use the usual procedure for evaluating algebraic expressions even when the values of the variables are decimal numbers.

If $x = -7.2$ and $y = 9.1$, then

$$x + y + 2 = -7.2 + 9.1 + 2 = 3.9$$

2. We use the Order of Operations to evaluate an expression that includes more than one operation.

For $x = 0.2$ and $y = 1.3$,

$$\begin{aligned} 3x + y^2 &= 3(0.2) + (1.3)^2 \\ &= 3(0.2) + 1.69 \\ &= 0.6 + 1.69 \\ &= 2.29 \end{aligned}$$

Copyright © Houghton Mifflin Company. All rights reserved.

B. Equations with Decimals

1. The Addition Property of Equations can be used to solve equations that contain decimal numbers.

$$x + 5.8 = 11.23$$
$$x + 5.8 - 5.8 = 11.23 - 5.8$$
$$x = 5.43$$

2. Some equations involving decimal numbers can be solved with the Multiplication Property of Equations.

$$\frac{x}{4.2} = 0.3$$
$$\frac{4.2}{1} \cdot \frac{x}{4.2} = 4.2(0.3)$$
$$x = 1.26$$

3. When an equation contains decimals, we can *clear the decimals* by multiplying both sides by a power of 10. The exponent on (or the number of zeros in) the power of 10 is the highest number of decimal places in the decimal numbers.

$$3.1x - 23 = -0.68$$

Multiply both sides (each term) by 100.

$$310x - 2{,}300 = -68$$
$$310x = 2{,}232$$
$$x = \frac{2{,}232}{310} = 7.2$$

Copyright © Houghton Mifflin Company. All rights reserved.

6.5 Exercises

A. Evaluating Algebraic Expressions

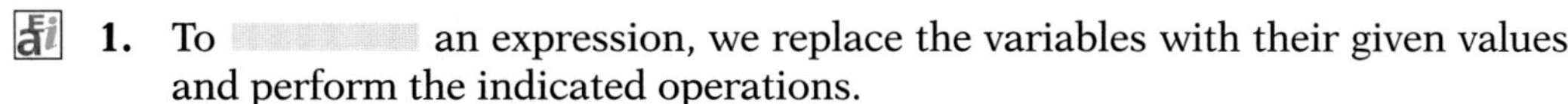

1. To ______ an expression, we replace the variables with their given values and perform the indicated operations.

2. Before evaluating an expression such as $3x - 7x$, we often ______ the expression.

In Exercises 3–6, evaluate the expression for the given values of the variables.

3. $a - b + c$; $a = 1.47, b = 2.09, c = 8.1$

4. $a + b - c$; $a = 1.1, b = -9.62, c = -1.1$

5. $x + y + z$; $x = 6.931, y = -4.96, z = -12.007$

6. $-x - y + z$; $x = -7.8, y = 2.936, z = -4.01$

In Exercises 7–14, evaluate the expression for the given value(s) of the variable(s).

7. $1.2(3a - 5)$ for $a = 2.7$

8. $4(3.5 - 0.6c)$ for $c = -1.4$

9. $a^2 + bc$ for $a = 0.6, b = -0.2, c = 3.1$

10. $(a - b)c^2$ for $a = 5.6, b = 12.3, c = 0.1$

11. $3.14r^2h$ for $r = 3.2, h = 10$

12. $6.28r(r + h)$ for $r = 4.1, h = 1.9$

13. Prt for $P = 500, r = 0.08, t = 1.5$

14. $P(1 + rt)$ for $P = 1{,}000, r = 0.1, t = 2.75$

In Exercises 15–18, evaluate the expression for the given values of the variables.

15. $x^2 + \frac{y}{x}$ for $x = -0.4$ and $y = -6.4$

16. $\frac{x - y}{x + y}$ for $x = 2.825$ and $y = -0.565$

Copyright © Houghton Mifflin Company. All rights reserved.

17. $\frac{ab}{c}$ for $a = -5.1$, $b = 3$, and $c = 3.4$

18. $\frac{2a}{b - c}$ for $a = 2.52$, $b = 5.9$, and $c = -6.7$

B. Equations with Decimals

19. We can multiply both sides of the equation $3.1x + 4 = 6.14$ by 100 to ________.

20. To clear decimals in an equation, multiply by a power of 10 with a(n) ________ that is equal to the greatest number of decimal places of a number in the equation.

In Exercises 21–24, solve the given equation.

21. $10 = x + 7.93$

22. $x - 60 = -31.2$

23. $n - 15.1 = 16.02$

24. $42.3 = 27.24 + m$

In Exercises 25–28, solve the given equation.

25. $\frac{x}{1.4} = 6$

26. $-5 = \frac{n}{0.6}$

27. $-1.5 = \frac{y}{0.2}$

28. $\frac{t}{7} = -1.1$

In Exercises 29–40, solve the given equation.

29. $13.8 = 6x$

30. $-5y = 8.5$

31. $-0.4n = -1.44$

32. $-0.96 = 1.2x$

33. $3x - 4.5 = 8.1$

34. $6y + 8.1 = 5.7$

35. $1.57 = 6.05 - 3.2n$

36. $7.88 = 4.1y - 6.47$

37. $1.7 - x = 3 - 0.8x$

38. $2x - 5.2 = 22 - 1.4x$

39. $2x - 6.08 = 21.2 - 4.2x$

40. $5x + 1.3 = 4.14 - 2.1x$

Copyright © Houghton Mifflin Company. All rights reserved.

C. Applications

41. *Ticket Purchase* TicketKing adds a service charge of $3.15 per ticket. If a person paid TicketKing $239.40 for 6 tickets to a baseball game, what was the actual price of 1 ticket?

42. *CD Sales* A band sold CDs for $2.75 more than tapes. If the band sold 19 tapes for $128.25, what was the price of a CD?

43. *Rental Car Charges* Suppose that a car rental company charges $19.90 per day and $.24 per mile. If your one-day rental cost was $77.98, for how many miles were you billed?

44. *Overtime* As a machine operator, you are paid $12.30 per hour for up to 40 hours per week and $18.45 per hour for overtime. If your pay for one week was $621.15, how many overtime hours did you work?

45. *Sewer Line* To connect to the city sewer system, a homeowner paid a tap-on fee of $1,500 plus a per-foot cost for installing the sewer line. If the total cost for 142 feet of sewer line was $2,472.70, what was the cost per foot for installation of the line?

46. *Medication* A doctor prescribed a certain medication to be administered as an initial injection of 12 milligrams followed by 6.5 milligrams per hour to be given intravenously. After how many hours had the patient received 44.5 milligrams of the medication?

Copyright © Houghton Mifflin Company. All rights reserved.

47. *Hourly Wages* A person earned \$2.35 per hour more as a dental assistant than at an evening job at a department store. In one week the person worked 50 hours of which 18 hours were at the department store. If the total pay for the week was \$430.20, what was the hourly pay at each job?

48. *Book Sales* The Lions Club sold 64 cookbooks of which 26 were mailed. The club charged an extra \$2.40 to mail a book. If the club collected a total of \$830.40, what was the price of a book?

Writing and Concept Extension

49. If you want to clear the decimals in an equation, how do you determine what to multiply by?

50. Explain how to easily simplify $100(4.3x + 7.05)$.

51. To clear the decimals in the equation $1.2x + 6 = 7.2$, you can multiply both sides of the equation by 10. Which is the correct result? Why?

(i) $12x + 6 = 72$ (ii) $12x + 60 = 72$

52. The following equations are equivalent. Explain how to begin solving each.

(i) $0.5x = 7$ (ii) $\frac{x}{2} = 7$

53. Suppose that you translate the statement "the value of n nickels is 35 cents" into an equation. Explain why each of the following is a correct equation.

(i) $0.05n = 0.35$ (ii) $5n = 35$

Copyright © Houghton Mifflin Company. All rights reserved.

6.6 Irrational Numbers and Square Roots

SUGGESTIONS FOR SUCCESS

Undoing is a favorite sport in mathematics.

We have seen that addition and subtraction are operations that undo each other; likewise, multiplication and division undo each other. Might there be an operation that undoes squaring? For example, if we square 3, we get 9. Is there an operation that we can perform on 9 to return to 3?

As you will see in this section, the answer is yes. You should find this topic easier to understand if you try to think of taking a square root as an operation that undoes squaring. You are already familiar with this simple idea, so relax and enjoy.

A *Irrational Numbers*

All the fractions we have worked with so far have been fractions in which both the numerator and the denominator were integers. The official name for such fractions is **rational numbers.**

Definition of a Rational Number

A **rational number** is a number that can be written in the form $\frac{a}{b}$, where a and b are integers and b is not 0.

Here are some examples of rational numbers.

$$\frac{5}{7} \qquad \frac{-2}{3} \qquad \frac{9}{-10} \qquad 4 \qquad 1\frac{3}{8}$$

The first three fractions in the list are rational numbers because their numerators and denominators are integers. You might be surprised to see that 4 and $1\frac{3}{8}$ are considered to be rational numbers. However, 4 can be written as $\frac{4}{1}$, and $1\frac{3}{8}$ can be written as $\frac{11}{8}$.

We know that we can find the decimal name of a rational number by dividing the numerator by the denominator. When we do that, there are two possible results.

1. The remainder is eventually 0. We call such decimals **terminating decimals.**
2. The remainders begin to repeat. We call such decimals **repeating decimals.**

Example 1

Determine whether the decimal names for the given fractions are terminating decimals or repeating decimals.

(a) $\frac{5}{8}$ **(b)** $\frac{5}{9}$ **(c)** $\frac{5}{11}$

Your Turn 1

Determine whether the decimal names for the given fractions are terminating decimals or repeating decimals.

(a) $\frac{1}{6}$ **(b)** $\frac{7}{20}$ **(c)** $\frac{1}{99}$

continued

Copyright © Houghton Mifflin Company. All rights reserved.

SOLUTION

Verify the decimal names in each part by dividing the numerator by the denominator.

(a) $\frac{5}{8} = 0.62500000000 \cdots$

The fourth decimal place is 0, and every decimal place thereafter is 0. Therefore, 0.625 is a terminating decimal.

(b) $\frac{5}{9} = 0.55555555555 \cdots$

The repeating pattern of 5s means that the decimal is a repeating decimal.

(c) $\frac{5}{11} = 0.454545454545 \cdots$

This time, the pattern consists of the block 45, which repeats over and over again. This is another repeating decimal.

NOTE Repeating decimals are often written with a bar over the repeating block. In Example 1(b), we can write $\frac{5}{9} = 0.\overline{5}$. In Example 1(c), we can write $\frac{5}{11} = 0.\overline{45}$. Some repeating blocks can be quite long. For example, $\frac{1}{7} = 0.\overline{142857}$.

Every rational number has a decimal name that is either a terminating decimal or a repeating decimal. Now think about this: Can you construct a decimal number that is neither terminating nor repeating? Look carefully at the following decimal number:

$$0.181181118111181111118 \cdots$$

The decimal is not terminating because it goes on forever. Furthermore, this number doesn't have a repeating block because the number of 1s keeps increasing. This *nonterminating, nonrepeating* decimal is an example of an **irrational number.** Simply said, an irrational number is a number that is not a rational number; that is, it can't be obtained by dividing one integer by another.

Taken together, the rational numbers and the irrational numbers form the **real numbers.**

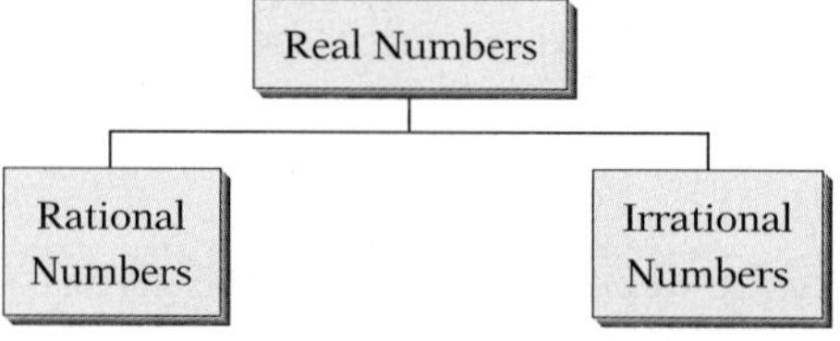

The figure indicates three facts:

1. Every rational number is also a real number.
2. Every irrational number is also a real number.
3. No number is both a rational number and an irrational number.

Copyright © Houghton Mifflin Company. All rights reserved.

B Square Roots of Perfect Squares

Squaring an integer means multiplying the integer by itself. The result is called a **perfect square.** Here are 15 perfect squares, which we obtain by squaring the integers 1 through 15.

$1^2 = 1$	$4^2 = 16$	$7^2 = 49$	$10^2 = 100$	$13^2 = 169$
$2^2 = 4$	$5^2 = 25$	$8^2 = 64$	$11^2 = 121$	$14^2 = 196$
$3^2 = 9$	$6^2 = 36$	$9^2 = 81$	$12^2 = 144$	$15^2 = 225$

We obtain the same perfect squares if we square the opposites of those integers. For example, $(-1)^2 = 1$, $(-2)^2 = 4$, $(-3)^2 = 9$, and so on. Note that perfect squares are never negative.

Being familiar with these perfect squares allows us to solve an equation such as $x^2 = 25$. In words, the equation is asking, "What number squared is 25?" We know that there are two such numbers, 5 and -5, because $5^2 = 25$ and $(-5)^2 = 25$. We say that 5 and -5 are **square roots** of 25.

Definition of Square Root

The number a is a **square root** of b if $a^2 = b$.

NOTE The definition of square root also gives us a way to check our answers. For example, we can prove that 3 and -3 are both square roots of 9 by showing that $3^2 = 9$ and $(-3)^2 = 9$. Note that a positive number always has two square roots, and they are opposites.

Here is an important detail: A negative number cannot have a square root. For example, -16 doesn't have a square root because there is no number that we can square to obtain -16.

Example 2

Find the square roots of these numbers.

(a) 49 **(b)** 0

(c) $\frac{4}{9}$ **(d)** -25

SOLUTION

(a) The square roots of 49 are 7 and -7 because $7^2 = 49$ and $(-7)^2 = 49$.

(b) The square root of 0 is 0 because $0^2 = 0$. Note that 0 has only one square root.

(c) The square roots of $\frac{4}{9}$ are $\frac{2}{3}$ and $-\frac{2}{3}$ because $\left(\frac{2}{3}\right)^2 = \frac{4}{9}$ and $\left(-\frac{2}{3}\right)^2 = \frac{4}{9}$.

(d) A negative number cannot have a square root.

Your Turn 2

Find the square roots of these numbers.

(a) 64 **(b)** $\frac{16}{25}$

Answers: (a) 8, -8; (b) $\frac{4}{5}$, $-\frac{4}{5}$

Copyright © Houghton Mifflin Company. All rights reserved.

> **LEARNING TIP**
> *Make sure that you know the difference between squaring a number and taking the square root of a number. The* square *of 4 is* $4^2 = 16$, *whereas the* square root *of 4 is* $\sqrt{4} = 2$.

Although every positive number has two square roots, usually we are most interested in the positive square root. We represent this number with a **radical symbol** $\sqrt{}$. For example, $\sqrt{25}$ means 5, not -5. In this expression, we call 25 the **radicand,** and the entire expression $\sqrt{25}$ is called a **radical.** Finding $\sqrt{25}$ is called "taking the square root of 25."

A radical symbol is also a grouping symbol. In expressions such as $-\sqrt{36}$ and $10 + \sqrt{9}$, we take the square root first.

Example 3

Evaluate these expressions.

(a) $\sqrt{36}$ **(b)** $-\sqrt{36}$ **(c)** $10 + \sqrt{9}$

(d) $\sqrt{25} - \sqrt{16}$ **(e)** $2\sqrt{81}$ **(f)** $5 + 2\sqrt{\frac{1}{4}}$

SOLUTION

(a) $\sqrt{36} = 6$

(b) $-\sqrt{36} = -6$ Take the square root of 36 and then write the opposite of the result.

(c) $10 + \sqrt{9} = 10 + 3 = 13$

(d) $\sqrt{25} - \sqrt{16} = 5 - 4 = 1$

(e) $2\sqrt{81} = 2 \cdot 9 = 18$

(f) $5 + 2\sqrt{\frac{1}{4}} = 5 + 2\left(\frac{1}{2}\right)$ $\sqrt{\frac{1}{4}} = \frac{1}{2}$ because $\left(\frac{1}{2}\right)^2 = \frac{1}{4}$.

$= 5 + 1 = 6$

Your Turn 3

Evaluate these expressions.

(a) $\sqrt{81}$

(b) $-\sqrt{4}$

(c) $5 - \sqrt{16}$

(d) $\sqrt{36} + \sqrt{49}$

(e) $8\sqrt{25}$

(f) $1 + 9\sqrt{\frac{1}{9}}$

Answers: (a) 9; (b) −2; (c) 1; (d) 13; (e) 40; (f) 4

KEYS TO THE CALCULATOR

Taking the square root of a number with a calculator is easily done with the $\boxed{\sqrt{}}$ key. Here are two ways to calculate $\sqrt{9}$.

Scientific calculator: **9** $\boxed{\sqrt{}}$ You usually don't have to press the $\boxed{=}$ key.

Graphing Calculator: $\boxed{\text{2nd}}$ $\boxed{\sqrt{}}$ **9** $\boxed{\text{ENTER}}$

Exercises

(a) $\sqrt{144}$ **(b)** $\sqrt{3249}$ **(c)** $\sqrt{841} - \sqrt{361}$ **(d)** $6\sqrt{225} - 2\sqrt{121}$

Copyright © Houghton Mifflin Company. All rights reserved.

C Square Roots and Irrational Numbers

Now we consider taking the square root of a number that is not a perfect square.

DEVELOPING THE CONCEPT

The Square Root of 3

To calculate $\sqrt{3}$, we must find a number whose square is 3. We can begin with some estimates.

Estimate	*Check*	
$\sqrt{3} \approx 1.7$	$(1.7)^2 = 2.89$	Not very close to 3.
$\sqrt{3} \approx 1.73$	$(1.73)^2 = 2.9929$	We're getting closer.
$\sqrt{3} \approx 1.732$	$(1.732)^2 = 2.999824$	Even closer!
$\sqrt{3} \approx 1.7321$	$(1.7321)^2 = 3.00017041$	1.7321 is too large.

Because $(1.732)^2$ is just under 3 and $(1.7321)^2$ is just over 3, we can conclude that $\sqrt{3}$ must lie between 1.7320 and 1.7321. However, we could continue this process forever and never find an exact decimal name for $\sqrt{3}$. The decimal representation is a nonterminating, nonrepeating decimal; that is, the decimal is an irrational number.

Although there is an arithmetic procedure for finding the square root of a number, hardly anyone uses it anymore. We can swiftly obtain square roots with a calculator and round off the results however we wish.

Example 4

Evaluate the following to the nearest hundredth:

(a) $\sqrt{18}$ **(b)** $4\sqrt{5}$ **(c)** $\sqrt{7} + \sqrt{2}$

SOLUTION

The following results were obtained with a calculator:

(a) $\sqrt{18} \approx 4.24$

(b) $4\sqrt{5} \approx 8.94$

(c) $\sqrt{7} + \sqrt{2} \approx 4.06$

Your Turn 4

Evaluate the following to the nearest hundredth:

(a) $\sqrt{23}$

(b) $3\sqrt{7}$

(c) $\sqrt{19} - \sqrt{6}$

Answers: (a) 4.80; (b) 7.94; (c) 1.91

NOTE Remember that a radical symbol is also a grouping symbol. In Example 4(c), your calculator finds the square roots of 7 and 2 and then adds the results. The expression $\sqrt{7} + \sqrt{2}$ is *not* the same as $\sqrt{7 + 2}$ in which we add 7 + 2 and then take the square root.

D Applications

In real-life situations, we can often describe the relationships among quantities with formulas or other equations that involve square roots.

Copyright © Houghton Mifflin Company. All rights reserved.

In Example 5, we use the fact that speed and braking distance are related. We measure the distance d (in feet) that a car travels from the point where the brakes are applied to the point where the car has come to a stop. Then we use the formula $s = 3.5\sqrt{d}$ to calculate the speed s (in miles per hour) that the car was traveling.

Example 5

At an accident scene, a police officer finds a skid mark that is 235 feet long. To the nearest whole number of miles per hour, how fast was the car traveling when the brakes were applied?

SOLUTION

$$s = 3.5\sqrt{d}$$
$$= 3.5\sqrt{235}$$
$$\approx 3.5(15.3)$$

We use the ≈ symbol because 15.3 is only an approximation of $\sqrt{235}$.

$$\approx 53.55$$

The car was traveling at about 54 miles per hour.

Your Turn 5

In Example 5, at what speed was the car traveling if the skid mark was 300 feet long?

Answer: 61 miles per hour

If wind resistance is ignored, gravity causes a free-falling object to drop at faster and faster speeds. The time t (in seconds) that an object takes to fall a certain distance d (in feet) is given by the formula $t = \sqrt{\frac{d}{16}}$.

Example 6

The Terminal Tower in Cleveland is 708 feet tall. If a sightseer drops his camera from the observation deck, how long (to the nearest tenth of a second) will the camera take to reach the ground? (Ignore wind resistance.)

SOLUTION

$$t = \sqrt{\frac{d}{16}}$$
$$= \sqrt{\frac{708}{16}}$$
$$= \sqrt{44.25}$$
$$\approx 6.65$$

The camera will take about 6.7 seconds to reach the ground.

Your Turn 6

Suppose that the events in Example 6 had occurred at Houston's Texas Commerce Tower, which is 1,002 feet high. Answer the same question.

Answer: 7.9 seconds

Copyright © Houghton Mifflin Company. All rights reserved.

6.6 Quick Reference

A. Irrational Numbers

1. A **rational number** is a number that can be written in the form $\frac{a}{b}$, where a and b are integers and b is not 0.

Rational numbers: $\frac{2}{7}$ $\frac{-5}{8}$ $\frac{8}{-3}$

10 is a rational number because it can be written $\frac{10}{1}$.

$2\frac{3}{4}$ is a rational number because it can be written $\frac{11}{4}$.

2. Every rational number has a decimal name that is either a terminating decimal or a repeating decimal.

$\frac{7}{8} = 0.875$ Terminating

$\frac{2}{3} = 0.666666\cdots$

$= 0.\overline{6}$ Repeating

3. An **irrational number** has a decimal name that is nonterminating and nonrepeating. Taken together, the rational numbers and the irrational numbers form the **real numbers.**

4. **(a)** Every rational number is also a real number.
   (b) Every irrational number is also a real number.
   (c) No number is both a rational number and an irrational number.

B. Square Roots of Perfect Squares

1. A **perfect square** is the result of multiplying an integer by itself.

Some perfect squares:

0, 1, 4, 9, 16, 25, . . .

2. The number a is a **square root** of b if $a^2 = b$.

The square roots of 16 are 4 and -4 because $4^2 = 16$ and $(-4)^2 = 16$.

3. A negative number does not have a square root.

No number can be squared to obtain -9. Therefore, -9 has no square roots.

4. Every positive number has two square roots, and they are opposites. The positive square root is denoted with the **radical symbol** $\sqrt{}$.

The square roots of 36 are 6 and -6, but $\sqrt{36} = 6$.

Copyright © Houghton Mifflin Company. All rights reserved.

5. An expression of the form $\sqrt{x}$ is called a **radical,** and x is called the **radicand.**

For the radical $\sqrt{100}$, the radicand is 100.

C. Square Roots and Irrational Numbers

1. If x is not a perfect square, then $\sqrt{x}$ is an irrational number that we can find (approximately) with a calculator.

$$\sqrt{10} \approx 3.16227766 \cdots$$

We usually round off to a meaningful number of decimal places.

Copyright © Houghton Mifflin Company. All rights reserved.

6.6 Exercises

A. Irrational Numbers

1. Fractions in which both the numerator and the denominator are integers are examples of ________ numbers.

2. A nonterminating, nonrepeating decimal is a(n) ________ number.

In Exercises 3–6, write the fraction as a terminating decimal.

3. $\frac{15}{4}$ **4.** $\frac{9}{8}$ **5.** $\frac{7}{16}$ **6.** $\frac{33}{25}$

In Exercises 7–12, write the fraction as a repeating decimal.

7. $\frac{7}{3}$ **8.** $\frac{2}{9}$ **9.** $\frac{5}{6}$

10. $\frac{16}{15}$ **11.** $\frac{35}{11}$ **12.** $\frac{8}{33}$

B. Square Roots of Perfect Squares

13. Because $(-3)^2 = 9$ and $3^2 = 9$, both -3 and 3 are called ________ of 9.

14. In $\sqrt{7}$, we call $\sqrt{\ }$ the ________ symbol and 7 the ________.

In Exercises 15–22, determine the square roots of the given number.

15. 16 **16.** 36 **17.** 81 **18.** 100

19. -4 **20.** -64 **21.** $\frac{1}{9}$ **22.** $\frac{49}{25}$

In Exercises 23–30, evaluate the radical.

23. $\sqrt{25}$ **24.** $\sqrt{49}$ **25.** $-\sqrt{100}$ **26.** $-\sqrt{16}$

27. $\sqrt{144}$ **28.** $\sqrt{121}$ **29.** $\sqrt{\frac{64}{25}}$ **30.** $\sqrt{\frac{4}{9}}$

Copyright © Houghton Mifflin Company. All rights reserved.

In Exercises 31–36, evaluate the radical expression.

31. $4\sqrt{9}$

32. $3\sqrt{4}$

33. $5\sqrt{36}$

34. $9\sqrt{100}$

35. $-2\sqrt{49}$

36. $-6\sqrt{16}$

In Exercises 37–42, evaluate the radical expression.

37. $\sqrt{64} + \sqrt{81}$

38. $\sqrt{49} - \sqrt{9}$

39. $2\sqrt{81} - 6\sqrt{9}$

40. $3\sqrt{49} + 2\sqrt{9}$

41. $\sqrt{16 + 9}$

42. $4 - \sqrt{36}$

C. Square Roots and Irrational Numbers

43. Numbers such as $\sqrt{5}$ and $\sqrt{3}$ are examples of ________ numbers.

44. Because a radical symbol is also a(n) ________ symbol, to evaluate $\sqrt{16 + 9}$, we add 16 and 9 before evaluating the square root.

In Exercises 45–48, use a calculator to evaluate the radical. Round answers to the nearest hundredth.

45. $\sqrt{45}$

46. $\sqrt{74}$

47. $\sqrt{3.7}$

48. $\sqrt{12.6}$

In Exercises 49–52, use a calculator to evaluate the radical expression. Round answers to the nearest hundredth.

49. $4\sqrt{7}$

50. $2\sqrt{2}$

51. $-3\sqrt{10}$

52. $-8\sqrt{3}$

In Exercises 53–56, use a calculator to evaluate the radical expression. Round answers to the nearest hundredth.

53. $\sqrt{5} + \sqrt{6}$

54. $\sqrt{10} - \sqrt{7}$

55. $-2\sqrt{15} + 5\sqrt{10}$

56. $\sqrt{12} + 2\sqrt{8}$

≈ *Estimation*

In Exercises 57–60, without using a calculator, determine two consecutive integers such that the value of the radical is between the integers.

57. $\sqrt{42}$

58. $\sqrt{115}$

59. $\sqrt{75}$

60. $\sqrt{150}$

Copyright © Houghton Mifflin Company. All rights reserved.

D. Applications

For Exercises 61–64, refer to the formulas given in Examples 5 and 6.

61. *Construction Accident* A construction worker dropped a hammer from a scaffold that was 196 feet above the ground. How many seconds did the hammer take to reach the ground?

62. *Water Balloon* If a prankster drops a water balloon from a window that is located 121 feet above the ground, how many seconds does the intended victim have to avoid being hit?

63. *Skid Marks* Following an accident, police investigators measured the skid marks made by one car. If the skid marks were 256 feet, what was the estimated speed of the car?

64. *Skid Marks* If a car left skid marks of 196 feet, by how much was the car exceeding the posted speed limit of 40 miles per hour?

65. *Estimated Salary* Suppose that a person's annual salary S can be estimated by $S = 30{,}000 + 5{,}000\sqrt{t}$, where t is the number of years that the person has worked for the company. What is the estimated salary for a person who has worked for the company for 9 years?

66. *Predicted Enrollment* Suppose that the enrollment E at a new campus of a community college can be predicted by $E = 2{,}500 + 400\sqrt{t}$, where t is the number of years after the new campus is opened. What is the predicted enrollment 4 years after the campus opens?

67. *Medication in the Blood Stream* Suppose that the amount M (in milligrams) of a certain medication in the blood stream can be estimated by $M = \dfrac{30}{\sqrt{h}}$, where h is the number of hours after the medication was injected. What is the predicted amount of medication in the blood stream 4 hours after an injection?

Copyright © Houghton Mifflin Company. All rights reserved.

68. *Reaction Time* Suppose that the reaction time R (in seconds) of a truck driver can be estimated by $R = 5 + \frac{8}{\sqrt{t}}$, where t is the number of thousands of miles that the trucker has driven. What is a trucker's estimated reaction time after 16,000 miles of driving?

Writing and Concept Extension

69. Explain the difference between "the square of 4" and "the square root of 4."

70. Explain the difference between $1.\overline{3}$ and 1.3.

In Exercises 71–74, evaluate the radical expression.

71. $(\sqrt{9})^2$ **72.** $\sqrt{9^2}$ **73.** $\sqrt{0.49}$ **74.** $\sqrt{5\frac{4}{9}}$

In Exercises 75 and 76, evaluate the expression for the given values of the variables.

75. $x + \sqrt{y}$ for $x = -1$ and $y = 16$ **76.** $\sqrt{a^2 + 3b}$ for $a = -8$ and $b = -5$

In Exercises 77–80, determine whether the equation is true or false.

77. $\sqrt{16 + 9} = \sqrt{16} + \sqrt{9}$ **78.** $\sqrt{100} - \sqrt{36} = \sqrt{100 - 36}$

79. $\sqrt{4} \cdot \sqrt{9} = \sqrt{4 \cdot 9}$ **80.** $\sqrt{\frac{64}{16}} = \frac{\sqrt{64}}{\sqrt{16}}$

Copyright © Houghton Mifflin Company. All rights reserved.

6.7 Geometry Formulas

A *The Pythagorean Theorem*
B *Circumference and Area of a Circle*
C *Spheres and Cylinders*

SUGGESTIONS FOR SUCCESS

We represent numbers such as three, five, and ten with the familiar symbols 3, 5, 10. However, other symbols, such as the Roman numerals III, V, and X, can be used. In this section, you will learn that the Greek letter π is used to represent a very important number. You will need to keep in mind that π is a specific number, not a variable.

A The Pythagorean Theorem

A **right triangle** is a triangle in which two sides form a right (90°) angle. As shown in the figure, the right angle is indicated by a small box.

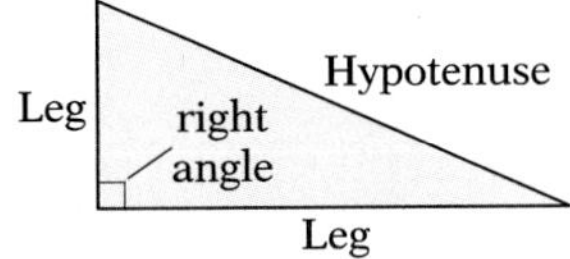

The two sides that form the right angle are called the **legs,** and the third side is called the **hypotenuse** (pronounced high-POT-en-oose).

A special relationship exists among the lengths of the three sides of a right triangle.

The Pythagorean Theorem

For a right triangle, if the lengths of the legs are a and b, and the length of the hypotenuse is c, then $c^2 = a^2 + b^2$.

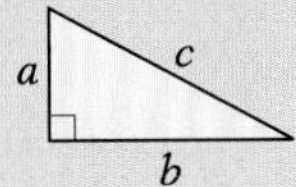

NOTE In mathematics, a *theorem* is a fact that can be proven to be true. The Pythagorean Theorem is named for Pythagoras, an early Greek philosopher. *Pythagorean* is pronounced Pi-thag-o-RE-an.

You can use a and b to represent the length of either leg. However, you must be able to identify the hypotenuse and label it as c. The hypotenuse is the longest side and lies opposite the right angle.

If a triangle is not a right triangle, then knowing the lengths of two sides will not be enough to find the length of the third side. However, for right triangles, if we know the lengths of any two sides, then we can use the Pythagorean Theorem to find the length of the third side.

Copyright © Houghton Mifflin Company. All rights reserved.

Example 1

Find the length of the hypotenuse of the given triangle.

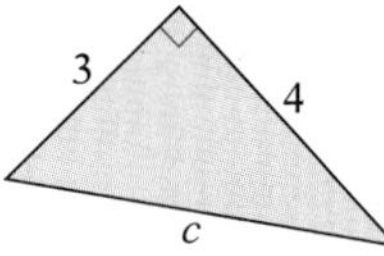

SOLUTION

We let $a = 3$ and $b = 4$.

$$\begin{aligned} c^2 &= a^2 + b^2 \\ &= 3^2 + 4^2 \\ &= 9 + 16 \\ &= 25 \end{aligned}$$

Because $c^2 = 25$, c is a square root of 25. We choose the positive square root because length is never negative.

$$c = \sqrt{25} = 5$$

Your Turn 1

Find the approximate length of the hypotenuse of the given triangle.

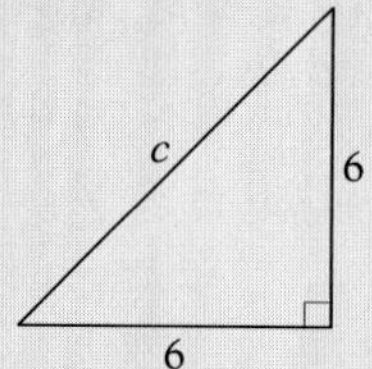

Answer: 8.49

As in Example 1, we need only simple arithmetic to find c when a and b are known. However, to find the length of a leg, we will always need to subtract a number from both sides in order to solve the equation. Example 2 illustrates this.

Example 2

One leg of a triangle is 8 feet, and the hypotenuse is 15 feet. Find the length of the other leg to the nearest tenth of a foot.

SOLUTION

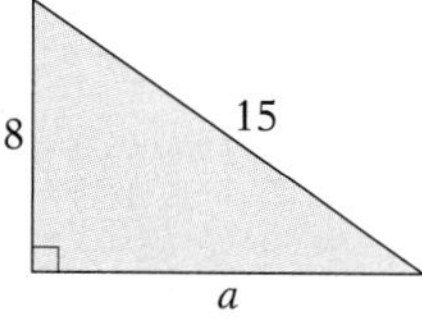

We let $b = 8$ and $c = 15$.

$$\begin{aligned} c^2 &= a^2 + b^2 \\ 15^2 &= a^2 + 8^2 \\ 225 &= a^2 + 64 \\ a^2 &= 225 - 64 = 161 && \text{Subtract 64 from both sides to isolate } a^2. \\ a &= \sqrt{161} \approx 12.7 \text{ feet} \end{aligned}$$

Your Turn 2

One leg of a triangle is 5 inches, and the hypotenuse is 12 inches. Find the length of the other leg to the nearest tenth of an inch.

Answer: 10.9 inches

Many real-life applications involve right triangles and can be solved with the Pythagorean Theorem. A key step in solving the problem is identifying the side of the triangle that is the hypotenuse.

Copyright © Houghton Mifflin Company. All rights reserved.

Example 3

A loading dock is 4 feet high. A ramp is to be built so that the bottom of the ramp is 20 feet from the base of the wall.

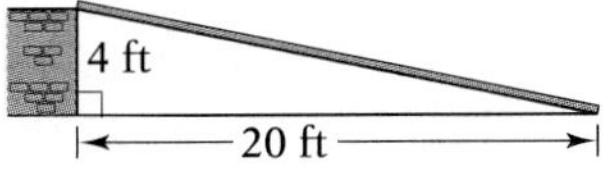

To the nearest tenth of a foot, how long should the ramp be?

SOLUTION

The wall, the ramp, and the ground form a right triangle, where the hypotenuse is the ramp.

$$\begin{aligned} c^2 &= a^2 + b^2 \\ &= 4^2 + 20^2 \\ &= 16 + 400 \\ &= 416 \\ c &= \sqrt{416} \approx 20.4 \text{ feet} \end{aligned}$$

The ramp is about 20.4 feet long.

Your Turn 3

A rectangular garden is 30 feet wide and 50 feet long. To the nearest tenth of a foot, how long is a path that cuts diagonally through the garden?

Answer: 58.3 feet

Example 4

A flag pole is supported by a guy wire that is 40 feet long. The guy wire is attached to the ground at a point that is 8 feet from the base of the pole. How high is the flag pole?

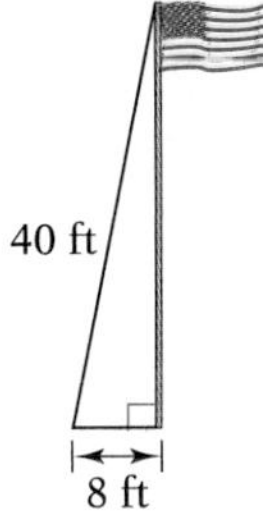

SOLUTION

In the figure, the guy wire is the hypotenuse, c. We let b represent the height of the pole.

$$\begin{aligned} c^2 &= a^2 + b^2 \\ 40^2 &= 8^2 + b^2 \\ 1{,}600 &= 64 + b^2 \\ b^2 &= 1{,}600 - 64 = 1{,}536 \\ b &= \sqrt{1{,}536} \approx 39.2 \end{aligned}$$

The pole is about 39.2 feet high.

Your Turn 4

A 10-foot framed wall is constructed on a foundation that is 2 feet high. The wall is braced with a board that is fastened at the ground 8 feet from the foundation. Approximately how long is the board?

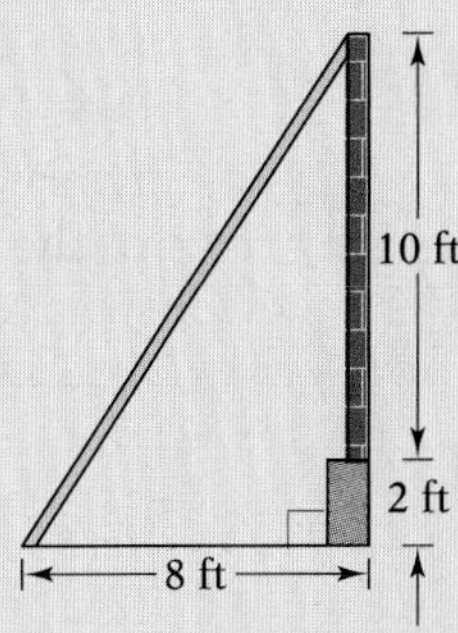

Answer: 14.4 feet

Copyright © Houghton Mifflin Company. All rights reserved.

We can also use the Pythagorean Theorem to test whether a given triangle is a right triangle.

Example 5

Forms for a foundation are to be set so that the corner at B is a right angle. The figure shows the lengths of the two forms.

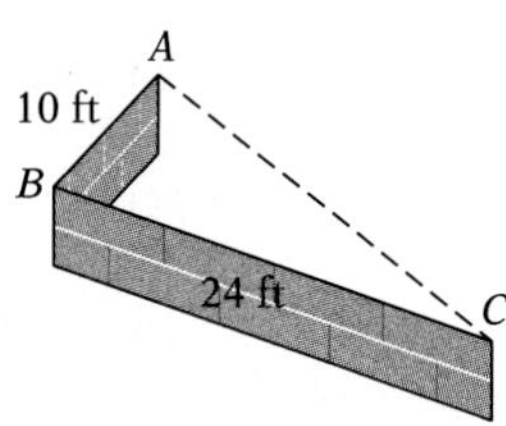

The supervisor measures the distance from A to C, and the distance is 27 feet. Are the forms set properly?

SOLUTION

If the corner at B is a right angle, then the triangle ABC is a right triangle, where the hypotenuse is 27 feet.

$$a^2 + b^2 = 10^2 + 24^2 = 100 + 576 = 676$$
$$c^2 = 27^2 = 729$$

Because $c^2 \neq a^2 + b^2$, the triangle ABC is not a right triangle. Therefore, the corner at B is not a right angle. The forms are not set correctly.

Your Turn 5

A person wants to build a rectangular frame that is 8 inches wide and 15 inches long. If the diagonal distance across the frame is 17 inches, are the corners of the frame right angles?

Answer: Yes

B Circumference and Area of a Circle

Every point of a circle is the same distance from a given point, called the **center** of the circle.

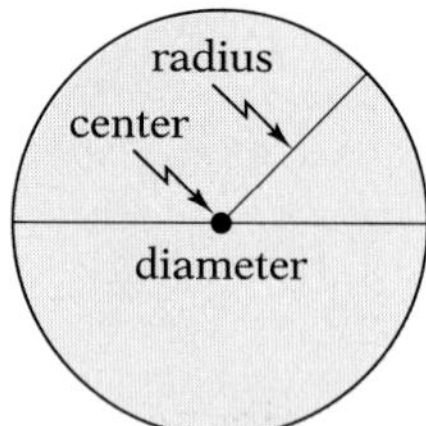

In the figure, the line segment that connects the center and a point of the circle is called the **radius,** usually denoted by r. The line segment that contains the center and whose endpoints are points of the circle is called the **diameter,** usually denoted by d. The diameter of a circle is twice as long as the radius: $d = 2r$.

NOTE The words *radius* and *diameter* can refer to the line segments themselves or to the lengths of those line segments.

Copyright © Houghton Mifflin Company. All rights reserved.

The distance around a circle is called the **circumference** of the circle. You can think of a circle formed by a string. If you cut the string and lay it out in a straight line, then the length of the string is the circumference of the circle.

The early Greeks who studied circles observed that the circumference is about 3 times the diameter. Through advances in mathematics and technology, we now know that the actual number is an irrational number called π, which is the Greek letter pi.

$$\pi \approx 3.141592653 \cdots$$

LEARNING TIP

The symbol π is not a variable. It is simply the name that we give to a specific number. Such symbols are often called constants.

The number π is perhaps the most well-known and most often used irrational number. When we use π in applications, we will round the decimal to 3.14. (Your calculator should have a $\boxed{\pi}$ key that can be used to enter π to any number of decimal places.)

The number π is a factor in the formulas for the circumference and the area of a circle.

The Circumference and Area of a Circle

For a circle with a diameter d and a radius r:

1. The circumference C is given by $C = \pi d$ or $C = 2\pi r$.
2. The area A is given by $A = \pi r^2$.

Example 6

Find the circumference and the area of the given circle.

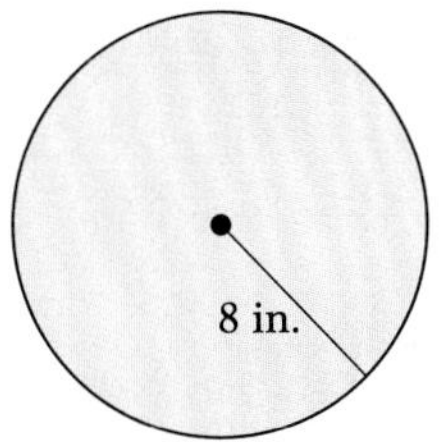

SOLUTION

$$C = 2\pi r = 2(3.14)(8) \qquad \pi \approx 3.14, r = 8$$
$$= 50.24 \text{ inches}$$
$$A = \pi r^2 = 3.14(8)^2 \qquad \pi \approx 3.14, r = 8$$
$$= 3.14(64)$$
$$= 200.96 \text{ square inches}$$

Your Turn 6

Find the circumference and the area of the given circle.

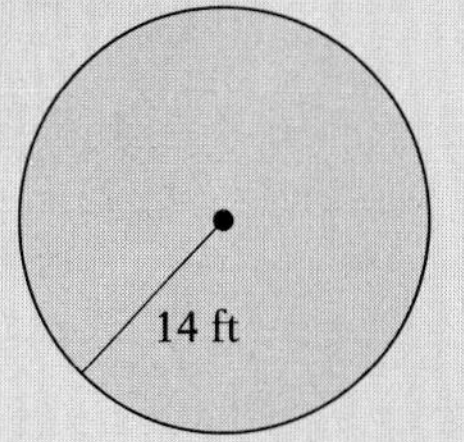

Answer: $C = 87.92$ feet; $A = 615.44$ square feet

NOTE Remember that 3.14 is only an approximate value of π. For simplicity we use equal symbols in our examples, but the results are really just rounded off approximations.

Copyright © Houghton Mifflin Company. All rights reserved.

Example 7

Suppose that the diameter of a bicycle wheel is 2.5 feet. How far does the bicycle travel after 5 revolutions of the wheel?

SOLUTION

For each revolution, the bicycle travels the circumference of the wheel.

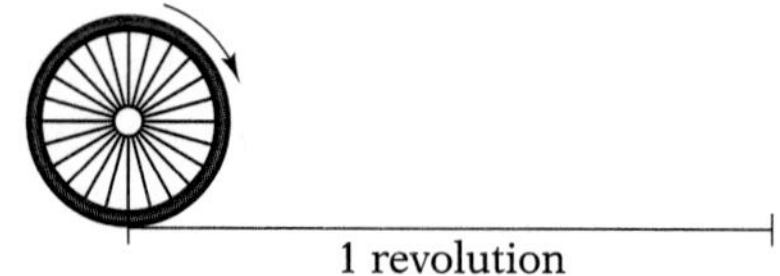

$$C = \pi d = 3.14(2.5) \quad \pi \approx 3.14,\ d = 2.5$$
$$= 7.85 \text{ feet}$$

After 5 revolutions, the bicycle travels

$$5(7.85) = 39.25 \text{ feet}$$

Your Turn 7

The wheels of a lawn mower have a radius of 4 inches. After 10 revolutions of the wheels, how far have you pushed the mower?

Answer: 251.2 inches

Example 8

A circular garden with a 50-foot radius is surrounded by a walkway of uniform width. If the total area of the garden and the walkway is 9,161 square feet, what is the area of the walkway?

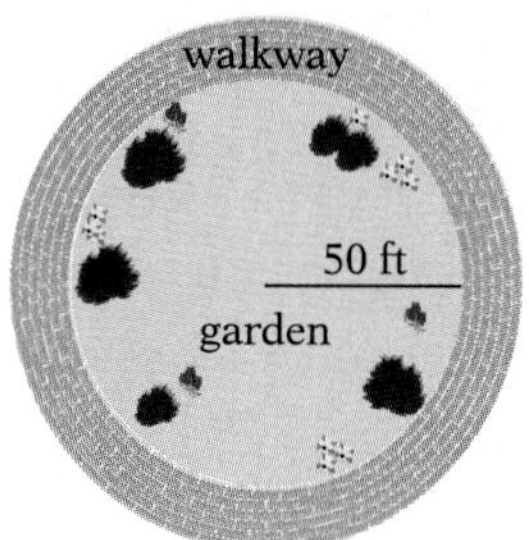

SOLUTION

First, we find the area of the garden.

$$A = \pi r^2 = 3.14(50)^2$$
$$= 3.14(2{,}500)$$
$$= 7{,}850 \text{ square feet}$$

The area of the walkway is the difference between the total area and the area of the garden.

$$9{,}161 - 7{,}850 = 1{,}311 \text{ square feet}$$

Your Turn 8

A washer is formed by drilling out a hole $\frac{1}{4}$ inch in diameter from a circular piece of metal 1 inch in diameter. What is the area of the washer?

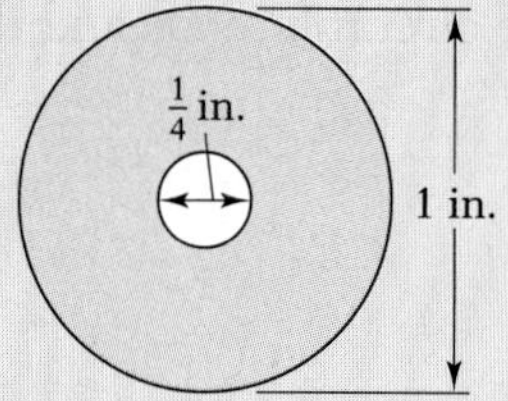

Answer: 0.736 square inch

C Spheres and Cylinders

The figures preceding Examples 9 and 10 show a sphere and a cylinder with the formulas for finding their volumes.

Copyright © Houghton Mifflin Company. All rights reserved.

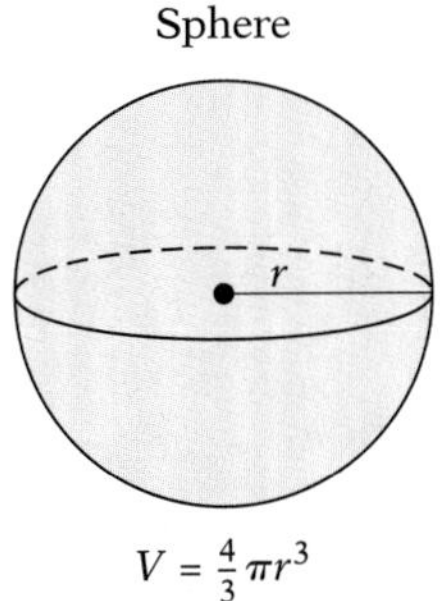

For the sphere formula, we will approximate $\frac{4}{3}$ with 1.33.

Example 9

The radius of a beach ball is about 9.55 inches. To the nearest whole number, what is its volume?

SOLUTION

$$V = \frac{4}{3}\pi r^3 = 1.33(3.14)(9.55)^3$$
$$= 1.33(3.14)(870.98)$$
$$= 3{,}637 \text{ cubic inches}$$

Your Turn 9

A person chewing bubble gum blows a perfectly spherical bubble with a $1\frac{1}{2}$-inch radius. What is the volume of air in the bubble?

Answer: 14.09 cubic inches

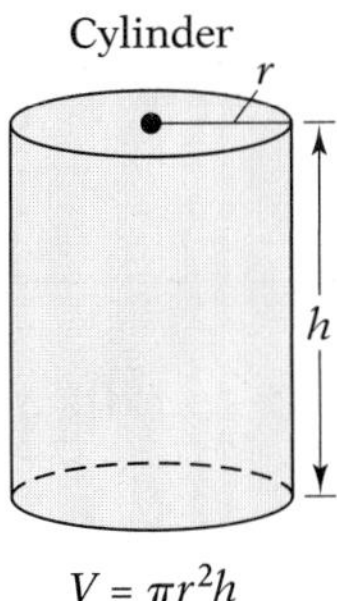

Example 10

Metal drums for storing toxic chemicals have a diameter of 2.6 feet and are 3.5 feet tall. How much liquid can the drums hold? (Round to the nearest cubic foot.)

SOLUTION

The radius of the drum is half the diameter: $r = 0.5(2.6) = 1.3$.

$$V = \pi r^2 h = 3.14(1.3)^2(3.5) \quad r = 1.3, h = 3.5$$
$$= 3.14(1.69)(3.5)$$
$$\approx 19 \text{ cubic feet}$$

Your Turn 10

The inner diameter of a drain pipe is 4 inches. How many cubic inches of water can a 10-foot length of this pipe hold?

Answer: 1,507.2 cubic inches

Copyright © Houghton Mifflin Company. All rights reserved.

6.7 Quick Reference

A. The Pythagorean Theorem

1. A **right triangle** is a triangle in which two sides, called **legs,** form a right angle. The side opposite the right angle is called the **hypotenuse.**

2. The Pythagorean Theorem

For a right triangle, if the lengths of the legs are a and b, and the length of the hypotenuse is c, then $c^2 = a^2 + b^2$.

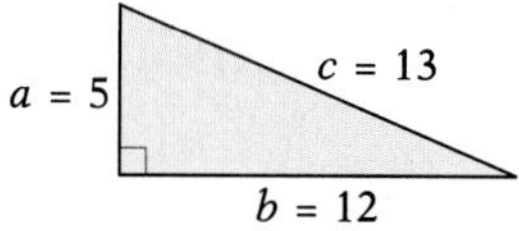

$$13^2 = 5^2 + 12^2$$

3. If we know the lengths of two sides of a right triangle, then we can calculate the length of the third side.

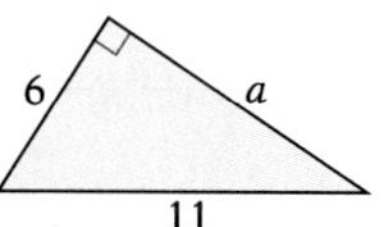

$$\begin{aligned} a^2 + 6^2 &= 11^2 \\ a^2 + 36 &= 121 \\ a^2 &= 121 - 36 \\ &= 85 \\ a &= \sqrt{85} \\ &\approx 9.22 \end{aligned}$$

4. We can also use the Pythagorean Theorem to test whether a given triangle is a right triangle.

$b = 9$, $a = 3$, $c = 10$

$$\begin{aligned} a^2 + b^2 &= 3^2 + 9^2 \\ &= 9 + 81 \\ &= 90 \\ c^2 &= 10^2 \\ &= 100 \end{aligned}$$

Because $c^2 \neq a^2 + b^2$, the triangle is not a right triangle.

B. Circumference and Area of a Circle

1. (a) Every point of a circle is the same distance from a given point, called the **center** of the circle.

Copyright © Houghton Mifflin Company. All rights reserved.

(b) The line segment that connects the center and a point of the circle is called the **radius.**

(c) The line segment that contains the center and whose endpoints are points of the circle is called the **diameter.**

(d) The diameter d of a circle is twice as long as the radius r: $d = 2r$.

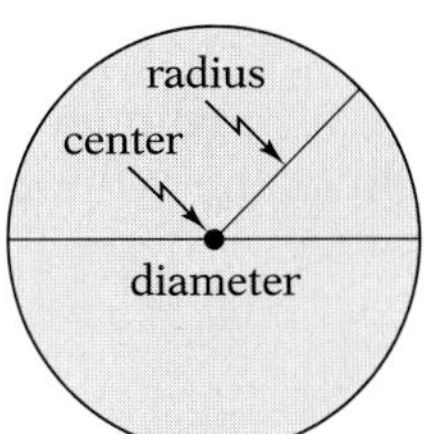

2. The **circumference** C of a circle is the distance around the circle. The circumference is π times as long as the diameter, where π is an irrational number approximately equal to 3.14.

3. For a circle with a diameter d and a radius r:

(a) The circumference C is given by $C = \pi d$ or $C = 2\pi r$.

(b) The area A is given by $A = \pi r^2$.

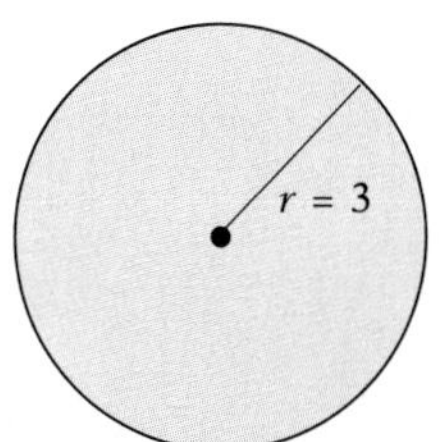

$$\begin{aligned} C &= \pi d \\ &= 3.14(6) \\ &= 18.84 \\ A &= \pi r^2 \\ &= 3.14(3)^2 \\ &= 3.14(9) \\ &= 28.26 \end{aligned}$$

C. Spheres and Cylinders

1. The volume V of a sphere is $V = \frac{4}{3}\pi r^3$.

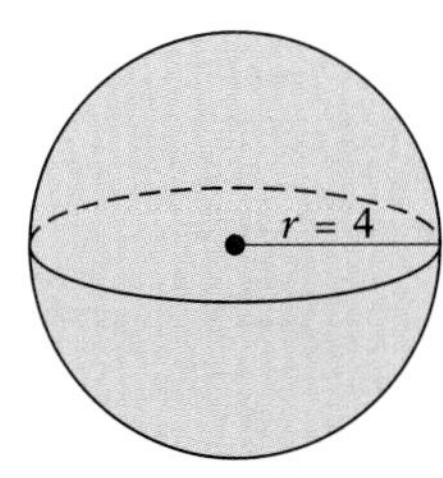

$$\begin{aligned} V &= \frac{4}{3}\pi r^3 \\ &= 1.33(3.14)(4)^3 \\ &= 1.33(3.14)(64) \\ &\approx 267.28 \end{aligned}$$

Copyright © Houghton Mifflin Company. All rights reserved.

2. The volume V of a cylinder is $V = \pi r^2 h$, where h is the height of the cylinder.

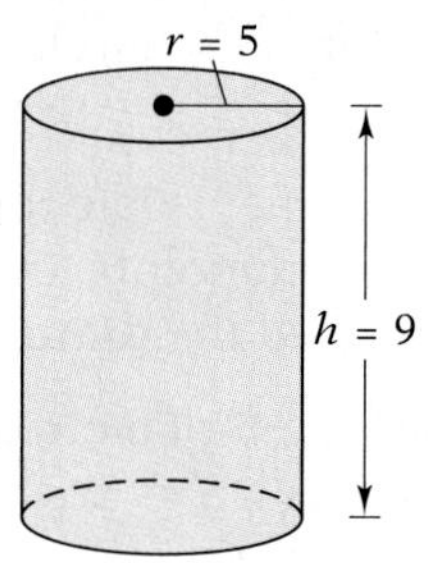

$$V = \pi r^2 h = 3.14(5)^2(9)$$
$$= 3.14(25)(9)$$
$$\approx 706.5$$

Copyright © Houghton Mifflin Company. All rights reserved.

6.7 Exercises

A. The Pythagorean Theorem

1. The two sides of a right triangle that form a right angle are called the ________, and the longest side is called the ________.

2. The ________ Theorem states the relationship among the lengths of the sides of a right triangle.

In Exercises 3–10, determine the length of the indicated side. Round all decimal answers to the nearest tenth.

3. 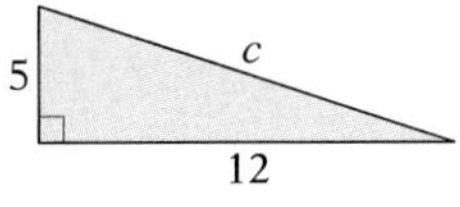

4. 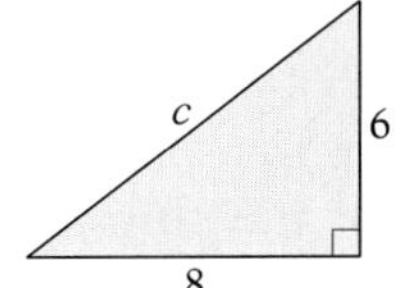

5. 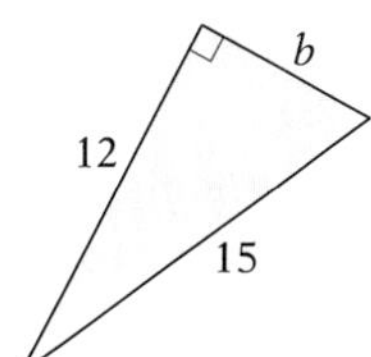

6. 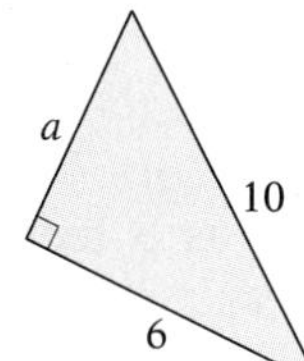

7. 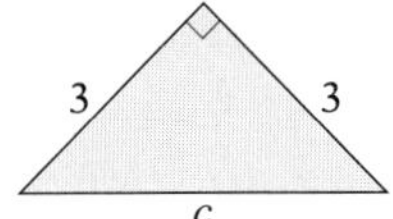

8. 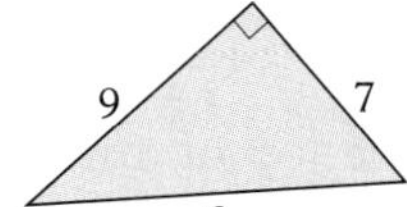

9. 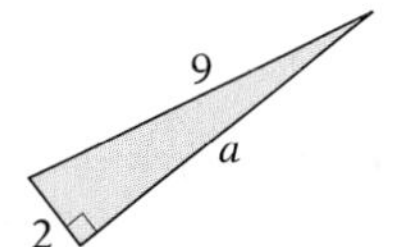

10. 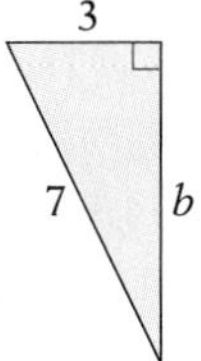

11. *Truck Distance* After leaving a truck stop, a trucker travels 72 miles west and then 30 miles north. Determine the distance d from the truck to the truck stop.

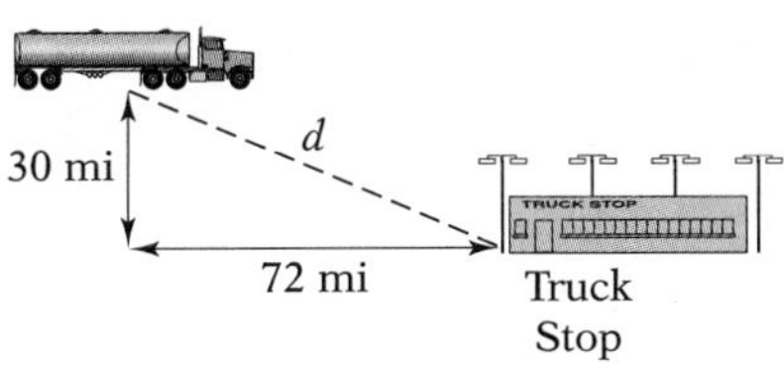

12. *Hiking* A hiker leaves camp and walks 3 miles south and then 4 miles east. How many miles is the hiker from camp?

Copyright © Houghton Mifflin Company. All rights reserved.

13. *Window Height* A 29-foot ladder rests on a window ledge. If the bottom of the ladder is 20 feet from the base of the building, how high is the window ledge above the ground?

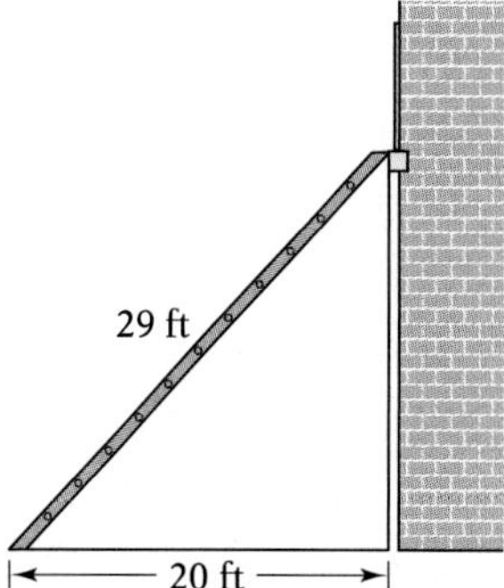

14. *Ship's Deck* A 10-foot ramp connects the deck of a ship to a point of a pier that is 8 feet from the ship. How high is the deck of the ship above the pier?

15. *Pole and Guy Wire* A 30-foot guy wire is attached to the top of a pole and is anchored at a point on the ground 15 feet from the base of the pole. To the nearest tenth of a foot, what is the height of the pole?

16. *Loading Platform* A 16-foot ramp extends from a loading platform to the ground. If the platform is 6 feet above the ground, to the nearest tenth of a foot, what is the distance from the base of the platform to the point where the ramp touches the ground?

17. *Little League Field* The distance between bases on a little league field is 60 feet. To the nearest tenth of a foot, what is the distance from home plate to second base?

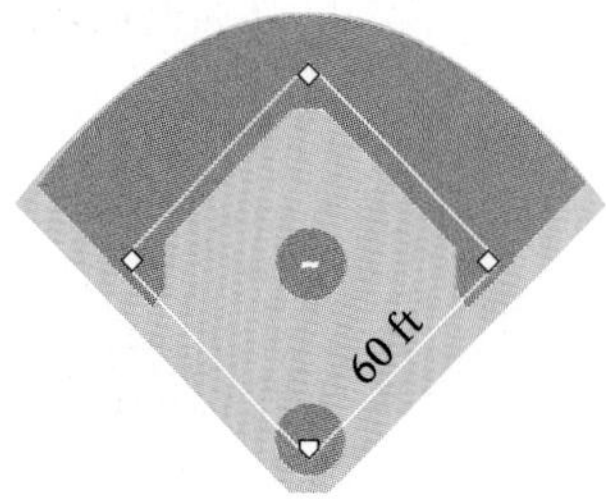

Copyright © Houghton Mifflin Company. All rights reserved.

18. *Baseball Field* The distance between bases on a major league baseball field is 90 feet. To the nearest tenth of a foot, what is the distance from third base to first base?

In Exercises 19–22, the lengths of the sides of a triangle are given. Determine whether the triangle is a right triangle.

19. 12, 13, 5

20. 13, 5, 9

21. 7, 14, 12

22. 10, 8, 6

The legs of a right triangle are said to be *perpendicular.* In general, perpendicular lines are lines that form a right angle.

23. *Retainer Wall* An 8-foot retainer wall is braced with a 10-foot pole that extends from the top of the wall to a point 6 feet from the base of the wall. Is the wall perpendicular to the ground?

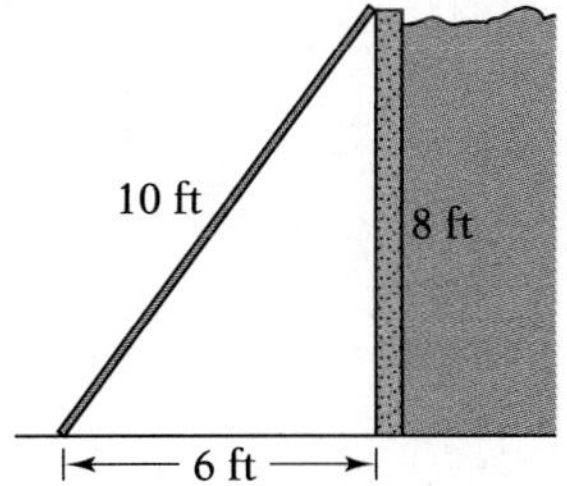

24. *Bookshelf Brace* A 17-inch piece of metal extends from a point 15 inches from the end of a bookshelf to a point on the side of the bookcase 8 inches below the end of the shelf. Are the shelf and the side of the bookcase perpendicular?

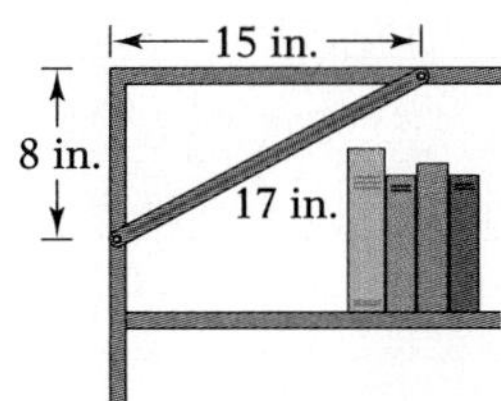

Copyright © Houghton Mifflin Company. All rights reserved.

25. *Ramp* An 8-foot ramp extends from a wall 5 feet above ground level to a point on the ground 5 feet from the base of the wall. Is the wall perpendicular to the ground?

26. *Pole and Guy Wire* A 20-foot guy wire connects the top of a 15-foot pole to a point on the ground 10 feet from the base of the pole. Is the pole perpendicular to the ground?

B. Circumference and Area of a Circle

27. The distance around a circle is called the ______ of the circle.

28. A line segment that connects the center and a point of a circle is called the ______.

In Exercises 29–36, determine the circumference and the area of the circle. Round your results to the nearest tenth.

29.

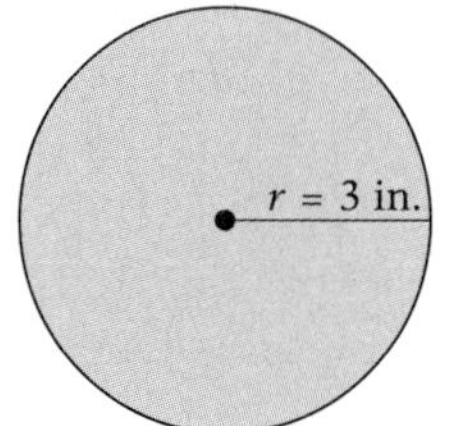

30.

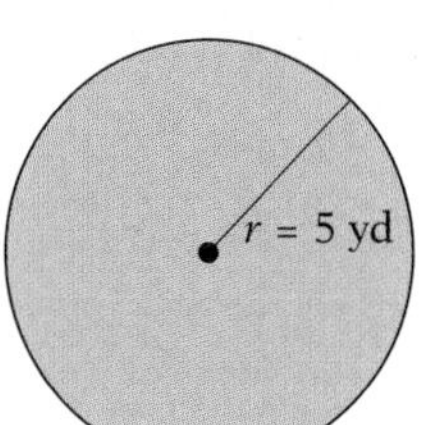

31.

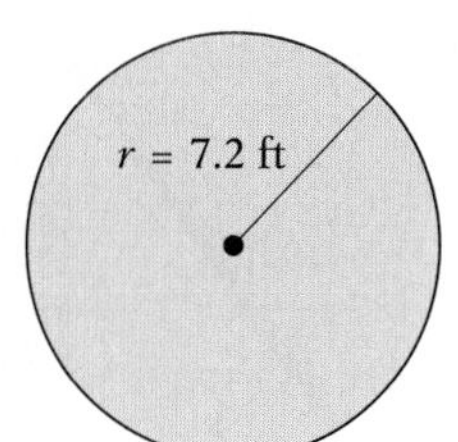

32.

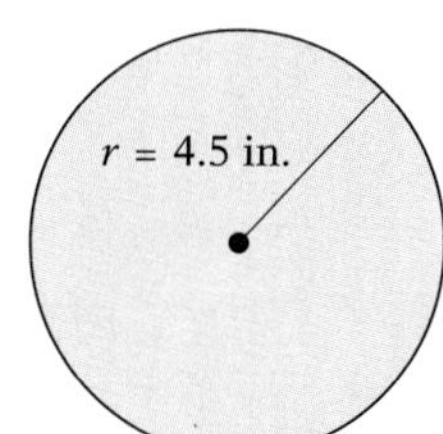

33.

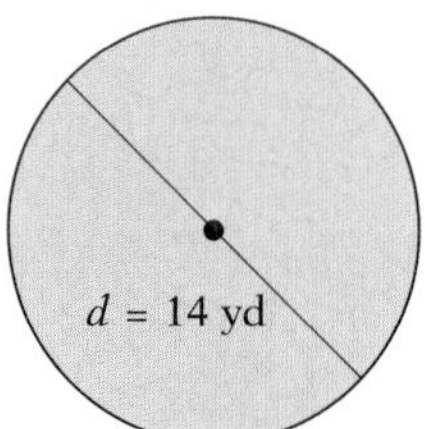

34.

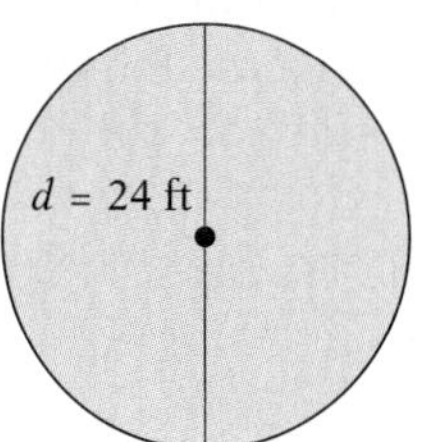

35.

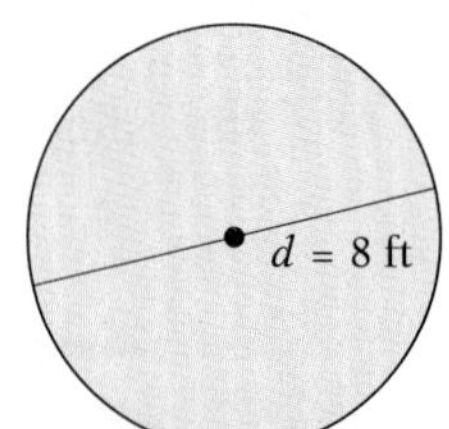

36.

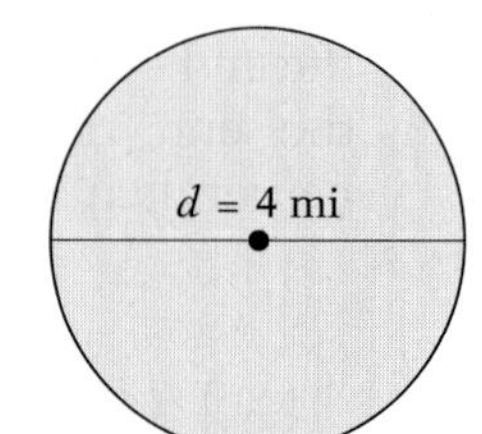

Copyright © Houghton Mifflin Company. All rights reserved.

37. *Distance Wheel* A utility worker uses a wheel with a radius of $\frac{3}{4}$ foot to measure the distance between two underground transformers. If the wheel makes 15 revolutions, to the nearest tenth of a foot, what is the distance?

38. *Bicycle Distance* A bicycle wheel with a radius of 1 foot makes 12 revolutions. To the nearest tenth of a foot, how many feet does the bicycle travel?

39. *Best Pizza Buy* In terms of dollars per square inch of area, which is the best buy, a 10-inch pizza for $12.75 or a 14-inch pizza for $21.55?

40. *Best Pizza Buy* In terms of dollars per square inch of area, which is the best buy, a 6-inch pizza for $5.95 or a 9-inch pizza for $14.35?

41. *Circumference of Earth* The radius of Earth is approximately 3,900 miles. To the nearest whole number, what is the distance around Earth at the equator?

42. *Seam Binding* To the nearest whole number, how many inches of seam binding are needed for the edge of a circular table cloth that is 36 inches in diameter?

Copyright © Houghton Mifflin Company. All rights reserved.

43. *Cake Recipe* A cake recipe makes enough batter for a rectangular pan that is 13 inches long and 9 inches wide. To the nearest tenth, by what number should the recipe be multiplied to create enough batter for three circular pans that are 8.5 inches in diameter?

44. *Table Cloth* A circular table cloth covers a circular table whose diameter is 30 inches. If the cloth extends 6 inches over the edge of the table, what is the area of the table cloth? (Round to the nearest whole number.)

In Exercises 45–48, determine the area of the shaded region. Round results to the nearest tenth.

45.

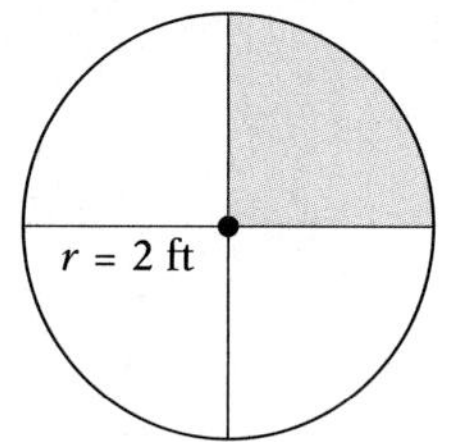

46.

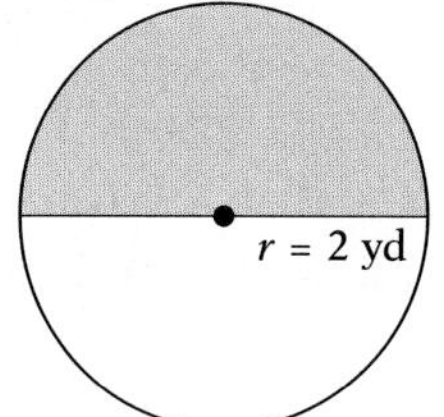

47.

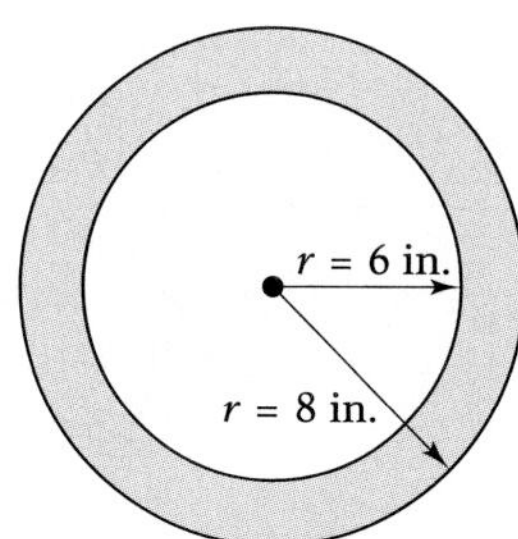

48.

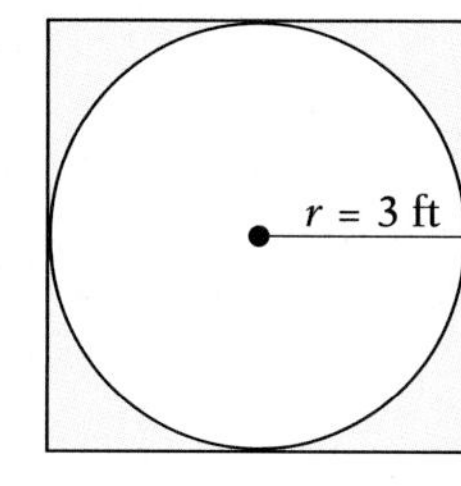

49. *Lobby Carpet* The lobby of an office building has a circular fountain with a radius of 3 feet. The fountain is surrounded by a carpeted area that is 2 feet wide. To the nearest tenth, how many square feet of carpet are needed to cover the area?

50. *Metal Work* A metal worker cuts a 3-inch diameter disk from the center of a 10-inch diameter metal disk. To the nearest tenth, what is the area of the remaining piece of metal?

Copyright © Houghton Mifflin Company. All rights reserved.

C. Spheres and Cylinders

51. The shape of a baseball is an example of a(n) ________.

52. The shape of a soup can is an example of a(n) ________.

In Exercises 53–60, determine the volume of the solid. Round results to the nearest tenth.

53.

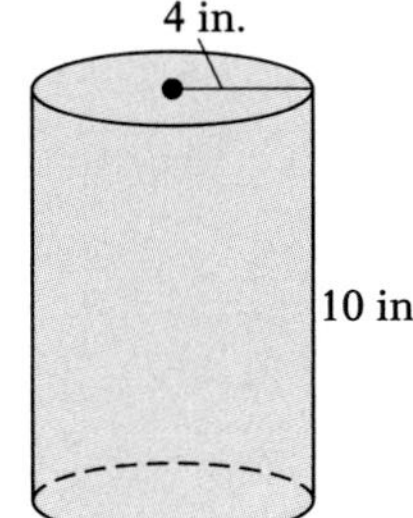

54.

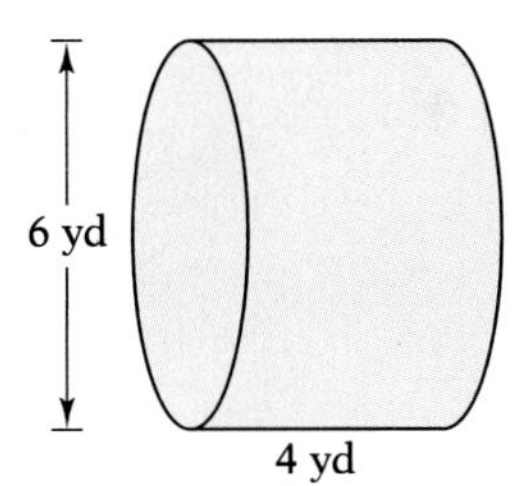

55.

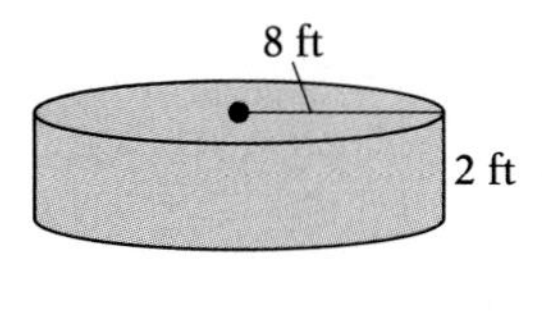

56.

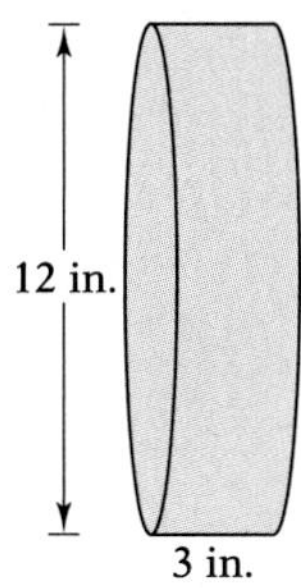

57.

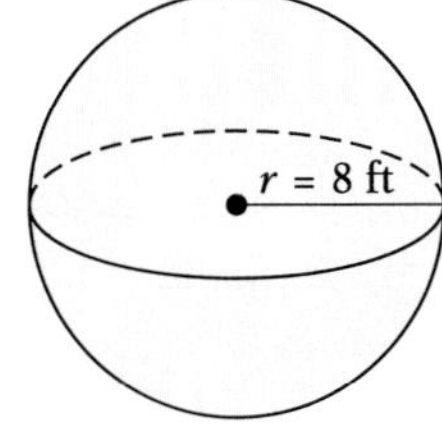

58.

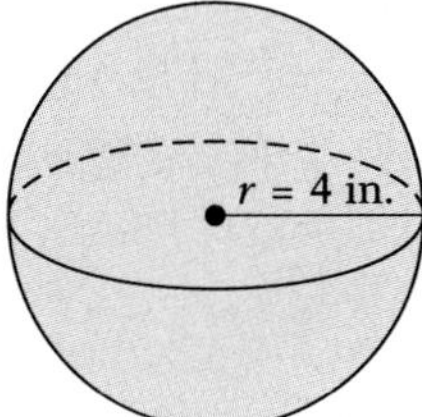

59.

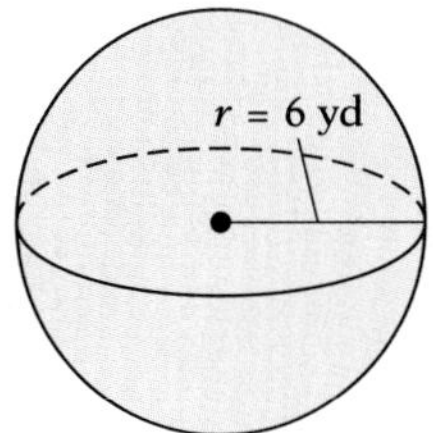

60.

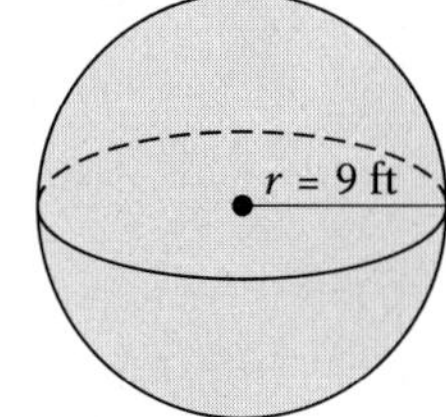

61. *Balloon Size* How many cubic inches of air are needed to fill a balloon so that its diameter is 10 inches?

62. *Baseball Volume* The radius of a baseball is about 1.43 inches. What is its volume?

63. *Soup Can Volume* The diameter of a soup can is 2.5 inches and the height is 4 inches. What is the volume of the can?

Copyright © Houghton Mifflin Company. All rights reserved.

64. *Pipe Capacity* In cubic feet, what is the capacity of a 6-inch pipe that is 5 feet long?

Writing and Concept Extension

65. Explain how to identify the hypotenuse of a right triangle.

66. How are the radius and diameter of a circle related?

In Exercises 67 and 68, determine the area of the figure. Round results to the nearest tenth.

67.

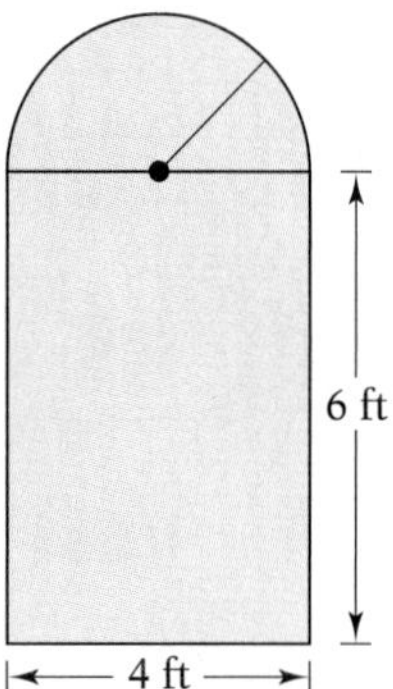

68.

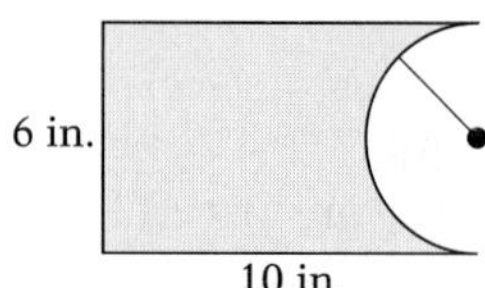

Copyright © Houghton Mifflin Company. All rights reserved.

CHAPTER 6 REVIEW EXERCISES

Section 6.1

1. One way to convert a fraction such as $\frac{3}{20}$ to a decimal is to write the denominator as a(n) ________ of 10.

2. In the number 3.472, the entry in the ________ column is 2, and the entry in the ________ column is 7.

3. Write 2.037 as a mixed number.

4. Write $\frac{7}{100}$ as a decimal.

5. The winning time in the Boston marathon was 2.176 hours. Write this number in words.

6. Write twenty and twelve thousandths as a decimal number.

7. Write 0.375 as a simplified fraction.

In Exercises 8 and 9, round the given number to the nearest (a) tenth and (b) hundredth.

8. 3.276

9. 1.983

10. Insert $<$ or $>$ to make the statement true.

    (a) 3.01 ____ 3.009

    (b) 8.1008 ____ 8.101

Section 6.2

11. Explain why 0.30 and 0.3 represent the same number.

In Exercises 12–17, perform the indicated operations.

12. $\begin{array}{r} 196.92 \\ +\ 214.03 \\ \hline \end{array}$

13. $\begin{array}{r} 307.6 \\ -\ 30.76 \\ \hline \end{array}$

14. $4.07 + 3.22 + 0.3$

15. $25.03 - 18.529$

16. $1.3 - 4.5$

17. $-6.7 - 0.6$

18. Simplify $2x + 4.3y + 7.1y + 5.6x$.

Copyright © Houghton Mifflin Company. All rights reserved.

19. *Sales Tax* Suppose that you purchased a jacket that was on sale for \$39.49. If you gave the clerk \$50 and received \$8.54 in correct change, what was the tax on the purchase?

20. *Grocery Purchase* At the grocery store, you purchased milk for \$2.69, cereal for \$3.19, juice for \$2.69, and a roast for \$8.94. What was your change from \$20?

Section 6.3

In Exercises 21–25, multiply.

21. $-5(0.3)$

22. $300(0.07)$

23. $\begin{array}{r} 20.38 \\ \times\ 0.16 \\ \hline \end{array}$

24. $\begin{array}{r} 9.7 \\ \times\ 0.004 \\ \hline \end{array}$

25. $-4.31(-0.22)$

26. To multiply by a power of 10, determine the number of zeros and shift the _______ that number of places to the right.

27. Multiply $8.4098 \times 1{,}000$.

28. Simplify $10(4.2x - 7)$.

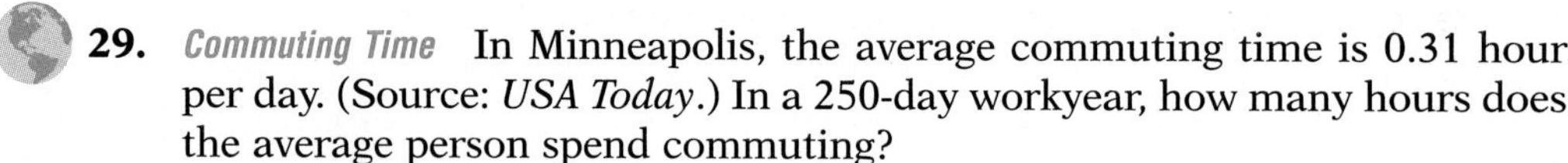

29. *Commuting Time* In Minneapolis, the average commuting time is 0.31 hour per day. (Source: *USA Today*.) In a 250-day workyear, how many hours does the average person spend commuting?

30. *Car Purchase* Suppose that you purchased a used car, priced at \$5,289, by paying \$350 down and \$241.35 per month for two years. How much more did you pay for the car than if you had paid the full amount at the time of the purchase?

Section 6.4

31. To round the result of a division to the nearest hundredth, carry out the division to _______ decimal places.

Copyright © Houghton Mifflin Company. All rights reserved.

In Exercises 32–36, divide to find the exact quotient.

32. $96 \div 30$

33. $\dfrac{4.5}{0.09}$

34. $273.78 \div 45$

35. $0.0288 \div 0.06$

36. $36.9 \div 1{,}000$

In Exercises 37 and 38, divide. Round the result as indicated.

37. $4.2 \div 9$, hundredth

38. $22.8 \div 1.4$, tenth

39. *Office Picnic* The 30 people who attended an office picnic ate 72 hot dogs. What was the average number of hot dogs that each person ate?

40. *State Populations* The population of Connecticut is 3.3 million, and the population of Wyoming is 0.5 million. To the nearest tenth, how many times as large is the population of Connecticut as that of Wyoming?

Section 6.5

In Exercises 41–43, evaluate the expression for the given values of the variables.

41. $a + b - c$ for $a = -5.2$, $b = 1.14$, and $c = -2.35$

Copyright © Houghton Mifflin Company. All rights reserved.

42. $x(y - 2x)$ for $x = 4.2$ and $y = 9.3$

43. $\dfrac{x + y}{y}$ for $x = -4.62$ and $y = 8.4$

44. To solve an equation such as $\dfrac{y}{1.3} = 7.1$, we use the ________ Property of Equations.

45. In which of the following can you clear decimals? Why?

(i) $5.3x - 12 = 19.8$

(ii) $5.3x - 12 + 19.8$

In Exercises 46–49, solve the given equation.

46. $x - 12.4 = 5.03$

47. $\dfrac{n}{0.3} = 0.15$

48. $x + 8.2 = 1 - 0.8$

49. $10x = 21.7 + 3.8x$

50. *Catering Cost* For a party, a caterer charged \$50 plus \$11.45 per person. If a customer paid the caterer \$347.70, how many people attended the party?

Section 6.6

51. When you divide to write the decimal name of a fraction, if the remainders begin to repeat, the result is called a(n) ________ decimal.

52. Explain the difference between 1.52 and $1.5\overline{2}$.

In Exercises 53 and 54, write the given number as a terminating or repeating decimal.

53. $\dfrac{11}{8}$

54. $\dfrac{3}{11}$

55. Numbers such as 9, 25, and 100 are called ________.

In Exercises 56–58, evaluate the expression.

56. $\sqrt{\dfrac{36}{25}}$

57. $7\sqrt{64}$

58. $6 - 2\sqrt{9}$

Copyright © Houghton Mifflin Company. All rights reserved.

59. Use a calculator to evaluate $3\sqrt{12} - 5\sqrt{2}$ to the nearest hundredth.

60. How many seconds does a diver take to reach the water from a platform 36 feet above the water? $\left(t = \sqrt{\frac{d}{16}}\right)$

Section 6.7

61. If two sides of a triangle form a 90° angle, we call the triangle a(n) ________ triangle.

62. In the figure, what is the length of the second leg?

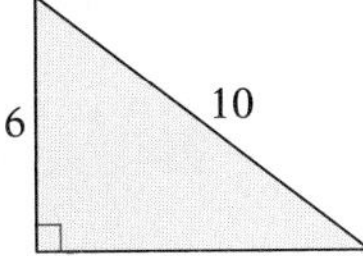

63. Determine whether the given triangle is a right triangle.

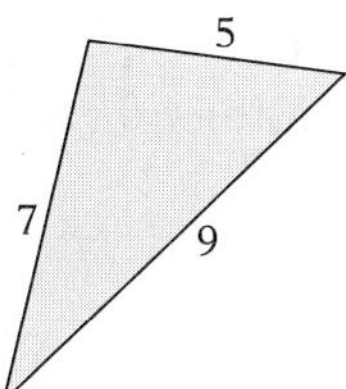

64. *Garden Path* A rectangular garden is 12 yards long and 9 yards wide. What is the length of a diagonal path through the garden?

65. A line segment that connects two points of a circle and contains the center is called a(n) ________.

66. Determine the area and circumference of a circle whose radius is 5 feet.

Copyright © Houghton Mifflin Company. All rights reserved.

67. *Pizza Cost* In terms of dollars per square inch of area, which is the best buy, an 8-inch pizza for \$7.95 or a 10-inch pizza for \$13.35?

68. In the figure, the two circles have the same center. To the nearest tenth, what is the area between the two circles?

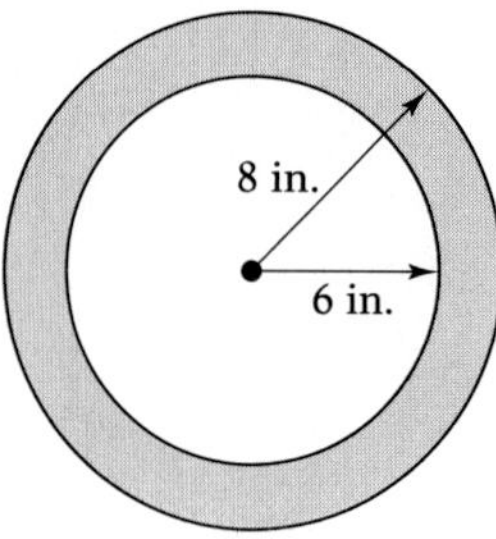

69. Determine the volume of a cylinder with radius 2 feet and height 5 feet.

70. Determine the volume of a sphere with diameter 6 inches.

Copyright © Houghton Mifflin Company. All rights reserved.

CHAPTER 6 TEST

1. In 0.3456, 5 is in the ________ column and 3 is in the ________ column.

2. When you divide to write the decimal name of a fraction, if the remainder is 0, the result is called a(n) ________ decimal.

3. The longest side of a right triangle is called the ________.

4. Write $4\frac{53}{1,000}$ as a decimal.

5. Round the given number to the nearest hundredth.

   (a) 1.4796

   (b) 3.0249

6. Arrange the numbers 1.0012, 1.012, and 1.001 from smallest to largest.

In Questions 7–17, perform the indicated operations.

7. $4.73 + 3.628$ **8.** $23.2 - 8.046$ **9.** $-12.3 + 6.7$ **10.** $4(-0.7)$

11. $403(0.016)$ **12.** 62.7×100 **13.** $61.18 \div 7$ **14.** $0.0216 \div 0.8$

15. $6.5 \div 0.07$ (Round the result to the nearest hundredth.)

16. $\sqrt{81} + \sqrt{9}$ **17.** $3\sqrt{49} - 2\sqrt{100}$

In Questions 18 and 19, solve the equation.

18. $1,000x = 41.78$ **19.** $4.2x - 5 = 3x + 4.6$

Copyright © Houghton Mifflin Company. All rights reserved.

20. Write the given number as a terminating or repeating decimal.

(a) $\frac{4}{3}$

(b) $\frac{3}{4}$

21. *Wages and Expenses* A child care worker's take home pay for one week was \$390.25. After paying \$140.40 for rent, \$53.10 for a car payment, and \$124.45 for groceries, how much money remained?

22. *Long Distance Charges* A long distance service charges \$.70 for the first 2 minutes of a call and \$.11 for each additional minute. What is the total charge for two calls of 16 minutes and 12 minutes?

23. In the figure, what is the length of the hypotenuse?

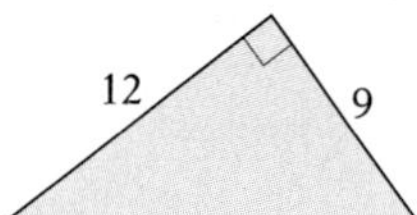

24. What is the volume of a sphere with radius 3 inches?

25. *Tricycle Wheel* A tricycle wheel with diameter 2 feet makes 10 revolutions. To the nearest tenth, how many feet does the tricycle travel?

Copyright © Houghton Mifflin Company. All rights reserved.

CHAPTER 7

Ratio and Proportion

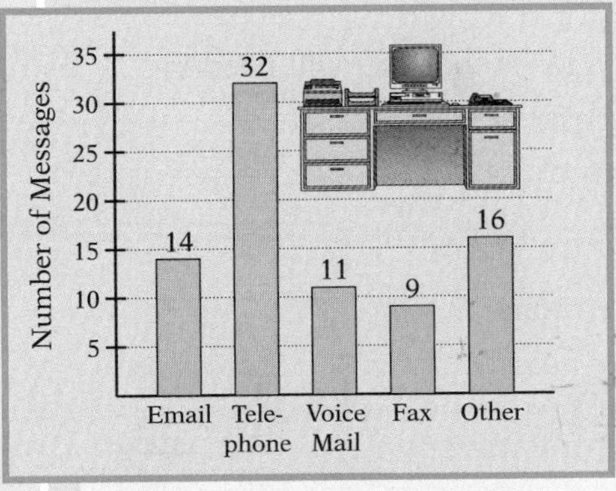

Corporate employees receive an average of 82 messages each day. The figure shows the forms of those messages. Of the 82 messages, 14 are by e-mail, a comparison that we can write in the form of a **ratio,** $\frac{14}{82}$. Similarly, the ratio of telephone messages to voice mail messages is $\frac{32}{11}$; this means that the employee receives nearly 3 times as many messages by phone as by voice mail.

Chapter Snapshot

In this chapter, you will learn that a **ratio** is a quotient that is used to compare two numbers with the *same* unit, whereas a **rate** is a quotient that is used to compare two numbers with *different* units. We will present ways to write, simplify, and interpret ratios and rates. In the next three sections, we discuss U.S. and metric units of length, volume, and weight, and we show how units can be converted within a unit system and between unit systems. We conclude with the topic of **proportions,** which are used to solve many real-life application problems. The last section is devoted specifically to geometric relationships in which proportions play an important role.

Copyright © Houghton Mifflin Company. All rights reserved.

For online resources, visit the web site **math.college.hmco.com/students** and follow the links to Hubbard/Robinson, *Prealgebra.*

Some Friendly Advice . . .

The time for registration for next semester is probably near. Do you know what courses you need to take? If not, talk to your instructor, adviser, or counselor. Asking your current instructor to suggest an instructor for your next course can be awkward. A better idea is to ask other students about their experiences and recommendations. Keep in mind, though, that each person learns in a different way, and each instructor teaches in a different way. Learning to be flexible as you adjust to new instructors and new approaches is an important part of your educational experience.

WARM-UP SKILLS

The following questions review concepts and skills that you will need in Chapter 7.

1. Simplify $\frac{80}{100}$.

In Exercises 2–4, perform the indicated operations.

2. $40 \cdot 700$

3. $96 \cdot \frac{1}{12} \cdot \frac{1}{3}$

4. $1\frac{1}{2} \div 3\frac{1}{4}$

5. Write $\frac{243}{100}$ as a decimal.

6. Solve $\frac{x}{21} = \frac{2}{3}$.

7. Solve $\frac{x - 4}{6} = \frac{1}{3}$.

8. What is the unit cost of potatoes if a 5-pound bag costs $1.90?

9. Determine the area of a rectangle that is 12 feet long and 10 feet wide.

10. Determine the volume of a rectangular box that is 4 feet long, 2 feet wide, and 1.5 feet deep.

Copyright © Houghton Mifflin Company. All rights reserved.

7.1 Ratios and Rates

A Ratios
B Rates

SUGGESTIONS FOR SUCCESS

Everything is relative.

Is a city of 100,000 people a large city? It probably would be to a person living in a rural town of 1,500 people, but it probably would not be to a person living in Chicago. Comparing two numbers is a natural thing to do because the numbers alone are not always meaningful.

Although we have previously used units, such as feet and pounds, we will talk a lot about units in this section and throughout this chapter. You will find the arithmetic easy, so pay particular attention to the examples and tips that we give you for working with units. They really are at the heart of the matter.

A Ratios

A **ratio** is a quotient that is used to compare two numbers with the same unit.

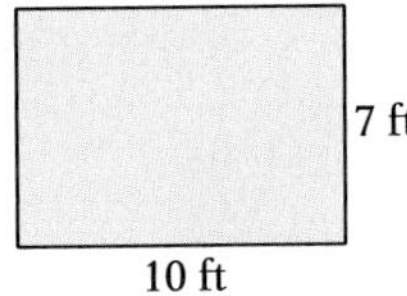

LEARNING TIP

You will need to read problems carefully to know whether $\frac{10}{7}$, for example, is supposed to be the ratio "10 to 7" or the fraction "ten sevenths."

In the rectangle, a comparison of the length to the width can be stated in words as "10 feet to 7 feet." Because the units are the same, we can express this comparison as a ratio, which we write as the quotient $\frac{10}{7}$. Note that the ratio does not include the units. We read the ratio as "10 to 7."

NOTE Another way to write this ratio is 10:7. However, the quotient form is easier to use, so we will not use the $a{:}b$ notation in this book.

When you write the ratio "a to b," make sure that a is in the numerator and b is in the denominator. As you will see in Example 1, starting with the ratio in words can help you write the ratio of numbers in the right order.

Example 1

In the preceding rectangle, what is the ratio of the width to the length?

SOLUTION

$$\frac{\text{Width}}{\text{Length}} = \frac{7 \text{ feet}}{10 \text{ feet}} = \frac{7}{10}$$

We are able to write the ratio because the width and the length have the same unit: feet. This allows us to "cancel" the units. A ratio is never written with units.

Your Turn 1

In the preceding rectangle, what is the ratio of the width to the perimeter?

Answer: $\frac{7}{34}$

Copyright © Houghton Mifflin Company. All rights reserved.

Example 2

A local Chamber of Commerce has 57 members, of whom 31 are men and 26 are women. What is the ratio of women to men?

SOLUTION

$$\frac{\text{Number of female members}}{\text{Number of male members}} = \frac{26}{31}$$

Your Turn 2

A coed soccer team has 16 players, of whom 7 are girls and 9 are boys. What is the ratio of boys to girls?

Answer: $\frac{9}{7}$

Example 3

A trip that takes 2 hours by plane takes 9 hours by car. Write two ratios that compare the trip times.

SOLUTION

$$\frac{\text{Time by plane}}{\text{Time by car}} = \frac{2 \text{ hours}}{9 \text{ hours}} = \frac{2}{9}$$

$$\frac{\text{Time by car}}{\text{Time by plane}} = \frac{9 \text{ hours}}{2 \text{ hours}} = \frac{9}{2}$$

Your Turn 3

An athlete can swim 100 yards in 51 seconds, and can run the same distance in 10 seconds. Write two ratios that compare those times.

Answer: $\frac{10}{51}$, $\frac{51}{10}$

Sometimes you may need to perform some arithmetic before you can compare two quantities.

Example 4

In the past year, a person has read 23 books, of which 9 were nonfiction. Write the ratio of fiction to nonfiction books that were read.

SOLUTION

The number of fiction books read is the difference between the total number of books read and the number that were nonfiction.

$$23 \text{ (total)} - 9 \text{ (nonfiction)} = 14 \text{ (fiction)}$$

Now we can write the required ratio.

$$\frac{\text{Number of fiction}}{\text{Number of nonfiction}} = \frac{14}{9}$$

Your Turn 4

In a 30-day period, rain fell on 17 days. What was the ratio of dry days to rainy days?

Answer: $\frac{13}{17}$

Suppose that a poll is taken to determine a political candidate's support. Of the 100 people polled, 60 are in favor of the candidate. Written as a ratio,

$$\frac{\text{number of supporters}}{\text{total number polled}} = \frac{60}{100}$$

This ratio can be simplified in the same way that we simplify any quotient.

$$\frac{60}{100} = \frac{3 \cdot \cancel{20}}{5 \cdot \cancel{20}} = \frac{3}{5}$$

The interpretation of this simplified ratio is that 3 out of every 5 people polled favor the candidate.

Copyright © Houghton Mifflin Company. All rights reserved.

Example 5

At a certain intersection, a traffic survey showed that 40 of 60 cars came to a complete stop. Compare the number of cars that did not come to a complete stop to the total number of cars in the survey.

SOLUTION

There were $60 - 40 = 20$ cars that did not come to a complete stop. The ratio is $\frac{20}{60} = \frac{1}{3}$. We can say that 1 out of every 3 cars did not come to a complete stop.

Your Turn 5

Of the 28 students who enrolled in a course, only 21 completed the course. Compare the number of students who dropped the course to the original enrollment.

Answer: $\frac{1}{4}$

Be aware that a simplified form of a ratio may not tell you what the original data were. Following the survey in Example 5, the evening news may report that 1 out of 3 cars did not obey the stop sign. However, this does not mean that only 3 cars were checked. There might have been 100 violators out of 300, or 25 out of 75, or any other ratio that is equivalent to $\frac{1}{3}$.

In all our examples so far, the ratios have involved comparisons of whole numbers. However, we can write ratios of any kind of numbers.

Example 6

Write and simplify the following ratios:

(a) 2.4 to 3.0

(b) $\frac{5}{6}$ to $\frac{2}{3}$

(c) $1\frac{1}{2}$ to $3\frac{1}{4}$

SOLUTION

(a) $\frac{2.4}{3.0} = \frac{24}{30}$ Multiply the numerator and the denominator by 10.

$= \frac{4 \cdot \cancel{6}}{5 \cdot \cancel{6}}$

$= \frac{4}{5}$

(b) $\frac{\frac{5}{6}}{\frac{2}{3}} = \frac{5}{6} \cdot \frac{3}{2}$ To divide, we multiply by the reciprocal.

$= \frac{5}{2 \cdot \cancel{3}} \cdot \frac{\cancel{3}}{2}$

$= \frac{5}{4}$

Your Turn 6

Write and simplify the following ratios.

(a) 0.8 to 4.0

(b) $1\frac{1}{5}$ to $\frac{3}{10}$

continued

Copyright © Houghton Mifflin Company. All rights reserved.

(c) $\dfrac{1\frac{1}{2}}{3\frac{1}{4}} = \dfrac{\frac{3}{2}}{\frac{13}{4}}$ Write the mixed numbers as improper fractions.

$= \dfrac{3}{2} \cdot \dfrac{4}{13}$ To divide, multiply by the reciprocal.

$= \dfrac{3}{\cancel{2}} \cdot \dfrac{\cancel{2} \cdot 2}{13}$

$= \dfrac{6}{13}$

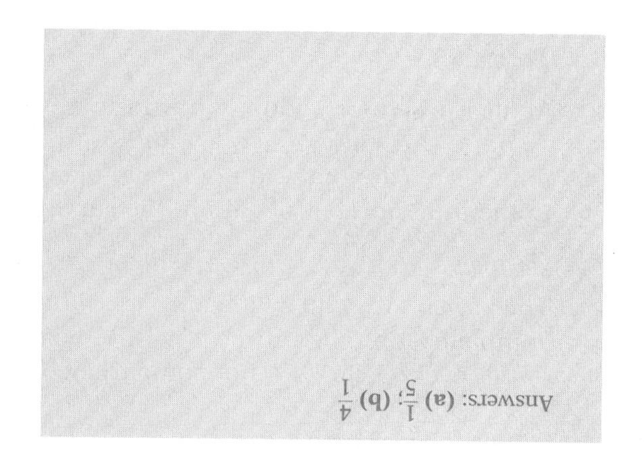

B Rates

A **rate** is a quotient that is used to compare two numbers that have different units.

A familiar example of a rate is speed, in which we compare distance to time. If you ride a bicycle 50 miles in 5 hours, then your rate is $\frac{50 \text{ miles}}{5 \text{ hours}}$ or, in simplified form, 10 miles per hour. This form is called the **unit rate** because the denominator of the quotient is 1.

Whereas ratios are written without units, the units for rates are different and must be retained. Here are two ways to write "10 miles per hour."

$$10 \frac{\text{miles}}{\text{hour}} \qquad 10 \text{ miles/hour}$$

The fraction bar and the slash symbol / are read *per,* which means "for each." In this example, you rode your bicycle 10 miles each hour. Rates involving other units of distance and time can also be written in these ways.

Example 7

A sprinter ran the 100-yard dash in 9.92 seconds. To the nearest hundredth, what was the sprinter's rate?

SOLUTION

$$\frac{100 \text{ yards}}{9.92 \text{ seconds}} \approx 10.08 \text{ yards per second}$$

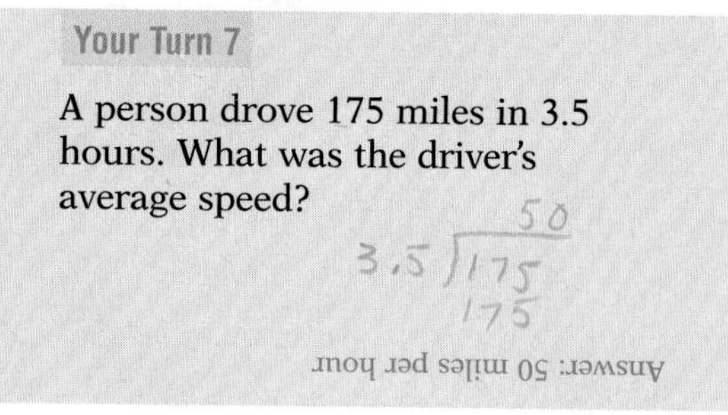

Your Turn 7

A person drove 175 miles in 3.5 hours. What was the driver's average speed?

Answer: 50 miles per hour

Rate does not always refer to speed. Any comparison of quantities that have different units is a rate.

Example 8

A 40-pound bag of fertilizer covers 5,000 square feet of lawn. What is the application rate in pounds per square foot?

SOLUTION

$$\frac{40 \text{ pounds}}{5{,}000 \text{ square feet}} = 0.008 \text{ pounds per square foot}$$

Your Turn 8

A drain pipe can empty a 1,000-gallon water tank in 25 minutes. What is the rate of flow of water through the pipe?

Answer: 40 gallons per minute

Copyright © Houghton Mifflin Company. All rights reserved.

Example 9

A college instructor earns an annual salary of $38,400. What are the instructor's monthly earnings?

SOLUTION

$$\frac{38{,}400 \text{ dollars}}{12 \text{ months}} = 3{,}200 \text{ dollars per month}$$

$$= \$3{,}200 \text{ per month}$$

Your Turn 9

The annual cost of health insurance is $4,080. What is the monthly cost?

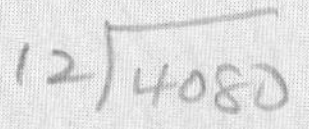

Answer: $340 per month

We first introduced the idea of the *unit cost* of an item in Section 6.4. We said that a unit cost is found by dividing the cost of the item by its weight or volume. We can also think of unit cost as a rate in the sense that it is a comparison of cost to weight or volume.

Example 10

In terms of unit cost, which ground beef package is the better buy?

SOLUTION

In each case, the unit cost is the ratio of the cost to the weight.

1-pound package: $\frac{.90 \text{ dollar}}{1 \text{ pound}} = \$.90 \text{ per pound}$

3-pound package: $\frac{2.73 \text{ dollars}}{3 \text{ pounds}} = \$.91 \text{ per pound}$

5-pound package: $\frac{4.40 \text{ dollars}}{5 \text{ pounds}} = \$.88 \text{ per pound}$

The 5-pound package is the better buy.

Your Turn 10

Dog food comes in 20-pound bags for $8.48 or 50-pound bags for $21.50. In terms of unit cost, which is the better buy?

Answer: 20-pound bag

Copyright © Houghton Mifflin Company. All rights reserved.

7.1 Quick Reference

A. Ratios

1. A **ratio** is a quotient that is used to compare two numbers with the same unit.	The ratio "2 to 5" can be written as $\frac{2}{5}$.
2. When you write "*a* to *b*" in quotient form, *a* is in the numerator, and *b* is in the denominator.	A town has 7 doctors and 2 dentists. The ratio of dentists to doctors is $\frac{2}{7}$.
3. Sometimes you may need to perform some arithmetic before you can compare two quantities.	In a 30-game season, a team won 19 games. The ratio of wins to losses is $\frac{\text{wins}}{\text{losses}} = \frac{19}{30 - 19} = \frac{19}{11}$.
4. Some ratios can be simplified in the same way that we simplify any quotient.	The ratio $\frac{35}{50}$ can be simplified as $\frac{\cancel{5} \cdot 7}{\cancel{5} \cdot 10} = \frac{7}{10}$.
5. We can write ratios of decimals, fractions, and mixed numbers.	$\frac{0.4}{1.2} = \frac{4}{12} = \frac{1}{3}$ $\frac{1\frac{1}{3}}{\frac{1}{2}} = \frac{\frac{4}{3}}{\frac{1}{2}} = \frac{4}{3} \cdot \frac{2}{1} = \frac{8}{3}$

B. Rates

1. A **rate** is a quotient that is used to compare two numbers with different units.	$\frac{\text{30 feet}}{\text{2 seconds}}$ is a rate because the units are different.
2. *Unit cost* is a rate that is found by dividing the cost of the item by its weight or volume.	The unit cost of an 18-ounce jar of peanut butter that sells for \$1.44 is $\frac{\text{1.44 dollars}}{\text{18 ounces}}$, or \$.08 per ounce.

Copyright © Houghton Mifflin Company. All rights reserved.

7.1 Exercises

A. Ratios

1. A(n) ______ is a quotient that is used to compare two numbers with the same unit.

2. We read the ratio $\frac{7}{3}$ as ______.

In Exercises 3–8, write the ratio as a fraction. Simplify if possible.

3. 5 to 3
4. 2 to 7
5. 6 to 9
6. 12 to 8
7. 15 to 3
8. 24 to 8

In Exercises 9–12, write the given ratio in words.

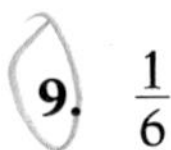

9. $\frac{1}{6}$
10. $\frac{3}{10}$
11. $\frac{8}{1}$
12. $\frac{41}{10}$

In Exercises 13 and 14, write the ratio of the length of the vertical leg to the length of the horizontal leg.

13.

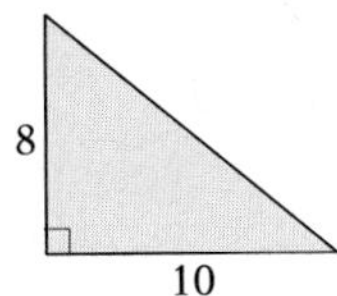

14.

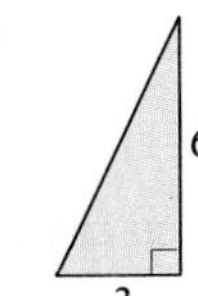

In Exercises 15–18, write the indicated ratio.

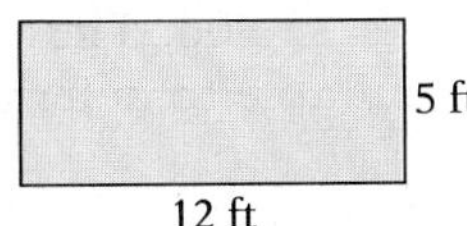

15. Length to width
16. Width to length
17. Width to perimeter
18. Length to perimeter

In Exercises 19–26, write the indicated ratio. Simplify the ratio, if possible.

19. *Commuting and Work Time* Each week a person spends 40 hours at work and 7 hours commuting to work. What is the ratio of the time spent commuting to the time spent working?

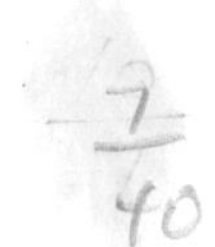

20. *Republicans and Democrats* Suppose that the Senate has 60 Republican senators and 40 Democratic senators. What is the ratio of Republicans to Democrats?

Copyright © Houghton Mifflin Company. All rights reserved.

21. *Acid Solution* A liquid solution consists of 12 ounces of acid and 20 ounces of water. What is the ratio of water to acid?

22. *Art Grades* In an art class, 10 students received A's and 15 students received B's. What is the ratio of the number of B's to the number of A's?

23. *Shopping Expenses* A survey of shoppers at a discount store found that 30 of 45 shoppers spent more than $50. What is the ratio of those who spent $50 or less to the total number of shoppers interviewed?

24. *Student Government* A student government organization has 26 elected representatives, of whom 14 are men. What is the ratio of men to women?

25. *Costs of Labor and Parts* The total cost of parts and labor for repairing a car was $210, of which $120 was for labor. What is the ratio of the cost of labor to the cost of parts?

26. *Win-Loss Ratio* Of the 30 games played during a season, a team lost 4 games. What is the ratio of wins to losses?

Household Budget Suppose that a person's monthly income is $2,800. The table shows the monthly expenditures. Use the information in the table to determine the ratios in Exercises 27 and 28.

Category	*Expenditure*
House payment	$800
Car payment	$280
Food	$460
Utilities	$240
Tax and insurance	$220
Miscellaneous	$400

27. **(a)** House payment to total income

(b) Utilities to car payment

Copyright © Houghton Mifflin Company. All rights reserved.

28. **(a)** Food and utilities to total income

(b) Remainder (amount left over after expenses have been paid) to total income

In Exercises 29–34, write the ratio as a simplified fraction.

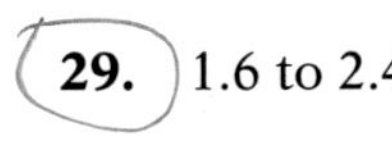

29. 1.6 to 2.4

30. 0.12 to 0.9

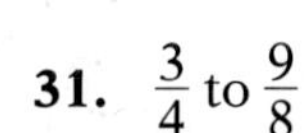

31. $\frac{3}{4}$ to $\frac{9}{8}$

32. $\frac{20}{9}$ to $\frac{3}{5}$

33. $2\frac{3}{4}$ to $1\frac{1}{2}$

34. $2\frac{2}{5}$ to $3\frac{1}{5}$

B. Rates

35. A(n) ________ is a quotient that is used to compare two numbers with different units.

36. A rate in which the denominator is 1 is called a(n) ________ rate.

In Exercises 37–46, determine the indicated rate.

37. ***Snail's Speed*** A snail can crawl approximately 13.2 inches in 5 minutes. (Source: *World Almanac.*) What is a snail's speed?

38. ***Boston Marathon*** The winning time for the Boston marathon (about 26 miles) was approximately $2\frac{1}{6}$ hours. (Source: *World Almanac.*) What was the runner's speed?

39. ***Hourly Wages*** A part-time clerk received $112.32 for 18 hours work. What was the rate of pay per hour?

40. ***Daily Wages*** A painter received $564.80 for 4 days of work. What was the rate of pay per day?

41. ***Cat Medication*** The dosage of a certain medication for a cat is based on the weight of the cat. If a 12-pound cat receives 32 milligrams of medication, what is the rate of medication per pound?

Copyright © Houghton Mifflin Company. All rights reserved.

42. *Scale Drawing* An architect uses $\frac{2}{3}$ inch to represent 18 feet. What is the scale in feet per inch?

43. *Tree Planting* A paper company planted 5,460 seedlings on 6.5 acres of land. At what rate per acre did they plant trees?

44. *Building Permits* A county issued 462 building permits in 30 days. What was the rate of permits issued per day?

45. *Water Flow* A water faucet can fill a 5-gallon container in 3 minutes. To the nearest tenth, at what rate per minute does water flow from the faucet?

46. *Fuel Consumption* A gas tank that holds 18 gallons was full at the beginning of a 3-hour trip. At the end of the trip 7 gallons remained. To the nearest tenth, at what rate did the vehicle use gasoline?

In Exercises 47 and 48, determine which size package is the best buy.

47. Orange juice concentrate

Size	*Price*
12 ounces	$1.39
8 ounces	$ .96

48. Vitamins

Quanity	*Price*
250 tablets	$9.87
160 tablets	$6.72

Calculator Exercises

49. *Child-care Costs* Women who make $10,000 per year spend approximately $2,459.08 each year for childcare. (Source: Bureau of Labor Statistics.) How much per dollar earned is spent for childcare?

Copyright © Houghton Mifflin Company. All rights reserved.

50. *Milk Consumption* Each year, Americans drink approximately 7.1 billion gallons of milk. (Source: U.S. Department of Agriculture.) Assuming a population of 0.29 billion, how many gallons of milk per person do Americans drink in a year?

Writing and Concept Extension

51. Explain why the ratio of 4 to 8 is the same as the ratio of 6 to 12.

52. Suppose that the ratio of two numbers is 1 to 1. What do you know about the numbers?

53. Suppose that the ratio of men to women in a composition class is 2 to 3. Identify the statement that is true. Explain why the other statement is not necessarily true.

(i) The class has 2 men and 3 women.

(ii) The class could have 12 men and 18 women.

54. Suppose that the ratio of unemployed adults to the entire population is the same for two states. Does that mean the two states have the same number of unemployed adults? Why?

55. Identify the ratio that is not the same as the other two.

(i) 1.5 to 4.5 (ii) $\frac{5}{6}$ to $\frac{5}{2}$ (iii) 4 to $1\frac{1}{3}$

Copyright © Houghton Mifflin Company. All rights reserved.

Exploring with Real-World Data: Collaborative Activities

Forms of Messages Corporate employees receive an average of 82 messages each day. The table describes the forms of those messages. (Source: *USA Today.*)

Form of Message	*Number of Messages*
Phone	32
E-mail	14
Voice mail	11
Fax	9
Other	16

Determine the following ratios:

56. Phone to e-mail

57. E-mail to fax

58. All messages to e-mail

59. Phone to all messages

Crime Victims According to the FBI, 634 people of every 100,000 are victims of violent crime. For a city with a population of 100,000, write the following ratios:

60. Number of crime victims to the entire population

61. Number of crime victims to the number who are not victims

62. Number of people who are not crime victims to the number who are victims

63. Number of people who are not crime victims to the entire population

Copyright © Houghton Mifflin Company. All rights reserved.

7.2 U.S. Units of Measure

A *Units of Length and Distance*
B *Units of Weight*
C *Units of Volume*
D *Units of Time and Rate*

SUGGESTIONS FOR SUCCESS

Knowing that the length of a rectangle is 8 is not very helpful. Do we mean 8 feet? 8 miles? 8 inches? 8 yards? A measure of length, weight, or volume is meaningful only if a unit of measure is given.

We remember the units we use most. One of the authors must look in a table every time to learn that there are 2 cups in a pint. Guess which one?

There are many weird units, such as rods, furlongs, fathoms, drams, and grains. Only specialists know about those. In this section, you should make it a point to memorize the familiar common units and their relationship to other units. This is not mathematics so much as just everyday practical knowledge.

A Units of Length and Distance

The following table shows the most common units of length and distance that are used in the United States. Also shown are the abbreviations of those units.

LEARNING TIP
Note that in. is the only abbreviation that has a period at the end (so that it doesn't look like the word in).

Common U.S. Units of Length and Distance

1 foot (ft) = 12 inches (in.)
1 yard (yd) = 3 feet = 36 inches
1 mile (mi) = 1,760 yards = 5,280 feet

Although any unit in the table can be used as a measure of length or distance, the unit that we choose ought to be meaningful.

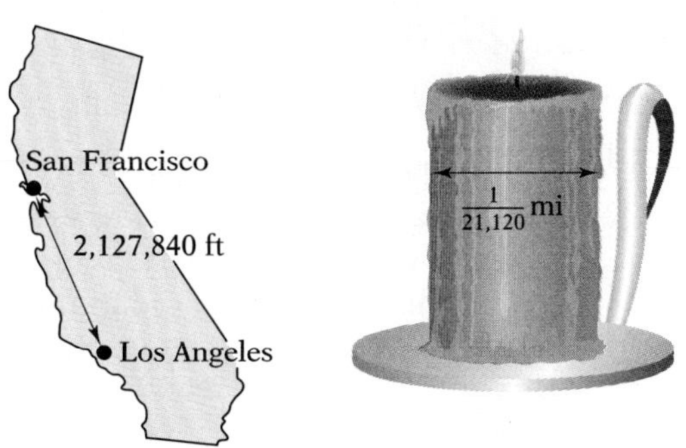

The measurements shown in the figures are silly. The more natural units are 403 miles (the distance from Los Angeles to San Francisco) and 3 inches (the diameter of the candle).

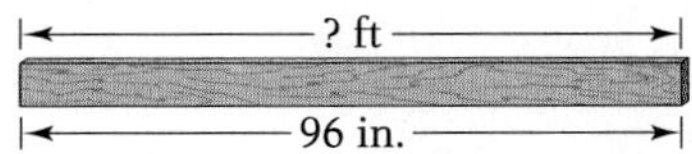

Suppose that you need a board that is 96 inches long. However, the length of lumber is given in feet. Therefore, you will need to convert 96 inches to feet. We often have a need to make such conversions from one unit to another.

Copyright © Houghton Mifflin Company. All rights reserved.

DEVELOPING THE CONCEPT

Conversion Factors

Consider the ratio $\frac{12\text{ inches}}{12\text{ inches}}$. Because the numerator and the denominator are the same, the value of the ratio is 1.

Because 1 foot = 12 inches, we can replace the 12 inches in the numerator of the ratio with 1 foot: $\frac{1\text{ foot}}{12\text{ inches}}$. Note, though, that the value of the ratio is still 1.

Now let's return to the 96-inch board that you need to buy. Observe what happens when we multiply 96 inches by $\frac{1\text{ foot}}{12\text{ inches}}$.

Write the given quantity as a fraction.

$$\frac{96\text{ inches}}{1} \cdot \frac{1\text{ foot}}{12\text{ inches}}$$

Multiplying a quantity by 1 does not change the value of the quantity.

$$= \frac{96\text{ }\cancel{\text{inches}}}{1} \cdot \frac{1\text{ foot}}{12\text{ }\cancel{\text{inches}}}$$

The in. units can be canceled.

$$= \frac{96}{12}\text{ feet} = 8\text{ feet}$$

We see that the 96-inch board is 8 feet long.

The ratio $\frac{1\text{ foot}}{12\text{ inches}}$ is an example of a **conversion factor,** which is a ratio whose value is 1. Multiplying a given quantity by a conversion factor changes the units without changing the value of the quantity.

Here are some guidelines for selecting a conversion factor.

Selecting a Conversion Factor

Select a conversion factor so that:

1. The unit that you want is in the numerator.
2. The unit that you are changing is in the denominator.

One way to get started is to write the conversion factor without numbers. Just write the units and verify that they will cancel as they are supposed to. Then you can go ahead and fill in the numbers and carry out the conversion. We will illustrate this technique in Example 1.

Example 1

In baseball, the distance between the bases is 90 feet. What is the distance in yards?

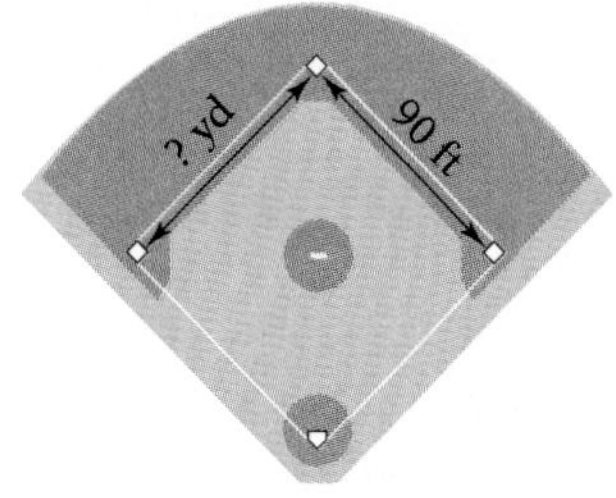

Your Turn 1

In softball, the distance between the bases is 20 yards. What is the distance in feet?

Copyright © Houghton Mifflin Company. All rights reserved.

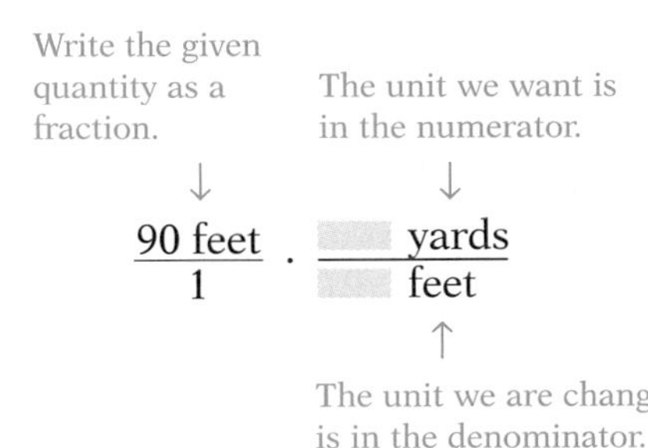

Now we can see that the units will cancel the way we want them to, so we can use 1 yard = 3 feet to fill in the numbers.

$$\frac{90 \text{ feet}}{1} \cdot \frac{1 \text{ yard}}{3 \text{ feet}} \qquad \text{Because 1 yard = 3 feet, } \frac{1 \text{ yard}}{3 \text{ feet}} = 1.$$

$$= \frac{90}{3} \text{ yards} \qquad \text{The foot units cancel.}$$

$$= 30 \text{ yards}$$

As you can tell from Example 1, writing the correct conversion factor is an essential step in the process. The trick is to cancel unwanted units so that you have the desired unit in the result.

In Example 1, we converted from a smaller unit (feet) to a larger unit (yards) by dividing. As shown in Example 2, converting from a larger unit to a smaller unit involves multiplication.

Although some conversions require multiplication and others require division, it is unnecessary, even unwise, to memorize the circumstances under which a particular operation is needed. Simply write the conversion factor so that the unit you are trying to change appears in the denominator. If you do that correctly, the operation you will need to perform will be apparent.

Example 2

Medical records often report a person's height in inches. What entry would be made for a person who is 6 feet tall?

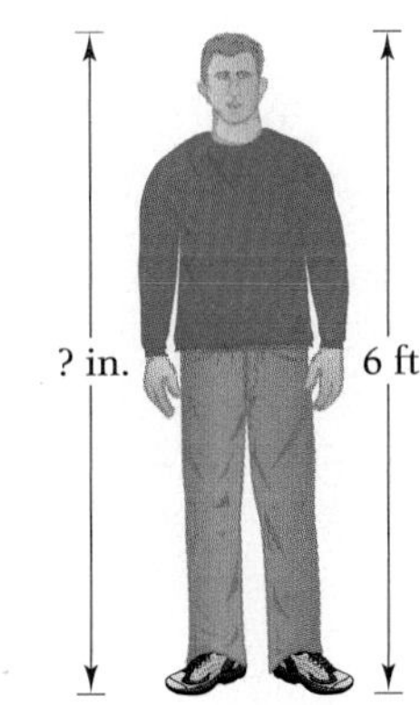

Your Turn 2

A kitchen counter is 90 inches long. What is the length of the counter in feet?

continued

Copyright © Houghton Mifflin Company. All rights reserved.

SOLUTION

Write the given quantity as a fraction. → $\frac{6 \text{ feet}}{1}$

The unit we want is in the numerator. → inches

$$\frac{6 \text{ feet}}{1} \cdot \frac{\square \text{ inches}}{\square \text{ feet}}$$

The unit we are changing is in the denominator. → feet

Now use the fact that 12 inches = 1 foot to fill in the numbers.

$$\frac{6 \text{ feet}}{1} \cdot \frac{12 \text{ inches}}{1 \text{ foot}}$$ Because 12 inches = 1 foot, $\frac{12 \text{ inches}}{1 \text{ foot}} = 1$.

$$= \frac{6 \cdot 12}{1} \text{ inches}$$ The foot units cancel.

$$= 72 \text{ inches}$$

B Units of Weight

The most commonly used U.S. units of weight are ounces, pounds, and tons. The ounce unit can be confusing because it is used for both volume (*fluid* ounce) and weight. Usually, you can tell which ounce unit is being used from the wording of a problem.

The following table gives the relationships among these units along with the abbreviations for the units.

Common U.S. Units of Weight

1 pound (lb) = 16 ounces (oz)

1 ton (T) = 2,000 pounds

The method for converting from one weight unit to another is exactly the same as that used for other kinds of units.

Example 3

Convert 20 ounces to pounds.

SOLUTION

$$\frac{20 \text{ ounces}}{1} \cdot \frac{1 \text{ pound}}{16 \text{ ounces}}$$ Because 1 pound = 16 ounces, $\frac{1 \text{ pound}}{16 \text{ ounces}} = 1$.

$$= \frac{20}{16} \text{ pounds}$$ The ounce units cancel.

$$= \frac{5}{4} \text{ pounds} = 1.25 \text{ pounds}$$

Your Turn 3

Convert $1\frac{1}{2}$ pounds to ounces.

Answer: 24 ounces

Copyright © Houghton Mifflin Company. All rights reserved.

Example 4

A person buys 80-pound bags of concrete mix. If his pickup truck can carry a maximum weight of 0.8 ton, what is the maximum number of bags that the person should buy?

SOLUTION

First, we find the maximum number of pounds that the truck can carry.

$$\frac{0.8 \text{ ton}}{1} \cdot \frac{2{,}000 \text{ pounds}}{1 \text{ ton}}$$ Because 2,000 pounds = 1 ton, $\frac{2{,}000 \text{ pounds}}{1 \text{ ton}} = 1.$

$$= 0.8(2{,}000) \text{ pounds}$$ The ton units cancel.

$$= 1{,}600 \text{ pounds}$$

Now we can determine the maximum number of bags that the truck can carry.

$$\frac{1{,}600 \text{ pounds}}{1} \cdot \frac{1 \text{ bag}}{80 \text{ pounds}}$$ Because 1 bag weighs 80 pounds, $\frac{1 \text{ bag}}{80 \text{ pounds}} = 1.$

$$= \frac{1{,}600}{80} \text{ bags}$$ The pound units cancel.

$$= 20 \text{ bags}$$

The person should buy no more than 20 bags.

Your Turn 4

In Example 4, if the concrete mix comes in 60-pound bags, what is the maximum (whole) number of bags that the person should buy?

Answer: 26

C *Units of Volume*

In Section 2.8, we gave the following unit conversions for volume:

1 cubic foot (cu ft) = 1,728 cubic inches (cu in.)

1 cubic yard (cu yd) = 27 cubic feet

The cubic units can also be written as in^3, ft^3, and yd^3.

Here are some other U.S. unit conversions that are commonly used for liquid volume, which is sometimes called *capacity.*

Common U.S. Units of Liquid Volume

1 cup = 8 fluid ounces (fl oz)

1 pint (pt) = 2 cups (cup)

1 quart (qt) = 2 pints (pt)

1 gallon (gal) = 4 quarts (qt)

Copyright © Houghton Mifflin Company. All rights reserved.

NOTE The unit *ounce* is also used for weight, so we use *fluid ounce* to indicate volume or capacity.

Example 5

In quarts, what is the capacity of a 55-gallon drum?

SOLUTION

$$\frac{55 \text{ gallons}}{1} \cdot \frac{4 \text{ quarts}}{1 \text{ gallon}}$$ Because 4 quarts = 1 gallon, $\frac{4 \text{ quarts}}{1 \text{ gallon}} = 1$.

$$= \frac{55 \cdot 4}{1} \text{ quarts}$$ The gallon units cancel.

$$= 220 \text{ quarts}$$

Your Turn 5

How many fluid ounces are in a quart of liquid?

Answer: 32

Example 6

A six-pack of a soft drink contains six 12-ounce cans. How many cups of liquid are there?

SOLUTION

The total number of ounces is

$$\frac{6 \text{ cans}}{1} \cdot \frac{12 \text{ ounces}}{1 \text{ can}} = 72 \text{ ounces}$$

$$\frac{72 \text{ ounces}}{1} \cdot \frac{1 \text{ cup}}{8 \text{ ounces}}$$ Because 1 cup = 8 ounces, $\frac{1 \text{ cup}}{8 \text{ ounces}} = 1$.

$$= \frac{72}{8} \text{ cups}$$ The ounce units cancel.

$$= 9 \text{ cups}$$

Your Turn 6

A half-gallon milk container holds 8 cups of milk. How many pints are in the container?

Answer: 4

Copyright © Houghton Mifflin Company. All rights reserved.

D Units of Time and Rate

Fortunately, the time units of seconds, minutes, hours, days, weeks, months, and years are used the world over. The following table summarizes the common time unit conversions:

Common Units of Time

1 minute (min) = 60 seconds (sec)

1 hour (hr) = 60 minutes (min)

1 day (day) = 24 hours (hr)

1 week (week) = 7 days (day)

1 year (yr) = 12 months (mo)

In science and engineering, we sometimes need to convert a rate, such as speed or liquid flow, from one system of units to another.

Example 7

A 1-gallon bottle can be filled from a kitchen faucet in 1 minute. What is the rate of liquid flow in quarts per hour?

SOLUTION

The given rate of flow is 1 gallon per minute. We must convert gallons to quarts and minutes to hours. First we convert gallons to quarts.

$$\frac{1 \text{ gallon}}{1 \text{ minute}} \cdot \frac{4 \text{ quarts}}{1 \text{ gallon}}$$ Because 4 quarts = 1 gallon, $\frac{4 \text{ quarts}}{1 \text{ gallon}} = 1$.

$$= \frac{4 \text{ quarts}}{1 \text{ minute}}$$ The gallon units cancel.

Now we convert minutes to hours.

$$\frac{4 \text{ quarts}}{1 \text{ minute}} \cdot \frac{60 \text{ minutes}}{1 \text{ hour}}$$ Because 60 minutes = 1 hour, $\frac{60 \text{ minutes}}{1 \text{ hour}} = 1$.

$$= \frac{(4 \cdot 60) \text{ quarts}}{1 \text{ hour}}$$ The minute units cancel.

$$= 240 \text{ quarts per hour}$$

Your Turn 7

A 1,000-gallon tank can be emptied in 2 hours. To the nearest whole number, what is the rate of flow in quarts per minute?

Answer: 33 quarts per minute

Copyright © Houghton Mifflin Company. All rights reserved.

A more efficient approach to Example 7 is to write both conversion factors in one step.

$$\frac{1\ \cancel{\text{gallon}}}{1\ \cancel{\text{minute}}} \cdot \frac{4\ \text{quarts}}{1\ \cancel{\text{gallon}}} \cdot \frac{60\ \cancel{\text{minutes}}}{1\ \text{hour}}$$

$$= \frac{(4 \cdot 60)\ \text{quarts}}{1\ \text{hour}}$$ The gallon and minute units cancel.

$$= 240\ \text{quarts per hour}$$

We will use this faster method in Example 8.

Example 8

If a car is traveling at 30 miles per hour (mph), what is the car's speed in feet per second?

SOLUTION

We have two unit conversions to make: miles to feet and hours to seconds. Observe how all of the conversion factors are arranged so that the unwanted units will cancel.

$$\frac{30\ \cancel{\text{miles}}}{1\ \cancel{\text{hour}}} \cdot \frac{5{,}280\ \text{feet}}{1\ \cancel{\text{mile}}} \cdot \frac{1\ \cancel{\text{hour}}}{60\ \cancel{\text{minutes}}} \cdot \frac{1\ \cancel{\text{minute}}}{60\ \text{seconds}}$$

$$= \frac{30(5{,}280)}{60(60)}\ \frac{\text{feet}}{\text{second}}$$

$$= \frac{158{,}400}{3{,}600}\ \frac{\text{feet}}{\text{second}}$$

$$= 44\ \text{feet per second}$$

Your Turn 8

Normal walking speed is about 6 feet per second. To the nearest tenth, what is this speed in miles per hour?

LEARNING TIP

Note that certain numerical factors can be canceled as well as the units. However, we want you to focus on the units. You can easily perform the arithmetic with a calculator.

Answer: 4.1 miles per hour

Example 9

A koala bear (which is not actually a bear) eats about 2.5 pounds of eucalyptus leaves per day. Does a koala eat more than or less than $\frac{1}{2}$ ton of eucalyptus leaves per year?

SOLUTION

We need to convert pounds to tons and days to years. Again, we arrange the conversion factors so that the unwanted factors cancel.

$$\frac{2.5\ \cancel{\text{pounds}}}{1\ \cancel{\text{day}}} \cdot \frac{1\ \text{ton}}{2{,}000\ \cancel{\text{pounds}}} \cdot \frac{365\ \cancel{\text{days}}}{1\ \text{year}}$$

$$= \frac{2.5(365)}{2{,}000}\ \frac{\text{ton}}{\text{year}}$$

$$\approx 0.46\ \text{ton per year}$$

A koala eats slightly less than $\frac{1}{2}$ ton per year.

Your Turn 9

If a person drinks three 12-ounce cans of a soft drink each day, to the nearest whole number, how many quarts does the person drink per week?

Answer: 8 quarts per week

Copyright © Houghton Mifflin Company. All rights reserved.

7.2 Quick Reference

A. Units of Length and Distance

1. In the United States, the common units of length and distance are inches (in.), feet (ft), yards (yd), and miles (mi).

2. A **conversion factor,** which is a ratio whose value is 1, is used to change the units of a quantity without changing its value.

We change 2.5 feet to inches as follows.

$$\frac{2.5 \text{ feet}}{1} \cdot \frac{12 \text{ inches}}{1 \text{ foot}}$$
$$= \frac{(2.5 \cdot 12)}{1} \text{ inches}$$
$$= 30 \text{ inches}$$

B. Units of Weight

1. The most commonly used U.S. units of weight are ounces (oz), pounds (lb), and tons (T).

2. Weight unit conversions can be made with conversion factors.

We change 60 ounces to pounds as follows:

$$\frac{60 \text{ ounces}}{1} \cdot \frac{1 \text{ pound}}{16 \text{ ounces}}$$
$$= \frac{60}{16} \text{ pounds}$$
$$= 3.75 \text{ pounds}$$

C. Units of Volume

1. The common U.S. units for volume (or *capacity*) are fluid ounces (fl oz), cups (cup), pints (pt), quarts (qt), and gallons (gal).

2. We use conversion factors in the usual way to convert volume units.

We change 50 quarts to gallons as follows:

$$\frac{50 \text{ quarts}}{1} \cdot \frac{1 \text{ gallon}}{4 \text{ quarts}}$$
$$= \frac{50}{4} \text{ gallons}$$
$$= 12.5 \text{ gallons}$$

Copyright © Houghton Mifflin Company. All rights reserved.

D. Units of Time and Rate

1. The time units of seconds (sec), minutes (min), hours (hr), days (day), weeks (week), months (mo), and years (yr) are used worldwide.

2. To convert a rate, such as speed or liquid flow, from one system of units to another, we may need more than one conversion factor.

We change 5 feet per second to yards per minute as follows:

$$\frac{5 \text{ feet}}{1 \text{ second}} \cdot \frac{60 \text{ seconds}}{1 \text{ minute}} \cdot \frac{1 \text{ yard}}{3 \text{ feet}}$$

$$= \frac{5 \cdot 60}{3} \text{ yards per minute}$$

$$= 100 \text{ yards per minute}$$

Copyright © Houghton Mifflin Company. All rights reserved.

7.2 Exercises

12 in = 1 ft

A. Units of Length and Distance

1. Ratios such as $\frac{1 \text{ yard}}{3 \text{ feet}}$ and $\frac{12 \text{ inches}}{1 \text{ foot}}$ are called ________.

2. In the United States, inches, feet, and miles are common units to measure ________.

In Exercises 3–12, convert the given measurement to the indicated units.

3. 2 feet = ____ inches

4. $\frac{2}{3}$ yard = ____ feet

5. 2 yards = ____ inches

6. $1\frac{1}{4}$ miles = ____ feet

7. 4 miles = ____ yards

8. 33 inches = ____ feet

9. 15 feet = ____ yards

10. 45 inches = ____ yards

11. 15,840 feet = ____ miles

12. 2,640 yards = ____ miles

13. *Ceiling Height* In most houses, the ceiling is 8 feet above the floor. What is the height of the ceiling in yards?

14. *Notebook Paper* The length of a sheet of notebook paper is 10 inches. What is the length in feet?

15. *Mt. Hubbard* The altitude of Mt. Hubbard in Alaska is 15,015 feet. To the nearest tenth, what is the altitude in miles?

16. *Empire State Building* The Empire State Building in New York is $416\frac{2}{3}$ yards high. What is the height of the building in feet?

17. *Sequoia Tree* The circumference of a large Sequoia tree is 998 inches. To the nearest whole number, what is the circumference in yards?

Copyright © Houghton Mifflin Company. All rights reserved.

18. *Yosemite Falls* Yosemite Falls is the highest waterfall in the United States. Its height is approximately 0.46 miles. To the nearest whole number, what is its height in feet?

B. Units of Weight

19. Ounces, pounds, and tons are common units of ______.

20. The conversion factor $\frac{16 \text{ ounces}}{1 \text{ pound}}$ can be used to convert ______ to ______.

In Exercises 21–28, convert the given measurement to the indicated units.

21. 3 tons = ____ pounds

22. 5 pounds = ____ ounces

23. $2\frac{1}{2}$ pounds = ____ ounces

24. $1\frac{1}{4}$ tons = ____ pounds

25. 48 ounces = ____ pounds

26. 10,000 pounds = ____ tons

27. 2,500 pounds = ____ tons

28. 36 ounces = ____ pounds

29. *Baby's Weight* At birth, a baby weighed 7 pounds 4 ounces. After 3 weeks, the baby had gained 2 pounds 12 ounces. What was the baby's new weight?

30. *Cat Care* A veterinarian recommended that a cat, whose weight was 15 pounds 8 ounces, be placed on a diet of low-calorie cat food. At the next checkup, the cat weighed 14 pounds 4 ounces. How much weight had the cat lost?

Copyright © Houghton Mifflin Company. All rights reserved.

31. *Produce Cost* A produce market charges $1.32 per pound for apples. What is the cost of 4 pounds 8 ounces of apples?

32. *Corn Seed* A seed store sells corn seed for $3.80 per pound. What is the cost of 1 pound 4 ounces of seed?

33. *Elevator Limit* A hotel elevator has a weight limit of 1 ton. If the average passenger weight is 165 pounds, what is the maximum (whole) number of people who should board the elevator?

34. *Hamburger Meat* A restaurant uses 6 ounces of ground meat to make a hamburger. How many pounds of meat are needed to make 30 hamburgers?

C. Units of Volume

35. We measure volume in ________ units.

36. We use the term *fluid ounce* to indicate volume or ________.

Copyright © Houghton Mifflin Company. All rights reserved.

In Exercises 37–48, convert the given measurement to the indicated units.

37. $3\frac{1}{2}$ gallons = ____ pints

38. 2 quarts = ____ fluid ounces

39. 3 quarts = ____ fluid ounces

40. $5\frac{3}{4}$ gallons = ____ quarts

41. $\frac{1}{2}$ cubic foot = ____ cubic inches

42. 2 cubic yards = ____ cubic feet

43. 48 fluid ounces = ____ pints

44. 10 quarts = ____ gallons

45. 128 fluid ounces = ____ cups

46. 26 pints = ____ gallons

47. 135 cubic feet = ____ cubic yards

48. 2,160 cubic inches = ____ cubic feet

49. *Aquarium Capacity* How many cubic inches of water are needed to fill an aquarium that is 10 inches wide, $2\frac{1}{5}$ feet long, and 15 inches deep?

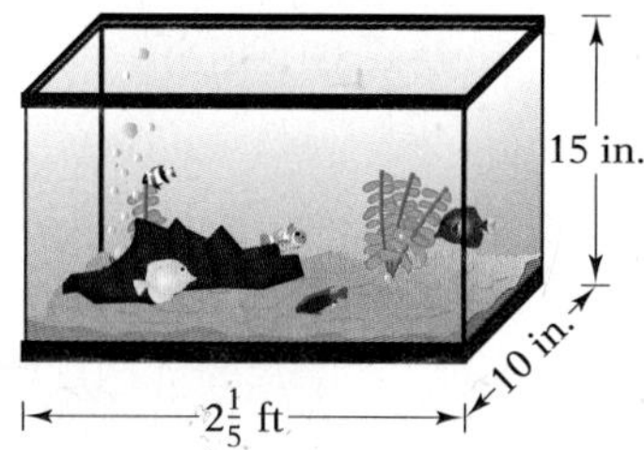

50. *Driveway Pavement* How many cubic yards of concrete are needed to pave a driveway that is 15 feet wide, 30 yards long, and 6 inches thick?

Copyright © Houghton Mifflin Company. All rights reserved.

51. *Milk Production* Each day a dairy farm processes eighty 15-gallon containers of milk. How many quarts of milk does the farm produce per day?

52. *Coffee Shop* Each day a coffee shop makes nine 2-gallon urns of coffee. How many 5-fluid-ounce cups of coffee can they serve?

D. Units of Time and Rate

53. In the table of time units, we didn't include the entry 1 month = 30 days. Why?

54. In the table of time units, we didn't include the entry 1 year = 365 days. Why?

In Exercises 55–64, convert the given measurement to the indicated units. Round to the nearest tenth if necessary.

55. 3 miles per hour = ____ feet per second

56. 40 feet per second = ____ miles per hour

57. 55 miles per hour = ____ yards per minute

58. 180 feet per minute = ____ miles per hour

59. 12 gallons per minute = ____ quart per second

60. 15 quarts per hour = ____ fluid ounce per second

61. 5 fluid ounces per second = ____ quarts per minute

62. 100 fluid ounces per minute = ____ gallons per hour

Copyright © Houghton Mifflin Company. All rights reserved.

63. 3 quarts per day = ▭ gallons per year

64. 100 gallons per year = ▭ quarts per month

65. *Cat's Speed* A domestic cat can run 30 miles per hour. What is the cat's speed in feet per second?

30 mph = ? ft/sec

66. *Kentucky Derby* A horse competing in the Kentucky Derby can run approximately 54.7 feet per second. To the nearest tenth, what is the horse's speed in miles per hour?

67. *Soda Dispenser* A dispenser can fill three 8-ounce cups with soda in 1 minute. At what rate in gallons per hour is the soda dispensed?

68. *Wright Brothers* The Wright brothers' plane flew 852 feet in 59 seconds. To the nearest tenth, what was the speed of the plane in miles per hour?

69. *Water Use* The average per-person use of water in the United States is 374 gallons per day. (Source: U.S. Geological Survey.) To the nearest tenth, what is the average in quarts per hour?

Copyright © Houghton Mifflin Company. All rights reserved.

70. *Waste Material* In the United States, the average amount of waste material generated is 4.3 pounds per day per person. (Source: EPA.) To the nearest tenth, what is the average in tons per year?

Writing and Concept Extension

71. *Recipe Units* The following units and conversions are frequently used in recipes:

16 tablespoons (Tsp) = 1 cup
3 teaspoons (tsp) = 1 tablespoon

The table gives common amounts used in recipes. Complete the table of equivalent measurements.

A Recipe Calls for	*Use*
$\frac{1}{3}$ cup	(a) ▒ tablespoons
12 tablespoons	(b) ▒ cup
2 ounces	(c) ▒ tablespoons
$1\frac{1}{3}$ tablespoons	(d) ▒ teaspoons

For Exercises 72–74, you may need to refresh your memory of certain area and volume formulas. If so, refer to the inside cover of the book.

72. *Fencing* A park manager needs to enclose a rectangular area that is $\frac{1}{2}$ mile long and 1,200 feet wide. The fence material and installation will cost $4.10 per yard. How much should the manager budget for the project?

73. *Garden Fence* To protect the plants from animals, a woman constructed a fence around her trapezoidal shaped vegetable garden shown in the figure at a cost of $.60 per yard.

(a) What was the total cost of the fence?

(b) How many square yards of garden space did she enclose?

(c) What was the cost per square yard to protect the garden?

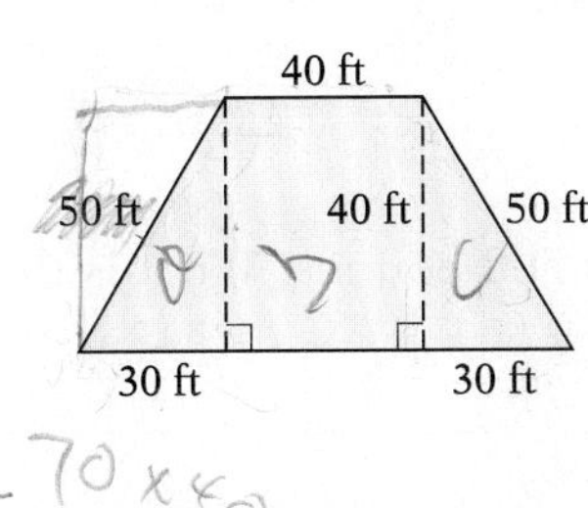

Copyright © Houghton Mifflin Company. All rights reserved.

74. *Ice Cream Container* A cylindrical plastic ice cream container is 9 inches high and has a diameter of 8 inches. What is the volume (in quarts) of the container? (*Hint:* Use the fact that 1 quart = 57.75 cubic inches.)

Exploring with Real-World Data: Collaborative Activities

Special Units of Measurement Other special units of measurement are used for certain situations. In Exercises 75–82, convert the given measurement to the indicated units.

75. *Horse Height* The height of a horse is sometimes measured in *hands.* A hand is 4 inches. What is the height, in feet, of a horse that is 15 hands high?

76. *Nautical Miles* Distance on an ocean is measured in *nautical miles.* One nautical mile is approximately 6,076 feet. If a hurricane is located 50 nautical miles from shore, how many land miles from shore is the storm?

77. *Leagues* A *league* is approximately 3 nautical miles. In the novel *20,000 Leagues Under the Sea,* how many nautical miles were traveled?

78. *Rods* Surveyors use a distance called a *rod,* which is $16\frac{1}{2}$ feet. If a surveyor measured a distance between two points as 20 rods, what is the distance in feet?

79. *River Depth* The depth of a river is measured in *fathoms.* A fathom is 6 feet. What is the depth in feet of a river whose depth is $2\frac{1}{2}$ fathoms?

80. *Print Size* Print is measured in *points.* One point is approximately 0.013 inch. What is the height, in inches, of 15-point type?

81. *Firewood* Firewood is measured in *cords.* A cord is 128 cubic feet. How many cubic feet does a load of $3\frac{1}{2}$ cords of wood occupy?

64

82. *Cotton* Cotton is measured in *bales.* A bale weighs 500 pounds. What is the weight, in tons, of 10 bales of cotton?

Copyright © Houghton Mifflin Company. All rights reserved.

7.3 Metric Units of Length and Distance

A *Introduction to the Metric System*
B *U.S. and Metric Unit Conversions*

SUGGESTIONS FOR SUCCESS

In 1999, *Time* magazine published a list of the 100 worst ideas of the century. Included was the plan to convert to the metric system in the United States.

However, because the rest of the world uses the metric system, it seems only a matter of time before the system is adopted in the United States. Scientists and engineers use it, as do international businesses. Even if you just travel to other countries, you will want to be familiar with the metric system.

Converting to the metric system in the United States will not be easy, but it is probably going to happen eventually. Knowing the metric system will put you ahead of the game.

A Introduction to the Metric System

If you think about the U.S. system of measurements, you will realize that there is no rhyme or reason to it. Why are there 12 inches in a foot rather than, say 10 or 23? Why are there 5,280 feet in a mile? In an unorganized way, this "Old English" system developed over the centuries, and it was carried over to our country when it was founded.

Meanwhile, England (and the rest of the world) has adopted the metric system. Like our decimal number system, the metric system is based on multiples of 10. Therefore, as you will see, converting between units is simply a matter of shifting decimal points.

In the metric system, the basic unit of length is the *meter* (m). A meter is about 3 inches longer than a yard.

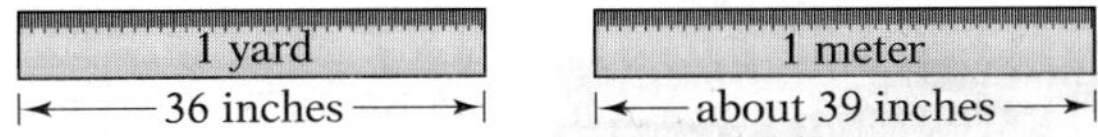

All metric length units contain the word *meter* preceded by a prefix that indicates a multiple or a fraction of a meter. The following table shows the metric units of length and distance and their abbreviations.

Metric Unit Prefixes for Length

Prefix	*Meaning*	*Metric Units of Length or Distance*
kilo	1,000	1 **kilo**meter (km) = 1,000 meters (m)
hecto	100	1 **hecto**meter (hm) = 100 meters (m)
deka	10	1 **deka**meter (dam) = 10 meters (m)
deci	$\frac{1}{10}$	1 **deci**meter (dm) = $\frac{1}{10}$ meter = 0.1 m
centi	$\frac{1}{100}$	1 **centi**meter (cm) = $\frac{1}{100}$ meter = 0.01 m
milli	$\frac{1}{1,000}$	1 **milli**meter (mm) = $\frac{1}{1,000}$ meter = 0.001 m

Copyright © Houghton Mifflin Company. All rights reserved.

Unless you use the metric system regularly, metric units will probably not seem natural to you. Try to relate the most commonly used metric units—meter, kilometer, centimeter, and millimeter—to what you know.

Because a meter is slightly longer than a yard, a length measured in meters is *less* than that same length measured in yards.

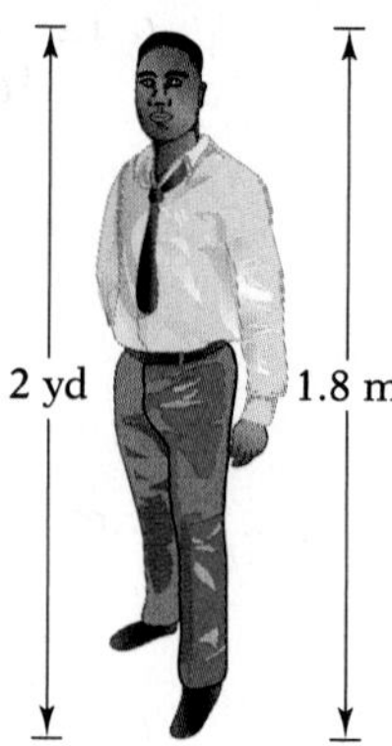

In the metric system, distances are measured in kilometers rather than in miles. A kilometer is about $\frac{6}{10}$ of a mile. This means that the number of kilometers between two points is *greater* than the number of miles.

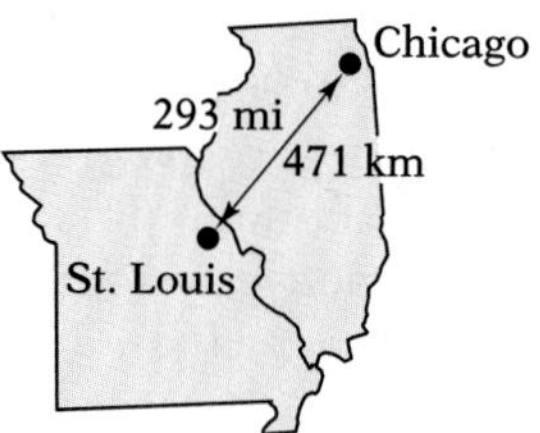

A centimeter is slightly less than $\frac{1}{2}$ inch. The figure shows that the diameter of a penny is about 2 centimeters.

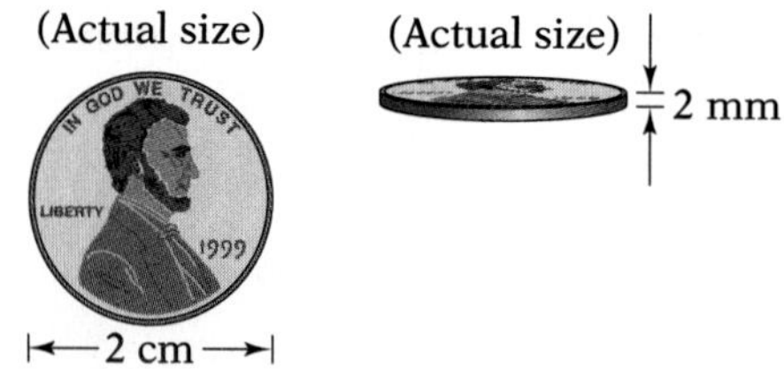

A millimeter is a very small measure that is often used for thickness. In the figure, we see that the thickness of a penny is about 2 millimeters.

The following table summarizes the relationships among these four most commonly used metric units of length:

Metric Units of Length and Distance

1 kilometer = 1,000 meters
1 meter = 100 centimeters
1 centimeter = 10 millimeters

Copyright © Houghton Mifflin Company. All rights reserved.

We can change one metric unit to another with conversion factors, as before. However, you will see that every conversion involves multiplying or dividing by some power of 10. This makes the arithmetic easy because we only need to shift the decimal point to the right or left.

Because metric-to-metric conversions involve only shifts of the decimal point, it will be tempting to bypass the use of conversion factors. However, this will invite errors because the conversion factors will tell you whether the decimal point needs to be shifted to the left or to the right.

Example 1

Convert 243 centimeters to meters.

SOLUTION

We set up the template so that the centimeter units will cancel.

$$\frac{243 \text{ cm}}{1} \cdot \frac{\quad \text{m}}{\quad \text{cm}}$$

$$\frac{243 \text{ cm}}{1} \cdot \frac{1 \text{ m}}{100 \text{ cm}}$$ Because 1 m = 100 cm, $\frac{1 \text{ m}}{100 \text{ cm}} = 1$.

$$= \frac{243}{100} \text{ m}$$ The cm units cancel.

$$= 2.43 \text{ m}$$ To divide by 100, we shift the decimal point 2 places to the left.

In the metric system, fractional parts are usually given as decimal numbers.

Your Turn 1

Convert 96 centimeters to meters.

Answer: 0.96 meter

Example 2

Make the following conversions:

(a) 56.9 centimeters to millimeters

(b) 213 millimeters to centimeters

SOLUTION

(a) $$\frac{56.9 \text{ cm}}{1} \cdot \frac{10 \text{ mm}}{1 \text{ cm}}$$ Because 1 cm = 10 mm, $\frac{10 \text{ mm}}{1 \text{ cm}} = 1$.

$$= \frac{56.9 \cdot 10}{1} \text{ mm}$$ The cm units cancel.

$$= 569 \text{ mm}$$ To multiply by 10, we shift the decimal point 1 place to the right.

(b) $$\frac{213 \text{ mm}}{1} \cdot \frac{1 \text{ cm}}{10 \text{ mm}}$$ Because 1 cm = 10 mm, $\frac{1 \text{ cm}}{10 \text{ mm}} = 1$.

$$= \frac{213}{10} \text{ cm}$$ The mm units cancel.

$$= 21.3 \text{ cm}$$ To divide by 10, we shift the decimal point 1 place to the left.

Your Turn 2

Make the following conversions.

(a) 998 millimeters to centimeters

(b) 1.27 centimeters to millimeters

Answers: (a) 99.8 cm; (b) 12.7 mm

Copyright © Houghton Mifflin Company. All rights reserved.

B U.S. and Metric Unit Conversions

When we use metric units for familiar lengths and distances, the measurements can seem odd to us. We might be used to the distance from home plate to the outfield fence being 340 feet, but fans at Montreal's Olympic Stadium would be accustomed to that same distance expressed as 102 meters. You may have a friend who lives in a town that is 200 miles from where you live, but a person living in Europe would think of the same distance as about 322 kilometers. To Americans, the numerical parts of metric measurements often seem too small or too large, but people from countries that use the metric system are just as baffled by the American system.

The following diagrams give some approximate values that can be used for conversion factors for converting between the U.S. and metric systems.

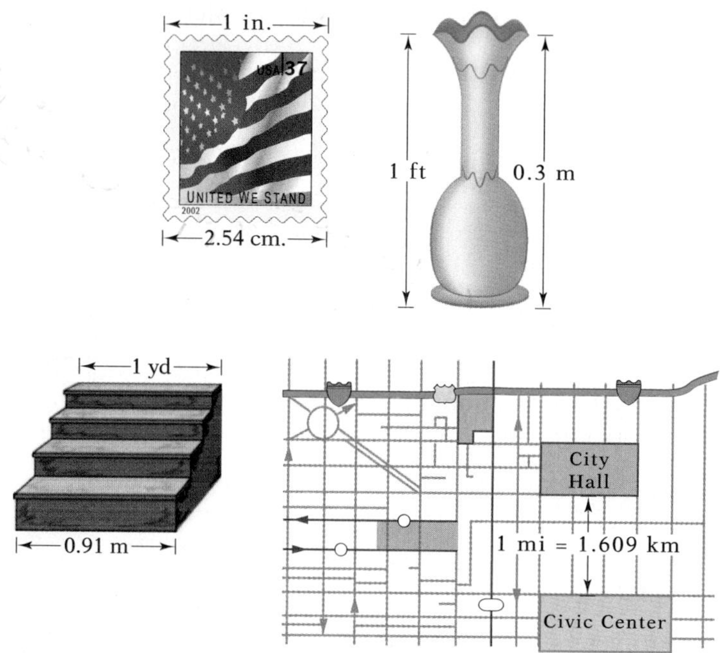

Because these values are only approximate, the solutions in the following examples and in the related exercises are also approximate.

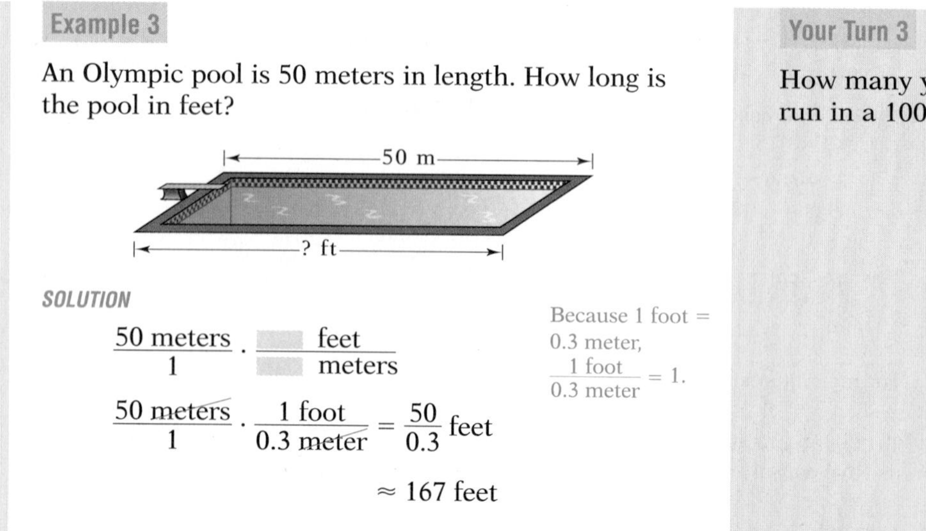

Example 3

An Olympic pool is 50 meters in length. How long is the pool in feet?

SOLUTION

$$\frac{50 \text{ meters}}{1} \cdot \frac{\boxed{} \text{ feet}}{\boxed{} \text{ meters}}$$

$$\frac{50 \text{ meters}}{1} \cdot \frac{1 \text{ foot}}{0.3 \text{ meter}} = \frac{50}{0.3} \text{ feet}$$

$$\approx 167 \text{ feet}$$

Because 1 foot = 0.3 meter, $\frac{1 \text{ foot}}{0.3 \text{ meter}} = 1$.

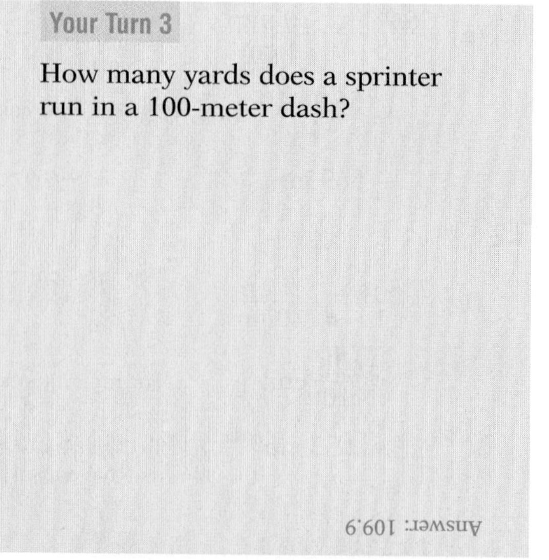

Your Turn 3

How many yards does a sprinter run in a 100-meter dash?

Answer: 109.9

Copyright © Houghton Mifflin Company. All rights reserved.

Example 4

In Canada, the distance from Vancouver to Winnipeg is 1,427 miles. What is this distance to the nearest kilometer?

SOLUTION

$$\frac{1{,}427 \text{ miles}}{1} \cdot \frac{1.609 \text{ kilometers}}{1 \text{ mile}}$$

1 mile = 1.609 kilometers

$$= (1{,}427)(1.609) \text{ kilometers}$$

$$\approx 2{,}296 \text{ kilometers}$$

Your Turn 4

The distance from New Orleans to Miami is 892 miles. What is this distance to the nearest kilometer?

Answer: 1,435 kilometers

Example 5

Standard copier paper is 8.5 inches wide and 11 inches high. To the nearest tenth, what are the dimensions in centimeters?

SOLUTION

For the conversion factors, we use the fact that 1 inch = 2.54 centimeters.

$$\frac{8.5 \text{ inches}}{1} \cdot \frac{2.54 \text{ centimeters}}{1 \text{ inch}}$$

$$= (8.5)(2.54) \text{ centimeters}$$

$$\approx 21.6 \text{ centimeters}$$

$$\frac{11 \text{ inches}}{1} \cdot \frac{2.54 \text{ centimeters}}{1 \text{ inch}}$$

$$= (11)(2.54) \text{ centimeters}$$

$$\approx 27.9 \text{ centimeters}$$

The width is 21.6 centimeters, and the height is 27.9 centimeters.

Your Turn 5

The width of a rectangle is 16 centimeters, and the length is 20 centimeters. To the nearest tenth, what are the dimensions of the rectangle in inches?

Answer: Width is 6.3 inches, length is 7.9 inches.

NOTE When you convert between U.S. and metric unit systems, answers can vary depending on the conversion factor that you use. For example, to convert feet to centimeters, converting feet to inches to centimeters or converting feet to meters to centimeters will likely result in answers that are somewhat different.

7.3 Quick Reference

A. Introduction to the Metric System

1. In the metric system, the basic unit of length and distance is the *meter,* which is about 3 inches longer than a yard. Prefixes such as *milli, centi,* and *kilo* indicate a multiple or fraction of a meter.

Copyright © Houghton Mifflin Company. All rights reserved.

2. Converting between metric units involves multiplying or dividing by powers of 10.

$$650 \text{ cm} = \frac{650 \text{ cm}}{1} \cdot \frac{1 \text{ m}}{100 \text{ cm}}$$
$$= \frac{650}{100} \text{ m}$$
$$= 6.5 \text{ m}$$

B. U.S. and Metric Unit Conversions

1. Using approximate conversion factors, we can convert between U.S. and metric units.

$$10 \text{ meters} = \frac{10 \text{ meters}}{1} \cdot \frac{1 \text{ yard}}{0.91 \text{ meter}}$$
$$= \frac{10}{0.91} \text{ yards}$$
$$\approx 11 \text{ yards}$$

2. Tables of U.S. and metric units of length can be found on the inside cover of the book.

Copyright © Houghton Mifflin Company. All rights reserved.

7.3 Exercises

A. Introduction to the Metric System

1. In the ________ system, common units of length are centimeter, meter, and kilometer.

2. The prefix *kilo* means ________, whereas the prefix ________ means $\frac{1}{1,000}$.

In Exercises 3–18, convert the given measurement to the indicated units.

3. 8 centimeters = ____ millimeters
4. 4.8 centimeters = ____ millimeters
5. 7.6 meters = ____ centimeters
6. 2 meters = ____ centimeters
7. 3 meters = ____ millimeters
8. 1.3 meters = ____ millimeters
9. 5.23 kilometers = ____ meters
10. 2 kilometers = ____ meters
11. 125 millimeters = ____ centimeters
12. 30 millimeters = ____ centimeters
13. 200 centimeters = ____ meters
14. 450 centimeters = ____ meters
15. 750 millimeters = ____ meter
16. 4,000 millimeters = ____ meters
17. 5,000 meters = ____ kilometers
18. 6,500 meters = ____ kilometers

19. *Dinner Plate Diameter* Suppose that the diameter of a dinner plate is 25 centimeters. What is the diameter in millimeters?

20. *Math Book Width* The width of your math book is approximately 215 millimeters. What is the width in centimeters?

21. *Man's Height* The height of an average adult male is 1.8 meters. What is his height in centimeters?

22. *Film Size* Some film is 35 millimeters wide. What is the width in centimeters?

23. *Nile River* The length of the Nile River is 6,695 kilometers. What is the length in meters?

Copyright © Houghton Mifflin Company. All rights reserved.

24. *Mount Blanc* The elevation of Mont Blanc in France is 4,807 meters. What is the elevation in kilometers?

B. U.S. and Metric Unit Conversions

25. One meter is slightly ____ than 1 yard.

26. One inch is approximately $2\frac{1}{2}$ ____ in the metric system.

In Exercises 27–36, convert the given metric measurement to U.S. units.

27. 20 centimeters = ____ inches

28. 10,000 meters = ____ miles

29. 50 millimeters = ____ inches

30. 45 centimeters = ____ feet

31. 3 kilometers = ____ yards

32. 10 meters = ____ yards

33. 750 millimeters = ____ feet

34. 8 kilometers = ____ miles

35. 1.5 kilometers = ____ feet

36. 0.8 meter = ____ inches

In Exercises 37–46, convert the given U.S. measurement to metric units.

37. 6 feet = ____ meters

38. 3 inches = ____ centimeters

39. 5 miles = ____ kilometers

40. 10 yards = ____ meters

41. $\frac{1}{4}$ inch = ____ millimeters

42. $\frac{4}{5}$ mile = ____ meters

43. 1 foot = ____ centimeters

44. 3,000 feet = ____ kilometer

45. 400 yards = ____ kilometer

46. 2 inches = ____ millimeters

Copyright © Houghton Mifflin Company. All rights reserved.

47. *Football Field Length* The length of a football field is 100 yards. What is the length in meters?

48. *Soccer Field Length* Suppose that the length of a soccer field is 100 meters. What is the length in yards?

49. *10-K Run* What is the length, in miles, of a 10-kilometer run?

50. *TV Screen Size* What is the size, in centimeters, of a 22-inch television screen?

51. *Matterhorn* The elevation of the Matterhorn in Switzerland is 4,478 meters. What is the elevation in feet?

52. *Paris to Berlin* The distance from Paris to Berlin is 548 miles. What is the distance in kilometers?

≈ Estimation

In Exercises 53–56, choose the best estimate of the value of the given measurement.

53.	20 kilometers	(i) 1.2 miles	(ii) 12 miles	(iii) 120 miles
54.	5 meters	(i) 50 yards	(ii) 0.5 yard	(iii) 5 yards
55.	6 miles	(i) 1 kilometer	(ii) 10 kilometers	(iii) 100 kilometers
56.	4 inches	(i) 10 centimeters	(ii) 100 centimeters	(iii) 1 centimeter

Calculator Exercises

In Exercises 57–60, convert the given measurement to the indicated units. Round to the nearest tenth if necessary. (*Hint:* 1 square foot = 144 square inches and 1 square yard = 9 square feet.)

57. 3.5 square feet = ▒ square inches

58. 5 square yards = ▒ square feet

59. 72 square feet = ▒ square yards

60. 360 square inches = ▒ square feet

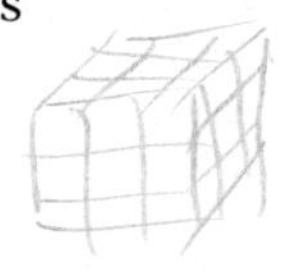

Copyright © Houghton Mifflin Company. All rights reserved.

61. *Carpeting a Room* How many square yards of carpet are needed for a room that is 18 feet wide and 25 feet long?

62. *Table Top Area* What is the area in square inches of the top of a round table with a diameter of $2\frac{1}{2}$ feet?

Writing and Concept Extension

63. Which is *not* a correct conversion factor? Why?

(i) $\dfrac{1 \text{ centimeter}}{10 \text{ millimeters}}$

(ii) $\dfrac{1 \text{ kilometer}}{100 \text{ meters}}$

(iii) $\dfrac{1 \text{ meter}}{100 \text{ centimeters}}$

64. Insert $<$ or $>$ to make the statement true.

(a) 1 yard ____ 1 meter

(b) 1 inch ____ 1 centimeter

(c) 1 mile ____ 1 kilometer

In the U.S. system, the area of land is often measured in acres. One acre is 43,560 square feet and 640 acres is 1 square mile. In Exercises 65–68, convert the given measurement to the indicated units.

65. $2\frac{1}{2}$ square miles = ____ acres

66. $\frac{1}{3}$ acre = ____ square feet

67. 32,670 square feet = ____ acre

68. 960 acres = ____ square miles

69. *Paved Area* An asphalt spreader paves a road that is 20 feet wide and 3 miles long. How many acres of area are covered?

70. *Seeding a Lawn* One bag of grass seed is needed for 5,000 square feet of lawn. To the nearest whole number, how many bags of seed should a homeowner purchase for $\frac{2}{3}$ acre?

Copyright © Houghton Mifflin Company. All rights reserved.

7.4 Metric Units of Volume and Weight

A Metric Units of Volume
B Metric Units of Weight
C U.S. and Metric Unit Conversions

SUGGESTIONS FOR SUCCESS

It is a sad fact that the illegal drug trade is helping us to become acquainted with metric units of weight. Grams and kilos (kilograms) are becoming part of the daily language of news reporting. Even our drug laws are framed in terms of these units.

As you begin your study of metric volume and weight units, try to keep the overall objective in mind: The entire metric system is based on multiples of 10. As was true with length and distance units, we can easily convert metric volume and weight units simply by shifting decimal points.

If you understand this basic concept of the metric system, then you will see its advantages over the crazy-quilt U.S. systems of measurements.

A Metric Units of Volume

When you buy gasoline in a country other than the United States, you are likely to buy *liters* rather than gallons. The liter (pronounced lee-ter) is the basic volume unit in the metric system, and it is equivalent to slightly more than 1 quart.

Unlike other metric units, the liter unit is commonly used even in the United States. We find 2-liter soft drink bottles in the grocery, and you may have a 3.5-liter engine in your automobile.

A liter is the volume of a cube that is 10 centimeters on each side.

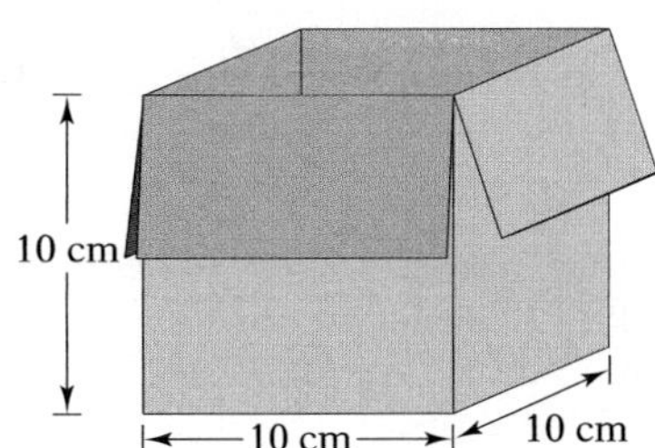

The same prefixes that are used with metric length units are used for multiples or fractions of liters.

Metric Unit Prefixes for Volume

Prefix	*Meaning*	*Metric Units of Volume*
kilo	1,000	1 **kilo**liter (kL) = 1,000 liters (L)
hecto	100	1 **hecto**liter (hL) = 100 liters (L)
deka	10	1 **deka**liter (daL) = 10 liters (L)
deci	$\frac{1}{10}$	1 **deci**liter (dL) = $\frac{1}{10}$ liter = 0.1 L
centi	$\frac{1}{100}$	1 **centi**liter (cL) = $\frac{1}{100}$ liter = 0.01 L
milli	$\frac{1}{1,000}$	1 **milli**liter (mL) = $\frac{1}{1,000}$ liter = 0.001 L

Copyright © Houghton Mifflin Company. All rights reserved.

In addition to the liter (L), the unit in the preceding table that is used most often is the milliliter (mL). A milliliter is the volume of a cube that is 1 centimeter on each side.

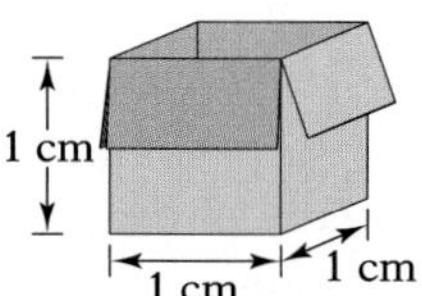

The volume of this cube is 1 cubic centimeter (cm^3). In chemistry and medicine, for example, you are more likely to see the abbreviation "cc" for cubic centimeters. Therefore, 1 mL, 1 cm^3, and 1 cc all mean the same thing.

$$\begin{aligned} 1 \text{ liter (L)} &= 1{,}000 \text{ milliliters (mL)} \\ &= 1{,}000 \text{ cubic centimeters (cm}^3\text{)} \\ &= 1{,}000 \text{ cubic centimeters (cc)} \end{aligned}$$

Soft drink bottles and cans usually give the volume (or capacity) in both fluid ounces and milliliters.

Example 1

An intravenous (IV) bag contains 120 cubic centimeters of a saline solution. What is the capacity of the bag in (a) milliliters and (b) liters?

Saline solution
120 cc

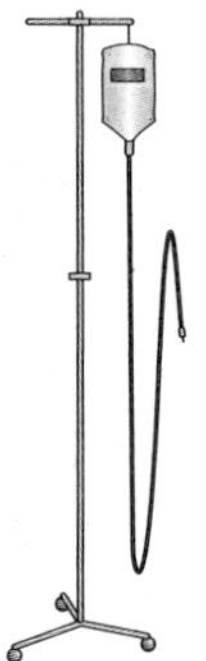

SOLUTION

(a) Because cubic centimeters (cc) and milliliters (mL) mean the same thing, the capacity of the bag is 120 milliliters.

Your Turn 1

A chemistry beaker contains 0.65 liter of a solution. How many milliliters of solution are in the beaker?

Copyright © Houghton Mifflin Company. All rights reserved.

(b) Because 1 liter = 1,000 cubic centimeters, we use the conversion factor $\frac{1 \text{ liter}}{1{,}000 \text{ cubic centimeters}} = 1$.

$$\frac{120 \text{ cc}}{1} \cdot \frac{1 \text{ liter}}{1{,}000 \text{ cc}} = \frac{120}{1{,}000} \text{ liter}$$
$$= 0.12 \text{ liter}$$

Example 2

A storage tank holds 3.6 kiloliters of kerosene. If 1,500 liters are drained from the tank, how many kiloliters remain?

SOLUTION

We convert the amount that was drained to kiloliters. Because 1 kiloliter = 1,000 liters, we use the conversion factor $\frac{1 \text{ kiloliter}}{1{,}000 \text{ liters}} = 1$.

$$\frac{1{,}500 \text{ liters}}{1} \cdot \frac{1 \text{ kiloliter}}{1{,}000 \text{ liters}} = \frac{1{,}500}{1{,}000} \text{ kiloliters}$$
$$= 1.5 \text{ kiloliters}$$

The amount of remaining kerosene is

$$3.6 - 1.5 = 2.1 \text{ kiloliters}$$

Your Turn 2

The capacity of a gasoline tank is 5 kiloliters. If the tank now contains 4.6 kiloliters, how many liters of gasoline are needed to fill the tank?

Answer: 400

B Metric Units of Weight

In the U.S. system of units, there is no relationship between weight and volume. You may be surprised to learn that weight and volume in the metric system are connected by water!

The basic unit of weight in the metric system is a gram (g), which is only a tiny fraction of an ounce. A gram is the weight of 1 cubic centimeter of water.

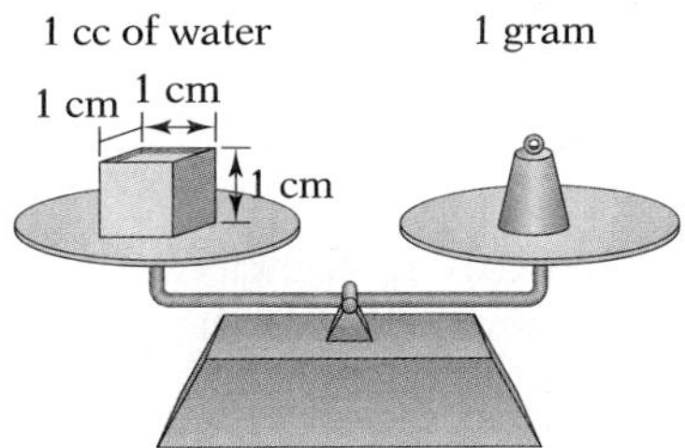

The same prefixes that are used for metric units of length and volume are used for metric units of weight.

NOTE Technically, a gram is the weight of 1 cubic centimeter of water whose temperature is 20°C. If you take chemistry, you will learn that the *density* of water decreases as its temperature increases from 4°C to 100°C. This means that 1 cubic centimeter of water at, say, 60°C will weigh slightly less than 1 gram. Nearly all measurement standards, both U.S. and metric, are based on specific temperatures.

Copyright © Houghton Mifflin Company. All rights reserved.

Metric Unit Prefixes for Weight

Prefix	*Meaning*	*Metric Units of Weight*
kilo	1,000	1 **kilo**gram (kg) = 1,000 grams (g)
hecto	100	1 **hecto**gram (hg) = 100 grams (g)
deka	10	1 **deka**gram (dag) = 10 grams (g)
deci	$\frac{1}{10}$	1 **deci**gram (dg) = $\frac{1}{10}$ gram = 0.1 g
centi	$\frac{1}{100}$	1 **centi**gram (cg) = $\frac{1}{100}$ gram = 0.01 g
milli	$\frac{1}{1,000}$	1 **milli**gram (mg) = $\frac{1}{1,000}$ gram = 0.001 g

In addition to the gram, the most commonly used metric weights are milligrams and kilograms.

NOTE In the United States, metric units of length and distance are among the hardest to get used to. However, we are becoming increasingly familiar with metric units of volume and weight because of their widespread use in the medical field. As we saw in Example 1, the volume of a liquid in an intravenous bag is typically measured in cubic centimeters. Pills are usually described by their weights in milligrams. Even if you don't plan to enter some medical field of study, an awareness of these metric units is important to your being a responsible health care consumer.

The following table summarizes some of the relationships that are often used to convert metric units of weight:

Metric Units of Weight

1 gram (g) = 1,000 milligrams (mg)
1 kilogram (kg) = 1,000 grams (g)
1 metric ton (t) = 1,000 kilograms (kg)

Example 3

Tagamet®, a heartburn medication, comes in 200-milligram tablets. In grams, what is the weight of 30 Tagamet tablets?

SOLUTION

We find the weight, in grams, of each tablet.

$$\frac{200 \text{ milligrams}}{1} \cdot \frac{1 \text{ gram}}{1{,}000 \text{ milligrams}} = \frac{200}{1{,}000} \text{ gram}$$
$$= 0.2 \text{ gram}$$

Now we multiply to determine the total weight of the 30 tablets.

$$30 \cdot 0.2 \text{ gram} = 6 \text{ grams}$$

Your Turn 3

With a prescription, Tagamet can also be purchased in 800-milligram tablets. In grams, what is the weight of 20 of these tablets?

Answer: 16 grams

Copyright © Houghton Mifflin Company. All rights reserved.

Example 4

A shipping container weighs 865 kilograms. What is the weight of the container in metric tons?

SOLUTION

$$\frac{865 \text{ kilograms}}{1} \cdot \frac{1 \text{ metric ton}}{1{,}000 \text{ kilograms}}$$

$$= \frac{865}{1{,}000} \text{ metric ton}$$

$$= 0.865 \text{ metric ton}$$

To divide by 1,000, shift the decimal point 3 places to the left.

Your Turn 4

A truck delivers 0.72 metric ton of gravel. What is the weight of the gravel in kilograms?

Answer: 720 kilograms

C U.S. and Metric Unit Conversions

Although the following diagrams do not provide every possible conversion between U.S. and metric units of volume and weight, they illustrate some of the more common ones:

Units of Volume

1 qt

0.946 L

1 gal

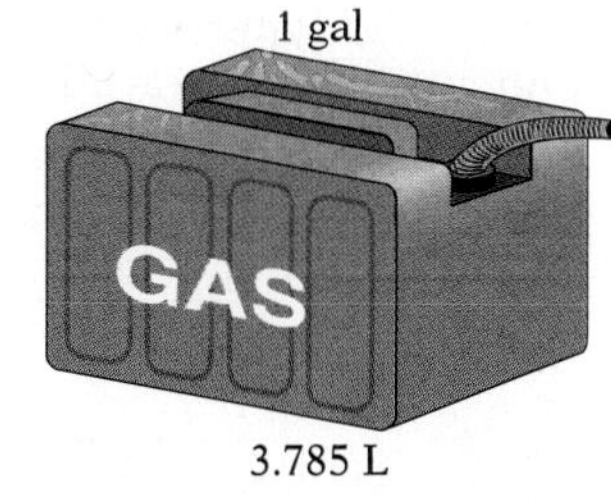

3.785 L

Units of Weight

1 oz

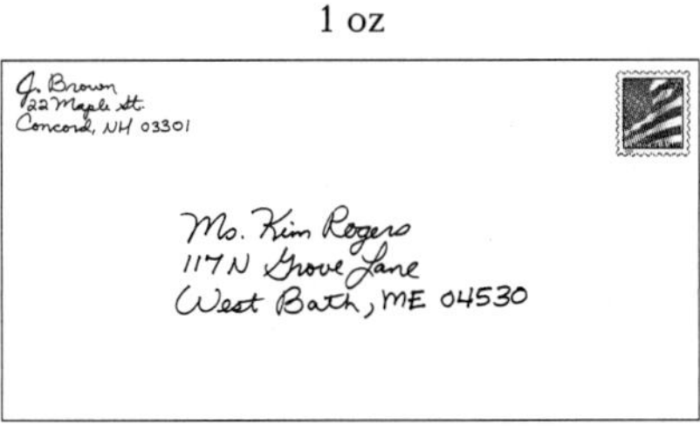

28.35 g

1 lb

0.454 kg

1 T

0.91 t

Copyright © Houghton Mifflin Company. All rights reserved.

NOTE Refer to the inside cover of this book for more complete listings of U.S. and metric units of length, weight, and volume.

Example 5

How many liters of gasoline would you need to buy to fill a 16-gallon tank?

SOLUTION

Because 1 gallon = 3.785 liters, we use the conversion factor $\frac{3.785 \text{ liters}}{1 \text{ gallon}} = 1$.

$$\frac{16 \text{ gallons}}{1} \cdot \frac{3.785 \text{ liters}}{1 \text{ gallon}} = 16(3.785) \text{ liters}$$
$$= 60.56 \text{ liters}$$

Your Turn 5

How many quarts of a soft drink does a 2-liter bottle contain?

Answer: 2.11

Recall that the "unit price" of an item is found by dividing the item's price by its weight, which can be given in any weight unit. However, if a unit price comparison between two items is to be made, then the weight units of the two items must be the same.

Example 6

In terms of unit price, which is the better buy, a 5-pound bag of flour for $.88 or a 5-kilogram bag for $1.65?

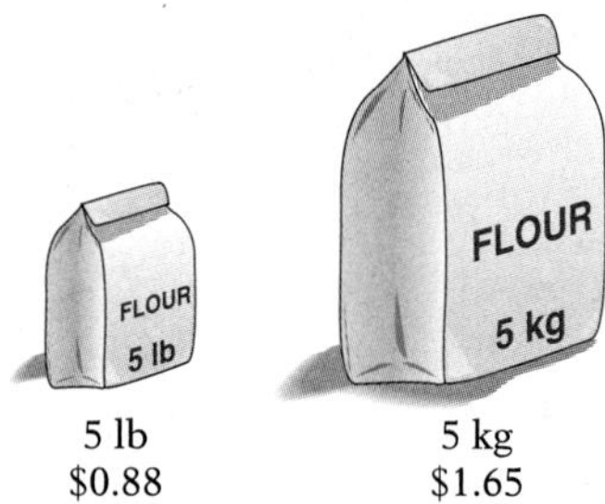

SOLUTION

To compare unit prices, we must know the two weights in the same units. We can convert both weights either to pounds or to kilograms. We use the fact that 1 lb ≈ 0.454 kilogram to convert to pounds.

$$\frac{5 \text{ kilograms}}{1} \cdot \frac{1 \text{ pound}}{0.454 \text{ kilogram}} \approx 11 \text{ pounds}$$

Now we can compare the unit prices.

5-pound bag: $\frac{\$.88}{5 \text{ pounds}} \approx \$.18$ per pound

5-kilogram bag: $\frac{\$1.65}{11 \text{ pounds}} \approx \$.15$ per pound

The 5-kilogram bag is the better buy.

Your Turn 6

Laundry detergent is available in a 96-ounce box or a 3-kilogram box. If both boxes cost $6, which is the better buy in terms of unit price?

Answer: 3-kilogram box

Copyright © Houghton Mifflin Company. All rights reserved.

Example 7

A conveyer can load a ship with 2 metric tons of coal per hour. To the nearest tenth, how many U.S. tons can be loaded per hour?

SOLUTION

We use the fact that 1 U.S. ton is equal to 0.91 metric ton.

$$\frac{2 \text{ metric tons}}{1 \text{ hour}} \cdot \frac{1 \text{ U.S. ton}}{0.91 \text{ metric ton}}$$

$$= \frac{2}{0.91} \text{ U.S. tons per hour}$$

$$\approx 2.2 \text{ U.S. tons per hour}$$

Your Turn 7

Large amounts of liquid fuel are sometimes measured by weight rather than volume. To the nearest tenth, what is the metric weight of 6 U.S. tons of jet fuel?

Answer: 5.5 metric tons

7.4 Quick Reference

A. Metric Units of Volume

1. The basic metric unit of volume is the liter (L), which is the volume of a cube that is 10 centimeters on each side. A kiloliter (kL) is 1,000 liters.

2. A milliliter (mL) is the volume of a cube that is 1 centimeter on each side. A liter is equal to 1,000 milliliters or 1,000 cubic centimeters (cc).

$$1{,}200 \text{ cc} = \frac{1{,}200 \text{ cc}}{1} \cdot \frac{1 \text{ L}}{1{,}000 \text{ cc}}$$

$$= \frac{1{,}200}{1{,}000} \text{ L} = 1.2 \text{ L}$$

B. Metric Units of Weight

1. The basic metric unit of weight is the gram (g), which is the weight of 1 cubic centimeter of water.

2. Other metric weights include kilogram (kg), milligram (mg), and metric ton (t). A kilogram is equal to 1,000 grams, and a metric ton is equal to 1,000 kilograms.

643 milligrams

$$= \frac{643 \text{ milligrams}}{1} \cdot \frac{1 \text{ gram}}{1{,}000 \text{ milligrams}}$$

$$= \frac{643}{1{,}000} \text{ gram}$$

$$= 0.643 \text{ gram}$$

Copyright © Houghton Mifflin Company. All rights reserved.

C. U.S. and Metric Unit Conversions

1. Conversion factors can be used to change between U.S. and metric units of volume.

$$10 \text{ liters} = \frac{10 \text{ liters}}{1} \cdot \frac{1 \text{ quart}}{0.946 \text{ liter}} \approx 10.6 \text{ quarts}$$

2. We can also use conversion factors to change between U.S. and metric units of weight.

$$2 \text{ pounds} = \frac{2 \text{ pounds}}{1} \cdot \frac{0.454 \text{ kilogram}}{1 \text{ pound}} \approx 0.908 \text{ kilogram}$$

Copyright © Houghton Mifflin Company. All rights reserved.

7.4 Exercises

A. Metric Units of Volume

1. The volume of a cube that is 10 centimeters on each side is called a(n) ______.

2. The abbreviations cc and mL represent the equal units ______ and ______, respectively.

In Exercises 3–10, convert the given measurements to the indicated units.

3. 0.25 liter = ______ milliliters

4. 1.5 kiloliters = ______ liters

5. 2 liters = ______ cubic centimeters

6. 0.1 liter = ______ cubic centimeters

7. 1,250 liters = ______ kiloliters

8. 6,500 milliliters = ______ liters

9. 800 milliliters = ______ liter

10. 400 cubic centimeters = ______ liter

11. *Car Engine Size* Suppose that a car has a 3.5-liter engine. What is the size of the engine in milliliters?

12. *Car Engine Size* Suppose that a car has a 2.1-liter engine. What is the size of the engine in cubic centimeters?

13. *Soft Drink Container* How many milliliters are in a $\frac{1}{4}$-liter bottle of a soft drink?

14. *Salad Dressing Bottle* How many liters of salad dressing are in two bottles of dressing that each contain 290 milliliters?

Copyright © Houghton Mifflin Company. All rights reserved.

B. Metric Units of Weight

15. In the metric system, the basic unit of weight is a(n) ________.

16. The conversion factor $\frac{1{,}000 \text{ grams}}{1 \text{ kilogram}}$ can be used to convert ________ to ________.

In Exercises 17–24, convert the given measurement to the indicated units.

17. 5 grams = ________ milligrams

18. 1.2 metric tons = ________ kilograms

19. 3.4 kilograms = ________ grams

20. 0.25 gram = ________ milligrams

21. 7,800 kilograms = ________ metric tons

22. 6,000 milligrams = ________ grams

23. 750 grams = ________ kilogram

24. 250 kilograms = ________ metric ton

25. *Calcium Supplement* One regular Tums® provides 200 milligrams of calcium, and one Tums EX® provides 0.3 gram of calcium. How many additional milligrams of calcium are in Tums EX than in regular Tums?

26. *Ibuprofen Tablet* An Advil® tablet contains 200 milligrams of ibuprofen. If a doctor recommends that a patient take 1.6 grams of ibuprofen per day, how many tablets should the patient take each day?

27. *Servings of Roast* How many 300-gram servings can be cut from a boneless roast that weighs 2.4 kilograms?

Copyright © Houghton Mifflin Company. All rights reserved.

28. *Case of Cheerios* A box of Cheerios® weighs 425 grams. What is the weight in kilograms of a case of 12 boxes of the cereal?

29. *Boxes of Nails* A packing case contains 600-gram boxes of nails. If a full packing case weighs 27 kilograms, how many boxes of nails does it contain?

30. *Granola Bars* A carton of 12 granola bars weighs 1.02 kilograms. What is the weight in grams of each granola bar?

C. U.S. and Metric Unit Conversions

In Exercises 31–38, convert the given measurement to the indicated units.

31. 3 gallons = ____ liters

32. 8 fluid ounces = ____ cubic centimeters

33. 1 cup = ____ milliliters

34. 6 quarts = ____ liters

35. 250 cubic centimeters = ____ cups

36. 4 liters = ____ gallons

37. $\frac{1}{2}$ liter = ____ pints

38. 10 cubic centimeters = ____ fluid ounce

Copyright © Houghton Mifflin Company. All rights reserved.

39. *Gas Tank Capacity* How many liters of gasoline will a 17-gallon tank hold?

40. *Milk Container* How many pints of milk are in a 4-liter container?

41. *Fresh Orange Juice* A store sells fresh-squeezed orange juice in 1-liter containers. The juice machine will squeeze 6 gallons of juice each hour. How many 1-liter bottles can be filled each hour?

42. *Servings of Soda* A person purchased twelve 2-liter bottles of soda for a party. How many 8-fluid-ounce cups can be served?

43. *Price of Gasoline* Suppose that gasoline costs $1.45 per gallon. What is the cost of gasoline in dollars per liter?

44. *Cost of Gasoline* Suppose that in France gasoline costs $1.80 per liter. What is the cost of gasoline in dollars per gallon?

In Exercises 45–52, convert the given measurement to the indicated units.

45. 12 ounces = ______ grams

46. 5 pounds = ______ kilograms

Copyright © Houghton Mifflin Company. All rights reserved.

47. 2 tons = ________ metric tons

48. 0.5 pound = ________ grams

49. 4 kilograms = ________ pounds

50. 400 grams = ________ ounces

51. 1.2 kilograms = ________ ounces

52. 2.5 metric tons = ________ tons

53. *Weight of Book* Your math book weighs approximately 3 pounds 4 ounces. What is the weight of the book in kilograms?

54. *Man's Weight* The recommended weight for a man who is 6 feet tall is 162 pounds. What is the recommended weight in kilograms?

55. *Bar of Soap* The weight of a bar of soap is 140 grams. What is the weight in pounds of an economy package of 12 bars of soap?

56. *Dishwasher Detergent* A box of dishwasher detergent that weighs 1.4 kilograms contains enough detergent for 25 loads of dishes. To the nearest tenth, how many ounces of detergent should be used in the dishwasher?

57. *Laundry Detergent* In terms of unit price, which is the better buy, a 4-pound box of laundry detergent for $5.79 or a 2-kilogram box for $6.17?

Copyright © Houghton Mifflin Company. All rights reserved.

58. *Cat Food* In terms of unit price, which is the better buy, a 3.5-pound bag of cat food for $4.85 or a 2.5-kilogram box for $7.99?

Estimation

In Exercises 59–62, choose the best estimate of the value of the given measurement.

59. 2 kilograms

(i) 2 pounds (ii) 5 pounds (iii) 1 pound

60. 2 pounds

(i) 4 kilograms (ii) 2 kilograms (iii) 1 kilogram

61. 12 pints

(i) 12 liters (ii) 3 liters (iii) 6 liters

62. 12 liters

(i) 12 quarts (ii) 4 quarts (iii) 12 gallons

Calculator Exercises

In Exercises 63–66, convert the given measurement to the indicated units. Round to the nearest tenth if necessary.

63. 50 milliliters per second = ▢ liters per minute

64. 100 milliliters per minute = ▢ liters per hour

65. 4 liters per day = ▢ gallons per year

Copyright © Houghton Mifflin Company. All rights reserved.

66. 100 gallons per year = ____ liters per month

67. *Glucose* A patient needs 0.5 liter of glucose in 8 hours. How many milliliters per hour should the patient receive?

68. *Saline Solution* A patient received 90 milliliters per hour of a saline solution. How many liters per day did the patient receive?

69. *Filling Gas Tank* An 18-gallon gas tank can be filled in 4 minutes. What is the rate of liquid flow in liters per minute?

70. *Milk Cost* If milk costs $1.79 per half-gallon, what is the price in cents per liter?

Writing and Concept Extension

71. Which is *not* a correct conversion factor? Why?

(i) $\frac{16\text{ ounces}}{1\text{ pound}}$ (ii) $\frac{1{,}000\text{ kilograms}}{1\text{ metric ton}}$ (iii) $\frac{1{,}000\text{ grams}}{1\text{ milligram}}$

72. Insert < or > to make the statement true.

(a) 1 kilogram ____ 1 pound

(b) 1 gram ____ 1 ounce

(c) 1 ton ____ 1 metric ton

Copyright © Houghton Mifflin Company. All rights reserved.

Exploring with Real-World Data: Collaborative Activities

Currency Exchange The table shows the numbers of units of currency of selected countries that are equal to 1 U.S. dollar.

British pound	0.64
Japanese yen	120.99
Mexican peso	7.91
French franc	5.84

73. How many Japanese yen does a traveler receive in exchange for $500?

74. A visitor to Mexico exchanges $200 for pesos. How many pesos does the visitor receive?

75. Suppose that dinner at a pub in Britain costs 15 pounds. To the nearest cent, what is the cost in U.S. dollars?

76. Suppose that in Paris a designer dress costs 1,200 francs. What is the price in U.S. dollars?

77. Suppose that in France a certain perfume costs 4 francs per gram. What is the price in U.S. dollars per ounce?

78. Suppose that in Japan gasoline costs 250 yen per liter. What is the cost in U.S. dollars per gallon?

79. A business traveler exchanges 250 British pounds for French francs. How many francs does the traveler receive?

80. How many British pounds does a tourist receive in exchange for 2,000 Japanese yen?

Copyright © Houghton Mifflin Company. All rights reserved.

7.5 Proportions

A *Introduction to Proportions*
B *Solving Proportions*
C *Applications*

SUGGESTIONS FOR SUCCESS

"Hat is to head as glove is to ______."

You have probably seen questions like this (they are called *analogies*) on standardized tests. In this section, we will consider similar questions that involve numbers: "2 is to 4 as 3 is to ______." In mathematics, we call such a statement a *proportion.*

Although we will present some methods for working with proportions, try to keep the concept simple by remembering that a proportion is just a comparison of numbers. As always in mathematics, if you are comfortable with the overall concept, then you will be much more successful with the details.

A Introduction to Proportions

Because $\frac{2}{4}$ and $\frac{3}{6}$ are both equal to $\frac{1}{2}$, we can write the following equation:

$$\frac{2}{4} = \frac{3}{6}$$

If we think of $\frac{2}{4}$ and $\frac{3}{6}$ as ratios, then the equation $\frac{2}{4} = \frac{3}{6}$ states that the two ratios are equal. We call such an equality of ratios a **proportion.**

Definition of a Proportion

A **proportion** is a statement that two ratios (or rates) are equal. The following is the general form of a proportion:

$$\frac{a}{b} = \frac{c}{d}, \qquad b \neq 0, d \neq 0$$

This proportion is read "*a* is to *b* as *c* is to *d*."

NOTE In the remainder of our discussion of proportions, we will assume that no denominator is 0.

Recall that a ratio is a comparison of two numbers. Therefore, in the proportion $\frac{2}{4} = \frac{3}{6}$, we are stating that 2 compares to 4 in the same way that 3 compares to 6. That's why we read this proportion as "2 is to 4 as 3 is to 6."

Here are some words that are commonly used with proportions.

Definition of Terms, Extremes, and Means

In the proportion $\frac{a}{b} = \frac{c}{d}$:

1. The numbers a, b, c, and d are called **terms.**
2. The numbers a and d are called the **extremes.**
3. The numbers b and c are called the **means.**

Copyright © Houghton Mifflin Company. All rights reserved.

Example 1

Consider the proportion $\frac{7}{10} = \frac{14}{20}$.

(a) What are the terms?

(b) What are the means?

(c) What are the extremes?

(d) How would you read the proportion?

SOLUTION

(a) The terms are 7, 10, 14, and 20.

(b) The means are 10 and 14.

$$\frac{7}{10} = \frac{14}{20}$$

(c) The extremes are 7 and 20.

$$\frac{7}{10} = \frac{14}{20}$$

(d) The proportion is read "7 is to 10 as 14 is to 20."

Your Turn 1

Consider the proportion $\frac{3}{8} = \frac{9}{24}$. Answer the questions in Example 1.

Answers: **(a)** 3, 8, 9, 24; **(b)** 8, 9; **(c)** 3, 24; **(d)** 3 is to 8 as 9 is to 24

Now we turn to an important property of proportions.

Means and Extremes

Suppose that $\frac{a}{b} = \frac{c}{d}$ is a true proportion.

To eliminate the fractions from this equation, we multiply both sides by the LCD, which in this case is bd.

$$\frac{\not{b}d}{1} \cdot \frac{a}{\not{b}} = \frac{b\not{d}}{1} \cdot \frac{c}{\not{d}}$$ On the left side, we cancel the b; on the right side, we cancel the d.

$$ad = bc$$

Observe that the numbers a and d in the product on the left side are the extremes of the proportion, and the numbers b and c in the product on the right side are the means. In words, the product of the extremes is equal to the product of the means.

Rather than multiplying both sides by the LCD, we can obtain the same result by multiplying across the proportion according to the following pattern.

$$\frac{a}{b} = \frac{c}{d}$$ ($b \cdot c \rightarrow bc$; $a \cdot d \rightarrow ad$)

$$ad = bc$$

This pattern shows why the products ad and bc are called **cross products.**

Copyright © Houghton Mifflin Company. All rights reserved.

Property of Cross Products

The proportion $\frac{a}{b} = \frac{c}{d}$ is true if and only if $ad = bc$.

In the Property of Cross Products, the phrase "if and only if" means that we can read the property in either order.

1. If $\frac{a}{b} = \frac{c}{d}$ is true, then the cross products are equal.
2. If the cross products are equal, then $\frac{a}{b} = \frac{c}{d}$ is true.

Example 2

Show whether the given proportion is true or false.

(a) $\frac{4}{15} = \frac{12}{45}$ **(b)** $\frac{5}{11} = \frac{6}{12}$

SOLUTION

(a) In the diagram, the arrows indicate the cross products.

$$\frac{4}{15} = \frac{12}{45} \qquad (15)(12) = 180 \qquad (4)(45) = 180$$

Because the cross products are equal, the proportion is true.

(b)

$$\frac{5}{11} = \frac{6}{12} \qquad (11)(6) = 66 \qquad (5)(12) = 60$$

The cross products are not equal, so the proportion is false.

Your Turn 2

Show whether the given proportion is true or false.

(a) $\frac{3}{11} = \frac{4}{20}$

(b) $\frac{18}{30} = \frac{9}{15}$

Answers: (a) False; (b) True

The Property of Cross Products leads to the following rules for proportions.

Properties of Proportions

If $\frac{a}{b} = \frac{c}{d}$ is a true proportion, then:

1. $\frac{b}{a} = \frac{d}{c}$ The two ratios can be inverted.
2. $\frac{d}{b} = \frac{c}{a}$ The extremes can be swapped.
3. $\frac{a}{c} = \frac{b}{d}$ The means can be swapped.

Copyright © Houghton Mifflin Company. All rights reserved.

Example 3

Write the proportion $\frac{1}{3} = \frac{4}{12}$ in three different ways.

SOLUTION

$$\frac{1}{3} = \frac{4}{12} \rightarrow \frac{3}{1} = \frac{12}{4}$$ Invert the ratios.

$$\frac{1}{3} = \frac{4}{12} \rightarrow \frac{12}{3} = \frac{4}{1}$$ Swap the extremes.

$$\frac{1}{3} = \frac{4}{12} \rightarrow \frac{1}{4} = \frac{3}{12}$$ Swap the means.

We can verify each of these by showing that the cross products are equal.

Your Turn 3

Write the proportion $\frac{2}{5} = \frac{4}{10}$ in three different ways.

Answer: $\frac{5}{2} = \frac{10}{4}$, $\frac{10}{5} = \frac{4}{2}$, $\frac{2}{4} = \frac{5}{10}$

B Solving Proportions

When at least one term of a proportion is a variable or a variable expression, we can *solve* the proportion to determine the unknown term.

Example 4

Solve the proportion $\frac{x}{3} = \frac{14}{21}$.

SOLUTION

Method 1

When the variable is in the numerator, we can simply clear the fractions in the usual way.

$$\frac{\cancel{3}}{1} \cdot \frac{x}{\cancel{3}} = \frac{\cancel{3}}{1} \cdot \frac{2 \cdot \cancel{7}}{\cancel{3} \cdot \cancel{7}}$$ Multiply both sides by 3 and simplify $\frac{14}{21}$.

$$x = 2$$

Method 2

For the proportion to be true, the cross products must be equal.

$$\frac{x}{3} = \frac{14}{21}$$ $(3)(14) = 42$; $x \cdot 21 = 21x$

$$21x = 42$$ The cross products are equal.

$$\frac{\cancel{21}x}{\cancel{21}} = \frac{42}{21}$$ Divide both sides by 21.

$$x = 2$$

Your Turn 4

Solve the proportion $\frac{6}{18} = \frac{x}{9}$.

Answer: 3

NOTE For any proportion, we can invert ratios or swap means or extremes. Therefore, if just one term of a proportion is a variable, we can always write the proportion with the variable in one of the numerators and apply Method 1. However, we will focus on Method 2 because it is efficient and because it works well when more than one term of a proportion is a variable.

Copyright © Houghton Mifflin Company. All rights reserved.

Example 5

Solve the proportion $\frac{100}{7} = \frac{4}{x}$.

SOLUTION

$100 \cdot x = 7 \cdot 4$ The cross products are equal.

$100x = 28$

$x = \frac{28}{100} = 0.28$ Divide both sides by 100.

Your Turn 5

Solve the proportion $\frac{40}{x} = \frac{1,000}{3}$.

Answer: 0.12

Example 6

Solve the proportion $\frac{6}{4} = \frac{x + 2}{2}$.

SOLUTION

$4(x + 2) = 6 \cdot 2$ The cross products are equal.

$4x + 8 = 12$ Distributive Property

$4x = 4$ Subtract 8 from both sides.

$x = 1$ Divide both sides by 4.

Your Turn 6

Solve the proportion $\frac{6}{x - 1} = \frac{3}{2}$.

Answer: 5

C Applications

Problems that involve ratios or rates can often be solved with proportions. The following examples are just samples of the wide variety of such problems that you might encounter in business, science, or in everyday life.

As always, the critical step in solving application problems is writing the equation (proportion) properly. As our examples show, using a table is an excellent way to organize information and to write proportions in the correct order.

Example 7

A woman has driven 165 miles in 3 hours. At the same rate, how long will it take her to drive another 275 miles?

SOLUTION

The following table summarizes the known and unknown information. Note that we let x represent the unknown number of hours.

	Miles	*Hours*
Known	165	3
Unknown	275	x

We can write the proportion directly from the table.

Your Turn 7

A man has driven 116 miles in 2 hours. At the same rate, how much farther will he travel in 3 hours?

continued

Copyright © Houghton Mifflin Company. All rights reserved.

Miles *Hours*

$$\frac{165}{275} = \frac{3}{x}$$

$165 \cdot x = 3 \cdot 275$ The cross products are equal.

$165x = 825$

$x = 5$ Divide both sides by 165.

The woman needs 5 hours to drive 275 miles.

NOTE The proportion that we wrote in Example 7 is not the only one that we could have used. If we think in terms of rate (speed), then the following proportion would work just as well:

$$\frac{165 \text{ miles}}{3 \text{ hours}} = \frac{275 \text{ miles}}{x \text{ hours}}$$

Example 8

If 6 cubic yards of topsoil cost \$135, what would be the cost of 10 cubic yards?

SOLUTION

	Cubic Yards	*Cost*
Known	6	\$135
Unknown	10	x

Again, we can write the proportion directly from the table.

Cubic Yards *Cost*

$$\frac{6}{10} = \frac{135}{x}$$

$6 \cdot x = 10 \cdot 135$ The cross products are equal.

$6x = 1{,}350$

$x = 225$ Divide both sides by 6.

The cost of 10 cubic yards would be \$225.

Your Turn 8

A 5-pound bag of onions costs \$3.50. At the same unit price, what would an 8-pound bag of onions cost?

Answer: \$5.60

Although we can use the Property of Cross Products to solve any proportion, there is an advantage to writing a proportion with the variable in the numerator. In Example 8, if we had written

$$\frac{x}{135} = \frac{10}{6}$$

then we could have simply multiplied both sides by 135 to solve for x.

$$\frac{\cancel{135}}{1} \cdot \frac{x}{\cancel{135}} = \frac{135}{1} \cdot \frac{10}{6}$$

$$x = \frac{135 \cdot 10}{6} = \frac{1{,}350}{6} = 225$$

Copyright © Houghton Mifflin Company. All rights reserved.

Example 9

The sales tax on a \$68 purchase is \$3.40. What would be the tax on a \$92 purchase?

SOLUTION

	Purchase	*Tax*
Known	\$68	\$3.40
Unknown	\$92	x

$$\begin{aligned} &\textit{Purchase} \quad \textit{Tax} \\ \frac{68}{92} &= \frac{3.40}{x} \\ 68 \cdot x &= (92)(3.40) && \text{The cross products are equal.} \\ 68x &= 312.80 \\ x &= 4.60 && \text{Divide both sides by 68.} \end{aligned}$$

The tax on a \$92 purchase is \$4.60.

Your Turn 9

One person paid \$1.84 in sales tax on a \$46 purchase. If another person paid a tax of \$2.88 on the purchase of an item, what was the cost of the item?

Answer: \$72

Example 10

On a map, 2 inches represent 300 kilometers. If the map distance between two cities is $4\frac{1}{3}$ inches, what is the actual distance between the cities?

SOLUTION

	Inches	*Kilometers*
Known	2	300
Unknown	$4\frac{1}{3}$	x

As we write the proportion, we use $\frac{13}{3}$ rather than $4\frac{1}{3}$.

$$\begin{aligned} &\textit{Inches} \quad \textit{Kilometers} \\ \frac{2}{\frac{13}{3}} &= \frac{300}{x} \\ 2 \cdot x &= \frac{13}{3} \cdot 300 && \text{The cross products are equal.} \\ 2x &= 1{,}300 \\ x &= 650 && \text{Divide both sides by 2.} \end{aligned}$$

The actual distance is 650 kilometers.

Your Turn 10

On an architect's scale drawing, a 150-foot side of a building is shown as a line 8 inches long. What line length would be used to represent a 100-foot side of the building?

Answer: $5\frac{1}{3}$ inches

Copyright © Houghton Mifflin Company. All rights reserved.

Example 11

A chemist has 40 cubic centimeters of a liquid solution, of which 15 cubic centimeters are acid. At the same rate, how much acid would be in a 90-cubic centimeter solution?

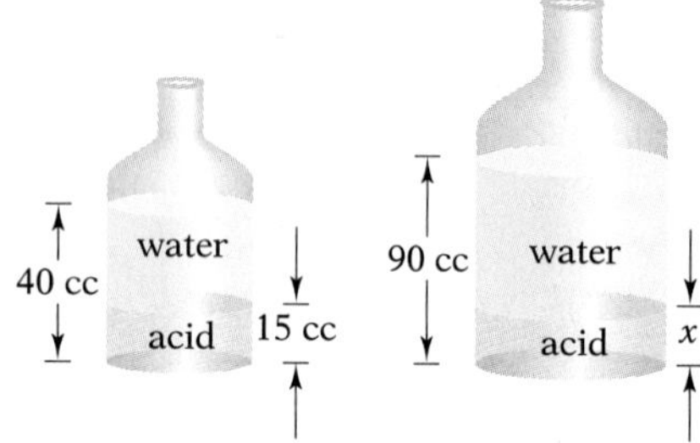

SOLUTION

	Total Solution	*Amount of Acid*
Known	40	15
Unknown	90	x

Total Solution / *Amount of Acid*

$$\frac{40}{90} = \frac{15}{x}$$

$$40 \cdot x = (90)(15) \quad \text{The cross products are equal.}$$

$$40x = 1{,}350$$

$$x = 33.75 \quad \text{Divide both sides by 40.}$$

There would be 33.75 cubic centimeters of acid in the solution of 90 cubic centimeters.

Your Turn 11

A 20-cubic centimeter liquid solution contains 3 cubic centimeters of alcohol. At the same rate, how much alcohol would be in a 70-cubic centimeter solution?

Answer: 10.5 cubic centimeters

Example 12

A survey of 100 registered voters showed that only 34 planned to vote in an upcoming election. According to the survey, if a community has 3,800 registered voters, how many are expected to vote?

SOLUTION

	Registered Voters	*Actual Voters*
Known	100	34
Unknown	3,800	x

Your Turn 12

A poll showed that 53 of every 100 people approved of the way the economy was being managed. If 800 people had been polled and the original poll was accurate, how many would have indicated approval?

Copyright © Houghton Mifflin Company. All rights reserved.

$$\frac{100}{3{,}800} = \frac{34}{x}$$

$$100x = (3{,}800)(34) = 129{,}200$$

$$x = 1{,}292$$

The survey predicts 1,292 actual voters.

7.5 Quick Reference

A. Introduction to Proportions

1. A **proportion** is a statement that two ratios (or rates) are equal. The following is the general form of a proportion:

$$\frac{a}{b} = \frac{c}{d}, \qquad b \neq 0,\ d \neq 0$$

This proportion is read "*a* is to *b* as *c* is to *d*."

True proportion:

$$\frac{6}{10} = \frac{12}{20}$$

This proportion is read "6 is to 10 as 12 is to 20."

2. In the proportion $\frac{a}{b} = \frac{c}{d}$:

(a) The numbers a, b, c, and d are called **terms.**

(b) The numbers a and d are called the **extremes.**

(c) The numbers b and c are called the **means.**

In the proportion $\frac{8}{3} = \frac{24}{9}$:

(a) Terms: 8, 3, 24, and 9

(b) Extremes: 8 and 9

(c) Means: 3 and 24

3. Property of Cross Products

The proportion $\frac{a}{b} = \frac{c}{d}$ is true if and only if $ad = bc$.
The products ad and bc are called **cross products.**

The proportion $\frac{14}{4} = \frac{7}{2}$ is true because the cross products, 14(2) and 4(7), are equal.

4. If $\frac{a}{b} = \frac{c}{d}$ is a true proportion, then:

(a) $\frac{b}{a} = \frac{d}{c}$ The two ratios can be inverted.

(b) $\frac{d}{b} = \frac{c}{a}$ The extremes can be swapped.

(c) $\frac{a}{c} = \frac{b}{d}$ The means can be swapped.

We can write $\frac{8}{x} = \frac{2}{5}$ as:

(a) $\frac{x}{8} = \frac{5}{2}$

(b) $\frac{5}{x} = \frac{2}{8}$

(c) $\frac{8}{2} = \frac{x}{5}$

Copyright © Houghton Mifflin Company. All rights reserved.

B. Solving Proportions

1. When at least one term of a proportion is a variable or a variable expression, we can *solve* the proportion to determine the unknown term. One method uses the fact that the cross products are equal.

$$\frac{x}{18} = \frac{5}{4}$$

$$x \cdot 4 = 18 \cdot 5$$

$$4x = 90$$

$$x = 22.5$$

Copyright © Houghton Mifflin Company. All rights reserved.

7.5 Exercises

A. Introduction to Proportions

1. A statement that two ratios are equal is called a ______.

2. In the proportion $\frac{y}{3} = \frac{7}{9}$, y and 9 are called the ______.

In Exercises 3–6, complete the table.

	Proportion	*Terms*	*Means*	*Extremes*	*In Words*
3.	$\frac{9}{6} = \frac{18}{12}$	**(a)** ______	**(b)** ______	**(c)** ______	**(d)** ______
4.	**(a)** ______	**(b)** ______	**(c)** ______	**(d)** ______	12 is to 8 as 6 is to 4.
5.	**(a)** ______	**(b)** ______	**(c)** ______	**(d)** ______	m is to n as 5 is to 3.
6.	$\frac{q}{9} = \frac{2}{p}$	**(a)** ______	**(b)** ______	**(c)** ______	**(d)** ______

In Exercises 7–14, determine whether the given proportion is true.

7. $\frac{11}{6} = \frac{10}{7}$ **8.** $\frac{18}{12} = \frac{12}{8}$ **9.** $\frac{14}{35} = \frac{6}{15}$ **10.** $\frac{5}{14} = \frac{7}{12}$

11. $\frac{1}{2.5} = \frac{0.2}{0.5}$ **12.** $\frac{1.5}{0.3} = \frac{1.2}{0.2}$ **13.** $\frac{\frac{3}{5}}{4} = \frac{12}{5}$ **14.** $\frac{\frac{3}{2}}{\frac{9}{4}} = \frac{4}{6}$

In Exercises 15–18, complete the table by writing the proportion in three different ways. (In Exercises 16–18, begin by writing the proportion in the first column.)

	Proportion	*Invert Ratios*	*Swap Extremes*	*Swap Means*
15.	$\frac{6}{15} = \frac{8}{20}$	**(a)** ______	**(b)** ______	**(c)** ______
16.	**(a)** ______	$\frac{6}{3} = \frac{2}{1}$	**(b)** ______	**(c)** ______
17.	**(a)** ______	**(b)** ______	**(c)** ______	$\frac{6}{11} = \frac{4}{y}$
18.	**(a)** ______	**(b)** ______	$\frac{x}{6} = \frac{9}{2}$	**(c)** ______

Copyright © Houghton Mifflin Company. All rights reserved.

B. Solving Proportions

19. When we determine the value of x in $\frac{x}{5} = \frac{3}{10}$, we say that we are ________ the proportion.

20. By ________ the ratios in $\frac{3}{x} = \frac{12}{7}$, we can write $\frac{x}{3} = \frac{7}{12}$.

In Exercises 21–38, solve the given proportion.

21. $\frac{x}{4} = \frac{7}{2}$

22. $\frac{2}{5} = \frac{8}{m}$

23. $\frac{6}{5} = \frac{3}{x}$

24. $\frac{4}{3} = \frac{n}{6}$

25. $\frac{4}{y} = \frac{4}{11}$

26. $\frac{10}{y} = \frac{20}{7}$

27. $\frac{10}{7} = \frac{100}{p}$

28. $\frac{1000}{7} = \frac{t}{0.01}$

29. $\frac{0.3}{1.5} = \frac{1}{x}$

30. $\frac{y}{10} = \frac{0.1}{0.01}$

31. $\frac{x+5}{3} = \frac{7}{6}$

32. $\frac{2}{15} = \frac{x-2}{5}$

33. $\frac{2}{3} = \frac{4}{x-5}$

34. $\frac{3}{x+1} = \frac{2}{5}$

35. $\frac{\frac{3}{5}}{m} = \frac{\frac{1}{2}}{10}$

36. $\frac{\frac{3}{4}}{\frac{5}{2}} = \frac{2}{x}$

37. $\frac{2\frac{1}{2}}{1\frac{2}{3}} = \frac{4\frac{1}{2}}{x}$

38. $\frac{2\frac{2}{3}}{2} = \frac{n}{2\frac{1}{2}}$

C. Applications

39. *Cough Medicine Dosage* The dose of a certain cough medicine for children is based on the weight of the child. If a child who weighs 40 pounds receives 25 milligrams of the medication, how many milligrams should be given to a 56-pound child?

Copyright © Houghton Mifflin Company. All rights reserved.

40. *Sodium Content of Soup* A 120-milliliter serving of chicken noodle soup contains 980 milligrams of sodium. How many milligrams of sodium are in a 200-milliliter serving?

41. *Miles Driven* For the first five months of a year, a sales representative drove 18,000 miles. Assuming that the same rate is maintained, how many miles would the person drive in a year?

42. *Weed Killer* A 40-pound bag of weed killer is recommended for 3,500 square feet of lawn. How many bags are needed for 14,875 square feet of lawn?

43. *Insecticide Solution* An insecticide solution is made up of 130 milliliters of insecticide and 2 liters of water. To maintain the same concentration, how many milliliters of insecticide should be added to 7 liters of water?

44. *Madrid to Paris Distance* On a European map, 2 centimeters represents 75 kilometers. If the map distance between Madrid and Paris is 14 centimeters, what is the actual distance in kilometers?

45. *Student-Faculty Ratio* An elementary school has 800 students and 44 faculty. If the number of students increases to 1,200, how many more faculty must be hired to maintain the same student-faculty ratio?

Copyright © Houghton Mifflin Company. All rights reserved.

46. *Hits and At-Bats* A baseball player had 9 hits in 32 at-bats. Assuming that he continues to hit at the same pace, how many hits should he expect after 128 at-bats?

47. *Soda Preference* A survey indicates that 2 of 5 people prefer diet soda to regular soda. Of 35 people in a class, how many would be expected to prefer regular soda?

48. *Defective Chips* An electronics manufacturer finds that 4 of every 1,500 chips are defective. How many defective chips are expected in a shipment of 7,500 chips?

49. *Disinfectant Solution* A hospital mixes 2 parts of disinfectant with 5 parts of water to make a cleaning solution. How many quarts of disinfectant should be mixed with 4 gallons of water?

50. *Weight Loss and Jogging* To lose 1 pound, you must burn 3,500 calories. If jogging for 30 minutes burns 300 calories, how many hours must you jog to lose 2 pounds?

51. *Investments* An investor earns \$520.80 of dividends on 420 shares of stock. How many additional shares of the same stock should the investor purchase to earn dividends of \$793.60?

Copyright © Houghton Mifflin Company. All rights reserved.

52. *Property Taxes* A home owner was charged \$3.70 property tax for each \$1,000 of the value of her home. What is the value of her property if the tax is \$462.50?

53. *Income Taxes* For each \$100 earned, \$38.20 goes to pay federal, state, and local taxes. (Source: Tax Foundation.) If a person earns \$32,000, how much goes for taxes?

54. *Loan Fee* A bank charges a processing fee of \$1.36 for each \$1,000 of a loan. What is the processing fee for a \$78,400 loan?

55. *Cat Food* Science Diet® cat food recommends feeding a 10-pound cat $\frac{3}{4}$ cup of food each day. How much should you feed a 15-pound cat?

56. *Hiking* A backpacker hiked $24\frac{3}{4}$ miles in 3 days. Assuming that the same pace is maintained, how many miles would the person hike in 8 days?

Copyright © Houghton Mifflin Company. All rights reserved.

57. *Recipe* A recipe for stew uses $3\frac{1}{2}$ pounds of meat and $1\frac{1}{3}$ cups of tomatoes. If the cook uses $8\frac{3}{4}$ pounds of meat, how many cups of tomatoes should he use?

58. *Potato Salad* If $2\frac{1}{4}$ pounds of potato salad cost $4.95, what is the cost of $5\frac{1}{2}$ pounds?

59. *Architect's Scale* An architect uses a scale of $1\frac{1}{2}$ inches for 20 feet. What length is represented by $3\frac{3}{8}$ inches?

60. *Benefits Packets* A clerk can assemble 7 benefit information packets in $\frac{1}{4}$ hour. How many packets could the clerk assemble in $1\frac{1}{2}$ hours?

61. *Fat Content* If 30 grams of ranch salad dressing contain 14 grams of fat, how many grams of fat are in a 50-gram serving?

62. *Juice Concentrate* If $1\frac{1}{2}$ pints of juice concentrate should be mixed with 3 quarts of water, how many quarts of water should be added to 1 quart of concentrate?

Copyright © Houghton Mifflin Company. All rights reserved.

Calculator Exercises

63. *TV Commercial Cost* Suppose that a 30-second commercial on a television show costs $565,000. At the same rate, what would a 45-second commercial cost?

64. *Time to Pay Taxes* In New Jersey, 3 hours and 1 minute of every 8 hours of work are spent earning money to pay taxes. (Source: Tax Foundation.) What amount of time is spent paying taxes in 60 hours of work?

65. *Swimming Pools* For every 10,000 residents of Florida, there are 894 residential swimming pools. (Source: Muskin Leisure Products.) If the population of Florida is 16.2 million, how many residential pools would you expect to find in the state?

66. *Birth Rate* In the United States, the number of births is 14.6 for every 1,000 people. (Source: Centers for Disease Control.) If the population of the United States is 287 million, how many births occurred during a year?

Writing and Concept Extension

67. Explain the difference between a ratio and a proportion.

68. Explain why the proportion $\frac{2}{y} = \frac{3}{4}$ is equivalent to the proportion $\frac{4}{y} = \frac{3}{2}$.

69. Which proportion is not equivalent to the proportion $\frac{5}{n} = \frac{3}{7}$?

(i) $\frac{7}{3} = \frac{n}{5}$ (ii) $\frac{7}{n} = \frac{3}{5}$ (iii) $\frac{5}{7} = \frac{n}{3}$

Copyright © Houghton Mifflin Company. All rights reserved.

70. An incumbent won an election by a 3 to 2 margin. Which statements are true?

(i) 3 of 5 voters cast a ballot for the candidate.

(ii) 2 of 5 voters cast a ballot for the opponent.

(iii) The opponent received 1 of 3 votes cast.

(iv) The opponent lost by a 2 to 3 margin.

Copyright © Houghton Mifflin Company. All rights reserved.

7.6 Proportions in Geometry

A Measures of Angles
B Similar Triangles
C Applications

SUGGESTIONS FOR SUCCESS

When you use a photocopier to enlarge or reduce an image, you don't want the image to come out squished or stretched out. The image, although larger or smaller, should have exactly the same shape as the original.

In this section, you will be working with similar triangles, which have the same shape but not necessarily the same size. You will learn that the sides of similar triangles are proportional.

A key to your success will be identifying the sides that correspond. Try this: Draw the triangles so that they have the same orientation and use three colors for the sides. The blue sides correspond, the red sides correspond, and the green sides correspond. This will give you a visual way to help you write the correct proportions.

A Measures of Angles

There are many geometric figures and relationships in which sides of figures and other line segments have lengths that are proportional. Knowing such proportions can be very helpful when, for example, we want to make a scale model or drawing of a large object. Architects, surveyors, and astronomers use proportions because their measurements are often difficult or impossible to make directly.

We begin our discussion with the meaning of a *degree,* which is the unit for measuring the size of an angle. Imagine a circle that has been divided into 360 equal arcs. Then half of the circle (*semicircle*) would have 180 equal arcs.

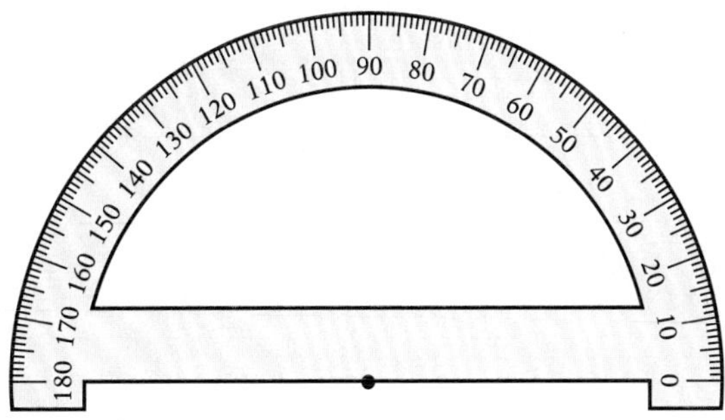

Now suppose that we place an angle so that its *vertex* (the point of intersection of the two sides) is at the center of the semicircle, and one side crosses the semicircle at the tick mark labeled 0.

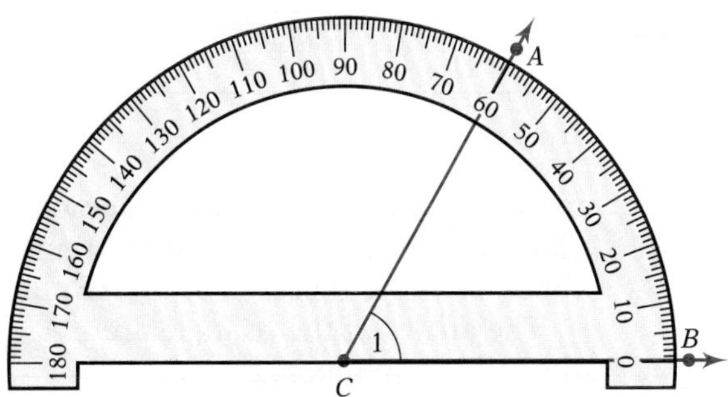

Copyright © Houghton Mifflin Company. All rights reserved.

The other side of the angle crosses the semicircle at the tick mark labeled 60. We say that the *measure* of the angle is 60 degrees or 60°.

There are several ways to name the angle in the preceding figure.

1. $\angle C$ (The letter C names the vertex of the angle.)
2. $\angle 1$ (Note that the number is placed inside the angle.)
3. $\angle ACB$ (In this notation, the middle letter names the vertex.)

When we use the symbols $\angle C$, $\angle 1$, or $\angle ACB$, we are naming the angle itself. To indicate the size of the angle, we use the letter m in front of the angle name. For example, in the preceding figure, $m\angle C = 60°$; we read this as "the measure of angle C is 60 degrees."

LEARNING TIP

You may be familiar with a protractor, which is a semicircular tool for measuring the sizes of angles. The tick marks on a protractor represent degrees.

Example 1

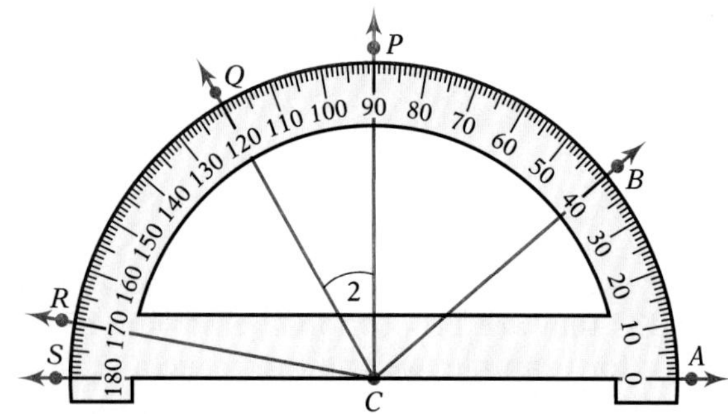

Using the figure, determine the following:

(a) $m\angle ACP$ **(b)** $m\angle ACQ$ **(c)** $m\angle 2$

SOLUTION

(a) $m\angle ACP = 90°$

Note that $\angle ACP$ is a *right* angle.

(b) $m\angle ACQ = 120°$

(c)
$$\begin{aligned} m\angle 2 &= m\angle ACQ - m\angle ACP \\ &= 120° - 90° \\ &= 30° \end{aligned}$$

Your Turn 1

Using the figure in Example 1, determine the following:

(a) $m\angle ACB$

(b) $m\angle ACR$

(c) $m\angle RCS$

Answers: (a) 40°; (b) 170°; (c) 10°

Just as we use different symbols to name an angle and to represent the size of that angle, we use different symbols to name the side of a triangle and to represent the length of that side.

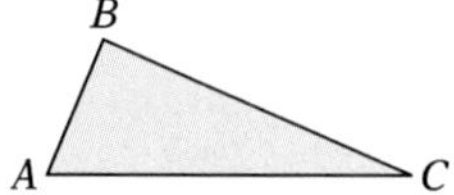

We represent the triangle shown in the figure with the symbol $\triangle ABC$. The sides are *line segments* that we represent with $\overline{AB}$, $\overline{BC}$, and $\overline{AC}$. The lengths of those sides are represented by AB, BC, and AC, respectively.

The angles of the triangle are $\angle A$, $\angle B$, and $\angle C$. Their measures are related by the following important rule:

Copyright © Houghton Mifflin Company. All rights reserved.

The Angles of a Triangle

For any triangle, the sum of the measures of the three angles is 180°.

For the triangle in the preceding figure, we can write

$$m\angle A + m\angle B + m\angle C = 180°$$

Example 2

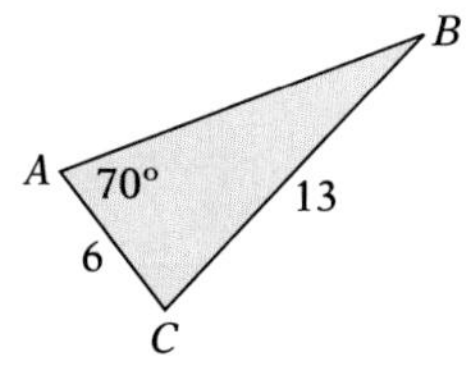

(a) What is BC?

(b) What is $m\angle A$?

(c) For which side is the length not given?

(d) For which angles are the measures not given?

SOLUTION

(a) BC refers to the length of side $\overline{BC}$.

$$BC = 13$$

(b) $m\angle A$ refers to the measure of $\angle A$.

$$m\angle A = 70°$$

(c) The length of $\overline{AB}$ is not given.

(d) The measures of $\angle B$ and $\angle C$ are not given.

Your Turn 2

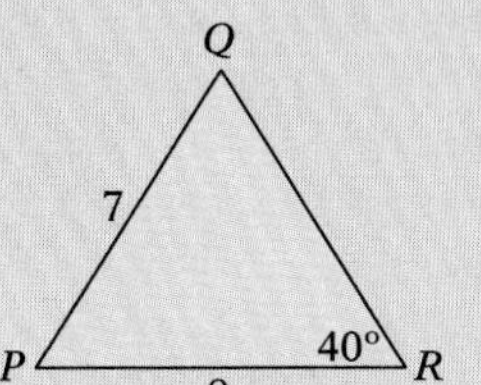

(a) What is PQ?

(b) What is $m\angle PRQ$?

(c) What side is not labeled?

(d) What angles are not labeled?

Answers: (a) 7; (b) 40°; (c) $\overline{QR}$; (d) $\angle P$, $\angle Q$

B Similar Triangles

Consider the following pair of triangles:

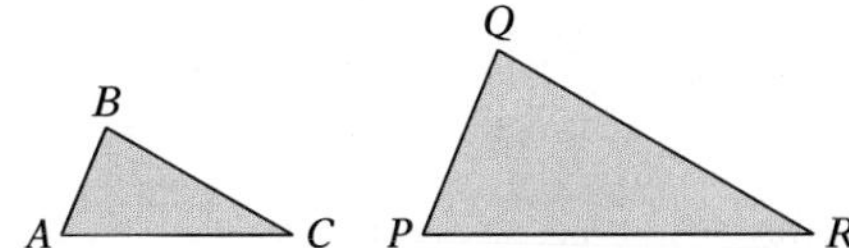

Although these two triangles are not the same size, they appear to have the same shape. We can think of using a photocopier to enlarge $\triangle ABC$ to $\triangle PQR$. Or we can think of $\triangle ABC$ as a reduction of $\triangle PQR$. Triangles with the same shape but not necessarily the same size are called **similar triangles.**

When we work with similar triangles, we refer to the *corresponding* sides and angles of the triangles. For the similar triangles in the preceding figure, the following sides and angles correspond.

Copyright © Houghton Mifflin Company. All rights reserved.

Corresponding Sides	*Corresponding Angles*
$\overline{AB}$ and $\overline{PQ}$	$\angle A$ and $\angle P$
$\overline{BC}$ and $\overline{QR}$	$\angle B$ and $\angle Q$
$\overline{AC}$ and $\overline{PR}$	$\angle C$ and $\angle R$

Now we can give a better description of similar triangles.

Similar Triangles

Two triangles are similar if and only if:

1. Their corresponding angles have the same measure.
2. Their corresponding sides are proportional.

NOTE The phrase "if and only if" means that if two triangles are similar, then properties (1) and (2) are true, and if properties (1) and (2) are true, then the triangles are similar.

The symbol ~ means "is similar to." If $\triangle ABC$ is similar to $\triangle PQR$, then we write $\triangle ABC \sim \triangle PQR$. The order of the letters is important because it indicates the corresponding angles and sides.

Example 3

In the figure, $\triangle EFG \sim \triangle RST$.

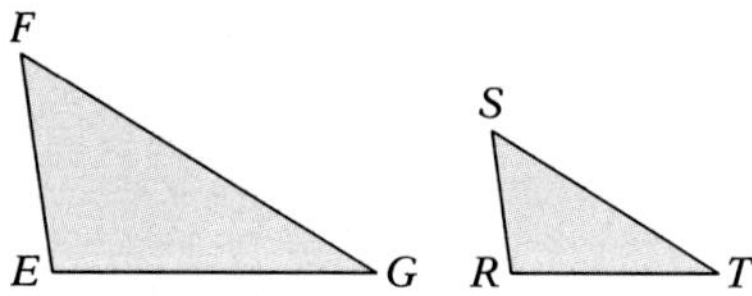

(a) How are the corresponding angles related?

(b) How are the corresponding sides related?

SOLUTION

(a) Because the triangles are similar, the corresponding angles have the same measures.

$$m\angle E = m\angle R$$
$$m\angle F = m\angle S$$
$$m\angle G = m\angle T$$

(b) Because the triangles are similar, the corresponding sides are proportional.

$$\frac{EF}{RS} = \frac{FG}{ST} = \frac{EG}{RT}$$

Your Turn 3

In the figure, $\triangle PQR \sim \triangle ABC$.

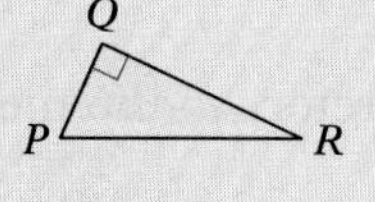

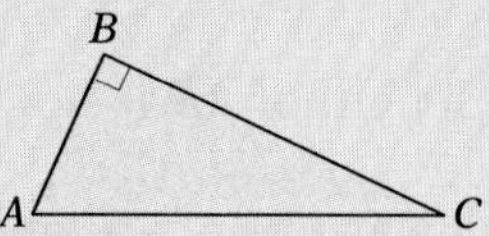

(a) How are the corresponding angles related?

(b) How are the corresponding sides related?

Answers: (a) $m\angle P = m\angle A$, $m\angle Q = m\angle B$, $m\angle R = m\angle C$; (b) $\frac{PQ}{AB} = \frac{QR}{BC} = \frac{PR}{AC}$

For the rest of our discussion of similar triangles, we will focus on the fact that corresponding sides are proportional. This property can sometimes be used to find unknown dimensions.

Copyright © Houghton Mifflin Company. All rights reserved.

Example 4

Given that $\triangle ABC \sim \triangle EFG$, find EG.

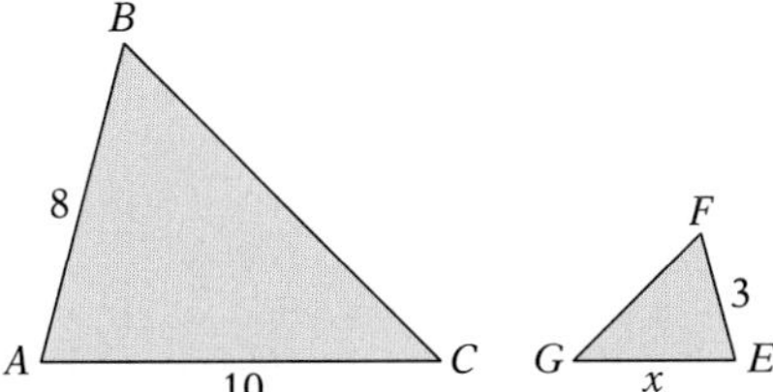

SOLUTION

Note that the triangles are not oriented in the same way, so we need to be careful about identifying the corresponding sides.

$\triangle ABC \sim \triangle EFG$ — $\overline{AB}$ corresponds to $\overline{EF}$.

$\triangle ABC \sim \triangle EFG$ — $\overline{AC}$ corresponds to $\overline{EG}$.

Because the triangles are similar, the corresponding sides are proportional. In particular,

$$\frac{AB}{EF} = \frac{AC}{EG}$$

$$\frac{8}{3} = \frac{10}{x}$$

$$8x = 30 \quad \text{The cross products are equal.}$$

$$x = \frac{30}{8} = \frac{15}{4} = 3\frac{3}{4}$$

Therefore, $EG = 3\frac{3}{4}$.

Your Turn 4

Given that $\triangle EFG \sim \triangle TSR$, find FG.

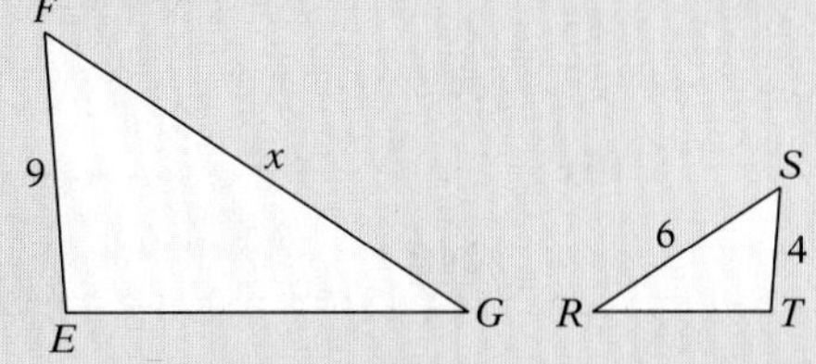

Answer: $13\frac{1}{2}$

Example 5

In the figure, it can be shown that if $\overline{PQ}$ is parallel to $\overline{AB}$, then $\triangle PQC \sim \triangle ABC$.

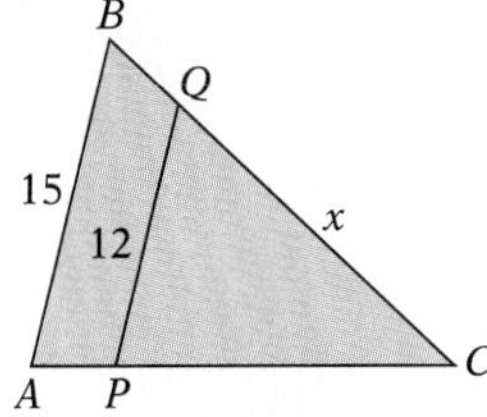

Given that $BC = 20$, find QC.

SOLUTION

Separating the triangles can sometimes help us to see the corresponding sides more easily.

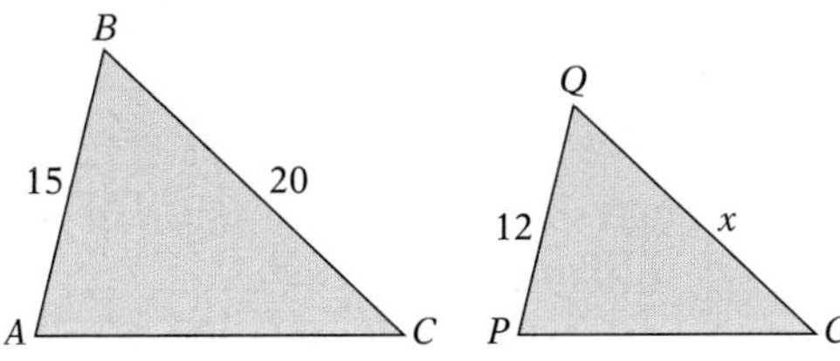

Your Turn 5

In the figure, $\overline{AB}$ is parallel to $\overline{EG}$. Given that $EF = 8$, find AB.

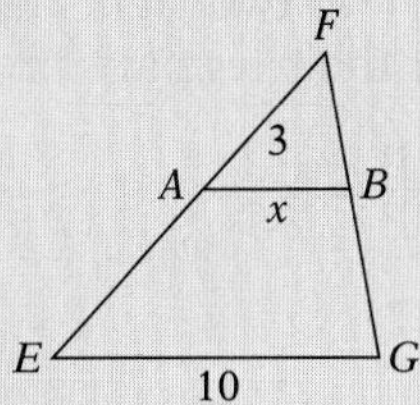

continued

Copyright © Houghton Mifflin Company. All rights reserved.

We use the corresponding sides to write a proportion.

$$\frac{AB}{PQ} = \frac{BC}{QC}$$

$$\frac{15}{12} = \frac{20}{x}$$

$$15x = 240 \quad \text{The cross products are equal.}$$

$$x = \frac{240}{15} = 16$$

Therefore, $QC = 16$.

Example 6

In the figure, $\triangle ABC$ is a right triangle, where $\angle B$ is the right angle. The line segment $\overline{BP}$ is an altitude of the triangle. It can be shown that $\triangle PAB \sim \triangle PBC$.

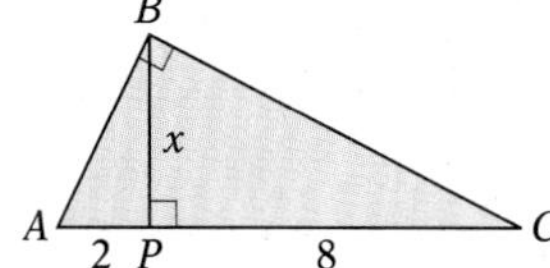

What is PB?

SOLUTION

Again, we can see the similarity of these two triangles if we separate them. Note how we rotate $\triangle PAB$ down so that it has the same orientation as $\triangle PBC$.

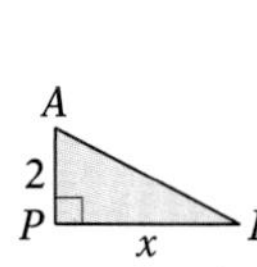

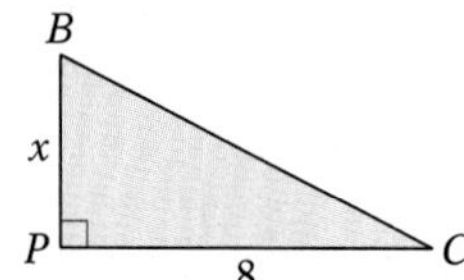

Because these two right triangles are similar, the lengths of their legs are proportional.

$$\frac{AP}{BP} = \frac{PB}{PC}$$

$$\frac{2}{x} = \frac{x}{8}$$

$$x^2 = 16 \quad \text{The cross products are equal.}$$

Therefore, $PB = 4$.

Your Turn 6

In the figure, $\triangle ABC$ is a right triangle. The line segment $\overline{AQ}$ is an altitude of the triangle. It can be shown that $\triangle BAQ \sim \triangle ACQ$. What is AQ?

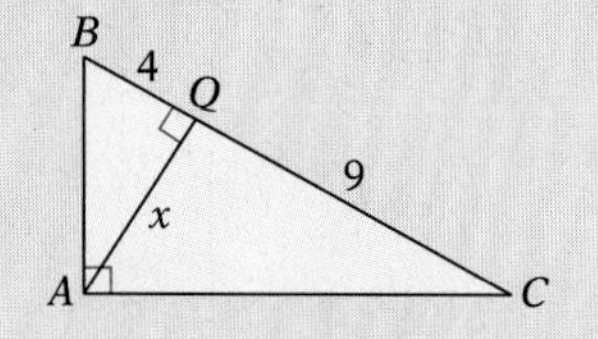

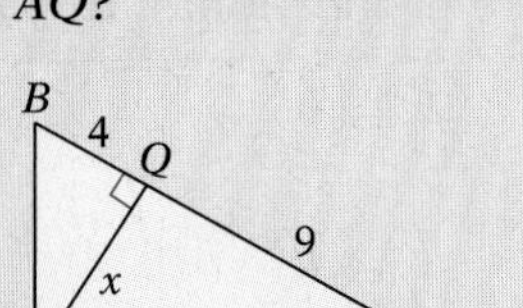

Answer: 6

C Applications

The proportional sides of similar triangles can be used to estimate lengths or distances that would otherwise be very difficult to measure.

Copyright © Houghton Mifflin Company. All rights reserved.

Example 7

A woman who is standing next to a tree observes that her shadow is 10 feet long and that the shadow of the tree is 40 feet long.

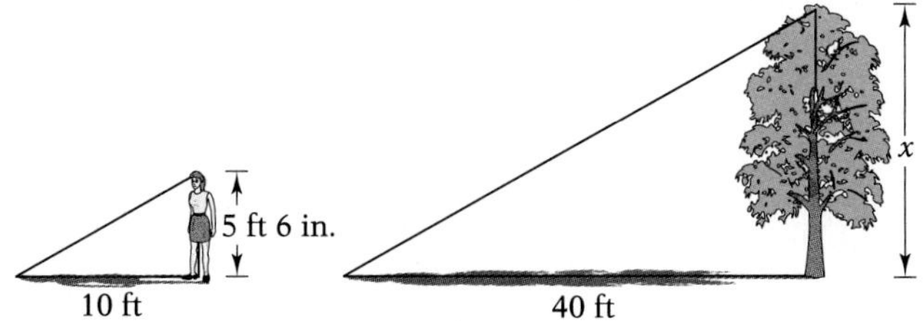

If the woman is 5 feet 6 inches tall, what is the height of the tree?

SOLUTION

The triangle formed by the woman and her shadow is similar to the triangle formed by the tree and its shadow. Using the fact that 5 feet 6 inches is the same as 5.5 ft, we write a proportion.

$$\frac{\text{Woman's height}}{\text{Tree's height}} = \frac{\text{Woman's shadow}}{\text{Tree's shadow}}$$

$$\frac{5.5}{x} = \frac{10}{40}$$

$$10x = 220$$ The cross products are equal.

$$x = 22$$

The height of the tree is 22 feet.

Your Turn 7

A Girl Scout sees that a 1-foot ruler casts a shadow that is 6 inches long. How high is a nearby flagpole whose shadow is 10 feet long?

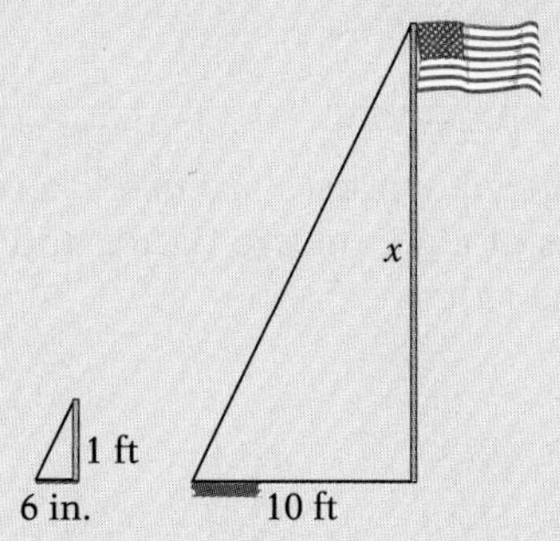

Answer: 20 feet

7.6 Quick Reference

A. Measures of Angles

1. Several methods are used to name an angle. The size of an angle is measured in *degrees*.

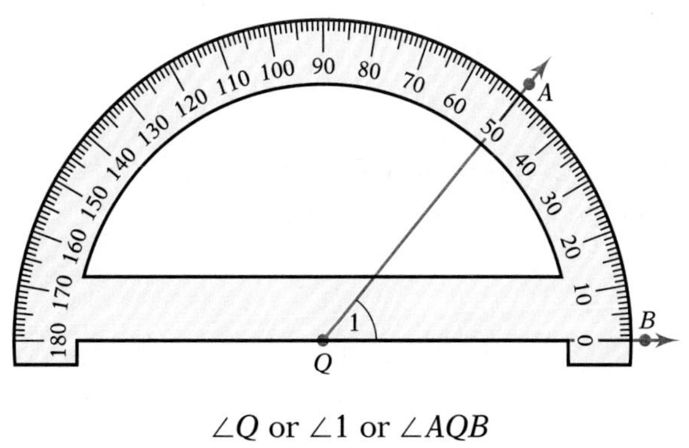

$\angle Q$ or $\angle 1$ or $\angle AQB$

$m\angle Q = 50°$

Copyright © Houghton Mifflin Company. All rights reserved.

2. For any triangle, the sum of the measures of the three angles is 180°.

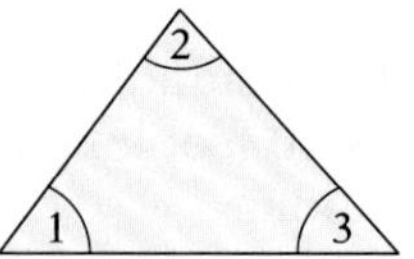

$m\angle 1 + m\angle 2 + m\angle 3 = 180°$

3. Different symbols are used to represent the sides of a geometric figure and the lengths of those sides.

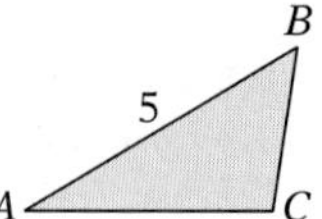

One side is $\overline{AB}$. The length of $\overline{AB}$ is $AB = 5$.

B. Similar Triangles

1. Two triangles are **similar triangles** if and only if:

(a) Their corresponding angles have the same measure.

(b) Their corresponding sides are proportional.

The symbol ~ means "is similar to."

$\triangle ABC \sim \triangle PQR$

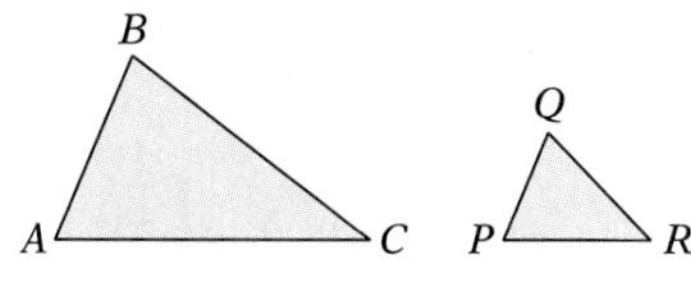

$m\angle A = m\angle P$

$m\angle B = m\angle Q$

$m\angle C = m\angle R$

$$\frac{AB}{PQ} = \frac{BC}{QR} = \frac{AC}{PR}$$

2. For similar triangles, we can sometimes use proportions to find the unknown lengths of sides.

$\triangle EFG \sim \triangle RST$

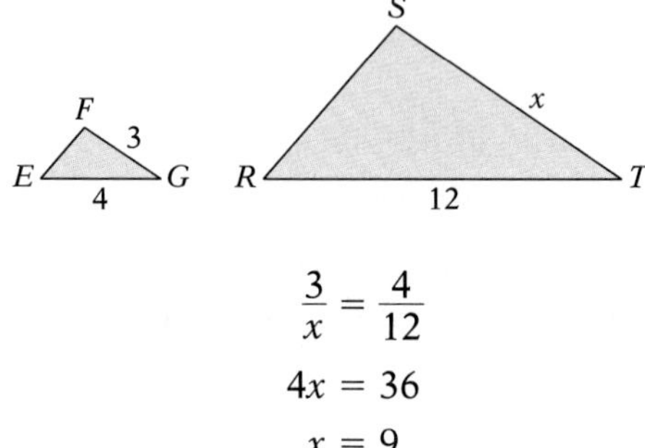

$$\frac{3}{x} = \frac{4}{12}$$

$4x = 36$

$x = 9$

Copyright © Houghton Mifflin Company. All rights reserved.

7.6 Exercises

A. Measures of Angles

1. The unit of measure for an angle is a(n) ________.

2. The ________ of an angle is the point of intersection of the two sides of an angle.

In Exercises 3 and 4, refer to the following figure to determine the measures of the given angles.

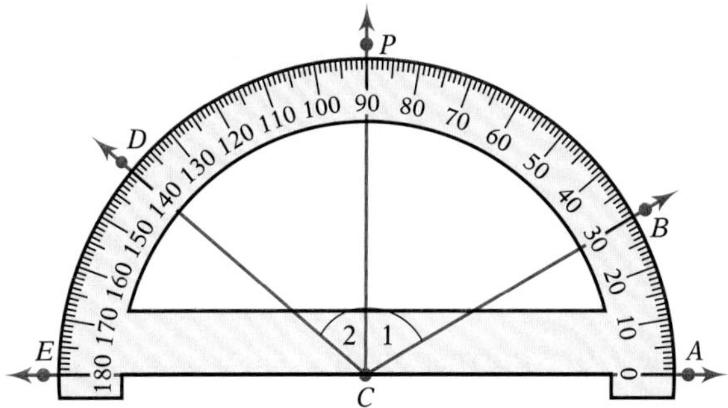

3. **(a)** $m\angle BCA$

   (b) $m\angle 2$

   (c) $m\angle DCE$

4. **(a)** $m\angle ACD$

   (b) $m\angle PCA$

   (c) $m\angle 1$

In Exercises 5–8, detemine the indicated value or identify the indicated parts of the given triangle.

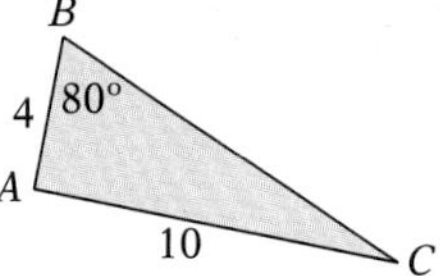

5. **(a)** AC

   (b) $m\angle B$

   (c) The side whose length is 4

   (d) The angles whose measures are not given

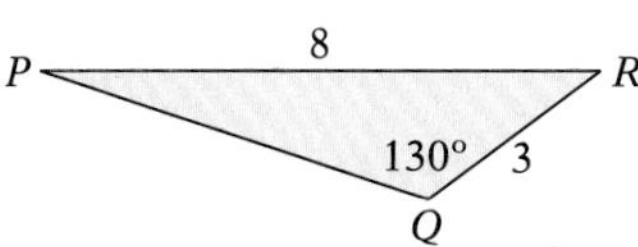

6. **(a)** $m\angle Q$

   (b) The angles whose measures are not given

   (c) The side whose length is 3

   (d) PR

Copyright © Houghton Mifflin Company. All rights reserved.

7. **(a)** $m\angle E$

(b) The sides whose lengths are not given

(c) $m\angle D$

(d) ED

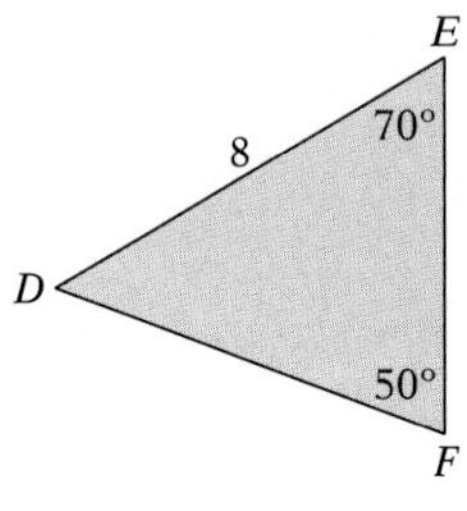

8. **(a)** $m\angle C$

(b) The sides whose lengths are not given

(c) CE

(d) $m\angle E$

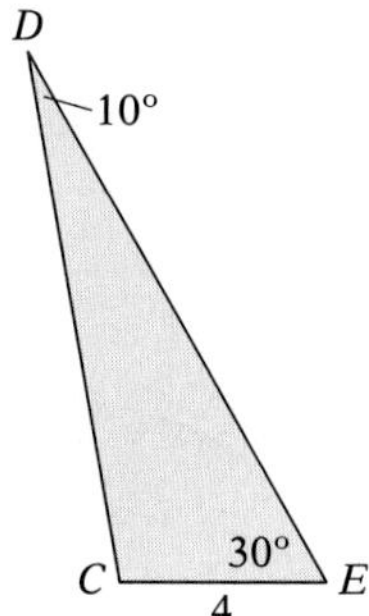

B. Similar Triangles

9. If two triangles are similar triangles, then the corresponding sides are ________.

10. If the corresponding angles of two triangles have the same measure, we say that the triangles are ________.

In Exercises 11–14, write the relationships among (a) the angles and (b) the sides.

11. $\triangle ABC \sim \triangle PQR$

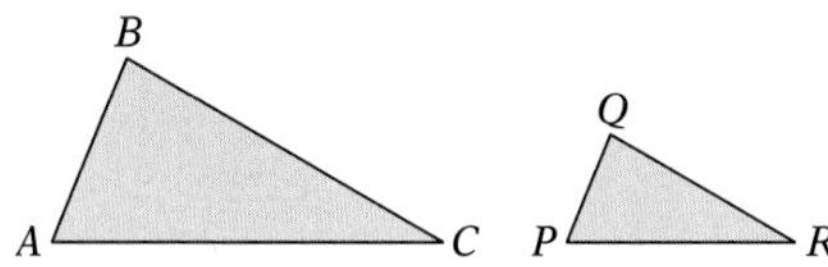

12. $\triangle CDE \sim \triangle RST$

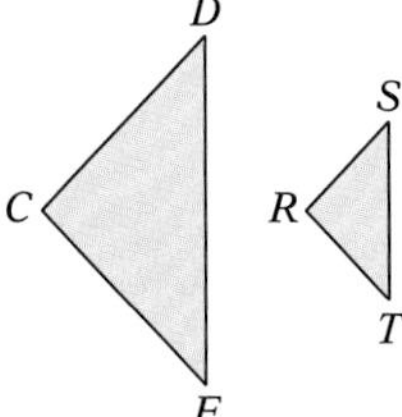

13. $\triangle ABC \sim \triangle DEF$

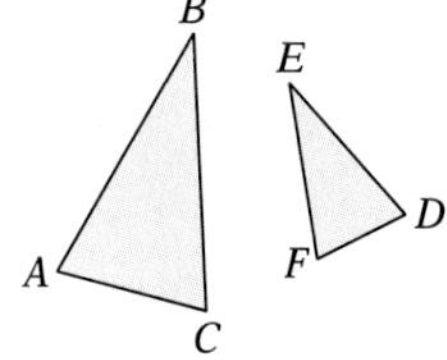

14. $\triangle RST \sim \triangle DFE$

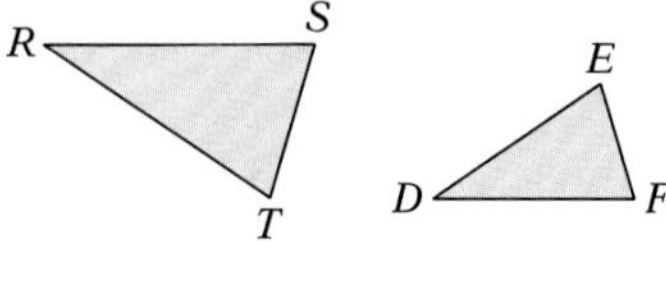

Copyright © Houghton Mifflin Company. All rights reserved.

In Exercises 15–22, determine the value of x.

15. $\triangle ABC \sim \triangle EFG$

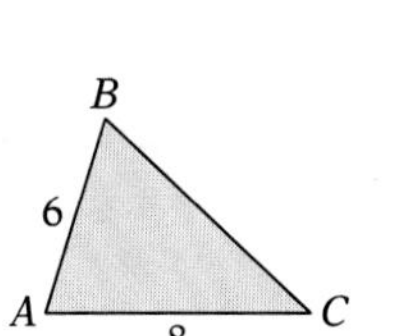

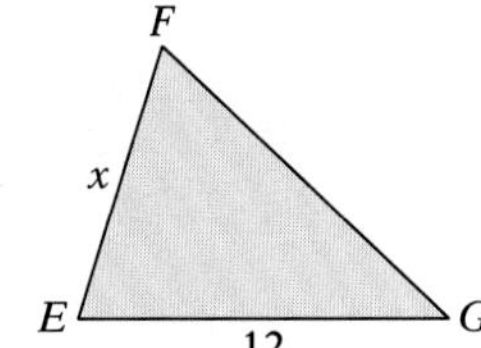

16. $\triangle ABC \sim \triangle PQR$

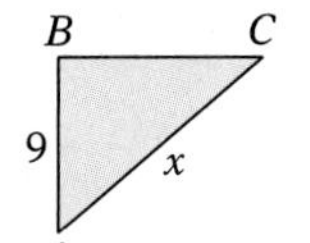

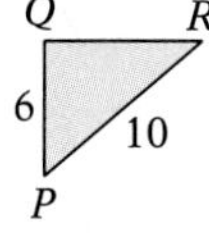

17. $\triangle PQR \sim \triangle ABC$

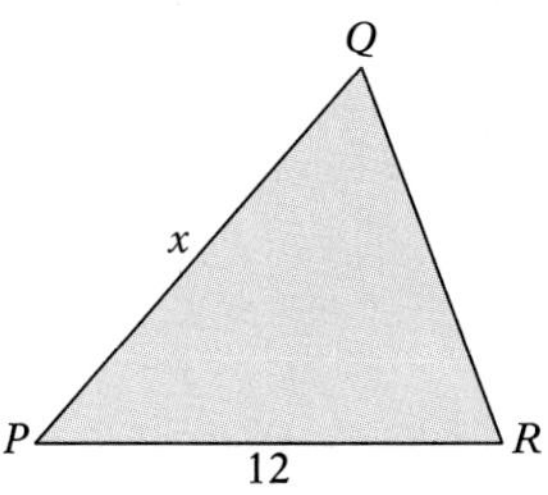

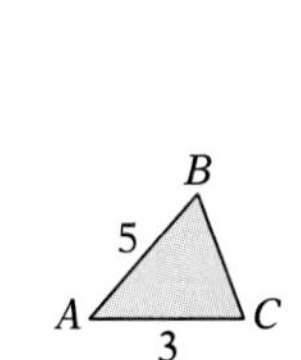

18. $\triangle RST \sim \triangle EFG$

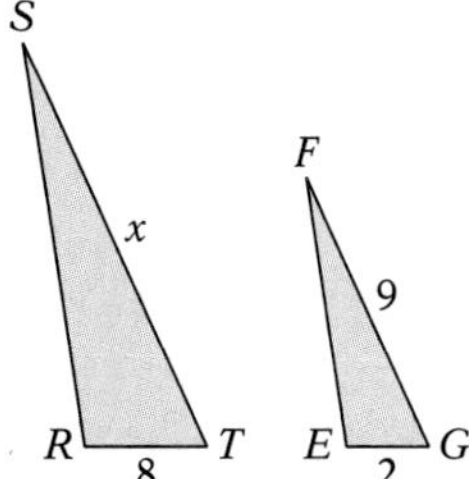

19. $\overline{PQ}$ is parallel to $\overline{CB}$.

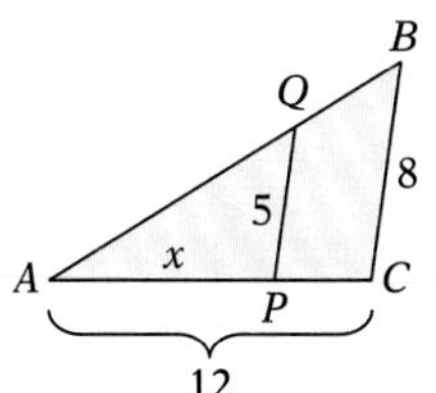

20. $\overline{EF}$ is parallel to $\overline{QR}$.

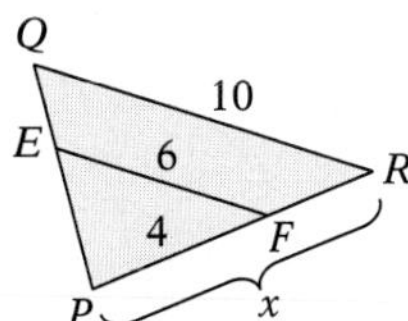

21.

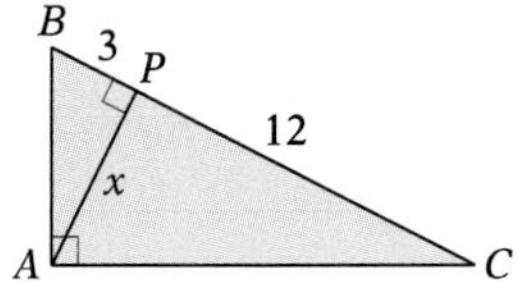

22.

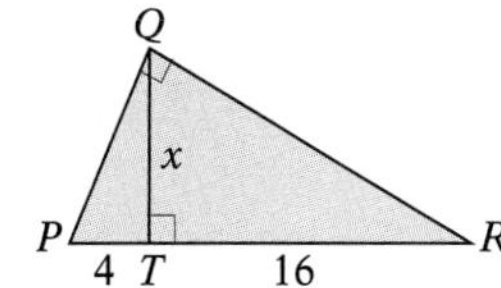

In Exercises 23–30, determine the values of x and y.

23. $\triangle ABC \sim \triangle PQR$

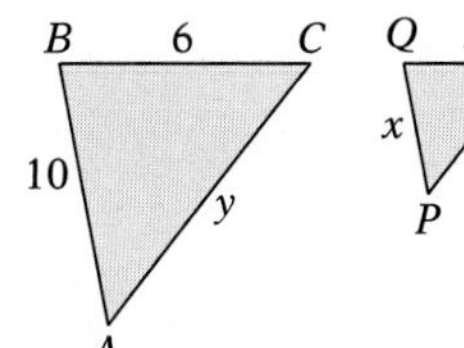

24. $\triangle RST \sim \triangle ABC$

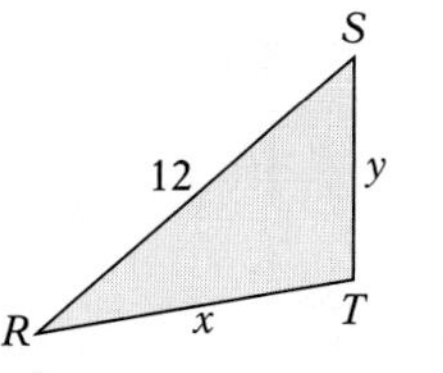

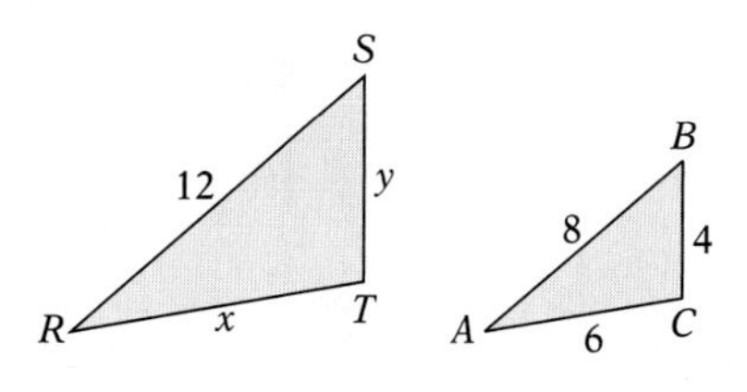

Copyright © Houghton Mifflin Company. All rights reserved.

25. $\triangle EFG \sim \triangle RST$

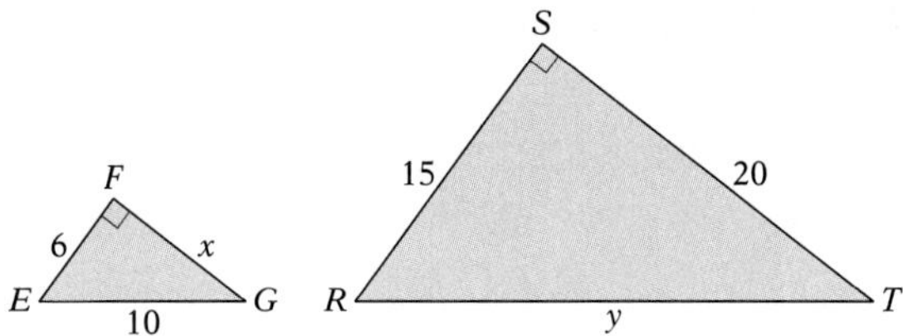

26. $\triangle PQR \sim \triangle EFG$

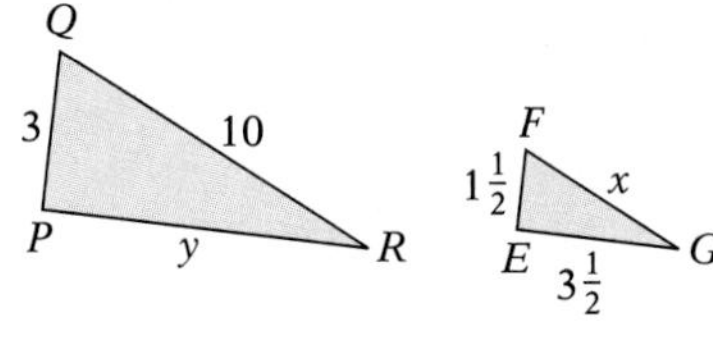

27. $\overline{PQ}$ is parallel to $\overline{AC}$.

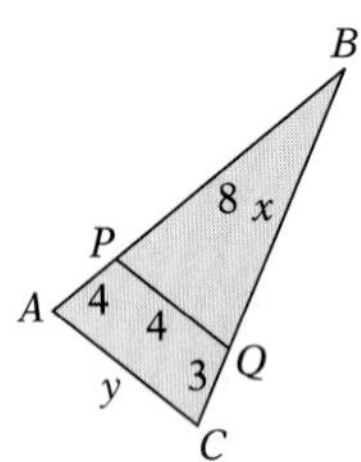

28. $\overline{MN}$ is parallel to $\overline{RS}$.

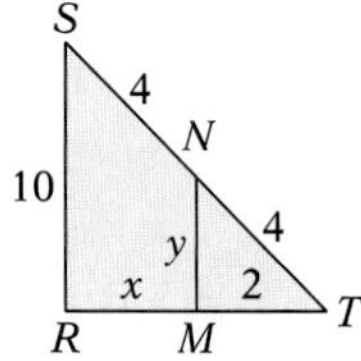

29. $\overline{AB}$ is parallel to $\overline{PR}$.

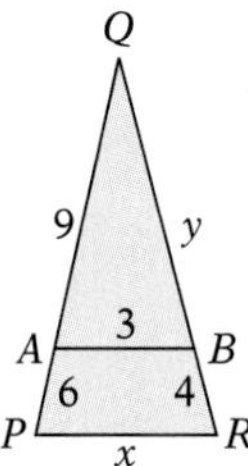

30. $\overline{RS}$ is parallel to $\overline{FG}$.

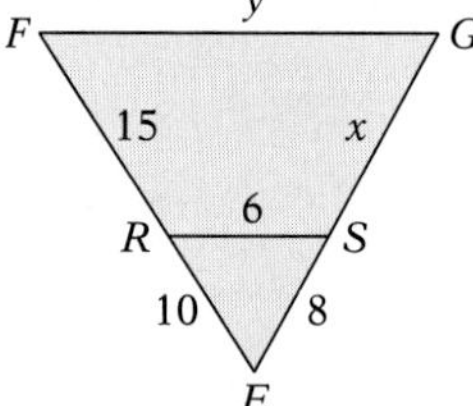

C. Applications

31. ***Flagpole Height*** A 10-foot flagpole is placed beside an existing taller flagpole. The shadows of the two flagpoles are 4 feet and 6.4 feet respectively. What is the height of the taller pole?

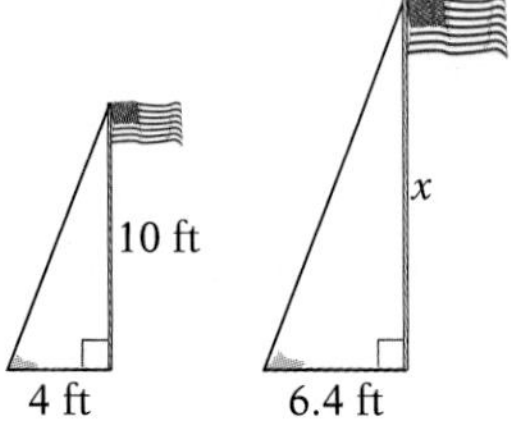

Copyright © Houghton Mifflin Company. All rights reserved.

32. *Height of a Tree* To determine the height of a tree, a forester places a yardstick beside the tree and measures the shadows cast by the yardstick and the tree. If the shadows are 15 inches and 20 feet respectively, what is the height of the tree?

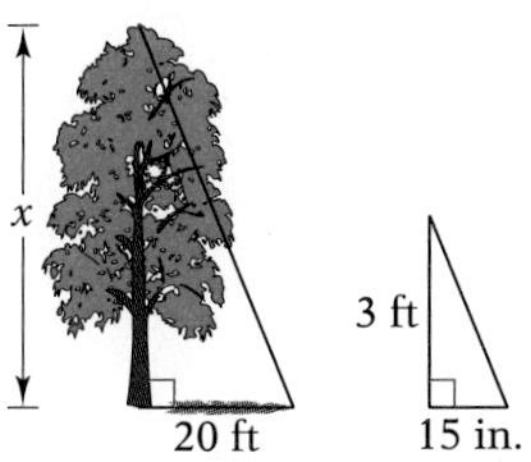

33. *Sailing Ship* A person is constructing a scale model of a sailing ship for display in a museum. The lengths of the sides of a triangular sail are 3, 4, and 5 inches. If the longest side of the actual sail is 25 feet, what are the lengths of the other sides?

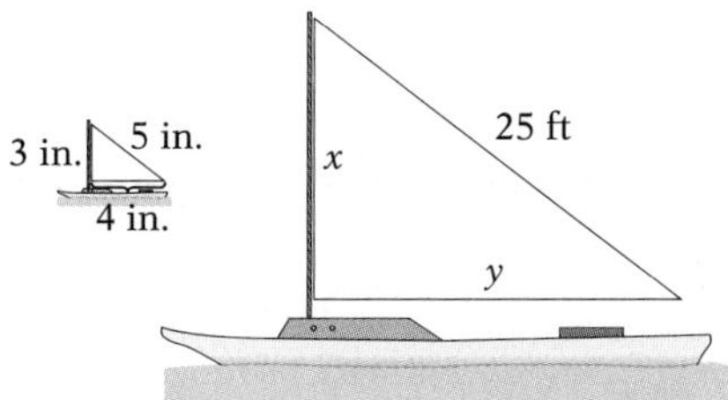

34. *Aircraft Wing* An engineer is constructing a scale model of a new aircraft wing design. The wing is triangular, and the lengths of the sides of the actual wing are 10, 25, and 27 feet. If the length of the shortest side of the model is 1 foot, what are the lengths of the other sides of the model?

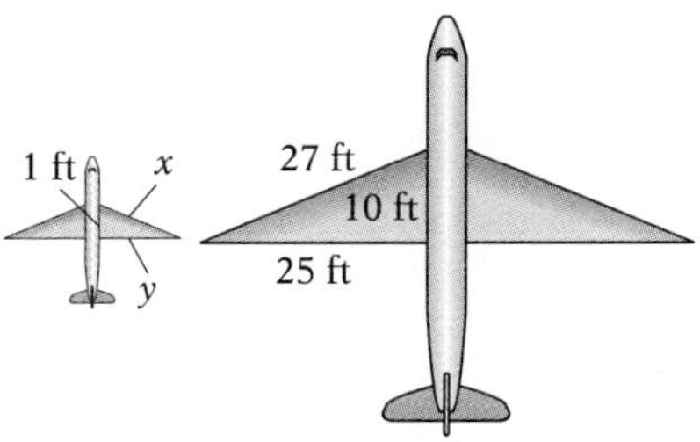

Writing and Concept Extension

35. What is the difference between the notation AB and $\overline{AB}$?

36. Suppose that in $\triangle ABC$ and $\triangle PQR$, $m\angle A = m\angle P = 50°$ and $m\angle B = m\angle Q = 60°$. What can you conclude about $\angle C$ and $\angle R$?

Copyright © Houghton Mifflin Company. All rights reserved.

In addition to triangles, other geometric figures can be similar, in which case their corresponding sides are proportional. For the similar figures in Exercises 37–40, determine x.

37.

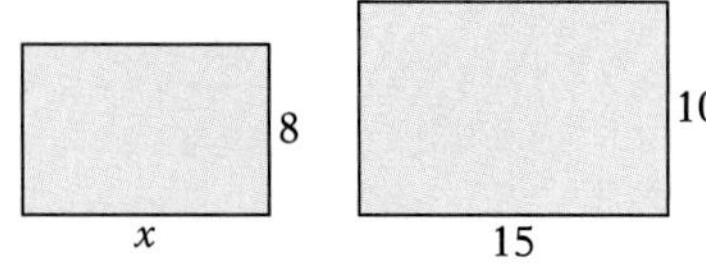

38.

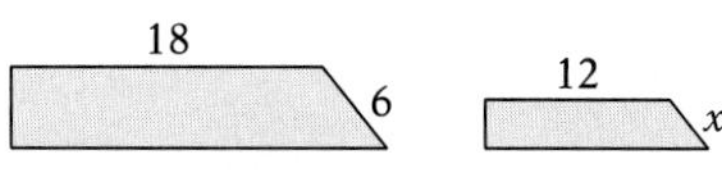

39.

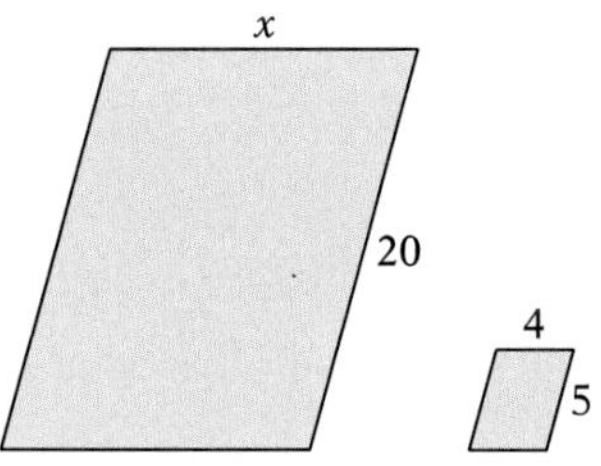

40.

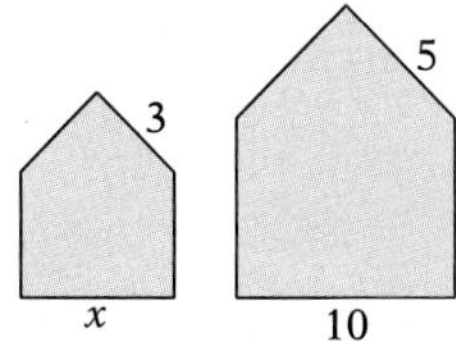

41. Suppose that two triangles are similar and that the ratio of the corresponding sides is 2. How are the perimeters of the triangles related?

Copyright © Houghton Mifflin Company. All rights reserved.

CHAPTER 7 REVIEW EXERCISES

Section 7.1

1. A quotient that compares two quantities with the same units is called a(n) ________.

In Exercises 2 and 3, write the ratio as a simplified fraction.

2. 15 to 12

3. 3.2 to 0.08

4. *Voter Survey* A survey of voters found that 45 of those surveyed voted for the incumbent and 35 voted for the challenger. What is the ratio of votes for the incumbent to votes for the challenger?

5. *Recipe* A recipe uses $1\frac{1}{2}$ cups of sugar and $2\frac{3}{4}$ cups of flour. What is the ratio of sugar to flour?

6. *Office Communications* An office worker received 24 memos in one day of which 14 were e-mails and the remainder were on paper. What was the ratio of e-mails to paper memos?

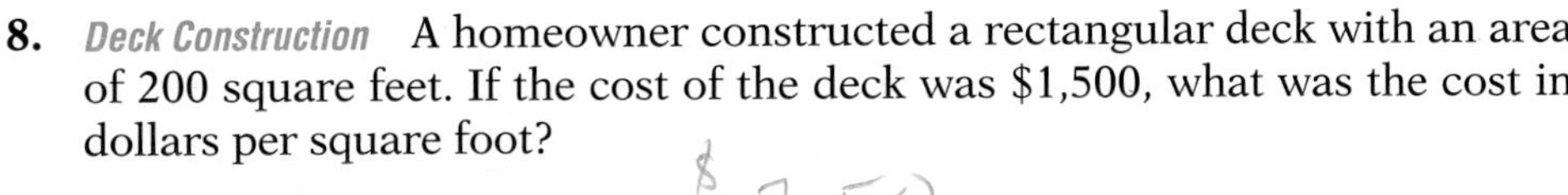

7. A unit rate is a rate in which the ________ is 1.

8. *Deck Construction* A homeowner constructed a rectangular deck with an area of 200 square feet. If the cost of the deck was \$1,500, what was the cost in dollars per square foot?

9. *Water Tank* With the drain open, a 300-gallon tank can be emptied in 12 minutes. At what rate does the water flow out of the tank?

10. *Coca-Cola®* Each year, Canadians drink 6.6 billion 8-ounce servings of Coca-Cola products. (Source: Coca-Cola Co.) Assuming a population of 0.031 billion, how many servings per person do Canadians drink each year?

Copyright © Houghton Mifflin Company. All rights reserved.

Section 7.2

11. A conversion factor is a ______ such as $\frac{3 \text{ feet}}{1 \text{ yard}}$.

In Exercises 12–18, convert the given measurement to the indicated units.

12. 52,800 feet = ______ miles

13. $1\frac{1}{3}$ yards = ______ inches

14. $2\frac{1}{4}$ pounds = ______ ounces

15. $3\frac{3}{4}$ quarts = ______ cups

16. 9 quarts = ______ gallons

17. 40 miles per hour = ______ feet per second

18. 24 gallons per minute = ______ quarts per second

19. *Maple Tree* A sugar maple tree is 65 feet high. What is the height in inches?

20. *Gravel* A contractor ordered 1 ton of gravel. A truck delivered $\frac{3}{4}$ ton. How many more pounds of gravel are needed?

Section 7.3

21. In the ______ system, the basic unit of length is the meter.

In Exercises 22–24, convert the given measurement to the indicated units.

22. 1.3 centimeters = ______ millimeters

23. 3,400 meters = ______ kilometers

24. 3.7 meters = ______ centimeters

25. *Long Jump* The men's world record for the long jump is 8.95 meters. What is the distance in centimeters?

Copyright © Houghton Mifflin Company. All rights reserved.

In Exercises 26–28, convert the given measurement to the indicated units.

26. 5,000 feet = ______ kilometers

27. 80 centimeters = ______ inches

28. 12 feet = ______ meters

29. *Hoover Dam* The height of the Hoover Dam is 221 meters. What is the height in feet?

30. *Film* The width of film is 35 millimeters. What is the width in inches?

Section 7.4

31. Area is measured in ______ units, whereas ______ is measured in cubic units.

In Exercises 32–37, convert the given measurement to the indicated units.

32. 1.2 liters = ______ cubic centimeters

33. 750 milliliters = ______ liter

34. 8.2 grams = ______ milligrams

35. 2.1 kilograms = ______ grams

36. 750 milliliters = ______ pints

37. 4 ounces = ______ grams

38. *Packages of Rice* A case of 800-gram packages of rice weighs 40 kilograms. How many packages of rice are in the case?

39. *Soda* How many liters of soda are in twelve 350-milliliter cans of soda?

40. *Cheese* Cheese costs $1.68 per kilogram. What is the cost in dollars per pound?

Copyright © Houghton Mifflin Company. All rights reserved.

Section 7.5

41. In the proportion $\frac{4}{n} = \frac{12}{11}$, n and 12 are called the ________.

42. The products of the means and the extremes of a proportion are called the ________.

In Exercises 43 and 44, determine whether the proportion is true.

43. $\frac{9}{17} = \frac{10}{18}$

44. $\frac{1\frac{1}{2}}{6} = \frac{3}{12}$

In Exercises 45–47, solve the given proportion.

45. $\frac{10}{3} = \frac{5}{n}$

46. $\frac{12}{x-1} = \frac{4}{21}$

47. $\frac{\frac{3}{4}}{6} = \frac{m}{2}$

48. *Flood Water* During a flood, the water level of a river rose 3 inches in 2 hours. Assuming that the rate continued, how many inches did the river rise in 9 hours?

49. *Copier Paper* An office copy room uses 15 cases of copier paper every 2 weeks. For how many weeks would 72 cases be expected to last?

50. *Stock Commission* A trader pays an \$8 fee for each 150 shares of stock traded. What is the fee for trading 325 shares?

Section 7.6

51. The point of intersection of the two sides of an angle is called the ________.

52. If two triangles are similar triangles, then the corresponding angles have the same ________.

53. What is the sum of the measures of the angles of a triangle?

Copyright © Houghton Mifflin Company. All rights reserved.

In Exercises 54–57, determine the indicated value or identify the indicated parts of the following triangle.

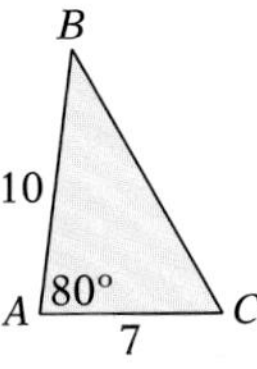

54. AB

55. $m\angle A$

56. The side whose length is 7.

57. The angles whose measures are not given

In Exercises 58 and 59, determine x.

58. $\overline{PQ}$ is parallel to $\overline{AC}$.

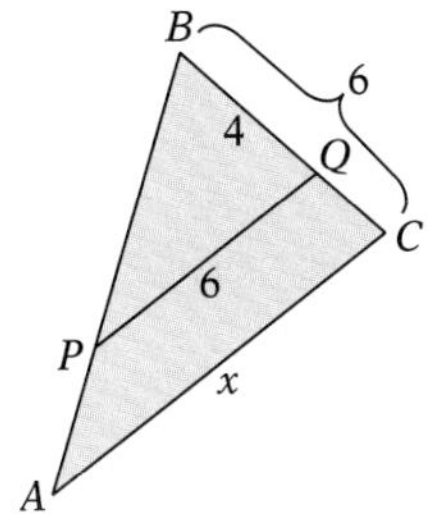

59. $\triangle ABC \sim \triangle PQR$

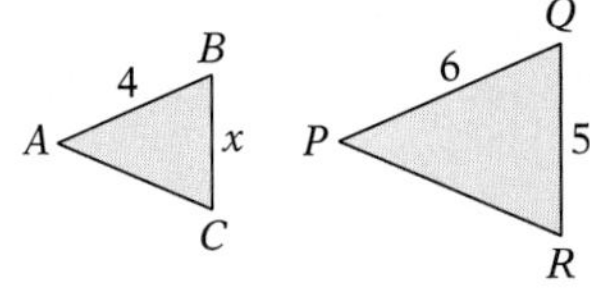

60. *Tower* A communication tower casts a shadow that is 50 feet long. If a yardstick casts a shadow that is 2 feet long, what is the height of the tower?

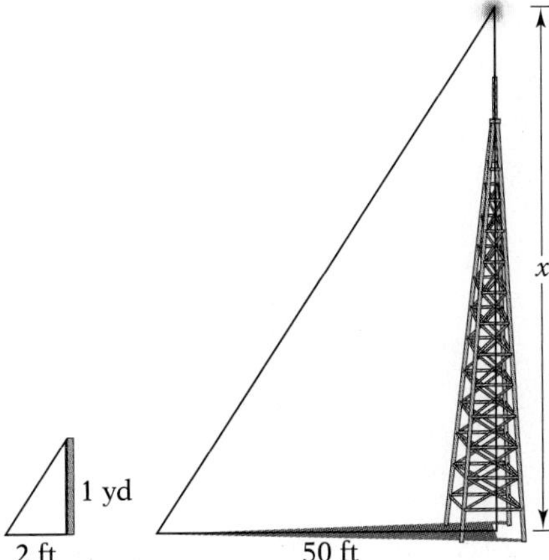

Copyright © Houghton Mifflin Company. All rights reserved.

CHAPTER 7 TEST

1. In the proportion $\frac{5}{9} = \frac{m}{12}$, 5, 9, m, and 12 are called the ______.

2. For two similar triangles, the measures of the corresponding angles are ______, and the corresponding sides are ______.

In Questions 3–12, convert the given measurement to the indicated units.

3. 27 inches = ______ yard

4. 6.4 meters = ______ centimeters

5. 5,000 meters = ______ miles

6. 2,300 milliliters = ______ liters

7. 10 liters = ______ quarts

8. $2\frac{1}{2}$ quarts = ______ cups

9. 1,600 grams = ______ kilograms

10. $\frac{3}{4}$ pound = ______ kilogram

11. 300 miles per hour = ______ feet per second

12. 36 ounces = ______ pounds

13. Write the ratio 36 to 27 as a simplified fraction.

14. Is the proportion $\frac{7.5}{50} = \frac{3}{20}$ true?

In Questions 15 and 16, solve the proportion.

15. $\frac{6}{a} = \frac{3}{2.5}$

16. $\frac{7}{3} = \frac{x+4}{6}$

Copyright © Houghton Mifflin Company. All rights reserved.

17. *Veterinarian* During one day, a veterinarian treated 34 pets of which 20 were dogs and the remainder were cats. What was the ratio of dogs to cats that were treated?

18. *Delivery Service* An overnight delivery service charges \$17 for a 3-pound package. What is the cost per pound?

19. *Honey bees* A beekeeper extracted 4.5 gallons of honey. How many pint jars can he fill with the honey?

20. *Distance* The distance from New York to Boston is 182 miles. What is the distance in kilometers?

21. *Cookies* A box of two dozen cookies weighs 1.2 kilograms. What is the weight in grams of one cookie?

22. *Candy* A confectioner charges \$4.90 per pound for chocolate creams. What is the cost of 2 pounds 4 ounces of chocolate creams? (Round to the nearest cent.)

23. *Printer* A printer will print 30 pages in 5 minutes. How many hours are needed to print 396 pages?

24. Determine BC and $m\angle A$.

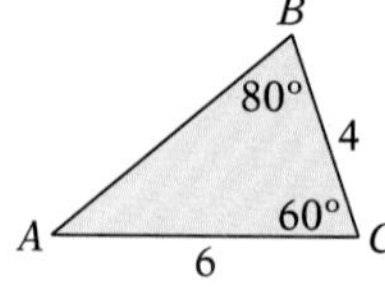

25. Determine x.
$\overline{EF}$ is parallel to $\overline{PR}$.

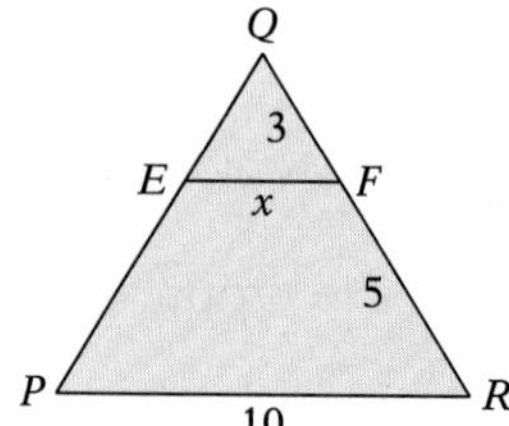

Copyright © Houghton Mifflin Company. All rights reserved.

CHAPTERS 6–7 CUMULATIVE TEST

1. Considered together, the rational and irrational numbers are called the ______ numbers.

2. **(a)** Write thirty and nine hundredths as a decimal number.

   (b) Write $\frac{77}{1,000}$ as a decimal number.

3. Round 2.973 to the nearest **(a)** tenth and **(b)** hundredth.

4. Arrange the numbers 0.04, 0.039, 0.041 from largest to smallest.

In Questions 5–8, perform the indicated operations.

5. $-5.2 - 3.47$

6. $-4.07 + 7.3$

7. $(4.2)(3.5)$

8. $0.049 \div 0.2$

In Questions 9 and 10, evaluate the expression.

9. $\sqrt{1} - \sqrt{25}$

10. $9 - 2\sqrt{\frac{9}{4}}$

11. Simplify $100(3.2x + 4.55)$.

In Questions 12 and 13, solve the equation.

12. $6.14 + n = 3.2$

13. $1.03 - 0.3x = 1.3$

14. Write the ratio of $\frac{3}{4}$ to 2 as a simplified fraction.

In Questions 15–20, convert the given measurement to the indicated units.

15. 976 millimeters = ______ meter

16. 2 feet = ______ centimeters

17. 9 pints = ______ gallons

18. $\frac{1}{5}$ liter = ______ fluid ounces

19. 4,750 pounds = ______ tons

20. 5 feet per second = ______ miles per hour

Copyright © Houghton Mifflin Company. All rights reserved.

21. Solve the proportion $\frac{5}{x + 3} = \frac{2}{3}$.

22. *Dining Out* A diner ordered salad for $3.85, steak for $14.75, and dessert for $2.95. If the diner gave the server $25 for the meal and tip, what was the tip?

23. *Trencher Rental* A tool rental company rents a trencher for $15.65 plus $3.25 per hour. What is the cost of renting the trencher for 6 hours?

24. *Football Record* Of the 11 football games that a college team played, the team won 8 games. What was the ratio of losses to games played?

25. *Wallpaper* If three rolls of wallpaper are needed for 180 square feet of wall space, how many rolls are needed for 1,380 square feet?

26. *Oatmeal* The weight of a box of oatmeal is 650 grams. What is the weight in ounces?

27. *Holland Tunnel* The Holland Tunnel in New York is 8,557 feet long. What is the length in miles?

28. The length of the hypotenuse of a right triangle is 13. If the length of one leg is 12, what is the length of the other leg?

29. Determine the circumference and the area of a circle whose radius is 4 feet.

30. Given that $\triangle ABC \sim \triangle RQP$, determine x.

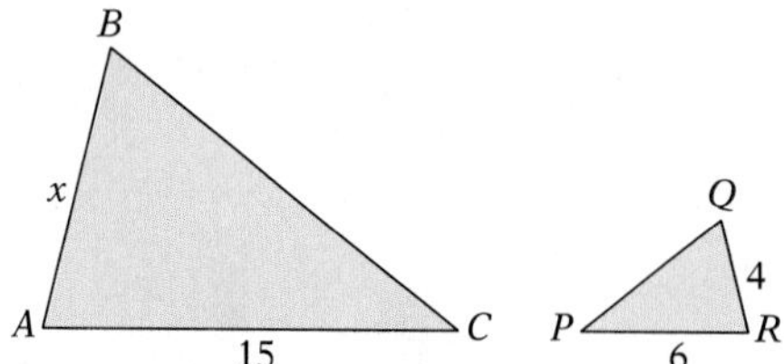

Copyright © Houghton Mifflin Company. All rights reserved.

CHAPTER 8

Percents

Soccer has become a popular sport for girls. In 2002, of the youngsters in the 11–12 age range who played soccer, 40 of every 100 were girls. The phrase "40 of every 100" can be written as the ratio $\frac{40}{100}$. A more common way to express the phrase is with a **percent:** 40%, which is read "40 percent."

Everywhere you look—television commercials and news reports, store signs, credit card bills, newspaper advertisements, bank statements—you will see percents. But what exactly do they mean?

A thorough understanding of percents will help you to make good decisions about your finances and to analyze and interpret information. Without this understanding, you could be vulnerable to being cheated and to making commitments that you might regret later.

Chapter Snapshot

In this chapter, we introduce the basic meaning of a percent and how to perform certain calculations with percents. The basic percent equation that we present in Section 8.2 is fundamental to all percent problems. Much of the chapter is devoted to practical applications of percents, including simple and compound interest, taxes, commissions, depreciation, markup, and discount. We conclude with the topic of percent increases and decreases.

Because percents play such an important role in our everyday lives, your mastery of the material in this chapter will serve you well in the real world.

Copyright © Houghton Mifflin Company. All rights reserved.

For online resources, visit the web site **math.college.hmco.com/students** and follow the links to Hubbard/Robinson, *Prealgebra*.

Some Friendly Advice . . .

There is a good chance that you are getting down toward the end of your syllabus, and your school term may be drawing to a close. If so, then you should be thinking now about your final exam.

If you have been following our earlier advice about reviewing on a daily basis, then you are already well on your way to preparing for the final. Even if you let things slide a little, it's not too late. The important thing is that you begin now.

To prepare for a final exam, your best resources are tests that you have taken. If a question was worth asking on those tests, then it is worth asking again on the final. The Quick References and your excellent notes will also serve you well. Make a list of things you have forgotten how to do, and start asking your instructor about them now. Above all, don't wait until the night before the exam!

WARM-UP SKILLS

The following questions review concepts and skills that you will need in Chapter 8.

1. Write the ratio of 37 to 100 as a fraction.
2. Write $\frac{9}{20}$ with a denominator of 100.
3. Reduce $\frac{135}{100}$.
4. Write $\frac{5}{8}$ as a terminating decimal.
5. Write $\frac{5}{6}$ as a repeating decimal.
6. Solve $0.3x = 18$.
7. In a sociology class, $\frac{3}{4}$ of the students were women. If 24 women were in the class, how many students were in the class?
8. In a 12-pound package of cookies, $\frac{2}{3}$ of the cookies were broken. How many pounds of broken cookies were in the package?
9. After \$124 was deducted for tax and insurance, a reporter received \$830. What were the reporter's earnings before the deductions?
10. The price per quart of milk rose from \$1.09 to \$1.23. By how many cents did the price increase?

Copyright © Houghton Mifflin Company. All rights reserved.

8.1 Introduction to Percents

SUGGESTIONS FOR SUCCESS

You will rarely go through an entire day without seeing or hearing some reference to a percent. Taxes, discounts, commissions on sales, mortgage rates, and results of political surveys are just a few examples of information that is usually presented as a percent.

The key to understanding this introductory section and the rest of the chapter is in the first sentence: A percent is a fraction whose denominator is 100. If you understand that one basic idea, then such tasks as writing a percent as a decimal will come naturally to you.

Much of the arithmetic involves multiplying or dividing by 100—that is, shifting the decimal point to the right or left. Put your calculator away. This stuff is easy.

A The Meaning of Percents

A **percent** is a fraction (or ratio) whose denominator is 100. We can think of the word *percent* as meaning "per 100," "out of 100," or "for every 100." The symbol for percent is %.

In the year 2000, the Senate of the United States consisted of 100 senators, of whom 55 were Republicans and 45 were Democrats.

The ratio of the number of Republicans to the total number of senators was $\frac{55}{100}$, and we say that 55% of the senators were Republicans. Similarly, the ratio of the number of Democrats to the total number of senators was $\frac{45}{100}$, which means that 45% of the senators were Democrats.

Any fraction whose denominator is 100 can easily be written as a percent. We simply drop the denominator and attach the percent symbol % to the numerator.

Example 1

Write the given fraction as a percent.

(a) $\frac{23}{100}$ **(b)** $\frac{99}{100}$ **(c)** $\frac{200}{100}$

SOLUTION

(a) $\frac{23}{100} = 23\%$ **(b)** $\frac{99}{100} = 99\%$

(c) $\frac{200}{100} = 200\%$

Your Turn 1

Write the given fraction as a percent.

(a) $\frac{39}{100}$ **(b)** $\frac{80}{100}$ **(c)** $\frac{150}{100}$

Answers: (a) 39%; (b) 80%; (c) 150%

In Example 1, the denominator of each fraction is 100, so we can quickly write the fraction as a percent. But how do we write a fraction as a percent if the denominator is not 100?

In some cases, we can easily rename a fraction so that its denominator is 100. In fact, with practice, you can do this mentally.

Copyright © Houghton Mifflin Company. All rights reserved.

Example 2

Write the given fraction as a percent.

(a) $\frac{30}{50}$ **(b)** $\frac{8}{25}$ **(c)** $\frac{7}{10}$

SOLUTION

In each case, we multiply the numerator and the denominator of the given fraction by a number chosen so that the resulting denominator will be 100.

(a) $\frac{30}{50} = \frac{30}{50} \cdot \frac{2}{2} = \frac{60}{100} = 60\%$

(b) $\frac{8}{25} = \frac{8}{25} \cdot \frac{4}{4} = \frac{32}{100} = 32\%$

(c) $\frac{7}{10} = \frac{7}{10} \cdot \frac{10}{10} = \frac{70}{100} = 70\%$

Your Turn 2

Write the given fraction as a percent.

(a) $\frac{7}{20}$ **(b)** $\frac{45}{50}$ **(c)** $\frac{15}{25}$

Answers: (a) 35%; (b) 90%; (c) 60%

Example 2 helps us to understand that we can talk about a percent even when the denominator of a ratio is not 100. The ratio in part (a) might have been the result of a poll of 50 people, 30 of whom favored a certain candidate for office. Even though 50 people were polled rather than 100, we can still say that 60% favored the candidate.

Similarly, the ratio in part (b) might represent a catcher's success in throwing out base stealers. The given ratio is based on 25 attempts, not 100, but we can still say that the catcher was successful 32% of the time.

As you work with percents such as 30%, which means 30 out of 100, it is important to keep in mind that the ratio of actual data could be $\frac{3}{10}$, $\frac{6}{20}$, $\frac{15}{50}$, or any other ratio that is equivalent to $\frac{30}{100}$.

We can reverse the process and write a percent as a fraction. Simply remove the % symbol and write the remaining number as the numerator. The denominator is 100.

Example 3

Write the given percent as a fraction.

(a) 57% **(b)** 91% **(c)** 123%

SOLUTION

(a) $57\% = \frac{57}{100}$

(b) $91\% = \frac{91}{100}$

(c) $123\% = \frac{123}{100}$

Your Turn 3

Write the given percent as a fraction.

(a) 19%

(b) 83%

(c) 137%

Answers: (a) $\frac{19}{100}$; (b) $\frac{83}{100}$; (c) $\frac{137}{100}$

After writing a percent as a fraction, you may find that you can simplify the fraction.

Copyright © Houghton Mifflin Company. All rights reserved.

Example 4

Write the given percent as a simplified fraction.

(a) 60% **(b)** 135%

(c) $33\frac{1}{3}\%$ **(d)** 2.5%

SOLUTION

(a) $60\% = \frac{60}{100} = \frac{3 \cdot \cancel{20}}{5 \cdot \cancel{20}} = \frac{3}{5}$

(b) $135\% = \frac{135}{100} = \frac{\cancel{5} \cdot 27}{\cancel{5} \cdot 20} = \frac{27}{20}$

(c) $33\frac{1}{3}\% = \frac{100}{3}\%$ Write the mixed number $33\frac{1}{3}$ as the improper fraction $\frac{100}{3}$.

$= \frac{\frac{100}{3}}{100}$ $\frac{\frac{100}{3}}{100}$ means $\frac{100}{3} \div 100$.

$= \frac{\cancel{100}}{3} \cdot \frac{1}{\cancel{100}}$ Multiply by the reciprocal of 100.

$= \frac{1}{3}$

(d) $2.5\% = \frac{2.5}{100}$

$= \frac{2.5}{100} \cdot \frac{10}{10}$ Multiply by $\frac{10}{10}$ to clear the decimal from the numerator.

$= \frac{25}{1{,}000}$

$= \frac{\cancel{25} \cdot 1}{\cancel{25} \cdot 40}$

$= \frac{1}{40}$

Your Turn 4

Write the given percent as a simplified fraction.

(a) 35%

(b) 225%

(c) $12\frac{1}{2}\%$

(d) 8.4%

Answers: (a) $\frac{7}{20}$; (b) $\frac{9}{4}$; (c) $\frac{1}{8}$; (d) $\frac{21}{250}$

B Percents and Decimals

When we perform arithmetic with percents, we usually want to write the percents as decimals. One method is first to write the percent as a fraction and then convert the fraction to a decimal.

$23\% = \frac{23}{100}$ To divide by 100, we shift the decimal point 2 places to the left.

$= 0.23$

Because this method always involves dividing by 100, we can skip the middle step and write the decimal directly. We just drop the percent symbol and shift the decimal point 2 places to the left.

Copyright © Houghton Mifflin Company. All rights reserved.

Example 5

Write the given percent as a decimal.

(a) 57% **(b)** 143% **(c)** 6% **(d)** 1.4%

SOLUTION

(a) 57% = 0.57. = 0.57

(b) 143% = 1.43. = 1.43

(c) 6% = 0.06. = 0.06

(d) 1.4% = 0.01.4 = 0.014

Your Turn 5

Write the given percent as a decimal.

(a) 28%

(b) 2%

(c) 110%

(d) 8.5%

Answers: (a) 0.28; (b) 0.02; (c) 1.10; (d) 0.085

Scientific calculators usually have a % key that converts a percent into a decimal. However, as shown in Example 5, you certainly don't need a calculator to shift the decimal point 2 places to the left.

We reverse the steps shown in Example 5 to write a decimal as a percent. That is, we move the decimal point 2 places to the right and attach the percent symbol.

Example 6

Write the given decimal as a percent.

(a) 0.72 **(b)** 1.5 **(c)** 0.007 **(d)** 0.1

SOLUTION

(a) 0.72 = 0.72.% = 72%

(b) 1.5 = 1.50.% = 150%

(c) 0.007 = 0.00.7% = 0.7%

(d) 0.1 = 0.10.% = 10%

Your Turn 6

Write the given decimal as a percent.

(a) 0.025

(b) 2.1

(c) 0.54

(d) 0.001

Answers: (a) 2.5%; (b) 210%; (c) 54%; (d) 0.1%

The following is a summary of the methods for converting between percents and decimals.

Converting Between Percents and Decimals

1. To convert from a percent to a decimal:
   (a) Drop the percent symbol.
   (b) Shift the decimal point 2 places to the left.
2. To convert from a decimal to a percent:
   (a) Move the decimal point 2 places to the right.
   (b) Attach the percent symbol.

Copyright © Houghton Mifflin Company. All rights reserved.

Writing a percent as a decimal involves dividing by 100, and writing a decimal as a percent involves multiplying by 100. Because these operations can be performed simply by shifting the decimal point, you should soon be able to make such conversions mentally.

C *Writing Fractions as Percents*

In Example 2, we were able to change a fraction to a percent by multiplying the numerator and the denominator by a number chosen so that the denominator of the resulting fraction would be 100. Because this method works well only in certain instances, we want to show you another method that can be used for all fraction-to-percent conversions.

The method is simply this: Write the fraction as a decimal, and then use the methods just presented to convert the decimal to a percent.

NOTE Recall that writing a fraction as a decimal involves dividing the numerator by the denominator. You can perform this division with a calculator or with long division. In the following examples, we show only the results.

Example 7

Write $\frac{3}{8}$ as a percent.

SOLUTION

First, we write $\frac{3}{8}$ as a decimal.

$$\frac{3}{8} = 0.375$$

Now we write 0.375 as a percent.

$$0.37.5 \rightarrow 37.5\%$$

Therefore, $\frac{3}{8} = 37.5\%$.

Your Turn 7

Write $\frac{1}{16}$ as a percent.

Answer: 6.25%

Example 8

Write $\frac{5}{6}$ as a percent.

SOLUTION

Method 1

$$\frac{5}{6} = 0.83333333 \cdots$$

$$\approx 0.8333$$

Now we write 0.8333 as a percent.

$$0.83.33 \rightarrow 83.33\%$$

Because we rounded off the decimal form, our answer is only approximate: $\frac{5}{6} \approx 83.33\%$.

Your Turn 8

Write $\frac{4}{9}$ (a) as a percent rounded to the nearest hundredth and (b) as an exact percent.

continued

Copyright © Houghton Mifflin Company. All rights reserved.

Method 2

When the decimal form is not a terminating decimal, we can use long division in order to obtain an exact answer.

$$\begin{array}{r} 0.83 \\ 6\overline{)5.00} \\ \underline{4\,8} \\ 20 \\ \underline{18} \\ 2 \end{array}$$

The result is $0.83\frac{2}{6} = 0.83\frac{1}{3} = 83\frac{1}{3}\%$.

LEARNING TIP

Make sure that you know whether answers are to be exact or approximate. You will need to use Method 2 if your answer must be exact. Otherwise, you can round off as in Method 1.

Answers: (a) 44.44%; (b) $44\frac{4}{9}\%$

Some fraction-to-percent conversions are so common that they are worth memorizing. For example, if the newspaper reports that 3 of every 5 people surveyed approve of the president's work, then, because $\frac{3}{5} = 60\%$, you know that the president enjoys a 60% approval rating.

The following table lists some of the more commonly used conversions.

Common Fraction and Decimal Equivalents

Halves	*Thirds*	*Fourths*	*Fifths*	*Eighths*	*Tenths*
$\frac{1}{2} = 50\%$	$\frac{1}{3} = 33\frac{1}{3}\%$	$\frac{1}{4} = 25\%$	$\frac{1}{5} = 20\%$	$\frac{1}{8} = 12\frac{1}{2}\%$	$\frac{1}{10} = 10\%$
	$\frac{2}{3} = 66\frac{2}{3}\%$	$\frac{3}{4} = 75\%$	$\frac{2}{5} = 40\%$	$\frac{3}{8} = 37\frac{1}{2}\%$	$\frac{3}{10} = 30\%$
			$\frac{3}{5} = 60\%$	$\frac{5}{8} = 62\frac{1}{2}\%$	$\frac{7}{10} = 70\%$
			$\frac{4}{5} = 80\%$	$\frac{7}{8} = 87\frac{1}{2}\%$	$\frac{9}{10} = 90\%$

In Example 9, you can practice the various methods that we have presented for converting a fraction to a percent.

Example 9

Write the given fraction as an exact percent.

(a) $\frac{13}{250}$ **(b)** $\frac{9}{8}$ **(c)** $\frac{9}{25}$ **(d)** $\frac{6}{7}$

SOLUTION

(a) $\frac{13}{250} = 0.052$

$= 5.2\%$

Your Turn 9

Write the given fraction as an exact percent.

(a) $\frac{36}{125}$

(b) $\frac{7}{4}$

(c) $\frac{11}{20}$

(d) $\frac{4}{15}$

Copyright © Houghton Mifflin Company. All rights reserved.

(b) $\frac{9}{8} = 1\frac{1}{8}$

$= 1 + \frac{1}{8}$

$= 100\% + 12\frac{1}{2}\%$ See the preceding table.

$= 112\frac{1}{2}\%$

(c) $\frac{9}{25} = \frac{9}{25} \cdot \frac{4}{4} = \frac{36}{100} = 36\%$

(d) The decimal name for $\frac{6}{7}$ is not a terminating decimal, so we use long division in order to obtain an exact answer.

$$\begin{array}{r} 0.85 \\ 7\overline{)6.00} \\ \underline{5.6} \\ 40 \\ \underline{35} \\ 5 \end{array}$$

Therefore,

$$\frac{6}{7} = 0.85\frac{5}{7} = 85\frac{5}{7}\%$$

Answers: **(a)** 28.8%; **(b)** 175%; **(c)** 55%; **(d)** $26\frac{2}{3}\%$

Example 10

In 2002, 62% of all personal computers sold cost less than $1,000. Write this percent (a) as a decimal and (b) as a simplified fraction.

SOLUTION

(a) $62\% = 0.62$

(b) $62\% = \frac{62}{100} = \frac{\cancel{2} \cdot 31}{\cancel{2} \cdot 50} = \frac{31}{50}$

Your Turn 10

A person's total monthly credit card charges are 38% of his monthly earnings. Write this percent

(a) as a decimal and

(b) as a simplified fraction.

Answers: **(a)** 0.38; **(b)** $\frac{19}{50}$

Copyright © Houghton Mifflin Company. All rights reserved.

Example 11

In a history class, 2 of every 5 students earned a B or better. What percentage of the class had an A or B for the course?

SOLUTION

The phrase "2 of every 5" translates to the ratio $\frac{2}{5}$.

$$\frac{2}{5} = 40\%$$ See the table of common conversions.

Therefore, 40% of the class had an A or B in the course.

Your Turn 11

A city census shows that 3 of every 8 people are under the age of 18. What percentage of the population of this town are under 18?

Answer: 37.5%

NOTE It is important to understand that knowing a percent does not mean knowing the actual number. In Example 11, we know that 40% of the students earned good grades, but we don't have enough information to know *how many* students earned those grades. We will take up that problem in the next section.

Suppose that a box contains 11 black marbles and 9 white marbles. Because there is a total of 20 marbles, the probability of randomly drawing a black marble from the box is $\frac{11}{20}$ or 55%. In general, the probability that an event will occur is the ratio of the number of ways in which the event *can* occur to the total number of possible outcomes. Of course, this ratio can be expressed as a percent.

Example 12

A box contains 16 red marbles, 14 blue marbles, and 20 green marbles.

(a) What is the probability that a blue marble will be drawn?

(b) What is the probability that either a red marble or a green marble will be drawn?

SOLUTION

First, we note that there are 16 + 14 + 20 = 50 ways in which a marble can be drawn.

(a) The event "blue marble drawn" can occur in 14 ways.

$$\frac{14}{50} = \frac{28}{100} = 28\%$$

There is a 28% probability that a blue marble will be drawn.

(b) The event "red marble drawn or green marble drawn" can occur in 16 + 20 = 36 ways.

$$\frac{36}{50} = \frac{72}{100} = 72\%$$

There is a 72% probability that either a red marble or a green marble will be drawn.

Your Turn 12

In Example 12,

(a) what is the probability that a green marble will be drawn?

(b) what is the probability that either a blue marble or a red marble will be drawn?

Answers: (a) 40% (b) 60%

Copyright © Houghton Mifflin Company. All rights reserved.

8.1 Quick Reference

A. The Meaning of Percents

1. A **percent** is a fraction (or ratio) in which the denominator is 100. The symbol for percent is %.

$$\frac{73}{100} = 73\%$$

2. If the denominator of a fraction is not 100, you may be able to rename the fraction so that the denominator is 100.

$$\frac{7}{25} = \frac{7}{25} \cdot \frac{4}{4} = \frac{28}{100} = 28\%$$

3. A percent can be written as a fraction by removing the % symbol and writing the remaining number as the numerator and 100 as the denominator.

$$29\% = \frac{29}{100}$$

4. After writing a percent as a fraction whose denominator is 100, you may be able to simplify the fraction.

$$52\% = \frac{52}{100} = \frac{4 \cdot 13}{4 \cdot 25} = \frac{13}{25}$$

B. Percents and Decimals

1. To convert from a percent to a decimal:

   (a) Drop the percent symbol.

   (b) Shift the decimal point 2 places to the left.

$$39\% = 0.39. = 0.39$$

2. To convert from a decimal to a percent:

   (a) Move the decimal point 2 places to the right.

   (b) Attach the percent symbol.

$$0.91 = 0.91.\% = 91\%$$

C. Writing Fractions as Percents

1. The general method for writing a fraction as a percent is to write the fraction as a decimal and then write the decimal as a percent.

$$\frac{7}{16} = 0.4375 = 43.75\%$$

2. If the decimal form of a fraction is a nonterminating decimal, you can round off the decimal to obtain an approximate percent.

$$\frac{5}{11} = 0.454545\cdots \approx 0.4545$$

$$0.4545 \rightarrow 45.45\%$$

Therefore, $\frac{5}{11} \approx 45.45\%$.

Copyright © Houghton Mifflin Company. All rights reserved.

3. If the decimal form is nonterminating, you can use long division to obtain an exact percent.

```
   0.45
11)5.00
   4 4
     60
     55
      5
```

Therefore, $\frac{5}{11} = 45\frac{5}{11}\%$.

Copyright © Houghton Mifflin Company. All rights reserved.

8.1 Exercises

A. The Meaning of Percents

1. A ratio in which the denominator is 100 is called a(n) ________.

2. To express a percent such as 43% as a fraction, write 43 as the ________ and 100 as the ________.

In Exercises 3–6, write the given fraction as a percent.

3. $\frac{36}{100}$ 4. $\frac{1}{100}$ 5. $\frac{140}{100}$ 6. $\frac{300}{100}$

In Exercises 7–10, write the given fraction as a percent.

7. $\frac{3}{10}$ 8. $\frac{6}{25}$ 9. $\frac{12}{5}$ 10. $\frac{21}{20}$

In Exercises 11–14, write the given percent as a fraction.

11. 79% 12. 51% 13. 147% 14. 209%

In Exercises 15–22, write the given percent as a simplified fraction.

15. 40% 16. 75% 17. 8% 18. 2%

19. 150% 20. 225% 21. $66\frac{2}{3}\%$ 22. 62.5%

B. Percents and Decimals

23. To perform arithmetic with percents, we usually write the percent as a(n) ________.

24. To write a percent as decimal, drop the ________ symbol and shift the decimal point 2 places to the ________.

In Exercises 25–34, write the given percent as a decimal.

25. 85% 26. 67% 27. 40% 28. 70%

29. 8% 30. 3% 31. 285% 32. 450%

33. 18.5% 34. 0.7%

In Exercises 35–46, write the given decimal as a percent.

35. 0.25 36. 0.42 37. 0.12 38. 0.87

39. 0.07 40. 0.05 41. 1.25 42. 2.4

43. 6.1 44. 0.6 45. 0.2 46. 0.005

Copyright © Houghton Mifflin Company. All rights reserved.

C. Writing Fractions as Percents

47. To write a fraction such as $\frac{7}{20}$ as a percent, rewrite the fraction so that the ________ is 100.

48. One method for writing a fraction as a percent is begun by writing the fraction as a(n) ________.

In Exercises 49–60, write the given fraction as an exact percent.

49. $\frac{5}{16}$ **50.** $\frac{3}{40}$ **51.** $\frac{1}{6}$

52. $\frac{5}{12}$ **53.** $\frac{1}{12}$ **54.** $\frac{8}{15}$

55. $\frac{6}{125}$ **56.** $\frac{7}{5}$ **57.** $\frac{14}{3}$

58. $\frac{11}{6}$ **59.** $\frac{27}{80}$ **60.** $\frac{29}{60}$

In Exercises 61–64, write the given fraction as a percent. Round to the nearest tenth of a percent.

61. $\frac{3}{7}$ **62.** $\frac{8}{9}$ **63.** $\frac{5}{14}$ **64.** $\frac{5}{11}$

In Exercises 65–68, write the given percent (a) as a decimal and (b) as a simplified fraction.

65. *Red Lights* According to the Federal Highway Administration, 26% of the people surveyed say that they see drivers go through red lights every day.

66. *Gas Grills* According to *Weber Grill Watch,* 58% of the grills sold in the United States are gas grills.

Copyright © Houghton Mifflin Company. All rights reserved.

67. *Air Travel Safety* The Marist Institute found that 45% of the people in the United States say that an airplane is the safest mode of transportation.

68. *Professional Football* The Harris Poll found that 28% of those surveyed said that professional football is their favorite sport.

69. Complete the table by writing the equivalent fractions, decimals, and percents.

Fraction	$\frac{1}{8}$						$\frac{7}{8}$	
Decimal			0.375			0.75		1
Percent		25%		50%	62.5%			

70. Complete the table by writing the equivalent fractions, decimals, and percents.

Fraction	$\frac{1}{10}$		$\frac{3}{10}$					$\frac{4}{5}$		1
Decimal		0.2			0.5				0.9	1
Percent				40%		60%	70%			

71. *Traffic Surveillance* The Federal Highway Administration reported that $\frac{7}{10}$ of the people surveyed say that they support the use of automated cameras to identify and ticket drivers who drive through red lights. What percent of those surveyed favor the use of cameras?

72. *Television Viewing* *USA Today* reported that $\frac{1}{5}$ of the people in the United States say that watching television is their favorite leisure activity. What percent of the population does this represent?

73. *Paper Recycling* *USA Today* reported that 2 of every 5 pounds of residential paper and paperboard are recycled. What percent of paper and paperboard is recycled?

Copyright © Houghton Mifflin Company. All rights reserved.

74. *Student Earnings* According to *Campus Concepts,* 1 of every 4 college students earns between \$200 and \$300 per month from a part-time job. What percent of students are in this category?

Calculator Exercises

In Exercises 75 and 76, write the fraction as a percent rounded to the nearest tenth.

75. $\frac{120}{347}$

76. $\frac{52}{21}$

Writing and Concept Extension

77. For which of the following fractions would writing the fraction with a denominator of 100 be convenient? Why?

(i) $\frac{3}{25}$ (ii) $\frac{3}{31}$ (iii) $\frac{3}{50}$ (iv) $\frac{3}{70}$

78. Identify the number that is not equivalent to the other two.

(a) $\frac{1}{4}\%$ 0.25 $\frac{0.25}{100}$

(b) $1\frac{1}{2}\%$ .015 $\frac{0.15}{100}$

(c) 175% 17.5 $1\frac{3}{4}$

79. *Population of Florida* The population of Florida is approximately 16 million, and the population of the United States is approximately 287 million. Write the ratio of the population of Florida to that of the United States. Then write the fraction as a percent.

80. *Television Viewing Time* The typical family in the United States watches television approximately 8 hours each day. Write the ratio of the number of hours of television viewing to the number of hours in a day. Then write the fraction as a percent.

Copyright © Houghton Mifflin Company. All rights reserved.

81. *Calories from Fat* A $\frac{1}{4}$-cup serving of mixed nuts contains a total of 220 calories, of which 160 calories are from fat. Write the ratio of calories from fat to the total number of calories. Then write the fraction as a percent.

82. *Reading Rate* Suppose that a book is 540 pages long. If you read 30 pages each weekday, 50 pages on Saturday, and 50 pages on Sunday, what percent of the book will you have completed at the end of the 7 days?

Exploring with Real-World Data: Collaborative Activities

Life in the Twenty-first Century A poll examined what Americans expect will happen in the twenty-first century. The pie graphs show the results of the poll. (Source: Pew Research Center.)

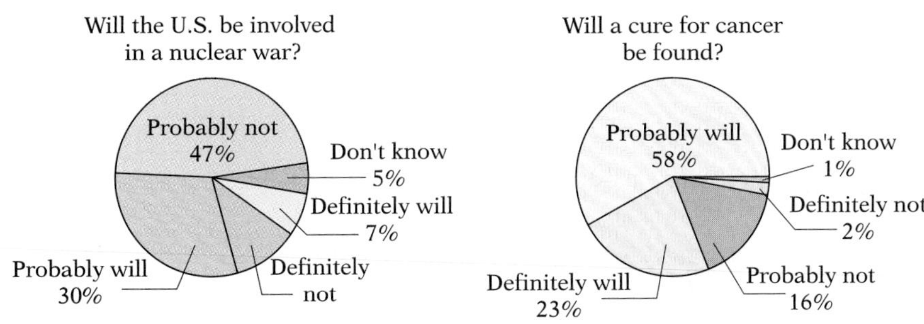

83. What *fraction* of the people believe that the United States will probably be involved in a nuclear war?

84. What percent of the people believe that the United States will definitely not be involved in a nuclear war?

85. What is the opinion of approximately $\frac{1}{2}$ of those surveyed about the possibility of a nuclear war?

Copyright © Houghton Mifflin Company. All rights reserved.

86. What percent of the people believe that the United States is at least unlikely to be involved in a nuclear war?

87. What is the opinion of approximately $\frac{1}{4}$ of those surveyed about the possibility of a cure for cancer?

88. What percent of the people believe that a cure for cancer is at least likely to be found?

89. What *fraction* of the people believe that a cure for cancer will probably not be found?

90. On the basis of the responses to the two events, would you classify Americans as optimistic or pessimistic about life in the twenty-first century.

Copyright © Houghton Mifflin Company. All rights reserved.

8.2 Solving Percent Problems

SUGGESTIONS FOR SUCCESS

A small word can have an important meaning.

A percent is often followed by the word "of." For example, you might read that 20% OF the taxpayers pay 80% OF the taxes collected. In mathematics, the word "of" means "times." Therefore, the way that you find a percent of some number is to multiply.

Knowing this simple fact is the key to your success with percent problems. Whenever you see the word "of," translate it to "times." This will help you to write and solve the basic percent equation that we present in this section.

A The Basic Percent Equation

The statement "50% of 20 is 10" includes three numbers: 50%, 20, and 10. In most percent problems, two of the numbers are known and the other number is unknown.

In this section, we will learn how to translate such statements into equations. Then we will discuss methods for solving percent equations when one of the three numbers is unknown and represented with a variable.

The Basic Percent Equation

Consider the following true sentence:

50% of 20 is 10.

Recall that we translate the word *of* as "times," and we translate the word *is* as "equals."

50% of 20 is 10.
↓ ↓
times equals
↓ ↓
$50\% \cdot 20 = 10$

We give names to the three numbers in this equation.

1. 50% is called the *percent* (P).
2. 20 is called the *base* (B).
3. 10 is called the *amount* (A).

Note that the *base* is the number that follows the word *of* (or the times symbol). The *amount* is the number that follows the word *is* (or the equals symbol).

Now we can write a formula that relates these words and symbols.

Percent · Base = Amount
$P \cdot B = A$

In a typical percent problem, one of the three numbers P, B, or A is unknown. The word *what* will indicate which number is unknown, and we will represent that number with x.

1. 30% of 60 is *what number?* $P = 30\%$, $B = 60$, and A is unknown.

$$30\% \cdot 60 = x$$

Copyright © Houghton Mifflin Company. All rights reserved.

2. 10% of *what number* is 6? $P = 10\%$, B is unknown, and $A = 6$.

$$10\% \cdot x = 6$$

3. *What percent* of 50 is 20? P is unknown, $B = 50$, and $A = 20$.

$$x \cdot 50 = 20$$

NOTE When the percent is unknown, we let x represent the percent expressed as a decimal number.

Your ability to translate the given information into an equation is crucial in solving percent problems.

Example 1

Letting x represent the unknown number, translate each of the following into a percent equation:

(a) 30% of what number is 15?

(b) What percent of 80 is 40?

(c) What is 12% of 70?

SOLUTION

We use the formula $P \cdot B = A$.

(a) $P \cdot B = A$ The percent and the amount are known.

$\downarrow \quad \downarrow \quad \downarrow$

$30\% \cdot x = 15$ The base is unknown.

(b) $P \cdot B = A$ The base and the amount are known.

$\downarrow \quad \downarrow \quad \downarrow$

$x \cdot 80 = 40$ The percent is unknown.

(c) $P \cdot B = A$ The percent and the base are known.

$\downarrow \quad \downarrow \quad \downarrow$

$12\% \cdot 70 = x$ The amount is unknown.

Your Turn 1

Letting x represent the unknown number, translate each of the following into a percent equation:

(a) What percent of 18 is 5?

(b) What is 26% of 90?

(c) 60% of what number is 112?

Answers: **(a)** $x \cdot 18 = 5$; **(b)** $26\% \cdot 90 = x$; **(c)** $60\% \cdot x = 112$

Example 1 shows that percent problems have three varieties, depending on the quantity that is unknown. However, all three varieties have the same form: $P \cdot B = A$.

B Solving Percent Equations

When we solve a percent equation, we need to change the percent to a fraction or a decimal. For familiar percents, you may find that converting the percent to a fraction is the easier method. (Refer to the table of common conversions in Section 8.1.) However, converting the percent to a decimal is generally the better approach.

Copyright © Houghton Mifflin Company. All rights reserved.

Here again are the three equations in Example 1, this time written in decimal form.

(a) $30\% \cdot x = 15 \rightarrow 0.30x = 15$
(b) $x \cdot 80 = 40 \quad \rightarrow 80x = 40$
(c) $12\% \cdot 70 = x \rightarrow 0.12(70) = x$

In part (c), we can solve for x simply by multiplying 0.12(70).

$$x = 0.12(70) = 8.40$$

NOTE In our examples and exercises, we will routinely round off percents and decimal numbers to the nearest hundredth.

In parts (a) and (b), we use the Multiplication Property of Equations to isolate the variable.

(a) $0.30x = 15$

$$\frac{\cancel{0.30}x}{\cancel{0.30}} = \frac{15}{0.30}$$

$$x = 50$$

(b) $80x = 40$

$$\frac{\cancel{80}x}{\cancel{80}} = \frac{40}{80}$$

$$x = \frac{1}{2} = 50\%$$

The following examples give you some practice with solving percent equations. Begin by identifying the percent (P), the base (B), and the amount (A). Then, using the formula $P \cdot B = A$, assign the variable to the quantity that is unknown. Be sure to write the percent as a decimal or a fraction.

Example 2

What is 63% of 120?

SOLUTION

$$\begin{array}{ccccc} P & \cdot & B & = & A \\ \downarrow & & \downarrow & & \downarrow \\ 0.63 & \cdot & 120 & = & x \end{array}$$

$$75.6 = x$$

63% of 120 is 75.6.

Your Turn 2

What is 18% of 90?

Answer: 16.2

Example 3

40% of what number is 98?

SOLUTION

$$\begin{array}{ccccc} P & \cdot & B & = & A \\ \downarrow & & \downarrow & & \downarrow \\ 0.40 & \cdot & x & = & 98 \end{array}$$

To isolate x, we divide both sides by 0.40.

$$\frac{\cancel{0.40}x}{\cancel{0.40}} = \frac{98}{0.40}$$

$$x = 245$$

40% of 245 is 98.

Your Turn 3

75% of what number is 9? (Work the problem with a decimal and with a fraction.)

continued

Copyright © Houghton Mifflin Company. All rights reserved.

Because 40% is $\frac{2}{5}$, we also can solve the problem with the fraction.

$$\frac{2}{5}x = 98$$

$$\frac{\cancel{5}}{\cancel{2}} \cdot \frac{\cancel{2}}{\cancel{5}}x = \frac{5}{\cancel{2}} \cdot \frac{\cancel{2} \cdot 49}{1}$$

$$x = 5 \cdot 49 = 245$$

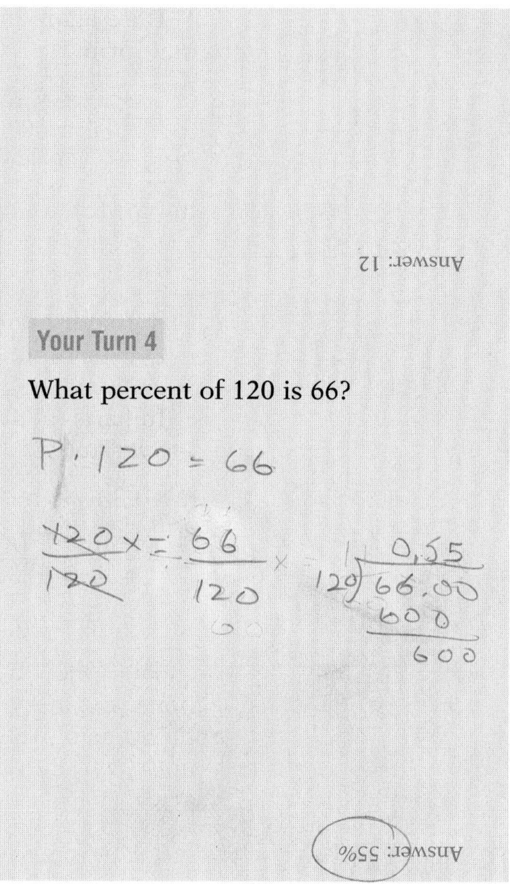

Answer: 12

Example 4

What percent of 116 is 145?

SOLUTION

$$\begin{array}{ccccc} P & \cdot & B & = & A \\ \downarrow & & \downarrow & & \downarrow \\ x & \cdot & 116 & = & 145 \end{array}$$

$$116x = 145$$

To isolate x, we divide both sides by 116.

$$\frac{\cancel{116}x}{\cancel{116}} = \frac{145}{116}$$

$$x = 1.25 = 125\%$$

125% of 116 is 145.

Your Turn 4

What percent of 120 is 66?

Answer: 55%

C Estimation

In their everyday use, percents are often used informally and sometimes do not need to be exact. For example, the phrase "1 of every 3" might be reported as 33%, even though the exact percent is $33\frac{1}{3}\%$. A 45% result of a statistical survey might be reported as accurate to ±5%, which means that the actual percent is between 45% − 5% = 40% and 45% + 5% = 50%.

Estimating percents can also be useful in making sure that your answers are reasonable. For example, to find 49% of 80, we can think of 49% as close to 50% or $\frac{1}{2}$. Because $\frac{1}{2}$ of 80 is 40, we know that 49% of 80 should be close to 40. Such checks can help you avoid errors that often occur in percent problems.

≈ **Example 5**

Approximate each of the following:

(a) What is 35% of 90?

(b) 8 is what percent of 17?

SOLUTION

(a) $\begin{array}{ccccc} P & \cdot & B & = & A \\ \downarrow & & \downarrow & & \downarrow \\ 0.35 & \cdot & 90 & = & x \end{array}$

35% is close to $33\frac{1}{3}\%$, or $\frac{1}{3}$. Because $\frac{1}{3}$ of 90 is 30, 35% of 90 is close to 30.

Your Turn 5

Approximate each of the following:

(a) What is 26% of 80?

(b) 14 is what percent of 30?

Copyright © Houghton Mifflin Company. All rights reserved.

(b) $P \cdot B = A$

$\downarrow \quad \downarrow \quad \downarrow$

$x \cdot 17 = 8$

$x = \frac{8}{17}$ Divide both sides by 17.

$\frac{8}{17}$ is close to $\frac{8}{16}$ or $\frac{1}{2}$.
Because $\frac{1}{2} = 50\%$, $P \approx 50\%$.
Therefore, 8 is about 50% of 17.

Answers: (a) 20; (b) 50%

Using the fact that $1\% = \frac{1}{100}$, or 0.01, can also be helpful in finding percents, estimating, or checking answers. To take 1% of a number, we simply move the decimal point 2 places to the left: 1% of 240 = 2.40.

Example 6

What is 3% of 500?

SOLUTION

$$1\% \text{ of } 500 = 5.00 \text{ or } 5$$

Therefore, 3% of 500 is $3 \cdot 5 = 15$.

Your Turn 6

What is 2% of 700?

Answer: 14

≈ **Example 7**

Approximately what is 5% of $29.95?

SOLUTION

$29.95 is close to $30.00.
1% of $30.00 = $0.30.
5% of $30.00 is 5 · $0.30 = $1.50.
Therefore, 5% of $29.95 is approximately $1.50.

Your Turn 7

Approximately what is 3% of $80.10?

Answer: $2.40

≈ **Example 8**

In order to purchase a home that costs $125,000, a couple is required to make a 19.8% down payment.

(a) Estimate the down payment.

(b) Determine the exact down payment.

SOLUTION

(a) 19.8% is close to 20%, or $\frac{1}{5}$.
$\frac{1}{5}$ of 125,000 is $\frac{125{,}000}{5} = 25{,}000$.
The estimated down payment is $25,000.

(b) $P \cdot B = A$

$\downarrow \quad \downarrow \quad \downarrow$

$0.198 \cdot 125{,}000 = x$

$24{,}750 = x$

The exact down payment is $24,750.

Your Turn 8

An auto dealer requires a 10.2% down payment on a new car that costs $26,000.

(a) Estimate the down payment.

(b) Determine the exact down payment.

Answers: (a) $2,600; (b) $2,652

Copyright © Houghton Mifflin Company. All rights reserved.

8.2 Quick Reference

A. The Basic Percent Equation

1. The basic percent equation is

$$\begin{array}{ccccc} \text{Percent} & \cdot & \text{Base} & = & \text{Amount} \\ \downarrow & & \downarrow & & \downarrow \\ P & \cdot & B & = & A \end{array}$$

$$\begin{array}{ccccc} 30\% & \text{of} & 60 & \text{is} & 18. \\ \downarrow & & \downarrow & & \downarrow \\ P & \cdot & B & = & A \end{array}$$

B. Solving Percent Equations

1. In a typical percent problem, you will be given two of the quantities P, B, and A, and the third quantity is unknown.

- What is 52% of 90?

$$P = 52\%,\ B = 90,\ A = ?$$
$$0.52 \cdot 90 = A$$
$$46.8 = A$$

- 30% of what number is 24?

$$P = 30\%,\ B = ?,\ A = 24$$
$$0.30 \cdot B = 24$$
$$B = \frac{24}{0.30} = 80$$

- What percent of 120 is 15?

$$P = ?,\ B = 120,\ A = 15$$
$$P \cdot 120 = 15$$
$$P = \frac{15}{120} = 0.125$$
$$P = 12.5\%$$

C. Estimation

1. We can estimate the answer to a percent problem by writing a percent as a fraction that is close in value or by writing a fraction as a percent that is close in value.

- What is 24% of 80?

24% is close to 25%, or $\frac{1}{4}$. Therefore, 24% of 80 is close to $\frac{1}{4}$ of 80, which is 20.

- 5 is what percent of 19?

$$P \cdot 19 = 5$$
$$P = \frac{5}{19} \approx \frac{5}{20} = \frac{1}{4}$$

$\frac{1}{4} = 25\%$, so 5 is about 25% of 19.

2. To take 1% of a number, we move the decimal point 2 places to the left. We can use this fact to estimate answers to percent problems.

What is 3% of 1,200?

1% of 1,200 is 12. So 3% of 1,200 is $3 \cdot 12 = 36$.

Copyright © Houghton Mifflin Company. All rights reserved.

8.2 Exercises

A. The Basic Percent Equation

1. In the statement "20% of 15 is 3", we call 15 the ________ and 3 the ________.

2. The basic percent equation $P \cdot B = A$ relates the ________, the ________, and the ________.

In Exercises 3–8, translate the statement into a percent equation. Let x represent the unknown number.

3. What number is 12% of 32?

4. Find 24% of 60.

5. 6 is what percent of 15?

6. What percent of 18 is 30?

7. 30 is 140% of what number?

8. 70% of what number is 21?

B. Solving Percent Equations

9. To solve a percent problem, we usually change the percent to a(n) ________ or a(n) ________.

10. To solve equations such as $0.4x = 24$ or $32x = 6$, we use the ________ Property of Equations.

In Exercises 11–16, fill in the blank.

11. 10% of 30 is ____.

12. 25% of 120 is ____.

13. 32% of 50 is ____.

14. 40% of 55 is ____.

15. 5% of 18 is ____.

16. 110% of 20 is ____.

In Exercises 17–22, determine the indicated number.

17. Find 44% of 75.

18. 50% of 44 is what number?

19. 20% of 35 is what number?

20. What number is 60% of 10?

21. What number is 150% of 30?

22. Find 2% of 24.

Copyright © Houghton Mifflin Company. All rights reserved.

In Exercises 23–28, fill in the blank.

23. ____% of 24 is 6.

24. ____% of 25 is 10.

25. ____% of 15 is 10.

26. ____% of 54 is 43.2.

27. ____% of 35 is 42.

28. ____% of 140 is 0.7.

In Exercises 29–34, determine the indicated percent.

29. What percent of 80 is 8?

30. 6.4 is what percent of 32?

31. What percent of 60 is 144?

32. 117 is what percent of 90?

33. 77.5 is what percent of 620?

34. What percent of 80 is 0.8?

In Exercises 35–40, fill in the blank.

35. 50% of ____ is 41.

36. 40% of ____ is 52.

37. 60% of ____ is 120.

38. 15% of ____ is 12.

39. 140% of ____ is 21.

40. 6.5% of ____ is 13.

In Exercises 41–46, determine the indicated number.

41. 20% of what number is 16?

42. 12 is 25% of what number?

43. 6 is 8% of what number?

44. 90% of what number is 81?

45. 45 is 125% of what number?

46. $2\frac{1}{2}$% of what number is 3.25?

≈ C. Estimation

47. Finding 1% or 10% of a number simply involves moving the ________ to the left.

Copyright © Houghton Mifflin Company. All rights reserved.

48. To check your answer when you find 24% of 60, you might ________ the answer by finding $\frac{1}{4}$ of 60.

In Exercises 49–52, choose the percent that is the best estimate of the given fraction.

49. $\frac{21}{30}$ **(A)** 50%

50. $\frac{23}{50}$ **(B)** 25%

51. $\frac{1}{9}$ **(D)** 66%

52. $\frac{9}{40}$ **(C)** 10%

In Exercises 53–56, estimate the amount.

53. 65% of 1,200

54. 23% of 360

55. 52% of 700

56. 77% of 1,600

In Exercises 57–60, estimate the percent.

57. 19 is what percent of 36?

58. What percent of 76 is 26?

59. What percent of 119 is 29?

60. 182 is what percent of 201?

In Exercises 61–68, determine the amount mentally.

61. 1% of 1,500

62. 1% of 940

63. 10% of 360

64. 10% of 63

65. 2% of 300

66. 6% of 90

67. 70% of 60

68. 30% of 800

Calculator Exercises

In Exercises 69–72, use a calculator to determine the indicated number.

69. What percent of 185 is 78?

70. Find 37% of 485.

71. 18% of what number is 1,280?

72. 75 is what percent of 160?

Copyright © Houghton Mifflin Company. All rights reserved.

Writing and Concept Extension

73. Suppose that you need to calculate $66\frac{2}{3}\%$ of 600. How should you express the percent to make the calculation easy?

74. Describe an easy way to calculate a 15% tip on a bill of $18.00.

75. Describe two ways to find 20% of a number.

76. Explain how to find:

(a) 10% of a number

(b) 1% of a number

In Exercises 77 and 78, determine the amount.

77. 50% of 50% of 140

78. 30% of 50% of 80

Copyright © Houghton Mifflin Company. All rights reserved.

8.3 Percent: General Applications

SUGGESTIONS FOR SUCCESS

Have you ever wondered, "When will I ever use this math?"

Studying percents gives you an excellent opportunity to answer that question. As you look at newspapers, news or business magazines, and television news reports, make a note of percentages that are mentioned. Can you interpret what they mean in the context of the report?

As you become skilled with percents, you will see how very useful this math is.

In 2002, there were 39,900 girls ages 11 and 12 who played organized soccer. This number of girls was 40% of all soccer players in that age group. For the 17 and 18 age group, the number of girls playing soccer was only 4,430. Interestingly, that number of girls still represented 38% of all soccer players in that age group. (Source: American Youth Soccer Organization.)

These statistics give us a percent (P) and an amount (A). This information allows us to calculate the base (B). In general, to use the general percent equation $P \cdot B = A$, we must know two of the three quantities.

The following examples are a sample of the many kinds of application problems that we encounter in everyday life:

Example 1

In a rural school of 950 students, 82% of the students ride a school bus. How many students ride the bus?

SOLUTION

Let x represent the number of students who ride the school bus.

82% of (total students) are (bus riders)

$$0.82 \cdot 950 = x$$

$$0.82(950) = x$$

$$779 = x$$

779 students ride the school bus.

Your Turn 1

If a driver has completed 58% of a 750-mile trip, how many miles have been driven?

Answer: 435

There are two important observations to be made in the solution of Example 1 and in all the other examples of this section. First, we write a very detailed description of what we are letting the variable represent. If we had written something like, "Let x = students," we would have created confusion from the beginning. All students? Students who ride a bus? Being clear on what the variable represents is essential.

Also, note the use of a "word equation" as an aid to writing the actual equation. A word equation is an intermediate step that gives you the opportunity to state the given information clearly. Then you can write the actual equation simply by replacing the words with numbers and symbols. This is an excellent technique for translating the given information into an equation that you can solve.

Copyright © Houghton Mifflin Company. All rights reserved.

Example 2

In a 10-kilometer charity run, 864 of the 1,200 entrants completed the run. What percent of the entrants crossed the finish line?

SOLUTION

Let x represent the percent (in decimal form) of the entrants who finished the run.

(Some %) of (entrants) were (finishers)

$$x \cdot 1{,}200 = 864$$

$$1{,}200x = 864$$

$$\frac{1{,}200x}{1{,}200} = \frac{864}{1{,}200} \quad \text{Divide both sides by 1,200.}$$

$$x = 0.72$$

72% of the entrants finished the run.

Your Turn 2

A person has read the first 234 pages of a 600-page book. What percent of the book has the person read?

Answer: 39%

Example 3

A bottle contains an acid and water solution, of which 12% is acid. If there are 9.6 cubic centimeters of acid in the bottle, what is the total amount of solution?

SOLUTION

Let x represent the total amount of solution in the bottle.

12% of (total solution) is (amount of acid)

$$0.12 \cdot x = 9.6$$

$$0.12x = 9.6$$

$$\frac{0.12x}{0.12} = \frac{9.6}{0.12} \quad \text{Divide both sides by 0.12.}$$

$$x = 80$$

There are 80 cubic centimeters of solution.

Your Turn 3

In a commercial area, 54 acres are planned for parking. How many acres are in the development if parking takes up 18% of the land?

Answer: 300

For Example 4 and Your Turn 4, refer to the soccer statistics given at the beginning of this section.

Example 4

What was the total number of boys and girls ages 11 and 12 who played soccer?

Your Turn 4

What was the total number of boys and girls ages 17 and 18 who played soccer?

Copyright © Houghton Mifflin Company. All rights reserved.

SOLUTION

From the beginning of the section, we know that 39,900 girls in the 11–12 age group represent 40% of all soccer players in that age group.

Let x represent the total number of soccer players ages 11 and 12.

$$\begin{array}{ccccc} 40\% & \text{of} & \text{(total players)} & \text{were} & \text{(girls)} \\ \downarrow & \downarrow & \downarrow & \downarrow & \downarrow \\ 0.40 & \cdot & x & = & 39{,}900 \end{array}$$

$$0.40x = 39{,}900$$

$$\frac{0.40x}{0.40} = \frac{39{,}900}{0.40} \quad \text{Divide both sides by 0.40.}$$

$$x = 99{,}750$$

The total number of players ages 11 and 12 was 99,750.

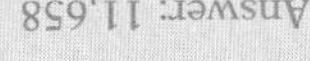

Example 5

The population of the United States is projected to be 337.5 million in the year 2030. In that year, 8.1 million people are projected to be 85 or older. (Source: U.S. Bureau of the Census.) What percent of the population in 2030 is expected to be 85 or older?

SOLUTION

Let x represent the percent (in decimal form) of the population that is expected to be 85 or older.

$$\begin{array}{ccccc} \text{(Some \%)} & \text{of} & \text{(total population)} & = & \text{(elderly population)} \\ \downarrow & \downarrow & \downarrow & \downarrow & \downarrow \\ x & \cdot & 337.5 & = & 8.1 \end{array}$$

$$337.5x = 8.1$$

$$\frac{337.5x}{337.5} = \frac{8.1}{337.5} \quad \text{Divide both sides by 337.5.}$$

$$x \approx 0.024$$

In 2030, 2.4% of the population is expected to be 85 or older.

Your Turn 5

The population of the United States is projected to be 391.3 million in the year 2050, and 18 million are projected to be 85 or older. What percent of the population is expected to be 85 or older?

Answer: 4.6%

Example 6

Of the 3,000 Internet service providers (ISP) in business in 1999, only 15% were expected to remain in the year 2002. (Source: Gartner Group.) How many ISPs were expected to be in business in 2002?

Your Turn 6

The number of ISPs in 1997 was 150% of the 3,000 ISPs in 1999. How many ISPs were there in 1997?

continued

Copyright © Houghton Mifflin Company. All rights reserved.

SOLUTION

Let x represent the number of ISPs expected to be in business in 2002.

$$\begin{array}{ccccc} 15\% & \text{of} & (\text{ISPs in 1999}) & = & (\text{ISPs in 2002}) \\ \downarrow & \downarrow & \downarrow & \downarrow & \downarrow \\ 0.15 & \cdot & 3{,}000 & = & x \end{array}$$

$$0.15(3{,}000) = x$$

$$450 = x$$

In 2002, only 450 Internet service providers were expected to be in business.

Copyright © Houghton Mifflin Company. All rights reserved.

8.3 Exercises

1. *Batting Average* A baseball player had 70 hits in 240 at-bats. What percent of the at-bats resulted in a hit?

2. *Test Score* A student correctly answered 32 of 36 test questions. What percent of the answers were correct?

3. *United Way Donations* The local United Way campaign announced that donations had reached $12,750, which was 85% of the goal for the campaign. What was the goal?

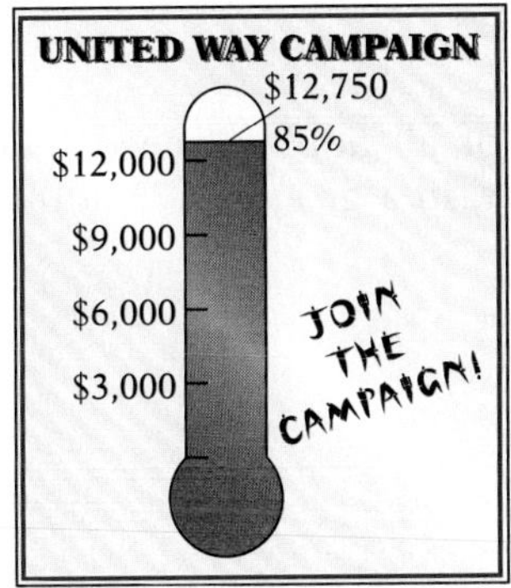

4. *Dog's Weight* At age 6 months, a dog weighed 59 pounds, which was 84% of its predicted adult weight. What was the predicted adult weight?

5. *Seasoning Mix* A seasoning mix contains 65% salt. How many ounces of salt are in 12 ounces of the mix?

Copyright © Houghton Mifflin Company. All rights reserved.

6. *Insecticide Solution* A solution contains 15% insecticide. How many liters of insecticide are there in 8 liters of the solution?

7. *Commuter Survey* A survey of 320 people found that 176 drove more than 10 miles to work. What percent of those surveyed drove more than 10 miles?

8. *Fitness Program* Of 80 people who entered a fitness program, 56 completed the program successfully. What percent of those who signed up stayed with the program to the end?

9. *Overweight Adults* *USA Today* reported that 76% of all adults exceed their recommended weight. On the basis of this report, in a group of 75 adults, how many do you expect to be overweight?

10. *Vacation Photos* According to Globus Travel, 40% of travelers rarely or never organize vacation photos. In a tour group of 65 people, how many do you expect *will* organize their vacation photos?

11. *Voter Turnout* For a general election, 702 people voted in a certain precinct. If the number represents a 36% turnout, how many voters are registered in the precinct?

Copyright © Houghton Mifflin Company. All rights reserved.

12. *Girl Scout Cookie Sales* Thin mints account for 26% of all Girl Scout cookie sales. (Source: Girl Scouts USA.) If a troop sold 91 boxes of thin mints, what was the total number of boxes of cookies that you would expect the troop to have sold?

13. *Team Wins* A baseball team won 5 of every 8 games played. What percent of the games did the team win?

14. *Registered Voters* A newspaper reported that 3 of every 5 registered voters in a certain district are Republicans. What percent of the registered voters are Republicans?

15. *Calorie Content* The number of calories in a certain brand of light ice cream is 60% of the number of calories in regular ice cream. If a serving of light ice cream contains 252 calories, how many calories are in a serving of regular ice cream?

16. *Youth Sports Donations* A community organization reports that 95% of the donations that it receives go to support youth sports. How much did the organization receive in donations if $8,493 was spent for youth sports?

17. *Graduation Rate* A study at a certain college found that 62% of the first-year students graduate within 5 years. If the college has 1,250 first-year students, how many are predicted to graduate within 5 years?

Copyright © Houghton Mifflin Company. All rights reserved.

18. *Land Allocation* A developer has 850 acres of land, of which 74% are designated for residential use and the remainder for commercial use. How many acres are designated for commercial use?

19. *Rent Expense* Each month a couple spends $656 of their $3,280 income for rent. What percent of the income is spent for rent?

20. *Company Employees* A small manufacturing company employs 78 workers and 12 managers. What percent of the employees are managers?

21. *Restaurant Bill* Suppose that a bill at a restaurant is $36.40. What is the total cost of the meal, including 15% for the tip?

22. *Annual Salary* A person whose annual salary is $35,320 receives a 4% raise. What is the annual salary after the raise?

23. *Vitamin Supplement* The label on a vitamin and mineral supplement states that the supplement contains 450 milligrams of calcium, which provides 30% of the recommended daily requirement of calcium. How many milligrams of calcium are recommended each day?

Supplement Facts
Serving Size 1 Tablet

Each Tablet Contains	% Daily Value
Calcium 450 mg	30%
Magnesium 40 mg	10%
Zinc 7.5 mg	50%
Copper 1 mg	50%
Manganese 1.8 mg	90%
Vitamin D 200 I.U.	50%

Copyright © Houghton Mifflin Company. All rights reserved.

24. *Food Expenditure* According to the American Farm Bureau, an average family spends 11% of its income on food. Assuming that the Farm Bureau estimate applies, if a family spends \$550 each month for food, what would be the family's annual income?

25. *Defective Calculators* In a shipment of calculators, $1\frac{1}{2}\%$ are defective. How many defective calculators would be expected in a shipment of 800 calculators?

26. *House Payment* A family spends 22% of its monthly income of \$2,640 for a house payment. What is the family's monthly house payment?

27. *Juice Drink* A fruit-flavored drink contains 12% pure fruit juice. How many quarts of the beverage can be made from 24 ounces of pure fruit juice? (Recall that 1 quart = 32 ounces.)

28. *Test Questions* A student answered 82% of all test questions correctly. If the student answered 9 questions incorrectly, how many questions were on the test?

Copyright © Houghton Mifflin Company. All rights reserved.

29. *Hockey Game Cost* The average cost for a family of four to attend a New York Rangers hockey game is \$347.71. The average cost for the National Hockey League is \$238.87. (Source: National Hockey League.) What percent of the average cost is the cost for the Rangers game?

30. *House of Representatives* In 2003, the U.S. House of Representatives had 221 Republicans, 212 Democrats, and 2 Independents. What percent of the members were Republicans?

31. *Fat Content of Milk* If milk contains 2% fat, how many ounces of fat are in $\frac{1}{2}$ gallon of milk? (1 gallon contains 128 ounces.)

32. *Employee Stock Plan* An employee contributes 6% of her salary to the Employee Stock Option Plan. If she made \$2,800 one month and the stock sold for \$35 per share, how many shares did she buy that month?

Writing and Concept Extension

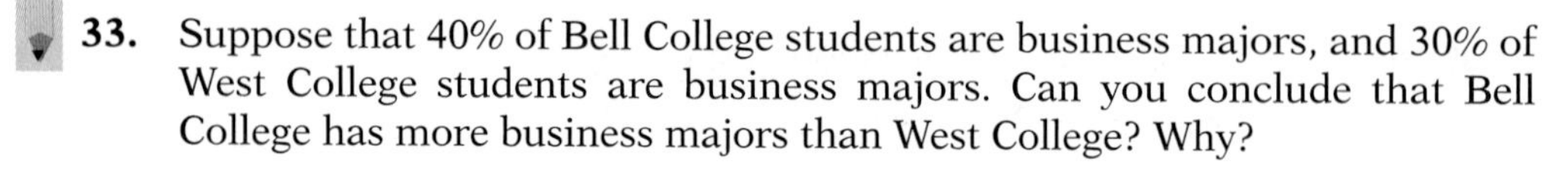

33. Suppose that 40% of Bell College students are business majors, and 30% of West College students are business majors. Can you conclude that Bell College has more business majors than West College? Why?

Copyright © Houghton Mifflin Company. All rights reserved.

34. Suppose that a company employs 340 women. Can you determine what percent of the employees are women? If not, what additional information do you need?

35. Suppose that an advertisement claims that 80% of all consumers who tried a new pain reliever thought that it was more effective than the brand they currently use.

(a) Can you determine the percent of consumers who did not find the medication to be more effective? If you can, what is it? If not, what additional information do you need?

(b) Can you determine the number of people who did not find the medication to be more effective? If you can, what is it? If not, what additional information do you need?

Exploring with Real-World Data: Collaborative Activities

Generated and Recycled Wastes The bar graph shows the number of pounds of certain types of wastes that each person in the United States generates and recycles each day. (Source: U.S. Environmental Protection Agency.)

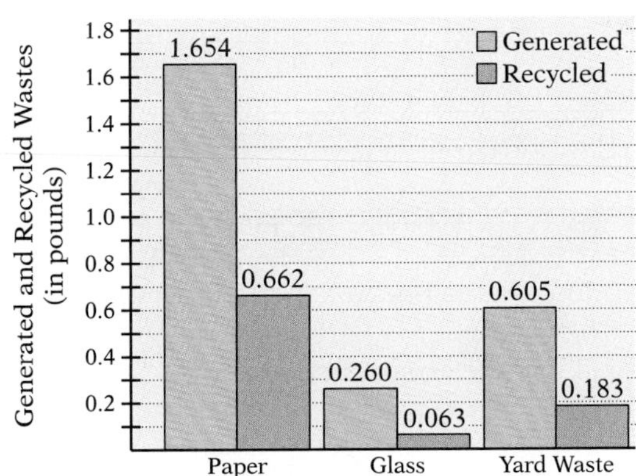

36. Determine the percent of waste paper that is recycled.

37. Determine the percent of yard waste that is recycled.

38. Determine the percent of waste glass that is recycled.

39. Suppose that you calculate the percent of waste paper that is *not* recycled. What is the relationship between that percent and the percent calculated in Exercise 36?

Copyright © Houghton Mifflin Company. All rights reserved.

For Exercises 40–43, consider the three types of waste shown in the bar graph.

40. What percent of the total generated waste is the amount of generated waste glass?

41. What percent of the total generated waste is the amount of generated waste paper?

42. What percent of the total recycled waste is the amount of recycled waste paper?

43. What percent of the total recycled waste is the amount of recycled yard waste?

Copyright © Houghton Mifflin Company. All rights reserved.

8.4 Percents and the Consumer

A *Simple Interest*
B *Compound Interest*
C *Taxes*

SUGGESTIONS FOR SUCCESS

As a consumer, you need to be careful about percents, which can be used to mislead you or cheat you out of your money. You will usually have to pay for the use of someone else's money, and you need to know exactly what you are agreeing to do.

As you study the material in this section, don't forget the basic percent equation given in Section 8.2: Percent · Base = Amount.

For interest, Percent is the interest rate, Base is the amount that you borrow (or invest), and Amount is the amount of interest.

For taxes, Percent is the tax rate, Base is the amount that is subject to tax, and Amount is the amount of tax you owe.

Although topics involving percent vary widely, the basic percent equation applies to all of them. Try writing the equation (in words) on a card and refer to it as you work examples and exercises.

A Simple Interest

Your neighbor may allow you to borrow his hammer, his lawn mower, or even his car and not expect anything other than your returning the item. However, when you borrow a significant amount of money, you will usually be required to pay for the use of it. The money that you borrow is called the **principal,** and the amount that you pay to borrow it is called the **interest.** The **loan payoff** (the amount that you must pay the lender) is the principal plus the interest.

As a borrower, you need to be exceedingly careful about the terms of a loan. Many people are victimized by unscrupulous lenders who mislead borrowers into interest obligations that they can never pay. Before you sign a loan agreement, make sure that you are clearly advised (by a lawyer, if necessary) about the commitment that you are making.

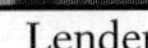

Of course, you can be the lender. When you invest money in your bank's savings account plan, a certificate of deposit, or any other account that pays interest, you are loaning money to the bank, and the bank pays you interest for the use of your money.

The amount of interest that you pay or receive depends on two quantities.

1. The *principal,* which is the amount that you borrow or invest
2. The *interest rate,* which is expressed as a percent

Simple interest is interest that is paid just once at the end of some specified period of time.

Copyright © Houghton Mifflin Company. All rights reserved.

Simple Interest for 1 Year

If the principal is P and the interest rate is r, then the simple interest I for 1 year can be found with the formula

$$I = Pr$$

NOTE Be careful about the letters we are using. In previous sections, we used the formula $P \cdot B = A$, where P represents percent. Here, we are using P to represent the principal and r to represent the interest rate, which is a percent.

Example 1

Suppose that you have borrowed \$2,000 at a simple interest rate of 6.5% for 1 year.

(a) How much interest must you pay?

(b) What is the loan payoff?

SOLUTION

(a) The principal P is \$2,000, and the interest rate r is 6.5% or, in decimal form, 0.065.

$$I = P \cdot r$$
$$= 2{,}000 \cdot 0.065$$
$$= 130$$

The interest is \$130.

(b) The loan payoff is the principal plus the interest: \$2,000 + \$130. The loan payoff is \$2,130.

Your Turn 1

To buy a car that costs \$21,000, you pay \$15,000 in cash and borrow the remainder at a simple interest rate of 9% for 1 year.

(a) How much interest must you pay?

(b) What is the loan payoff?

Answers: (a) \$540; (b) \$6,540

Example 2

A couple has received a \$5,000 wedding gift, and they decide to invest it in a certificate of deposit that pays 5.85% for 1 year. How much interest will they earn at the end of the year?

SOLUTION

The principal P is \$5,000 and the interest rate r is 5.85% or, in decimal form, 0.0585.

$$I = P \cdot r$$
$$= 5{,}000 \cdot 0.0585$$
$$= 292.5$$

The couple will earn \$292.50 in interest.

Your Turn 2

You have received a college grant of \$2,000, but you don't need it until next year. If you put the money in a savings account that pays 4% simple interest, how much interest will you earn for the year?

Answer: \$80

Copyright © Houghton Mifflin Company. All rights reserved.

The amount of interest that you pay or receive depends on the principal and the interest rate, but it also depends on the length of time that the principal is borrowed or invested. In Examples 1 and 2, the time was 1 year. For periods of time other than 1 year, we revise the interest formula as follows:

Simple Interest for *t* Years

If the principal is P, the interest rate is r, and the time period is t (in years), then the simple interest I can be found with the formula

$$I = Prt$$

NOTE For purposes of interest, 1 year is often regarded as 360 days. For a time t that is less than 1 year, we can use the conversion factor $\frac{1\text{ year}}{360\text{ days}}$. For example,

$$30\text{ days} = \frac{30\text{ days}}{1} \cdot \frac{1\text{ year}}{360\text{ days}} = \frac{30}{360}\text{ year} = \frac{1}{12}\text{ year}$$

Example 3

An accountant must make a $10,000 payment in 90 days. In the meantime, she decides to move the money into a short-term note that pays 4.8% simple annual interest. By doing so, how much interest will the accountant earn for her company?

SOLUTION

To use the formula $I = Prt$, we must express the time t in years.

$$\begin{aligned} 90\text{ days} &= \frac{90\text{ days}}{1} \cdot \frac{1\text{ year}}{360\text{ days}} \\ &= \frac{90}{360}\text{ year} \\ &= \frac{1}{4}\text{ year} \end{aligned}$$

The principal P is $10,000, and the interest rate r is 4.8%, or in decimal form, 0.048.

$$\begin{aligned} I &= P \cdot r \cdot t \\ &\quad\downarrow \quad \downarrow \quad \downarrow \\ &= 10{,}000 \cdot 0.048 \cdot \frac{1}{4} \\ &= 120 \end{aligned}$$

The accountant's decision will earn $120 for the company.

Your Turn 3

On January 1, you invest your holiday bonus of $1,500 in a 6-month certificate of deposit. How much interest will you have earned on July 1 if the certificate pays 5.2% simple annual interest?

Answer: $39

The simple interest formula $I = Prt$ involves four variables. If you know the value of any three of them, you can solve for the fourth one.

Copyright © Houghton Mifflin Company. All rights reserved.

Example 4

At the first of the year, a person borrowed $4,500 to pay the holiday bills. At the end of the year, the payoff for the loan was $4,702.50. What was the annual interest rate on the loan?

SOLUTION

The principal P was $4,500 and the time period t was 1 year.

The interest I is the difference between the loan payoff and the amount of the loan:

$$\$4{,}702.50 - \$4{,}500 = \$202.50$$

$$I = P \cdot r \cdot t$$

$$\downarrow \quad \downarrow \quad \downarrow \quad \downarrow$$

$$202.50 = 4{,}500 \cdot r \cdot 1$$

$$202.50 = 4{,}500r$$

$$\frac{202.50}{4{,}500} = r \qquad \text{Divide both sides by 4,500 to isolate } r.$$

$$0.045 = r$$

The interest rate was 0.045, or 4.5%.

Your Turn 4

A person wants an investment of $12,000 to grow to $12,840 in 1 year. What annual rate of interest is needed?

Answer: 7%

For simple interest, the number of times per year that interest is calculated has no bearing on the total amount of interest that will be paid. If the interest in Example 4 had been calculated quarterly rather than after 1 year, the quarterly interest amount would be $50.625. After 1 year, the interest would be 4($50.625) = $202.50, the same amount shown in Example 4.

B Compound Interest

Suppose that you agree to repay a $1,000 loan at the end of 2 years. Interest is calculated at the end of each year at an interest rate of 6%. How much interest will you pay? It depends on the fine print!

Interest Payments

One plan is to pay 6% interest each year on the *original principal.*

Year	*Principal*	*6% Interest*	
1	$1,000	$ 60	6% of $1,000 = $60.
2	$1,000	$ 60	6% of $1,000 = $60.
		$120	

With this plan, you will pay $120 in interest.

A second plan is to pay 6% interest on the *amount that you owe at the end of each year.*

Copyright © Houghton Mifflin Company. All rights reserved.

Year	*Principal*	*6% Interest*	
1	$1,000	$60.00	6% of $1,000 = $60.

Now you owe the original $1,000 plus the interest of $60. Therefore, your principal for the second year is $1,000 + $60 = $1,060.

Year	*Principal*	*6% Interest*	
2	$1,060	$63.60	6% of $1,060 = $63.60

Under this plan, you will pay $60.00 + $63.60 = $123.60 in interest.

The second plan is an example of **compound interest,** and, in this case, we say that the interest is *compounded annually.*

Interest can be compounded annually, monthly, daily, or for any other time period. In effect, you are paying interest on interest. This is costly for the borrower but good for the lender.

Using the formula $I = Prt$ is not practical if interest is compounded many times. For example, if you have money in a savings account that compounds interest daily, to find the interest for the year you would have to perform 360 calculations!

Fortunately, we have a formula that can be used, although a calculator is the only sensible way to use the formula.

Formula for Compound Interest

To calculate compound interest, use the formula

$$A = P\left(1 + \frac{r}{n}\right)^{nt}$$

where A = total value (principal plus interest)
P = principal
r = interest rate
t = time (in years)
n = number of times per year that interest is compounded

Try not to let the scary appearance of this formula put you off. The important part is to understand what the letters represent and to substitute values for the letters correctly. Then your calculator will make the rest of the work easy to do.

Example 5

At age 25, you decide to invest $10,000 in a safe fund that pays 6%, compounded monthly. Your plan is not to touch the investment until you retire at age 65. Substitute the given information into the compound interest formula. You do not need to attempt to evaluate the expression unless you want to.

Your Turn 5

(Work through the following Keys to the Calculator box before you attempt this problem.) Suppose that you invest $5,000 in a retirement fund that pays 5.5% interest, compounded quarterly (every 3 months). What will the value of the fund be after 8 years?

continued

Copyright © Houghton Mifflin Company. All rights reserved.

SOLUTION

$P = 10{,}000$ You invested \$10,000.

$r = 6\%$ or 0.06 The interest rate is 6%.

$t = 40$ years You will be age 65 in 40 years.

$n = 12$ Interest is compounded monthly or 12 times per year.

$$A = P\left(1 + \frac{r}{n}\right)^{nt}$$

$$= 10{,}000 \cdot \left(1 + \frac{0.06}{12}\right)^{12 \cdot 40}$$

Now let's see how the numerical expression in Example 5 can be evaluated. As always, we must follow the Order of Operations.

KEYS TO THE CALCULATOR

All the keystrokes that you need have been presented in previous chapters. The keystrokes shown are for a scientific calculator.

We begin with the expression inside the parentheses.

$$10{,}000 \cdot \left(1 + \frac{0.06}{12}\right)^{12 \cdot 40}$$

1 [+] **.06** [÷] **12** [=] Display: 1.005

Now we raise this result to the $12 \cdot 40 = 480$ power.

$$10{,}000 \cdot (1.005)^{480}$$

[x^y] **480** [=] Display: 10.95745367 . . .

Finally, we multiply this result by 10,000.

$$10{,}000 \cdot 10.95745367\ldots$$

[×] **10000** [=] Display: 109574.5367 . . .

As you can see, the plan outlined in Example 5 will result in an investment that is worth about \$109,575 at your retirement!

C Taxes

In effect, when you use a credit card, you are borrowing money. Unless you pay off the balance due each month, you will pay a hefty interest. Millions of people have serious financial troubles because they have allowed interest to eat into their earnings and savings.

As a consumer, you usually have a choice of whether to borrow money and pay interest. Unfortunately, such is not the case with taxes. You are required to pay taxes every time you fill your gas tank, make a long distance call, buy a home, or earn a living.

Copyright © Houghton Mifflin Company. All rights reserved.

A **sales tax** is a one-time tax that you pay for goods and services. Items that are taxed and the rate at which they are taxed vary from one locale to another.

Example 6

A 5% sales tax applies to the purchase of a computer that costs $1,500.

(a) What is the tax?

(b) What is the total cost of the computer?

SOLUTION

(a) (cost of computer) · (tax rate) = tax

$$1{,}500 \cdot 0.05 = x$$

$$1{,}500(0.05) = x$$

$$75 = x$$

The sales tax is $75.

(b) The total cost of the computer is

$$\$1{,}500 + \$75 = \$1{,}575.$$

Your Turn 6

The cost of a file cabinet is $180 plus a 4% sales tax.

(a) What is the tax?

(b) What is the total cost of the file cabinet?

Answers: (a) $7.20; (b) $187.20

Example 7

The sales tax on the purchase of a new car was $1,440. If the tax rate is 6%, what was the price of the car?

SOLUTION

(cost of car) · (tax rate) = tax

$$x \cdot 0.06 = 1{,}440$$

$$0.06x = 1{,}440$$ Divide both sides by 0.06 to isolate x.

$$x = \frac{1{,}440}{0.06} = 24{,}000$$

The price of the car was $24,000.

Your Turn 7

The total cost of a new pair of shoes includes a 5% sales tax of $4.10. What is the price of the shoes?

Answer: $82

Some have proposed replacing the complicated federal income tax with a simpler, federal sales tax. A flat rate of 18% has been suggested. Under such a plan, the tax on the new car in Example 7 would be $4,320! Nevertheless, the proposal has advantages as well as disadvantages, and there might come a time when you will be asked to vote on the issue. Your knowledge of percentages will be essential to making wise decisions.

Example 8

Including sales tax, the cost of a freezer is $752.40. If the freezer is advertised at $720, what is the sales tax rate?

Your Turn 8

A printer sells for $350, but your total bill, including sales tax, is $364. What is the sales tax rate?

continued

Copyright © Houghton Mifflin Company. All rights reserved.

SOLUTION

The sales tax is

$$\$752.40 - \$720.00 = \$32.40$$

$$\begin{array}{ccccc} (\text{cost of freezer}) & \cdot & (\text{tax rate}) & = & \text{tax} \\ \downarrow & & \downarrow & & \downarrow \\ 720 & \cdot & x & = & 32.40 \end{array}$$

$$720x = 32.40$$

$$x = \frac{32.40}{720} = 0.045 = 4.5\%$$

The sales tax rate is 4.5%.

Your earnings are subject to a federal tax, a state tax, possibly a local tax, a social security tax, and any number of other deductions. One estimate is that we work for the government from January through May of each year!

Example 9

Suppose that your annual salary is \$36,000. What is your monthly take-home pay if the following taxes are deducted?

Federal	28%
State	6%
Local	2%
Social security	6.5%

SOLUTION

Your monthly salary is

$$\$36{,}000 \div 12 = \$3{,}000$$

The total percent deductions are

$$28\% + 6\% + 2\% + 6.5\% = 42.5\%$$

$$\begin{array}{ccccc} (\text{salary}) & \cdot & (\text{tax rate total}) & = & \text{tax} \\ \downarrow & & \downarrow & & \downarrow \\ 3{,}000 & \cdot & 0.425 & = & x \end{array}$$

$$3{,}000(0.425) = x$$

$$1{,}275 = x$$

Your total taxes are \$1,275. Therefore, your take-home pay is

$$\$3{,}000 - \$1{,}275 = \$1{,}725$$

Your Turn 9

In Example 9, suppose that your employer also deducts 5% for the company's health plan. What is your monthly take-home pay?

Answer: \$1,575

NOTE In Example 9, we calculated the tax so that we could see how much tax is paid. We could have found the take-home pay directly. The percent of your take-home pay is 100% − 42.5% = 57.5%. Therefore, your take-home pay is 57.5% of \$3,000.

$$0.575 \cdot \$3{,}000 = \$1{,}725$$

Copyright © Houghton Mifflin Company. All rights reserved.

8.4 Exercises

A. Simple Interest

1. The amount that you earn from an investment (or pay to borrow money) is called the ________.

2. Interest paid just once at the end of a specified time period is called ________ interest.

For most loans or investments, compound interest is calculated rather than simple interest. However, in Examples 3–10, assume that the interest is *simple* interest and use the formula $I = Prt$.

3. *School Loan* To pay tuition, a student borrowed \$1,800 for 1 year. The interest rate was $8\frac{1}{2}\%$.

   (a) What was the interest for the year?

   (b) What was the payoff for the loan?

4. *Business Loan* To finance a new computer inventory system, a hardware store borrowed \$15,000 at $5\frac{1}{4}\%$ interest for 1 year.

   (a) What was the interest for the year?

   (b) What was the payoff for the loan?

5. *Home Improvement Savings* To help pay for remodeling their home next year, a couple invests \$12,000 in a money market fund at $7\frac{3}{4}\%$ interest for 1 year.

   (a) How much interest does the couple earn?

Copyright © Houghton Mifflin Company. All rights reserved.

(b) What is the value of the account at the end of the year?

6. *Vacation Savings* To pay for a vacation trip next year, a family invests \$900 at 8% for 1 year.

(a) How much interest does the couple earn?

(b) What is the value of the account at the end of the year?

7. *Credit Card Interest* A credit card company charges 18% annual interest. What is the interest for 1 month on an account balance of \$1,300?

8. *Construction Loan* A builder obtained a construction loan of \$90,000 at 9.5% annual interest. What was the interest for 2 months?

9. *Car Loan* Suppose that to buy a used car you borrow \$3,200 at 8% annual interest to be paid off at the end of 6 months. What is the loan payoff?

10. *Cash Advance* To pay for a trip to the beach, you obtain a cash advance of \$500 on your credit card. At an annual interest rate of 21%, how much do you owe if you wait 60 days to repay the advance?

Copyright © Houghton Mifflin Company. All rights reserved.

In Exercises 11 and 12, fill in the blank. Assume that interest is annual simple interest.

11.

I	P	r	t
$84	(a)	14%	9 months
(b)	$500	6.5%	18 months
$45	$3,000	(c)	60 days
$63	$300	7%	(d)

12.

I	P	r	t
(a)	$10,000	11%	180 days
$212.50	(b)	4.25%	3 months
$1,000	$5,000	10%	(c)
$6	$400	(d)	1 month

In Exercises 13–18, interest is assumed to be annual simple interest. Use the formula $I = Prt$.

13. *Tax Refund* A person invested a tax refund of $2,200 in a stock fund. One year later the account was worth $2,345. What was the interest rate?

14. *Loan Rate* After 60 days, the payoff for a $3,600 loan was $3,657. What rate of interest was charged on the loan?

15. *Inheritance* A person plans to invest part of an inheritance in a bond fund that pays 5% interest. How much should the person invest to receive $800 in interest in 1 year?

16. *Personal Loan* At the end of 1 month, a person owed $200 interest on a loan. If the bank charged 8% annual interest on the loan, how much money did the person borrow?

Copyright © Houghton Mifflin Company. All rights reserved.

17. *Investment Growth* A person invests $4,500 at 6% annual interest. After how many months will the investment be worth $4,635?

18. *Investment Value* How many months are required for an investment of $8,400 at 7% annual interest to grow to $8,841?

B. Compound Interest

19. Money that you borrow (or loan) is called the ________.

20. Interest that is paid on interest is called ________.

In Exercises 21–24, calculate the compound interest by using the simple interest formula $I = Prt$ repeatedly.

21. *Certificate of Deposit Value* A bank customer invests $30,000 in a certificate of deposit that pays 9% interest compounded monthly. What is the value of the account after 3 months?

22. *Flood Damage Repair* To repair flood damage, the owner of a coffee shop obtained a loan of $2,500 at 4% interest compounded monthly. If the owner pays off the loan at the end of 2 months, how much interest does she pay?

23. *Home Improvement Loan* To add a porch to the house, a homeowner borrowed $10,000 at 7% compounded semiannually (every 6 months). If he pays off the loan at the end of 1 year, how much interest does he pay?

Copyright © Houghton Mifflin Company. All rights reserved.

24. *Retirement Plan* An electrician invested \$7,000 in a retirement plan that paid 6% interest compounded semiannually (every 6 months). What was the value of the account after $1\frac{1}{2}$ years?

In Exercises 25–30, calculate the value A by using the compound interest formula $A = P\left(1 + \frac{r}{n}\right)^{nt}$.

25. *IRA Investment* Many people invest \$2,000 per year in an Individual Retirement Account (IRA). What of the value of \$2,000 invested at 8% compounded quarterly for 10 years?

26. *IRA* What is the value of \$2,000 invested in an IRA that pays 7.5% compounded monthly for 30 years?

27. *College Fund* Parents of a new baby invested \$5,000 in a college fund that paid 8.5% interest compounded semiannually. What is the value of the investment after 18 years?

28. *Education Investment* To provide for her 4-year-old child's education, a mother invested an inheritance of \$15,000 at 9% interest compounded monthly. What was the value of the investment 14 years later when the child began college?

29. *Certificate of Deposit* A bank advertises that a certificate of deposit pays 6% interest compounded daily. If a person invests \$10,000 in the CD, what is the value of the CD after 1 month? (Assume that 1 month = 30 days.)

Copyright © Houghton Mifflin Company. All rights reserved.

30. *Money Market Fund* A money market fund pays 4.5% interest compounded daily. What is the value of a $25,000 investment after 90 days?

C. Taxes

31. *Camera Purchase* A camera costs $132. The sales tax rate is 5%.

(a) What is the tax on the camera?

(b) What is the total cost of the camera?

32. *Hotel Tax* In a certain resort area the hotel tax is 14%.

(a) What is the tax for a 5-night stay at a resort that charges $124 per night?

(b) What is the total cost for the 5-night stay?

Copyright © Houghton Mifflin Company. All rights reserved.

33. *Take-Home Pay* Suppose that each week you make \$690, and your employer withholds 6.25% for social security, 18% for federal tax, and 4.5% for retirement. What is your annual take-home pay? (Assume that you work 50 weeks per year.)

34. *Import Tax* A traveler returning from Ireland had to pay a 12% import tax on 3 Waterford bowls that were valued at \$230 each. What total amount of tax did she pay?

35. *Income Tax for a Couple* A couple with an income of \$57,000 owed income tax of \$6,352.50 plus 28% of the amount of income over \$42,350. What was the total tax that the couple owed?

36. *Income Tax* A single taxpayer with an income of \$38,000 owed income tax of \$3,802.50 plus 28% of the amount of income over \$25,350. What was the total tax that the person owed?

37. *Tax on a Tiller* The sales tax on a tiller was \$70.80. If the sales tax rate is 6%, what was the price of the tiller?

38. *Auto Parts Sales* An auto parts store collected \$1,400 in sales tax for a week. If the sales tax rate is 4%, what were the store's sales for the week?

Copyright © Houghton Mifflin Company. All rights reserved.

39. *Library Upgrade Funding* A local sales tax referendum proposes to provide \$5.4 million for upgrading libraries. If 18% of the taxes proposed in the referendum are designated for library programs, how much tax is proposed by the referendum?

40. *Entertainment Tax* A state imposes a 6% sales tax and the city charges an 8% entertainment tax on tickets to professional sports events. If the total tax on a ticket is \$5.88, what is the price of the ticket?

41. *Tax on a Shirt* The total cost of a shirt priced at \$38 is \$39.71. What is the sales tax rate?

42. *Car Rental* A car rental company charges \$35 per day for a car. The total daily cost for a business traveler to rent a car was \$38.85. What tax rate was charged on the rental?

43. *Income Tax Rate* A person paid \$7,788 in taxes on an income of \$32,450. What percentage of the income was paid in taxes?

44. *Withholding Tax Rate* Suppose that you receive a monthly salary of \$2,250 of which \$607.50 is withheld for taxes. What percent of your salary is withheld for taxes?

Copyright © Houghton Mifflin Company. All rights reserved.

Writing and Concept Extension

45. In the formula $I = Prt$, explain what I, P, r, and t represent.

46. Suppose that you purchase a car for $15,000. The state sales tax rate is 4% and the city sales tax is 2.5%. Which of the following is a correct way to determine the tax on the car?

    (i) Add 4% and 2.5% to obtain 6.5%. Then find 6.5% of 15,000.

    (ii) Determine 4% of 15,000 and determine 2.5% of 15,000. Add the results.

A tax is usually charged on personal property, such as your home or a car. Rather than the tax's being based on the full value of the property, it is often charged on the *assessed value,* which is some percent of the full value.

47. *Property Tax* If 40% of the actual value of property is subject to property tax, what is the property value if $54,000 is subject to tax?

48. *Car Tax* Suppose that the value of a car is $16,000. If a county charges 3.6% tax on 40% of the value of the car, what is the tax on the car?

49. *Furniture Payment* Suppose that you buy living room furniture, and you have two payment options.

    (i) Pay the total cost of $4,350 at the time of the purchase.

    (ii) Pay nothing now and make no payments for 6 months. After 6 months, pay $4,000 plus interest based on a simple annual interest rate of 21%.

    Which is the best option?

Copyright © Houghton Mifflin Company. All rights reserved.

50. What is the difference between the interest on \$10,000 at 5% interest for 1 year:

(a) If the interest is simple interest or compounded monthly?

(b) If the interest is compounded monthly or compounded daily?

Copyright © Houghton Mifflin Company. All rights reserved.

8.5 Percents in Business

A *Commissions*
B *Depreciation*
C *Markup and Discount*

SUGGESTIONS FOR SUCCESS

Do you want to impress your boss and coworkers?

In a business environment, you will find that certain people seem particularly quick at estimating percents. At a meeting, someone might say, "Looks like we had about a 5% increase in sales last quarter." This kind of mental agility with percents always leaves a good impression.

As you work through the examples and exercises in this section, work on your estimating skills. Try to ballpark the answer before you do the arithmetic. Even if your estimate is not very good, making the attempt will give you a better understanding of the problem and the role that percents play in solving it.

A Commissions

When you are hired to perform a certain job, you are paid for your work. If you are paid an **hourly wage,** you receive a fixed amount for each hour that you work. A **salary** is a fixed amount that you are paid for some period of time, regardless of the number of hours that you spend on the job. Salaries are usually annual, but they can also be on a monthly basis or any other agreed-on work period.

Employees who work in sales are often paid a **commission,** which is a percent of the total sales made.

$$\text{Commission} = (\text{total sales}) \cdot (\text{commission rate})$$

NOTE A **royalty** is like a commission in that it is a percent of the total sales. Royalties are paid for certain creative work done, for example, by composers and authors.

Commissions and royalties are the least stable forms of income. A salesperson might work for several months to land an account but then lose the sale and the commission. Moreover, commissions are sometimes seasonal, such as in real estate sales. However, with a little luck and a lot of hard work, people can earn a very good living on commissions.

Example 1

A salesperson is paid an 8% commission. If her sales for the month were $40,800, what was her commission that month?

SOLUTION

$$\text{commission} = (\text{sales}) \cdot (\text{commission rate})$$

$$\downarrow \qquad \downarrow \qquad \downarrow$$

$$x = 40{,}800 \cdot 0.08$$

$$x = 40{,}800(0.08)$$

$$= 3{,}264$$

The commission for the month was $3,264.

Your Turn 1

At a certain department store, salespersons receive a 6% commission on all sales. How much commission is earned on the sale of a $700 refrigerator?

Answer: $42

Copyright © Houghton Mifflin Company. All rights reserved.

Example 2

A real estate agent received a commission of $8,700 on the sale of a new home. If the commission rate is 6%, what was the cost of the home?

SOLUTION

$$\begin{array}{ccccc} \text{commission} & = & (\text{sales}) & \cdot & (\text{commission rate}) \\ \downarrow & & \downarrow & & \downarrow \\ 8{,}700 & = & x & \cdot & 0.06 \end{array}$$

$$8{,}700 = 0.06x$$

$$\frac{8{,}700}{0.06} = x$$

$$145{,}000 = x$$

The cost of the home was $145,000.

Your Turn 2

A land auction company receives a 10% commission on all land sales. If the commission on a certain land sale was $3,250, what was the cost of the land?

Answer: $32,500

Example 3

In one month, a marketing representative for a newspaper received a $1,900 commission for selling $47,500 of advertising. What was the commission rate?

SOLUTION

$$\begin{array}{ccccc} \text{commission} & = & (\text{sales}) & \cdot & (\text{commission rate}) \\ \downarrow & & \downarrow & & \downarrow \\ 1{,}900 & = & 47{,}500 & \cdot & x \end{array}$$

$$1{,}900 = 47{,}500x$$

$$\frac{1{,}900}{47{,}500} = x$$

$$0.04 = x$$

The commission rate was 4%.

Your Turn 3

A real estate broker received a commission of $10,850 on the sale of a home that sold for $155,000. What was the commission rate?

Answer: 7%

Some employees receive a **base salary** plus a commission on sales. In such cases, the total monthly earnings are the monthly salary plus the commission on sales for that month.

Example 4

An employee is paid a salary of $600 per week plus a 7% commission on all monthly sales over $50,000. In April, the total sales were $80,000. How much did the employee earn that month?

SOLUTION

The commission is paid only on the sales over $50,000.

$$\$80{,}000 - \$50{,}000 = \$30{,}000$$

Your Turn 4

An employee receives an annual salary of $24,000 plus a 3% commission on all monthly sales over $10,000. If the sales for June were $18,000, what were the employee's earnings that month?

Copyright © Houghton Mifflin Company. All rights reserved.

$$\text{commission} = (\text{sales}) \cdot (\text{commission rate})$$

$$\downarrow \qquad \downarrow \qquad \downarrow$$

$$x = 30{,}000 \cdot 0.07$$

$$x = 30{,}000(0.07)$$

$$= 2{,}100$$

The commission was $2,100.

In addition, the employee received $600 per week or $2,400 for the month. The employee's earnings for the month were

$$\$2{,}400 + \$2{,}100 = \$4{,}500$$

B Depreciation

After a company purchases equipment, machinery, and other expensive items, those items begin to lose their value over time. This loss in value is called **depreciation.** Depreciation is a cost that can be written off as a deduction by the company for tax purposes.

There are several legal ways to calculate depreciation. One method (called *straight-line depreciation*) is to begin by estimating the useful life of the equipment. For example, if a machine has a useful life of 10 years, then we might say that the machine loses $\frac{1}{10}$, or 10%, of its original value each year.

Example 5

Suppose that a computer network costing $80,000 is estimated to have a useful life of 5 years. What is the depreciated value of the system after 3 years?

SOLUTION

Because the useful life is 5 years, the system loses $\frac{1}{5}$, or 20%, of its original value each year.

$$\text{Annual depreciation} = 0.20(80{,}000)$$

$$= 16{,}000$$

Year	*Depreciated Value*
0	$80,000
1	$80,000 − $16,000 = $64,000
2	$64,000 − $16,000 = $48,000
3	$48,000 − $16,000 = $32,000
4	$32,000 − $16,000 = $16,000
5	$16,000 − $16,000 = 0

The value at the end of each year is found by subtracting $16,000 from the value at the end of the previous year.

After 3 years, the depreciated value is $32,000. Note that the value is 0 after the useful life of 5 years.

Your Turn 5

A landscaper buys a truck for $40,000. What is the depreciated value of the truck after 4 years if its useful life is 10 years?

Answer: $24,000

Copyright © Houghton Mifflin Company. All rights reserved.

In straight-line depreciation, as in Example 5, we keep applying the annual depreciation rate to the *original* value. Another common method for calculating depreciation is to apply the annual depreciation rate to the *previous year's* value. For example, if the depreciation rate is 10%, then the value of the item in any given year is 90% of its value in the previous year. We use this method in Examples 6 and 7.

Example 6

In a rolling mill, thick aluminum ingots are rolled into thin sheets. The original cost of the roller was \$600,000. If the depreciation rate is 20%, what is the value of the roller after the second year?

SOLUTION

After the first year, the value of the roller is 100% − 20% = 80% of its original value.

depreciated value = 0.80(\$600,000) — The original value was \$600,000.
= \$480,000

After the second year, the value of the roller is 80% of the value in the preceding year.

depreciated value = 0.80(\$480,000) — The value at the end of the preceding year was \$480,000.
= \$384,000

Therefore, the roller's value after 2 years is \$384,000.

Your Turn 6

In Example 6, what is the value of the roller after the fourth year?

Answer: \$245,760

For this method of depreciation, called *exponential depreciation*, the following formula can be used to determine the depreciated value of an item after any given number of years.

Exponential Depreciation

For exponential depreciation, the depreciated (current) value V of an item is given by

$$V = P(1 - r)^t$$

where

P = the original value

r = the depreciation rate

t = the time in years

If we need to find the depreciated value of an item after 8 years, we could use the method of Example 6 to find the depreciated value after the first year, after the second year, after the third year, and so on through the 8 years. Obviously, this would be a tedious process. The formula above allows us to find the depreciated value after 8 years with just one calculation. In Example 7, we illustrate how to substitute into the formula. Then we learn how to perform the calculation.

Copyright © Houghton Mifflin Company. All rights reserved.

Example 7

Consider again the machine in Example 6. Suppose that you want to know the depreciated value after 8 years. Substitute the given information into the exponential depreciation formula. You do not need to attempt to evaluate the expression.

SOLUTION

The original value (P) was $600,000. The depreciation rate (r) is 20% or 0.20. The time (t) is 8 years.

$$V = P(1 - r)^t$$
$$= 600{,}000(1 - 0.20)^8$$

Your Turn 7

(Work through the following Keys to the Calculator box before you attempt this problem.) For the machine in Example 6, what is the depreciated value after 10 years?

Answer: $64,424.51

To evaluate the expression in Example 7, you will need a calculator.

KEYS TO THE CALCULATOR

The keystrokes that we give here are for a scientific calculator. Following the Order of Operations, we first calculate the quantity in parentheses.

$$600{,}000(1 - 0.20)^8$$

1 [−] **.20** [=] Display: .80

Then we raise that result to the 8th power.

$$600{,}000(0.80)^8$$

[x^y] **8** [=] Display: 0.16777216 . . .

Finally, we multiply that result by 600,000.

$$600{,}000(0.16777216\ldots)$$

[×] **600000** [=] Display: 100663.296 . . .

After 8 years, the depreciated value of the machine is about $100,663.

C Markup and Discount

In order to make a profit, a store owner buys an item at a **wholesale price** and then sells the item at a higher **retail price.** The difference between these prices is called the **markup.**

$$\text{markup} = \text{retail price} - \text{wholesale price}$$

or

$$\text{retail price} = \text{wholesale price} + \text{markup} \qquad (1)$$

Stores often calculate the markup as a percent of the wholesale price.

$$\text{markup} = (\text{markup rate}) \cdot (\text{wholesale price}) \qquad (2)$$

Copyright © Houghton Mifflin Company. All rights reserved.

We will use formulas (1) and (2) in the following examples.

Example 8

Suppose that a store marks up all its goods by 40% and the wholesale price of an item is $12.00.

(a) What is the markup on the item?

(b) What is the retail price of the item?

SOLUTION

(a) The markup is 40% of the wholesale price.

$$\text{markup} = (\text{markup rate}) \cdot (\text{wholesale price})$$
$$\downarrow \qquad\qquad \downarrow \qquad\qquad \downarrow$$
$$\text{markup} = 0.40 \cdot 12.00$$
$$= 4.80$$

The markup is $4.80.

(b) The retail price is the wholesale price plus the markup.

$$\text{retail price} = \text{wholesale price} + \text{markup}$$
$$\downarrow \qquad\qquad \downarrow \qquad\qquad \downarrow$$
$$\text{retail price} = 12.00 + 4.80$$
$$= 16.80$$

The retail price is $16.80.

Your Turn 8

A hardware store buys electric drills for $16.00 and marks them up by 45%.

(a) What is the markup on the drill?

(b) What is the retail price of the drill?

Answers: **(a)** $7.20; **(b)** $23.20

Example 9

Suppose that a store buys an item for $21 and marks up the item by $7. What is the markup rate for this item?

SOLUTION

Let r = the markup rate.
The wholesale price is $21 and the markup is $7.

$$\text{markup} = (\text{markup rate}) \cdot (\text{wholesale price})$$
$$\downarrow \qquad\qquad \downarrow \qquad\qquad \downarrow$$
$$7 = r \cdot 21$$
$$r = \frac{7}{21} = \frac{1}{3} = 33\frac{1}{3}\%$$

The markup rate is $33\frac{1}{3}\%$.

Your Turn 9

A grocer buys ear corn for 10 cents each and sells 4 ears for 98 cents. What is the markup rate?

Answer: 145%

If an item is not selling very well at its current retail price, the store owner may decide to offer a **discount,** which is a price reduction.

$$\text{new retail price} = \text{old retail price} - \text{discount} \qquad (3)$$

Copyright © Houghton Mifflin Company. All rights reserved.

Discounts are usually expressed as a percent of the original retail price.

$$\text{discount} = (\text{discount rate}) \cdot (\text{old retail price}) \qquad (4)$$

The next examples illustrate the use of formulas (3) and (4).

Example 10

A business displays a sign in its window: "All Goods 20% Off." Suppose that the original retail price of a sweater was $48.

(a) What is the discount?

(b) What is the new retail price of the sweater?

SOLUTION

(a) The discount rate is 20% and the old retail price was $48.

$$\text{discount} = (\text{discount rate}) \cdot (\text{old retail price})$$
$$\downarrow \qquad \downarrow \qquad \downarrow$$
$$\text{discount} = 0.20 \cdot 48$$
$$= 9.60$$

The discount is $9.60.

(b)
$$\text{new retail price} = \text{old retail price} - \text{discount}$$
$$\downarrow \qquad \downarrow \qquad \downarrow$$
$$\text{new retail price} = 48.00 - 9.60$$
$$= 38.40$$

The new retail price is $38.40.

Your Turn 10

Movie tickets cost $8, but senior citizens get a 10% discount.

(a) What is the discount?

(b) What is the cost of a senior citizen's ticket?

Answers: (a) $.80; (b) $7.20

Example 11

As a video club member, you get a 12% discount on all tapes that you buy. If you receive a $2.64 discount on a purchase, what was the original selling price of the tape?

SOLUTION

Let x represent the old retail price.

The discount rate is 12% and the discount is $2.64.

$$\text{discount} = (\text{discount rate}) \cdot (\text{old retail price})$$
$$\downarrow \qquad \downarrow \qquad \downarrow$$
$$2.64 = 0.12 \cdot x$$
$$\frac{2.64}{0.12} = x$$
$$22 = x$$

The price of the tape before the discount is $22.

Your Turn 11

A distributor gets a 2% discount if the invoice is paid within 10 days. If a $17 discount is taken on a particular purchase, what was the original cost?

Answer: $850

Copyright © Houghton Mifflin Company. All rights reserved.

Example 12

A freezer, originally priced at \$840, is on sale for \$705.60. What is the discount rate?

SOLUTION

The price discount is \$840.00 − \$705.60 = \$134.40.

$$\text{discount} = (\text{discount rate}) \cdot (\text{old retail price})$$

$$\downarrow \qquad\qquad \downarrow \qquad\qquad \downarrow$$

$$134.40 = r \cdot 840.00$$

$$\frac{134.40}{840.00} = r$$

$$0.16 = r$$

The discount rate is 16%.

Your Turn 12

For Fan Appreciation Night, hockey tickets normally costing \$22 are sold for \$18.70. What is the discount rate?

Answer: 15%

Copyright © Houghton Mifflin Company. All rights reserved.

8.5 Exercises

A. Commissions

1. If an employee is paid a percent of the total sales, we say that the employee receives a(n) ________.

2. A fixed amount that a person is paid regardless of the number of hours that the person worked is called a(n) ________.

3. *Furniture Sales* A furniture salesperson received an 8% commission on sales. What was the commission for selling a living room set for $2,400?

4. *Book Sales* A book sales representative received a 12% commission. If the representative made a sale of $34,000 to a college bookstore, what was the commission?

5. *Real Estate Commission* A real estate agent received a 7% commission on residential property sales and a 4% commission on commercial property sales. What was the total commission for an agent who sold $240,000 of residential property and $525,000 of commercial property?

6. *Used Car Sales* A used car sales agent received a 4% commission for sales. What was the commission for selling 3 cars for an average price of $5,260?

7. *Home Sale* If a house sold for $148,000, and the real estate agent received a commission of 8%, what did the seller receive from the sale of her home?

Copyright © Houghton Mifflin Company. All rights reserved.

8. *Antique Clock* The owner of an antique clock made arrangements for a dealer to sell the clock. The clock sold for $1,320, and the dealer received a 12% commission. What did the owner receive?

9. *Mud Tires* An employee of Al's Tire Sales earned a commission of $34 for selling a set of mud tires. If the commission rate was 4%, what was the price of the set of tires?

10. *Residential Property* A real estate agent received a $17,400 commission, which was 12% of the selling price of a home. For what price did the home sell?

11. *Magazine Sales* A telemarketer received a commission of 9.5% for selling magazines. One day he received a commission of $93.10. What were the total sales?

12. *Tapes and CDs* A musician received a royalty of 6.5% on sales of tapes and CDs. In the first year, the musician received $55,250. What were the sales for the year?

13. *Charitable Donations* A telephone solicitor obtained $5,600 in donations to a charity. If the solicitor received a commission of $448, what was the commission rate?

Copyright © Houghton Mifflin Company. All rights reserved.

14. *Jewelry Sales* A member of the Joy-Jewelry Club sold costume jewelry at parties that she hosted at her home. If she received a commission of $120.40 for selling $860 of jewelry, what was the commission rate?

15. *Consignment Shop* A consignment shop sells children's clothing on commission. The shop earned $112.80 in commissions for clothing sales of $470. What was the commission rate?

16. *Auction Company* An antiques auction company received a commission of $3,450 on sales of $230,000. What was the commission rate?

17. *Cosmetic Consultant* A Pace Department Store cosmetic consultant was paid $1,400 per month plus 3% of sales. What was the consultant's income from sales totaling $5,450?

18. *Photography Studio* Handy Photography Studio paid a salesperson $200 per week plus 15% of sales. What was the person's income if sales were $2,100?

19. *Stock Broker* Easy Trade paid brokers $2,100 per month plus $2\frac{1}{2}$% of stock sales over $250,000. If a broker's stock sales for a month were $440,000, what was the broker's income for the month?

20. *Pottery Sales* A potter paid a part-time clerk in his shop $180 per week plus 10% of sales over $400. What did the clerk earn for a week if sales were $1,200?

Copyright © Houghton Mifflin Company. All rights reserved.

B. Depreciation

21. The loss in value of an item over time is called ________.

22. One method of calculating ________ is to begin by estimating the useful life of the equipment.

In Exercises 23–26, depreciation is calculated according to the straight-line method, as in Example 5.

23. *Fitness Equipment* The original cost of fitness equipment at Get-in-Shape Gym was $18,000. If the estimated useful life of the equipment is 6 years, what is the depreciated value after 2 years?

24. *Forklift* Builders Supply purchased a forklift for $120,000. The forklift has an estimated useful life of 9 years. What is its depreciated value after 4 years?

25. *Tractor* A tractor that cost $32,000 has a useful life of 10 years. What is the depreciated value after 6 years?

26. *Fishing Boat* A fishing boat that costs $9,000 has a useful life of 12 years. What is the depreciated value after 8 years?

In Exercises 27–30, depreciation is calculated according to the exponential method, as in Examples 6 and 7.

27. *Motor Home* The original cost of a motor home was $52,000. If the depreciation rate is 15%, what is the value of the motor home after 2 years?

28. *Pickup Truck* A pickup truck, which cost $21,000, depreciates at the rate of 12%. What is the value after 3 years?

Copyright © Houghton Mifflin Company. All rights reserved.

29. *Copy Machine* Fast Copy purchased an office copy machine for $24,000. The copier depreciates at the rate of 18%. What is the value of the copier after 8 years?

30. *Airplane* The Etowah Construction Company purchased a corporate plane for $340,000. If the depreciation rate for the plane is 22%, what is the value after 6 years?

C. Markup and Discount

31. The price that a store owner pays for an item is called the ______ price, and the price that the owner charges for the item is called the ______ price.

32. The difference between the retail price and wholesale price is called the ______.

33. *Gas Grills* Stone-Mart buys a special shipment of gas grills for $140 each and uses a markup rate of 20%.

(a) What is the markup? **(b)** What is the retail price?

34. *Patio Furniture* Canton Outlet marks up furniture items by 35%. The wholesale price of a 5-piece patio set is $122.20.

(a) What is the markup? **(b)** What is the retail price?

Copyright © Houghton Mifflin Company. All rights reserved.

35. *Mountain Bike* Bike Authority marks up mountain bikes by 60%. The wholesale price of a Bluegoose mountain bike is $137.50.

(a) What is the markup?

(b) What is the retail price?

36. *Gym Set* Kids Play purchases 8-leg gym sets with a slide for $162.40. After a markup of 30%, what is the retail price?

37. *Glider Rockers* Furniture World sells glider rockers for $78. If the wholesale price is $52, what is the markup rate?

38. *Tents* Outdoor Adventure purchases Rocky Trail dome tents for $128 each and sells the tents for $158.72. What is the markup rate?

39. *Toothpaste* Green's Pharmacy purchases 8.2-ounce tubes of Fresh Toothpaste for 90 cents per tube and sells a tube for $1.53. What is the markup rate?

40. *Towels* Linen Closet sells bath towels for $6.84 each. If the wholesale price is $3.60, what is the markup rate?

Copyright © Houghton Mifflin Company. All rights reserved.

41. *Mountain Cabins* If you have a coupon, Dogwood Cabins offers a 20% discount on the rental of a mountain cabin. The regular weekly rate is $730.

(a) What is the discount? **(b)** What is the reduced weekly rate?

42. *Cat Food* PetsUSA advertises a 15% discount on a 21-pound bag of Super Diet cat food. The regular price is $9.80.

(a) What is the discount? **(b)** What is the reduced price?

43. *Air Conditioner* For one week only, Cool Zone sells a window air conditioner for 30% off the suggested retail price of $755.

(a) What is the discount? **(b)** What is the sale price?

44. *Fence Panels* For a special promotion, The Fence Post sells French Gothic fence panels for 12% off the regular price of $40.80.

(a) What is the discount? **(b)** What is the sale price?

45. *Alaska Cruise* A cruise line advertises a one-week Alaska cruise that is regularly $2,400 per person. For passengers who book early, the cost is only $1,950. What is the discount rate?

46. *Paint* Home Station features Hi-Gloss Bear Paint, regularly $21.97, for $18.67 per gallon. What is the discount rate?

47. *Computer System* CompuTown advertises a complete computer system for $660, which is $190 off the regular price. What is the discount rate?

Copyright © Houghton Mifflin Company. All rights reserved.

48. *Camcorder* Circuit Town sells a camcorder, regularly priced at $450, for only $387. What is the discount rate?

Writing and Concept Extension

49. Suppose that you have a coupon for 10% off any item at a certain department store. If that store advertises an item for 20% off and you use your coupon, which of the following would result in the same price?

(i) Use the coupon on the original price and then take 20% off the reduced price.

(ii) Take 20% off the original price and then use the coupon on the reduced price.

(iii) Take 30% off the original price.

50. A merchant marks up an item by 20% from the wholesale price of $100. When the item does not sell at that price, the merchant advertises the item at 20% off. Would the merchant then be selling the item for the wholesale price? Why?

51. *Appliance Sales* An appliance store paid a 4% commission for sales up to $3,000 and 5% on sales over $3,000. During one week, a salesperson sold $5,000 worth of merchandise. What was her commission?

52. *Boat Sales* A boat dealer paid a 10% commission on sales up to $50,000 and 12% commission on sales over $50,000. If a person's commission for the month was $9,200, what were the person's sales for the month?

53. *Car Sales* A car sales agent received a base salary of $1,500 per month plus a commission of 4% of sales. If the agent's income for the month was $2,700, what were the sales?

Copyright © Houghton Mifflin Company. All rights reserved.

8.6 Percent Increase and Decrease

A *Percent Increase*
B *Percent Decrease*

SUGGESTIONS FOR SUCCESS

As a greedy store owner, you decide to increase the prices on all your goods by 25%. Realizing that you no longer have any customers, you decrease those higher prices by 25%. So now your prices are what they were originally, right? Wrong! You will see why when you work through Example 8.

As in previous sections, we will give you a number of formulas that you can use for calculations involving percent increases and decreases. Again, we urge you to write them down on a card and keep them in front of you as you work the examples and exercises.

On the other hand, relying on formulas is not the same as knowing what you are doing. You will probably forget the formulas eventually. If you work on understanding the concept of a percent increase and decrease, then you won't even need the formulas.

A Percent Increase

One aspect of marketing is to convince consumers that they are getting more for their money. Typically, this is done by offering a percent increase in the product or a percent decrease in the cost. However, a careful consumer will want to know whether an increase in the amount of a product is quietly accompanied by an increase in the price. In fact, if the percent increase in price is greater than the percent increase in the product, you will actually be getting less for your money in terms of unit cost.

A Percent Increase in the Product

A candy bar maker produces Chocolate Gooie Bars that weigh 42 grams. To improve sales, the company has decided to market Chocolate Gooie Bars that are 10% larger.

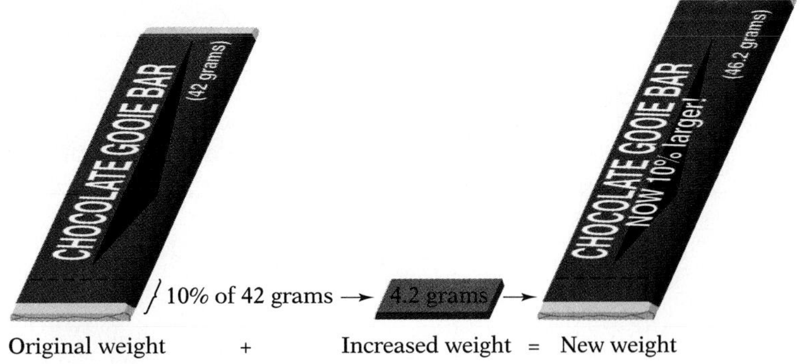

The increase in the weight is 10% of the original weight.

$$\begin{aligned} \text{increase} &= 10\% \text{ of the original weight} \\ &= 10\% \text{ of } 42 \text{ grams} \\ &= 0.10(42) \\ &= 4.2 \text{ grams} \end{aligned}$$

Copyright © Houghton Mifflin Company. All rights reserved.

The weight of the new candy bar is the original weight plus the increase in weight.

$$\text{new weight} = \text{original weight} + \text{increase in weight}$$
$$\downarrow \qquad\qquad \downarrow \qquad\qquad \downarrow$$
$$\text{new weight} = 42 \text{ grams} + 4.2 \text{ grams}$$
$$= 46.2 \text{ grams}$$

Note this important fact: A percent increase is a percent of the *original number.*

$$\text{increase} = (\text{percent increase}) \cdot (\text{original number})$$

Once we know the increase, we can determine the new number.

$$\text{new number} = (\text{old number}) + (\text{increase})$$

Example 1

Suppose that your salary last year was $36,000 and you received a 5% increase.

(a) What was your increase in salary?

(b) What is your salary this year?

SOLUTION

$$\text{original number} = 36{,}000$$
$$\text{percent increase} = 5\% = 0.05$$

(a) $\text{increase} = (\text{percent increase}) \cdot (\text{original number})$

$$\downarrow \qquad\qquad \downarrow \qquad\qquad \downarrow$$
$$\text{increase} = 0.05 \cdot 36{,}000$$
$$= 1{,}800$$

The salary increase was $1,800.

(b) Now we find the new salary.

$$\text{new salary} = \text{old salary} + \text{increase}$$
$$\downarrow \qquad\qquad \downarrow \qquad\qquad \downarrow$$
$$\text{new salary} = 36{,}000 + 1{,}800$$
$$= 37{,}800$$

Your salary this year is $37,800.

Your Turn 1

Last year, soft drinks in campus vending machines cost 50 cents. This year, the cost has been increased by 10%.

(a) What is the increase in cost?

(b) What is the new cost?

Answers: (a) 5¢; (b) 55¢

Example 2

In 1997, there were 3.5 million households in the United States with assets of at least $1 million. By 2005, that number is projected to increase to 5.6 million households. (Source: Affluent Market Institute.) What is the expected percent increase in the number of millionaire households?

Your Turn 2

From 1999 to 2005, the total charges to credit cards are projected to increase from $1.29 trillion to $2.16 trillion. (Source: *Nilson Report.*) What is the percent increase?

Copyright © Houghton Mifflin Company. All rights reserved.

SOLUTION

The numbers of households are given in millions, but we can solve the problem by using just 3.5 and 5.6.

First, we calculate the increase in the number of millionaire households.

$$\text{increase} = \text{new number} - \text{old number}$$
$$\downarrow \qquad\qquad \downarrow \qquad\qquad \downarrow$$
$$\text{increase} = 5.6 - 3.5 = 2.1$$

Now let x represent the percent increase.

$$\text{increase} = (\text{percent increase}) \cdot (\text{old number})$$
$$\downarrow \qquad\qquad \downarrow \qquad\qquad \downarrow$$
$$2.1 = x \cdot 3.5$$

$$2.1 = 3.5x$$

$$\frac{2.1}{3.5} = x \qquad \text{Divide both sides by 3.5.}$$

$$0.6 = x$$

$$60\% = x$$

The percent increase in millionaire households from 1997 to 2005 is 60%.

Answer: 67%

Example 3

A garden tractor that sold for \$2,000 last year now costs \$2,180. What is the percent increase in price?

SOLUTION

$$\text{new price} = \$2{,}180$$
$$\text{old price} = \$2{,}000$$
$$\text{increase} = \$2{,}180 - \$2{,}000 = \$180$$

We let x represent the percent increase.

$$\text{increase} = (\text{percent increase}) \cdot (\text{old number})$$
$$\downarrow \qquad\qquad \downarrow \qquad\qquad \downarrow$$
$$180 = x \cdot 2{,}000$$

$$180 = 2{,}000x$$

$$\frac{180}{2{,}000} = x \qquad \text{Divide both sides by 2,000.}$$

$$0.09 = x$$

$$9\% = x$$

The percent increase in price was 9%.

Your Turn 3

A company's profits increased from \$220,000 to \$246,400. What was the percent increase in profits?

Answer: 12%

We can combine the two formulas that precede Example 1 into one formula.

$$\begin{aligned}\text{new number} &= (\text{old number}) + (\text{increase})\\ &= (1) \cdot (\text{old number}) + (\text{percent increase}) \cdot (\text{old number})\\ &= (\text{old number})(1 + \text{percent increase}) && \text{Factor out “old number.”}\\ &= (\text{old number})(100\% + \text{percent increase}) && 1 = 100\%\end{aligned}$$

Copyright © Houghton Mifflin Company. All rights reserved.

Thus, to find the new number that results from a percent increase, an efficient method is to multiply the old number by the percent increase plus 100%.

new number = (percent increase + 100%) · (old number)

Example 4

This year, the enrollment at a certain college is 12% higher than last year. If last year's enrollment was 12,000, what is the enrollment this year?

SOLUTION

percent increase + 100% = 12% + 100% = 112%

old enrollment = 12,000

new enrollment = x

new number = (percent increase + 100%) · (old number)

$$x = 112\% \cdot 12{,}000$$
$$= 1.12(12{,}000)$$
$$= 13{,}440$$

This year's enrollment is 13,440.

Your Turn 4

A laundry detergent manufacturer markets a new carton with the label "Now 10% More!" If the old product was in a 90-ounce carton, how many ounces are in the new carton?

Answer: 99

B Percent Decrease

A percent decrease is an indication of how some number has decreased. You may read that unemployment has decreased by 1% or that crime has decreased by 6%.

Percent decreases are handled in much the same way as percent increases. The important thing to remember is that a percent decrease is a percent of the *original number.*

decrease = (percent decrease) · (original number)

Once we know the decrease, we can determine the new number.

new number = old number − decrease

Example 5

From 1980 to 2000, the population of North Dakota decreased by 1.6%. If the population in 1980 was 652,700, what was the population in 2000?

SOLUTION

percent decrease = 1.6% = 0.016

original (1980) population = 652,700

Your Turn 5

In 1914, a school year consisted of 204 days. By the year 2000, that number had decreased by 26%. To the nearest whole number, how many days were in a school year in 2000?

Copyright © Houghton Mifflin Company. All rights reserved.

First, we calculate the decrease in the population.

$$\text{decrease} = (\text{percent decrease}) \cdot (\text{original number})$$

$$\downarrow \qquad\qquad \downarrow \qquad\qquad \downarrow$$

$$\text{decrease} = 0.016 \cdot 652{,}700$$

$$\approx 10{,}443$$

The new number is the 2000 population.

$$\text{new number} = \text{old number} - \text{decrease}$$

$$\downarrow \qquad\qquad \downarrow \qquad\qquad \downarrow$$

$$\text{new number} \approx 652{,}700 - 10{,}443$$

$$\approx 642{,}257$$

The 2000 population was about 642,300.

When we know the original number and the percent decrease, a more efficient method for determining the new number is to use this formula.

$$\text{new number} = (100\% - \text{percent decrease}) \cdot (\text{old number})$$

In Example 5, 100% − (percent decrease) = 100% − 1.6% = 98.4%. Then the new (2000) population can be calculated as follows:

$$\text{new number} = (100\% - \text{percent decrease}) \cdot (\text{old number})$$

$$\downarrow \qquad\qquad \downarrow \qquad\qquad \downarrow$$

$$\text{new number} = 98.4\% \cdot 652{,}700$$

$$= 0.984(652{,}700)$$

$$\approx 642{,}300$$

Example 6

Insulation costs \$52, but a contractor gets a discount and pays only \$44.20. What is the percent decrease in price?

SOLUTION

$$\text{decrease} = 52.00 - 44.20 = 7.80$$

We let x represent the percent decrease.

$$\text{decrease} = (\text{percent decrease}) \cdot (\text{original number})$$

$$\downarrow \qquad\qquad \downarrow \qquad\qquad \downarrow$$

$$7.80 = x \cdot 52.00$$

$$7.80 = 52x$$

$$\frac{7.80}{52.00} = x \qquad \text{Divide both sides by 52.}$$

$$0.15 = x$$

$$15\% = x$$

The contractor's discount is 15%.

Your Turn 6

The rating for a certain TV show dropped from 16.0 to 13.6. What was the percent decrease in the rating?

Answer: 15%

Copyright © Houghton Mifflin Company. All rights reserved.

Example 7

A box spring and mattress set is reduced in price by 14%. If the set is on sale for \$464.40, what was the original price?

SOLUTION

The new price is 100% − 14% = 86% of the old price. Let x represent the old price.

new number = (100% − percent decrease) · (old number)

$$464.40 = 0.86 \cdot x$$

$$464.40 = 0.86x$$

$$\frac{464.40}{0.86} = x \qquad \text{Divide both sides by 0.86.}$$

$$540 = x$$

The original price of the set was \$540.

Your Turn 7

In 2000, there were 39 labor strikes. This number was 90% fewer than the number of strikes in 1970. (Source: Bureau of Labor Statistics.) How many strikes were there in 1970?

Answer: 390

You may be surprised to learn that a percent increase followed by an equal percent decrease does not result in the original number.

Example 8

A store owner buys an item for \$60 and marks it up by 20% for the first selling price. She then puts the item on sale at a 20% discount. What is the second selling price of the item?

SOLUTION

$$\begin{aligned} \text{first selling price} &= \$60 + 0.20(\$60) \\ &= \$60 + \$12 \\ &= \$72 \\ \text{second selling price} &= \$72 - 0.20(\$72) \\ &= \$72 - \$14.40 \\ &= \$57.60 \end{aligned}$$

Note that the second selling price is *not* \$60, the amount paid for the item. The store owner is actually losing money by selling the item for \$57.60.

Your Turn 8

For one week only, a store offers a 10% discount on an item that regularly sells for \$50. At the end of the week, the store marks up the discounted price by 10%. What is the new retail price?

Answer: \$49.50

Copyright © Houghton Mifflin Company. All rights reserved.

8.6 Exercises

A. Percent Increase

1. *Monthly Grocery Bill* After moving to a different state, a family found that their average monthly grocery bill increased by 15%. If the average monthly bill before the move was $620, what was the average bill after the move?

2. *Hurricanes* The number of hurricanes this season is expected to increase by 25% from the number last year. If there were 16 hurricanes last year, how many are expected this year?

3. *Office Rent* Office space in Manhattan rents for $85.30 per square foot. The cost of similar space in London is 30% higher. (Source: *USA Today.*) What is the cost per square foot for office space in London?

4. *Credit Cards* From 1999 to 2005, the number of credit cards is predicted to increase by 12.2%. The number of credit cards in 1999 was 1.4 million. (Source: *Nilson Report.*) How many credit cards are predicted to be in use in 2005?

5. *Motor Vehicle Production* From 1990 to 2000, the number of motor vehicles produced in the United States rose by 31%. In 1990, 9.78 million vehicles were produced. (Source: American Automobile Manufacturers Association.) How many vehicles were produced in 2000?

Copyright © Houghton Mifflin Company. All rights reserved.

6. *Sports Radio* In 1999, the number of all-sports radio stations was 256. That number increased by 32% in 2001. (Source: *USA Today.*) What was the number of all-sports radio stations in 2001?

7. *Checking Account Balance* Suppose that the balance in your checking account was $1,220. If you deposited $244, by what percent did the balance increase?

8. *Stock Price* The value of a share of stock rose from $36.60 to $42.09. What was the percent increase in the price?

9. *Athletic Shoes* Spending on athletic footwear increased from $11.6 billion in 1990 to $13.7 billion in 2000. (Source: National Sporting Goods Association.) What was the percent increase in spending?

10. *European Travel* From 1990 to 1999, the number of Americans traveling to Europe increased from 4.2 million to 6 million. (Source: Europe Travel Commission.) What was the percent increase?

11. *Singles* From 1980 to 2000, the number of adults who have never married rose from 32.3 million to 48.2 million. (Source: U.S. Bureau of Census.) What was the percent increase?

Copyright © Houghton Mifflin Company. All rights reserved.

12. *Tobacco Exports* The number of tons of tobacco exported from the United States declined from 223,000 in 1990 to 183,000 in 2000. (Source: Department of Agriculture.) What was the percent decrease?

13. *College Tuition* A college raised its tuition by 8%. If the current tuition is \$1,026, what was the tuition before the increase?

14. *Lost Luggage* The number of daily complaints about lost luggage that an airline received increased by 16% from the number last year. If the airline received 87 complaints per day this year, how many complaints per day did it receive last year?

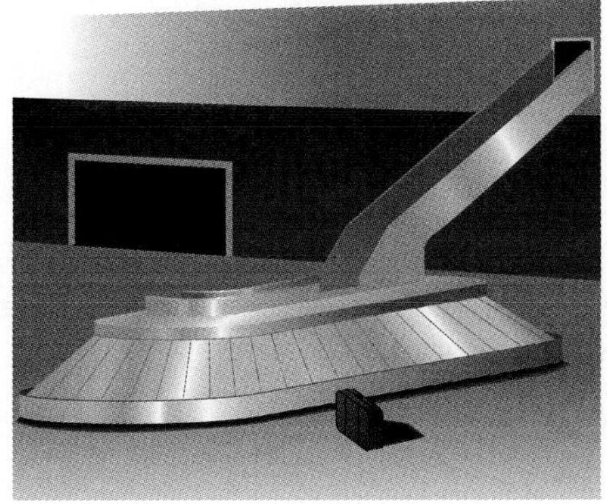

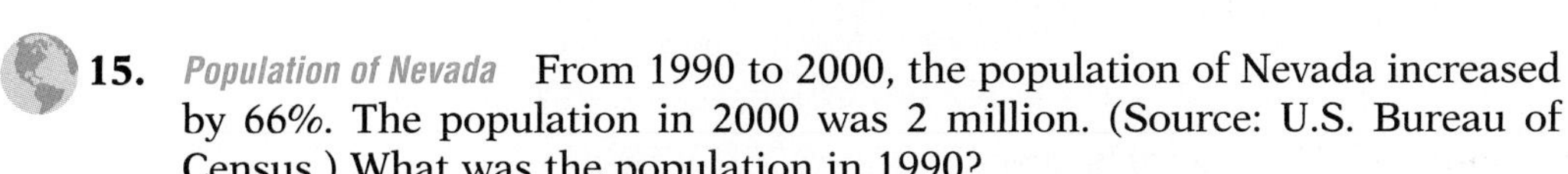

15. *Population of Nevada* From 1990 to 2000, the population of Nevada increased by 66%. The population in 2000 was 2 million. (Source: U.S. Bureau of Census.) What was the population in 1990?

16. *Salary Comparison* The average salary for a person with a bachelor's degree is \$40,500, which is 77% higher than the salary for a person with only a high school diploma. (Source: U.S. Bureau of Census.) What is the average salary for a person with a high school diploma?

Copyright © Houghton Mifflin Company. All rights reserved.

17. *College Education* In 2000, 3.3 million African American adults had 4 or more years of college education. This was a 74% increase from the number in 1990. (Source: U.S. Bureau of Census.) How many had 4 or more years of college education in 1990?

18. *Inmate Population* The number of people in state and federal prisons in 2000 was 1.4 million, which was an 86% increase over the number in 1990. (Source: U.S. Justice Department.) How many people were in state and federal prisons in 1990?

B. Percent Decrease

19. *Electric Bill* By adjusting the thermostat for the air conditioner, a person's summer electric bill decreased by 14% from last summer's bill of $780. What was this summer's bill?

20. *Dow Jones Average* The Dow Jones Industrial Average dropped 3% from its opening average of 11,300. What was the closing average?

21. *Baseball Attendance* Last year, the average attendance at a certain college's baseball games was 3,400. This year the attendance decreased by 9%. What was the attendance this year?

Copyright © Houghton Mifflin Company. All rights reserved.

22. *Voter Turnout* In a certain precinct, 1,250 people voted in the last election. The turnout at the next election is predicted to decrease by 14%. What is the predicted number of voters at that election?

23. *Cell Phone Cost* In 1998, the cost of a cell phone call was 33¢ per minute. In 2003, the cost was expected to decrease by 40%. (Source: Strategis Group.) What was the expected cost per minute in 2003?

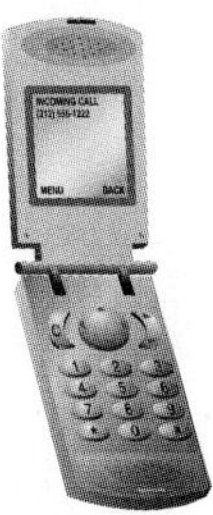

24. *Washington, D.C., Population* The population of Washington, D.C., declined by 6% from 1990 to 2000. The population in 1990 was 607,000. (Source: U.S. Bureau of Census.) What was the population in 2000?

25. *Weight Loss* Following a diet and exercise program, a person's weight was reduced from 190 pounds to the recommended weight of 146 pounds. What was the percent decrease?

26. *Heating Bill* By adding insulation, a home owner reduced the average monthly heating bill from $180 to $146. What was the percent decrease?

27. *Birth Rate* The number of babies born in the United States dropped from 4.2 million in 1990 to 3.9 million in 1999. (Source: *USA Today.*) What was the percent decrease?

Copyright © Houghton Mifflin Company. All rights reserved.

28. *Milk Consumption* In 1980, each person in the United States drank an average of 27.6 gallons of milk each year. By 2000, the average had dropped to 23.6 gallons. (Source: U.S. Department of Agriculture.) What was the percent decrease?

29. *Poverty Level Population* The total number of people below the poverty level dropped from 40 million in 1960 to 30 million in 2000. (Source: U.S. Bureau of Census.) What was the percent decrease?

30. *Skilled Nursing Facilities* From 1980 to 2000, the number of skilled nursing facilities increased from 5,155 to 15,313. (Source: U.S. Health Care Financing Administration.) What was the percent increase?

31. *Carpet Cleaning* A carpet cleaning service offered an introductory rate of \$35.70 per room for up to 5 rooms. If the price was 16% off the regular charge, what was the regular charge to clean the carpet in one room?

32. *Software* CompAmerica sold Instant Office software for \$19.95, which was 65% off the regular price. What was the original price?

Copyright © Houghton Mifflin Company. All rights reserved.

33. *Crime Rate* According to the FBI, the number of crimes per 100,000 population was 5,080, which was a drop of 3% from the previous year. What was the number of crimes per 100,000 population the previous year?

34. *PAC Contributions* From 1994 to 2000, the PAC contributions to House of Representative candidates increased by 48%. Contributions in 2000 were $196 million. (Source: U.S. Federal Elections Commission.) How much were the contributions in 1994?

35. *NBA Scoring* In 2001, NBA teams averaged 93.8 points per game, which was a 13% decrease from the 1990 average. (Source: NBA.) What was the 1990 average?

36. *Oil Imports* In 2000, the United States imported 3.5 million barrels of crude oil, which was an 80% increase from the number of barrels imported in 1980. (Source: U.S. Energy Information Administration.) How many barrels were imported in 1980?

Writing and Concept Extension

37. A person attempts to calculate the percent decrease from 50 to 40. The person writes $\frac{10}{40} = 0.25 = 25\%$. What mistake did the person make?

38. Is increasing a number by 10% and then increasing the result by 20% the same as increasing the original number by 30%? Why?

39. Suppose that you need to determine 40 increased by 150%. Which of the following is correct?

(i) $2.5 \cdot 40$ (ii) $1.5 \cdot 40$ (iii) $40 + 1.5 \cdot 40$

Copyright © Houghton Mifflin Company. All rights reserved.

Exploring with Real-World Data: Collaborative Activities

Population Trends The table shows the current and predicted populations (in billions) for selected countries and the world. (Source: U.S. Bureau of Census.)

	Present	*2050*
Germany	0.082	0.057
India	0.98	1.70
United States	0.27	0.39
World	5.9	9.3

40. What percent of the present world population is the population of India?

41. According to the prediction for 2050, what percent of the world population will be the population of India?

42. What is the predicted percent increase in the population for India?

43. What is the predicted percent decrease in the population for Germany?

44. What is the predicted percent increase in the population for the United States?

45. What is the predicted percent increase in the world population?

46. Suppose that the world population actually increases by 25% from the present to 2050. What would be the world population in 2050?

47. Suppose that the population of Germany actually decreases by 18% from the present to 2050. What would be the population (in millions) of Germany in 2050?

Copyright © Houghton Mifflin Company. All rights reserved.

CHAPTER 8 REVIEW EXERCISES

Section 8.1

1. The word ______ means "per 100" or "for every 100."

In Exercises 2–4, write the percent as a simplified fraction.

2. *Reading for Pleasure* *USA Today* reported that 30% of the people in the United States said that reading was their favorite leisure activity.

3. $3\frac{1}{4}\%$

4. 240%

5. Write 0.09 as a percent.

In Exercises 6 and 7, write the percent as a decimal.

6. 34%

7. 0.75%

In Exercises 8–10, write the fraction as a percent.

8. *Grills* According to *Weber Grill Watch*, $\frac{21}{50}$ of the grills sold in the United States were charcoal grills.

9. $\frac{7}{8}$

10. $\frac{11}{6}$

Section 8.2

11. In the basic percent equation $PB = A$, what do P, B, and A represent?

In Exercises 12–14, fill in the blank.

12. 7% of 21 is ______.

13. ______% of 45 is 9.

14. 3.5% of ______ is 14.

15. What number is 45% of 90?

16. 49 is what percent of 140?

17. 27 is 60% of what number?

Copyright © Houghton Mifflin Company. All rights reserved.

18. Explain how to find 1% of a number.

19. Find 10% of 540.

20. *Registered Voters* Suppose that a precinct has 1,825 registered voters, of which 949 are women. What percent of the registered voters are women?

Section 8.3

21. *Weight Loss* Of 60 people who participated in a weight loss study, 33 lost the target amount of weight. What percent of those in the study lost the target amount of weight?

22. *Work Force Composition* If 396 new employees represent 16% of a company's total work force, what is the total number of employees?

23. *Minority Students* In a school system, 34% of the students are minority students. If the system has 21,600 students, how many are minority students?

24. *First-Year Students* Of the 7,850 students enrolled at a certain college, 2,512 were first-year students. What percent of the total enrollment was first-year students?

25. *Bananas* Suppose that a produce manager expects to discard 15% of all produce due to spoilage. In a shipment of 80 pounds of bananas, how many pounds should the manager expect to discard?

Copyright © Houghton Mifflin Company. All rights reserved.

26. *Acid Solution* A metal drum contains a 12% acid solution. If the solution contains 1.8 liters of acid, how many liters of solution are in the drum?

27. *Recreational Fishing* In the United States, 29 million men and 4.1 million women fish as a form of recreation. (Source: U.S. Bureau of Census.) What percent of all those who fish are men?

28. *Parking Lot* In a 320-acre shopping complex, 37.5% of the land is used for parking. How many acres of parking are in the complex?

29. *Annual Salary* A person received a 7% raise, which resulted in an additional $2,940 per year. What was the annual salary before the raise?

30. *College Financial Aid* Of the 1,350 students at a community college, 64% received financial aid. How many students received financial aid?

Section 8.4

31. A loan payoff is the sum of the ________ and the ________.

Copyright © Houghton Mifflin Company. All rights reserved.

32. *Loan Payoff* Suppose that you borrowed \$2,800 at a simple interest rate of 8% for 1 year. What is the loan payoff?

33. *Investment Interest* A salesperson received a \$4,200 bonus and invested it at 7.5% annual simple interest for 8 months. How much interest did the person earn?

34. *Loan Rate* If the payoff for a 4-month loan of \$2,650 is \$2,703, what was the annual simple interest rate of the loan?

35. *Savings Account* A bank savings account paid 4% interest compounded semiannually. If you deposited \$1,000 in the account, what was the account balance at the end of 1 year?

36. *Retirement Account* Suppose that you invested \$2,000 in a retirement account that paid 6% compounded monthly. What was the value of the account in 8 years?

37. *Television Cost* A television was advertised for \$346 plus a 6% sales tax.

(a) What was the tax on the television?

(b) What was the total cost of the television?

38. *Luggage Price* The sales tax on a set of luggage was \$9.10. If the sales tax rate was 3.5%, what was the price of the luggage?

Copyright © Houghton Mifflin Company. All rights reserved.

39. *Sales Tax Rate* The total cost of a blouse priced at $27 was $27.81. What was the sales tax rate?

40. *Take-Home Pay* If you made $690 in a week but your employer withheld 14% for taxes, what was your take-home pay?

Section 8.5

41. *Ad Sales* A salesperson for a magazine received a salary of $2,000 per month plus an 8% commission on all advertisements sold. If he sold $17,000 worth of ads, what was his income for the month?

42. *Paper Shredder Price* An office supply salesperson received a commission of $29.05 for selling a heavy duty paper shredder. If the commission rate was 4%, what was the price of the shredder?

43. *Real Estate Commission* A home sold for $140,000 and the seller received $127,400. What was the real estate agent's commission rate?

44. *Pizza Oven* The original cost of a commercial pizza oven was $1,400. The estimated useful life of the oven was 8 years. If the straight-line depreciation method is used, what is the depreciated value after 3 years?

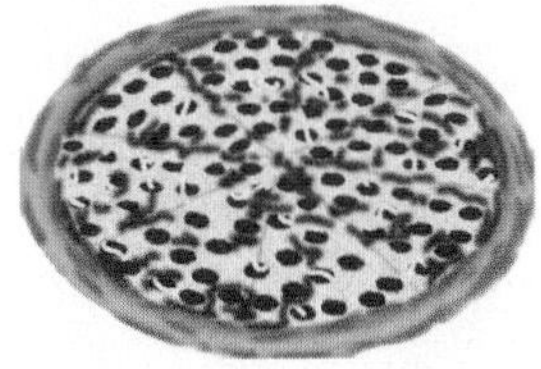

Copyright © Houghton Mifflin Company. All rights reserved.

45. *Lawn Tractor* A lawn tractor that cost $4,200 depreciated at a rate of 10%. If the exponential depreciation method is used, what is the depreciated value after 2 years?

46. The difference between an original retail price and a reduced retail price is called a(n) ______.

47. *Textbook Price* A college bookstore paid $56 for a history textbook. If the markup rate was 35%, what was the retail price of the book?

48. *Snow Tires* A tire dealer sold snow tires for $92.16. If the dealer bought the tires for $72, what was the markup rate?

49. *Lawn Mower* A hardware store advertised a lawn mower, originally priced at $142, at a discount of 45%. What was the sale price?

50. *Coffeemaker* A coffeemaker, originally priced at $39.40, was discounted to $29.55. What was the discount rate?

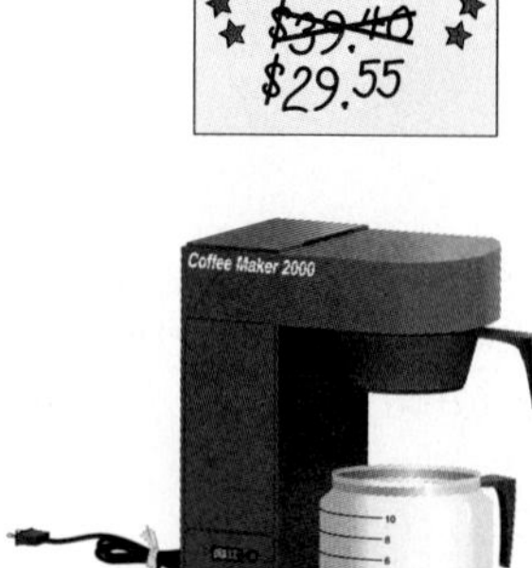

Copyright © Houghton Mifflin Company. All rights reserved.

Section 8.6

51. *Hourly Pay* A clerk made \$7.50 per hour. What was the clerk paid per hour after a 6% increase in pay?

52. *College Tuition* In 2000, the tuition for one year at a public college had increased 22% from the 1995 tuition of \$2,057. (Source: U.S. Center for Education Statistics.) What was the tuition for 2000?

53. *Pell Grants* The amount of Pell grant money available to college students rose from \$7.2 billion in 1998 to \$9.1 billion in 2001. (Source: U.S. Department of Education.) What was the percent increase?

54. *Consumer Debt* Consumer debt increased from \$902 million in 1994 to \$1,569 million in 2000. (Source: *USA Today.*) What was the percent increase?

55. *Work Force Size* As a result of a new contract, Russ Manufacturing increased its work force by 8% to 513 people. How many people did the company employ before the increase?

56. *Calorie Intake* A doctor instructed a patient to reduce his daily calorie intake by 35%. If the patient averaged 4,000 calories per day before the diet, what was the daily average for the diet?

Copyright © Houghton Mifflin Company. All rights reserved.

57. *Birthrate* From 1980 to 2000, the birthrate in the United States decreased by 8.8%. If the 1980 rate was 15.9 per 1,000 population, what was the 2000 rate per 1,000 population? (Source: U.S. National Center for Health Statistics.)

58. *Check Writing* From 1999 to 2005, the number of checks written is expected to decrease from 29.4 billion to 23.8 billion. (Source: *Nilson Report.*) What is the percent decrease?

59. *Loudspeakers* Circuit Town advertised bookshelf loudspeakers for $300 each. If the price was 40% off the regular price, what was the regular price?

60. *Air Compressor* As a grand opening special, Home Warehouse offered an air compressor for $259. If the price was 20% off the regular price, what was the regular price of the compressor?

Copyright © Houghton Mifflin Company. All rights reserved.

CHAPTER 8 TEST

1. To write a decimal as a percent, shift the decimal point 2 places to the ________ and attach the ________ symbol.

2. Explain how to find 10% of a number.

3. Suppose that you need to calculate $33\frac{1}{3}\%$ of 150. How should you express the percent to make the calculation easy?

4. Write the fraction as a percent.

   (a) $\frac{7}{20}$ **(b)** $\frac{2}{15}$

5. Write the percent as a simplified fraction and as a decimal.

   (a) 35% **(b)** 120%

6. Write the percent as a decimal.

   (a) 0.4% **(b)** $3\frac{3}{4}\%$

7. Write the decimal as a percent.

   (a) 0.6 **(b)** 1.4

8. Find 70% of 520.

9. What percent of 80 is 12?

10. 130% of what number is 78?

11. *Snack Mix* A snack mix contains 22% nuts. How many ounces of nuts are in 30 ounces of the mix?

12. *Hockey Team Record* A hockey team won 22 games and lost 10. What percent of the games did the team win?

Copyright © Houghton Mifflin Company. All rights reserved.

13. *Vitamin C* A vitamin tablet contains 48 milligrams of vitamin C, which is 80% of the recommended daily amount. How many milligrams of vitamin C are recommended each day?

14. *Business Loan* To expand the inventory in a craft shop, the manager borrows $6,200 for 90 days. If the bank charges 9% annual simple interest on the loan, what is the loan payoff?

15. *Bond Fund Investment* A bond fund pays 8% interest compounded monthly. After 2 months, what is the value of an investment of $2,000?

16. *Desk Cost* The total cost of a desk is $299.60. If the desk is advertised for $280, what is the sales tax rate?

17. *Take-Home Pay* If you earn $430, but your employer deducts 15% for federal tax and 6% for state tax, what is your take-home pay?

18. *Real Estate Commission* A real estate agent sells a house for $120,000. What is the agent's commission if the commission rate is 6%?

19. *Band Saw Depreciation* The straight-line depreciation rate for a band saw that costs $650 is 15%. What is the depreciated value after 2 years?

Copyright © Houghton Mifflin Company. All rights reserved.

20. *Golf Store Retail Price* A golf store marks up items by 45%. If the wholesale price of a putter is \$108, what is the retail price?

21. *Tennis Racket Discount* A tennis racket that regularly sells for \$120 is on sale for \$80. What is the discount rate?

22. *Construction Cost* Due to construction problems, the cost of a new building increased from \$3.2 million to \$4 million. What was the percent increase?

23. *National Parks Visitors* From 1995 to 2000, the number of visits to national parks increased by 7%. The number of visits in 1995 was 270 million. (Source: National Park Service.) What was the number of visits in 2000?

24. *Amusement Park Visitors* The average number of visitors to a new amusement park decreased by 18% after the opening month. If the daily average for the opening month was 1,200, what was the average after the opening month?

25. *Apple Price* The price per pound of apples decreased from \$1.40 to \$.98. What was the percent decrease?

Copyright © Houghton Mifflin Company. All rights reserved.

CHAPTER 9

Polynomials

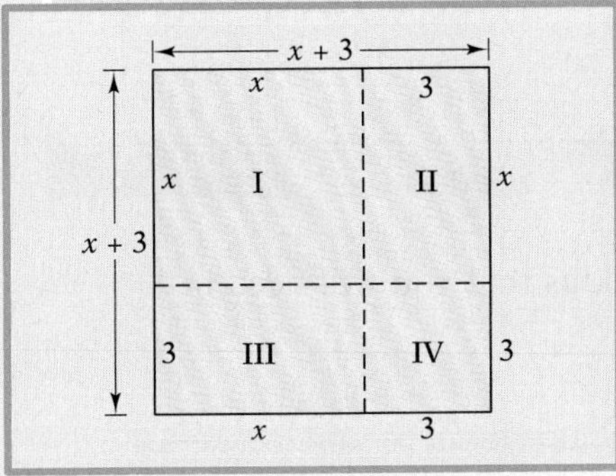

In the figure, the length of each side of the square is $x + 3$. Therefore, the area of the square is $(x + 3)(x + 3)$, or $(x + 3)^2$.

The area of the square also can be found by adding the areas of the four regions.

Region	*Area*
I	x^2
II	$3x$
III	$3x$
IV	3^2

Total area $= x^2 + 3x + 3x + 3^2 = x^2 + 6x + 9$

Because both methods result in the same area, we see that

$$(x + 3)^2 = (x + 3)(x + 3) = x^2 + 6x + 9$$

The expressions $x + 3$ and $x^2 + 6x + 9$ are examples of **polynomials.** With some visual help, we have seen the result of multiplying $x + 3$ by itself. Of course, we would like to be able to perform such an operation without having to draw a picture. In this chapter, you will learn how to do that.

Chapter Snapshot

After we discuss the meaning of a polynomial and other related words, we present methods for adding and subtracting polynomials. We then turn to some rules and techniques for multiplying expressions that involve exponents. You will use some of that information in our next topic, multiplying polynomials. Finally, we give you a start on the important skill of factoring polynomials.

For online resources visit the web site **math.college.hmco.com/students** and follow the links to Hubbard/Robinson, *Prealgebra.*

Copyright © Houghton Mifflin Company. All rights reserved.

Some Friendly Advice . . .

The prefix *pre-* means before, and so *pre-algebra* should mean "before algebra." Why, then, have we included algebra topics in this chapter and throughout the book?

To prepare you for algebra, our primary focus has been on integers, fractions, decimals, and percents—that is, the arithmetic that you need to know before algebra can make any sense. However, the topics of algebraic expressions, exponents, equations, and polynomials are at the heart of any study of beginning algebra. Therefore, previewing these topics now will make your life easier when you see them again.

A final note. If you have come this far, then you have probably overcome any fears that you might have had about mathematics. That's good! We wish you continued success.

WARM-UP SKILLS

The following questions review concepts and skills that you will need in Chapter 9:

1. Identify the terms in $5a + 2b + 9$.

In Exercises 2–4, simplify by combining like terms.

2. $6x + 10x$

3. $3a - 4a$

4. $5y + 2 - 7 + 6y$

5. Evaluate $7x + 8$ for $x = -3$.

6. Evaluate 4^3.

In Exercises 7 and 8, use the Distributive Property to remove the parentheses.

7. $4(x + 5)$

8. $-2(3n - 7)$

9. List the prime numbers that are less than 10.

10. Prime-factor 24.

Copyright © Houghton Mifflin Company. All rights reserved.

9.1 Adding and Subtracting Polynomials

SUGGESTIONS FOR SUCCESS

You won't be reading this unless your instructor has covered earlier algebra topics or you are working on your own.

We strongly encourage you to begin by reviewing Section 3.7 very carefully. Although we will review some of those topics, we will do so quickly. A thorough understanding of the vocabulary, concepts, and methods of Section 3.7 is essential to your success with this material.

A Introduction to Polynomials

Throughout this book, we have introduced you to some basic vocabulary and rules of algebra. In this chapter, we conclude with some essential skills that will help you bridge the gap to a real course in algebra.

Recall that the addends of an algebraic expression are called *terms*.

$$2x + 3y + 5$$

$\uparrow \quad \uparrow \quad \uparrow$

This expression has three terms: $2x$, $3y$, and 5.

Using the fact that $a - b = a + (-b)$, we can write any subtraction as an addition. We need to do this before we can identify terms.

$$3x^2 - 7x + 8 = 3x^2 + (-7x) + 8$$ The terms are $3x^2$, $-7x$, and 8.

Terms that have the same variable parts are called *like terms*. We use the Distributive Property to *combine* like terms.

$$5x + 3x = (5 + 3)x = 8x$$

Like terms

$$9y^2 + 2y + (-4y^2) = (9 - 4)y^2 + 2y = 5y^2 + 2y$$

Like terms

We give a special name to certain terms.

Definition of a Monomial

A **monomial** is a term whose exponents are whole numbers and that has no variable in the denominator.

These terms are monomials.	*These terms are not monomials.*	
$7x^2$	$7x^{-2}$	The exponent is not a whole number.
$\frac{y}{3}$	$\frac{y}{3x}$	There is a variable in the denominator.
$-2a^3x^4$		
9		

A sum or difference of monomials is also given a special name.

Definition of a Polynomial

A **polynomial** is a sum or difference of monomials.

Copyright © Houghton Mifflin Company. All rights reserved.

Here are some examples of polynomials. Note that the terms in each polynomial are monomials.

$$2x - 7 \qquad 3a^2 - 2a + 5 \qquad \frac{1}{2}x + \frac{1}{3} \qquad abc + 4$$

Example 1

Identify the terms of the given expression. Then decide whether the expression is a polynomial.

(a) $x + \frac{1}{x}$

(b) $a^2 + a - 1$

SOLUTION

(a) The terms are x and $\frac{1}{x}$.

Because $\frac{1}{x}$ is not a monomial, the expression is not a polynomial.

(b) $a^2 + a - 1 = a^2 + a + (-1)$
The terms are a^2, a, and -1, which are all monomials. Therefore, the expression is a polynomial.

Your Turn 1

Identify the terms of the given expression. Then decide whether the expression is a polynomial.

(a) $x^3 - 4x$

(b) $\frac{1}{2}x^2 - x + \frac{2}{3}$

Answers: (a) x^3, $-4x$; polynomial; (b) $\frac{1}{2}x^2$, $-x$, $\frac{2}{3}$; polynomial

We evaluate a polynomial in the same way that we evaluate any algebraic expression. Be sure to follow the Order of Operations.

Example 2

Evaluate $2x^2 - 9x + 4$ for $x = -2$.

SOLUTION

$$2x^2 \quad - \quad 9x \quad + 4$$
$$\downarrow \qquad\qquad \downarrow$$

$2(-2)^2 - 9(-2) + 4$	Replace each x with -2.
$= 2(4) - 9(-2) + 4$	Square the -2.
$= 8 + 18 + 4$	Then find the products.
$= 30$	Finally, find the sum.

Your Turn 2

Evaluate $x^3 + x - 4$ for $x = -1$.

Answer: -6

B Adding Polynomials

When two or more polynomials are to be added, the polynomials are usually enclosed in parentheses.

$(2x + 3) + (5x - 2)$	The sum of 2 polynomials
$(x^2 + 3x - 1) + (4x + 2) + (3x^2 - 8)$	The sum of 3 polynomials

To perform the addition we use the Associative Property of Addition to remove the parentheses. Then we use the Commutative Property of Addition to rearrange the terms so that like terms are together. Finally, we combine the like terms.

Copyright © Houghton Mifflin Company. All rights reserved.

Example 3

Add $(2x + 3) + (5x - 2)$.

SOLUTION

$(2x + 3) + (5x - 2) = 2x + 3 + 5x - 2$	Remove parentheses.
$= 2x + 5x + 3 - 2$	Rearrange terms.
$= (2x + 5x) + (3 - 2)$	Group like terms.
$= 7x + 1$	Combine like terms.

$(2x + 3) + (5x - 2) = 7x + 1$

Your Turn 3

Add $(-4a + 1) + (6a - 3)$.

Answer: $2a - 2$

If you are good at identifying like terms, then the first two steps in Example 3 can be omitted. Simply pick out the like terms and group them together.

Example 4

Add $(3x^2 - 5x + 4) + (x^2 + 2x + 6)$.

SOLUTION

$(3x^2 - 5x + 4) + (x^2 + 2x + 6)$

$= (3x^2 + 1x^2) + (-5x + 2x) + (4 + 6)$ We write x^2 as $1x^2$.

$= 4x^2 - 3x + 10$

Your Turn 4

Add $(x^2 - 3x - 5) + (2x^2 + 7x + 9)$.

Answer: $3x^2 + 4x + 4$

LEARNING TIP

Writing x^2, for example, as $1x^2$ can be helpful when you are combining like terms. Also, in the polynomial $3x^2 - 5x + 4$, the middle term is actually $-5x$. Write $3x^2 + (-5x) + 4$ until you are confident enough to keep track of the signs mentally.

We also can add polynomials vertically by writing the polynomials so that the like terms are lined up. Here again is the problem in Example 4, this time with the polynomials added vertically.

$$\begin{array}{r} 3x^2 - 5x + 4 \\ +\ 1x^2 + 2x + 6 \\ \hline 4x^2 - 3x + 10 \end{array}$$

The advantage of this method is that the like terms are already lined up in a column, and so they can be combined easily.

Example 5

Use the vertical format to find the sum of $x^3 + 2x - 1$ and $3x^2 - 8x + 7$.

SOLUTION

The first polynomial has an x^3 term, but the second polynomial does not. Also, the second polynomial has an x^2 term, but the first polynomial does not.

One way to account for these missing terms is to leave holes or spaces. Here is one way to write the problem.

$$\begin{array}{rrrr} 1x^3 & & +\ 2x & -\ 1 \\ + & 3x^2 & -\ 8x & +\ 7 \\ \hline \end{array}$$

Your Turn 5

Use the vertical format to find the sum of $2x^3 + 8x^2 - 6$ and $x^3 + 9x + 13$.

continued

Copyright © Houghton Mifflin Company. All rights reserved.

Another method is to write $0x^3$ and $0x^2$ just to fill in the columns.

$$\begin{array}{r} 1x^3 + 0x^2 + 2x - 1 \\ + \; 0x^3 + 3x^2 - 8x + 7 \\ \hline 1x^3 + 3x^2 - 6x + 6 \end{array}$$

The sum is $x^3 + 3x^2 - 6x + 6$.

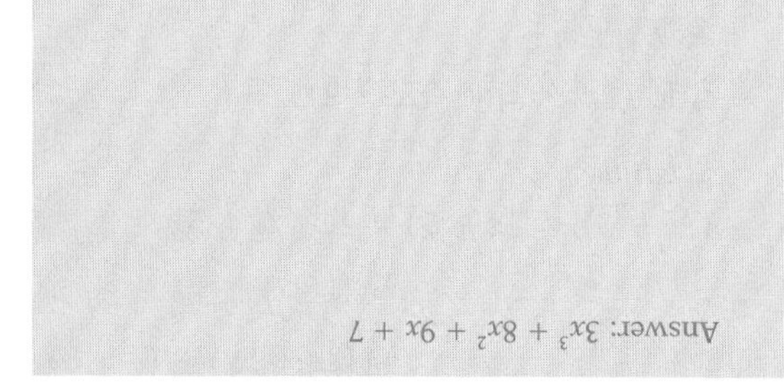

C Subtracting Polynomials

We subtract polynomials in the same way that we subtract numbers. If A and B represent polynomials, then $A - B = A + (-B)$. In words, we change the minus symbol to a plus symbol, and we change the second polynomial to its opposite.

Change the minus to a plus.

$$(7x + 2) - (3x - 4) = (7x + 2) + [-(3x - 4)]$$

Change $3x - 4$ to its opposite.

In Section 3.7, we found that we can take the opposite of an expression simply by changing the signs of the terms of the expression.

$$-(3x - 4) = -3x + 4$$

Therefore, we can write

$(7x + 2) - (3x - 4) = (7x + 2) + [-(3x - 4)]$	Change $-$ to $+$. Change $(3x - 4)$ to $-(3x - 4)$.
$= (7x + 2) + (-3x + 4)$	Change the signs in $(3x - 4)$.
$= (7x - 3x) + (2 + 4)$	Group like terms.
$= 4x + 6$	Combine like terms.

Here is a summary of the procedure for subtracting polynomials.

Subtracting Polynomials

To subtract polynomials, follow these steps.

1. Change the signs of the terms of the polynomial that you are subtracting.
2. Add the polynomials.

Example 6

Subtract $(15y - 8) - (9y + 6)$.

SOLUTION

$(15y - 8) - (9y + 6)$	
$= (15y - 8) + (-9y - 6)$	Change $-$ to $+$. Change $(9y + 6)$ to $(-9y - 6)$.
$= (15y - 9y) + (-8 - 6)$	Group like terms.
$= 6y + (-14)$	Combine like terms.
$= 6y - 14$	

Your Turn 6

Subtract $(4x + 7) - (9x - 1)$.

Answer: $-5x + 8$

Copyright © Houghton Mifflin Company. All rights reserved.

Example 7

Subtract $(-3x^2 + 4) - (x^2 + 5x - 1)$.

SOLUTION

$(-3x^2 + 4) - (x^2 + 5x - 1)$

$= (-3x^2 + 4) + (-x^2 - 5x + 1)$ — Change − to +. Change $(x^2 + 5x - 1)$ to $(-x^2 - 5x + 1)$.

$= (-3x^2 - 1x^2) + (-5x) + (4 + 1)$ — Group like terms.

$= -4x^2 - 5x + 5$ — Combine like terms.

Your Turn 7

Subtract $(5x^2 - 6x + 3) - (2x^2 + x - 6)$.

Answer: $3x^2 - 7x + 9$

In Example 7, the polynomial $x^2 + 5x - 1$ can be written $x^2 + 5x + (-1)$. When we change the signs, the polynomial becomes $-x^2 + (-5x) + 1 = -x^2 - 5x + 1$. We make these sign changes mentally without using the Definition of Subtraction each time.

As with addition, we can subtract in vertical columns. Remember, though, that you must change the signs of the terms of the polynomial that you are subtracting.

Example 8

Subtract $(8a^2 - 7a + 3) - (-2a^2 + 6a - 5)$.

SOLUTION

As we write the polynomials in vertical columns, we leave the first polynomial as is and change the signs of the terms of the second polynomial.

$$\begin{array}{r} 8a^2 - 7a + 3 \\ \underline{-\ -2a^2 + 6a - 5} \end{array} \rightarrow \begin{array}{r} 8a^2 - 7a + 3 \\ \underline{+\ +2a^2 - 6a + 5} \end{array}$$

Now we add the polynomials.

$$\begin{array}{r} 8a^2 - 7a + 3 \\ \underline{+\ 2a^2 - 6a + 5} \\ 10a^2 - 13a + 8 \end{array}$$

Your Turn 8

Use the vertical format to subtract $(3x^2 - 8)$ from $(6x^2 + 5x + 2)$.

Answer: $3x^2 + 5x + 10$

Example 9

A person drove the first $2x + 16$ miles of a 70-mile trip.

0 — 70

A — $2x + 16$ — B — ? — C

Write a polynomial that represents the remaining distance to be driven.

SOLUTION

The remaining distance is the distance from B to C, which is the total distance minus the distance from A to B.

Your Turn 9

A 12-foot board is cut into two pieces. The length of one piece is represented by $7x + 3$. Write a polynomial that represents the length of the other piece.

continued

Copyright © Houghton Mifflin Company. All rights reserved.

Remaining distance = total distance − distance from A to B

$$= 70 - (2x + 16)$$

$$= 70 + (-2x - 16) \quad \text{Change } - \text{ to } +. \text{ Change } (2x + 16) \text{ to } (-2x - 16).$$

$$= (70 - 16) + (-2x)$$

$$= 54 + (-2x)$$

$$= 54 - 2x$$

The remaining distance is $54 - 2x$.

Answer: $9 - 7x$

Example 10

Write a polynomial that represents the perimeter of the figure.

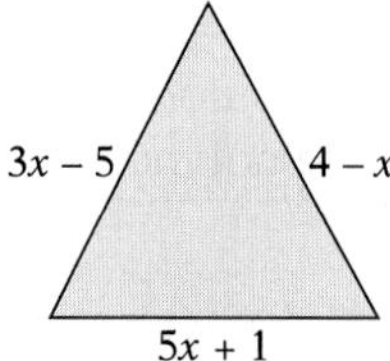

SOLUTION

The perimeter P is the sum of the lengths of the sides.

$$P = (3x - 5) + (4 - x) + (5x + 1)$$

$$= (3x - 1x + 5x) + (-5 + 4 + 1)$$

$$= 7x$$

Note that in this case the polynomial consists of just one term; that is, the polynomial is a monomial.

Your Turn 10

Write a polynomial that represents the perimeter of the figure.

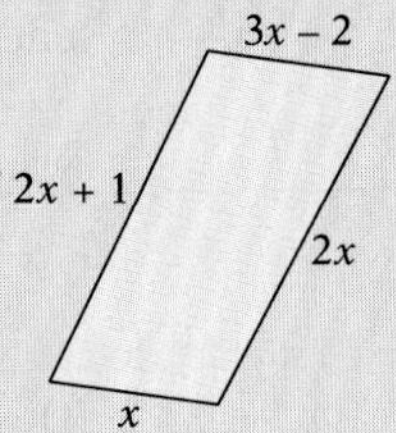

Answer: $8x - 1$

Example 11

A side of square A is represented by $2x + 1$. A side of a larger square B is represented by $3x - 1$. What is the difference in their perimeters?

SOLUTION

The perimeter P of a square is 4 times the length of its sides.

Square A: $P = 4(2x + 1) = 8x + 4$

Square B: $P = 4(3x - 1) = 12x - 4$

The difference between the perimeters is

$$(12x - 4) - (8x + 4)$$

$$= (12x - 4) + (-8x - 4)$$

$$= [12x + (-8x)] + [-4 + (-4)]$$

$$= 4x - 8$$

Your Turn 11

An *equilateral triangle* is a triangle whose sides are the same length. A side of an equilateral triangle A is represented by $x + 6$. A side of a larger equilateral triangle B is represented by $2x$. What is the difference in their perimeters?

Answer: $3x - 18$

Copyright © Houghton Mifflin Company. All rights reserved.

When you take algebra, you will spend a lot of time with polynomials. Of the two operations, addition is the easier because it requires only that we combine like terms. To subtract, we first need to change the signs of the terms of the polynomial that we are subtracting, but then we are right back to adding. Sometimes the big words are harder to learn than the operations.

9.1 Quick Reference

A. Introduction to Polynomials

1. The addends of an expression are called *terms*. Terms that have the same variable parts are called *like terms*.

Terms

$$x^2 + 5x + 3x + 4$$

Like terms

2. Using the fact that $a - b = a + (-b)$, we can write any subtraction as an addition. We need to do that before we can identify terms.

$5n - 7 = 5n + (-7)$
The terms are $5n$ and -7.

3. A **monomial** is a term whose exponents are whole numbers and that has no variable in the denominator. A **polynomial** is a sum or difference of monomials.

Monomials

$$3x^2 - 8x + 7$$

Polynomial

4. We evaluate a polynomial by replacing each occurrence of the variable with its given value.

If $x = 3$, then

$$\begin{aligned} &x^2 - 4x + 7 \\ &= (3)^2 - 4(3) + 7 \\ &= 9 - 4(3) + 7 \\ &= 9 - 12 + 7 \\ &= 4 \end{aligned}$$

B. Adding Polynomials

1. To add polynomials, we remove the parentheses and combine like terms.

$$\begin{aligned} &(2x^2 - 3x + 1) + (x^2 - 5) \\ &= (2x^2 + x^2) - 3x + (1 - 5) \\ &= 3x^2 - 3x - 4 \end{aligned}$$

2. The vertical format has the advantage of lining up like terms in columns.

$$\begin{array}{r} 4x^2 - 7x + 5 \\ +\ 3x^2 + 9x - 4 \\ \hline 7x^2 + 2x + 1 \end{array}$$

Copyright © Houghton Mifflin Company. All rights reserved.

C. Subtracting Polynomials

1. To subtract polynomials, follow these steps.

(a) Change the signs of the terms of the polynomial that you are subtracting.
(b) Add the polynomials.

$$(9x - 4) - (3x + 7)$$
$$= (9x - 4) + (-3x - 7)$$
$$= (9x - 3x) + (-4 - 7)$$
$$= 6x + (-11)$$
$$= 6x - 11$$

2. We can subtract polynomials with the vertical format. Remember to change the signs of the terms of the polynomial that you are subtracting.

Subtract.

$$\begin{array}{r} 6a^2 + 3a - 4 \\ -\quad 2a^2 - 5a + 1 \\ \hline \end{array}$$

$$\begin{array}{r} 6a^2 + 3a - 4 \\ + \ -2a^2 + 5a - 1 \\ \hline 4a^2 + 8a - 5 \end{array}$$

Copyright © Houghton Mifflin Company. All rights reserved.

9.1 Exercises

A. Introduction to Polynomials

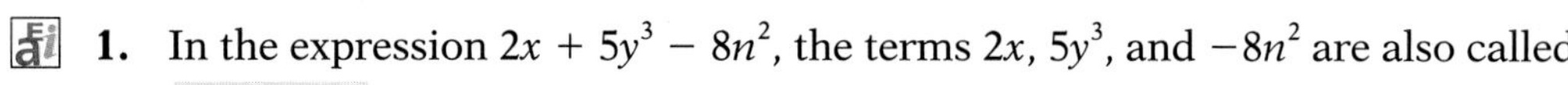

1. In the expression $2x + 5y^3 - 8n^2$, the terms $2x$, $5y^3$, and $-8n^2$ are also called _____.

2. The sum or difference of monomials is a(n) _____.

In Exercises 3–10, identify the terms of the expression and determine whether the expression is a polynomial.

3. $-\frac{2}{3}x$

4. $\frac{5b}{4}$

5. $2n - \frac{3}{4}$

6. $\frac{x}{4} - 5$

7. $\frac{3}{y^2} + 2y^2 - 4y$

8. $\frac{2}{a + 1} - 6a$

9. $2x^2 - 7x + 5$

10. $y^3 + 3y^2 - 5y - 2$

In Exercises 11–16, evaluate the polynomial for the given value of the variable.

11. $-3n + 5$ for $n = -2$

12. $-2y - 7$ for $y = -3$

13. $x^2 + 2x - 5$ for $x = 3$

14. $a^2 - 3a + 6$ for $a = 2$

15. $3y^2 - 5$ for $y = -2$

16. $2x^2 + 3x - 8$ for $x = -1$

B. Adding Polynomials

17. One advantage of the vertical method of adding polynomials is that _____ are aligned in columns.

18. Adding monomials is the same as _____ like terms.

In Exercises 19–28, add the polynomials.

19. $(5x - 6) + (2x + 9)$

20. $(3a + 1) + (-5a + 4)$

Copyright © Houghton Mifflin Company. All rights reserved.

21. $(2y^2 + 3y - 6) + (y^2 - 5y + 4)$

22. $(3x^2 - x + 4) + (2x^2 + 3x + 1)$

23. $(7x^2 + x + 4) + (x^2 + 2)$

24. $(9x^2 - 4x - 3) + (3x + 7)$

25. $(n^2 - n - 5) + (2n^3 - 2n^2 + 6)$

26. $(8x^3 + 4x + 6) + (x^3 + 2x^2 + 3)$

27. $(x^3 - 5x^2 + 4x + 6) + (4x^3 + 5x^2 - 3x - 6)$

28. $(2y^3 + y^2 + 3y - 9) + (5y^3 - y^2 - 2y + 9)$

29. What is the sum of $x^2 + 8x - 12$ and $3x^2 + 10$?

30. What is the sum of $-5b - 3$ and $b^2 + 6b$?

C. Subtracting Polynomials

31. To subtract two polynomials, change the minus symbol to a(n) ______ and change the second polynomial to its ______.

32. To take the opposite of an expression, we can change the ______ of the terms of the expression.

In Exercises 33–44, subtract the polynomials.

33. $(8x + 5) - (3x - 1)$

34. $(4x - 3) - (x + 5)$

35. $(5x^2 - 3x + 1) - (x^2 + x - 4)$

36. $(2a^2 + 4a - 1) - (a^2 - 2a - 6)$

37. $(5n^2 + n - 5) - (2n^2 - 4n + 5)$

38. $(y^2 - 3y - 2) - (4y^2 - 3y - 5)$

39. $(x^2 + 3) - (x + 3)$

40. $(2x - 5) - (x^2 + 3x)$

41. $(3y^3 + 2y^2 - y + 5) - (y^3 + 3y^2 + 5y - 3)$

42. $(4x^3 + x^2 - 7x + 1) - (2x^3 - x^2 - 6x + 1)$

43. $(2n^3 + 7n^2 - 5) - (7n^2 - 3n - 5)$

44. $(x^3 + 3x - 6) - (x^2 + 3x - 6)$

Copyright © Houghton Mifflin Company. All rights reserved.

45. Subtract $3x + 1$ from $4x + 1$.

46. Subtract $2m - 9$ from $m^2 + 3m - 7$.

47. What is $y^2 + 7y - 9$ minus $4y^2 - 6$?

48. What is $x^3 + 8x$ minus $x^2 + 8x + 2$?

Applications

The distance d (in feet) traveled by a falling object in t seconds is given by the formula $d = 16t^2$. Use this formula to answer the questions in Exercises 49 and 50.

49. *Height of Bridge* Suppose that a rock dropped from a bridge hit the water after 3 seconds. What was the height of the bridge?

50. *Scaffold Height* Four seconds after a worker dropped a hammer from a scaffold, the hammer landed on the ground. How high is the scaffold?

In Exercises 51 and 52, write the polynomial that represents the indicated length.

51. *Cable Length* A cable is cut into two pieces. The length of one of the pieces is given by $3x - 7$. If the entire length of the cable is 40 meters, what is the length of the second piece?

52. *Bolt of Cloth* The length of a bolt of cloth is 100 yards. A piece whose length is represented by $4x + 9$ is cut off. What is the length of the remaining cloth?

In Exercises 53 and 54, write the polynomial that represents the perimeter of the given figure.

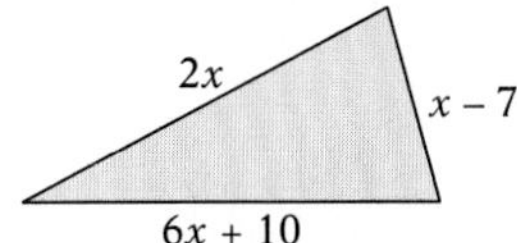

54.

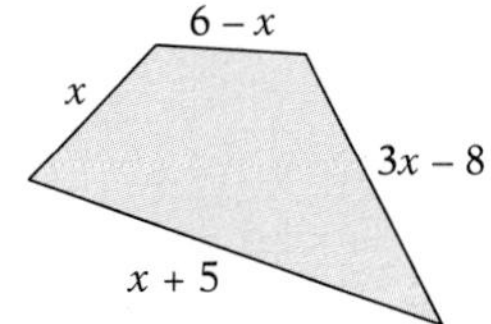

Copyright © Houghton Mifflin Company. All rights reserved.

Calculator Exercises

In Exercises 55–58, use a calculator to evaluate the polynomial for the given value of the variable.

55. $5x^2 - 6x + 2$ for $x = 3.2$

56. $-4x^2 + 7x + 10$ for $x = -2.7$

57. $x^3 + 2x^2 + 9x + 12$ for $x = -1.4$

58. $2x^3 - 6x^2 - 5x$ for $x = 7$

Writing and Concept Extension

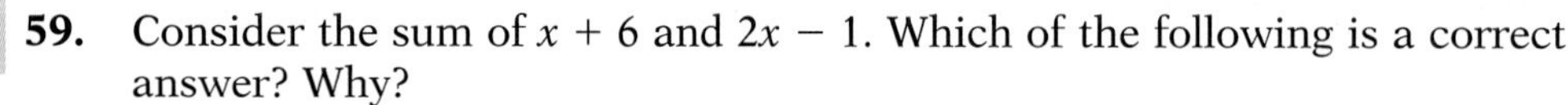

59. Consider the sum of $x + 6$ and $2x - 1$. Which of the following is a correct answer? Why?

(i) $5 + 3x$ (ii) $3x + 5$

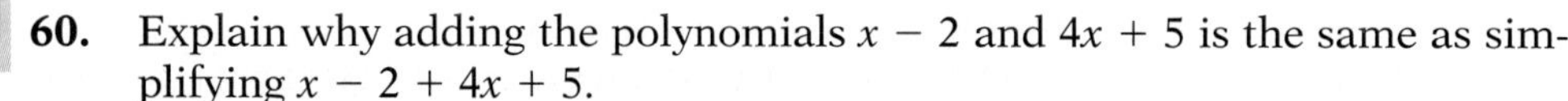

60. Explain why adding the polynomials $x - 2$ and $4x + 5$ is the same as simplifying $x - 2 + 4x + 5$.

In Exercises 61 and 62, perform the indicated operations.

61. $(3x^2 + x + 2) - (x^2 - 5) + (x^2 + 4x - 1)$

62. $(8y^2 - 10) + (y^2 - 3y - 7) - (2y^2 - 4y + 3)$

In Exercises 63 and 64, fill in the blank to make the statement true.

63. $(3x^2 + 2x + \text{(a)}) + (\text{(b)})x^2 - (\text{(c)}\ x + 3) = 7x^2 - 3x + 4$

64. $(x^2 + \text{(a)}\ x + 9) - (x + \text{(b)}) = x^2 + 5x + 2$

Copyright © Houghton Mifflin Company. All rights reserved.

9.2 Multiplication Rules for Exponents

SUGGESTIONS FOR SUCCESS

A good stenographer can use symbols to write down what we say as fast as we can say it. However, sometimes only the stenographer can read his or her own shorthand.

We need to be able to read and interpret the shorthand of mathematics, that is, the symbols used to represent numbers and words. To do this, sometimes we need to fill in the gaps.

A good example is the case of the missing 1s. Writing $1x$ instead of x is often helpful. Likewise, writing x as x^1 can be useful when you are working with exponents. You can even go all the way and write x as $1x^1$. This is not an uncool thing to do. Anything that helps you to understand the shorthand will improve your chances of success.

A The Product Rule for Exponents

In Section 2.4, we introduced the use of an *exponent* as a compact way of writing a repeated multiplication. The number being multiplied is called the *base,* and the exponent indicates the number of times that the base is a factor.

Expressions such as 3^5 and x^4 are called *exponential expressions*; they are read "3 to the fifth power" and "x to the fourth power."

Exponent ↓

$$3^5 = \underbrace{3 \cdot 3 \cdot 3 \cdot 3 \cdot 3}_{\text{3 is a factor 5 times.}} \qquad x^4 = \underbrace{x \cdot x \cdot x \cdot x}_{x \text{ is a factor 4 times.}}$$

↑ Base

The expressions 3^5 and x^4 are in *exponential* form. The expressions $3 \cdot 3 \cdot 3 \cdot 3 \cdot 3$ and $x \cdot x \cdot x \cdot x$ are in *expanded* form.

The Product Rule for Exponents

Suppose that we need to multiply two exponential expressions that have the same base. One way is to write the exponential expressions in expanded form and then just count the total number of factors.

$$\begin{aligned} a^3 \cdot a^5 &= (a \cdot a \cdot a)(a \cdot a \cdot a \cdot a \cdot a) \\ &= \underbrace{a \cdot a \cdot a \cdot a \cdot a \cdot a \cdot a \cdot a}_{a \text{ is a factor 8 times.}} \\ &= a^8 \end{aligned}$$

Do you see why this answer makes sense? The expression a^3 means that a is a factor 3 times, and the expression a^5 means that a is a factor 5 times. Therefore, in the result, a is a factor $3 + 5 = 8$ times. In other words, we could have just added the exponents.

$$a^3 \cdot a^5 = a^{3+5} = a^8$$

We summarize these results with the following rule.

Copyright © Houghton Mifflin Company. All rights reserved.

The Product Rule for Exponents

If m and n represent whole numbers and a represents some real number, then

$$a^m \cdot a^n = a^{m+n}$$

In words, to multiply exponential expressions *with the same base,* keep the base and add the exponents.

Example 1

Multiply $b^6 \cdot b^8$.

SOLUTION

$b^6 \cdot b^8 = b^{6+8}$ Product Rule for Exponents

$= b^{14}$

Your Turn 1

Multiply $x^5 \cdot x^{10}$.

Answer: x^{15}

We can use the Commutative and Associative Properties of Multiplication to rearrange and regroup factors.

Example 2

Multiply $4x^3 \cdot x^4$.

SOLUTION

$4x^3 \cdot x^4 = 4(x^3 \cdot x^4)$ Associative Property of Multiplication

$= 4x^{3+4}$ Product Rule for Exponents

$= 4x^7$

Your Turn 2

Multiply $2y^4 \cdot y^5$.

Answer: $2y^9$

Do you recall that a and a^1 mean the same thing? Writing the exponent 1 can be helpful as you perform operations with exponential expressions.

Example 3

Multiply $-4y^5 \cdot 3y$.

SOLUTION

$-4y^5 \cdot 3y = -4 \cdot 3 \cdot y^5 \cdot y$ Commutative Property of Multiplication

$= (-4 \cdot 3)(y^5 \cdot y^1)$ Associative Property of Multiplication

$= -12y^{5+1}$ Product Rule for Exponents

$= -12y^6$

Your Turn 3

Multiply $-2a^3 \cdot 5a$.

Answer: $-10a^4$

The Product Rule for Exponents applies to products of two or more exponential expressions with the same base.

Copyright © Houghton Mifflin Company. All rights reserved.

Example 4

Multiply: $5z^4 \cdot 6z^3 \cdot 2z$.

SOLUTION

$$\begin{aligned} 5z^4 \cdot 6z^3 \cdot 2z &= (5 \cdot 6 \cdot 2)(z^4 \cdot z^3 \cdot z^1) \\ &= 60z^{4+3+1} \quad \text{Product Rule for Exponents} \\ &= 60z^8 \end{aligned}$$

Your Turn 4

Multiply $3x^3 \cdot 4x^2 \cdot x$.

Answer: $12x^6$

B The Power to a Power Rule

Sometimes we need to raise an exponential expression to a power. For example, the expression $(a^3)^4$ is read "*a* to the third power raised to the fourth power." Once again, we can use the meaning of an exponent to perform the operation.

$$\begin{aligned} (a^3)^4 &= \underbrace{a^3 \cdot a^3 \cdot a^3 \cdot a^3}_{a^3 \text{ is a factor 4 times.}} \\ &= a^{3+3+3+3} \quad \text{Product Rule for Exponents} \\ &= a^{12} \end{aligned}$$

Note that the exponent 12 on the result is the product of the exponents 3 and 4 in the original expression. This suggests the following rule:

The Power to a Power Rule

If m and n represent whole numbers and a represents some real number, then

$$(a^m)^n = a^{mn}$$

In words, to raise a power to a power, keep the base and multiply the exponents.

Although the two rules that we have presented so far are not hard to use, you will need to think carefully about which rule to use in a given situation.

Multiplication: $x^2 \cdot x^6 = x^{2+6} = x^8$ *Add* the exponents.

Power to a Power: $(x^2)^6 = x^{2 \cdot 6} = x^{12}$ *Multiply* the exponents.

Make sure that you know exactly what operation you are performing and then apply the correct rule.

Example 5

Simplify $(x^5)^2$.

SOLUTION

$$\begin{aligned} (x^5)^2 &= x^{5 \cdot 2} \quad \text{Power to a Power Rule} \\ &= x^{10} \end{aligned}$$

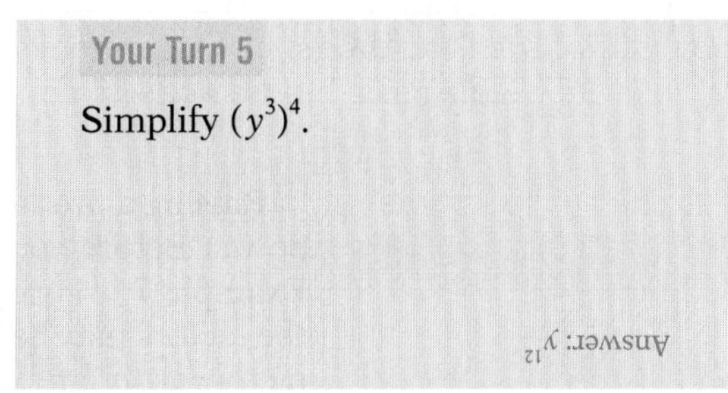

Your Turn 5

Simplify $(y^3)^4$.

Answer: y^{12}

Copyright © Houghton Mifflin Company. All rights reserved.

When an expression involves both exponents and multiplication, remember that the exponents have the higher priority in the Order of Operations.

Example 6

Simplify $(x^2)^3 \cdot (x^4)^2$.

SOLUTION

$$\begin{aligned}(x^2)^3 \cdot (x^4)^2 &= x^{2\cdot 3} \cdot x^{4\cdot 2} && \text{Power to a Power Rule}\\ &= x^6 \cdot x^8 \\ &= x^{6+8} && \text{Product Rule for Exponents}\\ &= x^{14}\end{aligned}$$

Your Turn 6

Simplify $(a^3)^2 \cdot (a^5)^3$.

Answer: a^{21}

C The Product to a Power Rule

The Power to a Power Rule can be extended to powers of products. For example, in the expression $(ab)^2$, we are raising the product ab to the second power.

$$\begin{aligned}(ab)^2 &= ab \cdot ab \\ &= a \cdot a \cdot b \cdot b && \text{Commutative Property of Multiplication}\\ &= (a \cdot a)(b \cdot b) && \text{Associative Property of Multiplication}\\ &= a^2b^2\end{aligned}$$

We see that raising ab to the second power is the same as raising both a and b to the second power. Here is the formal statement of the rule.

The Product to a Power Rule

If n represents a whole number and a and b represent real numbers, then

$$(ab)^n = a^n b^n$$

In words, to raise a product to a power, raise each factor to that power.

Example 7

Simplify $(3x)^4$.

SOLUTION

$$\begin{aligned}(3x)^4 &= 3^4 \cdot x^4 && \text{Product to a Power Rule}\\ &= 81x^4\end{aligned}$$

Your Turn 7

Simplify $(4b)^3$.

Answer: $64b^3$

Raising a product to a power involves raising *each factor* of the product to that power. Before you begin, it is a good idea to ask yourself what the factors are. Example 7 is straightforward because the factors are simply 3 and x. However, the factors may have exponents. For $(5y^3)^2$, the factors are 5 and y^3; for $(a^4b^2)^3$, the factors are a^4 and b^2. Identifying the factors gives you a blueprint for carrying out the operation.

Copyright © Houghton Mifflin Company. All rights reserved.

Example 8

Simplify $(2c^2)^3$.

SOLUTION

The factors of the product are 2 and c^2. Each factor is raised to the third power.

$(2c^2)^3 = (2)^3 \cdot (c^2)^3$ Product to a Power Rule

$= 8c^6$ Power to a Power Rule

Your Turn 8

Simplify $(3x^4)^2$.

Answer: $9x^8$

Example 9

Simplify $(x^3y^2)^5$.

SOLUTION

$(x^3y^2)^5 = (x^3)^5 \cdot (y^2)^5$ Product to a Power Rule

$= x^{15}y^{10}$

Your Turn 9

Simplify $(a^2b^4)^3$.

Answer: a^6b^{12}

The product can have any number of factors and the Product to a Power Rule still applies. Remember that writing the understood exponent of 1 can be helpful.

Example 10

Simplify $(4ab^2c^3)^2$.

SOLUTION

$(4ab^2c^3)^2 = (4)^2 \cdot (a^1)^2 \cdot (b^2)^2 \cdot (c^3)^2$ Product to a Power Rule

$= 16a^2b^4c^6$

Your Turn 10

Simplify $(3x^2yz^4)^3$.

Answer: $27x^6y^3z^{12}$

Our final example illustrates how all three of the exponent rules that we have presented might be used in a single problem.

Example 11

Simplify $(2a^2)^3 \cdot (3a)^2$.

SOLUTION

$(2a^2)^3 \cdot (3a)^2 = 2^3(a^2)^3 \cdot 3^2(a^1)^2$ Product to a Power Rule

$= 8a^6 \cdot 9a^2$ Power to a Power Rule

$= (8 \cdot 9)(a^6 \cdot a^2)$

$= 72a^8$ Product Rule for Exponents

Your Turn 11

Simplify $(4x^3)^2 \cdot (2x^2)^2$.

Answer: $64x^{10}$

Although Example 11 emphasizes the use of the exponent rules, our initial goal, in accordance with the Order of Operations, is to remove the parentheses. Because there are no operations that we can perform inside the parentheses, we turn our attention to the exponents outside the parentheses. That is why the Product to a Power Rule is used first. Even though a problem might involve only exponents, the Order of Operations continues to guide us in the sequence of steps.

Copyright © Houghton Mifflin Company. All rights reserved.

9.2 Quick Reference

A. The Product Rule for Exponents

1. If m and n represent whole numbers and a represents some real number, then

$$a^m \cdot a^n = a^{m+n}$$

In words, to multiply exponential expressions *with the same base,* keep the base and add the exponents.

$x^3 \cdot x^5 = x^{3+5} = x^8$

$3x \cdot 2x^4 = (3 \cdot 2)(x^1 \cdot x^4)$
$= 6x^5$

B. The Power to a Power Rule

1. If m and n represent whole numbers and a represents some real number, then

$$(a^m)^n = a^{mn}$$

In words, to raise a power to a power, keep the base and multiply the exponents.

$(x^3)^4 = x^{3 \cdot 4} = x^{12}$

$(x^2)^5 \cdot (x^3)^4 = x^{10} \cdot x^{12}$
$= x^{22}$

C. The Product to a Power Rule

1. If n represents a whole number and a and b represent real numbers, then

$$(ab)^n = a^n b^n$$

In words, to raise a product to a power, raise each factor to that power.

$(2x)^3 = 2^3 \cdot x^3 = 8x^3$

$(3x^3)^2 = 3^2 \cdot (x^3)^2 = 9x^6$

$(2x^2y^4)^3 = 2^3 \cdot (x^2)^3 \cdot (y^4)^3$
$= 8x^6y^{12}$

Copyright © Houghton Mifflin Company. All rights reserved.

9.2 Exercises

A. The Product Rule for Exponents

1. To multiply exponential expressions with the same base, keep the ______ and ______ the exponents.

2. We use a(n) ______ to indicate repeated multiplication.

In Exercises 3–22, multiply.

3. $x^4 \cdot x^2$
4. $y^3 \cdot y^7$
5. $c \cdot c^3$
6. $n^2 \cdot n$
7. $y^5 \cdot y^9$
8. $x^6 \cdot x^9$
9. $y \cdot 6y$
10. $n^2 \cdot 3n$
11. $3x^5 \cdot x^2$
12. $2a^3 \cdot a^2$
13. $(-4x^2) \cdot (2x^4)$
14. $(7y) \cdot (-6y^4)$
15. $(-5a^4) \cdot (-3a^3)$
16. $(6b) \cdot (2b)$
17. $(3x^5) \cdot (-8x^3)$
18. $(10n^3) \cdot (-5n^6)$
19. $w^2 \cdot 3w \cdot 7w^5$
20. $-6z^3 \cdot 2z^2 \cdot 5z^7$
21. $-5x \cdot 3x \cdot 2x$
22. $x \cdot x^8 \cdot x^5$

B. The Power to a Power Rule

23. To raise a power to a power, keep the ______ and ______ the exponents.

24. According to the ______ Rule, to simplify $(y^3)^2$, we multiply the exponents.

In Exercises 25–30, simplify.

25. $(x^7)^3$
26. $(y^2)^4$
27. $(a^2)^{10}$
28. $(n^6)^5$
29. $(b^4)^6$
30. $(x^4)^3$

In Exercises 31–36, simplify.

31. $(y^3)^4 \cdot (y^5)^2$
32. $(x^2)^5 \cdot (x^4)^6$
33. $(x^4)^5 \cdot (x^3)^2$
34. $(n^6)^2 \cdot (n^8)^3$
35. $a^6 \cdot (a^2)^7$
36. $b^3 \cdot (b^4)^6$

C. The Product to a Power Rule

37. To raise a product to a power, raise each ______ to that power.

38. To simplify an expression such as $(3y)^2$, we apply the ______ Rule.

Copyright © Houghton Mifflin Company. All rights reserved.

In Exercises 39–50, simplify.

39. $(2n)^3$
40. $(5y)^2$
41. $(6x^3)^2$
42. $(3a^4)^3$

43. $(a^3b^2)^5$
44. $(m^6n^3)^4$
45. $(xy^4)^3$
46. $(a^4b)^7$

47. $(7x^6y^4)^2$
48. $(2x^2y)^5$
49. $(3ab^5c^2)^4$
50. $(5x^4yz^3)^3$

In Exercises 51–54, simplify.

51. $(n^3)^5 \cdot (7n^6)^2$
52. $(4x^4)^3 \cdot (x^2)^6$
53. $(3y^4)^2 \cdot (2y)^3$
54. $(4a^5)^2 \cdot (2a^2)^3$

In Exercises 55–58, write an expression for the area of the geometric figure.

55.
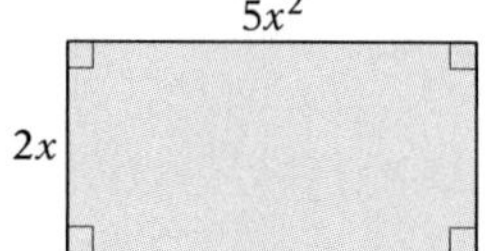

56.
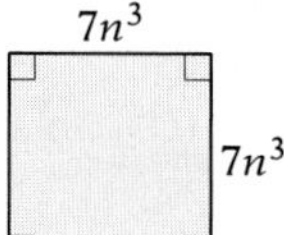

57.
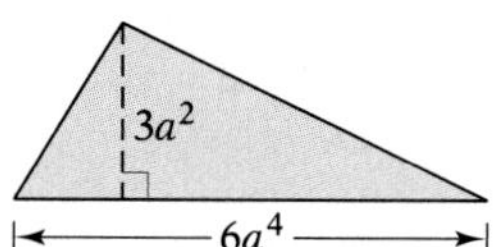

58.
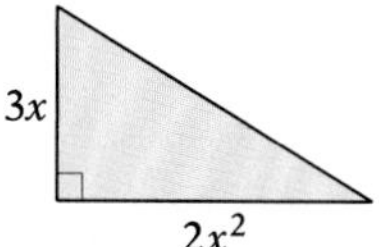

Writing and Concept Extension

59. Explain why you can use the Product Rule for Exponents to multiply $x^4 \cdot x^6$ but not $a^4 \cdot b^4$.

60. Compare the methods for simplifying $n^3 \cdot n^2$ and $(n^3)^2$.

In Exercises 61–64, multiply.

61. $6x^2y^7 \cdot 2xy^4$
62. $-5a^8b^2 \cdot 4ab^3$

63. $-m^2n^6 \cdot 8m^8n^2$
64. $-3x^8y^6 \cdot (-7xy^{10})$

In Exercises 65 and 66, perform the indicated operations.

65. **(a)** $x^3 \cdot x^3$ **(b)** $x^3 + x^3$

66. **(a)** $5x^2 \cdot 3x^2$ **(b)** $5x^2 + 3x^2$

Copyright © Houghton Mifflin Company. All rights reserved.

9.3 Multiplying Polynomials

A *Multiplying by a Monomial*
B *Multiplying Binomials*
C *Multiplying Larger Polynomials*

SUGGESTIONS FOR SUCCESS

As you make these last-minute preparations for a course in algebra, ask yourself whether you know the Distributive Property backward and forward, literally: $a(b + c) = ab + ac$ and $ab + ac = a(b + c)$.

We use this property to remove parentheses and combine like terms, to multiply polynomials (as in this section), and to factor expressions (as in the next section). The Distributive Property plays a huge role in the basic tasks of algebra.

If you enter a real course in algebra knowing how to use the Distributive Property, your success is guaranteed.

A Multiplying by a Monomial

When we multiply $3x^2$ by $2x^3$, we are multiplying two monomials. We learned how to do that in the preceding section. In this section, we extend the concept to multiplying polynomials.

A **binomial** is a polynomial with two terms. Suppose that we need to multiply the monomial $3x$ times the binomial $4x + 5$. The Distributive Property tells us how to perform the operation.

$$a\ (b + c) = a \cdot b + a \cdot c \quad \text{Distributive Property}$$
$$\downarrow\ \downarrow\ \ \downarrow\quad \downarrow\ \ \downarrow\quad \downarrow\ \ \downarrow$$
$$3x(4x + 5) = 3x \cdot 4x + 3x \cdot 5 \quad \text{Distributive Property}$$
$$= 12x^2 + 15x$$

NOTE Observe that each term of the binomial is a monomial. The Distributive Property tells us to multiply each monomial inside the parentheses by the monomial in front of the parentheses. This means that multiplying a binomial by a monomial simply comes down to multiplying monomials.

Recall that the Distributive Property can also be written

$$a(b - c) = ab - ac$$

This allows us to multiply polynomials that contain minus symbols.

Example 1

Multiply $4x^3(2x^2 - 3x)$.

SOLUTION

$$a\ (b - c) = a \cdot b - a \cdot c \quad \text{Distributive Property}$$
$$\downarrow\ \downarrow\quad \downarrow\quad \downarrow\ \ \downarrow\quad \downarrow\ \ \downarrow$$
$$4x^3(2x^2 - 3x) = 4x^3 \cdot 2x^2 - 4x^3 \cdot 3x^1 \quad \text{Distributive Property}$$
$$= 8x^5 - 12x^4 \quad \text{Product Rule for Exponents}$$

Your Turn 1

Multiply $3b^2(2b - 5b^3)$.

Answer: $6b^3 - 15b^5$

In Example 1, both the monomial and the binomial involve exponents. We assume that you can use the Product Rule for Exponents to add the exponents

Copyright © Houghton Mifflin Company. All rights reserved.

mentally, and we don't explicitly show that step in the solution. Note, again, that writing the exponent 1 can be helpful in remembering to add the exponents and in avoiding careless errors.

Example 2

Multiply $-3n^2(2n^3 + n)$.

SOLUTION

$$\begin{array}{ccc} a\,(b\,+c) & = & a \cdot b + a \cdot c \\ \downarrow\ \downarrow\ \ \downarrow & & \downarrow\ \ \downarrow\ \ \ \downarrow\ \ \downarrow \\ -3n^2(2n^3 + n) & = & -3n^2 \cdot 2n^3 + (-3n^2) \cdot 1n \\ & = & -6n^5 + (-3n^3) \\ & = & -6n^5 - 3n^3 \quad A + (-B) = A - B \end{array}$$

Your Turn 2

Multiply $-5y(y^4 + 3y^2)$.

Answer: $-5y^5 - 15y^3$

A **trinomial** is a polynomial with three terms. The Distributive Property can be extended to products involving trinomials or, for that matter, polynomials with any number of terms.

Example 3

Multiply $5a(3a^2 - 4a + 2)$.

SOLUTION

$$\begin{aligned} 5a(3a^2 - 4a + 2) &= 5a \cdot 3a^2 - 5a \cdot 4a + 5a \cdot 2 \\ &= 15a^3 - 20a^2 + 10a \end{aligned}$$

Your Turn 3

Multiply $2x(7x^2 + 3x - 5)$.

Answer: $14x^3 + 6x^2 - 10x$

B Multiplying Binomials

The Distributive Property also applies to products of two binomials.

Multiplying Binomials

Consider the product $(x + 3)(x + 7)$ and compare it to the expression $a(b + c)$.

$$\begin{array}{c} a\quad (b + c) \\ \downarrow\quad \downarrow\quad \downarrow \\ (x + 3)(x + 7) \end{array} \quad \text{Note that } a \text{ is } (x + 3).$$

With these assignments of a, b, and c, we can use the Distributive Property to carry out the multiplication.

$$\begin{array}{ccc} a\,(b + c) & = & a \cdot b + a \cdot c \\ \downarrow\ \ \downarrow\ \ \downarrow & & \downarrow\ \ \ \downarrow\ \ \ \downarrow\ \ \ \downarrow \\ (x + 3)(x + 7) & = & (x + 3) \cdot x + (x + 3) \cdot 7 \\ & = & x(x + 3) + 7(x + 3) \quad \text{Commutative Property of Multiplication} \\ & = & x^2 + 3x + 7x + 21 \quad \text{Distributive Property applied twice.} \end{array}$$

Copyright © Houghton Mifflin Company. All rights reserved.

We can combine the like terms $3x$ and $7x$. However, before we do that, observe that the same four terms in the result can be obtained by multiplying each term in the second binomial by each term in the first binomial.

$$(x + 3)(x + 7) = x \cdot x + x \cdot 7 + 3 \cdot x + 3 \cdot 7$$

In other words, an efficient way to multiply two binomials is to multiply each term of one binomial by each term of the other binomial.

We write the simplified result by combining like terms.

$$x^2 + 10x + 21$$

Example 4

Multiply $(5a + 4)(2a - 3)$.

SOLUTION

$$(5a + 4)(2a - 3)$$

$$= 5a \cdot 2a - 5a \cdot 3 + 4 \cdot 2a - 4 \cdot 3$$

$$= 10a^2 - 15a + 8a - 12$$

$$= 10a^2 - 7a - 12 \quad \text{Combine like terms.}$$

Your Turn 4

Multiply $(3x + 2)(x - 1)$.

Answer: $3x^2 - x - 2$

The most common errors that occur in multiplying polynomials are sign errors. Until you are confident about managing signs mentally, you might want to rewrite subtractions as additions.

Example 5

Multiply $(c - 2)(4c - 3)$.

SOLUTION

$$(c - 2)(4c - 3) = [c + (-2)][4c + (-3)]$$

$$= c(4c) + c(-3) + (-2)(4c) + (-2)(-3)$$

$$= 4c^2 + (-3c) + (-8c) + 6$$

$$= 4c^2 + (-11c) + 6 \quad \text{Combine like terms.}$$

$$= 4c^2 - 11c + 6$$

Your Turn 5

Multiply $(2x - 5)(x - 4)$.

Answer: $2x^2 - 13x + 20$

A special case of multiplying a binomial is squaring a binomial, that is, raising the binomial to the second power. There is nothing new about this operation as long as you remember that squaring a quantity means multiplying the quantity by itself.

Copyright © Houghton Mifflin Company. All rights reserved.

Example 6

Simplify.

(a) $(n + 5)^2$ **(b)** $(x - y)^2$

SOLUTION

(a) $(n + 5)^2 = (n + 5)(n + 5)$

$= n \cdot n + n \cdot 5 + 5 \cdot n + 5 \cdot 5$

$= n^2 + 5n + 5n + 25$

$= n^2 + 10n + 25$

(b) Again, we rewrite the differences as sums.

$(x - y)^2 = [x + (-y)]^2$ $A - B = A + (-B)$

$= [x + (-y)][x + (-y)]$

$= x \cdot x + x \cdot (-y) + (-y) \cdot x + (-y)(-y)$

$= x^2 + (-xy) + (-xy) + y^2$

$= x^2 + (-2xy) + y^2$ Combine like terms.

$= x^2 - 2xy + y^2$ $A + (-B) = A - B$

Your Turn 6

Simplify.

(a) $(x + 4)^2$

(b) $(a - b)^2$

Answers: (a) $x^2 + 8x + 16$; (b) $a^2 - 2ab + b^2$

NOTE Squaring a binomial, as in Example 6, can easily be confused with squaring a product.

$(ab)^2 = a^2b^2$ is true. $(a + b)^2 = a^2 + b^2$ is *not* true!

To perform $(a + b)^2$, you should write $(a + b)^2 = (a + b)(a + b)$ to remind yourself that you are multiplying two binomials.

C Multiplying Larger Polynomials

The methods that we used for multiplying two binomials can be extended to multiplying any two polynomials, no matter how large they are. Here is a summary of the general method.

Multiplying Two Polynomials

1. Multiply each term of one polynomial by each term of the other polynomial.
2. Combine like terms.

We illustrate this method with the product of a binomial and a trinomial.

Copyright © Houghton Mifflin Company. All rights reserved.

Example 7

Multiply $(x - 2)(x^2 + 3x - 9)$.

SOLUTION

We continue to change differences to sums. With practice, you will be able to omit those steps and manage the signs mentally.

$(x - 2)(x^2 + 3x - 9)$

$= [x + (-2)][x^2 + 3x + (-9)]$

$= x \cdot x^2 + x \cdot 3x + x \cdot (-9) + (-2) \cdot x^2 + (-2) \cdot 3x + (-2) \cdot (-9)$

$= x^3 + 3x^2 + (-9x) + (-2x^2) + (-6x) + 18$

$= x^3 + 1x^2 + (-15x) + 18$ Combine like terms.

$= x^3 + x^2 - 15x + 18$

Your Turn 7

Multiply $(x + 3)(x^2 - 2x + 1)$.

Answer: $x^3 + x^2 - 5x + 3$

In Example 7, we had to search through a long string of terms to find the like terms. We can make that process easier by arranging the polynomials in a vertical format so that like terms are written in columns.

In Example 8, we repeat the product in Example 7, this time with the vertical format. Observe how the like terms are arranged in columns.

Example 8

Use the vertical format to multiply

$$(x - 2)(x^2 + 3x - 9)$$

SOLUTION

As we write the product in the vertical format, we change differences to sums.

$$\begin{array}{rrrr} & x^2 + & 3x & + (-9) \\ & & x & + (-2) \\ \hline & -2x^2 + & (-6x) + & 18 \\ x^3 + & 3x^2 + & (-9x) & \\ \hline x^3 + & 1x^2 + & (-15x) + & 18 \end{array}$$

Multiply $x^2 + 3x + (-9)$ by -2.

Multiply $x^2 + 3x + (-9)$ by x.

Add down the columns.

The result is $x^3 + x^2 - 15x + 18$.

Your Turn 8

Use the vertical format to multiply

$$(x + 4)(x^2 - 5x - 2)$$

Answer: $x^3 - x^2 - 22x - 8$

We can use the same methods to multiply a trinomial by a trinomial or to multiply any polynomial by any other polynomial. Of course, the larger the polynomials, the more terms we obtain when we multiply. However, often some of those terms can be combined.

Multiplying polynomials is a basic skill in algebra. Mastering the skill now will give you a big advantage in later courses.

Copyright © Houghton Mifflin Company. All rights reserved.

9.3 Quick Reference

A. Multiplying by a Monomial

1. A **binomial** is a polynomial with two terms. To multiply a binomial by a monomial, we use the Distributive Property.

$$a(b + c) = ab + ac$$

or

$$a(b - c) = ab - ac$$

$7x(2x + 3)$
$= 7x \cdot 2x + 7x \cdot 3$
$= 14x^2 + 21x$

$2y^3(y^2 - 4y)$
$= 2y^3 \cdot y^2 - 2y^3 \cdot 4y$
$= 2y^5 - 8y^4$

2. A **trinomial** is a polynomial with three terms. The Distributive Property can be extended to products involving trinomials or polynomials of any number of terms.

$4x(2x^2 + 3x - 5)$
$= 4x(2x^2) + 4x(3x) - 4x(5)$
$= 8x^3 + 12x^2 - 20x$

B. Multiplying Binomials

1. To multiply two binomials, we multiply each term of one binomial by each term of the other binomial. After you have performed the multiplication, you may find that there are like terms that can be combined.

$(x + 2)(x + 5)$
$= x(x) + x(5) + 2(x) + 2(5)$
$= x^2 + 5x + 2x + 10$
$= x^2 + 7x + 10$

2. Squaring a binomial means multiplying the binomial by itself.

$(x + 3)^2$
$= (x + 3)(x + 3)$
$= x(x) + x(3) + 3(x) + 3(3)$
$= x^2 + 3x + 3x + 9$
$= x^2 + 6x + 9$

C. Multiplying Larger Polynomials

1. To multiply any two polynomials:

(a) Multiply each term of one polynomial by each term of the other polynomial.
(b) Combine like terms.

The advantage of the vertical format is that like terms can be aligned in columns.

$$\begin{array}{r} x^2 + 2x + 3 \\ x + 4 \\ \hline 4x^2 + 8x + 12 \\ x^3 + 2x^2 + 3x \phantom{{}+12} \\ \hline x^3 + 6x^2 + 11x + 12 \end{array}$$

Copyright © Houghton Mifflin Company. All rights reserved.

9.3 Exercises

A. Multiplying by a Monomial

1. A polynomial that has three terms is called a(n) ________.

2. When we write $x(x + 3) = x \cdot x + x \cdot 3$, we are applying the ________ Property.

In Exercises 3–20, multiply.

3. $2x(4x + 5)$

4. $5b(7b - 2)$

5. $7y(2y - 7)$

6. $3x(6x + 1)$

7. $-8x(2x - 1)$

8. $-6x(2x + 1)$

9. $-3a(5a + 2)$

10. $-n(8n - 9)$

11. $y^4(y^2 + 2y)$

12. $a^5(a^3 - 6a^2)$

13. $2x^2(x^3 + 5x)$

14. $4x^3(x^2 - 2x)$

15. $-5n^4(n^4 + 3)$

16. $-3x^2(x^3 - 1)$

17. $3b(2b^2 - b + 4)$

18. $6y(y^2 + 5y - 1)$

19. $2x^2(x^2 - 3x - 6)$

20. $5n^2(4n^2 - 2n + 7)$

B. Multiplying Binomials

21. A polynomial that has two terms is called a(n) ________.

22. Raising a binomial to the second power, as in $(x + 5)^2$, is called ________ the binomial.

In Exercises 23–34, multiply.

23. $(x + 4)(x + 3)$

24. $(x + 1)(x - 6)$

25. $(y - 2)(y + 5)$

Copyright © Houghton Mifflin Company. All rights reserved.

26. $(x - 7)(x - 3)$

27. $(z + 7)(z - 7)$

28. $(a + 1)(a - 1)$

29. $(3x - 1)(x + 4)$

30. $(2n + 3)(n + 5)$

31. $(2b - 1)(3b - 5)$

32. $(5x + 2)(4x - 3)$

33. $(7x + 2)(2x - 3)$

34. $(2a - 5)(6a - 1)$

In Exercises 35–42, simplify.

35. $(a + 3)^2$

36. $(x - 1)^2$

37. $(n - 2)^2$

38. $(y + 7)^2$

39. $(2x + 3)^2$

40. $(5x - 2)^2$

41. $(6x - 1)^2$

42. $(4b + 7)^2$

C. Multiplying Larger Polynomials

In Exercises 43–50, multiply.

43. $(x + 5)(x^2 - 2x + 4)$

44. $(x - 6)(x^2 + x + 7)$

45. $(x - 2)(2x^2 + 5x + 3)$

46. $(x + 4)(3x^2 - x - 1)$

47. $(2x - 1)(x^2 + 3x + 1)$

48. $(3x + 1)(x^2 + 4x - 5)$

Copyright © Houghton Mifflin Company. All rights reserved.

49. $(3x - 2)(9x^2 + 6x + 4)$

50. $(3y + 1)(9y^2 - 3y + 1)$

In Exercises 51 and 52, write an expression for the area of the given geometric figure.

51.

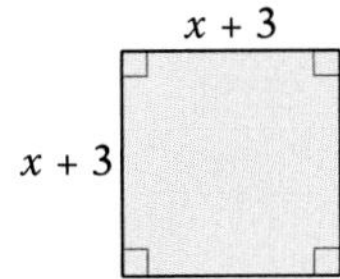

52.

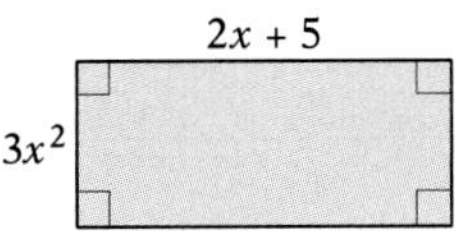

In Exercises 53 and 54, write an expression for the area of the shaded region.

53.

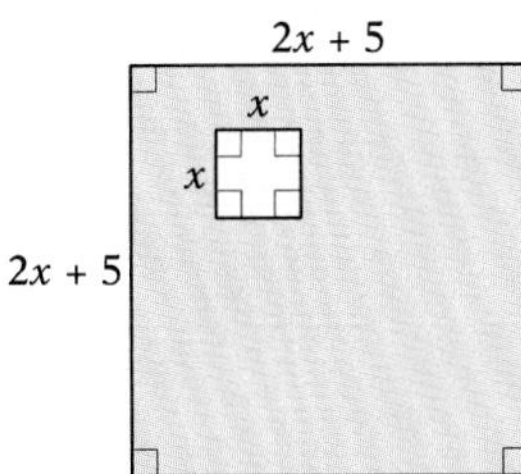

54.

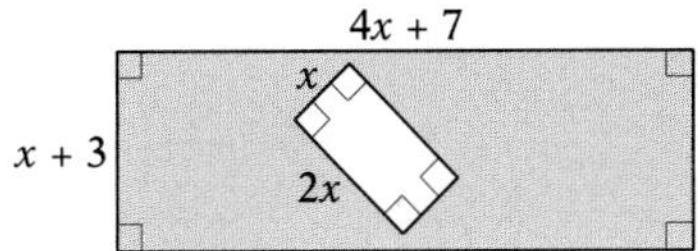

Writing and Concept Extension

55. What is an advantage of using the vertical format when you multiply polynomials?

56. Identify the statement that is true. Explain why the other two statements are false.

(i) $(x + 3)^2 = x^2 + 9$

(ii) $(x + 3)(x - 2) = x^2 - 6$

(iii) $(x + 3)(x - 3) = x^2 - 9$

In Exercises 57–62, multiply.

57. $(2x^2 - 5)(3x^2 + 4x + 2)$

58. $(x^2 + 1)(x^2 + 2x - 4)$

Copyright © Houghton Mifflin Company. All rights reserved.

59. $(x^2 + 3x - 1)(x^2 - x + 2)$

60. $(x^2 - 3x - 1)(x^2 + x - 4)$

61. $(2x^2 + x + 3)(x^2 - 6x - 2)$

62. $(x^2 - x - 6)(4x^2 + 3x + 2)$

In Exercises 63–66, perform the indicated operations.

63. $(x + 5)^2 + (x - 5)^2$

64. $(x + 3)(x - 3) + (x - 1)^2$

65. $x(x - 2)^2$

66. $3x(x + 4)(x - 4)$

Copyright © Houghton Mifflin Company. All rights reserved.

9.4 Factoring Polynomials

A *Greatest Common Factors*
B *Factoring Polynomials*

SUGGESTIONS FOR SUCCESS

The context in which a word is used often determines its meaning.

For instance, *face* may mean the front of someone's head or the part of a clock with the numbers. You may use the word *factor* to mean a condition that brings about a result. For example, the weather may be a factor in your decision to go somewhere.

However, in mathematics the word *factor* has a different meaning. Make sure that you think about the mathematical meaning of words rather than meanings from common usage.

LEARNING TIP

A good way to begin this section is with a review of Section 5.2. You will need that information in order to be successful with this topic.

A Greatest Common Factors

In Section 5.2, we defined a *prime number* as any whole number greater than 1 that is divisible only by itself and 1. We also showed how a number can be *prime-factored,* that is, written as a product of prime numbers.

Example 1 reviews the method of prime-factoring.

Example 1

Prime-factor 18 and 30.

SOLUTION

$$18 = 2 \cdot 9 = 2 \cdot 3 \cdot 3$$

$$30 = 2 \cdot 15 = 2 \cdot 3 \cdot 5$$

Your Turn 1

Prime-factor 24 and 42.

Answer: $2 \cdot 2 \cdot 2 \cdot 3$, $2 \cdot 3 \cdot 7$

In Example 1, we see that 2 and 3 are factors of both 18 and 30. We say that 2 and 3 are *common factors* of 18 and 30. If we multiply these factors, then another way to write 18 and 30 as products is as follows.

$$18 = 2 \cdot 3 \cdot 3 = 6 \cdot 3$$

$$30 = 2 \cdot 3 \cdot 5 = 6 \cdot 5$$

Now we can see that the factor 6 appears in both products, so 6 is also a common factor of 18 and 30. In fact, 6 is the largest common factor that we can find. Therefore, we call 6 the **greatest common factor (GCF)** of 18 and 30.

Definition of Greatest Common Factor

The **greatest common factor (GCF)** of two numbers is the largest integer that is a factor of both numbers. If the two numbers have no common prime factors, then the GCF is 1.

You can often find the GCF of small numbers by inspection. However, the following procedure gives you a systematic way of determining a GCF:

Copyright © Houghton Mifflin Company. All rights reserved.

Finding the GCF of Two or More Numbers

To find the GCF of two or more numbers, follow these steps.

1. Write each number as a product of prime factors.
2. Identify the common prime factors.
3. The GCF is the product of the common prime factors.

Example 2

Find the GCF of 20 and 36.

SOLUTION

First, we prime-factor 20 and 36.

$$20 = 2 \cdot 10 = 2 \cdot 2 \cdot 5$$

$$36 = 2 \cdot 18 = 2 \cdot 2 \cdot 9 = 2 \cdot 2 \cdot 3 \cdot 3$$

Now we see that 2 appears twice in the factorizations of 20 and 36. Therefore, the GCF is $2 \cdot 2 = 4$.

Your Turn 2

Find the GCF of 27 and 45.

Answer: 9

We can find the GCF of monomials in much the same way. Consider the monomials x^3 and x^5.

$x^3 = x \cdot x \cdot x$ — x is a factor 3 times.

$x^5 = x \cdot x \cdot x \cdot x \cdot x$ — x is a factor 5 times.

Because x is a factor 3 times in both x^3 and x^5, the GCF is x^3. If you think that the variable part of the GCF will always be the variable with the smallest exponent, you are right!

Finding the GCF of Two or More Monomials

To find the GCF of two or more monomials, follow these steps.

1. Use the preceding method to find the GCF of the numerical parts.
2. For each variable part, find the smallest exponent and write the variable with that exponent.
3. The GCF of the monomials is the product of the results in steps 1 and 2.

Example 3

Find the GCF of a^4, a^2, and a^5.

SOLUTION

The smallest exponent on a is 2. Therefore, the GCF is a^2.

Your Turn 3

Find the GCF of x^5, x^7, and x^3.

Answer: x^3

Copyright © Houghton Mifflin Company. All rights reserved.

Example 4

Find the GCF of $12x^4$ and $20x^3$.

SOLUTION

First, we find the GCF of the numerical parts.

$$12 = 2 \cdot 6 = 2 \cdot 2 \cdot 3$$

$$20 = 2 \cdot 10 = 2 \cdot 2 \cdot 5$$

The GCF is $2 \cdot 2 = 4$.

For the variable parts, we see that the smallest exponent on x is 3.

The GCF of the monomials is $4x^3$.

Your Turn 4

Find the GCF of $9y^3$ and $15y^2$.

Answer: $3y^2$

Example 5

Find the GCF of $6y^3$, $9y^2$, and $15y$.

SOLUTION

$$6 = 2 \cdot 3$$

$$9 = 3 \cdot 3$$

$$15 = 5 \cdot 3$$

The GCF of the numerical parts is 3. The smallest exponent on y is 1. (Remember that $y = y^1$.)

The GCF of the monomials is $3y$.

Your Turn 5

Find the GCF of $8x^2$, $12x$, and $16x^4$.

Answer: $4x$

Example 6

Find the GCF of $9a^2b^5$ and $16a^4b^3$.

SOLUTION

Because 9 and 16 have no common prime factors, the GCF of the numerical parts is 1.

The smallest exponent on a is 2, and the smallest exponent on b is 3.

The GCF of the monomials is a^2b^3.

Your Turn 6

Find the GCF of $2x^5y$ and $6x^2y^3$.

Answer: $2x^2y$

B Factoring Polynomials

When we multiply a monomial times a polynomial we use the Distributive Property.

$$a(b + c) = a \cdot b + a \cdot c$$

If we swap the sides of this equation, we have another version of the Distributive Property.

$$a \cdot b + a \cdot c = a(b + c)$$

Common factor

Copyright © Houghton Mifflin Company. All rights reserved.

We use this version of the Distributive Property to write a polynomial as a product. We call this process factoring out the greatest common factor (GCF).

Factoring out a GCF from a Polynomial

1. Determine the GCF of the terms of the polynomial.
2. Write each term as a product with the GCF as one factor.
3. To factor the polynomial,
   a. Write the GCF as one factor.
   b. For the other factor, write (in parentheses) what is left of the terms after the GCF has been removed.

Example 7

Factor $3x^2 + 8x$.

SOLUTION

The GCF of the two terms is x.

$$3x^2 + 8x = x \cdot 3x + x \cdot 8$$
$$= x(3x + 8) \quad \text{Distributive Property}$$

Your Turn 7

Factor $5y^2 - 7y$.

Answer: $y(5y - 7)$

NOTE Because factoring is the reverse of multiplying, you can always check the results of a factoring problem by multiplying the factors to verify that we obtain the original expression. For Example 7,

$$x(3x + 8) = 3x^2 + 8x$$

Example 8

Factor $5a^2 + 15a - 10$.

SOLUTION

The GCF is 5.

$$5a^2 + 15a - 10 = 5 \cdot a^2 + 5 \cdot 3a - 5 \cdot 2$$
$$= 5(a^2 + 3a - 2)$$

Your Turn 8

Factor $6x^2 - 2x + 12$.

Answer: $2(3x^2 - x + 6)$

Example 9

Factor $4x^5 + 6x^3$.

SOLUTION

The GCF is $2x^3$.

$$4x^5 + 6x^3 = 2x^3 \cdot 2x^2 + 2x^3 \cdot 3$$
$$= 2x^3(2x^2 + 3)$$

Your Turn 9

Factor $14x^3 - 7x^2$.

Answer: $7x^2(2x - 1)$

Copyright © Houghton Mifflin Company. All rights reserved.

NOTE Keep in mind that not all polynomials can be factored. For instance, the terms of the polynomial $x^2 + 7$ have no common factor, so we can't use our current methods to factor the expression. For now, we will just say that the polynomial cannot be factored.

Example 10

Factor $9y^2 - y$.

SOLUTION

The GCF is y.
You might think that when we remove y from the second term, we have nothing left. Remember, though, that $y = y \cdot 1$.

$$\begin{aligned} 9y^2 - y &= y \cdot 9y - y \cdot 1 \\ &= y(9y - 1) \end{aligned}$$

Your Turn 10

Factor $2x^2 + 2x$.

LEARNING TIP

Get in the habit of checking answers by multiplying the factors. Doing so will help you to avoid the kind of error described in Example 10.

Answer: $2x(x + 1)$

Example 11

Factor $-4x + 12a - 20b$.

SOLUTION

Method 1

Think of the GCF as 4.

$$\begin{aligned} &-4x + 12a - 20b \\ &= 4 \cdot (-1x) + 4 \cdot 3a - 4 \cdot 5b \\ &= 4(-x + 3a - 5b) \end{aligned}$$

Method 2

Think of the GCF as -4.

$$\begin{aligned} &-4x + 12a - 20b \\ &= (-4) \cdot x + (-4) \cdot (-3a) + (-4) \cdot 5b \\ &= -4(x - 3a + 5b) \end{aligned}$$

When the numerical part of the first term is negative, Method 2 is often the preferred way to factor the expression.

Your Turn 11

Use the two methods shown in Example 11 to factor $-3a - 9b + 15c$.

Answer: $3(-a - 3b + 5c)$, $-3(a + 3b - 5c)$

In algebra, there are many different ways to factor an expression. For example, although we can't factor $x^2 - 4$ with the methods discussed in this section, you will learn that $x^2 - 4$ can be written as a product. However, all factoring methods begin with factoring out the GCF, so this section should give you a good head start toward the very important skill of factoring.

Copyright © Houghton Mifflin Company. All rights reserved.

9.4 Quick Reference

A. Greatest Common Factors

1. The **greatest common factor (GCF)** of two numbers is the largest integer that is a factor of both numbers. If the two numbers have no common prime factors, then the GCF is 1.

   Because $16 = 8 \cdot 2$ and $24 = 8 \cdot 3$, the GCF of 16 and 24 is 8.

2. To find the GCF of two or more numbers, follow these steps.
   (a) Write each number as a product of prime factors.
   (b) Identify the common prime factors.
   (c) The GCF is the product of the common prime factors.

   Find the GCF of 24 and 40.

   $$24 = 2 \cdot 2 \cdot 2 \cdot 3$$
   $$40 = 2 \cdot 2 \cdot 2 \cdot 5$$

   The common prime factors are $2 \cdot 2 \cdot 2$.

   The GCF of 24 and 40 is $2 \cdot 2 \cdot 2 = 8$.

3. To find the GCF of two or more monomials, follow these steps.
   (a) Use the preceding method to find the GCF of the numerical parts.

   Find the GCF of $15x^2$ and $9x^4$.

   $$15 = 3 \cdot 5 \qquad 9 = 3 \cdot 3$$

   The GCF of the numerical parts is 3.

   (b) For each variable part, find the smallest exponent and write the variable with that exponent.

   The smallest exponent on x is 2.

   (c) The GCF of the monomials is the product of the results in steps 1 and 2.

   The GCF is $3x^2$.

B. Factoring Polynomials

1. To **factor** a polynomial, we use the Distributive Property.

   $$a \cdot b + a \cdot c = a(b + c)$$

   Factor $5x^3 + 10x^2$.

   The GCF of $5x^3$ and $10x^2$ is $5x^2$.

   $$5x^3 + 10x^2 = 5x^2(x + 2)$$

2. When we factor a polynomial, we sometimes need to use the rule $x = x \cdot 1$.

   $$4x^3 + x = 4x^3 + 1x$$
   $$= x(4x^2 + 1)$$

3. When the first term of a polynomial is negative, we usually factor out a negative common factor.

   $$-2x^2 - 4x = -2x(x + 2)$$

4. You can check your factoring results by multiplying the factors to verify that you obtain the original expression.

   $$3ab + 2a = a(3b + 2)$$

   Check: $a(3b + 2) = 3ab + 2a$

Copyright © Houghton Mifflin Company. All rights reserved.

9.4 Exercises

A. Greatest Common Factors

1. The largest integer that is a factor of 8 and 12 is 4. We call 4 the _____ of 8 and 12.

2. Numbers such as 3 and 11 that are divisible only by themselves and 1 are called _____ numbers.

In Exercises 3–12, determine the greatest common factor.

3. 20, 35
4. 21, 30
5. 9, 18
6. 27, 36
7. 30, 40, 60
8. 8, 24, 32
9. x^3, x^2, x^6
10. a^5, a^6, a^4
11. y^3, y^7, y^5
12. b^6, b^{10}, b^7

In Exercises 13–20, determine the greatest common factor.

13. $20y^4, 24y^7$
14. $9x^3, 12x^2$
15. $4x^4, 8x^6$
16. $16y^2, 8y^4$
17. $10x^3, 6x^4, 12x^6$
18. $10a^6, 20a^4, 5a^8$
19. $3b^4, 9b, 12b^3$
20. $9y^2, 18y^5, 36y$

In Exercises 21–24, determine the greatest common factor.

21. $3a^2b^6, 4a^4b^3$
22. $15x^3y^6, 9x^7y^8$
23. $2x^4y^5, 12x^2y^3$
24. $12m^5n, 16mn$

B. Factoring Polynomials

25. The process of writing the polynomial $6a + 4$ as the product $2(3a + 2)$ is called _____ the polynomial.

26. When we factor the polynomial $5x + 5y$, we use the _____ Property to write $5(x + y)$.

In Exercises 27–30, fill in the blank.

27. $18x - 3 =$ _____ $(6x - 1)$
28. $a^2 + 5a =$ _____ $(a + 5)$
29. $2y^2 + 7y = y($ _____ $)$
30. $9b + 12a = 3($ _____ $)$

Copyright © Houghton Mifflin Company. All rights reserved.

In Exercises 31–52, factor.

31. $x^2 + 3x$

32. $y^2 - 8y$

33. $5a + 15$

34. $3x + 12$

35. $6x + 6$

36. $9a - 36$

37. $12a - 20b$

38. $42x + 12y$

39. $2y^2 - 12y$

40. $7b^3 - 21b^2$

41. $9x^2 - 3x$

42. $56n^2 + 8n$

43. $3x^2 + 12x - 3$

44. $14x^2 - 7x - 28$

45. $8x^2 + 4x - 28$

46. $15x^2 - 20x - 5$

47. $9m + 18n - 12$

48. $10x + 25y - 10z$

49. $-3a - 6b + 15c$

50. $-15x + 5y + 10$

51. $-4x - 4y + 4$

52. $-18a - 6b - 6$

In Exercises 53–58, factor.

53. $2a^3 - 5a^2$

54. $x^5 + x^3$

55. $b^5 + 6b^3$

56. $a^8 - 10a^5$

57. $14y^4 - 21y^2$

58. $15x^3 + 5x$

Copyright © Houghton Mifflin Company. All rights reserved.

Writing and Concept Extension

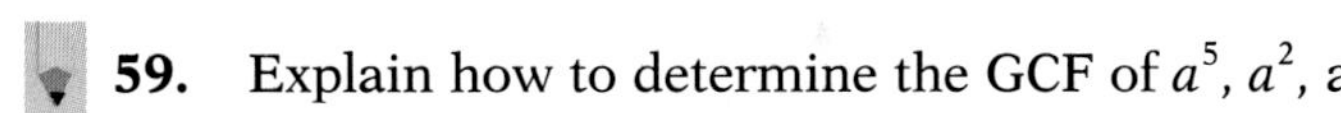

59. Explain how to determine the GCF of a^5, a^2, and a^4.

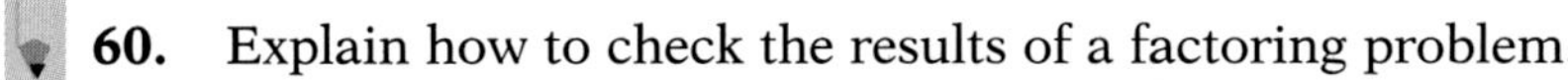

60. Explain how to check the results of a factoring problem.

61. The area of the rectangle is represented by $6x^2 + 10x$. Write a polynomial that represents the length.

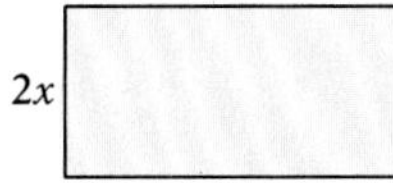

62. The area of the right triangle is represented by $\frac{1}{2}x^2 + \frac{3}{2}x$. Write a polynomial that represents the length of the base.

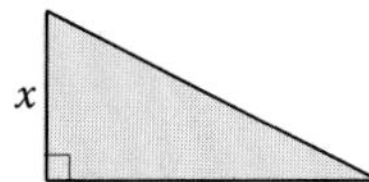

Copyright © Houghton Mifflin Company. All rights reserved.

CHAPTER 9 REVIEW EXERCISES

Section 9.1

1. The terms of a polynomial are called ________.

2. Identify the terms of the expression and determine whether the expression is a polynomial.

   (i) $\frac{2}{5}y^2 - 3y + 1$ (ii) $\frac{5}{x^2} + 7x - 9$

3. Evaluate $x^2 + 5x + 7$ for $x = -3$.

4. To add polynomials, we remove parentheses and combine ________.

In Exercises 5 and 6, add the polynomials.

5. $(x^2 - 3x + 1) + (2x^2 + x + 1)$

6. $(2y^2 + 4y - 5) + (5y^2 - 4y + 5)$

In Exercises 7–9, subtract the polynomials.

7. $(4x + 1) - (x - 3)$

8. $(4a^2 - 3a) - (a^2 - a + 2)$

9. $(8y^3 + 2y + 1) - (6y^3 - y^2 + 2y)$

10. *Hiking* A hiker walked the first $x + 2$ miles of a 10-mile trail. Write an expression for the remaining distance to be walked.

Section 9.2

11. According to the ________ Rule for Exponents, to multiply $y^2 \cdot y^3$, we add the exponents.

In Exercises 12–15, multiply.

12. $n^4 \cdot n^7$

13. $-5x^3 \cdot 4x$

14. $6c \cdot 5c$

15. $z^3 \cdot z^7 \cdot z^2$

Copyright © Houghton Mifflin Company. All rights reserved.

In Exercises 16 and 17, simplify.

16. $(y^4)^6$

17. $(x^2)^4 \cdot (x^3)^2$

In Exercises 18–20, simplify.

18. $(3x)^4$

19. $(a^3b^5)^2$

20. $(2x^3y)^5$

Section 9.3

In Exercises 21–24, multiply.

21. $2x(x - 5)$

22. $-5x(3x - 7)$

23. $2a^2(3a + 10)$

24. $4n(2n^2 + 3n - 6)$

In Exercises 25–28, multiply.

25. $(x + 8)(x - 6)$

26. $(b - 5)(b + 5)$

27. $(3a - 2)(5a - 1)$

28. $(y + 6)^2$

In Exercises 29 and 30, multiply.

29. $(x + 3)(x^2 - x - 5)$

30. $(3x - 2)(2x^2 + x - 4)$

Section 9.4

31. We call 7 the ________ of 14 and 21.

In Exercises 32–35, determine the greatest common factor.

32. 8, 20

33. x^4, x, x^3

34. $9a^6, 15a^2$

35. $12x^3y^5, 8x^2y^7$

36. To factor a polynomial, we write the polynomial as a(n) ________.

In Exercises 37–40, factor.

37. $10x - 25$

38. $2y^2 + y$

39. $x^3 - x^2 + 4x$

40. $8a + 4b - 16$

Copyright © Houghton Mifflin Company. All rights reserved.

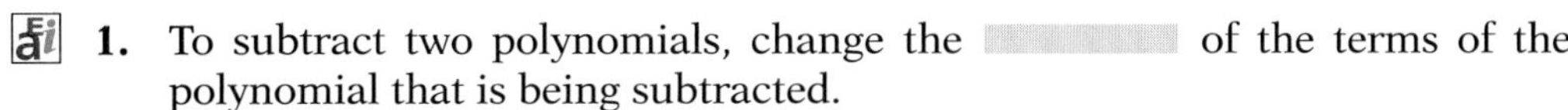

CHAPTER 9 TEST

1. To subtract two polynomials, change the ________ of the terms of the polynomial that is being subtracted.

2. To simplify $x^5 \cdot x^2$, we ________ the exponents, whereas to simplify $(x^5)^2$, we ________ the exponents.

3. Evaluate $x^2 - 5x + 6$ for $x = 2$.

In Questions 4–7, perform the indicated operation.

4. $(7n^2 + 4n + 1) + (n^2 - 3n - 1)$

5. $(y^3 + 3y - 2) + (y^2 + 2)$

6. $(6x^2 + x - 5) - (2x^2 + 3x - 2)$

7. $(x^3 + 2x^2 - x - 1) - (x^2 - x + 1)$

In Questions 8–13, use the properties of exponents to simplify the expression.

8. $y^6 \cdot y^3$

9. $(4x^5) \cdot (-5x^2)$

10. $(n^3)^7$

11. $(x^2)^7 \cdot (x^3)^2$

12. $(5y)^3$

13. $(m^4n^7)^2$

In Questions 14–18, multiply.

14. $-7b(2b - 3)$

15. $5x^3(2x^2 + x - 6)$

16. $(2x + 7)(4x - 3)$

17. $(n - 8)^2$

18. $(x - 4)(x^2 - 5x - 2)$

In Questions 19 and 20, determine the greatest common factor.

19. 24, 40

20. $6x^5, 3x^8$

In Questions 21–23, factor.

21. $24y + 6$

22. $10a^2 + 8a$

23. $14x^2 + 7x - 28$

Copyright © Houghton Mifflin Company. All rights reserved.

CHAPTERS 8–9 CUMULATIVE TEST

1. Write the percent as a simplified fraction and as a decimal.

(a) 80% **(b)** 0.5% **(c)** 160%

2. Write the fraction as a percent.

(a) $\frac{7}{40}$ **(b)** $\frac{2}{3}$

3. Write the decimal as a percent.

(a) 0.12 **(b)** 3.2 **(c)** 0.008

4. What is 7% of 12?

5. 21 is what percent of 60?

6. 20% of what number is 2.4?

7. *Pay Raise* A person making $2,600 per month received a 6% raise. What was the new monthly salary?

8. *Allergic Reaction* Of the 140 patients who received a new medication, 21 experienced an allergic reaction. What percent of the patients had an allergic reaction?

9. *Simple Interest* What is the simple interest on a $2,500 loan for 6 months at a 14% annual interest rate?

10. *Take-Home Pay* If you earned $680, but 21% was deducted for taxes and 3% was deducted for the employee savings program, what was your take-home pay?

Copyright © Houghton Mifflin Company. All rights reserved.

11. *Ticket Sale Commission* Ticket King received a 9% commission on all tickets sold. What was the commission on the sale of 4 football tickets that cost $45 each?

12. *Electric Range* A self-cleaning electric range sells for $350. If the wholesale price is $218.75, what is the markup rate?

13. *Electric Mixer* An electric mixer priced at $220 is discounted by 14%. What is the sale price?

14. *Sightseeing Boat* The average number of passengers per day on a sightseeing boat in June increased 15% from the number in May. If the average number of passengers per day in May was 80, what was the average in June?

15. *Dress Sale* A dress was advertised for $85.80, which was 22% off the regular price. What was the regular price?

16. Evaluate $3x^2 + 2x$ for $x = -1$.

In Questions 17 and 18, perform the indicated operation.

17. $(x^2 + 6x - 8) + (x^3 - 2x^2 + 9)$

18. $(4x^2 + 7x - 4) - (3x^2 - 5x + 2)$

19. According to the ________ Rule, to simplify $(a^5)^3$, we ________ the exponents.

20. Multiply.

(a) $a^7 \cdot a^2$

(b) $7y^3 \cdot 5y^4$

21. Simplify.

(a) $(z^6)^5$

(b) $(y^7)^3 \cdot y^9$

22. Simplify.

(a) $(a^5b)^3$

(b) $(2x^2y^6)^4$

In Questions 23–25, multiply.

23. **(a)** $8x(2x - 5)$

(b) $y^3(y^2 + 4)$

24. $(4x + 1)(x + 3)$

25. $(a + 7)^2$

Copyright © Houghton Mifflin Company. All rights reserved.

26. Determine the GCF.

(a) b^4, b^8, b^6

(b) $6n^3, 9n^2$

In Questions 27–29, factor.

27. $27x + 45$

28. $3x^2 - 7x$

29. $6x^2 - 10x + 8$

Copyright © Houghton Mifflin Company. All rights reserved.

Answers to Exercises

CHAPTER 1

Section 1.1 (page 11)

1. number line

3. 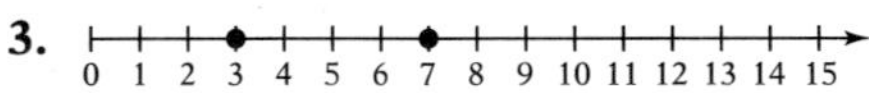

5. 0 1 2 3 4 5 6 7 8 9 10 11 12 13 14 15

7. < **9.** > **11.** < **13.** >
15. 41, 32, 21, 20, 12, 3
17. 101, 117, 210, 213, 245, 312
19. 4 **21.** 5 **23.** standard, expanded
25. 500 + 30 + 6 **27.** 40,000 + 600 + 90
29. 700,000 + 90,000 + 6,000
31. 10,000,000 + 6,000,000 + 200,000 + 30 + 7
33. 643 **35.** 14,006 **37.** 620,049 **39.** 600,062,513
41. Fifty-four **43.** Four hundred seven
45. Eight thousand fourteen
47. Three million two hundred ninety thousand seven hundred
49. Five thousand two hundred eighty
51. Five hundred seventy-nine thousand eight hundred thirty-three
53. 93 **55.** 602 **57.** 990 **59.** 42,000,008 **61.** 462
63. 50,455,000 **65.** 2,357
67. The number 12 is located 3 units to the right of 9 and the number 6 is located 3 units to the left of 9.
69. 7 **71.** 100

Section 1.2 (page 21)

1. 0 **3.** 800 **5.** 4,380 **7.** 16,000 **9.** 94,000
11. 540,000 **13.** 8,000,000 **15.** July
17. June and August **19.** 200,154 < 227,818
21. 500,000 **23.** McKinley, St. Elias, and Hubbard
25. 18,008 > 15,015
27. The distance between Earth and the Sun, because that distance is more than a million, whereas the number of students at a college is much less than a million.
29. **(a)** 30,000 **(b)** 30,000 **(c)** 30,000 **(d)** 30,000
30. 832,000 **31.** Teacher aides
32. Teacher aides and systems analysts
33. Nearest thousand **34.** 832,000 < 893,000
35. Answers will vary.

Section 1.3 (page 31)

1. sum, addends **3.** 6 + 5 **5.** $x + 4$ **7.** 3 + 2
9. 7 + 4 **11.** 36 + 9 **13.** variable
15. Commutative **17.** 6 **19.** z **21.** 0 **23.** 9
25. z **27.** 2 + 9 **29.** $n + c$ **31.** 3 + (8 + 7)
33. $(4 + x) + y$
35. Commutative Property of Addition
37. Addition Property of 0
39. Associative Property of Addition
41. Two numbers can be added in either order and the sum is the same.
43. **(a)** Commutative Property of Addition
(b) Commutative Property of Addition
45. $x + (4 + 5)$ **47.** $b + (9 + 2)$

Section 1.4 (page 43)

1. Commutative **3.** 79 **5.** 9,697 **7.** 97 **9.** 987
11. 1,404 **13.** 9,077 **15.** 6,290 **17.** 213
19. 21,886 **21.** subtrahend, minuend **23.** 12 − 3
25. $x - 9$ **27.** 11 − 2 **29.** 110 − 28
31. 1,830 − 1,497 **33.** 23 **35.** 2,372 **37.** 18
39. 108 **41.** 129 **43.** 3,192 **45.** 750 **47.** rounding
49. 400 **51.** 2,000 **53.** 1,222 **55.** 657 **57.** 11
59. 62,169 **61.** 741,857 **63.** \$815 **65.** \$355
67. 1,219 **69.** \$1,649 **71.** \$211 **73.** 2,757
75. **(a)** \$1,783 **(b)** \$1,288 **(c)** \$931 **(d)** \$1,083
77. Group the addends so that the sums are 10:

$$(3 + 7) + (6 + 4) + (8 + 2)$$

79. Because $n = n + 0$, $n - 0 = n$. **81.** 6,977
83. \$495 **85.** 24 million **86.** 4,489,322 **87.** 19
88. 151 **89.** 49 acres **90.** 3,521 **91.** 1,221 acres
92. 1,079 acres

Section 1.5 (page 53)

1. 273,115 **3.** 107,605 **5.** 13,770,000
7. 8,680,000 **9.** \$224 million **11.** \$187 million
13. \$1,389 million **15.** 9 **17.** 12,646 **19.** 73,688
21. perimeter **23.** 16 feet **25.** 22 yards
27. 18 inches **29.** 95 feet **31.** 46 yards **33.** 1,050
35. 324
37. No, the sum of the same four whole numbers cannot be 6.
39. 4 inches long and 1 inch wide or 3 inches long and 2 inches wide
41. 24 inches **42.** (ii)

Copyright © Houghton Mifflin Company. All rights reserved.

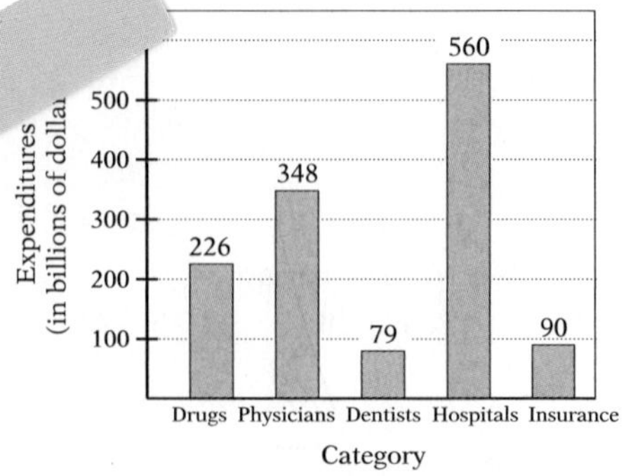

44. \$1,303 billion **45.** \$908 billion
46. Hospitals and dentists **47.** \$14 billion
48. \$427 billion **49.** \$395 billion
50. \$269 billion
51. No, we do not have enough information to calculate average charges.

Section 1.6 *(page 65)*

1. variable, numerical **3.** 10 **5.** 4 **7.** 16 **9.** 5
11. 12 **13.** 5 **15.** 18 **17.** 8 **19.** 2,262 **21.** 800
23. 163 **25.** 66 **27.** $b + 12$ **29.** $11 + y$
31. $z + 12$ **33.** $9 + x$ **35.** equation **37.** Yes
39. No **41.** 9 **43.** 8 **45.** 2 **47.** 4 **49.** 4
51. 14 **53.** 5 **55.** 10 **57.** 10 **59.** 4 **61.** 3
63. The Commutative Property of Addition states that $x + 5 = 5 + x$ for any number.
65. Any number **67.** $x + 9$ **69.** $b + 11$

Review Exercises *(page 69)*

1. natural, whole **2.** inequality

3. 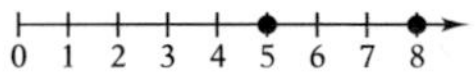

4. $>$ **5.** $<$ **6.** 139, 132, 21, 12 **7.** 7
8. 3 hundreds **9.** $7{,}000 + 40 + 5$ **10.** 563
11. Three hundred thirty-four thousand five hundred sixty-three
12. 3,006 **13.** estimate **14.** tens **15.** 7,000
16. 4,000,000 **17.** 82,000 **18.** 400
19. United States **20.** Japan **21.** $1{,}670 > 936$
22. $936 < 1{,}410$ **23.** \$2,000, \$1,000, \$1,000
24. \$1,700, \$1,400, \$900 **25.** sum, addends
26. $n + 8$ **27.** $32 + 5$
28. A variable is a symbol that represents an unknown number.
29. $a + (7 + b)$ **30.** $n + m$ **31.** y
32. Commutative Property of Addition
33. Addition Property of 0
34. Associative Property of Addition
35. 5,091 **36.** 448
37. Round the numbers to 200 and 500 and add to obtain the estimate of 700.
38. subtrahend, minuend **39.** $n - 7$ **40.** 206
41. 2,606 **42.** \$1,734 **43.** 1,210,000 **44.** 59,638
45. 39,000,000 **46.** 1,000,000 **47.** 16,018,929
48. 21,291,584 **49.** 6,338,808 **50.** 3,599,636
51. perimeter **52.** triangle **53.** 26 feet
54. 294 yards **55.** variable **56.** 11 **57.** 8 **58.** 89
59. $11 + a$ **60.** equation **61.** 9 **62.** 8 **63.** 0
64. 6

Chapter Test *(page 75)*

1. [1.3, 1.4] addends, minuend, subtrahend
2. [1.1] **(a)** Seventy-two million one hundred thirty-six thousand [1.2] **(b)** 70,000,000
3. [1.1] 3,610 **4.** [1.1] **(a)** $<$ [1.1] **(b)** $>$
5. [1.3] **(a)** (ii) [1.3] **(b)** (i)
6. [1.4] 1,300 **7.** [1.4] 2,068
8. [1.3, 1.4] addition, subtraction
9. [1.6] $5 + n$, variable expression; $30 - 17$, numerical expression
10. [1.4] 1,443 **11.** [1.3] $\$123 + \$15$ **12.** [1.4] 281
13. [1.4] \$24
14. [1.4] **(a)** 190 million [1.4] **(b)** 22 million
15. [1.2] Adult paperback **16.** [1.2] $116 < 270$
17. [1.5] \$560 million **18.** [1.5] \$212 million
19. [1.5] 212 **20.** [1.5] 56 inches **21.** [1.6] $x + 10$
22. [1.6] 16 **23.** [1.6] 8 **24.** [1.6] 6

CHAPTER 2

Section 2.1 *(page 87)*

1. factors **3.** 5, 11 **5.** 7, x **7.** 2, a, b **9.** $9 \cdot 7$
11. $3 \cdot 5$ **13.** $4y$ **15.** $58 \cdot 6$ **17.** $12 \cdot 3$ **19.** $2x$
21. Commutative **23.** $12 \cdot 5$ **25.** $8 \cdot c$
27. $(5 \cdot 6) \cdot 9$ **29.** $(7 \cdot 3)b$ **31.** 0 **33.** 7
35. Multiplication Property of 1
37. Commutative Property of Multiplication
39. multiples **41.** 800 **43.** 320 **45.** 497,000
47. 4,800 **49.** 21,000 **51.** 200 **53.** \$5,200
55. \$240 **57.** 1,250 **59.** 140 **61.** 6,500 **63.** \$320
65. \$288 **67.** \$1,200
69. The product 6(13) means $13 + 13 + 13 + 13 + 13 + 13$.
71. 2, 6 **73.** 2, 10 **75.** 0

Section 2.2 *(page 99)*

1. Distributive **3.** $5(2) + 5(7) = 45$
5. $4(20) + 4(3) = 92$ **7.** $6(300) + 6(7) = 1{,}842$
9. 469 **11.** 19,236 **13.** 3,526 **15.** 810
17. 36,504 **19.** 84,915 **21.** 359,840 **23.** 205,623
25. 583,500 **27.** 63,000 **29.** 6,000,000
31. 256,743,240 **33.** 743,341,458 **35.** 700

Copyright © Houghton Mifflin Company. All rights reserved.

37. 834 **39.** 6,316 **41.** 1,925 miles **43.** 380
45. $18 billion **47.** $384 **49.** $13,624 **51.** 9,463
53. 645 **55.** 3,499,186
57. (i) The second step involves multiplication by 0.
59. $553

61.

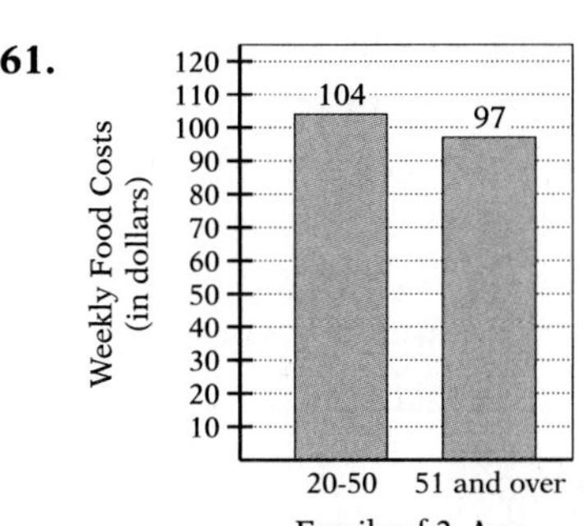

62.

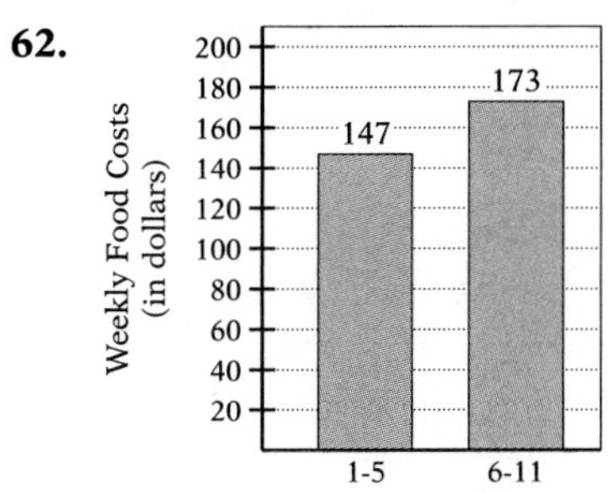

63. $7,644 **64.** $692 **65.** $28 **66.** $1,352
67. No **68.** (ii)

Section 2.3 *(page 117)*

1. divisor, dividend **3.** **(a)** $15 \div 3, 5$ **(b)** 15 **(c)** 3
5. **(a)** $\frac{8}{4}, 2$ **(b)** 8 **(c)** 4 **7.** 7, 7 **9.** 4, 4
11. undefined **13.** 0 **15.** 1 **17.** 52
19. undefined **21.** (iii) **23.** 31 **25.** 68 **27.** 58
29. 71 **31.** 307 **33.** divisible **35.** 8 R 4
37. 422 R 2 **39.** 61 R 20 **41.** 70 R 21 **43.** 463 R 8
45. 405 **47.** 105 **49.** 10; 4 ounces **51.** 154
53. 154 **55.** $6 **57.** 7; 4 ounces **59.** 17
61. $4,683 **63.** 38,637 **65.** 744 **67.** 5,118
69. 647 **71.** Both are the same: $15 \div 3$.
73. $24 \div (6 \div 2) = 8$
75. 8 people per square mile
76. 602 people per square mile
77. 869 people per square mile
78. 75 people per square mile

79.

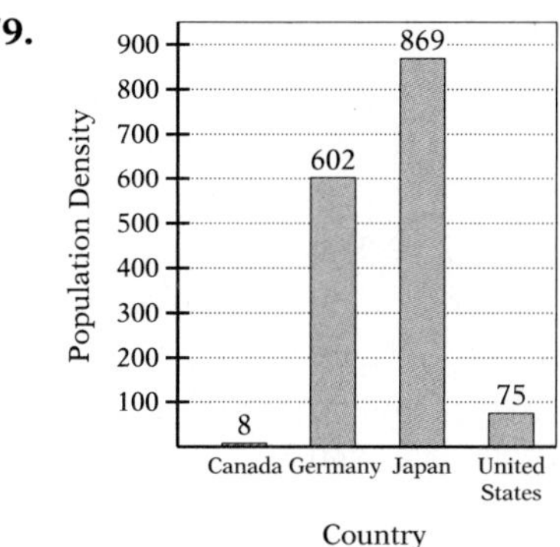

80. 9 **81.** 8 **82.** 27 **83.** 4 **84.** 108

Section 2.4 *(page 129)*

1. exponent, base
3. Base 5, exponent 3; 5 cubed
5. Base 7, exponent 5; 7 to the fifth power
7. 7^3 **9.** 8^4 **11.** 3^9 **13.** 15^3 **15.** $4 \cdot 4 \cdot 4$
17. $7 \cdot 7 \cdot 7 \cdot 7 \cdot 7$ **19.** 64 **21.** 729 **23.** 1
25. 1,000,000 **27.** 100,000,000 **29.** 10^4 **31.** 10^9
33. 1 **35.** 0 **37.** 1 **39.** 1 **41.** 3,375
43. 1,073,741,824 **45.** 1 **47.** 19 R 6 **49.** 105
51. 7,720 **53.** 81 **55.** 3,562
57. Read 7^2 as "7 to the second power" or "7 squared."
59. (iv)

Section 2.5 *(page 139)*

1. grouping symbols **3.** 4 **5.** 14 **7.** 5 **9.** 8
11. 72 **13.** 7 **15.** 40 **17.** 1 **19.** 45 **21.** 53
23. 34 **25.** 9 **27.** 30 **29.** 0 **31.** 1 **33.** 1,026
35. 23 **37.** $4 \cdot 8 - 5 = 27$ **39.** $\frac{10}{7-2} = 2$
41. $8^2 - 7 \cdot 4 = 36$ **43.** 89 **45.** 3 **47.** $193
49. Add the numbers and divide by 2. **51.** 5,000
53. $3 \cdot (5 + 2) = 21$ **55.** 6,952,000 **56.** 640,000
57. 1,975,000 **58.** domestic trucks **59.** 1990
60. 8,927,000 **61.** 1990 **62.** 15,691,000

Section 2.6 *(page 149)*

1. simplifying **3.** $15x$ **5.** $8a$ **7.** $10ax$ **9.** $12xy$
11. $5x + 30$ **13.** $14 + 2a$ **15.** $14x + 7$
17. $56a + 48$ **19.** Operations **21.** 16 **23.** 24
25. 150 **27.** 6 **29.** 3 **31.** 1 **33.** 12 **35.** 10
37. 17 **39.** 5 **41.** expression, equation **43.** 8
45. 1 **47.** 8 **49.** 0 **51.** 81 **53.** 0 **55.** 7
57. 10 **59.** 4 **61.** 3
63. (ii) because the quantity in parentheses is a sum
65. Any number divided by 1 is the number.
67. 15 **69.** **(a)** $3x + 2xy$ **(b)** $10ax + 6bx$

Copyright © Houghton Mifflin Company. All rights reserved.

)

are inches **5.** 81 square miles

11. 48 square inches

ies **15.** \$18 **17.** 21 yards

0 feet **23.** 2,544 square feet

27. 30 square feet **29.** 56 square yards

31 square inches **33.** 152 square feet

35. 210 square feet **37.** 316,800 square feet

39. \$2,400 **41.** \$70,720 **43.** 360 feet

45. The length and width must be expressed in the same units.

47. **(a)** Increased by 4 inches
(b) Increased by 7 square inches

Section 2.8 *(page 173)*

1. volume **3.** 42 cubic inches **5.** 152 cubic yards
7. 570 cubic feet **9.** 600 cubic inches
11. 1,216 cubic inches **13.** \$5,600
15. cubic, square **17.** 82 square inches
19. 1,712 square inches **21.** 16 **23.** 363,780
25. The area is measured in square units, and the volume in cubic units.
27. **(a)** Increased by 61 cubic inches **(b)** Increased by 54 square inches

Review Exercises *(page 177)*

1. product, factors **2.** 7, a, b
3. Multiplication Property of 0
4. Multiplication Property of 1
5. Associative Property of Multiplication
6. Commutative Property of Multiplication
7. Multiply 5 by 3 and place three 0s after the result.
8. 68,000 **9.** 2,400 **10.** 2,100 **11.** Distributive
12. $2 \cdot 400 + 2 \cdot 7$ **13.** 90,000 **14.** 29,376
15. 22,952 **16.** 1,627,920 **17.** 145,285
18. 350,000,000 **19.** \$606 **20.** \$2,184
21. quotient, divisor, dividend **22.** Multiply 8 by 4.
23. **(a)** 0 **(b)** undefined **24.** **(a)** 1 **(b)** 17 **25.** 52
26. 504 **27.** 57 R 1 **28.** 87 R 58 **29.** 156
30. 9; 16 **31.** exponential, exponent, base **32.** y^5
33. $6 \cdot 6 \cdot 6 \cdot 6$; 6 to the fourth power **34.** 7^3 **35.** 32
36. 1 **37.** 10,000 **38.** 1 **39.** 225 **40.** 10^7
41. Order of Operations **42.** grouping **43.** 8
44. 23 **45.** 7 **46.** 32 **47.** 2 **48.** $6 \cdot 4 - 3 = 21$
49. $\dfrac{20}{9-5} = 5$ **50.** 18 **51.** Distributive **52.** $15ab$
53. $12x + 28$ **54.** 280 **55.** 5 **56.** 6 **57.** 4
58. 0 **59.** 42 **60.** 5 **61.** area **62.** formula
63. Divide by 36. **64.** Multiply by 12.
65. 49 square yards **66.** 6 feet **67.** \$1,320
68. 175 feet **69.** 65 square miles **70.** 10 square feet
71. volume **72.** rectangular **73.** $V = s^3$
74. **(a)** 125 cubic inches **(b)** 150 square inches
75. 15,552 cubic inches **76.** 3,888 square inches
77. 14,720 **78.** 130 cubic inches **79.** 504
80. \$18,720

Chapter Test *(page 183)*

1. [2.1] factors
2. [2.3] (i) because division by 0 is not defined.
3. [2.1] **(a)** (i) [2.1] **(b)** (ii) **4.** [2.1] 3,500
5. [2.2] 1,209,600 **6.** [2.2] 5,840,282 **7.** [2.3] 504
8. [2.3] 100 R 77 **9.** [2.4] 7^4; base 7, exponent 4
10. [2.4] 64 **11.** [2.4] 1 **12.** [2.5] 1
13. [2.5] 17 **14.** [2.5] $2(7 + 9) = 32$ **15.** [2.6] $15a$
16. [2.6] $30x + 35$ **17.** [2.6] 31 **18.** [2.6] 9
19. [2.6] 7 **20.** [2.2] 9,600 **21.** [2.3] \$16
22. [2.5] 89 **23.** [2.7] 40 square feet
24. [2.7] 75 square inches
25. [2.8] **(a)** 200 cubic feet [2.8] **(b)** 210 square feet

CHAPTER 3

Section 3.1 *(page 195)*

1. integers **3.** $<$ **5.** $>$ **7.** $<$ **9.** $>$
11. $-5, -3, -1, 1, 3$ **13.** $-7, -4, -3, 0, 2$
15. 1 **17.** -6 **19.** -4 **21.** opposites **23.** 3
25. -22 **27.** 7
29. Negative 4 minus 3
31. Negative 6 plus x
33. Negative 3 minus negative 4
35. The opposite of negative 5
37. 15 **39.** y **41.** -8 **43.** absolute value **45.** 7
47. 12 **49.** 9 **51.** -5 **53.** $=$ **55.** $<$ **57.** $>$
59. $-|2|, -1, |0|, -(-4), |-5|$
61. $-|4|, -3, -|-2|, |2|, |-6|$
63. 535, -8 **65.** 350, -128
67. Both a and $-a$ are the same distance from 0.
69. If n represents a negative number, then $-n$ represents a positive number.
71. -3 and 3 **72.** 40°F, -27°F
73. $-27, -25, -17, 1, 17, 40$

74.

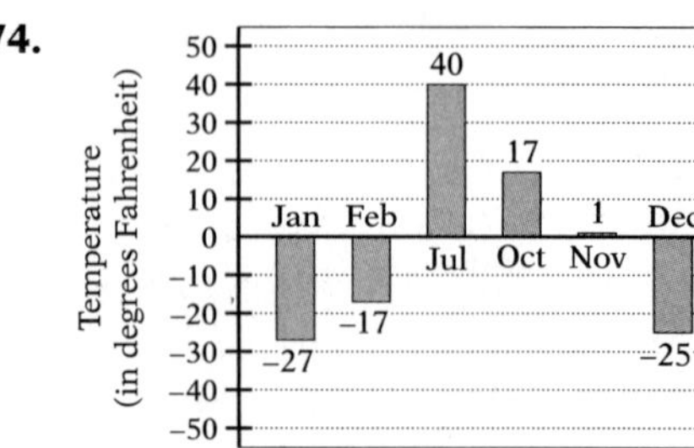

75. $-17 > -25$ **76.** $-27 < -17$
77. February and October **78.** July and January
79. November

Copyright © Houghton Mifflin Company. All rights reserved.

Section 3.2 *(page 205)*

1. absolute values **3.** −9 **5.** −16 **7.** −7 **9.** 4
11. 6 **13.** −6 **15.** −20 **17.** 9 **19.** −4 **21.** 0
23. −1 **25.** 7 **27.** −5 **29.** −7 **31.** 2 **33.** −15
35. 0 **37.** −28 **39.** −5 **41.** 11 **43.** −12
45. opposites **47.** −8 **49.** −7 **51.** −6 **53.** −20
55. 6 **57.** 3 **59.** 4 **61.** 6 **63.** D **65.** A **67.** 64
69. −985 **71.** 640 + (−320) + (−70) + 200 = \$450
73. −3 + 1 + (−4) + (−1) + 2 + (−3) = −8; 8 pounds lost
75. Begin at −2. To add −4, move to the left 4 units. To add 4, move to the right 4 units.
77. We must agree that −0 = 0. **79.** 4
81. **(a)** Positive **(b)** Negative

Section 3.3 *(page 215)*

1. minus, plus, opposite **3.** 11 **5.** 2 **7.** −13
9. 5 **11.** −3 **13.** −2 **15.** −5 **17.** 14 **19.** 7
21. 7 **23.** −20 **25.** −5 **27.** 21 **29.** 11 **31.** −5
33. −15 **35.** 0 **37.** −7 **39.** −17
41. Order of Operations **43.** −1 **45.** 1 **47.** −4
49. 10 **51.** −8 **53.** −16 **55.** B **57.** C **59.** 261
61. −409 **63.** 73 − (−5) = 78°F
65. 730 − 875 = −145 **67.** −15 **69.** −11 **71.** 20
73. −11 **75.** −1 **77.** The values are opposites.
79. −10 **81.** 9 **83.** 40 **84.** 8
85. San Diego and Tennessee; Baltimore and Buffalo
86. Seattle and Tennessee; San Diego and Baltimore; Tennessee and Buffalo
87. 34 **88.** −44

Section 3.4 *(page 227)*

1. positive **3.** −30 **5.** 63 **7.** 40 **9.** −56 **11.** 0
13. 48 **15.** −500 **17.** 120 **19.** −36 **21.** 70
23. 0 **25.** −16 **27.** −280 **29.** even **31.** 36
33. −24 **35.** −60 **37.** 48 **39.** −32 **41.** −27
43. 1 **45.** opposite **47.** $-1x$ **49.** $-a$
51. grouping **53.** 30 **55.** −12 **57.** −30 **59.** 9
61. −44 **63.** 9 **65.** D **67.** C **69.** 51,136
71. −279,552 **73.** 1,400 **75.** −64 **77.** −11
79. −15 **81.** 8 **83.** One of the numbers is 0.
85. −4, 2 **87.** −1, −3

Section 3.5 *(page 237)*

1. multiplication **3.** 4 **5.** −5 **7.** −5 **9.** −12
11. undefined **13.** 3 **15.** −10 **17.** 1 **19.** −8
21. 0 **23.** −4 **25.** 7 **27.** grouping **29.** 1
31. −4 **33.** 3 **35.** 1 **37.** −2 **39.** −4 **41.** 3
43. C **45.** B **47.** 65 **49.** −465 **51.** −1°F
53. Raleigh **55.** −28 **57.** 8 **59.** −1 **61.** −12
63. 21 **65.** 7 **67.** −6 **69.** Multiply 4 by 6.
71. −2 **73.** **(a)** Negative **(b)** Positive
75. 18,602 feet **77.** 420 feet **79.** 9,209 feet
81. 22,000 feet

Section 3.6 *(page 247)*

1. evaluate **3.** −18 **5.** −6 **7.** 2 **9.** −4
11. −36 **13.** 28 **15.** −14 **17.** −56 **19.** −72
21. −5 **23.** 12 **25.** −1 **27.** **(a)** −5 **(b)** 5
29. **(a)** 1 **(b)** −1 **31.** 5 **33.** 7 **35.** solving
37. 0 **39.** −5 **41.** 4 **43.** −7 **45.** 12 **47.** 3
49. −7 **51.** 7 **53.** 4 **55.** 3 **57.** −2 **59.** −4
61. 3 **63.** −1 **65.** 9 **67.** −36 **69.** −9 **71.** −7
73. The opposite of what number is −2? **75.** −7, 7
77. 1 **79.** **(a)** Even **(b)** Odd

Section 3.7 *(page 259)*

1. simplify **3.** $-10x$ **5.** $42a$ **7.** $-36x$
9. Distributive **11.** $-5x - 40$ **13.** $-14x - 49$
15. $8x - 24y$ **17.** $-9a + 45$ **19.** $-10m + 12n$
21. $4x + 4y + 12$ **23.** $-3m - 3n + 12$
25. $-21a - 56b - 7c$ **27.** $-6x + 15y - 21$
29. opposite **31.** $-5 - y$ **33.** $3x - 8$ **35.** $-5 + 3b$
37. $-2a + 5$ **39.** terms **41.** $12x$ **43.** $-a$
45. $-20x$ **47.** $5a + 8b$ **49.** $-4n$ **51.** $9y$
53. $12x - 5$ **55.** $6x - 4$ **57.** $5n + 2$
59. $-3x + 4y + 3$ **61.** $a - 4$ **63.** $2n$ **65.** $4x - 27$
67. $2a - 6$ **69.** $7x - 29$ **71.** $3x + 3$
73. Change subtractions to additions. The terms are the addends of the expression.
75. Yes **77.** Yes **79.** No

Review Exercises *(page 263)*

1. origin **2.** **(a)** < **(b)** < **3.** −3
4. **(a)** −9 **(b)** 12 **(c)** 0
5. The opposite of *c* minus 5
6. **(a)** 7 **(b)** x **7.** **(a)** 9 **(b)** −6
8. $-4, -|-3|, |0|, -(-5), |-7|$
9. Both numbers are 3 units from the origin.
10. −97, 305 **11.** negative **12.** −16 **13.** −7
14. 6 **15.** −5 **16.** **(a)** −5 **(b)** (−10) **17.** −4
18. 0 **19.** 4
20. 380 + (−60) + (−18) + (−24) = \$278
21. Change the minus symbol to a plus symbol and write the opposite of −3: −5 + 3.
22. −6 **23.** 9 **24.** −12 **25.** −8 **26.** 16
27. −22 **28.** −1 **29.** −9 **30.** 7,362 feet
31. positive, negative **32.** 30 **33.** −16 **34.** −70
35. −48 **36.** **(a)** 1 **(b)** −32 **37.** **(a)** $-1c$ **(b)** $-n$
38. 31 **39.** 3 **40.** 27 **41.** undefined **42.** −4
43. 0 **44.** −1 **45.** 10 **46.** 4 **47.** 3 **48.** 2
49. −3 **50.** −3°F **51.** 2 **52.** −3 **53.** 0 **54.** −6
55. **(a)** −6 **(b)** 6 **(c)** 17 **56.** −10 **57.** 6 **58.** 4
59. 4 **60.** −10 **61.** $15x$ **62.** $-3x - 12$
63. $-8x + 32$ **64.** $6a - 12b + 3$
65. opposite, difference **66.** $-3m + 8n$ **67.** $-4n$
68. −5 **69.** $-14a + 4$ **70.** $-11x + 22$

Copyright © Houghton Mifflin Company. All rights reserved.

Chapter Test (page 267)

1. [3.1] **(a)** > [3.1] **(b)** <
2. [3.1] $-|-7|, -2, -|0|, -(-1), |3|$
3. [3.2] **(a)** −15 [3.2] **(b)** 2
4. [3.3] **(a)** −18 [3.3] **(b)** 3
5. [3.4] **(a)** −48 [3.4] **(b)** 14
6. [3.5] **(a)** 4 [3.5] **(b)** −9
7. [3.2] −1 **8.** [3.3] −7 **9.** [3.4] −120
10. [3.4] $-1m, -x$ **11.** [3.5] undefined, 0
12. [3.6] 5 **13.** [3.6] −9
14. [3.4] If n is even, the number is positive. If n is odd, the number is negative.
15. [3.4] −27 **16.** [3.5] 2 **17.** [3.5] −3
18. [3.6] 0 **19.** [3.6] −50 **20.** [3.6] −12°F
21. [3.7] Multiplication **22.** [3.7] $4x + 12y - 20$
23. [3.7] $4b - 9$ **24.** [3.7] $x + 8$
25. [3.7] $-a - 10b$

Cumulative Test, Chapters 1–3 (page 269)

1. [1.1] **(a)** Nine thousand, five hundred thirty-four [1.2] **(b)** 10,000
2. [1.1] 1,070
3. [1.3] **(a)** 0 [2.1] **(b)** 5 [2.1] **(c)** 1 [1.3] **(d)** 7
4. [1.4] 1,594 **5.** [1.4] 1,549 **6.** [2.2] 534,204
7. [2.3] 264 **8.** [2.3] (i) is undefined and (ii) is 0.
9. [1.5, 2.7] 30 yards, 36 square yards
10. [2.4] base, exponent **11.** [2.3] **(a)** 5 [2.3] **(b)** 20
12. [2.8] **(a)** 2,328 [2.8] **(b)** 4,320
13. [3.1] $-7, -|-3|, -0, -(-3), |-7|$
14. [3.2] **(a)** −15 [3.2] **(b)** 8
15. [3.3] **(a)** −18 [3.3] **(b)** 2
16. [3.4] **(a)** −90 [3.4] **(b)** 18
17. [3.5] **(a)** 3 [3.5] **(b)** −8
18. [3.3] 5 **19.** [3.4] **(a)** −1 [3.4] **(b)** 1
20. [3.6] 7 **21.** [2.6] 0 **22.** [1.6] 3
23. [3.4] 6 **24.** [3.4] 8 **25.** [3.5] 5
26. [3.7] $-3x + y$ **27.** [3.7] $10x - 25$
28. [3.7] Like terms are numbers or terms that have the same variable parts.
29. [3.7] $x + 3$ **30.** [3.7] $-9x$

CHAPTER 4

Section 4.1 (page 279)

1. Addition **3.** Yes **5.** Yes **7.** No **9.** −4
11. 7 **13.** −7 **15.** 5 **17.** 4 **19.** −7 **21.** 14
23. 2 **25.** −3 **27.** 4 **29.** −8 **31.** Multiplication
33. −5 **35.** 2 **37.** −3 **39.** 9 **41.** 0 **43.** 7
45. 0 **47.** −6 **49.** 3 **51.** −904 **53.** −155
55. **(a)** −3 **(b)** −5 **57.** **(a)** 3 **(b)** 15
59. Subtracting a number is the same as adding the opposite of the number.
61. To isolate the variable, we must add the opposite of −3 so that the sum of the terms is 0: $-3 + x + 3 = 0 + x = x$.
63. All numbers **65.** 5 **67.** 5

Section 4.2 (page 289)

1. linear, variable **3.** −7 **5.** −5 **7.** −5 **9.** 2
11. −1 **13.** 4 **15.** 2 **17.** 0 **19.** add, right **21.** 6
23. 3 **25.** 2 **27.** −5 **29.** 0 **31.** −5 **33.** simplify
35. 0 **37.** 2 **39.** −3 **41.** 0 **43.** 3 **45.** 5
47. −4 **49.** 2 **51.** −2 **53.** −10 **55.** −5 **57.** −1
59. 7 **61.** 0 **63.** 22 **65.** −89
67. (i); it is not an equation.
69. No solution

Section 4.3 (page 299)

1. $n + 7$ **3.** $-3 + y$ **5.** $b - 5$ **7.** $2z$ **9.** $2n + 3$
11. $\frac{1}{2}n - 8$ **13.** $3(n + 7)$ **15.** $n - 5 = 8$
17. $3(n + 3) = -21$ **19.** $5n - 6 = -1$
21. $x + 8 = -4$; −12 **23.** $2x - 7 = -9$; −1
25. $2[x - (-6)] = 20$; 4 **27.** $2x - 3 = 5x$; −1
29. $5x - 1 = 7 - 3x$; 1 **31.** 8 feet, 5 feet
33. 5 inches **35.** 52 **37.** 60 feet, 120 feet **39.** 5
41. 12, 15 **43.** 4 feet, 6 feet **45.** $775 **47.** 10
49. Translate a phrase into an expression and a sentence into an equation.
51. 7, 14

Section 4.4 (page 311)

1. variables **3.** $x + y = 7$ **5.** $x = 2y$ **7.** $y = x - 4$
9. ordered pair **11.** (−6, 4), (1, −3)
13. (5, −1), (0, −3) **15.** (3, 0), (−1, 4)
17. (0, 0), (−1, −3) **19.** solution
21. (−1, 8), (3, 4), (9, −2) **23.** (−2, 12), (0, 9), (6, 0)
25. (−1, −3), (0, −1), (3, 5) **27.** (4, −4), (−2, 2), (−5, 5)

29.

x	y	(x, y)
−6	4	(−6, 4)
0	2	(0, 2)
3	1	(3, 1)
6	0	(6, 0)

31.

x	y	(x, y)
−4	−3	(−4, −3)
2	0	(2, 0)
0	−1	(0, −1)
6	2	(6, 2)

Copyright © Houghton Mifflin Company. All rights reserved.

33.

x	y	(x, y)
0	0	(0, 0)
−1	−2	(−1, −2)
1	2	(1, 2)
6	12	(6, 12)

35.

x	y	(x, y)
−2	−1	(−2, −1)
4	5	(4, 5)
6	7	(6, 7)
−4	−3	(−4, −3)

37.

x	y	(x, y)
−2	3	(−2, 3)
2	−5	(2, −5)
1	−3	(1, −3)
0	−1	(0, −1)

39.

x	y	(x, y)
−2	6	(−2, 6)
0	2	(0, 2)
2	−2	(2, −2)
3	−4	(3, −4)

41.

x	y	(x, y)
−2	−5	(−2, −5)
−1	−4	(−1, −4)
2	−1	(2, −1)
3	0	(3, 0)

43.

x	y	(x, y)
−4	5	(−4, 5)
−1	2	(−1, 2)
0	1	(0, 1)
4	−3	(4, −3)

45. (12, −43) **47.** (16, 15), (48, 70)

49. The first number is the value of x and the second is the value of y.

51. (−3, 9), (3, 9), (−2, 4), (2, 4); (−5, 25), (5, 25)

Section 4.5 *(page 321)*

1. rectangular coordinate system

3.

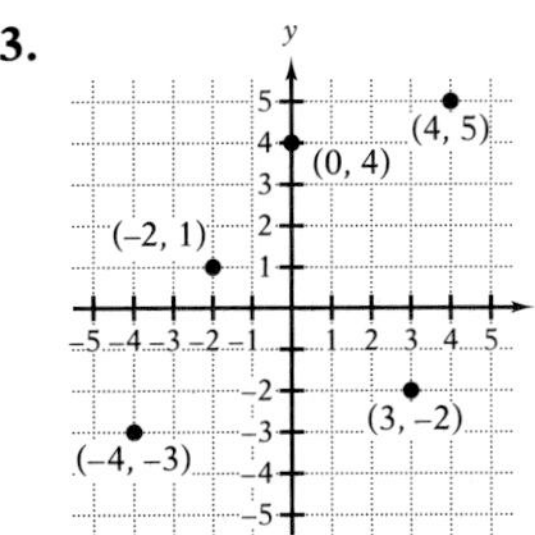

5.

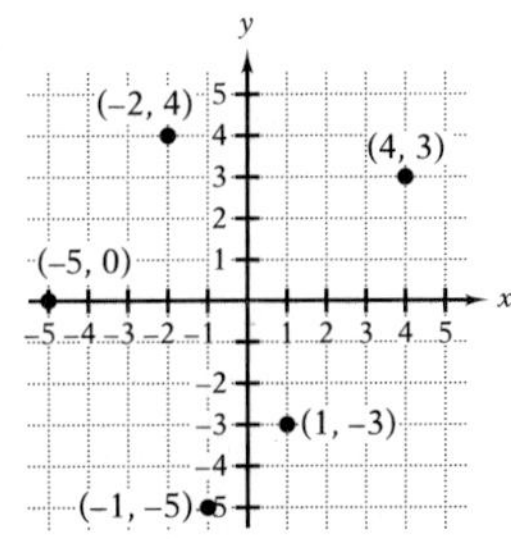

7. II **9.** y-axis **11.** III **13.** IV **15.** x-axis **17.** I

19. $A(2, 4)$, $B(5, 0)$, $C(4, -4)$, $D(0, -2)$, $E(-4, -3)$, $F(-5, 1)$

21. $A(-2, -4)$, $B(-3, 0)$, $C(-1, 2)$, $D(0, 3)$, $E(5, 1)$, $F(2, -2)$

23. (a) (2000, 3,900), (2003, 4,100), (2008, 4,300), (2013, 4,600)

(b)

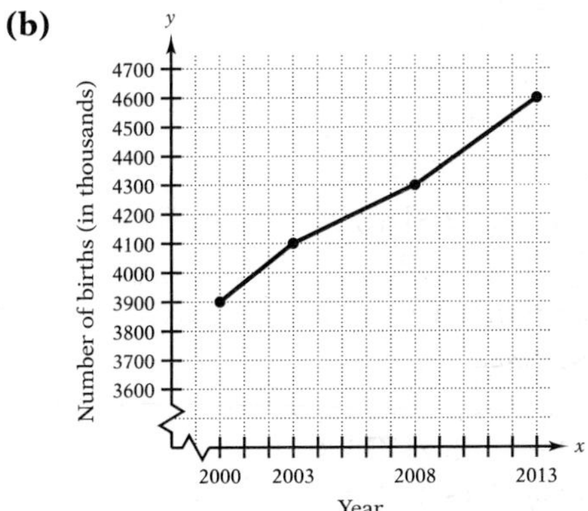

(c) Increasing

25. (a) (1996, 20), (1998, 45), (2000, 70), (2001, 107)

(b)

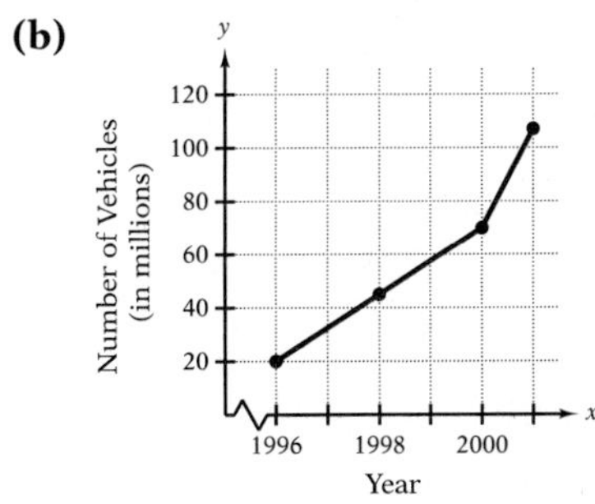

(c) Increasing

27. The second coordinate is 0.

29. The order in which the numbers are written is different.

31. The x- and y-coordinates must be equal: $a = b$.

33. I, IV

Copyright © Houghton Mifflin Company. All rights reserved.

35. III, IV

37. (3, 1)

Section 4.6 *(page 333)*

1. solutions

3. (0, 3), (−3, 0), (1, 4)

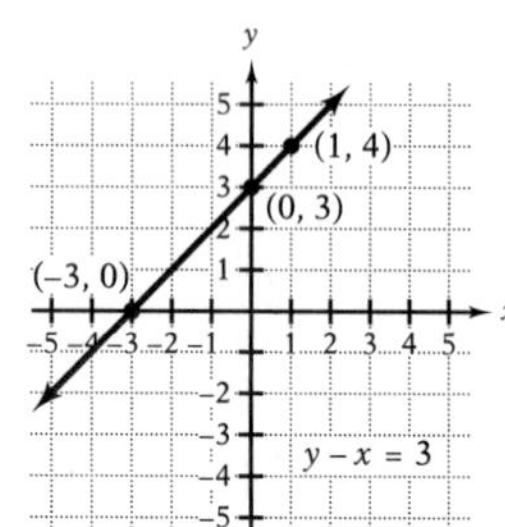

5. (−1, 4), (0, 2), (2, −2)

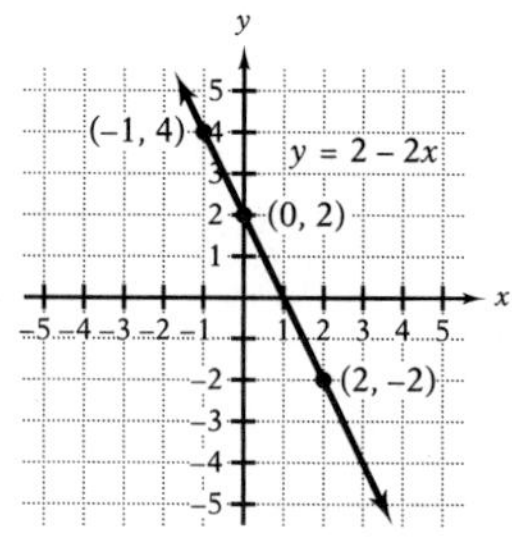

7.

x	y	(x, y)
−2	1	(−2, 1)
0	0	(0, 0)
4	−2	(4, −2)

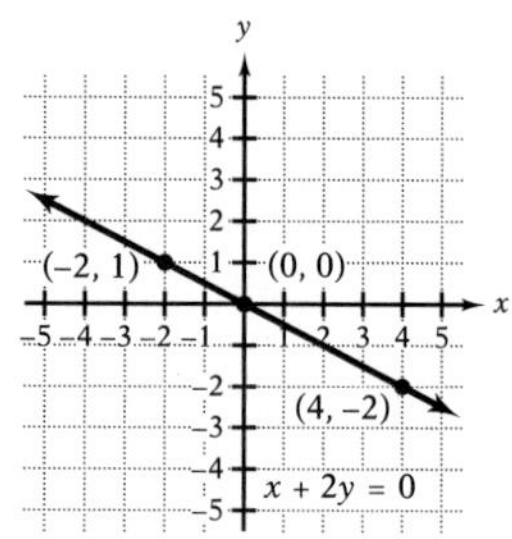

9.

x	y	(x, y)
−2	−5	(−2, −5)
0	−1	(0, −1)
3	5	(3, 5)

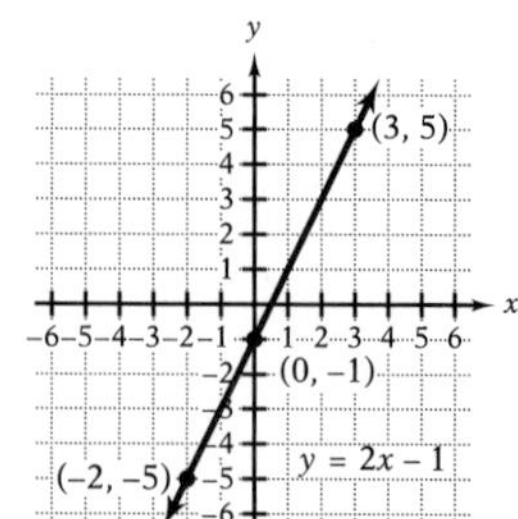

11.

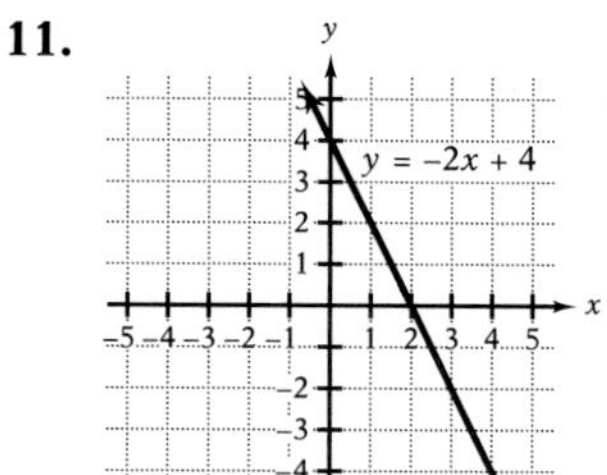

13.

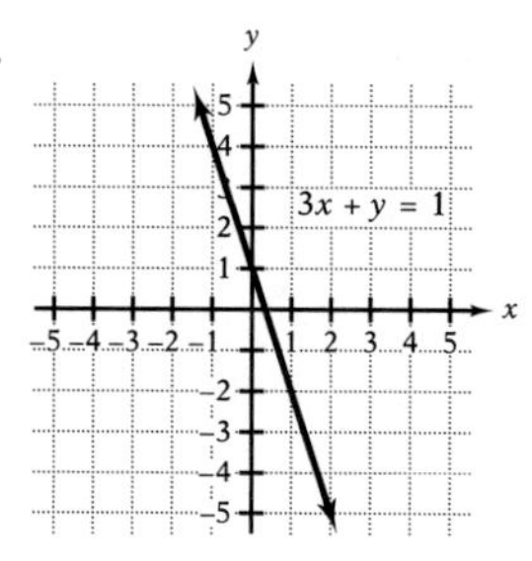

15.

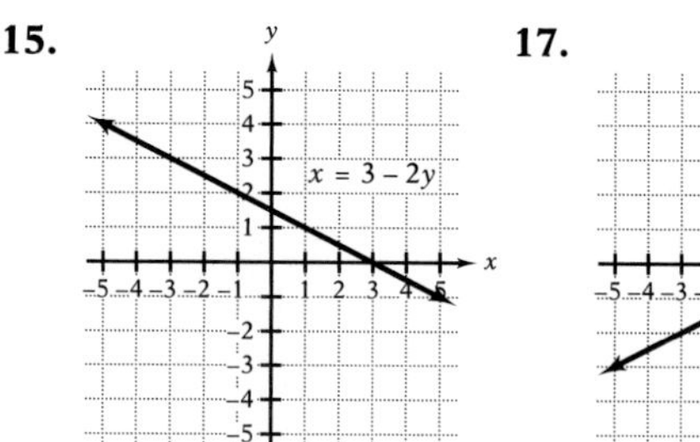

17.

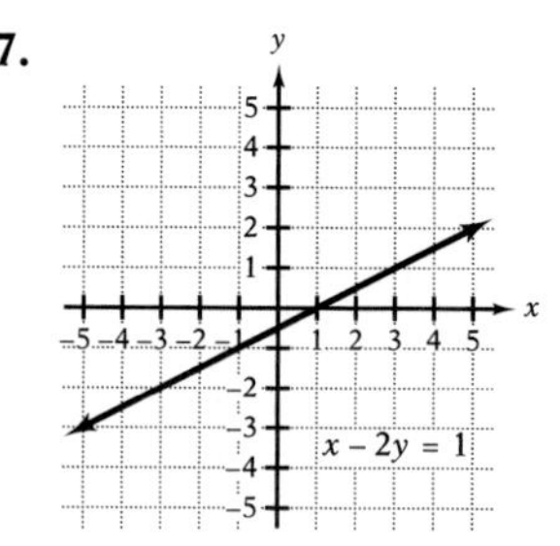

19. **21.**

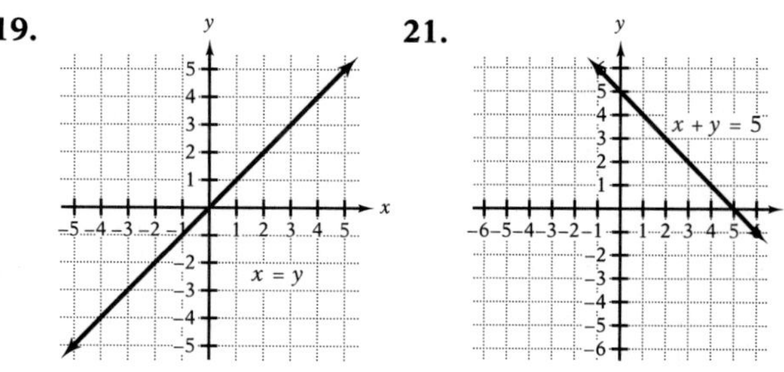

23. y-coordinate

25. $x = 2$

27. $x = -1$

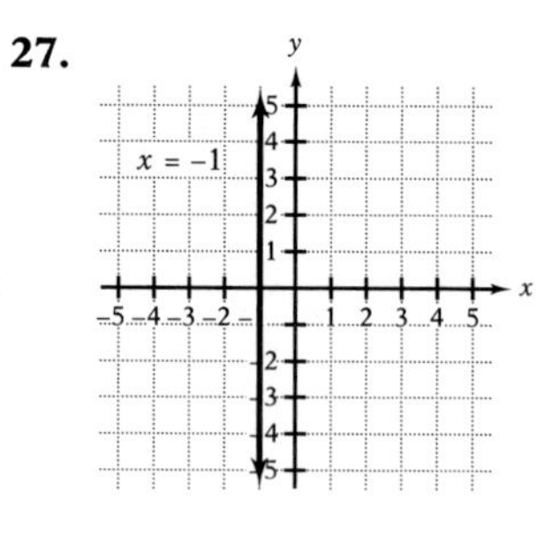

29. **31.**

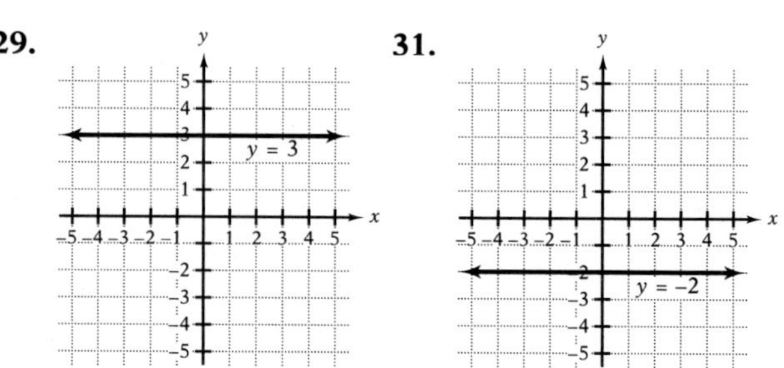

Copyright © Houghton Mifflin Company. All rights reserved.

33. (a)

T	D	(T, D)
1	60	(1, 60)
2	120	(2, 120)
3	180	(3, 180)
4	240	(4, 240)
5	300	(5, 300)

(b)

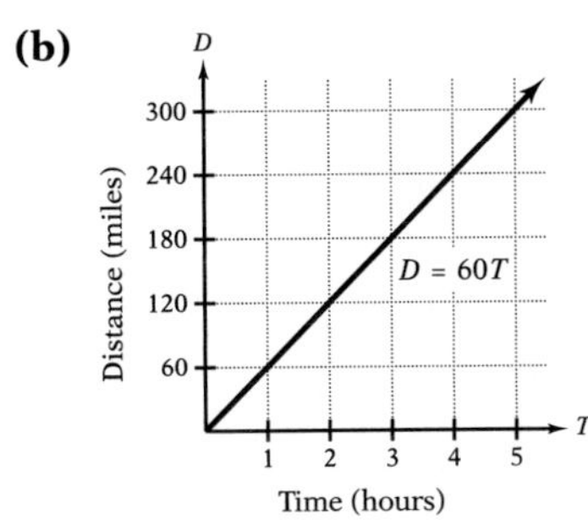

(c) 240 miles
(d) 2 hours
(e) When the time is 0, the car hasn't started yet, so the distance traveled is 0.
(f) Neither time nor distance can be negative.

35. Determining three points helps us check for errors.
37. The graph of the equation is a straight line.
39. (2, 0), (0, −3)

41.

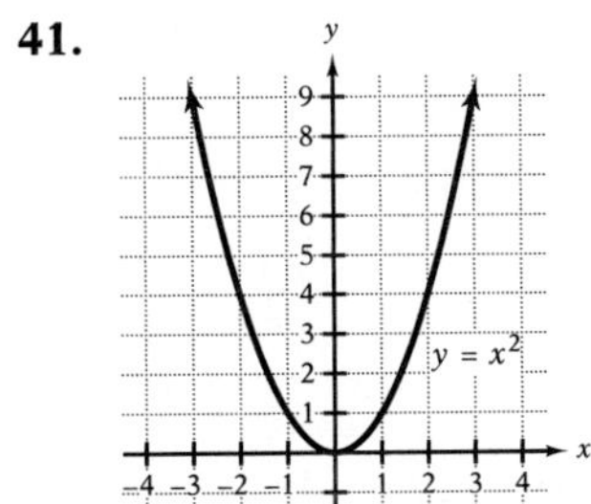

Review Exercises *(page 339)*

2. Addition **3.** Yes **4.** −8 **5.** 4 **6.** 3
7. Multiplication **8.** −5 **9.** 0 **10.** −7
11. An expression does not contain an equality symbol and an equation does.
12. simplifying **13.** −6 **14.** 5 **15.** 4 **16.** 9
17. 3 **18.** 12 **19.** 2 **20.** −8 **21.** $x + (-5)$
22. $3(5 - x)$ **23.** $x - 5 = -14$ **24.** 2 **25.** −4
26. 11 inches, 7 inches **27.** 50 feet **28.** 5 hours
29. −6, 2 **30.** 8 feet, 17 feet **31.** $x - y = 7$
32. ordered pair **33.** Yes **34.** No **35.** Yes
36. (0, 6), (5, 3) **37.** (−4, 10), (2, 4) **38.** (3, 7), (−5, 7)

39.

x	y	(x, y)
−3	2	(−3, 2)
0	1	(0, 1)
3	0	(3, 0)
6	−1	(6, −1)

40.

x	y	(x, y)
−2	−4	(−2, −4)
0	0	(0, 0)
1	2	(1, 2)
3	6	(3, 6)

41. plotting **42.** y-axis **43.** x-axis **44.** Origin

45.

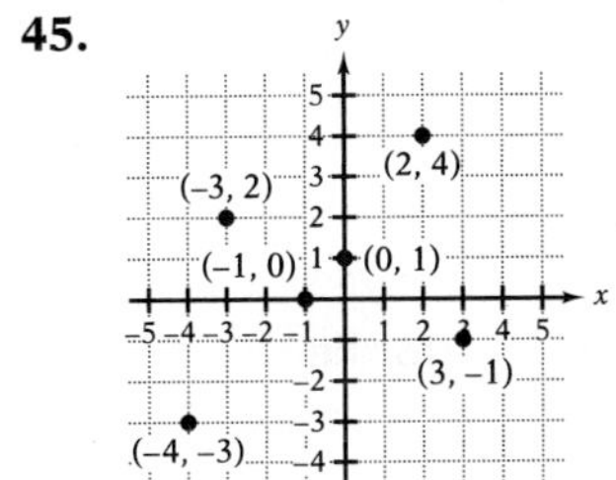

46. III **47.** IV **48.** y-axis **49.** x-axis
50. $A(0, -2)$, $B(2, -4)$, $C(3, 5)$, $D(-1, 4)$, $E(-5, 0)$, $F(-2, -1)$
51. Any number can be chosen for x.
52. (−3, 2), (0, 1), (6, −1)

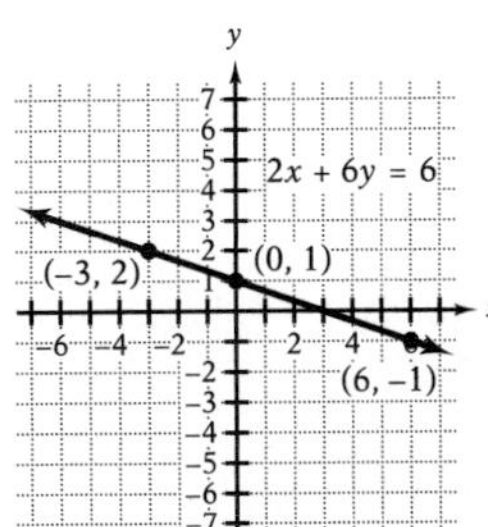

Copyright © Houghton Mifflin Company. All rights reserved.

53.

x	y	(x, y)
0	−4	(0, −4)
5	0	(5, 0)
−5	−8	(−5, −8)

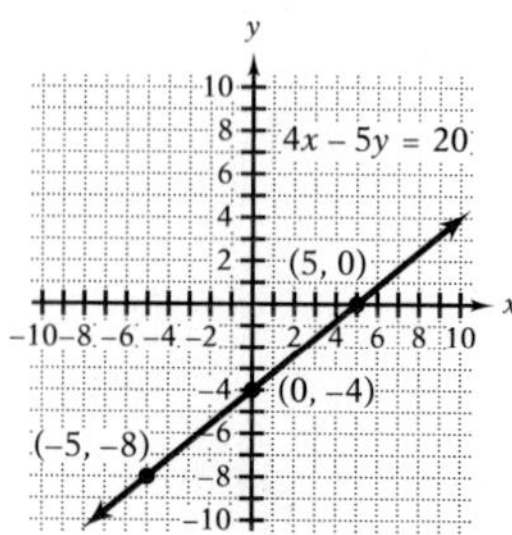

54.

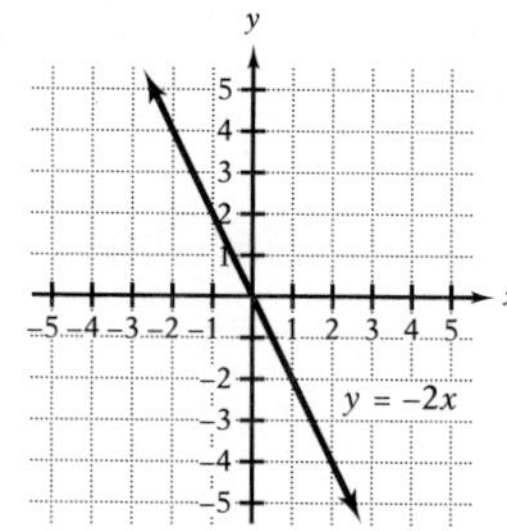

55.

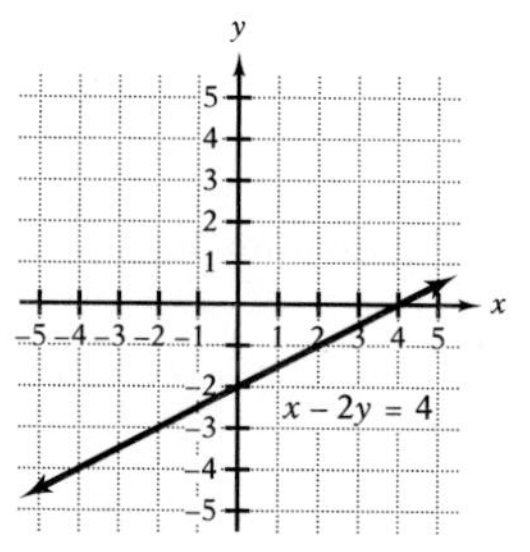

56.

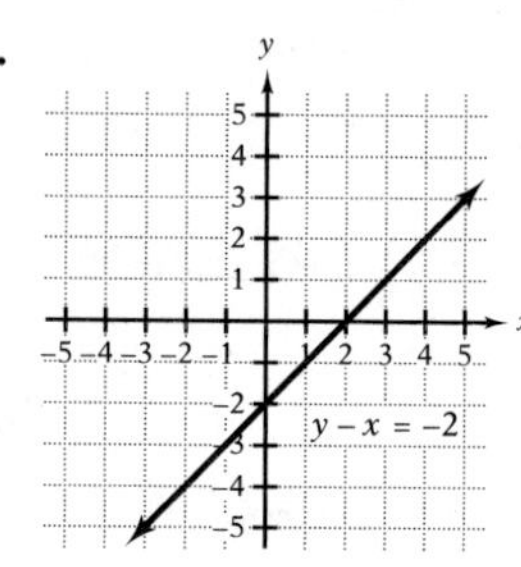

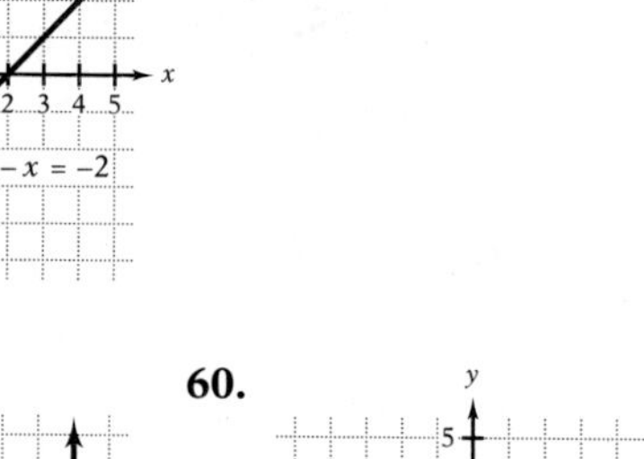

57. x-coordinate
58. y-coordinate

59.

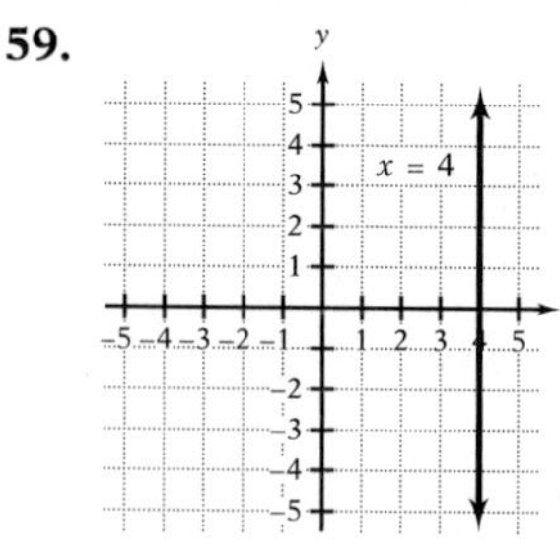

60.

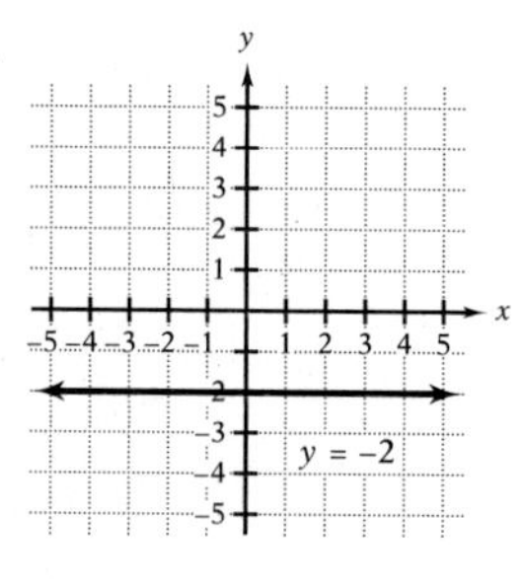

Chapter Test *(page 345)*

1. [4.1] (i) is an expression and (ii) is an equation.
2. [4.1] multiply, divide
3. [4.1] Multiplication, Addition **4.** [4.1] −10
5. [4.1] 2 **6.** [4.1] −5 **7.** [4.2] 4 **8.** [4.2] 0
9. [4.2] 8 **10.** [4.3] 7 **11.** [4.3] 4
12. [4.3] 6 feet, 9 feet **13.** [4.5] quadrants
14. [4.5] x-axis **15.** [4.5] II **16.** [4.5] IV
17. [4.5] y-axis
18. [4.5] $A(0, -3), B(4, -2), C(-2, 3)$
19. [4.6] horizontal, vertical **20.** [4.4] (6, 0), (0, −8)
21. [4.4] 6
22. [4.6] (−1, 3), (0, 6), (−3, −3)

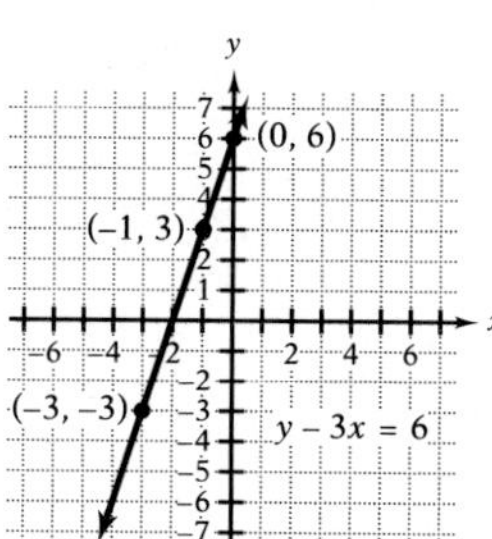

23. [4.6]

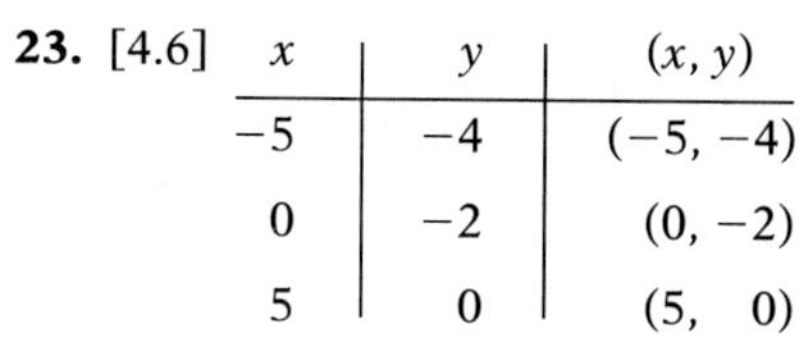

x	y	(x, y)
−5	−4	(−5, −4)
0	−2	(0, −2)
5	0	(5, 0)

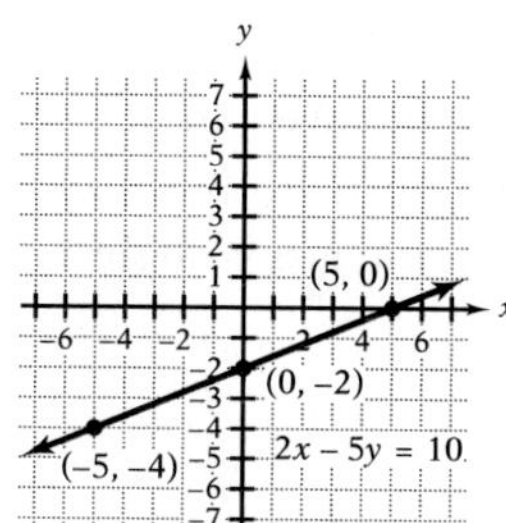

24. [4.6]

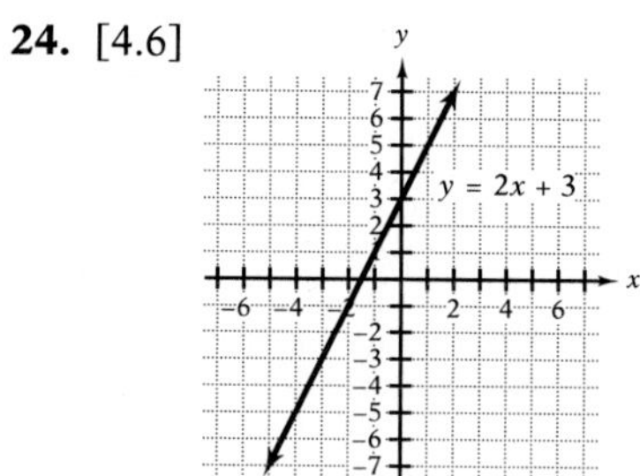

CHAPTER 5

Section 5.1 *(page 359)*

1. fraction, denominator, numerator **3.** $\frac{2}{3}$ **5.** $\frac{3}{4}$
7. $\frac{3}{6}$ **9.** $\frac{2}{3}$ **11.** $\frac{8}{6}$ **13.** −1 **15.** 0 **17.** $\frac{4}{1}$
19. $\frac{-12}{1}$ **21. (a)** $\frac{5}{9}, \frac{12}{19}$ **(b)** $\frac{12}{7}, \frac{14}{11}$
23. (a) $\frac{9}{20}, \frac{6}{17}$ **(b)** $\frac{11}{4}, \frac{8}{4}$
25. numerators, denominators **27.** $-\frac{5}{12}, \frac{5}{-12}$

Copyright © Houghton Mifflin Company. All rights reserved.

29. $-\frac{x}{3}, \frac{-x}{3}$ **31.** $\frac{9}{10}$ **33.** $\frac{3}{8}$ **35.** $-\frac{10}{21}$ **37.** $\frac{12}{77}$

39. -1 **41.** $\frac{3x}{4y}$ **43.** $-\frac{15a}{14b}$ **45.** $\frac{15}{4}$ **47.** $-\frac{35}{12}$

49. $-\frac{16}{13}$ **51.** $\frac{4n}{15}$ **53.** x **55.** $-n$ **57.** a **59.** $\frac{18}{31}$

61. $\frac{59}{129}$ **63.** $\frac{3}{8}$ cup **65.** $\frac{15}{32}$ square inch

67. For a proper fraction, the numerator is less than the denominator. For an improper fraction, the numerator is greater than or equal to the denominator.

69. $\frac{35}{36}$ **71.** $\frac{45}{16}$ **75.** $\frac{7}{16}$ **76.** (i) **77.** (iii)

78. $\frac{6}{25} < \frac{33}{50}$ **79.** $\frac{6}{25} > \frac{1}{10}$ **80.** 198 **81.** 72

Section 5.2 *(page 373)*

1. divisible **3. (a)** 11, 23 **(b)** 9, 27 **5.** $7 \cdot 3 \cdot 2$

7. $2^4 \cdot 3$ **9.** $2^3 \cdot 3^2$ **11.** $2^3 \cdot 3^2 \cdot 5$ **13.** $3^3 \cdot 11$

15. simplified, reduced **17.** $\frac{3}{5}$ **19.** $\frac{4}{7}$ **21.** $\frac{5}{3}$

23. $\frac{3}{7}$ **25.** $\frac{1}{2}$ **27.** 3 **29.** $-\frac{11}{5}$ **31.** $\frac{9}{2}$ **33.** $\frac{7}{4}$

35. $-\frac{5}{9}$ **37.** $\frac{3}{8}$ **39.** $\frac{1}{2}$ **41.** $\frac{3}{2}$ **43.** $\frac{1}{5}$ **45.** $\frac{5}{3b}$

47. $\frac{4n}{3}$ **49.** $\frac{1}{3b}$ **51.** $\frac{3x}{5}$ **53.** $\frac{5b^2}{6a}$ **55.** $\frac{9}{4xy^2}$

57. $3^4 \cdot 5^2$ **59.** $\frac{14}{15}$ **61.** $\frac{2}{3}$ **63.** $\frac{2}{5}$ **65.** $\frac{1}{10}$

67. $\frac{181}{266}$ **69.** $\frac{3}{20}$ **71.** $\frac{25}{48}$

73. The number is written as a product of prime numbers.

75. Because the numbers 760 and 995 end in 0 or 5, the numbers are divisible by 5. The number 5,552 does not end in 0 or 5, so it is not divisible by 5.

76. A prime number must be larger than 1. So 1 is not prime. The only factor of 1 is 1, so it is not a composite number.

77. $-\frac{-3}{-4}$

78. Positive factors: 1, 2, 3, 4, 6, 9, 12, 18, 36; prime factors: 2, 3

Section 5.3 *(page 387)*

1. simplify **3.** $\frac{7}{20}$ **5.** $-\frac{1}{4}$ **7.** 8 **9.** $-\frac{12}{25}$ **11.** -5

13. 1 **15.** $\frac{7}{6}$ **17.** $\frac{1}{3b}$ **19.** $\frac{3a}{7}$ **21.** reciprocal

23. $\frac{11}{6}$ **25.** $-\frac{1}{10}$ **27.** $\frac{3b}{a}$ **29.** $\frac{5}{6}$ **31.** $-\frac{9}{8}$ **33.** $\frac{2x}{5y}$

35. $\frac{3}{2}$ **37.** $\frac{4}{15}$ **39.** 6 **41.** $-\frac{5}{12}$ **43.** $\frac{32}{21}$ **45.** $-\frac{3}{7}$

47. 6 **49.** $\frac{3x}{y}$ **51.** $-\frac{3a}{2}$ **53.** 15 inches **55.** 40

57. 24 **59.** 48 **61.** 1,120 **63.** 360

65. 10 divided by $\frac{1}{2}$ means 10 multiplied by 2, which is 20.

67. (ii) **69.** 75

Section 5.4 *(page 399)*

1. renaming **3.** 9 **5.** 7 **7.** 6 **9.** 35 **11.** -20

13. -21 **15.** $4a$ **17.** $7c$ **19.** xy **21.** multiple

23. 10 **25.** 20 **27.** 90 **29.** 840 **31.** 210

33. 900 **35.** $12ab$ **37.** $6xy$ **39.** $10a^2b$

41. numerators **43.** 14 **45.** 24 **47.** $<$ **49.** $<$

51. $>$ **53.** $\frac{2}{5}, \frac{1}{2}, \frac{3}{4}$ **55.** $\frac{4}{9}, \frac{2}{3}, \frac{5}{6}$ **57.** B **59.** D

61. 60 **63.** 90

65. The LCM is the product of m and n. **67.** $12x^2y^2$

69. 28 **71.** All categories **72.** Sometimes

73. $\frac{21}{100}, \frac{33}{100}, \frac{46}{100}$ **74.** $\frac{30}{100}, \frac{34}{100}, \frac{36}{100}$ **75.** $\frac{23}{50} > \frac{33}{100}$

76. $\frac{3}{10} < \frac{9}{25}$ **77.** Often **78.** Sometimes, Rarely

Section 5.5 *(page 411)*

1. numerators, denominator **3.** $\frac{6}{7}$ **5.** $\frac{1}{3}$ **7.** $-\frac{4}{y}$

9. $\frac{1}{2}$ **11.** $-\frac{4}{5}$ **13.** $\frac{3}{4}$ **15.** $-\frac{2}{3}$

17. least common denominator **19.** $\frac{11}{12}$ **21.** $\frac{17}{12}$

23. $\frac{7}{6}$ **25.** $\frac{1}{15}$ **27.** $\frac{17}{5}$ **29.** $-\frac{1}{5}$ **31.** $\frac{8}{15}$ **33.** $-\frac{3}{20}$

35. $\frac{11}{12}$ **37.** $-\frac{1}{6}$ **39.** $\frac{2}{3}$ **41.** $-\frac{7}{24}$ **43.** $\frac{3x+4}{4x}$

45. $\frac{4c-3b}{bc}$ **47.** $\frac{4a-1}{a}$ **49.** $\frac{13}{14}$ **51.** $\frac{10}{3}$ **53.** $\frac{1}{30}$

55. $-\frac{5}{9}$ **57.** $-\frac{13}{90}$ **59.** $\frac{9}{16}$ **61.** $\frac{3}{20}$ **63.** $\frac{7}{24}$

65. $\frac{9}{40}$; 31 **67.** $\frac{24}{35}$ **69.** $1,440

71. The LCD is the product of the two denominators.

72. The factored form makes simplifying easier.

73. Addition and subtraction

74. Multiplication and division **75.** $\frac{1}{a}$ **76.** $\frac{2a}{5}$

77. $\frac{5}{6}$ **78.** $\frac{2}{3}$ **79.** $\frac{1}{150}$ **80.** $\frac{1}{6}$ **81.** $\frac{7}{20}$ **82.** $\frac{11}{30}$

83. $\frac{1}{12}$ **84.** $\frac{1}{10}$ **85.** $\frac{9}{10}$ **86.** $\frac{11}{12}$

Section 5.6 *(page 427)*

1. mixed **3.** $5\frac{4}{9}$ **5.** $-3\frac{2}{3}$ **7.** $3 + \frac{5}{8}$ **9.** $-2 - \frac{3}{4}$

11. $\frac{11}{4}$ **13.** $\frac{25}{6}$ **15.** $1\frac{7}{8}$ **17.** $4\frac{1}{4}$ **19.** improper

21. 8 **23.** $17\frac{1}{2}$ **25.** $7\frac{1}{7}$ **27.** $2\frac{18}{23}$ **29.** $1\frac{2}{7}$ **31.** $7\frac{1}{2}$

Copyright © Houghton Mifflin Company. All rights reserved.

33. whole, fractions **35.** $6\frac{3}{5}$ **37.** $2\frac{1}{2}$ **39.** $1\frac{2}{5}$

41. $3\frac{1}{4}$ **43.** $\frac{8}{9}$ **45.** $15\frac{1}{6}$ **47.** $3\frac{19}{20}$ **49.** $11\frac{7}{24}$

51. $\frac{17}{20}$ **53.** 3 **55.** $42\frac{1}{6}$ **57.** $14\frac{2}{3}$ **59.** $2\frac{7}{24}$

61. $56\frac{1}{2}$ **63.** $30\frac{3}{4}$ **65.** $119\frac{1}{6}$

67. The notation $5\frac{1}{2} = 5 + \frac{1}{2}$ indicates addition, whereas $5 \cdot \frac{1}{2}$ indicates multiplication.

69. The fraction $\frac{4}{3}$ is an improper fraction.

71. $7\frac{9}{16}$ **73.** $1\frac{1}{5}$ **75.** 7 million **76.** $2\frac{1}{5}$ million

77. $6\frac{9}{10}$ million **78.** 2 million

Section 5.7 *(page 441)*

1. divide, reciprocal **3.** $\frac{12}{5}$ **5.** $-\frac{1}{2}$ **7.** -35

9. -12 **11.** $\frac{8}{5}$ **13.** -8 **15.** -8 **17.** $-\frac{4}{3}$

19. -4 **21.** -12 **23.** $\frac{4}{9}$ **25.** 8 **27.** 6 **29.** clear

31. $-\frac{1}{2}$ **33.** $\frac{1}{2}$ **35.** $-\frac{5}{24}$ **37.** 3 **39.** $\frac{13}{2}$ **41.** $\frac{16}{3}$

43. 7 **45.** $-\frac{3}{2}$ **47.** $\frac{9}{2}$ **49.** $\frac{3}{2}$ **51.** 36 **53.** 6

55. 0 **57.** 300 miles **59.** 48

61. Multiply both sides by 4 or divide both sides by $\frac{1}{4}$.

63. -4 **65.** -1 **67.** 12

Section 5.8 *(page 453)*

1. 35 square feet **3.** $\frac{5}{12}$ square yard

5. $7\frac{1}{2}$ square inches **7.** 6 miles **9.** 12 yards

11. 28 square inches **13.** $2\frac{1}{4}$ feet **15.** $1\frac{3}{4}$ miles

17. 24 square inches **19.** 3 feet **21.** 6 yards

23. \$72 **25.** \$539 **27.** \$2,130 **29.** 30°C

31. -13°F **33.** $1\frac{1}{2}$ hours **35.** 24 miles per hour

37. A formula is an equation with more than one variable.

39. 56 square feet **41.** 91 **43.** 74

45. 156 square inches

Review Exercises *(page 459)*

1. numerator, denominator

2. Both fractions are equivalent to 1.

3. **(a)** 0 **(b)** $2x$ **4.** improper, proper **5.** $-\frac{3}{8}, \frac{3}{-8}$

6. $-\frac{35}{12}$ **7.** $-\frac{32}{3}$ **8.** $\frac{5a}{9b}$ **9.** y **10.** $\frac{15}{26}$ **11.** prime

12. $2^4 \cdot 5$ **13.** $3^2 \cdot 5^2 \cdot 11$ **14.** $\frac{8}{5}$ **15.** $\frac{7}{9}$ **16.** $\frac{1}{2}$

17. $\frac{2}{5}$ **18.** $\frac{3a}{4}$ **19.** $\frac{3}{5}$ **20.** $\frac{5}{8}$ **21.** $-\frac{27}{4}$ **22.** $\frac{1}{3}$

23. $\frac{4y}{3}$ **24.** multiply, reciprocal

25. **(a)** $-\frac{7}{15}$ **(b)** $\frac{1}{9}$ **26.** $-\frac{9}{77}$ **27.** $\frac{3}{32}$ **28.** $\frac{4a}{3}$

29. $\frac{1}{2}$ **30.** 12 **31.** renaming **32.** 30 **33.** -15

34. ab **35.** multiples **36.** 60 **37.** 210 **38.** $6x^2y$

39. $>$ **40.** 120 **41.** $\frac{1}{2}$ **42.** $-\frac{2}{5}$

43. denominator, numerators

44. $\frac{1}{20}$ **45.** $\frac{8}{9}$ **46.** $\frac{2}{3}$ **47.** $\frac{17}{18}$ **48.** $\frac{4x}{7}$

49. $\frac{m-12}{2}$ **50.** **(a)** $\frac{13}{30}$ **(b)** 85 **51.** mixed

52. $\frac{365}{7}$ **53.** $2\frac{5}{12}$ **54.** $2\frac{2}{3}$ **55.** $\frac{4}{15}$ **56.** $6\frac{1}{2}$

57. $1\frac{1}{2}$ **58.** $10\frac{23}{36}$ **59.** $5\frac{5}{8}$ tablespoons **60.** $56\frac{1}{4}$

61. Multiply both sides by the lowest common denominator.

62. $\frac{5}{2}$ **63.** -24 **64.** $-\frac{9}{2}$ **65.** 12 **66.** $\frac{1}{8}$ **67.** $\frac{4}{3}$

68. $-\frac{2}{5}$ **69.** 6 **70.** 90 **71.** formula

72. $1\frac{1}{5}$ square yards **73.** 7 inches **74.** 2 feet

75. \$48 **76.** \$9,300 **77.** 5°F **78.** 30°C **79.** $2\frac{1}{3}$

80. $62\frac{1}{2}$ miles per hour

Chapter Test *(page 465)*

1. [5.1] denominator, numerator, reciprocal

2. [5.1] **(a)** 0 [5.1] **(b)** $3y$ **3.** [5.2] $2 \cdot 3^2 \cdot 5$

4. [5.2] **(a)** $\frac{4}{7}$ [5.2] **(b)** $\frac{6}{5}$ [5.2] **(c)** $\frac{4}{3y}$

5. [5.4] **(a)** 12 [5.4] **(b)** 15 **6.** [5.4] 210

7. [5.6] **(a)** $\frac{19}{8}$ [5.6] **(b)** $7\frac{1}{2}$ **8.** [5.4] $<$

9. [5.3] $\frac{2}{9}$ **10.** [5.3] $\frac{9}{2}$ **11.** [5.3] $-\frac{9}{20}$ **12.** [5.3] $\frac{1}{3}$

13. [5.5] $\frac{2}{3}$ **14.** [5.5] $\frac{5}{16}$ **15.** [5.5] $\frac{5x+2}{x}$

16. [5.5] $\frac{13}{10}$ **17.** [5.6] $\frac{8}{3}$ **18.** [5.6] $4\frac{1}{8}$

19. [5.6] $2\frac{5}{6}$ **20.** [5.7] $-\frac{14}{3}$ **21.** [5.7] $\frac{17}{2}$

22. [5.1] $\frac{3}{4}$ **23.** [5.4] 90 **24.** [5.5] $\frac{1}{8}$

25. [5.8] $7\frac{1}{2}$ square feet **26.** [5.8] \$17

Copyright © Houghton Mifflin Company. All rights reserved.

Cumulative Test, Chapters 4–5 *(page 467)*

1. [5.3] reciprocal **2.** [5.6] mixed, improper

3. [5.2] $\frac{3}{4}$ **4.** [5.3] $\frac{14}{3}$ **5.** [5.3] $\frac{1}{4}$ **6.** [5.3] $-\frac{1}{6}$

7. [5.5] $\frac{2}{3}$ **8.** [5.5] $\frac{1}{2}$ **9.** [5.5] $\frac{13}{30}$ **10.** [5.6] $\frac{4}{3}$

11. [5.6] $2\frac{1}{3}$ **12.** [5.6] $8\frac{1}{6}$ **13.** [4.2] -3

14. [5.7] -35 **15.** [4.2] 2 **16.** [4.2] 3 **17.** [5.7] $\frac{8}{7}$

18. [5.4] $\frac{5}{9}, \frac{4}{7}, \frac{5}{6}$ **19.** [5.3] 16 **20.** [5.5] $\frac{4}{15}$

21. [5.8] 3 square yards

22. [5.8] 48 miles per hour **23.** [4.3] 7

24. [4.3] 21 **25.** [4.5] x-coordinate, y-coordinate

26. [4.5] **(a)** x-axis [4.5] **(b)** III

27. [4.5] $A(-2, 4), B(1, -1), C(-5, 0), D(3, 2), E(0, 2), F(-4, -3)$

28. [4.6] $(2, 3), (0, 0), (-2, -3)$

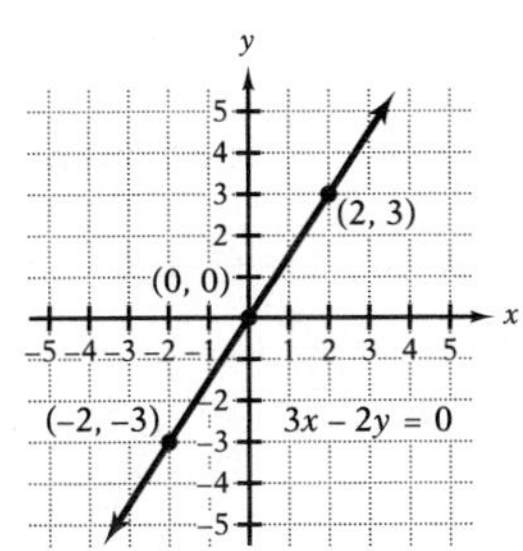

29. [4.6]

x	y	(x, y)
-3	3	$(-3, 3)$
0	1	$(0, 1)$
6	-3	$(6, -3)$

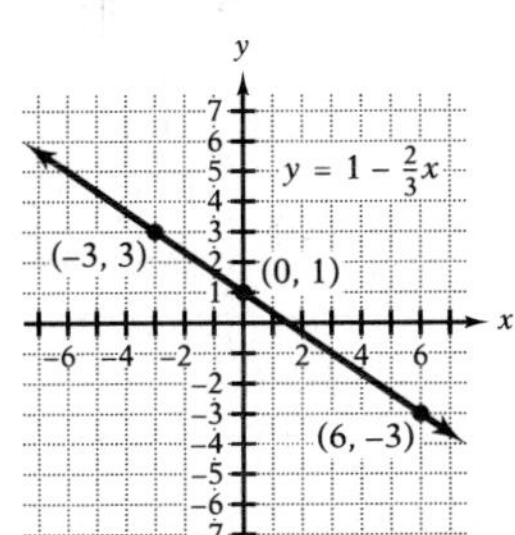

30. [4.6]

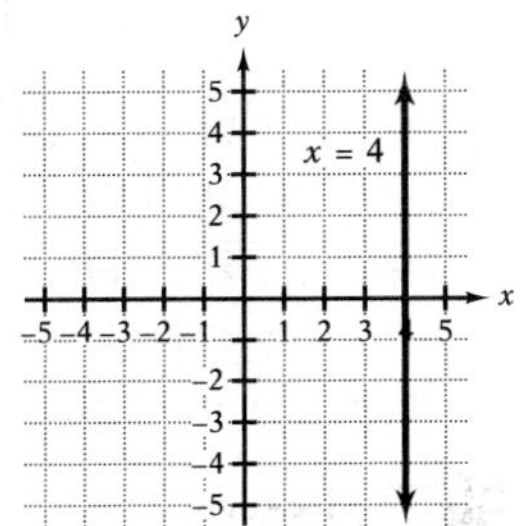

CHAPTER 6

Section 6.1 *(page 483)*

1. decimal point **3.** hundredths **5.** ten thousandths

7. $\frac{43}{100}$ **9.** $15\frac{9}{1{,}000}$ **11.** 0.7 **13.** 18.003

15. Forty-five hundredths

17. Six and forty-five thousandths

19. Fifty-two and twenty-seven hundredths

21. Two and two hundred eighty-three thousandths

23. 0.24 **25.** 11.6 **27.** 0.554 **29.** 0.17 **31.** $\frac{4}{5}$

33. $\frac{3}{50}$ **35.** 0.35 **37.** 0.038 **39.** digits **41.** 5.4

43. 6.40 **45.** 36.00 **47.** 35, 34.7

49. 3,000, 3,456.461

51. **(a)** 2 **(b)** 2.4 **(c)** 2.43 **(d)** 2.435

53. **(a)** 2 **(b)** 2.1 **(c)** 2.08 **(d)** 2.083

55. order **57.** $<$ **59.** $>$

61. 0.069, 0.07, 0.0701, 0.071

63. 1.0201, 1.02, 1.019, 1.009

65. A comma separates the whole number place values in groups of three. A period separates the whole number place values and the fractional place values.

67. $<$ **69.** 100.44, 27.9, 15.5, 4.805, 2.79

70. Oxygen and carbon **71.** $2.79 < 4.805$

72. $27.9 > 15.5$ **73.** 28, 3, 16, 5, 100 **74.** Nitrogen

75. Carbon **76.** (iii)

Section 6.2 *(page 493)*

1. decimal points **3.** 15.74 **5.** 97.691 **7.** 7.739

9. 36.077 **11.** 1.33 **13.** 19.661 **15.** 0.47

17. 3.04 **19.** 0.93 **21.** 22.493 **23.** 3.651

25. 34.686 **27.** 2.74 **29.** -16.12 **31.** -8.11

33. 17.05 **35.** -4.333 **37.** -3.106 **39.** round

41. 1 **43.** 0.3 **45.** $5.5x$ **47.** $2.3c$ **49.** $7.6y$

51. $-1.7a - 5.2b$ **53.** 14.5 feet **55.** \$39.51

57. \$411.81 **59.** \$271 **61.** \$16.41 **63.** 4,442.8036

65. 1.918997 **67.** (i), (ii) **69.** \$988.41

71. **(a)** 1.5 inches **(b)** $3x$ inches **73.** 33.89

74. 64.57 inches **75.** 5.36 **76.** 7.64 **77.** 51.2

78. 59.1 **79.** 13.6 **80.** 43.3

Section 6.3 *(page 505)*

1. whole **3.** 0.18 **5.** 0.014 **7.** 0.094 **9.** 0.016

11. -0.37 **13.** 36 **15.** -1.5 **17.** 1.72

19. 0.28008 **21.** 54 **23.** 31.08 **25.** -27

27. -22.25 **29.** 34.56 **31.** 9.6936 **33.** 106.4

35. exponent **37.** 1,430.6 **39.** 85,830 **41.** 378

43. 678,300 **45.** $456x + 320$ **47.** $3{,}247n + 35$

49. 240 **51.** 4 **53.** 4.698 billion **55.** 2.35

57. \$17.98 **59.** \$327.05 **61.** \$329

63. 18,415.83794 **65.** 27.8649 **67.** 6.74

Copyright © Houghton Mifflin Company. All rights reserved.

69. 0.00246 **71.** 69.92
73. Multiply 2 by 4. Write the answer with 3 decimal places. **75.** 5.73 × 100 **77.** 0.345 **78.** 2.519
79. 17.633 **80.** 0.789 **81.** 474,698,000
82. 173,635,000 **83.** 94.9 **84.** 603.71

Section 6.4 *(page 519)*

1. divisor, dividend **3.** 7.4 **5.** 12.56 **7.** 1.31
9. 0.005 **11.** 1.8 **13.** 2.46 **15.** one, right **17.** 40
19. 400 **21.** 639 **23.** 500 **25.** 0.2 **27.** 0.5
29. 1.7 **31.** 6.04 **33.** 494.3 **35.** 15.17 **37.** round
39. 0.4362 **41.** 0.3986 **43.** 0.0004 **45.** 0.05698
47. B **49.** D **51.** 16-ounce box **53.** 3.1
55. $1.12 **57.** $84.25 **59.** 3.13 **61.** 56.89
63. 4.11 **65.** 0.347 **67.** 3.8 **69.** 399.42 **71.** 0.42
73. **(a)** Move the decimal point 3 places to the right and obtain 3,570.
(b) Move the decimal point 3 places to the left and obtain 0.00357.
75. 6 ÷ 100 **77.** 5.6-pound box **78.** 11.25
79. 3,032.25 **80.** 1,368 **81.** 1,061.58 **82.** 4.36
83. 4.43 **84.** 0.86 **85.** 81.38

Section 6.5 *(page 531)*

1. evaluate **3.** 7.48 **5.** −10.036 **7.** 3.72
9. −0.26 **11.** 321.536 **13.** 60 **15.** 16.16
17. −4.5 **19.** clear the decimals **21.** 2.07
23. 31.12 **25.** 8.4 **27.** −0.3 **29.** 2.3 **31.** 3.6
33. 4.2 **35.** 1.4 **37.** −6.5 **39.** 4.4 **41.** $36.75
43. 242 **45.** $6.85 **47.** Store: $7.10; assistant: $9.45
49. Determine the highest number of decimal places in all the decimals. The number indicates the number of zeros in the power of 10 by which you multiply.
51. (ii); each term must be multiplied by 10.
53. They each express the same amount. In (i) the values are expressed in dollars. In (ii) the values are expressed in cents.

Section 6.6 *(page 543)*

1. rational **3.** 3.75 **5.** 0.4375 **7.** $2.\overline{3}$ **9.** $0.8\overline{3}$
11. $3.\overline{18}$ **13.** square roots **15.** −4, 4 **17.** −9, 9
19. no square root **21.** $-\frac{1}{3}, \frac{1}{3}$ **23.** 5 **25.** −10
27. 12 **29.** $\frac{8}{5}$ **31.** 12 **33.** 30 **35.** −14 **37.** 17
39. 0 **41.** 13 **43.** irrational **45.** 6.71 **47.** 1.92
49. 10.58 **51.** −9.49 **53.** 4.69 **55.** 8.07 **57.** 6, 7
59. 8, 9 **61.** 3.5 **63.** 56 miles per hour
65. $45,000 **67.** 15 milligrams
69. The square of 4 means 4^2. The square root of 4 means the number whose square is 4.
71. 9 **73.** 0.7 **75.** 3 **77.** False **79.** True

Section 6.7 *(page 557)*

1. legs, hypotenuse **3.** 13 **5.** 9 **7.** 4.2 **9.** 8.8
11. 78 miles **13.** 21 feet **15.** 26.0 feet
17. 84.9 feet **19.** Yes **21.** No **23.** Yes **25.** No
27. circumference
29. C = 18.8 inches, A = 28.3 square inches
31. C = 45.2 feet, A = 162.8 square feet
33. C = 44.0 yards, A = 153.9 square yards
35. C = 25.1 feet, A = 50.2 square feet
37. 70.7 feet **39.** 14-inch pizza **41.** 24,492 miles
43. 1.5 **45.** 3.1 square feet **47.** 87.9 square inches
49. 50.2 **51.** sphere **53.** 502.4 cubic inches
55. 401.9 cubic feet **57.** 2,138.2 cubic feet
59. 902.1 cubic yards **61.** 522.0
63. 19.6 cubic inches
65. The hypotenuse is the longest side and is opposite the right angle.
67. 30.3 square feet

Review Exercises *(page 565)*

1. power **2.** thousandths, hundredths **3.** $2\frac{37}{1{,}000}$
4. 0.07
5. Two and one hundred seventy-six thousandths
6. 20.012 **7.** $\frac{3}{8}$ **8.** **(a)** 3.3 **(b)** 3.28
9. **(a)** 2.0 **(b)** 1.98 **10.** **(a)** $>$ **(b)** $<$
11. Because $0.30 = \frac{30}{100} = \frac{3}{10} = 0.3$ **12.** 410.95
13. 276.84 **14.** 7.59 **15.** 6.501 **16.** −3.2
17. −7.3 **18.** $-3.6x + 2.8y$ **19.** $1.97 **20.** $2.49
21. −1.5 **22.** 21 **23.** 3.2608 **24.** 0.0388
25. 0.9482 **26.** decimal point **27.** 8,409.8
28. $42x - 70$ **29.** 77.5 **30.** $853.40 **31.** three
32. 3.2 **33.** 50 **34.** 6.084 **35.** 0.48 **36.** 0.0369
37. 0.47 **38.** 16.3 **39.** 2.4 **40.** 6.6 **41.** −1.71
42. 3.78 **43.** 0.45 **44.** Multiplication
45. (i) We can clear decimals in an equation, not an expression.
46. 17.43 **47.** 0.045 **48.** −8 **49.** 3.5 **50.** 26
51. repeating
52. The notation $1.5\overline{2}$ means the 2 repeats: 1.522222 · · · . The notation 1.52 means that every decimal place after 2 is 0.
53. 1.375 **54.** $0.\overline{27}$ **55.** perfect squares **56.** $\frac{6}{5}$
57. 56 **58.** 0 **59.** 3.32 **60.** 1.5 **61.** right **62.** 8
63. No **64.** 15 yards **65.** diameter
66. A = 78.5 square feet, C = 31.4 feet
67. 8-inch pizza **68.** 87.9 square inches
69. 62.8 cubic feet **70.** 112.8 cubic inches

Copyright © Houghton Mifflin Company. All rights reserved.

Chapter Test *(page 571)*

1. [6.1] thousandths, tenths **2.** [6.6] terminating
3. [6.7] hypotenuse **4.** [6.1] 4.053
5. [6.1] **(a)** 1.48 [6.1] **(b)** 3.02
6. [6.1] 1.001, 1.0012, 1.012 **7.** [6.2] 8.358
8. [6.2] 15.154 **9.** [6.2] −5.6 **10.** [6.3] −2.8
11. [6.3] 6.448 **12.** [6.3] 6,270 **13.** [6.4] 8.74
14. [6.4] 0.027 **15.** [6.4] 92.86 **16.** [6.6] 12
17. [6.6] 1 **18.** [6.5] 0.04178 **19.** [6.5] 8
20. [6.6] **(a)** $1.\overline{3}$ **(b)** 0.75 **21.** [6.2] $72.30
22. [6.3] $4.04 **23.** [6.7] 15
24. [6.7] 112.8 cubic inches **25.** [6.7] 62.8

CHAPTER 7

Section 7.1 *(page 581)*

1. ratio **3.** $\frac{5}{3}$ **5.** $\frac{2}{3}$ **7.** $\frac{5}{1}$ **9.** 1 to 6 **11.** 8 to 1
13. $\frac{4}{5}$ **15.** $\frac{12}{5}$ **17.** $\frac{5}{34}$ **19.** $\frac{7}{40}$ **21.** $\frac{5}{3}$ **23.** $\frac{1}{3}$
25. $\frac{4}{3}$ **27. (a)** $\frac{2}{7}$ **(b)** $\frac{6}{7}$ **29.** $\frac{2}{3}$ **31.** $\frac{2}{3}$ **33.** $\frac{11}{6}$
35. rate **37.** 2.64 inches per minute
39. $6.24 per hour **41.** $2\frac{2}{3}$ milligrams per pound
43. 840 seedlings per acre **45.** 1.7 gallons per minute
47. 12 ounces **49.** 24.6 cents per dollar
51. Both $\frac{4}{8}$ and $\frac{6}{12}$ are equal to $\frac{1}{2}$.
53. (ii) is true. We do not know the number of people in the class so (i) is not necessarily true.
55. (iii) **56.** $\frac{16}{7}$ **57.** $\frac{14}{9}$ **58.** $\frac{41}{7}$ **59.** $\frac{16}{41}$
60. $\frac{317}{50{,}000}$ **61.** $\frac{317}{49{,}683}$ **62.** $\frac{49{,}683}{317}$ **63.** $\frac{49{,}683}{50{,}000}$

Section 7.2 *(page 597)*

1. conversion factors **3.** 24 **5.** 72 **7.** 7,040
9. 5 **11.** 3 **13.** $2\frac{2}{3}$ yards **15.** 2.8 miles
17. 28 yards **19.** weight **21.** 6,000 **23.** 40 **25.** 3
27. $1\frac{1}{4}$ **29.** 10 pounds **31.** $5.94 **33.** 12
35. cubic **37.** 28 **39.** 96 **41.** 864 **43.** 3 **45.** 16
47. 5 **49.** 3,960 **51.** 4,800
53. Not all months have 30 days. **55.** 4.4 **57.** 1613.3
59. 0.8 **61.** 9.4 **63.** 273.8 **65.** 44 feet per second
67. 11.25 gallons per hour **69.** 62.3 quarts per hour
71. (a) $5\frac{1}{3}$ **(b)** $\frac{3}{4}$ **(c)** 4 **(d)** 4
73. (a) $48 **(b)** 311.1 **(c)** $.15
75. 5 feet **76.** 57.5 **77.** 60,000 **78.** 330 feet
79. 15 feet **80.** 0.195 inch **81.** 448 **82.** $2\frac{1}{2}$ tons

Section 7.3 *(page 611)*

1. metric **3.** 80 **5.** 760 **7.** 3,000 **9.** 5,230
11. 12.5 **13.** 2 **15.** 0.75 **17.** 5
19. 250 millimeters **21.** 180 centimeters
23. 6,695,000 meters **25.** more **27.** 7.9 **29.** 2.0
31. 3,296.7 **33.** 2.5 **35.** 4,922.3 **37.** 1.8
39. 8.0 **41.** 6.4 **43.** 30.5 **45.** 0.4 **47.** 91 meters
49. 6.2 miles **51.** 14,926.7 feet **53.** (ii) **55.** (ii)
57. 504 **59.** 8 **61.** 50
63. (ii); One kilometer is 1,000 meters, not 100 meters.
65. 1,600 **67.** 0.75 **69.** 7.3

Section 7.4 *(page 623)*

1. liter **3.** 250 **5.** 2,000 **7.** 1.25 **9.** 0.8
11. 3,500 milliliters **13.** 250 **15.** gram **17.** 5,000
19. 3,400 **21.** 7.8 **23.** 0.75 **25.** 100 **27.** 8
29. 45 **31.** 11.4 **33.** 236.5 **35.** 1.1 **37.** 1.1
39. 64.3 **41.** 22.7 **43.** $0.38 **45.** 340.2 **47.** 1.82
49. 8.8 **51.** 42.3 **53.** 1.5 kilograms
55. 3.7 pounds **57.** 2-kilogram box **59.** (ii)
61. (iii) **63.** 3 **65.** 385.8 **67.** 62.5
69. 17.0 liters per minute
71. (iii); One gram is 1,000 milligrams. **73.** 60,495
74. 1,582 **75.** $23.44 **76.** $205.48
77. $19.42 per ounce **78.** $7.82 per gallon
79. 2,281.25 **80.** 10.58

Section 7.5 *(page 641)*

1. proportion
3. (a) 9, 6, 18, 12 **(b)** 6, 18 **(c)** 9, 12 **(d)** 9 is to 6 as 18 is to 12.
5. (a) $\frac{m}{n} = \frac{5}{3}$ **(b)** $m, n, 5, 3$ **(c)** $n, 5$ **(d)** $m, 3$
7. False **9.** True **11.** True **13.** False
15. (a) $\frac{15}{6} = \frac{20}{8}$ **(b)** $\frac{20}{15} = \frac{8}{6}$ **(c)** $\frac{6}{8} = \frac{15}{20}$
17. (a) $\frac{6}{4} = \frac{11}{y}$ **(b)** $\frac{4}{6} = \frac{y}{11}$ **(c)** $\frac{y}{4} = \frac{11}{6}$
19. solving **21.** 14 **23.** 2.5 **25.** 11 **27.** 70
29. 5 **31.** $-\frac{3}{2}$ **33.** 11 **35.** 12 **37.** 3 **39.** 35
41. 43,200 **43.** 455 **45.** 22 **47.** 21 **49.** 6.4
51. 220 **53.** $12,224 **55.** $1\frac{1}{8}$ cups of food
57. $3\frac{1}{3}$ **59.** 45 feet **61.** $23\frac{1}{3}$ **63.** $847,500
65. 1,448,280
67. A ratio uses a quotient to compare two numbers. A proportion states that two ratios are equal.
69. (iii)

Copyright © Houghton Mifflin Company. All rights reserved.

Section 7.6 (page 657)

1. degree **3.** **(a)** 30° **(b)** 50° **(c)** 40°
5. **(a)** 10 **(b)** 80° **(c)** $\overline{AB}$ **(d)** $\angle A, \angle C$
7. **(a)** 70° **(b)** $\overline{DF}, \overline{EF}$ **(c)** 60° **(d)** 8
9. proportional
11. **(a)** $m\angle A = m\angle P, m\angle B = m\angle Q, m\angle C = m\angle R$
(b) $\frac{AB}{PQ} = \frac{BC}{QR} = \frac{AC}{PR}$
13. **(a)** $m\angle A = m\angle D, m\angle B = m\angle E, m\angle C = m\angle F$
(b) $\frac{AB}{DE} = \frac{BC}{EF} = \frac{AC}{DF}$
15. 9 **17.** 20 **19.** 7.5 **21.** 6 **23.** $x = 5, y = 10$
25. $x = 8, y = 25$ **27.** $x = 6, y = 6$ **29.** $x = 5, y = 6$
31. 16 feet **33.** 15 feet, 20 feet
35. AB represents the length of line segment $\overline{AB}$.
37. 12 **39.** 16
41. The perimeter of the larger triangle is twice the perimeter of the smaller triangle.

Review Exercises (page 663)

1. ratio **2.** $\frac{5}{4}$ **3.** $\frac{40}{1}$ **4.** $\frac{9}{7}$ **5.** $\frac{6}{11}$ **6.** $\frac{7}{5}$
7. denominator **8.** $7.50 per square foot
9. 25 gallons per minute **10.** 212.9 **11.** ratio
12. 10 **13.** 48 **14.** 36 **15.** 15 **16.** 2.25 **17.** $58\frac{2}{3}$
18. 1.6 **19.** 780 inches **20.** 500 **21.** metric
22. 13 **23.** 3.4 **24.** 370 **25.** 895 centimeters
26. 1.5 **27.** 31.5 **28.** 3.6 **29.** 736.7 feet
30. 1.4 inches **31.** square, volume **32.** 1,200
33. 0.75 **34.** 8,200 **35.** 2,100 **36.** 1.6 **37.** 113.4
38. 50 **39.** 4.2 **40.** $.76 per pound **41.** means
42. cross products **43.** False **44.** True **45.** 1.5
46. 64 **47.** $\frac{1}{4}$ **48.** $13\frac{1}{2}$ **49.** 9.6 **50.** $17.33
51. vertex **52.** measure **53.** 180° **54.** 10
55. 80° **56.** $\overline{AC}$ **57.** $\angle B, \angle C$ **58.** 9 **59.** $\frac{10}{3}$
60. 75 feet

Chapter Test (page 669)

1. [7.5] terms **2.** [7.6] equal, proportional
3. [7.2] $\frac{3}{4}$ **4.** [7.3] 640 **5.** [7.3] 3.1 **6.** [7.4] 2.3
7. [7.4] 10.6 **8.** [7.2] 10 **9.** [7.4] 1.6
10. [7.4] 0.3 **11.** [7.2] 440 **12.** [7.2] $2\frac{1}{4}$
13. [7.1] $\frac{4}{3}$ **14.** [7.5] Yes **15.** [7.5] 5
16. [7.5] 10 **17.** [7.1] $\frac{10}{7}$
18. [7.1] $5.67 per pound **19.** [7.2] 36
20. [7.3] 292.8 kilometers **21.** [7.4] 50 grams
22. [7.2] $11.03 **23.** [7.5] 1.1 **24.** [7.6] 4, 40°
25. [7.6] 3.75

Chapters 6–7 Cumulative Test (page 671)

1. [6.6] real **2.** [6.1] **(a)** 30.09; [6.1] **(b)** 0.077
3. [6.1] **(a)** 3.0; [6.1] **(b)** 2.97
4. [6.1] 0.041, 0.04, 0.039 **5.** [6.2] −8.67
6. [6.2] 3.23 **7.** [6.3] 14.7 **8.** [6.4] 0.245
9. [6.6] −4 **10.** [6.6] 6 **11.** [6.3] $320x + 455$
12. [6.5] −2.94 **13.** [6.5] −0.9 **14.** [7.1] $\frac{3}{8}$
15. [7.3] 0.976 **16.** [7.3] 60.96 **17.** [7.2] $1\frac{1}{8}$
18. [7.4] 6.8 **19.** [7.2] $2\frac{3}{8}$ **20.** [7.2] 3.4
21. [7.5] 4.5 **22.** [6.2] $3.45 **23.** [6.3] $35.15
24. [7.1] $\frac{3}{11}$ **25.** [7.5] 23 **26.** [7.4] 22.9 ounces
27. [7.2] 1.6 miles **28.** [6.7] 5
29. [6.7] $C = 25.1$ feet, $A = 50.2$ square feet
30. [7.6] 10

CHAPTER 8

Section 8.1 (page 685)

1. percent **3.** 36% **5.** 140% **7.** 30% **9.** 240%
11. $\frac{79}{100}$ **13.** $\frac{147}{100}$ **15.** $\frac{2}{5}$ **17.** $\frac{2}{25}$ **19.** $\frac{3}{2}$ **21.** $\frac{2}{3}$
23. decimal **25.** 0.85 **27.** 0.4 **29.** 0.08 **31.** 2.85
33. 0.185 **35.** 25% **37.** 12% **39.** 7% **41.** 125%
43. 610% **45.** 20% **47.** denominator **49.** 31.25%
51. $16\frac{2}{3}\%$ **53.** $8\frac{1}{3}\%$ **55.** 4.8% **57.** $466\frac{2}{3}\%$
59. $33\frac{3}{4}\%$ **61.** 42.9% **63.** 35.7%
65. **(a)** 0.26 **(b)** $\frac{13}{50}$ **67.** **(a)** 0.45 **(b)** $\frac{9}{20}$
69.

Fraction	$\frac{1}{8}$	$\frac{1}{4}$	$\frac{3}{8}$	$\frac{1}{2}$	$\frac{5}{8}$	$\frac{3}{4}$	$\frac{7}{8}$	1
Decimal	0.125	0.25	0.375	0.5	0.625	0.75	0.875	1
Percent	12.5%	25%	37.5%	50%	62.5%	75%	87.5%	100%

71. 70% **73.** 40% **75.** 34.6%
77. (i) and (iii) because their denominators are factors of 100 so they are easy to rewrite
79. $\frac{16}{287}$; 5.6% **81.** $\frac{160}{220}$; 72.7% **83.** $\frac{3}{10}$ **84.** 11%
85. will probably not happen **86.** 58%
87. will definitely happen **88.** 81% **89.** $\frac{4}{25}$
90. optimistic

Section 8.2 (page 697)

1. base, amount **3.** $12\% \cdot 32 = x$ **5.** $x \cdot 15 = 6$
7. $140\% \cdot x = 30$ **9.** fraction, decimal **11.** 3

Copyright © Houghton Mifflin Company. All rights reserved.

13. 16 **15.** 0.9 **17.** 33 **19.** 7 **21.** 45 **23.** 25
25. $66\frac{2}{3}$ **27.** 120 **29.** 10% **31.** 240% **33.** 12.5%
35. 82 **37.** 200 **39.** 15 **41.** 80 **43.** 75 **45.** 36
47. decimal point **49.** D **51.** C **53.** 800 **55.** 350
57. 50% **59.** 25% **61.** 15 **63.** 36 **65.** 6 **67.** 42
69. 42.16% **71.** 7,111.11
73. Express the percent as the fraction $\frac{2}{3}$.
75. Divide the number by 5. Find 10% of the number and double the result.
77. 35

Section 8.3 *(page 705)*

1. 29.2% **3.** $15,000 **5.** 7.8 **7.** 55% **9.** 57
11. 1,950 **13.** 62.5% **15.** 420 **17.** 775 **19.** 20%
21. $41.86 **23.** 1,500 **25.** 12 **27.** 6.25
29. 145.6% **31.** 1.28
33. No. You need to know the total number of students at each college.
35. **(a)** Yes. The percent is 100% − 80% = 20%. **(b)** No. You need to know the total number of people who tried the medication.
36. 40.0% **37.** 30.2% **38.** 24.2%
39. The sum is 100%. **40.** 10.3% **41.** 65.7%
42. 72.9% **43.** 20.2%

Section 8.4 *(page 721)*

1. interest **3.** **(a)** $153 **(b)** $1,953
5. **(a)** $930 **(b)** $12,930 **7.** $19.50 **9.** $3,328
11. **(a)** $800 **(b)** $48.75 **(c)** 9% **(d)** 3 years
13. 6.6% **15.** $16,000 **17.** 6 **19.** principal
21. $30,680.08 **23.** $712.25 **25.** $4,416.08
27. $22,372.18 **29.** $10,050.12
31. **(a)** $6.60 **(b)** $138.60
33. $24,581.25 **35.** $10,454.50 **37.** $1,180
39. $30 million **41.** 4.5% **43.** 24%
45. I represents the interest, P the principal, r the interest rate, and t the time in years.
47. $135,000 **49.** (i)

Section 8.5 *(page 739)*

1. commission **3.** $192 **5.** $37,800 **7.** $136,160
9. $850 **11.** $980 **13.** 8% **15.** 24%
17. $1,563.50 **19.** $6,850 **21.** depreciation
23. $12,000 **25.** $12,800 **27.** $37,570
29. $4,905.94 **31.** wholesale, retail
33. **(a)** $28 **(b)** $168 **35.** **(a)** $82.50 **(b)** $220
37. 50% **39.** 70% **41.** **(a)** $146 **(b)** $584
43. **(a)** $226.50 **(b)** $528.50 **45.** 18.75% **47.** 22.4%
49. (i) and (ii) result in the same price.
51. $220 **53.** $30,000

Section 8.6 *(page 753)*

1. $713 **3.** $110.89 **5.** 12.81 million **7.** 20%
9. 18.1% **11.** 49.2% **13.** $950 **15.** 1.2 million
17. 1.9 million **19.** $670.80 **21.** 3,094 **23.** 19.8¢
25. 23.2% **27.** 7.1% **29.** 25% **31.** $42.50
33. 5,237 **35.** 107.8 points
37. The person should have divided by 50, not by 40.
39. (i) and (iii) are correct. **40.** 16.6% **41.** 18.3%
42. 73.5% **43.** 30.5% **44.** 44.4% **45.** 57.6%
46. 7.4 billion **47.** 67.24 million

Review Exercises *(page 761)*

1. percent **2.** $\frac{3}{10}$ **3.** $\frac{13}{400}$ **4.** $\frac{12}{5}$ **5.** 9% **6.** 0.34
7. 0.0075 **8.** 42% **9.** 87.5% **10.** $183\frac{1}{3}$%
11. P represents the percent, B the base, and A the amount.
12. 1.47 **13.** 20 **14.** 400 **15.** 40.5
16. 35% **17.** 45
18. Move the decimal point 2 places to the left.
19. 54 **20.** 52% **21.** 55% **22.** 2,475
23. 7,344 **24.** 32% **25.** 12 **26.** 15
27. 87.6% **28.** 120 **29.** $42,000 **30.** 864
31. principal, interest **32.** $3,024 **33.** $210
34. 6% **35.** $1,040.40 **36.** $3,228.29
37. **(a)** $20.76 **(b)** $366.76 **38.** $260 **39.** 3%
40. $593.40 **41.** $3,360 **42.** $726.25 **43.** 9%
44. $875 **45.** $3,402 **46.** discount **47.** $75.60
48. 28% **49.** $78.10 **50.** 25% **51.** $7.95
52. $2,509.54 **53.** 26.4% **54.** 73.9% **55.** 475
56. 2,600 calories **57.** 14.5 **58.** 19% **59.** $500
60. $323.75

Chapter Test *(page 769)*

1. [8.1] right, percent
2. [8.2] Move the decimal point 1 place to the left.
3. [8.2] Express the percent as the fraction $\frac{1}{3}$.
4. [8.1] **(a)** 35% [8.1] **(b)** $13\frac{1}{3}$%
5. [8.1] **(a)** $\frac{7}{20}$, 0.35 [8.1] **(b)** $\frac{6}{5}$, 1.20
6. [8.1] **(a)** 0.004 [8.1] **(b)** 0.0375
7. [8.1] **(a)** 60% [8.1] **(b)** 140% **8.** [8.2] 364
9. [8.2] 15% **10.** [8.2] 60 **11.** [8.3] 6.6
12. [8.3] 68.75% **13.** [8.3] 60 **14.** [8.4] $6,339.50
15. [8.4] $2,026.76 **16.** [8.4] 7% **17.** [8.4] $339.70
18. [8.5] $7,200 **19.** [8.5] $455 **20.** [8.5] $156.60
21. [8.5] $33\frac{1}{3}$% **22.** [8.6] 25% **23.** [8.6] 289 million
24. [8.6] 984 **25.** [8.6] 30%

Copyright © Houghton Mifflin Company. All rights reserved.

CHAPTER 9

Section 9.1 (page 783)

1. monomials **3.** $-\frac{2}{3}x$; polynomial

5. $2n$, $-\frac{3}{4}$; polynomial

7. $\frac{3}{y^2}$, $2y^2$, $-4y$; not a polynomial

9. $2x^2$, $-7x$, 5; polynomial **11.** 11 **13.** 10 **15.** 7
17. like terms **19.** $7x + 3$ **21.** $3y^2 - 2y - 2$
23. $8x^2 + x + 6$ **25.** $2n^3 - n^2 - n + 1$ **27.** $5x^3 + x$
29. $4x^2 + 8x - 2$ **31.** plus, opposite **33.** $5x + 6$
35. $4x^2 - 4x + 5$ **37.** $3n^2 + 5n - 10$ **39.** $x^2 - x$
41. $2y^3 - y^2 - 6y + 8$ **43.** $2n^3 + 3n$ **45.** x
47. $-3y^2 + 7y - 3$ **49.** 144 feet **51.** $47 - 3x$
53. $9x + 3$ **55.** 34 **57.** 0.576
59. Both are correct because addition is commutative.
61. $3x^2 + 5x + 6$ **63. (a)** 1 **(b)** 4 **(c)** 5

Section 9.2 (page 793)

1. base, add **3.** x^6 **5.** c^4 **7.** y^{14} **9.** $6y^2$ **11.** $3x^7$
13. $-8x^6$ **15.** $15a^7$ **17.** $-24x^8$ **19.** $21w^8$
21. $-30x^3$ **23.** base, multiply **25.** x^{21} **27.** a^{20}
29. b^{24} **31.** y^{22} **33.** x^{26} **35.** a^{20} **37.** factor
39. $8n^3$ **41.** $36x^6$ **43.** $a^{15}b^{10}$ **45.** x^3y^{12} **47.** $49x^{12}y^8$
49. $81a^4b^{20}c^8$ **51.** $49n^{27}$ **53.** $72y^{11}$ **55.** $10x^3$ **57.** $9a^6$
59. To apply the Product Rule for Exponents, the bases must be the same.
61. $12x^3y^{11}$ **63.** $-8m^{10}n^8$ **65. (a)** x^6 **(b)** $2x^3$

Section 9.3 (page 801)

1. trinomial **3.** $8x^2 + 10x$ **5.** $14y^2 - 49y$
7. $-16x^2 + 8x$ **9.** $15a^2 - 6a$ **11.** $y^6 + 2y^5$
13. $2x^5 + 10x^3$ **15.** $-5n^8 - 15n^4$ **17.** $6b^3 - 3b^2 + 12b$
19. $2x^4 - 6x^3 - 12x^2$ **21.** binomial **23.** $x^2 + 7x + 12$
25. $y^2 + 3y - 10$ **27.** $z^2 - 49$ **29.** $3x^2 + 11x - 4$
31. $6b^2 - 13b + 5$ **33.** $14x^2 - 17x - 6$ **35.** $a^2 + 6a + 9$
37. $n^2 - 4n + 4$ **39.** $4x^2 + 12x + 9$
41. $36x^2 - 12x + 1$ **43.** $x^3 + 3x^2 - 6x + 20$
45. $2x^3 + x^2 - 7x - 6$ **47.** $2x^3 + 5x^2 - x - 1$
49. $27x^3 - 8$ **51.** $x^2 + 6x + 9$ **53.** $3x^2 + 20x + 25$
55. Lining up the like terms is easy.
57. $6x^4 + 8x^3 - 11x^2 - 20x - 10$
59. $x^4 + 2x^3 - 2x^2 + 7x - 2$
61. $2x^4 - 11x^3 - 7x^2 - 20x - 6$
63. $2x^2 + 50$ **65.** $x^3 - 4x^2 + 4x$

Section 9.4 (page 811)

1. greatest common factor **3.** 5 **5.** 9 **7.** 10
9. x^2 **11.** y^3 **13.** $4y^4$ **15.** $4x^4$ **17.** $2x^3$ **19.** $3b$
21. a^2b^3 **23.** $2x^2y^3$ **25.** factoring **27.** 3
29. $2y + 7$ **31.** $x(x + 3)$ **33.** $5(a + 3)$ **35.** $6(x + 1)$
37. $4(3a - 5b)$ **39.** $2y(y - 6)$ **41.** $3x(3x - 1)$
43. $3(x^2 + 4x - 1)$ **45.** $4(2x^2 + x - 7)$
47. $3(3m + 6n - 4)$ **49.** $-3(a + 2b - 5c)$
51. $-4(x + y - 1)$ **53.** $a^2(2a - 5)$ **55.** $b^3(b^2 + 6)$
57. $7y^2(2y^2 - 3)$
59. Determine the smallest exponent on a: 2. The GCF is a^2.
61. $3x + 5$

Review Exercises (page 815)

1. monomials
2. (i) $\frac{2}{5}y^2$, $-3y$, 1; polynomial **(ii)** $\frac{5}{x^2}$, $7x$, -9; not a polynomial
3. 1 **4.** like terms **5.** $3x^2 - 2x + 2$ **6.** $7y^2$
7. $3x + 4$ **8.** $3a^2 - 2a - 2$ **9.** $2y^3 + y^2 + 1$
10. $8 - x$ **11.** Product **12.** n^{11} **13.** $-20x^4$
14. $30c^2$ **15.** z^{12} **16.** y^{24} **17.** x^{14} **18.** $81x^4$
19. a^6b^{10} **20.** $32x^{15}y^5$ **21.** $2x^2 - 10x$
22. $-15x^2 + 35x$ **23.** $6a^3 + 20a^2$
24. $8n^3 + 12n^2 - 24n$ **25.** $x^2 + 2x - 48$ **26.** $b^2 - 25$
27. $15a^2 - 13a + 2$ **28.** $y^2 + 12y + 36$
29. $x^3 + 2x^2 - 8x - 15$ **30.** $6x^3 - x^2 - 14x + 8$
31. greatest common factor **32.** 4 **33.** x **34.** $3a^2$
35. $4x^2y^5$ **36.** product **37.** $5(2x - 5)$ **38.** $y(2y + 1)$
39. $x(x^2 - x + 4)$ **40.** $4(2a + b - 4)$

Chapter Test (page 817)

1. [9.1] signs **2.** [9.2] add, multiply **3.** [9.1] 0
4. [9.1] $8n^2 + n$ **5.** [9.1] $y^3 + y^2 + 3y$
6. [9.1] $4x^2 - 2x - 3$ **7.** [9.1] $x^3 + x^2 - 2$
8. [9.2] y^9 **9.** [9.2] $-20x^7$ **10.** [9.2] n^{21}
11. [9.2] x^{20} **12.** [9.2] $125y^3$ **13.** [9.2] m^8n^{14}
14. [9.3] $-14b^2 + 21b$ **15.** [9.3] $10x^5 + 5x^4 - 30x^3$
16. [9.3] $8x^2 + 22x - 21$ **17.** [9.3] $n^2 - 16n + 64$
18. [9.3] $x^3 - 9x^2 + 18x + 8$ **19.** [9.4] 8
20. [9.4] $3x^5$ **21.** [9.4] $6(4y + 1)$
22. [9.4] $2a(5a + 4)$ **23.** [9.4] $7(2x^2 + x - 4)$

Chapters 8–9 Cumulative Test (page 819)

1. [8.1] **(a)** $\frac{4}{5}$, 0.8 **(b)** $\frac{1}{200}$, 0.005 **(c)** $\frac{8}{5}$, 1.6
2. [8.1] **(a)** 17.5% **(b)** $66\frac{2}{3}$%
3. [8.1] **(a)** 12% **(b)** 320% **(c)** 0.8%
4. [8.2] 0.84 **5.** [8.2] 35% **6.** [8.2] 12
7. [8.3] \$2,756 **8.** [8.3] 15% **9.** [8.4] \$175
10. [8.4] \$516.80 **11.** [8.5] \$16.20 **12.** [8.5] 60%
13. [8.5] \$189.20 **14.** [8.6] 92 **15.** [8.6] \$110
16. [9.1] 1 **17.** [9.1] $x^3 - x^2 + 6x + 1$
18. [9.1] $x^2 + 12x - 6$
19. [9.2] Power to a Power, multiply
20. [9.2] **(a)** a^9 **(b)** $35y^7$

Copyright © Houghton Mifflin Company. All rights reserved.

21. [9.2] **(a)** z^{30} **(b)** y^{30}
22. [9.2] **(a)** $a^{15}b^3$ **(b)** $16x^8y^{24}$
23. [9.3] **(a)** $16x^2 - 40x$ **(b)** $y^5 + 4y^3$
24. [9.3] $4x^2 + 13x + 3$ **25.** [9.3] $a^2 + 14a + 49$
26. [9.4] **(a)** b^4 **(b)** $3n^2$ **27.** [9.4] $9(3x + 5)$
28. [9.4] $x(3x - 7)$ **29.** [9.4] $2(3x^2 - 5x + 4)$

Copyright © Houghton Mifflin Company. All rights reserved.

Answers to Selected Features

CHAPTER 1

KEYS TO THE CALCULATOR

Section 1.4 *(page 35)*

(a) 59 **(b)** 787 **(c)** 769 **(d)** 8,938

Section 1.4 *(page 39)*

(a) 19 **(b)** 536 **(c)** 2,001 **(d)** 54,279

SELECTED YOUR TURNS

Section 1.4

Your Turn 6. **(a)**

Given Number		*Rounded Number*
111	→	100
159	→	200
290	→	300
		600

(b)

Given Number		*Rounded Number*
1,198	→	1,200
653	→	− 700
		500

Your Turn 7.

Books Sold		*Rounded Number*
395	→	400
276	→	300
144	→	100
815		800

(a) Approximately 800 textbooks were sold.
(b) Exactly 815 textbooks were sold.

Your Turn 8. $14 - 5 = 9$

The chat room had 9 more women than men.

Section 1.5

Your Turn 1.

$$\begin{array}{r} 893{,}000 \\ +\ 355{,}000 \\ \hline 1{,}248{,}000 \end{array}$$

In 2005, the expected total number of systems analysts and computer engineers is 1,248,000.

Your Turn 2.

$$\begin{array}{r} 114 \\ -\ 63 \\ \hline 51 \end{array}$$

The first-place team won 51 more games than the last-place team.

Your Turn 3. **(a)** $12 + 12 + 12 + 12 = 48$
The perimeter of the square is 48 inches.
(b) $5 + 12 + 13 = 30$
The perimeter of the triangle is 30 yards.
(c) $7 + 2 + 3 + 12 = 24$
The perimeter of the polygon is 24 feet.

Section 1.6

Your Turn 1. **(a)** $3 + 15 + x$
↓
$3 + 15 + 9 = 27$
(b) $x - 6$
↓
$9 - 6 = 3$

Your Turn 2. **(a)** $m - n$
↓ ↓
$9 - 4 = 5$
(b) $n + m + 1$
↓ ↓
$4 + 9 + 1 = 14$

Your Turn 3. $a + 28 + b$
↓ ↓
$37 + 28 + 86$

$$\begin{array}{r} 37 \\ 28 \\ +\ 86 \\ \hline 151 \end{array}$$

Your Turn 4. $(y + 5) + 8$
$= y + (5 + 8)$
$= y + 13$

Your Turn 5. **(a)**

Left Side	Right Side
8 + 1	(2 + 3) + 3
	5 + 3
9	8

Because $9 \neq 8$, the equation is false.

(b)

Left Side	Right Side
6 + (5 + 4)	7 + 8
6 + 9	
15	15

Because $15 = 15$, the equation is true.

Your Turn 6. The sum of some number and 4 is 12. Because $8 + 4 = 12$, the solution is 8.

Copyright © Houghton Mifflin Company. All rights reserved.

Your Turn 7. **(a)** *Subtraction* / *Addition*

$y - 5 = 3$ $\quad$ $3 + 5 = y$

From the related addition, y is 8.

(b) *Subtraction* / *Addition*

$15 - a = 9$ $\quad$ $9 + a = 15$

From the related addition, a must be 6.

Your Turn 8. $3 + y = 6 + 2$
$3 + y = 8$
Because $3 + 5 = 8$, the solution is 5.

Your Turn 9. $2 + (4 + x) = 11$
$(2 + 4) + x = 11$
$6 + x = 11$
The solution is 5 because $6 + 5 = 11$.

CHAPTER 2

WARM-UP SKILLS *(page 78)*

1. 36 **2.** 300 **3.** 15 **4.** 312 **5.** 15 **6.** 39
7. (a) Commutative Property of Addition
(b) Associative Property of Addition
8. 12 inches **9.** 31 **10.** 8

KEYS TO THE CALCULATOR

Section 2.2 *(page 95)*

(a) 3,496 **(b)** 56,448 **(c)** 720,054 **(d)** 1,465,284

Section 2.3 *(page 111)*

(a) 26 **(b)** 52 **(c)** 748 **(d)** 354

Section 2.4 *(page 126)*

(a) 25 **(b)** 1,728 **(c)** 729 **(d)** 10,556,001

Section 2.5 *(page 133)*

(a) 19 **(b)** 49 **(c)** 25 **(d)** 9

SELECTED YOUR TURNS

Section 2.1

Your Turn 6. $6 \cdot 70 = 6 \cdot (7 \cdot 10) = (6 \cdot 7) \cdot 10 = 42 \cdot 10 = 420$

Your Turn 7. $9 \cdot 500 = 9 \cdot (5 \cdot 100) = (9 \cdot 5) \cdot 100 = 45 \cdot 100 = 4{,}500$

Your Turn 8. $5 \cdot 400 = 5 \cdot 4 \cdot 100 = 20 \cdot 100 = 2{,}000$

Section 2.2

Your Turn 1. $4(7 + 9) = 4(7) + 4(9) = 28 + 36 = 64$

Your Turn 4.

```
  37
× 42
  74
148
1554
```

Your Turn 5.

```
   891
 × 346
  5346
 3564
2673
308286
```

Your Turn 6.

Given Number		*Rounded Number*
79	→	80
92	→	90

Multiply $80(90) = 7{,}200$.

Your Turn 7.

	Hours	× *Hourly pay*	= *Total*
Regular	40	\$ 9	\$360
Overtime	6	\$18	\$108
Total pay			\$468

Your Turn 8. The total number of chairs that were set up was $22(36) = 792$. Therefore, $796 - 792 = 4$ people had to stand.

Your Turn 9. The increase per hour is 9,600. Because a day has 24 hours, the daily increase is $24(9{,}600) = 230{,}400$.

Section 2.3

Your Turn 4.

```
   83
4)332
  32
   12
   12
    0
```

Your Turn 5.

```
    241
36)8676
   72
   147
   144
     36
     36
      0
```

Your Turn 6.

```
     40
92)3685
   368
     05
      0
      5
```

$3{,}685 \div 92 = 40$ R 5

Your Turn 7.

```
  5
6)32
  30
   2
```

Five family members can have a 6-ounce serving, and 2 ounces will be left over.

Copyright © Houghton Mifflin Company. All rights reserved.

Your Turn 8.

$$\begin{array}{r} 3200 \\ 12\overline{)38400} \\ \underline{36} \\ 24 \\ \underline{24} \\ 000 \\ \underline{0} \\ 0 \end{array}$$

The monthly salary is \$3,200 and you need \$3,000. So you should be able to save \$200 each month.

Section 2.4

Your Turn 5. **(a)** $3^3 = 3 \cdot 3 \cdot 3 = 27$
(b) Because all the factors are 0, the product is 0.
(c) $9^2 = 9 \cdot 9 = 81$
(d) $2^5 = 2 \cdot 2 \cdot 2 \cdot 2 \cdot 2 = 32$

Your Turn 8. $10 \cdot 10 \cdot 10 \cdot 10 = 10^4$
The answer is 1 followed by four 0s: 10,000.

Section 2.5

Your Turn 1. $6 + 24 \div 8 = 6 + 3 = 9$

Your Turn 2. $15 - 2(5 - 3) = 15 - 2(2) = 15 - 4 = 11$

Your Turn 3. $2^3 + \frac{8}{5 - 3} = 8 + \frac{8}{2} = 8 + 4 = 12$

Your Turn 4. $2[(3 + 9) \div 4] = 2[12 \div 4]$
$= 2[3]$
$= 6$

Your Turn 5. 4
4 + ( )
4 + ( )( )
4 + (5)(3)

Your Turn 6. $\frac{82 + 85 + 91 + 73 + 94 + 79}{6} = \frac{504}{6} = 84$

Your Turn 7. The number of charter schools would be 422, 276, and 196.
$\frac{422 + 276 + 196}{3} = \frac{894}{3} = 298$

Section 2.6

Your Turn 1. $6(3y) = (6 \cdot 3)y = 18y$

Your Turn 2. $(8a)(3b) = 8 \cdot a \cdot 3 \cdot b = 8 \cdot 3 \cdot a \cdot b$
$= (8 \cdot 3)(a \cdot b) = 24ab$

Your Turn 3. $9(y + 2) = 9(y) + 9(2) = 9y + 18$

Your Turn 4. $2(7a + 4) = 2(7a) + 2(4) =$
$(2 \cdot 7)a + 2(4) = 14a + 8$

Your Turn 5. $2(3x) = 2(3 \cdot 3) = 2(9) = 18$

Your Turn 6. **(a)** $6ab = 6(3)(2) = 36$
(b) $\frac{ab}{6} = \frac{3(2)}{6} = \frac{6}{6} = 1$

Your Turn 7. $x - yz = 10 - 4(1) = 10 - 4 = 6$

Your Turn 8. **(a)** $(a + b)^2 = (4 + 2)^2 = 6^2 = 36$
(b) $a^2 + 2ab + b^2 = 4^2 + 2(4)(2) + 2^2 =$
$16 + 16 + 4 = 36$

Your Turn 9. Because $8(7) = 56$, the solution is 7.

Your Turn 10. $4(2y) = 3 \cdot 8$
$(4 \cdot 2)y = 24$
$8y = 24$
The solution is 3 because $8(3) = 24$.

Your Turn 11. The related multiplication is $9 \cdot 2 = x$. Therefore, $x = 18$.

Your Turn 12. The related multiplication is $27 = 3x$. The solution is 9 because $3 \cdot 9 = 27$.

Your Turn 13. **(a)** Ask, "For what whole number x does $x \cdot x = 100$?" The solution is 10 because $10 \cdot 10 = 100$.
(b) Ask, "For what whole number z does $z \cdot z \cdot z = 27$?" The solution is 3 because $3 \cdot 3 \cdot 3 = 27$.

Section 2.7

Your Turn 1. $A = s^2 = 9^2 = 81$ square inches

Your Turn 2. $A = s^2$
$49 = s^2$
$7 = s$
The length of a side is 7 inches.

Your Turn 3. $A = LW$
$A = (42)(30) = 1{,}260$ square inches

Your Turn 4. $A = LW$
$63 = 9W$
$7 = W$
The width of the picture is 7 inches.

Your Turn 5.

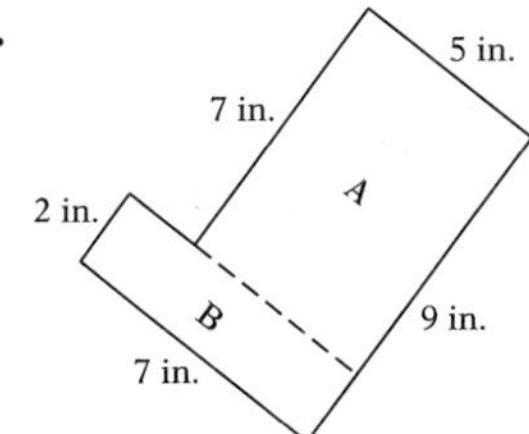

Total area = area A + area B
$= 5 \cdot 7 + 2 \cdot 7 = 35 + 14 = 49$ square inches

Your Turn 6. The length of a side is 1 foot, or 12 inches. The area in square inches is $A = s^2 = 12^2 = 144$.

Your Turn 7. The width is 36 inches, or 1 yard.
$A = LW = 5(1) = 5$ square yards

Section 2.8

Your Turn 1. $V = LWH = 8(1)(3) = 24$ cubic feet

Your Turn 2. The area of the top is $A = LW = 8(6) = 48$. The surface area without the top is $236 - 48 = 188$ square feet.

Your Turn 3. Original surface area = 96
New surface area with $s = 3$:
$S = 6s^2 = 6(3)^2 = 6 \cdot 9 = 54$
The surface area is reduced by $96 - 54 = 42$ square inches.

Copyright © Houghton Mifflin Company. All rights reserved.

CHAPTER 3

WARM-UP SKILLS *(page 186)*

1. **(a)** $<$ **(b)** $>$ **2.** $x + 14$ **3.** $10a$ **4.** 20
5. $6x + 21$ **6.** 7 **7.** 125 **8.** 2 **9.** 92 **10.** 2

KEYS TO THE CALCULATOR

Section 3.2 *(page 202)*

(a) -13 **(b)** -140 **(c)** 491 **(d)** -145

Section 3.3 *(page 212)*

(a) -42 **(b)** 52 **(c)** 45 **(d)** -17

Section 3.4 *(page 221)*

(a) $-3{,}570$ **(b)** -768 **(c)** 1,026 **(d)** $-213{,}504$

Section 3.5 *(page 232)*

(a) -15 **(b)** 19 **(c)** 0 **(d)** undefined **(e)** -12

SELECTED YOUR TURNS

Section 3.1

Your Turn 1. **(a)**

-1 0

0 is to the right of -1. Therefore $0 > -1$.

(b)

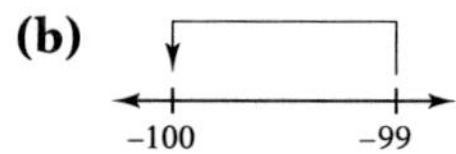

-100 is to the left of -99. Therefore $-100 < -99$.

Your Turn 2.

-7 -6 -5 -4 -3 -2 -1 0 1 2 3 4 5 6 7

Start with the leftmost number. The order is $-7, -1, 0, 2, 5$.

Your Turn 3.

-5 -4 -3 -2 -1 0 1 2

Count 4 units to the right. The number is 0.

Your Turn 8. **(a)** $|0| < |-1|$ because $0 < 1$.
(b) $|-3| > |2|$ because $3 > 2$.
(c) $|8| = |-8|$ because $8 = 8$.

Your Turn 9.

Expression	*Value*
$-\|4\|$	-4
-1	-1
$-(-3)$	3
$\|-4\|$	4

The order is $-|4|, -1, -(-3), |-4|$.

Section 3.2

Your Turn 1.

Move 3 units to the left.
-3 -2 -1 0 1 2
Start at 1.

$1 + (-3) = -2$

Your Turn 2.

Move 3 units to the right.
-3 -2 -1 0 1 2
Start at -2.

$-2 + 3 = 1$

Your Turn 3.

Move 1 unit to the left.
-5 -4 -3 -2 -1 0 1
Start at -2.

$-2 + (-1) = -3$

Your Turn 7. $9 + (-4) + (-5) + 2 = 5 + (-5) + 2 = 0 + 2 = 2$

Section 3.3

Your Turn 1. $11 - 8 = 11 + (-8) = 3$
Your Turn 2. $-8 - 6 = -8 + (-6) = -14$
Your Turn 3. $8 - (-2) = 8 + 2 = 10$
Your Turn 4. $-10 - (-4) = -10 + 4 = -6$
Your Turn 5. **(a)** $6 - (-3) = 6 + 3 = 9$
(b) $-7 - 5 = -7 + (-5) = -12$
(c) $-2 - (-8) = -2 + 8 = 6$
(d) $4 - 12 = 4 + (-12) = -8$
Your Turn 6. $-5 - 3 - (-8) + 4$
$= -5 + (-3) + 8 + 4$
$= [-5 + (-3)] + (8 + 4)$
$= -8 + 12 = 4$
Your Turn 7. $4 - 3^2 - 5 = 4 - 9 - 5$
$= 4 + (-9) + (-5) = -10$

Section 3.4

Your Turn 1. $3(-7) = -7 + (-7) + (-7) = -21$
Your Turn 2. $-6(3) = 3(-6) = -6 + (-6) + (-6) = -18$
Your Turn 5. $5 \cdot 3 \cdot 1 \cdot 2 = 30$. Four factors are negative so the product is positive. $-5(-3)(-1)(-2) = 30$
Your Turn 6. $(-2)^4 = (-2)(-2)(-2)(-2) = 16$
Your Turn 8. **(a)** $8 - (-2)(5) = 8 - (-10)$
$= 8 + 10 = 18$
(b) $-3[5 + (-4)(2)] = -3[5 + (-8)] = -3[-3] = 9$
(c) $3(-2)^2 + 5(-4) = 3(4) + 5(-4) = 12 + (-20) = -8$

Copyright © Houghton Mifflin Company. All rights reserved.

Section 3.5

Your Turn 2. $\dfrac{2(3)}{7-1} = \dfrac{6}{6} = 1$

Your Turn 3. $\dfrac{-3(-1) - 4(2)}{-15 \div 3} = \dfrac{3-8}{-5} = \dfrac{-5}{-5} = 1$

Your Turn 4. $\dfrac{6-4^2}{2(-1)} = \dfrac{6-16}{-2} = \dfrac{-10}{-2} = 5$

Your Turn 5. The temperatures would be −7, −9, 8, 5, and −2°F. Calculate the average.

$$\frac{-7 + (-9) + 8 + 5 + (-2)}{5} = \frac{13 + (-18)}{5} =$$

$$\frac{-5}{5} = -1°\text{F}$$

Your Turn 6.

$$\frac{0 + (-1) + (-2) + (-3) + (-2) + (-1) + (-5)}{7}$$

$$= \frac{-14}{7} = -2$$

Your Turn 7.

$$\frac{[4(-2) + 6(-1) + 18(0) + 10(+1) + 2(+2)]}{40}$$

$$= \frac{[-8 + (-6) + 0 + 10 + 4]}{40}$$

$$= \frac{[-14 + 14]}{40} = \frac{0}{40} = 0$$

Section 3.6

Your Turn 1. $a + (-4) + b = -3 + (-4) + (-10) = -17$

Your Turn 2. $a - 1 - b = -6 - 1 - 4$
$= -6 + (-1) + (-4) = -11$

Your Turn 3. $5ab = 5(-2)(-3) = 30$

Your Turn 4. $\dfrac{a}{3b} = \dfrac{12}{3(-1)} = \dfrac{12}{-3} = -4$

Your Turn 5. $2a^4 = 2(-3)^4 = 2(81) = 162$

Your Turn 6. **(a)** $-|c| = -|-2| = -2$
(b) $-(-c) = -[-(-2)] = -[2] = -2$
(c) $-|-c| = -|-(-2)| = -|2| = -2$

Your Turn 7. $|-a| + |b| = |-4| + |-1| = 4 + 1 = 5$

Your Turn 8. **(a)**

$-5x + 3$	13
$-5(-2) + 3$	13
$10 + 3$	13
13	13

Yes, −2 is a solution.

(b)

$-5x + 3$	13
$-5(-4) + 3$	13
$20 + 3$	13
23	13

No, −4 is not a solution.

Your Turn 9. **(a)** If the opposite of y is −6, then y must be 6. The solution is 6.
(b) Because $-(-a) = a$ and $-|3| = -3$, we can write the equation as $-3 = a$. The solution is −3.

Your Turn 10. Ask, "What number added to −2 gives −5?" The solution is −3 because $-3 + (-2) = -5$.

Your Turn 11. Because $x - (-3)$ can be written $x + 3$, we can write the equation as $x + 3 = 7$. The solution is 4 because $4 + 3 = 7$.

Your Turn 12. Ask, "What number multiplied by −2 gives −12?" The solution is 6 because $-2(6) = -12$.

Your Turn 13. $\dfrac{x}{-6} = 3$ means $3(-6) = x$.
Therefore, the solution is −18.

Your Turn 14. $\dfrac{-32}{x} = -8$ means $-8x = -32$.
Because $-8(4) = -32$, the solution is 4.

Section 3.7

Your Turn 1. **(a)** $5(-4x) = [5(-4)]x = -20x$
(b) $(8a)(-2) = (-2)(8a) = [(-2)(8)]a = -16a$

Your Turn 2. **(a)** $-2(y + 9) = -2(y) + (-2)(9) = -2y + (-18) = -2y - 18$
(b) $-4(2a + 5b) = -4(2a) + (-4)(5b) = -8a + (-20b) = -8a - 20b$

Your Turn 3. **(a)** $8(x - 2) = 8(x) - 8(2) = 8x - 16$
(b) $-4(a - 5) = -4(a) - (-4)(5) = -4a - (-20) = -4a + 20$

Your Turn 4. $2(x - y + 8) = 2(x) - 2(y) + 2(8) = 2x - 2y + 16$

Your Turn 5. **(a)** $-(y + 5) = -1(y + 5)$
$= -1(y) + (-1)(5) = -1y + (-5) = -y - 5$
(b) $-(-a + 3) = -1(-1a + 3) = -1(-1a) + (-1)(3) = 1a + (-3) = a - 3$

Your Turn 6. $-(2 - a) = -1(2 - a)$
$= -1(2) - (-1)(a)$
$= -2 - (-a) = -2 + a$

Your Turn 8. **(a)** $9x - 5y = 9x + (-5y)$
The terms are $9x$ and $-5y$.
(b) $-x + 8y - 2 = -x + 8y + (-2)$
The terms are $-x$, $8y$, and −2.

Your Turn 9. **(a)** $7b + 3b = (7 + 3)b = 10b$
(b) Because $5x$ and 5 are not like terms, they cannot be combined.
(c) $-a + 4a = -1a + 4a = (-1 + 4)a = 3a$
(d) $y - 10y = 1y + (-10y)$
$= [1 + (-10)]y = -9y$

Your Turn 10. **(a)** $y - 3y + 8y = 1y - 3y + 8y = (1 - 3 + 8)y = 6y$
(b) $x + y - 3x + 2y$
$= (1x - 3x) + (1y + 2y)$
$= (1 - 3)x + (1 + 2)y = -2x + 3y$

Copyright © Houghton Mifflin Company. All rights reserved.

Your Turn 11. **(a)** $-2 + (y - 4) = -2 + y - 4 = y - 6$
(b) $(7 - 2x) - 3 = 7 - 2x - 3 = 4 - 2x$

Your Turn 12. **(a)** $-(3a + 4) + 2a - 1$
$= -3a - 4 + 2a - 1 = -1a - 5$
$= -a - 5$
(b) $-3(x + 1) + 4(x - 2)$
$= -3x - 3 + 4x - 8$
$= (-3x + 4x) + [-3 + (-8)]$
$= 1x + (-11) = x - 11$

CHAPTER 4

WARM-UP SKILLS *(page 272)*

1. **(a)** x; **(b)** y **2.** $8x - 3$ **3.** $5 + y$ **4.** 17 **5.** 3 **6.** 5 **7.** 7 **8.** $3 + 7 = 10$ **9.** $2 - 5 = -3$ **10.** $\frac{24}{6} = 4$

SELECTED YOUR TURNS

Section 4.1

Your Turn 1.
$$\begin{aligned} x - 3 &= 3x + 1 \\ -2 - 3 &= 3(-2) + 1 \\ -5 &= -6 + 1 \\ -5 &= -5 \quad \text{True} \end{aligned}$$
Yes, -2 is a solution.

Your Turn 2.
$$\begin{aligned} 8 - x &= 2x - 3 \\ 8 - 4 &= 2(4) - 3 \\ 4 &= 8 - 3 \\ 4 &= 5 \quad \text{False} \end{aligned}$$
No, 4 is not a solution.

Your Turn 3.
$$\begin{aligned} x + 2 &= 9 \\ x + 2 - 2 &= 9 - 2 \\ x + 0 &= 7 \\ x &= 7 \end{aligned}$$

Your Turn 4.
$$\begin{aligned} y - 6 &= 8 \\ y - 6 + 6 &= 8 + 6 \\ y + 0 &= 14 \\ y &= 14 \end{aligned}$$

Your Turn 5. **(a)**
$$\begin{aligned} 15 &= x + 9 \\ 15 - 9 &= x + 9 - 9 \\ 6 &= x + 0 \\ 6 &= x \end{aligned}$$
(b)
$$\begin{aligned} 21 &= c - 4 \\ 21 + 4 &= c - 4 + 4 \\ 25 &= c + 0 \\ 25 &= c \end{aligned}$$

Your Turn 6.
$$\begin{aligned} -2y + 7 + 3y &= 9 - 5 \\ y + 7 &= 4 \\ y + 7 - 7 &= 4 - 7 \\ y + 0 &= -3 \\ y &= -3 \end{aligned}$$

Your Turn 7.
$$\begin{aligned} 6x &= 18 \\ \frac{6x}{6} &= \frac{18}{6} \\ x &= 3 \end{aligned}$$

Your Turn 8. **(a)**
$$\begin{aligned} -2x &= -14 \\ \frac{-2x}{-2} &= \frac{-14}{-2} \\ x &= 7 \end{aligned}$$
(b)
$$\begin{aligned} -4 &= -y \\ \frac{-4}{-1} &= \frac{-y}{-1} \\ 4 &= y \end{aligned}$$

Your Turn 9.
$$\begin{aligned} 4x + 3x &= 2(3) + 8 \\ 7x &= 6 + 8 \\ 7x &= 14 \\ \frac{7x}{7} &= \frac{14}{7} \\ x &= 2 \end{aligned}$$

Section 4.2

Your Turn 1.
$$\begin{aligned} -4x + 9 &= -7 \\ -4x + 9 - 9 &= -7 - 9 \\ -4x &= -16 \\ \frac{-4x}{-4} &= \frac{-16}{-4} \\ x &= 4 \end{aligned}$$

Your Turn 2.
$$\begin{aligned} -19 &= 5x - 4 \\ -19 + 4 &= 5x - 4 + 4 \\ -15 &= 5x \\ \frac{-15}{5} &= \frac{5x}{5} \\ -3 &= x \end{aligned}$$

Your Turn 3.
$$\begin{aligned} 6y - 1 &= 2y + 11 \\ 6y - 1 + 1 &= 2y + 11 + 1 \\ 6y &= 2y + 12 \\ 6y - 2y &= 2y - 2y + 12 \\ 4y &= 12 \\ \frac{4y}{4} &= \frac{12}{4} \\ y &= 3 \end{aligned}$$

Copyright © Houghton Mifflin Company. All rights reserved.

Your Turn 4.
$$3(x - 5) = 6$$
$$3x - 15 = 6$$
$$3x - 15 + 15 = 6 + 15$$
$$3x = 21$$
$$\frac{3x}{3} = \frac{21}{3}$$
$$x = 7$$

Your Turn 5.
$$-2x + 4(x + 3) = 4$$
$$-2x + 4x + 12 = 4$$
$$2x + 12 = 4$$
$$2x + 12 - 12 = 4 - 12$$
$$2x = -8$$
$$\frac{2x}{2} = \frac{-8}{2}$$
$$x = -4$$

Your Turn 6.
$$-(x - 2) = 5$$
$$-x + 2 = 5$$
$$-x + 2 - 2 = 5 - 2$$
$$-x = 3$$
$$\frac{-x}{-1} = \frac{3}{-1}$$
$$x = -3$$

Section 4.3

Your Turn 1. **(a)**

$(\ ) - (\ )$	The difference between
$x - (\ )$	a number
$x - 7$	and 7
$x - 7 =$	is
$x - 7 = -4$	-4.

(b)

$(\ ) + (\ )$	Adding
$(\ ) + 6$	6 to
$2x + 6$	twice a number
$2x + 6 =$	results in
$2x + 6 = 8$	8.

(c)

$2 \cdot$	Twice
$2 \cdot (\ \ + \ \ )$	the sum of
$2 \cdot (x + \ \ )$	a number
$2 \cdot (x + 3)$	and 3
$2 \cdot (x + 3) =$	is
$2 \cdot (x + 3) = -5$	-5.

Your Turn 2. Let x = the number.

$x + 5$	5 more than a number
$x + 5 =$	is
$x + 5 = (\ ) - 1$	1 less than
$x + 5 = 3x - 1$	3 times the number.

Solve the equation.
$$x + 5 = 3x - 1$$
$$x + 5 - 5 = 3x - 1 - 5$$
$$x = 3x - 6$$
$$x - 3x = 3x - 3x - 6$$
$$-2x = -6$$
$$\frac{-2x}{-2} = \frac{-6}{-2}$$
$$x = 3$$

Your Turn 3.

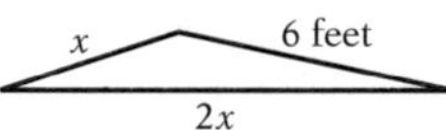

Let x = the length of one side of the triangle. Then $2x$ = the length of another side of the triangle. The perimeter is 21.
$$x + 2x + 6 = 21$$
$$3x + 6 = 21$$
$$3x + 6 - 6 = 21 - 6$$
$$3x = 15$$
$$\frac{3x}{3} = \frac{15}{3}$$
$$x = 5$$
The length of the shortest side is 5 feet.

Your Turn 4. Let x = the number of days that the car was rented. Then $30x$ is the daily charge for x days. The total charge is the flat rate plus the daily charge.
$$30x + 25 = 115$$
Solve the equation.
$$30x + 25 = 115$$
$$30x + 25 - 25 = 115 - 25$$
$$30x = 90$$
$$\frac{30x}{30} = \frac{90}{30}$$
$$x = 3$$
The car was rented for 3 days.

Copyright © Houghton Mifflin Company. All rights reserved.

Your Turn 5. 12 feet; x, $2x$

Let x = the length of the shorter piece.
Then $2x$ = the length of the other piece.
The total length is 12 feet.

$$x + 2x = 12$$

Solve the equation.

$$x + 2x = 12$$
$$3x = 12$$
$$\frac{3x}{3} = \frac{12}{3}$$
$$x = 4$$

The length of the shorter piece is 4 feet.

Section 4.4

Your Turn 1. Let x = the smaller number and y = the larger number. Then $y = x + 4$.

Your Turn 2. **(a)** $y = x + 2$

$3 = 1 + 2$

$3 = 3$ True

Yes, (1, 3) is a solution.

(b) $y = x + 2$

$1 = -1 + 2$

$1 = 1$ True

Yes, (−1, 1) is a solution.

(c) $y = x + 2$

$2 = -2 + 2$

$2 = 0$ False

No, (−2, 2) is not a solution.

Your Turn 3. **(a)**
$$x - 2y = 5$$
$$x - 2(0) = 5$$
$$x - 0 = 5$$
$$x = 5$$

(5, 0)

(b)
$$x - 2y = 5$$
$$3 - 2y = 5$$
$$3 - 3 - 2y = 5 - 3$$
$$-2y = 2$$
$$\frac{-2y}{-2} = \frac{2}{-2}$$
$$y = -1$$

(3, −1)

(c)
$$x - 2y = 5$$
$$x - 2(-2) = 5$$
$$x + 4 = 5$$
$$x + 4 - 4 = 5 - 4$$
$$x = 1$$

(1, −2)

Your Turn 4.

For $x = -1$

$$2x + y = 7$$
$$2(-1) + y = 7$$
$$-2 + y = 7$$
$$-2 + 2 + y = 7 + 2$$
$$y = 9$$

For $x = 0$

$$2x + y = 7$$
$$2(0) + y = 7$$
$$0 + y = 7$$
$$y = 7$$

For $y = 5$

$$2x + y = 7$$
$$2x + 5 = 7$$
$$2x + 5 - 5 = 7 - 5$$
$$2x = 2$$
$$\frac{2x}{2} = \frac{2}{2}$$
$$x = 1$$

For $y = -1$

$$2x + y = 7$$
$$2x + (-1) = 7$$
$$2x - 1 = 7$$
$$2x - 1 + 1 = 7 + 1$$
$$2x = 8$$
$$\frac{2x}{2} = \frac{8}{2}$$
$$x = 4$$

x	y	(x, y)
−1	9	(−1, 9)
0	7	(0, 7)
1	5	(1, 5)
4	−1	(4, −1)

Section 4.5

Your Turn 1.

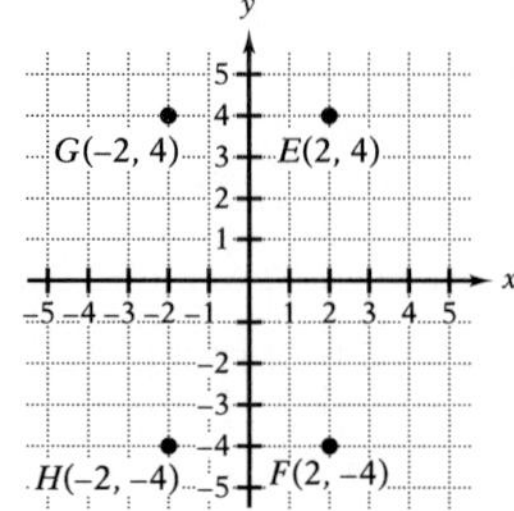

Your Turn 3.

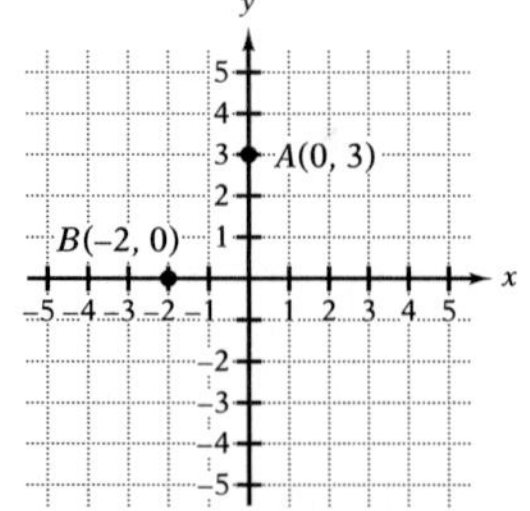

Copyright © Houghton Mifflin Company. All rights reserved.

Section 4.6

Your Turn 1.

For $x = -5$	$y = -5 + 3 = -2$	$(-5, -2)$
For $x = 1$	$y = 1 + 3 = 4$	$(1, 4)$
For $x = 2$	$y = 2 + 3 = 5$	$(2, 5)$

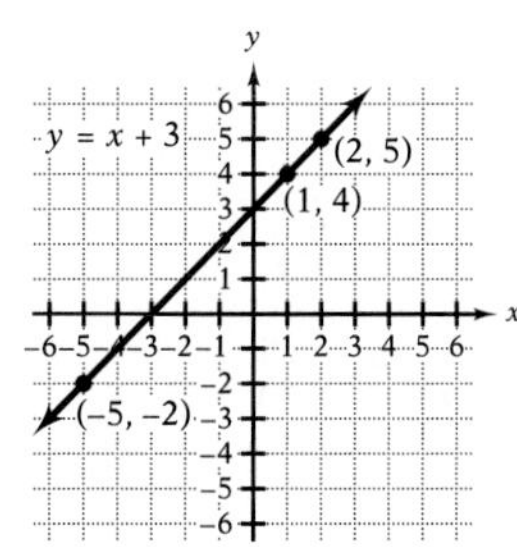

Your Turn 2.

For $x = 0$	*For $y = 0$*	*For $y = 1$*
$0 + 2y = 4$	$x + 2(0) = 4$	$x + 2(1) = 4$
$2y = 4$	$x + 0 = 4$	$x + 2 = 4$
$y = 2$	$x = 4$	$x = 2$
$(0, 2)$	$(4, 0)$	$(2, 1)$

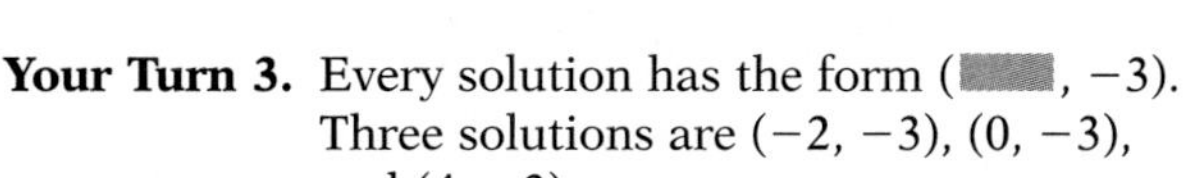

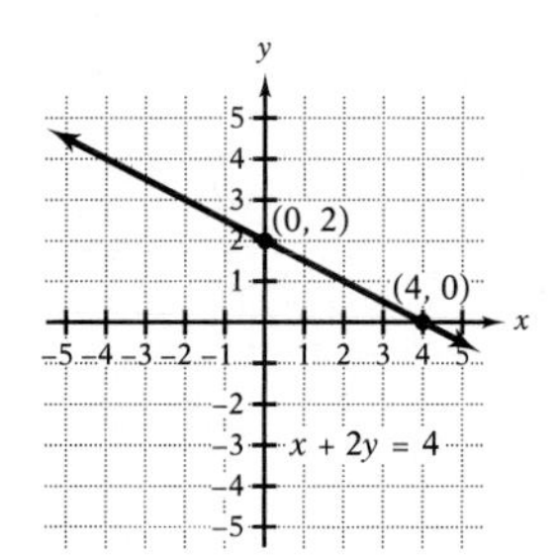

Your Turn 3. Every solution has the form (▬, -3). Three solutions are $(-2, -3)$, $(0, -3)$, and $(4, -3)$.

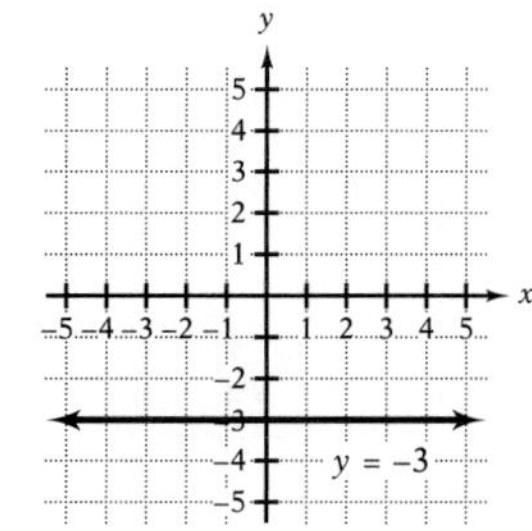

Your Turn 4. Every solution has the form (-2, ▬). Three solutions are $(-2, -1)$, $(-2, 0)$, and $(-2, 3)$.

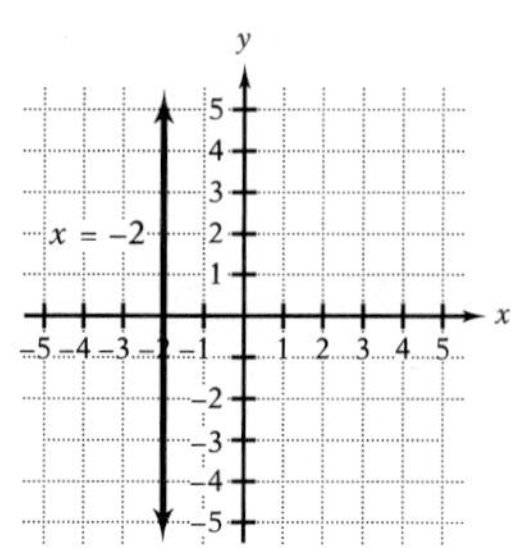

Your Turn 5. The point of intersection has the coordinates $(-3, -2)$, which is the solution of both equations.

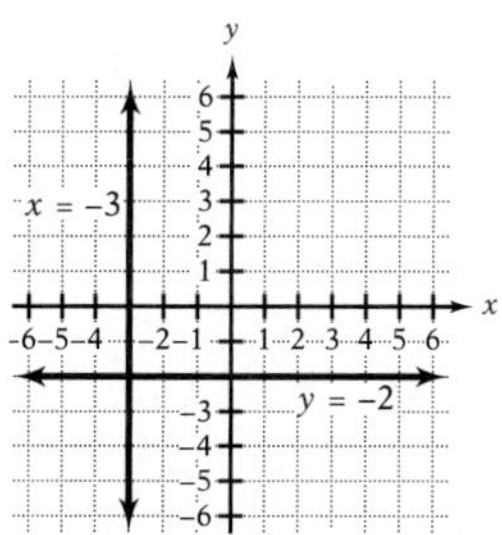

CHAPTER 5

WARM-UP SKILLS *(page 348)*

1. 2 **2.** 0 **3.** 30 **4.** 10 **5.** 120 **6.** 5 R 3 **7.** $8x$ **8.** 20 **9.** -7

10. Perimeter: 40 inches; area: 96 square inches

KEYS TO THE CALCULATOR

Section 5.2 *(page 368)*

(a) $\frac{3}{4}$ **(b)** $\frac{6}{5}$ **(c)** $\frac{7}{4}$ **(d)** $\frac{2}{9}$

Section 5.3 *(page 384)*

(a) $\frac{7}{6}$ **(b)** 10 **(c)** $\frac{3}{7}$ **(d)** $\frac{1}{8}$

Section 5.5 *(page 407)*

(a) $\frac{4}{5}$ **(b)** $\frac{7}{16}$ **(c)** $\frac{5}{8}$ **(d)** $\frac{19}{24}$

Section 5.6 *(page 420)*

Box 1 **(a)** $8\frac{2}{5}$ **(b)** $\frac{4}{9}$ **(c)** $1\frac{1}{2}$ **(d)** $1\frac{8}{15}$

Box 2 **(a)** $5\frac{11}{40}$ **(b)** $2\frac{9}{16}$ **(c)** $1\frac{2}{7}$ **(d)** $9\frac{5}{6}$

Copyright © Houghton Mifflin Company. All rights reserved.

SELECTED YOUR TURNS

Section 5.1

Your Turn 4. **(a)** Proper. The numerator is less than the denominator. **(b)** Improper. The numerator is greater than the denominator.

Your Turn 5. **(a)** $\frac{7}{2} \cdot \frac{5}{6} = \frac{7 \cdot 5}{2 \cdot 6} = \frac{35}{12}$

(b) $\frac{4}{x} \cdot \frac{y}{3} = \frac{4 \cdot y}{x \cdot 3} = \frac{4y}{3x}$

Your Turn 6. $4 \cdot \frac{7}{3} = \frac{4}{1} \cdot \frac{7}{3} = \frac{4 \cdot 7}{1 \cdot 3} = \frac{28}{3}$

Your Turn 7. $\frac{2}{3} \cdot (-8) = \frac{2}{3} \cdot \frac{-8}{1} = \frac{2 \cdot (-8)}{3 \cdot 1} = \frac{-16}{3}$

Your Turn 9. $\frac{5}{2} \cdot \frac{3}{-2} = \frac{5 \cdot 3}{2 \cdot (-2)} = \frac{15}{-4} = -\frac{15}{4} = \frac{-15}{4}$

Your Turn 10. $-\frac{1}{6} \cdot \frac{5}{-3} = \frac{-1}{6} \cdot \frac{5}{-3}$

$= \frac{-1 \cdot 5}{6 \cdot (-3)} = \frac{-5}{-18} = \frac{5}{18}$

Your Turn 11. **(a)** $-\frac{1}{2}(-2x) = \left(-\frac{1}{2} \cdot (-2)\right)x$

$= \left(\frac{-1}{2} \cdot \frac{-2}{1}\right)x = \frac{2}{2}x = 1x = x$

(b) $\frac{3}{4}\left(\frac{4}{3}a\right) = \left(\frac{3}{4} \cdot \frac{4}{3}\right)a = \frac{12}{12}a = 1a = a$

Your Turn 12. **(a)** The slice for 1–2 hours indicates 3 students are in this category. The fraction is $\frac{3}{22}$.

(b) The number of students who study more than 2 hours is $4 + 6 + 9 = 19$. The fraction is $\frac{19}{22}$.

Section 5.2

Your Turn 1. **(a)** Prime because 13 is divisible only by itself and 1 **(b)** Composite because 18 is divisible by 6 and 3

Your Turn 2. **(b)** $24 = 8 \cdot 3 = 2 \cdot 2 \cdot 2 \cdot 3 = 2^3 \cdot 3$

Your Turn 3. **(a)** $180 = 18 \cdot 10 = 2 \cdot 9 \cdot 2 \cdot 5 = 2 \cdot 2 \cdot 3 \cdot 3 \cdot 5 = 2^2 \cdot 3^2 \cdot 5$

(b) $735 = 5 \cdot 147 = 5 \cdot 3 \cdot 49 = 5 \cdot 3 \cdot 7 \cdot 7 = 5 \cdot 3 \cdot 7^2$

Your Turn 4. $\frac{8}{10} = \frac{\cancel{2} \cdot 4}{\cancel{2} \cdot 5} = \frac{4}{5}$

Your Turn 5. **(a)** $\frac{12}{18} = \frac{\cancel{2} \cdot 2 \cdot \cancel{3}}{\cancel{2} \cdot 3 \cdot \cancel{3}} = \frac{2}{3}$

(b) $\frac{18}{45} = \frac{2 \cdot \cancel{3} \cdot \cancel{3}}{5 \cdot \cancel{3} \cdot \cancel{3}} = \frac{2}{5}$

Your Turn 6. **(a)** $\frac{6}{90} = \frac{\cancel{2} \cdot \cancel{3}}{\cancel{2} \cdot \cancel{3} \cdot 3 \cdot 5} = \frac{1}{15}$

(b) $\frac{27}{9} = \frac{\cancel{3} \cdot \cancel{3} \cdot 3}{\cancel{3} \cdot \cancel{3}} = 3$

Your Turn 7. $\frac{8}{-20} = -\frac{8}{20} = -\frac{\cancel{2} \cdot \cancel{2} \cdot 2}{\cancel{2} \cdot \cancel{2} \cdot 5} = -\frac{2}{5}$

Your Turn 8. $\frac{-9}{-6} = \frac{9}{6} = \frac{3 \cdot \cancel{3}}{2 \cdot \cancel{3}} = \frac{3}{2}$

Your Turn 9. $\frac{12a}{3abc} = \frac{2 \cdot 2 \cdot \cancel{3} \cdot \cancel{a}}{\cancel{3} \cdot \cancel{a} \cdot b \cdot c} = \frac{4}{bc}$

Your Turn 10. $\frac{20x^3y}{6xy^2z} = \frac{\cancel{2} \cdot 2 \cdot 5 \cdot \cancel{x} \cdot x \cdot x \cdot \cancel{y}}{\cancel{2} \cdot 3 \cdot \cancel{x} \cdot \cancel{y} \cdot y \cdot z}$

$= \frac{2 \cdot 5 \cdot x \cdot x}{3 \cdot y \cdot z} = \frac{10x^2}{3yz}$

Your Turn 11. Total customers: 90
Number in the 17–50 group: 75
$\frac{75}{90} = \frac{\cancel{15} \cdot 5}{\cancel{15} \cdot 6} = \frac{5}{6}$

Section 5.3

Your Turn 1. $\frac{3}{2} \cdot \frac{6}{7} = \frac{3}{2} \cdot \frac{2 \cdot 3}{7} = \frac{3 \cdot \cancel{2} \cdot 3}{\cancel{2} \cdot 7} = \frac{3 \cdot 3}{7} = \frac{9}{7}$

Your Turn 2. $\frac{14}{15} \cdot \frac{10}{7} = \frac{2 \cdot \cancel{7}}{3 \cdot \cancel{5}} \cdot \frac{2 \cdot \cancel{5}}{\cancel{7}} = \frac{4}{3}$

Your Turn 3. $\frac{3}{6y} \cdot \frac{4xy}{9} = \frac{\cancel{3}}{\cancel{2} \cdot \cancel{3} \cdot \cancel{y}} \cdot \frac{\cancel{2} \cdot 2 \cdot x \cdot \cancel{y}}{3 \cdot 3} = \frac{2x}{9}$

Your Turn 4. $\frac{5}{6} \cdot \frac{3}{4} \cdot \frac{2}{3} = \frac{5}{2 \cdot \cancel{3}} \cdot \frac{\cancel{3}}{2 \cdot \cancel{2}} \cdot \frac{\cancel{2}}{3} = \frac{5}{12}$

Your Turn 5. **(a)** $\frac{2}{-3}\left(-\frac{3}{4}\right) = \frac{\cancel{2}}{\cancel{-3}} \cdot \frac{\cancel{-3}}{\cancel{2} \cdot 2} = \frac{1}{2}$

(b) $20 \cdot \frac{3}{5} = \frac{20}{1} \cdot \frac{3}{5} = \frac{2 \cdot 2 \cdot \cancel{5}}{1} \cdot \frac{3}{\cancel{5}} = 12$

Your Turn 6. **(c)** $\frac{8}{-1} = -8$ **(d)** $\frac{1}{1} = 1$

Your Turn 7. **(a)** $\frac{6}{7} \div 5 = \frac{6}{7} \div \frac{5}{1} = \frac{6}{7} \cdot \frac{1}{5} = \frac{6}{35}$

(b) $\frac{8}{3} \div \frac{1}{4} = \frac{8}{3} \cdot \frac{4}{1} = \frac{32}{3}$

(c) $3 \div \frac{2}{3} = \frac{3}{1} \cdot \frac{3}{2} = \frac{9}{2}$

Your Turn 8. **(a)** $-\frac{1}{3} \div (-2) = -\frac{1}{3}\left(-\frac{1}{2}\right) = \frac{1}{6}$

(b) $8 \div \left(-\frac{3}{2}\right) = \frac{8}{1}\left(-\frac{2}{3}\right) = -\frac{16}{3}$

(c) $-\frac{1}{9} \div \left(-\frac{3}{2}\right) = -\frac{1}{9}\left(-\frac{2}{3}\right) = \frac{2}{27}$

Your Turn 9. **(a)** $-\frac{3}{4} \div \frac{9}{2} = -\frac{3}{4} \cdot \frac{2}{9}$

$= -\frac{\cancel{3}}{2 \cdot \cancel{2}} \cdot \frac{\cancel{2}}{\cancel{3} \cdot 3} = -\frac{1}{6}$

(b) $16 \div \left(-\frac{12}{5}\right) = \frac{16}{1} \cdot \left(-\frac{5}{12}\right)$

$= \frac{\cancel{2} \cdot \cancel{2} \cdot 2 \cdot 2}{1} \cdot \left(-\frac{5}{\cancel{2} \cdot \cancel{2} \cdot 3}\right) = -\frac{20}{3}$

(c) $-\frac{1}{2} \div \left(-\frac{5}{6}\right) = -\frac{1}{2} \cdot \left(-\frac{6}{5}\right)$

$= -\frac{1}{\cancel{2}} \cdot \left(-\frac{\cancel{2} \cdot 3}{5}\right) = \frac{3}{5}$

(d) $\frac{12}{5} \div 6 = \frac{12}{5} \cdot \frac{1}{6}$

$= \frac{2 \cdot \cancel{2} \cdot \cancel{3}}{5} \cdot \frac{1}{\cancel{2} \cdot \cancel{3}} = \frac{2}{5}$

Your Turn 10. $\frac{1}{2} \cdot \frac{3}{4} \cdot 120 = 45$

Section 5.4

Your Turn 1. $\frac{4}{7} = \frac{4 \cdot 2}{7 \cdot 2} = \frac{8}{14}$

Copyright © Houghton Mifflin Company. All rights reserved.

Your Turn 2. $\frac{2}{5} = \frac{2 \cdot 10}{5 \cdot 10} = \frac{20}{50}$; 20

Your Turn 3. $\frac{2}{-9} = \frac{-2}{9} = \frac{-2 \cdot 4}{9 \cdot 4} = \frac{-8}{36}$; -8

Your Turn 4. $\frac{a}{2} = \frac{a \cdot z}{2 \cdot z} = \frac{az}{2z}$; az

Your Turn 5. Multiples of 9 are 18, 27, 36, ... Because 36 is divisible by 4, the LCM of 4 and 9 is 36.

Your Turn 6. $10 = 2 \cdot 5$
$42 = 6 \cdot 7 = 2 \cdot 3 \cdot 7$
LCM: $2 \cdot 3 \cdot 5 \cdot 7 = 210$

Your Turn 7. $8 = 2 \cdot 2 \cdot 2$
$12 = 2 \cdot 2 \cdot 3$
LCM: $2 \cdot 2 \cdot 2 \cdot 3 = 24$

Your Turn 8. $6 = 2 \cdot 3$
$9 = 3 \cdot 3$
$60 = 6 \cdot 10 = 2 \cdot 2 \cdot 3 \cdot 5$
LCM: $2 \cdot 2 \cdot 3 \cdot 3 \cdot 5 = 180$

Your Turn 9. $2x^2 = 2 \cdot x \cdot x$
$3xy = 3 \cdot x \cdot y$
LCM: $2 \cdot 3 \cdot x \cdot x \cdot y = 6x^2y$

Your Turn 10. $9 = 3 \cdot 3$
$12 = 4 \cdot 3 = 2 \cdot 2 \cdot 3$
LCD: $2 \cdot 2 \cdot 3 \cdot 3 = 36$

Your Turn 11. The LCD is 36.

$$\frac{5}{9} = \frac{5 \cdot 4}{9 \cdot 4} = \frac{20}{36}$$

$$\frac{7}{12} = \frac{7 \cdot 3}{12 \cdot 3} = \frac{21}{36}$$

Because $20 < 21$, $\frac{20}{36} < \frac{21}{36}$. So $\frac{5}{9} < \frac{7}{12}$.

Section 5.5

Your Turn 1. $\frac{4}{5} + \frac{2}{5} = \frac{4 + 2}{5} = \frac{6}{5}$

Your Turn 2. $\frac{7}{y} + \frac{1}{y} = \frac{7 + 1}{y} = \frac{8}{y}$

Your Turn 3. $\frac{11}{9} - \frac{8}{9} = \frac{11 - 8}{9} = \frac{3}{9} = \frac{1 \cdot \cancel{3}}{3 \cdot \cancel{3}} = \frac{1}{3}$

Your Turn 4. $\frac{4}{7} - \frac{9}{7} = \frac{4 - 9}{7} = \frac{4 + (-9)}{7} = \frac{-5}{7} = -\frac{5}{7}$

Your Turn 5. $\frac{1}{5} - \left(-\frac{9}{5}\right) = \frac{1}{5} + \frac{9}{5} = \frac{10}{5} = 2$

Your Turn 6. Because 9 is a multiple of 3, the LCD is 9.

$$\frac{4}{9} + \frac{2}{3} = \frac{4}{9} + \frac{2 \cdot 3}{3 \cdot 3} = \frac{4}{9} + \frac{6}{9} = \frac{10}{9}$$

Your Turn 7. The LCD is the product of 2 and 7, which is 14.

$$\frac{3}{2} + \frac{2}{7} = \frac{3 \cdot 7}{2 \cdot 7} + \frac{2 \cdot 2}{7 \cdot 2} = \frac{21}{14} + \frac{4}{14} = \frac{25}{14}$$

Your Turn 8. Because 15 is a multiple of 5, the LCD is 15.

$$\frac{2}{15} + \frac{1}{5} = \frac{2}{15} + \frac{1 \cdot 3}{5 \cdot 3} = \frac{2}{15} + \frac{3}{15} = \frac{5}{15} = \frac{1 \cdot 5}{3 \cdot 5} = \frac{1}{3}$$

Your Turn 9. $3 - \frac{15}{4} = \frac{3}{1} - \frac{15}{4} = \frac{3 \cdot 4}{1 \cdot 4} - \frac{15}{4}$
$= \frac{12}{4} - \frac{15}{4} = \frac{-3}{4} = -\frac{3}{4}$

Your Turn 10. $\frac{2}{3} - \frac{1}{6} + \frac{1}{7} = \frac{2 \cdot 14}{3 \cdot 14} - \frac{1 \cdot 7}{6 \cdot 7} + \frac{1 \cdot 6}{7 \cdot 6} =$
$\frac{28}{42} - \frac{7}{42} + \frac{6}{42} = \frac{28 - 7 + 6}{42} = \frac{27}{42} = \frac{9}{14}$

Your Turn 11. **(a)** $\frac{1}{a} + \frac{3}{5} = \frac{1 \cdot 5}{a \cdot 5} + \frac{3 \cdot a}{5 \cdot a} = \frac{5}{5a} + \frac{3a}{5a}$
$= \frac{5 + 3a}{5a}$

(b) $\frac{5a}{b} - 2 = \frac{5a}{b} - \frac{2}{1} = \frac{5a}{b} - \frac{2 \cdot b}{1 \cdot b}$
$= \frac{5a}{b} - \frac{2b}{b} = \frac{5a - 2b}{b}$

Section 5.6

Your Turn 1. $4\frac{3}{8} = 4 + \frac{3}{8} = \frac{4}{1} + \frac{3}{8} = \frac{4 \cdot 8}{1 \cdot 8} + \frac{3}{8}$
$= \frac{32}{8} + \frac{3}{8} = \frac{35}{8}$

Your Turn 2. Multiply the 8 in the denominator by the whole number 2. Add the result 16 to the numerator 5. The result 21 is the numerator of the improper fraction.

$$2\frac{5}{8} = \frac{21}{8}$$

Your Turn 3.

$$\begin{array}{r} 2 \\ 9\overline{)20} \\ \underline{18} \\ 2 \end{array}$$

$$\frac{20}{9} = 2 + \frac{2}{9} = 2\frac{2}{9}$$

Your Turn 4. $2\frac{2}{5} \cdot 1\frac{7}{8} = \frac{12}{5} \cdot \frac{15}{8} = \frac{\cancel{2} \cdot \cancel{2} \cdot 3}{\cancel{5}} \cdot \frac{3 \cdot \cancel{5}}{\cancel{2} \cdot \cancel{2} \cdot 2}$
$= \frac{9}{2} = 4\frac{1}{2}$

Your Turn 5. $1\frac{2}{7} \div \frac{3}{14} = \frac{9}{7} \div \frac{3}{14} = \frac{9}{7} \cdot \frac{14}{3}$
$= \frac{3 \cdot \cancel{3}}{\cancel{7}} \cdot \frac{2 \cdot \cancel{7}}{\cancel{3}} = \frac{6}{1} = 6$

Your Turn 6. **(a)** $1\frac{3}{8} + 4\frac{1}{4} = \frac{11}{8} + \frac{17}{4} = \frac{11}{8} + \frac{17 \cdot 2}{4 \cdot 2}$
$= \frac{11}{8} + \frac{34}{8} = \frac{45}{8} = 5\frac{5}{8}$

(b) $5\frac{1}{2} - 1\frac{3}{4} = \frac{11}{2} - \frac{7}{4} = \frac{11 \cdot 2}{2 \cdot 2} - \frac{7}{4}$
$= \frac{22}{4} - \frac{7}{4} = \frac{15}{4} = 3\frac{3}{4}$

Your Turn 7.

$$\begin{array}{llll} 2\frac{5}{18} & = 2 + \frac{5}{18} & = 2 + \frac{5}{18} & = 2 + \frac{5}{18} \\ 7\frac{8}{9} & = 7 + \frac{8}{9} & = 7 + \frac{8 \cdot 2}{9 \cdot 2} & = \underline{7 + \frac{16}{18}} \\ & & & \ 9 + \frac{21}{18} \end{array}$$

Simplify. $\frac{21}{18} = \frac{\cancel{3} \cdot 7}{2 \cdot \cancel{3} \cdot 3} = \frac{7}{6} = 1 + \frac{1}{6}$

Finally, write $9 + \frac{21}{18} = 9 + 1 + \frac{1}{6} = 10 + \frac{1}{6} = 10\frac{1}{6}$.

Copyright © Houghton Mifflin Company. All rights reserved.

Your Turn 8.

$$\begin{array}{rcr} 4\frac{5}{8} = & & 4 + \frac{5}{8} \\ -1\frac{3}{8} = & & -1 - \frac{3}{8} \\ \hline & & 3 + \frac{2}{8} \end{array}$$

Simplify. $3 + \frac{2}{8} = 3 + \frac{1}{4} = 3\frac{1}{4}$

Your Turn 9.

$$\begin{array}{lr} 8\frac{3}{10} = 8 + \frac{3}{10} = 7 + 1 + \frac{3}{10} = 7 + \frac{10}{10} + \frac{3}{10} = & 7 + \frac{13}{10} \\ -5\frac{7}{10} & = -5 - \frac{7}{10} \\ \hline & 2 + \frac{6}{10} \end{array}$$

Simplify. $2 + \frac{6}{10} = 2 + \frac{3}{5} = 2\frac{3}{5}$

Your Turn 10.

$$\begin{array}{l} 9\frac{2}{3} = \quad 9 + \frac{2 \cdot 5}{3 \cdot 5} = \quad 9 + \frac{10}{15} \\ -6\frac{4}{5} = -6 - \frac{4 \cdot 3}{5 \cdot 3} = -6 - \frac{12}{15} \end{array}$$

Now borrow.

$$\begin{array}{lr} 9 + \frac{10}{15} = 8 + 1 + \frac{10}{15} = 8 + \frac{15}{15} + \frac{10}{15} = & 8 + \frac{25}{15} \\ -6 - \frac{12}{15} & = -6 - \frac{12}{15} \\ \hline & 2 + \frac{13}{15} = 2\frac{13}{15} \end{array}$$

Section 5.7

Your Turn 1.

$$\begin{aligned} 7y &= 3 \\ \frac{\cancel{7} \cdot y}{\cancel{7}} &= \frac{3}{7} \\ y &= \frac{3}{7} \end{aligned}$$

Your Turn 2.

$$\begin{aligned} \frac{y}{5} &= 4 \\ \frac{y}{\cancel{5}} \cdot \frac{\cancel{5}}{1} &= 4 \cdot 5 \\ y &= 20 \end{aligned}$$

Your Turn 3.

$$\begin{aligned} -6 &= -\frac{x}{3} \\ -6 &= \frac{x}{-3} \\ -6(-3) &= \frac{x}{\cancel{-3}} \cdot \frac{\cancel{-3}}{1} \\ 18 &= x \end{aligned}$$

Your Turn 4.

$$\begin{aligned} \frac{1}{5}x &= 2 \\ 5 \cdot \left(\frac{1}{5}x\right) &= 5(2) \\ \left(5 \cdot \frac{1}{5}\right)x &= 10 \\ 1x &= 10 \\ x &= 10 \end{aligned}$$

Your Turn 5.

$$\begin{aligned} \frac{7}{3}x &= \frac{5}{6} \\ \frac{3}{7}\left(\frac{7}{3}x\right) &= \frac{3}{7} \cdot \frac{5}{6} \\ \left(\frac{3}{7} \cdot \frac{7}{3}\right)x &= \frac{\cancel{3}}{7} \cdot \frac{5}{2 \cdot \cancel{3}} \\ x &= \frac{5}{14} \end{aligned}$$

Your Turn 6.

$$\begin{aligned} \frac{3}{5}x - 2 &= 7 \\ \frac{3}{5}x - 2 + 2 &= 7 + 2 \\ \frac{3}{5}x &= 9 \\ \frac{5}{3}\left(\frac{3}{5}x\right) &= \frac{5}{3} \cdot \frac{9}{1} \\ \left(\frac{5}{3} \cdot \frac{3}{5}\right)x &= \frac{5}{\cancel{3}} \cdot \frac{\cancel{3} \cdot 3}{1} \\ x &= 15 \end{aligned}$$

Your Turn 7.

$$\begin{aligned} y + \frac{1}{2} &= \frac{5}{4} \\ y + \frac{1}{2} - \frac{1}{2} &= \frac{5}{4} - \frac{1}{2} \\ y + 0 &= \frac{5}{4} - \frac{2}{4} \\ y &= \frac{3}{4} \end{aligned}$$

Your Turn 8.

$$\begin{aligned} y + \frac{1}{2} &= \frac{5}{4} \\ 4\left(y + \frac{1}{2}\right) &= 4\left(\frac{5}{4}\right) \\ 4y + \frac{4}{1} \cdot \frac{1}{2} &= \frac{\cancel{4}}{1} \cdot \frac{5}{\cancel{4}} \\ 4y + 2 &= 5 \\ 4y + 2 - 2 &= 5 - 2 \\ 4y &= 3 \\ y &= \frac{3}{4} \end{aligned}$$

Your Turn 9.

$$\begin{aligned} 2 + \frac{2}{3}x &= \frac{5}{6} \\ 6\left(2 + \frac{2}{3}x\right) &= 6\left(\frac{5}{6}\right) \\ 6(2) + 6\left(\frac{2}{3}x\right) &= 6\left(\frac{5}{6}\right) \\ 12 + \left(\frac{2 \cdot \cancel{3}}{1} \cdot \frac{2}{\cancel{3}}\right)x &= \frac{\cancel{6}}{1} \cdot \frac{5}{\cancel{6}} \\ 12 + 4x &= 5 \\ 12 + 4x - 12 &= 5 - 12 \\ 4x &= -7 \\ \frac{\cancel{4} \cdot x}{\cancel{4}} &= \frac{-7}{4} \\ x &= -\frac{7}{4} \end{aligned}$$

Copyright © Houghton Mifflin Company. All rights reserved.

Your Turn 10.
$$\frac{1}{2} + \frac{1}{3}x = \frac{1}{4}x$$
$$12\left(\frac{1}{2}\right) + 12\left(\frac{1}{3}x\right) = 12\left(\frac{1}{4}x\right)$$
$$\frac{12}{1}\cdot\frac{1}{2} + \left(\frac{12}{1}\cdot\frac{1}{3}\right)x = \left(\frac{12}{1}\cdot\frac{1}{4}\right)x$$
$$6 + 4x = 3x$$
$$6 + 4x - 4x = 3x - 4x$$
$$6 = -1x$$
$$\frac{6}{-1} = \frac{\cancel{-1}\cdot x}{\cancel{-1}}$$
$$-6 = x$$

Your Turn 11. Let x = total pages in the paper.
Number of pages written in week 1: $\frac{1}{2}x$

Number of pages written in week 2: $\frac{3}{7}x$

$$\frac{1}{2}x + \frac{3}{7}x = 26$$
$$14\cdot\frac{1}{2}x + 14\cdot\frac{3}{7}x = 14\cdot 26$$
$$\left(\frac{14}{1}\cdot\frac{1}{2}\right)x + \left(\frac{14}{1}\cdot\frac{3}{7}\right)x = 364$$
$$7x + 6x = 364$$
$$13x = 364$$
$$x = 28$$

The completed paper was 28 pages long.

Section 5.8

Your Turn 1. $A = \frac{1}{2}bh = \frac{1}{2}\cdot 7\cdot 3$
$$= \frac{1}{2}\cdot\frac{7}{1}\cdot\frac{3}{1}$$
$$= \frac{21}{2} = 10\frac{1}{2} \text{ square inches}$$

Your Turn 2. $A = bh = \frac{3}{2}\cdot\frac{4}{7}$
$$= \frac{3}{\cancel{2}}\cdot\frac{\cancel{2}\cdot 2}{7}$$
$$= \frac{6}{7} \text{ square foot}$$

Your Turn 3. $A = \frac{1}{2}h(b_1 + b_2) = \frac{1}{2}\cdot\frac{2}{3}\cdot(2 + 1)$
$$= \frac{1}{\cancel{2}}\cdot\frac{\cancel{2}}{\cancel{3}}\cdot\frac{\cancel{3}}{1} = 1 \text{ square yard}$$

Your Turn 4.
$$\frac{1}{2}bh = A$$
$$\frac{1}{2}\cdot b\cdot\frac{8}{9} = \frac{1}{3}$$
$$\frac{1}{2}\cdot\frac{8}{9}\cdot b = \frac{1}{3}$$
$$\frac{1}{\cancel{2}}\cdot\frac{\cancel{2}\cdot 2\cdot 2}{3\cdot 3}\cdot b = \frac{1}{3}$$
$$\frac{4}{9}b = \frac{1}{3}$$
$$\frac{9}{4}\cdot\frac{4}{9}b = \frac{9}{4}\cdot\frac{1}{3}$$
$$\frac{9}{4}\cdot\frac{4}{9}b = \frac{3\cdot\cancel{3}}{2\cdot 2}\cdot\frac{1}{\cancel{3}}$$
$$b = \frac{3}{4} \text{ foot}$$

Your Turn 5. $P = 48$, $R = 200$
$$R - W = P$$
$$200 - W = 48$$
$$200 - W - 200 = 48 - 200$$
$$-W = -152$$
$$\frac{\cancel{-1}\cdot W}{\cancel{-1}} = \frac{-152}{-1}$$
$$W = \$152$$

Your Turn 6. $D = 150$, $S = 1{,}420$
$$P - D = S$$
$$P - 150 = 1{,}420$$
$$P - 150 + 150 = 1{,}420 + 150$$
$$P = \$1{,}570$$

Your Turn 7.
$$\frac{9}{5}C + 32 = F$$
$$\frac{9}{5}C + 32 = 95$$
$$\frac{9}{5}C + 32 - 32 = 95 - 32$$
$$\frac{9}{5}C = 63$$
$$\frac{5}{9}\cdot\frac{9}{5}C = \frac{5}{9}(63)$$
$$\left(\frac{5}{9}\cdot\frac{9}{5}\right)C = \frac{5}{\cancel{3}\cdot\cancel{3}}\cdot\frac{\cancel{3}\cdot\cancel{3}\cdot 7}{1}$$
$$1C = 35$$
$$C = 35°C$$

Your Turn 8. $D = 135$, $R = 45$
$$RT = D$$
$$45\cdot T = 135$$
$$\frac{\cancel{45}\cdot T}{\cancel{45}} = \frac{135}{45}$$
$$T = 3 \text{ hours}$$

Copyright © Houghton Mifflin Company. All rights reserved.

CHAPTER 6

WARM-UP SKILLS (page 472)

1. 23,000 **2.** 16 **3.** $\frac{37}{100}$ **4.** $3\frac{2}{5}$ **5.** 1,274
6. 1,776 **7.** 17,787 **8.** 7 **9.** 25 **10.** 1,500

KEYS TO THE CALCULATOR

Section 6.2 *(page 489)*

(a) −181.08 **(b)** −3.62 **(c)** −0.52 **(d)** 0.934

Section 6.3 *(page 501)*

(a) 1,036.9584 **(b)** 0.05368 **(c)** 0.00144
(d) 28,550.673

Section 6.4 *(page 515)*

(a) 2.83 **(b)** −3.262 **(c)** 612.8571 **(d)** 0.4

Section 6.6 *(page 538)*

(a) 12 **(b)** 57 **(c)** 10 **(d)** 68

SELECTED YOUR TURNS

Section 6.1

Your Turn 4. $0.14 = \frac{14}{100} = \frac{\cancel{2} \cdot 7}{\cancel{2} \cdot 50} = \frac{7}{50}$

Your Turn 5. $\frac{3}{50} = \frac{3 \cdot 2}{50 \cdot 2} = \frac{6}{100} = 0.06$

Section 6.2

Your Turn 1. **(a)**
$$\begin{array}{r} 6.23 \\ +\ 14.80 \\ \hline 21.03 \end{array}$$

(b)
$$\begin{array}{r} 24.000 \\ +\ 9.531 \\ \hline 33.531 \end{array}$$

Your Turn 2.
$$\begin{array}{r} 84.16 \\ -\ 19.03 \\ \hline 65.13 \end{array}$$

Your Turn 3. **(a)**
$$\begin{array}{r} 9.72 \\ -\ 1.60 \\ \hline 8.12 \end{array}$$

$1.6 + (-9.72) = -8.12$

(b) $8 - (-11.3) = 8 + 11.3$

$$\begin{array}{r} 8.0 \\ +\ 11.3 \\ \hline 19.3 \end{array}$$

$8 - (-11.3) = 19.3$

Your Turn 4.
$$\begin{array}{rcr} 50.25 & \rightarrow & 50 \\ -\ 29.4 & \rightarrow & -\ 30 \\ \hline & & 20 \end{array}$$

Your Turn 5.
$$\begin{array}{rcr} 10.18 & \rightarrow & 10 \\ +\ 0.49 & \rightarrow & +\ 0.5 \\ \hline & & 10.5 \end{array}$$

Your Turn 6. $8.4y - 2.3y = (8.4 - 2.3)y = 6.1y$

Your Turn 7. $2.4x + 5y - 3.4y + 8x =$
$(2.4 + 8)x + (5 - 3.4)y = 10.4x + 1.6y$

Your Turn 8. The selling price is $7.65 more than your cost of $19.23.

$$\begin{array}{r} \$19.23 \\ +\ \$\ 7.65 \\ \hline \$26.88 \end{array}$$

The selling price is $26.88.

Your Turn 9. Find the total length of the two sides that are given.

$$\begin{array}{r} 18.3 \\ +\ 12.9 \\ \hline 31.2 \end{array}$$

The length of the third side is the difference between the perimeter and the total lengths of the other two sides.

$$\begin{array}{r} 40.8 \\ -\ 31.2 \\ \hline 9.6 \end{array}$$

The length of the third side is 9.6 inches.

Section 6.3

Your Turn 1.
$$\begin{array}{r} 2.87 \\ \times\ \ 41.5 \\ \hline 1435 \\ 287\ \\ 1148\ \ \\ \hline 119.105 \end{array}$$

Your Turn 2. Because $8 \cdot 3 = 24$ and the answer has 3 decimal places, $8 \times 0.003 = 0.024$.

Your Turn 4.

(a) Shift the decimal point 4 places to the right.

$$5.47 \times 10^4 = 54{,}700$$

(b) Shift the decimal point 2 places to the right.

$$1.32867 \times 100 = 132.867$$

Your Turn 5. **(a)** $10(0.1 + 2.7x)$
$= 10(0.1) + 10(2.7x)$
$= 1 + 27x$

(b) $1{,}000(5.86x - 0.027)$
$= 1{,}000(5.86x) - 1{,}000(0.027)$
$= 5{,}860x - 27$

Your Turn 6. $9.1 \approx 9$ and $5.92 \approx 6$. So $9.1 \times 5.92 \approx 9 \times 6 = 54$.

Your Turn 7. $0.1 \times 287 = 28.7$ million

Your Turn 8. $1.5 \times \$16.78 = \$25.17$

Copyright © Houghton Mifflin Company. All rights reserved.

Section 6.4

Your Turn 1.

$$\begin{array}{r} 3.588 \\ 51\overline{)183.000} \\ \underline{153} \\ 30\,0 \\ \underline{25\,5} \\ 4\,50 \\ \underline{4\,08} \\ 420 \\ \underline{408} \\ 12 \end{array}$$

$183 \div 51 \approx 3.59$

Your Turn 2.

$$\begin{array}{r} 3.2875 \\ 4\overline{)13.1500} \\ \underline{12} \\ 1\,1 \\ \underline{8} \\ 35 \\ \underline{32} \\ 30 \\ \underline{28} \\ 20 \\ \underline{20} \\ 0 \end{array}$$

$13.15 \div 4 \approx 3.288$

Your Turn 3. Move the decimal point 2 places to the right.

$$\begin{array}{r} 1.688 \\ 162\overline{)273.500} \\ \underline{162} \\ 111\,5 \\ \underline{97\,2} \\ 14\,30 \\ \underline{12\,96} \\ 1\,340 \\ \underline{1\,296} \\ 44 \end{array}$$

$2.735 \div 1.62 \approx 1.69$

Your Turn 4. Move the decimal point 2 places to the right.

$$\begin{array}{r} 2.051 \\ 39\overline{)80.000} \\ \underline{78} \\ 2\,00 \\ \underline{1\,95} \\ 50 \\ \underline{39} \\ 11 \end{array}$$

$0.8 \div 0.39 \approx 2.05$

Your Turn 5.

(a) Move the decimal point 2 places to the left.

$$\frac{8.1}{10^2} = 0.081$$

(b) Move the decimal point 4 places to the left.

$$\frac{643.9}{10{,}000} = 0.06439$$

Your Turn 6. $25.04 \approx 25$ and $4.83 \approx 5$.

$$25.04 \div 4.83 \approx 25 \div 5 = 5$$

Your Turn 7.

Course	*Grade · Hours*
English	4 · 3 = 12
Prealgebra	4 · 3 = 12
First-year seminar	4 · 2 = 8
History	2 · 3 = 6
Total	38

The total number of credit hours was 11. The GPA is

$$\frac{38}{11} \approx 3.45$$

Your Turn 8. To divide the cost by 10, move the decimal point one place to the left.

Section 6.5

Your Turn 1.

$$\begin{aligned} x + y - z &= 1.52 + (-2.7) - (-3.846) \\ &= 1.52 + (-2.7) + 3.846 = 2.666 \end{aligned}$$

$$\begin{array}{r} 1.520 \\ \underline{+\ 3.846} \\ 5.366 \\ \underline{-\ 2.700} \\ 2.666 \end{array}$$

Your Turn 2.

$$\begin{aligned} 2.4(x - y) &= 2.4(9.21 - 3.76) \\ &= 2.4(5.45) = 13.08 \end{aligned}$$

Your Turn 3. $\frac{x}{y} + 3y = \frac{25.2}{8.4} + 3(8.4) = 3 + 25.2 = 28.2$

Your Turn 4.

$$\begin{aligned} x + 6.3 &= 10.8 \\ x + 6.3 - 6.3 &= 10.8 - 6.3 \\ x &= 4.5 \end{aligned}$$

Your Turn 5.

$$\begin{aligned} \frac{x}{0.06} &= 9.1 \\ \left(\frac{0.06}{1}\right)\left(\frac{x}{0.06}\right) &= (0.06)(9.1) \\ x &= 0.546 \end{aligned}$$

Your Turn 6.

$$\begin{aligned} 0.03x &= 0.12 \\ 100(0.03x) &= 100(0.12) \\ 3x &= 12 \\ x &= \frac{12}{3} = 4 \end{aligned}$$

Copyright © Houghton Mifflin Company. All rights reserved.

Your Turn 7. $1.7x + 17.214 = 6.23x$

$$1{,}000(1.7x) + 1{,}000(17.214) = 1{,}000(6.23x)$$
$$1{,}700x + 17{,}214 = 6{,}230x$$
$$1{,}700x + 17{,}214 - 1{,}700x = 6{,}230x - 1{,}700x$$
$$17{,}214 = 4{,}530x$$
$$x = \frac{17{,}214}{4{,}530} = 3.8$$

Your Turn 8. Let x = the original price. Then the sale price is $x - 2.65$. The cost of a dozen wrenches is $12(x - 2.65)$.

$$12(x - 2.65) = 111$$
$$12x - 31.8 = 111$$
$$12x - 31.8 + 31.8 = 111 + 31.8$$
$$12x = 142.8$$
$$x = \frac{142.8}{12} = \$11.90$$

Your Turn 9. Let h = the number of overtime hours.
Regular pay: 40($11.33) = $453.20
Overtime pay: $\$22.66h$

$$453.20 + 22.66h = 521.18$$
$$22.66h = 67.98$$
$$h = 3$$

Your Turn 10. Let h = the number of hours after the first hour. Then $7.2h$ = the distance traveled after the first hour.

$$9.6 + 7.2h = 31.2$$
$$7.2h = 21.6$$
$$h = \frac{21.6}{7.2} = 3$$

The cyclist rode for 4 hours.

Section 6.6

Your Turn 1.

(a) $\frac{1}{6} = 0.16666666\cdots$; repeating

(b) $\frac{7}{20} = 0.35000000\cdots = 0.35$; terminating

(c) $\frac{1}{99} = 0.01010101\cdots$; repeating

Your Turn 2.

(a) The square roots of 64 are 8 and -8 because $8^2 = 64$ and $(-8)^2 = 64$.

(b) The square roots of $\frac{16}{25}$ are $\frac{4}{5}$ and $-\frac{4}{5}$ because $\left(\frac{4}{5}\right)^2 = \frac{16}{25}$ and $\left(-\frac{4}{5}\right)^2 = \frac{16}{25}$.

Your Turn 3.

(c) $5 - \sqrt{16} = 5 - 4 = 1$

(d) $\sqrt{36} + \sqrt{49} = 6 + 7 = 13$

(e) $8\sqrt{25} = 8 \cdot 5 = 40$

(f) $1 + 9\sqrt{\frac{1}{9}} = 1 + \frac{9}{1} \cdot \frac{1}{3} = 1 + 3 = 4$

Your Turn 5. $s = 3.5\sqrt{d} = 3.5\sqrt{300} \approx 61$ miles per hour

Your Turn 6. $t = \sqrt{\frac{d}{16}} = \sqrt{\frac{1{,}002}{16}} \approx 7.9$ seconds

Section 6.7

Your Turn 1. $c^2 = a^2 + b^2 = 6^2 + 6^2 = 36 + 36 = 72$

$c = \sqrt{72} \approx 8.49$

Your Turn 2.

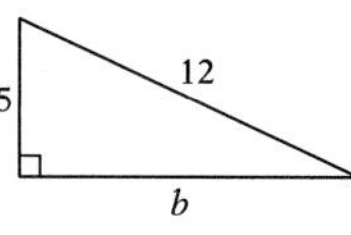

$$c^2 = a^2 + b^2$$
$$12^2 = 5^2 + b^2$$
$$144 = 25 + b^2$$
$$b^2 = 144 - 25 = 119$$
$$b = \sqrt{119} \approx 10.9 \text{ inches}$$

Your Turn 3.

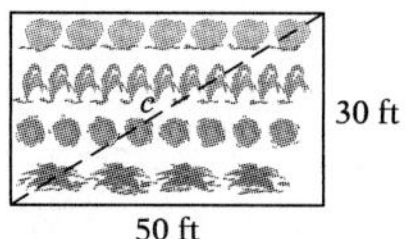

$$c^2 = a^2 + b^2 = 30^2 + 50^2 = 900 + 2{,}500 = 3{,}400$$
$$c = \sqrt{3{,}400} \approx 58.3 \text{ feet}$$

Your Turn 4. Let c = the length of the board. The lengths of the legs are 8 and 10 + 2 or 12.

$$c^2 = a^2 + b^2 = 8^2 + 12^2 = 64 + 144 = 208$$
$$c = \sqrt{208} \approx 14.4 \text{ feet}$$

Your Turn 5.

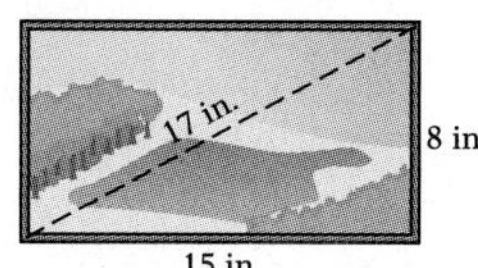

The hypotenuse is 17.

$$a^2 + b^2 = 8^2 + 15^2 = 64 + 225 = 289$$
$$c^2 = 17^2 = 289$$

Yes, because $c^2 = a^2 + b^2$, the corners are right angles.

Your Turn 6. $C = 2\pi r = 2(3.14)14 = 87.92$ feet

$A = \pi r^2 = (3.14)14^2 = 615.44$ square feet

Your Turn 7. $C = 2\pi r = 2(3.14)4 = 25.12$ inches
After 10 revolutions, you pushed the mower $10(25.12) = 251.2$ inches.

Your Turn 8. The radius of the circular piece is 0.5 inch and the area is $A = \pi r^2 = 3.14(0.5)^2 = 0.785$ square inch.
The radius of the hole is 0.125 inch and the area is $A = \pi r^2 = 3.14(0.125)^2 = 0.049$ square inch.
The area of the washer is $0.785 - 0.049 = 0.736$ square inch.

Copyright © Houghton Mifflin Company. All rights reserved.

Your Turn 9.

$V = \frac{4}{3}\pi r^3 = (1.33)(3.14)(1.5)^3 = 14.09$ cubic inches

Your Turn 10. The radius of the pipe is half the diameter: $r = 2$ inches. The length is 10 feet, which is 120 inches.

$V = \pi r^2 h = (3.14)(2)^2(120) = 1{,}507.2$ cubic inches

CHAPTER 7

WARM-UP SKILLS *(page 574)*

1. $\frac{4}{5}$ **2.** 28,000 **3.** $\frac{8}{3}$ **4.** $\frac{6}{13}$ **5.** 2.43 **6.** 14
7. 6 **8.** $.38 per pound **9.** 120 square feet
10. 12 cubic feet

SELECTED YOUR TURNS

Section 7.1

Your Turn 1. $P = 2L + 2W$

$P = 2 \cdot 10 + 2 \cdot 7 = 20 + 14 = 34$

$\frac{\text{Width}}{\text{Perimeter}} = \frac{7\text{ feet}}{34\text{ feet}} = \frac{7}{34}$

Your Turn 2. $\frac{\text{Boys}}{\text{Girls}} = \frac{9}{7}$

Your Turn 3. $\frac{\text{Time running}}{\text{Time swimming}} = \frac{10\text{ seconds}}{51\text{ seconds}} = \frac{10}{51}$

$\frac{\text{Time swimming}}{\text{Time running}} = \frac{51\text{ seconds}}{10\text{ seconds}} = \frac{51}{10}$

Your Turn 4. The number of dry days is $30 - 17 = 13$.

$\frac{\text{Dry days}}{\text{Rainy days}} = \frac{13}{17}$

Your Turn 5. The number of students who dropped the course is $28 - 21 = 7$.

$\frac{\text{Number who dropped}}{\text{Original enrollment}} = \frac{7}{28} = \frac{1}{4}$

Your Turn 6. **(a)** $\frac{0.8}{4.0} = \frac{8}{40} = \frac{1}{5}$

(b) $\frac{1\frac{1}{5}}{\frac{3}{10}} = \frac{\frac{6}{5}}{\frac{3}{10}} = \frac{6}{5} \cdot \frac{10}{3} = \frac{4}{1}$

Your Turn 7. $\frac{175\text{ miles}}{3.5\text{ hours}} = 50$ miles per hour

Your Turn 8. $\frac{1{,}000\text{ gallons}}{25\text{ minutes}} = 40$ gallons per minute

Your Turn 9. $\frac{4{,}080\text{ dollars}}{12\text{ months}} = \$340$ per month

Your Turn 10.

20-pound bag: $\frac{8.48\text{ dollars}}{20\text{ pounds}} = \$.42$ per pound

50-pound bag: $\frac{21.50\text{ dollars}}{50\text{ pounds}} = \$.43$ per pound

The 20-pound bag is the better buy.

Section 7.2

Your Turn 1. $\frac{20\text{ yards}}{1} \cdot \frac{3\text{ feet}}{1\text{ yard}} = 60$ feet

Your Turn 2. $\frac{90\text{ inches}}{1} \cdot \frac{1\text{ foot}}{12\text{ inches}} = 7\frac{1}{2}$ feet

Your Turn 3.

$\frac{1\frac{1}{2}\text{ pounds}}{1} \cdot \frac{16\text{ ounces}}{1\text{ pound}} = \frac{3}{2} \cdot \frac{16}{1}$ ounces $= 24$ ounces

Your Turn 4. $\frac{1{,}600\text{ pounds}}{1} \cdot \frac{1\text{ bag}}{60\text{ pounds}} = 26\frac{2}{3}$

The person should buy 26 bags.

Your Turn 5.

$\frac{1\text{ quart}}{1} \cdot \frac{2\text{ pints}}{1\text{ quart}} \cdot \frac{2\text{ cups}}{1\text{ pint}} \cdot \frac{8\text{ ounces}}{1\text{ cup}} = 32$ ounces

Your Turn 6. $\frac{8\text{ cups}}{1} \cdot \frac{1\text{ pint}}{2\text{ cups}} = 4$ pints

Your Turn 7. $\frac{1{,}000\text{ gallons}}{2\text{ hours}} \cdot \frac{4\text{ quarts}}{1\text{ gallon}} \cdot \frac{1\text{ hour}}{60\text{ minutes}} =$

33 quarts per minute

Your Turn 8.

$\frac{6\text{ feet}}{1\text{ second}} \cdot \frac{1\text{ mile}}{5{,}280\text{ feet}} \cdot \frac{60\text{ seconds}}{1\text{ minute}} \cdot \frac{60\text{ minutes}}{1\text{ hour}} \approx$

4.1 miles per hour

Your Turn 9.

$\frac{36\text{ ounces}}{1\text{ day}} \cdot \frac{7\text{ days}}{1\text{ week}} \cdot \frac{1\text{ quart}}{32\text{ ounces}} = \frac{36 \cdot 7\text{ quarts}}{32\text{ weeks}}$

≈ 8 quarts per week

Section 7.3

Your Turn 1.

$\frac{96\text{ centimeters}}{1} \cdot \frac{1\text{ meter}}{100\text{ centimeters}} = 0.96$ meter

Your Turn 2.

(a) $\frac{998\text{ millimeters}}{1} \cdot \frac{1\text{ centimeter}}{10\text{ millimeters}} = 99.8$ centimeters

(b) $\frac{1.27\text{ centimeters}}{1} \cdot \frac{10\text{ millimeters}}{1\text{ centimeter}} = 12.7$ millimeters

Your Turn 3. $\frac{100\text{ meters}}{1} \cdot \frac{1\text{ yard}}{0.91\text{ meter}} \approx 109.9$ yards

Your Turn 4.

$\frac{892\text{ miles}}{1} \cdot \frac{1.609\text{ kilometers}}{1\text{ mile}} \approx 1{,}435$ kilometers

Your Turn 5.

$\frac{16\text{ centimeters}}{1} \cdot \frac{1\text{ inch}}{2.54\text{ centimeters}} = \frac{16}{2.54}$ inches

≈ 6.3 inches

$\frac{20\text{ centimeters}}{1} \cdot \frac{1\text{ inch}}{2.54\text{ centimeters}} = \frac{20}{2.54}$ inches

≈ 7.9 inches

Copyright © Houghton Mifflin Company. All rights reserved.

Section 7.4

Your Turn 1.

$$\frac{0.65 \text{ liter}}{1} \cdot \frac{1{,}000 \text{ milliliters}}{1 \text{ liter}} = 650 \text{ milliliters}$$

Your Turn 2. The amount of gasoline needed to fill the tank is $5.0 - 4.6 = 0.4$ kiloliter.

$$\frac{0.4 \text{ kiloliter}}{1} \cdot \frac{1{,}000 \text{ liters}}{1 \text{ kiloliter}} = 400 \text{ liters}$$

Your Turn 3. 20 tablets weigh $20 \cdot 800 =$ 16,000 milligrams.

$$\frac{16{,}000 \text{ milligrams}}{1} \cdot \frac{1 \text{ gram}}{1{,}000 \text{ milligrams}} = 16 \text{ grams}$$

Your Turn 4.

$$\frac{0.72 \text{ metric ton}}{1} \cdot \frac{1{,}000 \text{ kilograms}}{1 \text{ metric ton}} = 720 \text{ kilograms}$$

Your Turn 5. $\frac{2 \text{ liters}}{1} \cdot \frac{1 \text{ quart}}{0.946 \text{ liter}} \approx 2.11 \text{ quarts}$

Your Turn 6. Convert both weights to the same units. We convert both to pounds.

	Weight	*Unit Price*
96-oz box	$\frac{96 \text{ oz}}{1} \cdot \frac{1 \text{ lb}}{16 \text{ oz}} = 6 \text{ lb}$	$\frac{\$6}{6 \text{ lb}} = \$1 \text{ per lb}$
3-kg box	$\frac{3 \text{ kg}}{1} \cdot \frac{1 \text{ lb}}{0.454 \text{ kg}} \approx 6.6 \text{ lb}$	$\frac{\$6}{6.6 \text{ lb}} \approx \$.91 \text{ per lb}$

The 3-kilogram box is the better buy.

Your Turn 7. $\frac{6 \text{ U.S. tons}}{1} \cdot \frac{0.91 \text{ metric ton}}{1 \text{ U.S. ton}}$

$= 6 \cdot 0.91$ metric tons

≈ 5.5 metric tons

Section 7.5

Your Turn 2. (a)

$$\begin{aligned} \frac{3}{11} &= \frac{4}{20} \\ 3 \cdot 20 &= 11 \cdot 4 \\ 60 &= 44 \quad \text{False} \end{aligned}$$

(b)

$$\begin{aligned} \frac{18}{30} &= \frac{9}{15} \\ 18 \cdot 15 &= 30 \cdot 9 \\ 270 &= 270 \quad \text{True} \end{aligned}$$

Your Turn 4.

$$\begin{aligned} \frac{6}{18} &= \frac{x}{9} \\ 18x &= 54 \\ \frac{18x}{18} &= \frac{54}{18} \\ x &= 3 \end{aligned}$$

Your Turn 5.

$$\begin{aligned} \frac{40}{x} &= \frac{1{,}000}{3} \\ 1{,}000x &= 120 \\ x &= 0.12 \end{aligned}$$

Your Turn 6.

$$\begin{aligned} \frac{6}{x-1} &= \frac{3}{2} \\ 3(x-1) &= 12 \\ 3x - 3 &= 12 \\ 3x &= 15 \\ x &= 5 \end{aligned}$$

Your Turn 7. Let x = the additional miles traveled.

	Miles	Hours
Known	116	2
Unknown	$x + 116$	3

$$\begin{aligned} \frac{116}{x + 116} &= \frac{2}{3} \\ 2(x + 116) &= 3 \cdot 116 \\ 2x + 232 &= 348 \\ 2x &= 116 \\ x &= 58 \text{ miles} \end{aligned}$$

Your Turn 8.

	Pounds	Cost
Known	5	\$3.50
Unknown	8	x

$$\begin{aligned} \frac{5}{8} &= \frac{3.50}{x} \\ 5x &= 8(3.50) \\ 5x &= 28 \\ x &= \$5.60 \end{aligned}$$

Your Turn 9.

	Purchase	Tax
Known	\$46	\$1.84
Unknown	x	\$2.88

$$\begin{aligned} \frac{46}{x} &= \frac{1.84}{2.88} \\ 1.84x &= 46(2.88) \\ 1.84x &= 132.48 \\ x &= \$72 \end{aligned}$$

Your Turn 10.

	Inches	Feet
Known	8	150
Unknown	x	100

$$\begin{aligned} \frac{8}{x} &= \frac{150}{100} \\ 150x &= 800 \\ x &= 5\tfrac{1}{3} \text{ inches} \end{aligned}$$

Copyright © Houghton Mifflin Company. All rights reserved.

Your Turn 11.

	Total Solution	Amount of Alcohol
Known	20	3
Unknown	70	x

$$\frac{20}{70} = \frac{3}{x}$$
$$20x = 70 \cdot 3$$
$$20x = 210$$
$$x = 10.5 \text{ cc}$$

Your Turn 12.

	Number Polled	Number Who Approve
Known	100	53
Unknown	800	x

$$\frac{100}{800} = \frac{53}{x}$$
$$100x = (800)(53) = 42{,}400$$
$$x = 424$$

Section 7.6

Your Turn 1. **(c)** $m\angle RCS = 180° - m\angle ACR = 180° - 170° = 10°$

Your Turn 4. $\frac{EF}{ST} = \frac{FG}{RS}$

$$\frac{9}{4} = \frac{x}{6}$$
$$4x = 54$$
$$x = 13\tfrac{1}{2}$$

Your Turn 5. $\frac{AF}{EF} = \frac{AB}{EG}$

$$\frac{3}{8} = \frac{x}{10}$$
$$8x = 30$$
$$x = 3\tfrac{3}{4}$$

Your Turn 6. $\frac{QB}{QA} = \frac{QA}{QC}$

$$\frac{4}{x} = \frac{x}{9}$$
$$x^2 = 36$$
$$x = 6$$

Your Turn 7. $\frac{x}{1} = \frac{10}{\frac{1}{2}}$

$$\tfrac{1}{2}x = 10$$
$$x = 20 \text{ feet}$$

CHAPTER 8

WARM-UP SKILLS (page 674)

1. $\frac{37}{100}$ **2.** $\frac{45}{100}$ **3.** $\frac{27}{20}$ **4.** 0.625 **5.** $0.8\overline{3}$ **6.** 60
7. 32 **8.** 8 **9.** \$954 **10.** 14

SELECTED YOUR TURNS

Section 8.1

Your Turn 2. **(a)** $\frac{7}{20} = \frac{7}{20} \cdot \frac{5}{5} = \frac{35}{100} = 35\%$

(b) $\frac{45}{50} = \frac{45}{50} \cdot \frac{2}{2} = \frac{90}{100} = 90\%$

(c) $\frac{15}{25} = \frac{15}{25} \cdot \frac{4}{4} = \frac{60}{100} = 60\%$

Your Turn 4. **(a)** $35\% = \frac{35}{100} = \frac{\cancel{5} \cdot 7}{\cancel{5} \cdot 20} = \frac{7}{20}$

(b) $225\% = \frac{225}{100} = \frac{9 \cdot \cancel{25}}{4 \cdot \cancel{25}} = \frac{9}{4}$

(c) $12\tfrac{1}{2}\% = \frac{25}{2}\% = \frac{\frac{25}{2}}{100} = \frac{25}{2} \cdot \frac{1}{100} = \frac{1}{8}$

(d) $8.4\% = \frac{8.4}{100} = \frac{8.4}{100} \cdot \frac{10}{10} = \frac{84}{1{,}000} = \frac{\cancel{4} \cdot 21}{\cancel{4} \cdot 250} = \frac{21}{250}$

Your Turn 5. **(a)** $28\% = 0.28. = 0.28$

(b) $2\% = 0.02. = 0.02$

(c) $110\% = 1.10. = 1.10$

(d) $8.5\% = 0.08.5 = 0.085$

Your Turn 6. **(a)** $0.025 = 0.02.5\% = 2.5\%$

(b) $2.1 = 2.10.\% = 210\%$

(c) $0.54 = 0.54.\% = 54\%$

(d) $0.001 = 0.00.1\% = 0.1\%$

Your Turn 7. $\frac{1}{16} = 0.0625 = 6.25\%$

Your Turn 8. **(a)** $\frac{4}{9} = 0.44444 \cdots \approx 44.44\%$

(b) $44\tfrac{4}{9}\%$

Your Turn 9. **(a)** $\frac{36}{125} = 0.288 = 28.8\%$

(b) $\frac{7}{4} = 1.75 = 175\%$

(c) $\frac{11}{20} = 0.55 = 55\%$

(d) $\frac{4}{15} = 0.26666 \cdots = 26\tfrac{2}{3}\%$

Your Turn 10. **(b)** $38\% = \frac{38}{100} = \frac{\cancel{2} \cdot 19}{\cancel{2} \cdot 50} = \frac{19}{50}$

Copyright © Houghton Mifflin Company. All rights reserved.

Your Turn 11. $\frac{3}{8} = 0.375 = 37.5\%$

Your Turn 12. **(a)** $\frac{20}{50} = \frac{40}{100} = 40\%$

(b) $\frac{16 + 14}{50} = \frac{30}{50} = \frac{60}{100} = 60\%$

Section 8.2

Your Turn 2. $0.18 \cdot 90 = x$

$16.2 = x$

Your Turn 3. $0.75 \cdot x = 9$

$\frac{\cancel{0.75}x}{\cancel{0.75}} = \frac{9}{0.75}$

$x = 12$

75% is $\frac{3}{4}$ as a fraction.

$\frac{3}{4}x = 9$

$\frac{\cancel{4}}{3} \cdot \frac{3}{\cancel{4}}x = \frac{4}{3} \cdot \frac{9}{1}$

$x = 12$

Your Turn 4. $x \cdot 120 = 66$

$\frac{\cancel{120}x}{\cancel{120}} = \frac{66}{120}$

$x = 0.55 = 55\%$

Your Turn 5. **(a)** $0.26 \cdot 80 = A$

26% is close to 25%, or $\frac{1}{4}$.

$\frac{1}{4}$ of 80 is 20.

(b) $P \cdot 30 = 14$

$P = \frac{14}{30}$

$\frac{14}{30}$ is close to $\frac{15}{30}$, or $\frac{1}{2}$, and $\frac{1}{2} = 50\%$.

Your Turn 6. 1% of 700 = 7. So, 2% of 700 is $2 \cdot 7 = 14$.

Your Turn 7. \$80.10 is close to \$80. 1% of 80 is 0.8. So 3% of \$80.10 is approximately \$2.40.

Your Turn 8. **(a)** 10.2% is close to 10%. 10% of \$26,000 is \$2,600.

(b) $0.102 \cdot 26{,}000 = \$2{,}652$

Section 8.3

Your Turn 1. Let x = the number of miles driven.

$0.58 \cdot 750 = x$

$435 = x$

The driver has traveled 435 miles.

Your Turn 2. Let x = the percent (in decimal form) of the book that has been read.

$x \cdot 600 = 234$

$\frac{\cancel{600}x}{\cancel{600}} = \frac{234}{600}$

$x = 0.39 = 39\%$

The person has read 39% of the book.

Your Turn 3. Let x = the number of acres in the development.

$0.18 \cdot x = 54$

$\frac{\cancel{0.18}x}{\cancel{0.18}} = \frac{54}{0.18}$

$x = 300$

The development has 300 acres.

Your Turn 4. Let x = the number of players ages 17 and 18.

$0.38 \cdot x = 4{,}430$

$\frac{\cancel{0.38}x}{\cancel{0.38}} = \frac{4{,}430}{0.38}$

$x = 11{,}658$

11,658 boys and girls played soccer.

Your Turn 5. Let x = the percent of the population that is 85 or older.

$x \cdot 391.3 = 18$

$x = \frac{18}{391.3} = 0.046$

4.6% of the population is expected to be 85 or older.

Your Turn 6. Let x = the number of ISPs in 1997.

$1.5 \cdot 3{,}000 = x$

$4{,}500 = x$

In 1997, the number of ISPs was 4,500.

Section 8.4

Your Turn 1. The amount borrowed is \$21,000 − \$15,000 = \$6,000.

(a) $I = Pr$

$I = 6{,}000 \cdot 0.09$

$I = 540$

The interest is \$540.

(b) The loan payoff is \$6,000 + \$540 = \$6,540.

Your Turn 2. $I = Pr$

$I = 2{,}000 \cdot 0.04$

$I = 80$

You will earn \$80 in interest.

Your Turn 3. 6 months = $\frac{1}{2}$ year

$I = Prt$

$I = 1{,}500 \cdot 0.052 \cdot \frac{1}{2}$

$I = 39$

You will earn \$39 in interest.

Copyright © Houghton Mifflin Company. All rights reserved.

Your Turn 4. $P = 12{,}000$, $I = 12{,}840 - 12{,}000 = 840$, $t = 1$

$$I = Prt$$
$$840 = 12{,}000 \cdot r \cdot 1$$
$$840 = 12{,}000r$$
$$\frac{840}{12{,}000} = r$$
$$r = 0.07$$

The person needs to earn 7% interest.

Your Turn 5. $P = 5{,}000$, $r = 0.055$, $t = 8$, $n = 4$

$$A = P\left(1 + \frac{r}{n}\right)^{nt} = 5{,}000\left(1 + \frac{0.055}{4}\right)^{4 \cdot 8} = \$7{,}740.30$$

Your Turn 6. **(a)** $180 \cdot 0.04 = 7.20$
The tax is \$7.20.
(b) $180 + 7.20 = 187.20$
The total cost is \$187.20.

Your Turn 7. Let x = the price of the shoes.
The tax is $0.05 \cdot x$.

$$0.05x = 4.10$$
$$\frac{\cancel{0.05}x}{\cancel{0.05}} = \frac{4.10}{0.05}$$
$$x = 82$$

The price of the shoes was \$82.

Your Turn 8. The sales tax is \$364 − \$350 = \$14.
Let x = the sales tax rate.

$$350 \cdot x = 14$$
$$\frac{\cancel{350}x}{\cancel{350}} = \frac{14}{350}$$
$$x = 0.04$$

The sales tax rate is 4%.

Your Turn 9. The total deductions are 42.5% + 5% = 47.5%. Let x = the tax.

$$3{,}000 \cdot 0.475 = x$$
$$1{,}425 = x$$

Your take-home pay is \$3,000 − \$1,425 = \$1,575.

Section 8.5

Your Turn 1. Let x = the commission.

$$x = 700 \cdot 0.06 = 42$$

The commission is \$42.

Your Turn 2. Let x = the cost of the land.

$$0.10 \cdot x = 3{,}250$$
$$\frac{\cancel{0.1}x}{\cancel{0.1}} = \frac{3{,}250}{0.1}$$
$$x = 32{,}500$$

The cost of the land was \$32,500.

Your Turn 3. Let x = the commission rate.

$$155{,}000 \cdot x = 10{,}850$$
$$\frac{\cancel{155{,}000}x}{\cancel{155{,}000}} = \frac{10{,}850}{155{,}000}$$
$$x = 0.07$$

The commission rate was 7%.

Your Turn 4. The commission is paid on \$18,000 − \$10,000 = \$8,000. Let x = the commission.

$$x = \$8{,}000 \cdot 0.03 = \$240$$

The monthly salary is \$24,000 ÷ 12 or \$2,000. The employee's monthly earnings were \$2,000 + \$240 = \$2,240.

Your Turn 5. The truck loses $\frac{1}{10}$ of its value each year.

$$\text{annual depreciation} = \frac{1}{10}(\$40{,}000) = \$4{,}000$$
$$\text{loss of value after 4 years} = \$4{,}000 \cdot 4 = \$16{,}000$$

The depreciated value after 4 years is \$40,000 − \$16,000 = \$24,000.

Your Turn 6. After the third year the value of the roller is 0.80(\$384,000) = \$307,200.
The value after the fourth year is 0.80(\$307,200) = \$245,760.

Your Turn 7. $V = P(1 - r)^t = 600{,}000(1 - 0.20)^{10} = 64{,}424.51$
The depreciated value after 10 years is \$64,424.51.

Your Turn 8. **(a)** markup = 0.45 · \$16 = \$7.20
(b) retail price = \$16 + \$7.20 = \$23.20

Your Turn 9. The retail price for an ear of corn is $\frac{98¢}{4} = 24.5¢$.
The markup is 24.5¢ − 10¢ = 14.5¢.
Let r = markup rate.

$$r \cdot 10 = 14.5$$
$$\frac{\cancel{10}r}{\cancel{10}} = \frac{14.5}{10}$$
$$r = 1.45$$

The markup rate is 145%.

Your Turn 10. **(a)** The discount is 0.10 · \$8 = \$.80.
(b) The cost of a ticket is \$8 − \$.80 = \$7.20.

Your Turn 11. Let x = the original cost.

$$0.02 \cdot x = 17$$
$$\frac{\cancel{0.02}x}{\cancel{0.02}} = \frac{17}{0.02}$$
$$x = 850$$

The cost was \$850.

Copyright © Houghton Mifflin Company. All rights reserved.

Your Turn 12. The price discount is

$$22.00 - 18.70 = 3.30$$
$$3.30 = r \cdot 22.00$$
$$\frac{3.30}{22.00} = r$$
$$0.15 = r$$

The discount rate is 15%.

Section 8.6

Your Turn 1. **(a)** The increase in cost is $0.10 \cdot 50¢ = 5¢$.
(b) The new cost is $50¢ + 5¢ = 55¢$.

Your Turn 2. increase = \$2.16 trillion − \$1.29 trillion = \$.87 trillion
Let x = the percent increase.

$$x \cdot 1.29 = 0.87$$
$$\frac{\cancel{1.29}x}{\cancel{1.29}} = \frac{0.87}{1.29}$$
$$x \approx 0.67$$

The percent increase is 67%.

Your Turn 3. increase = 246,400 − 220,000 = 26,400
Let x = the percent increase.

$$x \cdot 220{,}000 = 26{,}400$$
$$\frac{\cancel{220{,}000}x}{\cancel{220{,}000}} = \frac{26{,}400}{220{,}000}$$
$$x = 0.12$$

The percent increase was 12%.

Your Turn 4. percent increase + 100% = 10% + 100% = 110%
The new carton contains $110\% \cdot 90 = 1.10\,(90) = 99$ ounces.

Your Turn 5. decrease = $0.26 \cdot 204 \approx 53$
The number of days in the school year in 2000 was 204 − 53 = 151.

Your Turn 6. decrease = 16.0 − 13.6 = 2.4
Let x = the percent decrease.

$$x \cdot 16 = 2.4$$
$$\frac{\cancel{16}x}{\cancel{16}} = \frac{2.4}{16}$$
$$x = 0.15$$

The percent decrease was 15%.

Your Turn 7. The number in 2000 was 100% − 90%, or 10%, of the number in 1970.
Let x = the number of strikes in 1970.

$$0.10 \cdot x = 39$$
$$\frac{\cancel{0.10}x}{\cancel{0.10}} = \frac{39}{0.10}$$
$$x = 390$$

The number of strikes in 1970 was 390.

Your Turn 8. sale price = $\$50 - 0.10 \cdot \$50 = \$50 - \$5 = \$45$
The new retail price is $\$45 + 0.10 \cdot \$45 = \$45 + \$4.50 = \$49.50$.

CHAPTER 9

WARM-UP SKILLS *(page 774)*

1. $5a, 2b, 9$ **2.** $16x$ **3.** $-a$ **4.** $11y - 5$ **5.** -13 **6.** 64 **7.** $4x + 20$ **8.** $-6n + 14$ **9.** 2, 3, 5, 7 **10.** $2^3 \cdot 3$

SELECTED YOUR TURNS

Section 9.1

Your Turn 1.
(a) $x^3 + (-4x)$; terms: x^3, $-4x$; polynomial
(b) $\frac{1}{2}x^2 + (-x) + \frac{2}{3}$; terms: $\frac{1}{2}x^2$, $-x$, $\frac{2}{3}$; polynomial

Your Turn 2. $x^3 + x - 4 = (-1)^3 + (-1) - 4 = -6$

Your Turn 3. $(-4a + 1) + (6a - 3)$
$= -4a + 1 + 6a - 3$
$= (-4a + 6a) + (1 - 3) = 2a - 2$

Your Turn 4. $(x^2 - 3x - 5) + (2x^2 + 7x + 9)$
$= x^2 - 3x - 5 + 2x^2 + 7x + 9$
$= (x^2 + 2x^2) + (-3x + 7x) + (-5 + 9)$
$= 3x^2 + 4x + 4$

Your Turn 5.

$$\begin{array}{r} 2x^3 + 8x^2 - 6 \\ + \; x^3 + 9x + 13 \\ \hline 3x^3 + 8x^2 + 9x + 7 \end{array}$$

Your Turn 6. $(4x + 7) - (9x - 1)$
$= (4x + 7) + (-9x + 1)$
$= (4x - 9x) + (7 + 1)$
$= -5x + 8$

Your Turn 7. $(5x^2 - 6x + 3) - (2x^2 + x - 6)$
$= (5x^2 - 6x + 3) + (-2x^2 - x + 6)$
$= (5x^2 - 2x^2) + (-6x - x) + (3 + 6)$
$= 3x^2 - 7x + 9$

Your Turn 8.

$$\begin{array}{r} 6x^2 + 5x + 2 \\ - \; 3x^2 - 8 \\ \hline \end{array} \rightarrow \begin{array}{r} 6x^2 + 5x + 2 \\ + \; -3x^2 + 8 \\ \hline 3x^2 + 5x + 10 \end{array}$$

Your Turn 9. $12 - (7x + 3)$
$= 12 + (-7x - 3)$
$= (12 - 3) + (-7x)$
$= 9 - 7x$

Your Turn 10. $(2x + 1) + (3x - 2) + 2x + x$
$= (2x + 3x + 2x + x) + (1 - 2)$
$= 8x - 1$

Your Turn 11.
Triangle A: $P = 3(x + 6) = 3x + 18$
Triangle B: $P = 3(2x) = 6x$

$$6x - (3x + 18) = 6x + (-3x - 18)$$
$$= 3x - 18$$

Copyright © Houghton Mifflin Company. All rights reserved.

Section 9.2

Your Turn 1. $x^5 \cdot x^{10} = x^{5+10} = x^{15}$

Your Turn 2. $2y^4 \cdot y^5 = 2(y^4 \cdot y^5) = 2y^{4+5} = 2y^9$

Your Turn 3.

$$-2a^3 \cdot 5a = (-2 \cdot 5)(a^3 \cdot a^1) = -10a^{3+1} = -10a^4$$

Your Turn 4.

$$3x^3 \cdot 4x^2 \cdot x = (3 \cdot 4)(x^3 \cdot x^2 \cdot x^1) = 12x^{3+2+1} = 12x^6$$

Your Turn 5. $(y^3)^4 = y^{3 \cdot 4} = y^{12}$

Your Turn 6. $(a^3)^2 \cdot (a^5)^3 = a^6 \cdot a^{15} = a^{6+15} = a^{21}$

Your Turn 7. $(4b)^3 = 4^3 \cdot b^3 = 64b^3$

Your Turn 8. $(3x^4)^2 = 3^2 \cdot (x^4)^2 = 9x^{4 \cdot 2} = 9x^8$

Your Turn 9. $(a^2b^4)^3 = (a^2)^3 \cdot (b^4)^3 = a^{2 \cdot 3}b^{4 \cdot 3} = a^6b^{12}$

Your Turn 10. $(3x^2yz^4)^3 = 3^3 \cdot (x^2)^3 \cdot (y^1)^3 \cdot (z^4)^3 = 27x^{2 \cdot 3}y^{1 \cdot 3}z^{4 \cdot 3} = 27x^6y^3z^{12}$

Your Turn 11.

$$(4x^3)^2 \cdot (2x^2)^2 = 4^2(x^3)^2 \cdot 2^2(x^2)^2 = 16x^{3 \cdot 2} \cdot 4x^{2 \cdot 2} = 16x^6 \cdot 4x^4 = 64x^{6+4} = 64x^{10}$$

Section 9.3

Your Turn 1.

$$3b^2(2b - 5b^3) = 3b^2 \cdot 2b^1 - 3b^2 \cdot 5b^3 = 6b^3 - 15b^5$$

Your Turn 2.

$$\begin{aligned}-5y(y^4 + 3y^2) &= -5y^1 \cdot y^4 + (-5y^1) \cdot 3y^2 \\ &= -5y^5 + (-15y^3) \\ &= -5y^5 - 15y^3\end{aligned}$$

Your Turn 3.

$$\begin{aligned}2x(7x^2 + 3x - 5) &= 2x \cdot 7x^2 + 2x \cdot 3x - 2x \cdot 5 \\ &= 14x^3 + 6x^2 - 10x\end{aligned}$$

Your Turn 4.

$$\begin{aligned}(3x + 2)(x - 1) &= 3x \cdot x - 3x \cdot 1 + 2 \cdot x - 2 \cdot 1 \\ &= 3x^2 - 3x + 2x - 2 \\ &= 3x^2 - x - 2\end{aligned}$$

Your Turn 5.

$$\begin{aligned}(2x - 5)(x - 4) &= [2x + (-5)](x - 4) \\ &= 2x \cdot x - 2x \cdot 4 + (-5) \cdot x - (-5) \cdot 4 \\ &= 2x^2 - 8x + (-5x) - (-20) \\ &= 2x^2 + (-13x) + 20 \\ &= 2x^2 - 13x + 20\end{aligned}$$

Your Turn 6.

(a)
$$\begin{aligned}(x + 4)^2 &= (x + 4)(x + 4) \\ &= x \cdot x + x \cdot 4 + 4 \cdot x + 4 \cdot 4 \\ &= x^2 + 4x + 4x + 16 \\ &= x^2 + 8x + 16\end{aligned}$$

(b)
$$\begin{aligned}(a - b)^2 &= (a - b)(a - b) \\ &= [a + (-b)](a - b) \\ &= a \cdot a - a \cdot b + (-b) \cdot a - (-b) \cdot b \\ &= a^2 - ab + (-ab) - (-b^2) \\ &= a^2 - 2ab + b^2\end{aligned}$$

Your Turn 7.

$$\begin{aligned}&(x + 3)(x^2 - 2x + 1) \\ &= x \cdot x^2 - x \cdot 2x + x \cdot 1 + 3 \cdot x^2 - 3 \cdot 2x + 3 \cdot 1 \\ &= x^3 - 2x^2 + x + 3x^2 - 6x + 3 \\ &= x^3 + x^2 - 5x + 3\end{aligned}$$

Your Turn 8.

$$\begin{array}{r} x^2 - 5x - 2 \\ x + 4 \\ \hline 4x^2 - 20x - 8 \\ x^3 - 5x^2 - 2x \phantom{{}- 8} \\ \hline x^3 - x^2 - 22x - 8 \end{array}$$

Section 9.4

Your Turn 1. $24 = 8 \cdot 3 = 2 \cdot 2 \cdot 2 \cdot 3$

$42 = 6 \cdot 7 = 2 \cdot 3 \cdot 7$

Your Turn 2. $27 = 3 \cdot 9$

$45 = 5 \cdot 9$

The GCF is 9.

Your Turn 3. The smallest exponent is 3. Therefore, the GCF is x^3.

Your Turn 4. $9y^3 = 3 \cdot 3 \cdot y^3$

$15y^2 = 3 \cdot 5 \cdot y^2$

The GCF is $3y^2$.

Your Turn 5. $8x^2 = 4 \cdot 2 \cdot x^2$

$12x = 4 \cdot 3 \cdot x$

$16x^4 = 4 \cdot 4 \cdot x^4$

The GCF is $4x$.

Your Turn 6. $2x^5y = 2 \cdot x^5 \cdot y$

$6x^2y^3 = 2 \cdot 3 \cdot x^2 \cdot y^3$

The GCF is $2x^2y$.

Your Turn 7. $5y^2 - 7y = y \cdot 5y - y \cdot 7 = y(5y - 7)$

Your Turn 8. $6x^2 - 2x + 12 = 2 \cdot 3x^2 - 2 \cdot x + 2 \cdot 6 = 2(3x^2 - x + 6)$

Your Turn 9. $14x^3 - 7x^2 = 7x^2 \cdot 2x - 7x^2 \cdot 1 = 7x^2(2x - 1)$

Your Turn 10. $2x^2 + 2x = 2x \cdot x + 2x \cdot 1 = 2x(x + 1)$

Your Turn 11. *Method 1*

$$\begin{aligned}-3a - 9b + 15c &= 3 \cdot (-1a) - 3 \cdot 3b + 3 \cdot 5c \\ &= 3(-a - 3b + 5c)\end{aligned}$$

Method 2

$$\begin{aligned}-3a - 9b + 15c &= -3 \cdot a + (-3) \cdot 3b + (-3) \cdot (-5c) \\ &= -3(a + 3b - 5c)\end{aligned}$$

Copyright © Houghton Mifflin Company. All rights reserved.

Index